STOCHASTIC QUANTIZATION

STOCHASTIC QUANTIZATION

Editors

P.H. Damgaard and H. Hüffel

Cern, Geneva, Switzerland

World Scientific

Singapore • New Jersey • Hong Kong

Published by

World Scientific Publishing Co. Pte. Ltd.

5 Toh Tuck Link, Singapore 596224

USA office: 27 Warren Street, Suite 401-402, Hackensack, NJ 07601

UK office: 57 Shelton Street, Covent Garden, London WC2H 9HE

British Library Cataloguing-in-Publication Data
A catalogue record for this book is available from the British Library.

The editors and publisher would like to thank all authors and the following publishers for their permission to reproduce the reprinted papers found in this volume:

American Physical Society (*Phys. Rev. & Phys. Rev. Lett.*); D. Riedel Publishing Co. (*Lett. Math. Phys.*); Kyoto University (*Prog. Theor. Phys.*); North Holland (*Nucl. Phys. & Phys. Lett.*); Plenum Publishing Corp. (*J. Stat. Phys.*).

To those who have not granted us permission before publication, we have taken the liberty to reproduce their articles without consent. We shall however acknowledge them in future editions of this work.

STOCHASTIC QUANTIZATION

ISBN-13 978-9971-5-0254-6
ISBN-10 9971-5-0254-2
ISBN-13 978-9971-5-0298-0 (pbk)
ISBN-10 9971-5-0298-4 (pbk)

PREFACE

The growing interest in Parisi and Wu's method of Stochastic Quantization has brought about the need to have some of the most important papers in the field available in compact form. This book contains a selection of research papers centred around this subject, and a set of introductory chapters which provide background material and a guide to the reprinted literature.

In making this selection, our goal has been to include as broad a spectrum as possible. Stochastic Quantization has found applications in a diverse set of physics problems, and we felt that this fact should be reflected in our choice of reprinted papers. The introductory chapters serve to explain some of the basic concepts behind this quantization method, as well as to present a link between the different areas of field theory/statistical mechanics covered in the reprinted papers.

We would like to express our gratitude to the authors and publishers of the journals for their kind permission to reprint the selected papers. We also thank Bo-Sture Skagerstam for encouragement, and Maurice Jacob and the CERN Theory Division for support.

Geneva, January 1987

Poul Henrik Damgaard
Helmuth Hüffel

CONTENTS

REPRINTED PAPERS

Gauge Fields

Fermions

Numerical Applications

INTRODUCTORY CHAPTERS

1. INTRODUCTION

Stochastic Quantization, in the sense in which we will use it in this book, is a comparatively new quantization method, which was proposed in 1981 by Parisi and Wu [R1]*). As it relies strongly on concepts from statistical mechanics and probability theory, it seems appropriate to recall briefly some of the underlying basic aspects.

As a starting point we consider the analogy between classical statistical mechanics and Euclidean quantum field theory. This is based on the fact that the path integral measure $D\phi \exp(-S_E[\phi])$ can be identified with the Boltzmann probability distribution of a statistical system. Here $S_E[\phi]$ denotes the action of the field theory under study, Wick rotated to D-dimensional Euclidean space. As a consequence of this analogy, we can view Euclidean Green functions as the correlation functions of a statistical system in thermal equilibrium.

In Stochastic Quantization this formal analogy is carried one step further, and one allows the system to start out of equilibrium at an arbitrary initial state. The system is then slowly forced into equilibrium, with its stationary probability distribution given by the Euclidean path integral measure. Euclidean Green functions may finally be computed as the equilibrium limits of thermal correlation functions.

---

*) References of the form [Rnn] refer to reprinted papers.

The relaxation towards thermal equilibrium occurs in a new additional coordinate, a 'fictitious' time with respect to which all fields evolve. But before getting to that point, we must introduce the notion from probability theory of a stochastic process [1].

Consider first an object X, which is defined by a set of possible values (its 'range') and a probability distribution P(x) over this set. This object X is called a *stochastic variable.* A particularly interesting case arises when the range I is continuous. With a probability density $P(x) \geq 0$, normalized to unity on its range, the probability that X takes a value between x and $x + dx$ is simply given by P(x)dx.

Given X, we can construct a new stochastic variable by any well-defined function f acting on X. Thus $Y = f(X)$ is also a well-defined stochastic variable. Consider now in particular a Y which is also a function of an additional real parameter t: $Y(t) = f(t, X)$. This is called a *stochastic process.* It is a function of a real parameter t (the 'time') and a stochastic variable X.

Of particular importance are so-called *Markov processes.* These are stochastic processes Y(t) where the conditional probability that Y takes the value y_k at t_k is uniquely determined by the value y_{k-1} at t_{k-1}; in other words, it does not depend on the value y_m at any earlier time $t_m < t_{k-1}$.

A well-known *physical* process which is Markovian in character is that of Brownian motion. If we let $\vec{p}$ denote the momentum vector of a particle undergoing Brownian motion, then we can assume that it will be governed by the equation

$$d\vec{p}/dt = -\gamma\vec{p}(t) + \vec{\eta}(t) \tag{1}$$

which was first discussed in detail by Langevin [2]. Here γ gives the friction coefficient of the medium (fluid or gas) in which the particle is immersed; $\vec{\eta}(t)$ is a specific Markov process, usually called Gaussian white noise, which represents collectively the reaction from the background medium. It makes the particle undergo random and sudden changes of momentum, whereas the friction term in Eq. (1) acts as to damp all motion.

The Gaussian white noise $\vec{\eta}(t)$ derives its name from the fact that its correlations can be obtained as functional averages with Gaussian distribution:

$$\langle \ldots \rangle = \int D\eta(t)(\ldots) \exp\left[-\tfrac{1}{4}\int d\tau\, \eta^2(\tau)^2\right] \Big/ \int D\eta(t) \exp\left[-\tfrac{1}{4}\int d\tau\, \eta(\tau)^2\right] \,, \quad (2)$$

so that, for instance,

$$\langle \eta(t) \rangle = 0, \qquad \langle \eta(t_1)\eta(t_2) \rangle = 2\delta(t - t'), \text{ etc.} \quad (3)$$

Interestingly, for large times the two terms on the r.h.s of Eq. (1) balance each other, and a particular *equilibrium distribution* for $\vec{p}$ is obtained. This can be seen most easily by writing the Fokker–Planck [3] equation associated with the Langevin equation (1):

$$\dot{P}(t) = (\partial/\partial\vec{p})[(\partial/\partial\vec{p}) + \gamma\vec{p}]\, P(t) \,, \quad (4)$$

where

$$P(t) = \int D\eta\, \delta[p - p(t)] \exp\left[-\tfrac{1}{4}\int d\tau\, \eta(\tau)^2\right] \quad (5)$$

is the probability distribution. The equilibrium solution $\dot{P}(t) = 0$ to Eq. (4) is easily found to be (up to a normalization)

$$P(t \rightarrow \infty) = \exp\left(-\gamma/2\, \vec{p}^{\,2}\right] \,. \quad (6)$$

With $\gamma = 1/(mkT)$, this is recognized as the familiar Maxwell distribution.

2. STOCHASTIC QUANTIZATION OF FIELD THEORIES

Consider a scalar field theory, whose Euclidean action is given by

$$S[\phi] = \int d^Dx\, [\tfrac{1}{2}(\partial_\mu\phi)(\partial_\mu\phi) + \tfrac{1}{2}m^2\phi^2 + (\lambda/3!)\,\phi^3] \;. \tag{7}$$

We supplement the field $\phi(x)$ with an extra coordinate t, the fictitious time, such that $\phi(x) \rightarrow \phi(x, t)$, $x = (x_0, x_1, \ldots, x_{D-1})$. The fictitious time t should of course not be confused with the Euclidean 'time' x_0.

The t-evolution of $\phi(x, t)$ is assumed to be given by the Langevin equation:

$$(\partial/\partial t)\,\phi(x, t) = -\delta S[\phi]/\delta\phi(x)|_{\phi(x) = \phi(x,t)} + \eta(x, t) \;, \tag{8}$$

which in this case reads

$$(\partial/\partial t)\,\phi(x, t) = (\partial^2 - m^2)\,\phi(x, t) - \tfrac{1}{2}\lambda\phi(x,t)^2 + \eta(x, t) \;. \tag{9}$$

Given some initial condition for $\phi(x, t)$ at $t = t_0$, one solves the Langevin equation (9) by standard analytic methods. To begin with, let us denote the solution by $\phi^{[\eta]}(x, t)$ in order to indicate explicitly the dependence on the noise field $\eta(x, t)$. [We shall later omit this little complication, and simply keep in mind that the solution to any Langevin equation will always be a functional of η (x, t).] Correlation functions are defined, as before, by performing Gaussian averages over $\eta(x, t)$. Explicitly,

$$\langle\phi^{[\eta]}(x_1, t_1) \ldots \phi^{[\eta]}(x_n, t_n)\rangle = (1/Z_0) \int D\eta\, \phi^{[\eta]}(x_1, t_1) \ldots \phi^{[\eta]}(x_n, t_n) \exp\left[-\tfrac{1}{4} \int d^D x\, d\tau\, \eta(x, t)^2\right] \tag{10}$$

with

$$Z_0 = \int D\eta \exp\left[-\tfrac{1}{4} \int d^D x\, d\tau\, \eta(x, t)^2\right] . \tag{11}$$

The central assertion of Stochastic Quantization is that in the limit $t \to \infty$ equilibrium is reached, and that the (equal-time) correlation functions of $\phi^{[\eta]}$ become identical to the corresponding quantum Green functions, i.e.

$$\lim_{t\to\infty} \langle\phi^{[\eta]}(x_1, t_1) \ldots \phi^{[\eta]}(x_n, t_n)\rangle = \langle\phi(x_1) \ldots \phi(x_n)\rangle \tag{12}$$

Note that the l.h.s. contains a *stochastic* η-average, whereas the r.h.s. represents the standard field theoretic *vacuum expectation value* in Euclidean space!

This equivalence can be shown formally by a Fokker–Planck analysis similar to the one used in the simple case of Brownian motion [R1]. Alternatively, in the weak-coupling region $\lambda \ll 1$, a perturbation theory treatment can be carried through in detail. This is most easily done by first transforming the Langevin equation (9) into a (stochastic) integral equation. This integral equation may then solved by iteration up to any given order in the coupling constant λ.

It is convenient to Fourier transform the Langevin equation into momentum space, and first determine the Green function $G(k, t)$ which obeys

$$\begin{aligned} [(\partial/\partial t) + k^2 + m^2]\, G(k, t) &= \delta(t) , \qquad t \geq 0 \\ G(k, t) &= 0 , \qquad t < 0 \end{aligned} \tag{13}$$

i.e. $$G(k, t) = \theta(t) \exp\left[-(k^2 + m^2)t\right] \; . \tag{14}$$

The solution to Eq. (9) can then be written

$$\phi(k, t) = \int_0^t \exp\left[-(k^2 + m^2)(t - \tau)\right] \{\eta(k, \tau) - \tfrac{1}{2}\lambda \int [d^D p d^D q/(2\pi)^D] \times \times \phi(p, \tau)\, \phi(q, \tau)\, \delta^D(k - p - q)\} \; , \tag{15}$$

which for $\lambda \ll 1$ can be solved by iteration. This can be illustrated graphically as

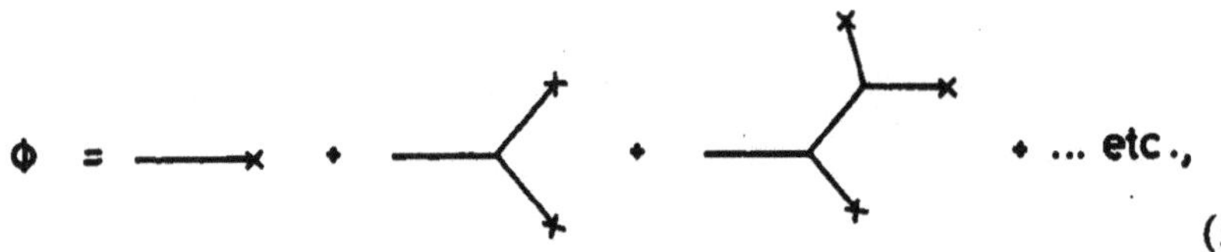

(16)

where we denote the Green function G by a line, and η by a cross. Appropriate integrations are implicit.

We can then study the n-point function $\langle \phi^{[\eta]}(x_1,t_1) \ldots \phi^{[\eta]}(x_n,t_n) \rangle$ by substituting the graphical expansion equation (16). Taking the Gaussian average means that only diagrams with an *even* number of crosses (noise fields) will survive. In these diagrams the crosses are joined in all possible ways. We will call such graphs *stochastic diagrams.*

The easiest example is the lowest-order contribution to the scalar propagator:

$$\begin{aligned} \langle \phi^{[\eta]}(k, t)\, \phi^{[\eta]}(k', t) \rangle &= \int_0^t d\tau \int_0^t d\tau' \exp\left[-(k^2 + m^2)(t - \tau)\right] \\ &\quad \exp\left[-(k'^2 + m^2)(t - \tau')\right] \langle \eta(k, \tau)\, \eta(k', \tau') \rangle \qquad (17) \\ &= (2\pi)^D \delta^D(k + k')(1/(k^2 + m^2))\{1 - \exp[-2t(k^2 + m^2)]\} \; . \end{aligned}$$

We see that indeed when $t \to \infty$ the standard Euclidean field theory expression is obtained.

Generally, to any ordinary Feynman diagram there corresponds a number of stochastic diagrams with the same topology. It can be shown that in the equilibrium limit these stochastic diagrams exactly sum up to the corresponding Feynman diagram [4; R3, R4]. For different methods of proving the equivalence between stochastic quantization and standard quantization methods, see refs. [R2, R15, R17; 5–8].

Consider now the stochastic quantization of gauge fields. It is well known that in the conventional approach we need to introduce a gauge-fixing term in the action, and in general Faddeev–Popov ghost fields as well. Moreover, in the non-Abelian case, it can be shown in a non-perturbative manner that standard gauge-fixing procedures may still imply ambiguities in the path integral. This is known as the Gribov problem.

It is a remarkable fact that within the stochastic quantization scheme, *gauge-invariant* observables can be computed without fixing the gauge. For reasons of simplicity, let us first see how this works in the case of a free Maxwell field. With the standard action

$$S_E[A_\mu] = \tfrac{1}{4}\int d^Dx\, F_{\mu\nu}(x)\, F_{\mu\nu}(x) \tag{18}$$

and $F_{\mu\nu}(x) \equiv \partial_\mu A_\nu(x) - \partial_\nu A_\mu(x)$, the Langevin equation reads

$$(\partial/\partial t)\, A_\mu(x, t) = \partial_\nu F_{\nu\mu}(x, t) + \eta_\mu(x, t)\ . \tag{19}$$

The white-noise correlations are straightforward generalizations of the scalar ones, namely

$$\langle \eta_\mu(x, t)\rangle = 0; \qquad \langle \eta_\mu(x, t)\, \eta_\nu(x', t')\rangle = 2\delta_{\mu\nu}\, \delta^D(x - x')\, \delta(t - t'), \text{ etc.} \tag{20}$$

In momentum space the Langevin equation becomes

$$(\partial/\partial t)\, A_\mu(k,t) = -k^2\, T_{\mu\nu} A_\nu(k,t) + \eta_\mu(k,t) \;, \tag{21}$$

where we have introduced the transverse projection operator $T_{\mu\nu}$:

$$T_{\mu\nu} \equiv \delta_{\mu\nu} - k_\mu k_\nu/k^2 \;; \qquad T_{\mu\varrho} T_{\varrho\nu} = T_{\mu\nu} \;. \tag{22}$$

(We also have the longitudinal projection operator $L_{\mu\nu} = \delta_{\mu\nu} - T_{\mu\nu} = k_\mu k_\nu/k^2$.) Now introducing, as in the scalar case, a Green function $G_{\mu\nu}$ which obeys

$$\begin{aligned} &[(\partial/\partial t)\, \delta_{\mu\nu} + T_{\mu\nu}(k)\, k^2]\, G_{\nu\varrho}(k,t) = \delta_{\mu\varrho}\, \delta(t) \;, \quad t \geq 0 \;, \\ &G_{\nu\varrho}(k,t) = 0, \quad t < 0 \;, \end{aligned} \tag{23}$$

we find

$$\begin{aligned} G_{\mu\nu}(k,t) &= \theta(t)[\exp(e^{-k^2 T\, t})]_{\mu\nu} \\ &= \theta(t)\, [T_{\mu\nu}(k) e^{-k^2 t} + L_{\mu\nu}(k)] \;, \end{aligned} \tag{24}$$

as can easily be checked by using the properties of the projection operator $T_{\mu\nu}$.

The solution for $A_\mu(k,t)$ then reads:

$$\begin{aligned} A_\mu(k,t) &= \int_0^t d\tau\, [e^{-k^2(t-\tau)}\, T_{\mu\nu}(k) + L_{\mu\nu}(k)]\eta_\nu(k,t) \\ &\quad + [T_{\mu\nu}(k) e^{-k^2 t} + L_{\mu\nu}(k)]\, A_\nu(k,0) \;, \end{aligned} \tag{25}$$

with arbitrary initial condition $A_\mu(k,0)$.

Let us at this point mention the important fact that the longitudinal mode of the gauge field A_μ is *not* exponentially damped. This follows from

$$L_{\mu\nu}(k)\,(\partial/\partial t)\,A_\nu(k,t) = L_{\mu\nu}(k)\,\eta_\nu(k,t) \ , \tag{26}$$

which shows that the longitudinal mode undergoes unbounded diffusion. As a consequence [we put $A_\mu(k, t = 0) = 0$ for simplicity],

$$\langle A_\mu(k,t)\,A_\nu(k',t)\rangle = \{[T_{\mu\nu}(k)/k^2]\,(1 - e^{-2k^2 t}) + L_{\mu\nu}\,2t\}\,(2\pi)^D\,\delta(k+k') \ , \tag{27}$$

which shows that we cannot quite identify the large time-limit of the two-point function with the standard propagator! This should not come as a big surprise, though, since we did not fix the gauge, and since the two-point function in any case is not a gauge-invariant object. We also see that (as a consequence of our chosen initial conditions) the transverse part of $\langle A_\mu A_\nu\rangle$ approaches the usual Landau-gauge propagator as $t \to \infty$. However, a linearly divergent longitudinal contribution remains.

Stochastic diagrams will in general contain potentially divergent contributions. Remarkably, these terms all cancel each other when one computes gauge-invariant objects [R1]. This has later been checked by perturbation theory in several cases.

The non-Abelian case only presents a slight complication. The Langevin equation now reads for an SU(N) gauge group:

$$(\partial/\partial t)\,A_\mu^a(x,t) = D_\nu^{ab}\,F_{\nu\mu}^b(k,t) + \eta_\mu^a(k,t) \ , \tag{28}$$

where

$$F_{\mu\nu}^a(x) = \partial_\mu\,A_\nu^a(x) - \partial_\nu\,A_\mu^a(x) - g f^{abc}\,A_\mu^b(x)\,A_\nu^c(x) \tag{29}$$

and

$$D_\mu^{ab} = \partial_\mu\,\delta^{ab} - g f^{abc}\,A_\mu^c(x) \tag{30}$$

is the covariant derivative. The coefficients f^{abc} are the structure constants of the SU(N) algebra.

When calculating stochastic diagrams, contributions from transverse lines sum up to the conventional Landau gauge expectations. But here also, longitudinal lines will give rise to potentially infinite terms. It should be pointed out that *finite* terms can arise, as well, from these longitudinal lines. The non-trivial assertion of Ref. [R1] is that for gauge-invariant quantities not only do the infinite terms drop out, but also the finite parts arrange themselves to give just those contributions which are conventionally associated with Faddeev–Popov ghosts. For one of the explicit verifications of this assertion see, for example, Ref. [R7].

The fact that we do not need to fix the gauge in stochastic quantization is, of course, related to the fact that the Langevin equation contains the invertible operator $\partial_t \delta_{\mu\nu} + T_{\mu\nu}$, in contrast to just $T_{\mu\nu}$, which normally appears in perturbation theory.

Interestingly, an alternative formulation for the stochastic quantization of gauge fields is possible. This is commonly referred to as *stochastic gauge-fixing* [R5, R6, R8; 9, 10]. It offers several advantages, both at the technical and at the conceptual level. For example, within the stochastically gauge-fixed scheme, it can be shown explicitly that all individual stochastic diagrams contributing to some gauge-invariant quantity remain finite as $t \to \infty$. Moreover, a detailed analysis of the stochastic evolution of the gauge field A_μ in 'gauge parameter space' becomes possible; this question has even been studied numerically [11; R36].

The basic idea behind stochastic gauge-fixing (sketched below in the Abelian case) is to transform the field A_μ by a *t-dependent* gauge transformation into a new field B_μ:

$$B_\mu (k, t) = A_\mu (k, t) + ik_\mu \Lambda (k, t) \ . \tag{31}$$

Obviously the Langevin equation for B_μ will look different; one finds

$$(\partial/\partial t)\, B_\mu(k,t) = -k^2\, T_{\mu\nu}(k)\, B_\nu(k,t) + ik_\mu\, (\partial/\partial t)\, \Lambda(k,t) + \eta_\mu(k,t) \;, \quad (32)$$

where on the r.h.s. the previous transverse projection operator is replaced by an invertible matrix. In fact, the simplest choice [R5] for $\Lambda(k, t)$ is one satisfying

$$(\partial/\partial t)\, \Lambda(k,t) = \alpha^{-1} ik_\mu\, B_\mu(k,t) \;, \quad (33)$$

so that

$$(\partial/\partial t)\, B_\mu(k,t) = -k^2 (T_{\mu\nu} + \alpha^{-1} L_{\mu\nu})\, B_\nu(k,t) + \eta_\mu(k,t) \quad (34)$$

and

$$\lim_{t\to\infty} \langle B_\mu(k,t)\, B_\nu(k',t)\rangle = (2\pi)^D \delta^D(k+k')(1/k^2)\,(T_{\mu\nu} + \alpha\, L_{\mu\nu}) \;, \quad (35)$$

which we recognize as the familiar propagator in a covariant α-gauge.

Pursuing a similar procedure for the Yang–Mills case, we arrive at

$$(\partial/\partial t)\, B^a_\mu(k,t) = D^{ab}_\nu\, F^b_{\nu\mu}(k,t) + \alpha^{-1} D^{ab}_\mu\, \partial_\nu B^b_\nu(k,t) + \eta^a_\mu(k,t) \quad (36)$$

which, besides the usual interaction terms [see Eq. (28)], induces a new vertex. It is a three-gluon coupling which, unlike the standard three-gluon interaction, is symmetric under the interchange of *two* lines only.

If one computes a gauge-invariant observable, the contributions from the various stochastic diagrams containing just the ordinary vertices sum up to the corresponding Feynman diagrams in the covariant α-gauge. Similarly, the extra vertices give rise to new stochastic diagrams which nicely sum up to the usual ghost contributions.

Fermion fields can be treated in stochastic quantization as well [R9, R10, R20; 12]. Since non-positive definite operators (e.g. the Dirac operator) appear, some special care is, however, required. One solution to this problem lies in judiciously choosing a 'bosonized' version of the Langevin equation.

We work with a Euclidean Clifford algebra representation $\{\gamma_\mu, \gamma_\nu\} = -2\,\delta_{\mu\nu}$. The free Euclidean action can then be chosen as

$$S[\psi, \bar{\psi}] = -i \int d^D x\, \bar{\psi}(x)(\partial\!\!\!/ + im)\, \psi(x) \ , \tag{37}$$

and we use our freedom to introduce a kernel K(x,y) into the Langevin equation:

$$(\partial/\partial t)\, \psi(x, t) = -\int d^D y\, K(x, y)\, [\delta S/\delta\bar{\psi}(y, t)] + \theta(x, t) \tag{38}$$

(and similarly with its conjugate equation), with stochastic noise fields of a Grassmann nature:

$$\begin{aligned} \langle\theta(x, t)\rangle &= \langle\bar{\theta}(x, t)\rangle = 0 \ , \\ \langle\theta(x, t)\, \bar{\theta}(x', t')\rangle &= 2K(x, x')\, \delta(t - t') \ . \end{aligned} \tag{39}$$

The fact that these noise fields are fermionic means that all anticommutators involving these fields only vanish.

Whereas kernels in scalar cases generally have to be positive definite (since otherwise the white-noise measure becomes ill-defined), this is not the case for fermions. Now the kernel, rather, should be chosen in such a way as to cancel precisely the negative eigenvalues from $\delta S/\delta\bar{\psi}$. The simplest choice for such a kernel is [R10, R20]

$$K(x, y) = (i\partial\!\!\!/_x + m)\, \delta^D(x - y) \ , \tag{40}$$

which leads to

$$(\partial/\partial t)\,\psi(x,t) = (\partial^2 - m^2)\,\psi(x,t) + \theta(x,t)\ , \tag{41}$$

and similarly for $\bar{\psi}(x, t)$. All information about the fermionic nature of the system is now in effect contained in the white-noise distribution.

It is not difficult to stochastically quantize interacting theories with fermions (such as electrodynamics) either. For some explicit calculations—see, for example, Refs. [R10] and [R20]—supersymmetric theories can be treated as well [R20; 13]; this has particularly interesting consequences in the case of stochastic *regularization,* which will be discussed below.

But supersymmetry is related to stochastic quantization also at a completely different and more fundamental level [R12–R14, R16; 13]. This connection with supersymmetry is related to the so-called Nicolai map [14] for supersymmetric field theories: if we consider a supersymmetric theory and integrate out the fermion fields from the path integral, a non-trivial determinant arises. The Nicolai map is a specific transformation of the bosonic field(s) of the action: $T[\phi] = \tilde{\phi}$, whose Jacobian determinant exactly cancels the fermionic determinant, *and* transforms the purely bosonic part of the action into a free, Gaussian, bosonic action.

The Langevin equation for a scalar field coordinate may be viewed as a particular example of a Nicolai map. It is instructive to see this by considering the argument in *reverse.*

To begin, let us consider the partition function for the Langevin dynamics,

$$Z = \int D\eta \exp\left[-\tfrac{1}{4}\int d^Dx\, dt\, \eta(x,t)^2\right] \tag{42}$$

and perform the variable transformation $\eta \rightarrow \phi$, where ϕ is the scalar field of the Langevin equation. Clearly,

$$Z = \int D\phi \det [\delta\eta/\delta\phi] \exp \{-\tfrac{1}{4}\int d^Dx\, dt\, [\dot{\phi} + (\delta S/\delta\psi)]^2\} \ . \tag{43}$$

Next we use the well-known fermionic path-integral representation of a determinant, omit total time derivatives inside the integral sign in the exponent, and thus end up with

$$Z = \int D\phi\, D\psi\, D\bar{\psi} \exp (-\tfrac{1}{4}\int d^Dx\, dt\, S_{FP}) \ , \tag{44}$$

where the 'effective action' S_{FP} reads

$$S_{FP} = \int d^Dx\, dt\, \{\tfrac{1}{2}\dot{\phi}^2 + \tfrac{1}{8}(\delta S/\delta\phi)^2 - \bar{\psi}[\partial/\partial t + \tfrac{1}{2}(\delta S/\delta\phi)]\psi\} \ . \tag{45}$$

The action S_{FP} of Eq. (45) is indeed invariant under a supersymmetry transformation! This is simply the inverse of the Nicolai map, which would start with Eqs. (44) and (45), and end up with Eq. (42)—the free bosonic action. (In general, for an arbitrary supersymmetric action the Nicolai map will not be of the simple local form of a Langevin-type equation).

Since the effective action (45) is supersymmetric, it can be cast into a form which is manifestly so by means of superfields. This can be used to give a rather elegant equivalence proof to standard path-integral quantization [R15, R17; 8]. The dimensional reduction (from D + 1 dimensions to the D-dimensional Euclidean field theory) of the fictitious time t in the equilibrium limit can be shown to be a manifestation of a more general phenomenon, investigated first by Parisi and Sourlas [R12] in a related, but different, context.

The presence of the extra dimension of the fictitious time provides an opportunity for an interesting generalization of stochastic quantization. This is stochastic *regularization*, a new regularization scheme for quantum field theory. By using the freedom to modify the

Langevin dynamics in the extra time dimension [R20; 15], stochastic regularization can be made to manifestly preserve all symmetries of the underlying field theory, although its application to non-Abelian gauge theories is a problematic issue. Another stochastic regularization scheme manages to preserve internal symmetries even while modifying the Langevin dynamics in the Euclidean space–time dimensions [R23]. This scheme can also be used in connection with non-Abelian gauge theories.

The stochastic regularization which modifies the t-evolution, works by changing the stochastic process $\eta(x, t)$ into a specific *non-Markovian* process. This is accomplished by changing, for example, the noise–noise correlations to

$$\langle \eta(x, t)\, \eta(x', t') \rangle = 2\, \delta^D(x - x')\, K(t - t') \ , \tag{46}$$

where K is a 'memory kernel', which can represent a smeared-out δ-function. As the 'dynamics' in the Euclidean space–time dimensions remains unchanged, the noise–noise correlations remain invariant under all the t-independent symmetries which left the original Langevin system invariant. Explicitly, the following series of kernels have been proposed [R20]:

$$K_\Lambda^m(t) = (1/2m!)\, \Lambda^2 (\Lambda^2 |t|)^m \exp(-\Lambda^2 |t|) \tag{47}$$

or [15]

$$K_\sigma(t) = \tfrac{1}{2} \sigma |t|^{\sigma - 1} \ , \qquad \sigma > 0 \ . \tag{48}$$

The unregularized theory is obviously obtained in the limits $\Lambda \to \infty$, $\sigma \to 0$, respectively.

As a consequence of the more complicated noise–noise correlations, the fictitious time integrations in the stochastic diagrams become much more involved, and the final result can in fact be shown to be regularized in the ultraviolet for appropriately chosen values of Λ, m, or σ in Eqs. (47) and (48)—all depending on the degrees of divergence in the theories under study.

To give the easiest example, consider [m = 0 in Eq. (47)]

$$K(t) = \tfrac{1}{2}\Lambda^2 \exp(-\Lambda^2|t|) \ , \tag{49}$$

which leads (for $t \geq t'$) to the following two-point function:

$$\langle\phi(k,t)\,\phi(k',t')\rangle = (2\pi)^D\delta^D(k+k')[\Lambda^2/(k^4-\Lambda^4)]\{-(\Lambda^2/k^2)\exp[-k^2(t-t')] + \exp[-\Lambda^2(t-t')]\} + \ldots \ . \tag{50}$$

In the $t \to \infty$ limit this becomes

$$\lim_{t\to\infty}\langle\phi(k,t)\,\phi(k',t)\rangle = (2\pi)^D\delta^D(k+k')(1/k^2)[\Lambda^2/(\Lambda^2+k^2)] \ , \tag{51}$$

which clearly shows an improved ultraviolet behaviour. Without going into further details, let us remark that in the interacting case a Fokker–Planck analysis seems rather complicated. Generally, no equilibrium distribution can be given, and a study of, for example, Ward Identities still remains an open issue.

Next we discuss the covariant derivative regularization scheme [R23] which, in contrast to the previous scheme, is purely Markovian.

The basic idea is to generalize, in a covariant way, the noise structure of the gauge-field Langevin equation to

$$\dot{A}^{a}_{\mu} = -(\delta S/\delta A^{a}_{\mu}) + \int d^4y\, K^{ab}(x,y)\,\eta^{b}_{\mu}(y,t) \ , \tag{52}$$

where K is a function of the covariant Laplacian Δ

$$\Delta^{ab}(x,y) = \int d^4z\, D^{ac}_{\mu}(x,z)\, D^{cb}_{\mu}(z,y) \ . \tag{53}$$

with

$$D_\mu^{ab}(x, y) = D_\mu^{ab}(x)\delta^4(x - y) \ . \tag{54}$$

There is a large class of possible regulator functions, such as

$$K_n^{ab}(x, y) = \{[1 - (\Delta/\Lambda^2)]^{-n}\}^{ab}(x, y) \tag{55}$$

which approaches unity when $\Lambda^2 \to \infty$. We note that gauge covariance of the Langevin equation is guaranteed by general construction; this scheme is convenient for a formulation in terms of *regularized* Schwinger–Dyson equations.

The Langevin equation is ideally suited for *numerical* simulations of (quantum) field theories, at least to the extent in which ergodicity is obeyed [R33]. To solve the Langevin equation numerically, one first discretizes the 'time'. The simplest discretization reads, for a scalar theory,

$$\phi(x, t + \delta t) - \phi(x, t) = -[\delta S/d\phi(x, t)]\,\delta t + \sqrt{\delta t}\,\eta(x, t) \ , \tag{56}$$

where the $\sqrt{\delta t}$ in front of the noise field has been inserted for convenience. It makes the noise–noise correlation of the simple form $\langle\eta(x, t)\eta(x', t')\rangle = 2\delta_{xx'}\delta_{tt'}$, if we also discretize ordinary Euclidean space–time (in order to regulate short-distance divergences from the field theory). The specific discretization (56) is, of course, by no means unique, but it is straightforward to verify that, indeed, as $\delta t \to 0$ one recovers the continuous-time Langevin equation.

Keeping δt *finite* means that the approach to equilibrium can be different (even jeopardized), and that the equilibrium distribution *itself* in general will be different [16, 17; R38]. Equivalence proofs to standard quantization techniques then take on a rather special meaning, since the resulting equilibrium correlation functions will be δt-dependent. Many other features of stochastic quantization are disturbed as well. For instance, it is no longer true that kernels do not modify the equilibrium distribution either. On the other hand, all these consequences of the discretization procedure may be turned to one's advantage. For example, kernels can be introduced which optimize the convergence towards equilibrium [18]. Also, clearly if δt is chosen small enough, the effects of discretization should be irrelevant. The only difficulty lies in *a priori* quantifying what 'small enough' δt means. This can only be done on the basis of observed dynamics of the theory in question.

To illustrate these problems, consider the simplest possible case—a free scalar theory with discretized action $S[\phi] = \frac{1}{2}\Sigma_{x,y}\, \phi(x)\, M(x,y)\, \phi(y)$, with some 'operator' $M(x,y)$ [16]. The simplest Langevin equation (56) then reads

$$\phi(x, t + \delta t) - \phi(x, t) = - \sum_y M(x, y)\, \phi(y)\, \delta t + \sqrt{\delta t}\, \eta(x, t) \ , \tag{57}$$

which can be solved exactly. However, it is amusing to introduce first a kernel $K(x, y) = M^{-1}(x, y)$ in order to make the Langevin equation manifestly trivial. The solution for, for example, the two-point function is then easily found to be

$$\begin{aligned} \langle \phi(x, t_n)\, \phi(x', t_n) \rangle &= 2\delta t\, M^{-1}(x, x') \sum_{j=0}^{n-1} (1 - \delta t)^{n-j-1} \sum_{i=0}^{n-1} (1 - \delta t)^{n-i-1}\, \delta_{ij} \\ &= 2\delta t M^{-1}(x, x')(1 - \delta t)^{-2(n-1)} \sum_{j=0}^{n-1} (1 - \delta t)^{-2j} \\ &= M^{-1}(x, x')\, (1/(1 - \tfrac{1}{2}\delta t))\, [1 - (1 - \delta t)^{2n}] \ , \end{aligned} \tag{58}$$

so that even in the $n \to \infty$ (infinite time) limit, we do *not* recover the 'exact' answer $M^{-1}(x, x')$. It is also clear that the approach to equilibrium is no longer completely trivial; clearly δt should be chosen 'small' (in this case at least $\delta t < 2$) in order to reach any equilibrium correlation function at all.Corresponding problems are found in more realistic (interacting) cases, but since they are similar in principle, we shall not enter into a detailed description here.

Particularly interesting problems occur in the case of gauge theories, which of course are related to the internal invariances associated with these theories [17; R37, R38]. The discussion looks a little different from our earlier treatment of continuum gauge theories, since on a discrete Euclidean space-time it is more convenient to work with elements of the gauge *group,* rather than elements of the gauge *algebra.* Let us restrict ourselves here to unitary groups. Clearly one problem that must be solved is how to set up a Langevin equation for group elements [19, 20; R35], and in particular, how to ensure that the stochastic diffusion remains on the group manifold, i.e. for instance preserves the unitarity constraint. One way of doing this is to write the discrete Langevin equation as

$$U(t + \delta t) = U(t) \exp [iW[U(t), \eta(t)] \;, \qquad (59)$$

where $W[U(t), \eta(t)]$ is Hermitian:

$$W[U(t), \eta(t)] = -i\delta t \,\nabla S[U(t)] + \sqrt{\delta t}\, \eta(t) \;, \qquad (60)$$

where $S[U(t)]$ is the gauge group action and ∇ is a Lie derivative. One can easily show that Eqs. (59) and (60) preserve the unitarity constraint $U(t)U^{+}(t) = 1$. In this way standard lattice gauge theories can be treated by Langevin simulations.

Fermions cannot easily be treated by direct numerical methods, but once the theories have been put in a form where the fermion fields occur only bilinearly (e.g. by the introduction of auxiliary fields), they can be explicitly integrated out of the path integral. This leaves an effective action which often can be rewritten in effectively bosonic form. Such effective actions can then be treated by the techniques mentioned above [17, 21, 22; R38].

Concerning numerical applications of the Langevin equation, one interesting possibility which has been investigated in some detail is a simulation of actions which are *complex* [23, 24]. It should be emphasized that such actions may arise in several physical problems, where all computed observables are manifestly *real*. Complex actions cannot be given a probabilistic interpretation, and hence cannot be treated straightforwardly by, for example, standard Monte Carlo techniques.

Some simple examples of complex actions can be treated analytically. Consider, for example, the 'zero-dimensional' field theory defined by [R28]

$$S(x) = \tfrac{1}{2}\sigma x^2 + \tfrac{1}{4}\lambda x^4 \ . \tag{61}$$

Assume that λ is real (and positive), but that σ is allowed to be complex. Naïvely one might imagine the x^4 term to (almost) always enforce convergence for sufficiently long times, irrespective of the magnitude or complex phase of σ. This, however, is too simplified. When $\sigma \in \mathbb{C}$, the general solution to the Langevin equation will also diffuse into the complex plane and a more detailed stability analysis is required. Nevertheless, this is an example of a rather well-behaved diffusion problem [R28, R29; 25]. Serious problems can, however, arise in cases of complex actions which can attain one or more zeros [26]. In this case ergodicity can effectively be lost for discretized Langevin equations, since the stochastic path can be trapped for infinitely long times inside sub-regions of the available phase space. This obviously makes numerical simulations of such complex actions, based on discretized Langevin equations, highly problematical.

In our presentation here of the most basic aspects of stochastic quantization, we have intentionally tried to follow the rule of not providing any details about any alternative formulations [27–33] of the Parisi–Wu scheme. These can be found in the literature listed below and, to some extent, among the reprinted papers.

We are aware of the fact that stochastic quantization represents an active research field and that, therefore, in addition to our discussion and reprint selection, new results may already have been obtained.

3. A GUIDE TO THE REPRINTED PAPERS

The first selected paper is the one of Parisi and Wu [R1] in which the concept of stochastic quantization is first introduced. The paper discusses all the basic features of this new scheme for the case of a self-interacting scalar field. Subsequently, gauge theories are treated, and it is pointed out that no gauge fixing is needed. This paper discusses several of the most important ideas behind the stochastic quantization method, some of which have later found their explicit verification in terms of, e.g. perturbation theory.

Stochastic quantization becomes particularly transparent in a perturbative analysis. Apart from working out several examples in detail, equivalence proofs relating stochastic quantization to the more standard methods are established. The perturbative analysis can either be carried out in a functional formulation [R2] or in an iterative scheme based on the Langevin equation [R3, R4]. The concept of stochastic diagrams is studied in detail.

The idea of stochastic gauge fixing is first introduced in [R5]. It is studied in further detail in [R6], where a formal connection to the standard path-integral measure (including ghosts) of Yang–Mills theories is established. In [R8] the $t \to \infty$ limit of stochastically gauge-fixed theores is used to set up a new renormalizable and ghost-free perturbation theory for non-Abelian gauge fields.

In [R7] it is verified, in a specific example of perturbation theory, that indeed ghost effects are reproduced in stochastic quantization of Yang–Mills theories. The calculations are performed within the original scheme of [R1].

Quantization of fermions requires some care in the stochastic formulation. Two such approaches are presented. In [R9], convergence of the stochastic process to equilibrium is assured by a non-vanishing fermion mass term. In [R10] (as well as in [R20]), a non-trivial kernel operator effectively 'bosonizes' the Langevin equation, leaving (in the free case) all the fermionic nature of the theory contained in the correlations among noise fields.

Stochastic quantization of gravity [R11] is based on an obvious generalization of the stochastic formulation of gauge theories. Moreover, by exploiting the principle of general covariance with respect to field redefinitions, a one parameter family of Langevin equations is shown to emerge; a preferred value of this parameter is pointed out.

The intrinsic connection of stochastic quantization to supersymmetry has been widely discussed in the literature. Its basic aspects are noted in [R12] within the context of critical phenomena associated with spin systems in random external fields; they are discussed in further depth in [R13]. Immediate consequences of the underlying supersymmetric structure of stochastic quantization are presented in [R14] and [R16]. New equivalence proofs of stochastic quantization to the standard quantization procedure can be found in [R15] and [R17].

The Parisi–Wu scheme of stochastic quantization can be formulated in somewhat more general terms by relying on a canonical description [R18, R19]. In this case a Gibbs distribution emerges in the equilibrium.

As mentioned in Chapter 2, it is one of the appealing features of stochastic quantization to imply new regularization schemes [R20, R23], which manifestly preserve all symmetries of the unregularized action. We note various applications, e.g. for scalar electrodynamics [R21], for quantum chromodynamics [R23], or for the evaluation of critical exponents [R22].

From a mathematical point of view, the regularization and renormalization program for the stochastic quantization of field theories represents a delicate issue. In [R24] a rigorous non-perturbative analysis of a self-interacting scalar field in two dimensions is carried out in detail.

One of the surprising applications of stochastic quantization lies in the field of large-N physics. Specifically it offers an extremely simple derivation [R25, R26] of the 'quenched momentum prescription'. Furthermore, if the quenching prescription is extended to incorporate the fictitious time direction as well, then a novel interpretation of a 'master field' emerges [R25].

Another interest in stochastic quantization arises when complex actions are considered. Analytical as well as numerical studies are carried out in [R27] to [R29]. In [R30] the complex Langevin equation is successfully tested on a more physical problem such as SU(3) lattice gauge theory in the presence of a chemical potential.

Stochastic quantization directly in Minkowski space is suggested in [R31] and [R32] by appropriately generalizing the original Parisi–Wu approach; also in this case the stochastic process becomes complex.

We have already mentioned several numerical applications of stochastic quantization. Basic features of such a program can be found in [R33]. Various aspects of numerical investigations of gauge theories are represented by [R34] to [R38].

REFERENCES

[1] N.G. van Kampen, Stochastic processes in physics and chemistry (North-Holland Publ. Co., Amsterdam, 1981).
C.W. Gardiner, Handbook of stochastic methods (Springer Verlag, Heidelberg, 1983).
L. Arnold, Stochastic differential equations (Wiley & Sons, Inc., New York, 1974).

[2] P. Langevin, C.R. Acad. Sci. (Paris) **146** (1908) 530.

[3] A.D. Fokker, Ann. Phys. **43** (1914) 810.
M. Planck, Sitzber. Pr. Akad. Wiss. p. 324 (1917).

[4] C. De Dominicis, Lett. Nuovo Cimento **12** (1975) 567.

[5] H. Nakazato, M. Namiki, I. Ohba and K. Okano, Prog. Theor. Phys. **70** (1983) 298.
Y. Nakano, Prog. Theor. Phys. **69** (1983) 361.

[6] G. Aldazabal, E. Dagotto, A. Gonzáles-Arroyo and N. Parga, Phys. Lett. **125B** (1983) 305.
G. Marchesini, Nucl. Phys. **B239** (1984) 135.

[7] S. Araki and Y. Nakano, Prog. Theor. Phys. **71** (1984) 1074.

[8] E. Gozzi, Phys. Lett. **143B** (1984) 183.

[9] R.F. Alvarez-Estrada and A. Muñoz Sudupe, Phys. Lett. **164B** (1985) 102 and **166B** (1986) 186.

[10] M.S. Chan and M.B. Halpern, Phys. Rev. **D33** (1986) 540.

[11] E. Seiler, I.O. Stamatescu and D. Zwanziger, Nucl. Phys. **B239** (1984) 177.
E. Seiler, Acta Phys. Austriaca, Suppl. **26** (1984) 259.

[12] K. Ishikawa, Nucl. Phys. **B241** (1984) 589.

[13] E. Gozzi, Phys. Rev. **D30** (1984) 1218.

[14] H. Nicolai, Phys. Lett. **89B** (1980) 341; Nucl. Phys. **B176** (1980) 419.

[15] J. Alfaro, Nucl. Phys. **B253** (1985) 464.

[16] O.C. Martin, S.W. Otto and J.W. Flower, Nucl. Phys. **B264** (1986) 89.

[17] K.G. Wilson, Phys. Rev. **D32** (1985) 2736.

[18] G. Parisi, *in* Progress in gauge field theory, eds. G. 't Hooft et al. (Plenum Publ. Corp., New York, 1984).

[19] I.T. Drummond, S. Duane and R.R. Horgan, Nucl. Phys. **B220** [FS8] (1983) 119.

[20] M.B. Halpern, Nucl. Phys. **B228** (1983) 173.
A. Guha and S.-C. Lee, Phys. Rev. **D27** (1983) 2412.
J. Alfaro and B. Sakita, *in* Topical Symp. on High-Energy Physics, eds. T. Eguchi and Y. Yamaguchi (World Scientific, Singapore, 1983).
G.G. Batrouni, H. Kawai and P. Rossi, Cornell Univ. preprint CLNS–85/712 (1985).

[21] G.G. Batrouni, Phys. Rev. **D33** (1986) 1815.
A. Kronfeld, Phys. Lett. **172B** (1986) 93.

[22] M. Fukugita, Y. Oyanagi and A. Ukawa, Tsukuba Univ. preprint UTHEP (1985) 152.
M. Fukugita and A. Ukawa, Kyoto Univ. preprint RIFP 642 (1986).

[23] J.R. Klauder, *in* Recent developments in high-energy physics, eds. H. Mitter and C.B. Lang (Springer, 1983).

[24] J.R. Klauder, Phys. Rev. **A29** (1984) 2036.

[25] J. Ambjorn and S.-K. Yang, Nucl. Phys. **B275** [FS17] (1986) 18.

[26] J. Flower, S.W. Otto and S. Callahan, Phys. Rev. **D34** (1986) 598.

J. Ambjorn, M. Flensburg and C. Peterson, Nucl. Phys. **B275** [FS17] (1986) 375.

[27] D.J. Callaway and A. Rahman, Phys. Rev. **D28** (1983) 1506.
M. Creutz, Phys. Rev. Lett. **50** (1983) 1411.
A. Strominger, Ann. Phys. (NY) **146** (1983) 419.
A. Iwazaki, Phys. Lett. **141B** (1984) 342.
D.J. Callaway, Phys. Lett. **145B** (1984) 363.

[28] V. de Alfaro, S. Fubini and G. Furlan, Phys. Lett. **105B** (1981) 462; Nuovo Cimento **74A** (1983) 365.
E. Gozzi, Phys. Lett. **130B** (1983) 183.

[29] S. Chaturvedi, A.K. Kapoor and V. Srinivasan, Phys. Lett. **157B** (1985) 400.

[30] B. McClain, A. Niemi and C. Taylor, Ann. Phys. (NY) **140** (1982) 232.
A. Niemi and L. Wijewardhana, Ann. Phys. (NY) **140** (1982) 247.
B. McClain, A. Niemi, C. Taylor and L. Wijewardhana, Phys. Rev. Lett. **49** (1982) 252; Nucl. Phys. **B217** (1983) 430.

[31] M. Namiki and Y. Yamanaka, Prog. Theor. Phys. **69** (1983) 1764.

[32] E. Nelson, Phys. Rev. **150** (1966) 1079.

[33] F. Guerra and P. Ruggiero, Phys. Rev. Lett. **31** (1973) 1022.
F. Guerra and M.I. Loffredo, Lett. Nuovo Cimento **27** (1980) 41.
S.C. Lim, Lett. J. Math. Phys. **7** (1983) 469.

REPRINTED PAPERS

The Classic

Vol. XXIV No. 4 SCIENTIA SINICA April 1981

PERTURBATION THEORY WITHOUT GAUGE FIXING

G. Parisi

(*Institute of Theoretical Physics, Academia Sinica; Laboratori Nazionale, INFN, Frascati, Italy*)

and Wu Yongshi (吴泳时)

(*Institute of Theoretical Physics, Academia Sinica*)

Received July 7, 1980.

Abstract

We propose to formulate the perturbative expansion for field theory starting from the Langevin equation which describes the approach to equilibrium. We show that this formulation can be applied to gauge theories to compute gauge invariant quantities without fixing the gauge. A very simple example is worked out in detail. We also discuss the speed of approaching to equilibrium of the solution of the Langevin equation in the framework of perturbation theory.

I. Introduction

In the standard perturbative approach to gauge theories, it is necessary to introduce a gauge fixing term. In the Abelian case this introduction does not give serious problems (if one uses a linear gauge condition), while in non-abelian gauge theories it is necessary to introduce the Faddeev-Popov ghost[1]. However, this procedure breaks down in a nonperturbative approach if one uses a covariant gauge: the gauge condition does not fix uniquely the gauge for large gauge potentials and this phenomenon goes under the name of Gribov ambiguity[2].

In lattice gauge theories the situation is very different: no gauge fixing is needed and owing to the compactness of the gauge group, all quantities which are not gauge invariant are zero[3].

The aim of our work is to construct a new perturbation theory for continuum gauge theories without introducing a gauge fixing. The quantization method we propose has the advantage of working also in the non-perturbative region independently of the Gribov ambiguity. Moreover, we think that it is better to respect as far as possible the symmetry of the problem at all stages in the computation. In our case we would still find that quantities which transform homogeneously under a gauge transformation (e.g. a charged field $\phi:\delta\phi = i\alpha\phi$) must have vanishing expectation values. Therefore, quantities such as the propagator for a charged field become zero when the group charge e is different from zero. This discontinuity in the behaviour of non-gauge-invariant quantities as a function of e will be reflected in the presence of divergences in their perturbative expansion. However, the perturbative expansion for gauge invariant quantities is obviously free of these divergences.

Our approach is based on the Langevin equation of non-equilibrium statistical mechanics[4]. As it will be clear later, the Langevin equation (a stochastic evolution equation) is strongly connected to the Monte Carlo procedure which is used to do computer simulation in gauge theories[5]1). Now it is interesting to study in the framework of perturbation theory the speed of approaching to equilibrium of these random constructive procedures, as we shall do in this paper.

We first recall the general properties of the Langevin equation in Sec. II, and write down the diagrammatic rules in Sec. III. Then, in Sec. IV we show how to obtain the correct results in a case where standard perturbation theory cannot be used. In Sec. V we write the Langevin equation and the diagrammatical rules for gauge theories, and present a simple computation to show how the correct result is obtained. The arguments for the correctness of our new perturbative expansion at all orders are given in Sec. VI, and finally Sec. VII is devoted to the presentation of our conclusions.

II. Langevin Equation

Let us consider, for definiteness, an Euclidean scalar field theory. Usually we want to compute the correlation function, e.g., $\langle\phi(x)\phi(y)\rangle$, where the bracket denotes the statistical expectation value at a temperature T:

$$\langle\phi(x)\phi(y)\rangle = \frac{\int d[\phi]\phi(x)\phi(y)\exp\{-\beta V(\phi)\}}{\int d[\phi]\exp\{-\beta V(\phi)\}}, \tag{2.1}$$

where $\beta = 1/kT$.

It may be convenient to generalize the problem. We can consider that the field ϕ is also a function of a time $t(0<t<\infty)$, and it is coupled with a heat reservoir at temperature T. It will reach the equilibrium distribution for large time t. If we know the evolution equation of the field $\phi(x,t)$, we can use it to compute the large time behaviour of $\phi(x,t)$, and consequently the equilibrium distribution and the correlation function Eq. (2.1). In other words, we have the freedom to assign to the field $\phi(x,t)$ any time evolution equation so as to reach equilibrium for large times.

The simplest equation we can write is the so-called Langevin equation[4],

$$\frac{\partial\phi(x,t)}{\partial t} = -\frac{\delta V}{\delta\phi(x,t)} + \eta(x,t), \tag{2.2}$$

where $\eta(x,t)$ are Gaussian random variables:

$$\langle\eta(x,t)\rangle = 0,$$

1) In certain cases the Monte Carlo procedure can be considered as a time-discretized Langevin equation, where the discretization is done in such a way as to preserve the same asymptotic limit for large times.

$$\langle\eta(x,t)\eta(y,t')\rangle = 2\beta^{-1}\delta(x-y)\delta(t-t')$$

$$= \frac{\int d[\eta]\eta(x,t)\eta(y,t')\exp\left\{-\frac{\beta}{2}\int d^D x dt \eta^2(x,t)\right\}}{\int d[\eta]\exp\left\{-\frac{\beta}{2}\int d^D x dt \eta^2(x,t)\right\}}, \tag{2.3}$$

$$\langle\eta_1\eta_2\eta_3\eta_4\rangle_c = \langle\eta_1\eta_2\eta_3\eta_4\rangle - \langle\eta_1\eta_2\rangle\langle\eta_3\eta_4\rangle - \langle\eta_1\eta_3\rangle\langle\eta_2\eta_4\rangle - \langle\eta_1\eta_4\rangle\langle\eta_2\eta_3\rangle = 0, \tag{2.4}$$

where $\eta_i = \eta(x,t_i)(i=1,2,3,4)$. If we impose a boundary condition at $t=0$ (in the following we will assume $\phi(x,0)=0$ and $V(0)=0$), the solution of Eq. (2.2) is uniquely given in terms of η; let us call it $\phi^\eta(x,t)$. The stochastic correlation function are defined by $\langle\phi^\eta(x,t)\phi^\eta(x',t')\rangle$, where the bracket indicates the mean value over η. (In the rest of the paper we shall write $\phi(x,t)$ at the place of $\phi^\eta(x,t)$.)[1] It is a well-known theorem of statistical mechanics[4] that when t goes to infinity,

$$\langle\phi(x,t)\phi(x',t)\rangle \to \langle\phi(x)\phi(x')\rangle, \tag{2.5}$$

i.e. the equal time non-equilibrium correlation functions tend to the equilibrium ones for large times.

We note that the probability distribution $P(\phi,t)$ satisfies the Fokker-Planck equation, as is proved in [4], (we set $\beta=1$)

$$\frac{d}{dt}P(\phi,t) = \frac{\delta^2 P}{\delta\phi(x)^2} + \frac{\delta}{\delta\phi(x)}\left(P\frac{\delta V}{\delta\phi(x)}\right)$$

$$= -2\exp\left(-\frac{1}{2}V\right)\hat{H}\left[P(\phi,t)\exp\left(\frac{1}{2}V\right)\right], \tag{2.6}$$

where

$$\hat{H} = -\frac{1}{2}\frac{\delta^2}{\delta\phi(x)^2} + U, \quad U = \frac{1}{8}\left(\frac{\delta V}{\delta\phi}\right)^2 - \frac{1}{4}\frac{\delta^2 V}{\delta\phi^2}. \tag{2.7}$$

Eq. (2.5) can be proved in many ways. In the following we present a simple proof in the case in which ϕ is defined only on one point, i.e. it is a number q, not a function. In this case we have

$$\hat{H} = \frac{1}{2}P^2 + U(q). \tag{2.8}$$

If $V(q)$ increases fast at infinity, $\hat{H}$ has a discrete spectrum. Let us denote by $\psi_n(q)$ and λ_n its eigenvectors and eigenvalues:

$$\hat{H}\psi_n(q) = \lambda_n\psi_n(q), \quad (\lambda_{i+1} > \lambda_i). \tag{2.9}$$

We can write the correlation functions at equal times as

$$\langle q(t)^K\rangle = \sum_{n=0}^{\infty} c_n \exp(-2\lambda_n t)\int dq q^K \psi_n(q)\psi_0(q). \tag{2.10}$$

1) We write the Langevin equation using a simple, but not mathematically correct notation. The rigorous notation uses the Ito differential calculus[4].

It is very easy to verify that $\exp\left(-\frac{1}{2}V(q)\right)$ is an eigenvector of $\hat{H}$ with eigenvalue zero; $\exp\left(-\frac{1}{2}V(q)\right)$ is also the ground state of $\hat{H}$, because it is a function without zeros. Therefore, for large times we have

$$\langle q(t)^K\rangle = \frac{\int dq q^K \exp(-V(q))}{\int dq \exp(-V(q))} + O[\exp(-2\lambda_1 t)], \tag{2.11}$$

i.e. for large times we reach the equilibrium distribution $\exp(-V(q))$, and the corrections are exponentially small. It will be useful to note that the exponent λ_1 can be written as the eigenvalue of a Schrödinger operator. This implies that λ_1 must be a continuous function of V.

The same argument can be done for the general case. The Hamiltonian now is

$$H = \int d^D x \left\{\frac{1}{2}\pi(x)^2 + U(\phi(x))\right\}, \quad [\pi(x), \phi(y)] = -i\delta(x-y), \tag{2.12}$$

and

$$\psi_0[\phi] = \exp\left\{-\frac{1}{2}\int d^D x V(\phi(x))\right\} \tag{2.13}$$

is the solution of the following Schrödinger functional equation with $\lambda_0 = 0$:

$$\left\{-\frac{1}{2}\frac{\delta^2}{\delta\phi(x)^2} + U(\phi(x))\right\}\psi_0[\phi] = \lambda_0\psi_0[\phi]. \tag{2.14}$$

It is possible to consider a more general Langevin equation:

$$\dot{\phi}(x,t) = -\int d^D y M(x,y)\frac{\delta V}{\delta\phi(y)} + \eta(x,t). \tag{2.15}$$

If

$$\langle\eta(x,t)\eta(y,t')\rangle = 2M(x,y)\delta(t-t') \tag{2.16}$$

and M is a positive matrix, one finds the same conclusions of the previous case.

These results imply that the functional integral formulation of field theories can be replaced by a parabolic nonlinear stochastic equation[1]. The Langevin equation can be used both in perturbative theory or in a nonperturbative framework. In this paper we will show how to use the Langevin equation for constructing a perturbative expansion in cases where the standard perturbative approach must be modified.

III. Diagrams

Let us study the diagrammatic approach to the solution of the Langevin equation. We will consider for simplicity the case in which,

$$V(\phi) = \int d^D x\left\{\frac{1}{2}(\partial_\mu\phi)^2 + \frac{1}{2}m^2\phi^2 + \frac{1}{3}g\phi^3\right\}. \tag{3.1}$$

---

1) In Ref. [6] it is shown how to use an elliptic nonlinear stochastic equation instead of a parabolic one.

The Langevin equation with $M(x,y) = \delta(x-y)$ is

$$\left.\begin{aligned} \dot{\phi} &= \partial^2\phi - m^2\phi + g\phi^2 + \eta, \\ \langle \eta(x,t)\eta(x',t')\rangle &= 2\delta(x-x')\delta(t-t'). \end{aligned}\right\} \tag{3.2}$$

Let us first study the case $g=0$. The solution of Eq. (3.2) can be written as

$$\phi(x,t) = \int_0^t d\tau \int d^D y G(x-y, t-\tau)\eta(y,\tau), \tag{3.3}$$

where $G(x,t)$ is the retarded Green function which satisfies

$$\left.\begin{aligned} &\frac{\partial}{\partial t} G(x,t) = (\partial^2 - m^2)G(x,t) + \delta(x)\delta(t), \\ &G(x,t) = 0. \quad (\text{for } t < 0) \end{aligned}\right\} \tag{3.4}$$

It is obvious that

$$G(x,t) = \int \frac{d^D k}{(2\pi)^D} \exp\{-t(k^2+m^2) + ik\cdot x\}\theta(t). \tag{3.5}$$

Eq. (3.3) implies that $\phi(x,t)$ is a Gaussian stochastic variable, being the linear combination of Gaussian variables. Now the correlation function

$$\langle \phi(x,t)\phi(x',t)\rangle = D(x-x';t,t')$$

can be easily computed. We find

$$D(x-x';t,t') = 2\int_0^\infty d\tau \int d^D y G(x-y,t-\tau)G(x'-y,t'-\tau). \tag{3.6}$$

In the momentum space we have, for $t' < t$,

$$D(k;t,t') = \frac{\exp[-(k^2+m^2)(t-t')]}{k^2+m^2}\{1 - \exp[-2(k^2+m^2)t']\}. \tag{3.7}$$

When $t' \to \infty$, the second term can be neglected. At equal times ($t = t' \to \infty$) we obtain the equilibrium result $1/(k^2+m^2)$.

For $g \neq 0$, we can write

$$\phi(x,t) = \int_0^t d\tau \int d^D y G(x-y,t-\tau)[\eta(y,\tau) + g\phi^2(y,\tau)]. \tag{3.8}$$

If we denote G by a line, η by a cross and ϕ by a point, assign a factor g to each three-line vertex and integrate over the times and coordinates of all crosses and vertices, then we obtain by iterating Eq. (3.8)

$\phi =$ 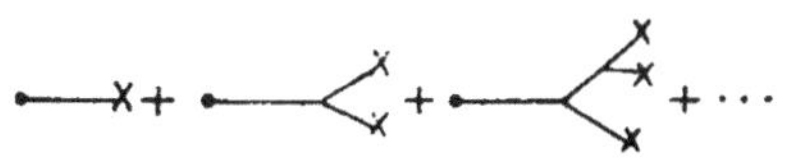 (3.9)

The mean over η is zero, if two crosses do not coincide. So we get up to the order

g^2 (if we neglect tadpole-like diagrams),

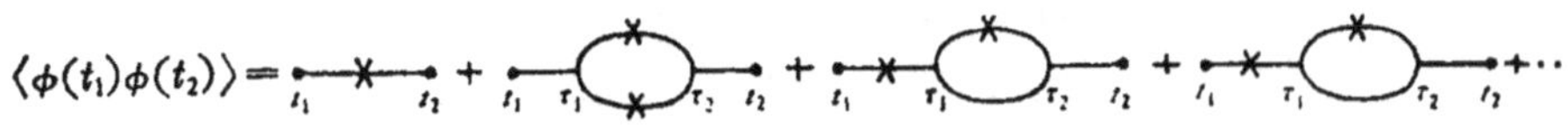

(a) (b) (c) (d) (3.10)

The diagram (a) gives the free propagator Eq. (3.6) or (3.7), and the contributions of the diagrams (b), (c) and (d) are respectively:

$$b = g^2 \int \frac{d^D k_1}{(2\pi)^D} \int_0^{t_1} d\tau_1 \int_0^{t_2} d\tau_2 G(k; t_1 - \tau_1) G(k; t_2 - \tau_2)$$

$$\times D(k_1; \tau_1, \tau_2) D(k - k_1; \tau_1, \tau_2), \tag{3.11}$$

$$c + d = g^2 \int \frac{d^D k_1}{(2\pi)^D} \int_0^{t_1} d\tau_1 \int_0^{t_2} d\tau_2 \{ D(k - k_1; \tau_1, \tau_2)$$

$$\times [D(k; t_1, \tau_1) G(k_1; \tau_2 - \tau_1) G(k; t_2 - \tau_2)$$

$$+ D(k; t_2, \tau_2) G(k_1; \tau_1 - \tau_2) G(k; t_1 - \tau_1)]$$

$$+ \text{terms obtained by } k_1 \rightleftarrows k - k_1 \}. \tag{3.12}$$

A simple computation shows that we recover the correct equilibrium result at equal large times ($t_1 = t_2 \to \infty$):

$$b = g^2 \frac{d^D k_1}{(2\pi)^D} \frac{1}{(k^2 + m^2)(k_1^2 + m^2)(k_2^2 + m^2)(k^2 + k_1^2 + k_2^2 + 3m^2)}, \tag{3.13}$$

$$c + d = g^2 \int \frac{d^D k_1}{(2\pi)^D} \left(\frac{1}{k_1^2 + m^2} + \frac{1}{k_2^2 + m^2} \right) \frac{1}{(k^2 + m^2)^2 (k^2 + k_1^2 + k_2^2 + 3m^2)}, \tag{3.14}$$

$$b + c + d = g^2 \int \frac{d^D k_1}{(2\pi)^D} \frac{1}{(k_1^2 + m^2)(k_2^2 + m^2)(k^2 + m^2)^2}, \tag{3.15}$$

where $k_2 = k - k_1$. This is an example showing explicitly the validity of Eq. (2.5) to the order g^2. That it is generally true for all orders follow from the general consideration of the previous section or, alternatively, from a diagrammatical proof given by De Dominicis for the case (3.1)[7].

IV. A Simple Example

In this section we will present a simple example of a more general phenomenon. Let us consider the potential

$$V(\boldsymbol{q}) = -\frac{1}{2} \mu^2 \boldsymbol{q}^2 + \frac{1}{4} g (\boldsymbol{q}^2)^2, \tag{4.1}$$

where $\boldsymbol{q}$ is an n-dimensional vector, and $\boldsymbol{q}^2 = \sum_{i=1}^{n} (q^i)^2$.

If we want to compute

$$\langle q^2\rangle \propto \int d[q]q^2 \exp\{-V(q)\} \tag{4.2}$$

in perturbation theory in g, we have first to find the minimum of $V(q)$. The minimum happens at $|q|^2 = \mu^2/g$, but owing to the $O(n)$ symmetry of the potential $V(q)$, it is not an isolated minimum. If we choose the minimum

$$q = q_0 \equiv \left(\sqrt{\frac{\mu^2}{g}}, 0, \cdots, 0\right),$$

we can write $q = q_0 + \tilde{q}$ and develop the exponent in powers of $\tilde{q}$, but we shall get divergent intergals. Indeed, let us write

$$q = \left(\sqrt{\frac{\mu^2}{g}} + \tilde{q}_L, q_T\right), \tag{4.3}$$

where q_T is an $(n-1)$-dimensional vector. We find

$$\begin{aligned} V(q) &= \mu^2\tilde{q}_L^2 + \sqrt{\mu^2 g}\,\tilde{q}_L(\tilde{q}_L^2 + q_T^2) \\ &\quad + \frac{1}{4}g(\tilde{q}_L^2 + q_T^2)^2 - \frac{1}{4g}\mu^4, \end{aligned} \tag{4.4}$$

and the integration over q_T is not damped for $g = 0$. If we add a regularizing term hq_T^2 and send h to zero, we would get the wrong result at the first order in g. The correct result can always be obtained by doing the nonlinear transformation to the variable $r = (q^2)^{1/2}$ and the set of angular coordinates of the $(n-1)$-dimensional sphere,

$$\langle q^2\rangle = \frac{\int dr r^{N-1} r^2 \exp\left\{-\left(\frac{g}{4}r^4 - \frac{\mu^2}{2}r^2\right)\right\}}{\int dr r^{N-1} \exp\left\{-\left(\frac{g}{4}r^4 - \frac{\mu^2}{2}r^2\right)\right\}}. \tag{4.5}$$

We want to show that from the Langevin equation one can get the correct result without having to do the nonlinear transformation. Let us consider the first order in g. We know that we must have

$$\langle q^2\rangle = \frac{\mu^2}{g} + A(n-1) + B. \tag{4.6}$$

If $n = 1$, we obtain automatically the correct result, so we can only compute the term proportional to $(n-1)$. Now the Langevin equations are

$$\left.\begin{aligned} \dot{q}_T &= \eta_T + O(g^{1/2}), \\ \dot{\tilde{q}}_L &= -2\mu^2\tilde{q}_L - \sqrt{\mu^2 g}(3\tilde{q}_L^2 + q_T^2) + O(g) + \eta_L. \end{aligned}\right\} \tag{4.7}$$

An easy computation shows that

$$\langle q_T^i(t)q_T^j(t')\rangle = 2\delta_{ij}\min(t,t'). \tag{4.8}$$

In order to get $\langle q^2\rangle$, we must compute $\tilde{q}_L$ at the order $g^{1/2}$. The only diagram (for the term proportional to $(n-1)$) is

$\tilde{q}_L(t) \approx$ 

$$= \mu g^{1/2} \int_0^t dt' \langle q_T^2(t') \rangle \exp[-2\mu^2(t-t')]$$
$$= -(n-1)\sqrt{g/\mu^2}(t - 1/2\mu^2), \tag{4.9}$$

where the dashed line stands for the transverse propagator, and we have neglected terms which vanish when $t \to \infty$. If we compute the terms proportional to $(n-1)$ in

$$\langle q^2 \rangle = \frac{\mu^2}{g} + 2\sqrt{\frac{\mu^2}{g}}\langle \tilde{q}_L \rangle + \langle \tilde{q}_L^2 \rangle + \langle q_T^2 \rangle, \tag{4.10}$$

we find they are

$$-2(n-1)\left(t - \frac{1}{2\mu^2}\right) + 2(n-1)t = \frac{n-1}{\mu^2}. \tag{4.11}$$

From this equation we see that the terms proportional to t coming from $\langle \tilde{q}_L \rangle$ and $\langle q_T^2 \rangle$ cancel each other, and the finite contribution gives the correct result as can be easily checked from Eq. (4.5) by the saddle point method.

Of course, we can also apply the nonlinear transformations to the Langevin equation. Let us consider the case $n = 2$. It can be shown that if we go to the variables $r(t)$ and $\theta(t)$, the Langevin equation

$$\dot{q} = -\frac{\partial V(q)}{\partial q} + \eta \tag{4.12}$$

becomes[1)]

$$\left.\begin{aligned} \dot{r} &= -\frac{d}{dr}[V(r) + \ln r] + \eta_r, \\ \dot{\theta} &= \eta_\theta, \end{aligned}\right\} \tag{4.13}$$

where

$$\langle \eta_r(t)\eta_r(t') \rangle = 2\delta(t-t'), \quad \langle \eta_\theta(t)\eta_\theta(t') \rangle = \frac{2}{r^2}\delta(t-t').$$

The evolution for r is decoupled from that for θ, and the perturbative expansion in g is uniform at all times. However, we do not need at any rate to do explicitly the nonlinear transformation. The solution of the Langevin equation is finite at all t (t plays the role of a regulator), and we get the correct results automatically. Of course, if we compute $\langle q_T^2 \rangle = \langle r^2 \sin^2\theta \rangle$ we would get divergent results in perturbation theory:

$$\langle q_T^2 \rangle \simeq t + t^2 g + \cdots \tag{4.14}$$

indicating that the equilibrium is reached only for times $t \gg 1/g$. Therefore, the

1) If one is not careful in doing the nonlinear transformation, one would miss the $\ln r$ term.

only distinction is between quantities which go fast to equilibrium ($\lambda_1 \sim O(1)$) and those which go slowly ($\lambda_1 \sim O(1/g)$); only the first ones can be correctly computed in perturbation theory. For example, we obtain

$$\langle q \rangle \simeq O\,(\exp(-tg)). \tag{4.15}$$

The existence of an equation for r being decoupled from the equation for θ is a general consequence of the symmetry of the problem, and it implies that the approach to equilibrium must be fast for symmetric invariant quantities.

Indeed, let us consider the Fokker-Planck equation for $\psi = P(q,t)\exp\left(\frac{1}{2}V(q)\right)$. Using the radial variables r and θ we can see that the "S-wave" component of ψ,

$$\psi_S(r,t) = \int d\theta \psi(r,\theta,t) \tag{4.16}$$

satisfies the radial equation

$$\frac{\partial}{\partial t}\psi_S(r,t) = \hat{H}_S\psi_S(r,t). \tag{4.17}$$

The eigenvalue λ_1 of $\hat{H}_S$ controls the approach to equilibrium of radially symmetric quantities. By general theorems λ_1 is a continuously differentiable function of the coupling constant, so that we expect that the approach to equilibrium must be uniform in g and time.

As the reader can see, this approach has the virtue of giving automatically the correct results in perturbation theory without worrying about the nonlinear transformations. This is in contrast with the approach consisting in adding to V a symmetry breaking term hq_T^2 and firstly expanding in g and later sending h to zero.

V. Gauge Theories

Now we proceed to consider a gauge theory, the Euclidean Hamiltonian being

$$V = \int d^D x \left\{(D_\mu\phi^+)(D_\mu\phi) + \frac{1}{2}\,\mathrm{Tr}F_{\mu\nu}^2\right\}, \tag{5.1}$$

where

$$D_\mu\phi = (\partial_\mu - ieA_\mu)\phi, \quad F_{\mu\nu} = \partial_\mu A_\nu - \partial_\nu A_\mu - ie[A_\mu, A_\nu]$$

$$A_\mu = A_\mu^a\tau_a, \quad F_{\mu\nu} = F_{\mu\nu}^a\tau_a, \quad \mathrm{Tr}(\tau_a\tau_b) = \frac{1}{2}\,\delta_{ab}.$$

The associate Langevin equations are

$$\begin{cases} \dot{\phi} = D^2\phi + \eta_\phi, \quad \dot{\phi}^+ = D^2\phi^+ + \eta_\phi^+, \\ \dot{A}_\mu = D_\nu F_{\nu\mu} + J_\mu + \eta_\mu, \end{cases} \tag{5.2}$$

where

$$J_\mu = J_\mu^a\tau_a, \quad J_\mu^a = ie\phi^+\tau_a\partial_\mu\phi + e^2\phi^+\{\tau_a, A_\mu\}\phi,$$

$$\left.\begin{aligned} \langle\eta_\phi(x,t)\eta_\phi^+(x',t')\rangle &= 2\delta(x-x')\delta(t-t'), \\ \langle\eta_\mu(x,t)\eta_\nu(x',t')\rangle &= 2\delta_{\mu\nu}\delta(x-x')\delta(t-t')C_2, \end{aligned}\right\} \tag{5.3}$$

and $C_2 = \delta^{ab}\tau_a\tau_b$ is the second Casimir operator, which is a multiple of the unit matrix.

Let us first consider the free Abelian case in which,

$$\dot{A}_\mu = \partial^2 A_\mu - \partial_\mu\partial_\nu A_\nu + \eta_\mu. \tag{5.4}$$

After being imposed the boundary condition

$$A_\mu(x,t)|_{t=0} = 0, \tag{5.5}$$

the solution of the Eq. (5.4) for $t > 0$ (in the momentum space) is,

$$A_\mu(k,t) = \int_0^t dt' G_{\mu\nu}(k,t-t')\eta_\nu(k,t'), \tag{5.6}$$

where the retarded Green function defined only for $t > t'$, is

$$G_{\mu\nu}(k,t-t') = \left(\delta_{\mu\nu} - \frac{k_\mu k_\nu}{k^2}\right)\exp[-k^2(t-t')] + \frac{k_\mu k_\nu}{k^2}. \tag{5.7}$$

The expectation value of $A_\mu(x,t)A_\nu(y,t')$ in the momentum space is

$$D_{\mu\nu}(k;t,t') = \left(\delta_{\mu\nu} - \frac{k_\mu k_\nu}{k^2}\right)\frac{1}{k^2}\{\exp[-k^2|t-t'|] - \exp[-k^2(t+t')]\} + \frac{k_\mu k_\nu}{k^2} 2\min(t,t'). \tag{5.8}$$

For large equal times ($t = t' \to \infty$), we find

$$D_{\mu\nu}(k;t,t) = \left(\delta_{\mu\nu} - \frac{k_\mu k_\nu}{k^2}\right)\frac{1}{k^2} + 2t\frac{k_\mu k_\nu}{k^2}. \tag{5.9}$$

Diagrammatically, Eq. (5.7) is the (retarded) propagator without any cross and Eq. (5.8) is the propagator with a cross.

We note that Eq. (5.9) is just the usual Feynman propagator in the Landau gauge plus a longitudinal term which is divergent as the time. Indeed, we can write

$$A_\mu(x,t) = A_\mu^T(x,t) + \partial_\mu\alpha(x,t), \tag{5.10}$$

where $A_\mu^T(x,t)$ satisfies

$$\partial_\mu A_\mu^T(x,t) = 0. \tag{5.11}$$

$A_\mu^T(x,t)$ is gauge invariant and $\alpha(x,t)$ looks like a gauge transformation. Then we find in the momentum space,

$$\left.\begin{aligned} \langle A_\mu^T(k,t)A_\nu^T(-k,t)\rangle &= \left(\delta_{\mu\nu} - \frac{k_\mu k_\nu}{k^2}\right)\frac{1}{k^2}, \\ \langle \alpha(k,t)\alpha(-k,t)\rangle &= \frac{2}{k^2}t. \end{aligned}\right\} \tag{5.12}$$

This means that the evolution for gauge invariant quantities is fast, while the system undergoes a random walk in the gauge parameter space.

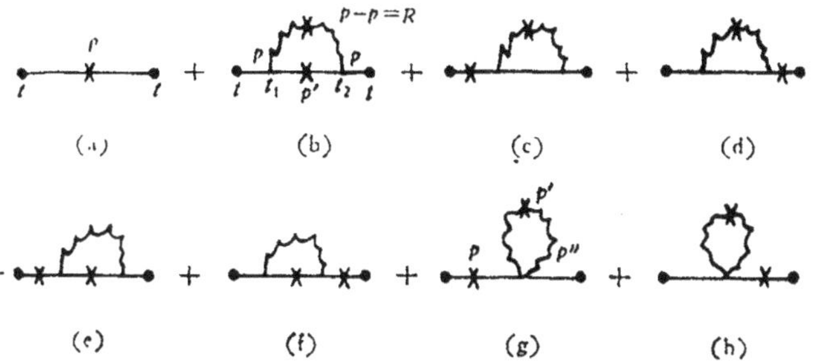

Fig. 1

Now let us consider the $\phi^+\phi$ propagator in the Abelian theory at equal times. The diagrammatical rules for scalar QED are very similar to those in Sec. III. In addition to the propagators given by Eqs. (3.7), (5.7) and (5.8), what we should add here are only the rules for the $A_\mu\phi^+\phi$ and $A_\mu A_\nu\phi^+\phi$ vertices, but they are the same as in usual Feynman diagrams. We shall do the computation up to the order e^2, keeping only the terms which survive at large t. The diagrams are shown in Fig. 1, where the straight line stands for the ϕ field, and the wavy line the photon. Their contributions at equal times t are given as follows for large t (after being integrated over the times, t_1 and t_2, of the vertices):

$$a = 2\int_0^t dt' \exp(-2p^2t') = \frac{1}{p^2},$$

$$b = e^2\int \frac{d^Dp'}{(2\pi)^D}\frac{1}{p^2p'^2k^2(p^2+p'^2+k^2)}\left[(p+p')^2 - \frac{(p^2-p'^2)^2}{k^2}\right]$$
$$+ e^2\int \frac{d^Dp'}{(2\pi)^D}\frac{1}{p^2p'^2(p^2+p'^2)}\left[2t - \frac{2}{p^2+p'^2} - \frac{1}{p^2}\right]\frac{(p^2-p'^2)^2}{k^2}, \tag{5.13}$$

$$c + e = d + f = \frac{e^2}{2}\int \frac{d^Dp'}{(2\pi)^D}\frac{1}{p^2}\left(\frac{1}{p'^2}+\frac{1}{k^2}\right)\frac{1}{p^2+p'^2+k^2}\left[(p+p')^2 - \frac{(p^2-p'^2)^2}{k^2}\right]$$
$$+ \frac{e^2}{2}\int \frac{d^Dp'}{(2\pi)^D}\frac{1}{p^4(p^2+p'^2)}\left[\frac{1}{p'^2} + 2t - \frac{2}{p^2+p'^2} - \frac{1}{p^2}\right]\frac{(p^2-p'^2)^2}{k^2}, \tag{5.14}$$

$$g + h = -3e^2\int \frac{d^Dp'}{(2\pi)^D}\frac{1}{p^4p'^2} - e^2\int \frac{d^Dp'}{(2\pi)^D}\frac{1}{p^4}\left[2t - \frac{1}{p^2}\right]. \tag{5.15}$$

Here $k = p - p'$; the first and the second integrals represent respectively the contributions of the transverse and longitudinal part of the A_μ field.

If we sum all the contributions of the transverse part, we find the usual result in the Landau gauge. For the contribution of the longitudinal part, at large times the term proportional to t is equal to the variation of the equilibrium $\phi^+\phi$ propagator induced by adding the gauge term $t(k_\mu k_\nu/k^2)$ to the equilibrium $A_\mu A_\nu$ propagator. Indeed, the main contribution to the diagrams comes from the region of integration where $t - t_1, t - t_2$ are of order 1, so that at the leading order in t we can substitute t for t_1, t_2. In this way we lose terms of $O(1)$ when $t \to \infty$, but we compute correctly the terms of order t. This argument can be generalized to the leading order in t at fixed e^2. Using the standard theorem on the variation of Green functions under a gauge transformation[6], we get

$$\langle\phi^+(x,t)\phi(y,t)\rangle \underset{t\to\infty}{\sim} \langle\phi^+(x)\phi(y)\rangle_f \exp[-e^2 t\omega(x-y)], \tag{5.16}$$

where $\omega(x) \propto 1/|x|^{D-2}$ is the Fourier transform of $1/k^2$, and $\langle\phi^+(x)\phi(y)\rangle_f$ is the free propagator. Eq. (5.16) implies that for large times the charged field propagator is very near to zero, but this happens only for times t greater than $1/e^2\omega(x-y)$. In the asymptotic limit $t\to\infty$, the sum of the leading terms in t for each order in e^2 goes to zero. However, a completely different result would be obtained considering only a finite number of terms in the perturbation expansion.

Let us consider a gauge invariant quantity. The simplest one is $\phi^+(x,t)\phi(x,t)$. The contribution proportional to t is obviously zero, because it corresponds to a gauge transformation. This can be explicitly checked from Eqs. (5.13)—(5.15) by using dimensional regularization. As for the remaining finite terms, after some algebraic operation and having eliminated terms odd in $k=p-p'$, we find they are equal to the well-known contribution in the Landau gauge plus the following term,

$$2\int \frac{d^D p}{(2\pi)^D}\frac{d^D p'}{(2\pi)^D}\frac{p^2-p'^2}{p^2p'^2(p^2+p'^2)(p-p')^2}. \tag{5.17}$$

Being odd under the exchange $p\longleftrightarrow p'$, this term is equal to zero, as required by consistency.

We have seen that also in gauge theories, at least in this simple example, the correct results for gauge invariant quantities can be obtained in our approach with the Langevin equation by expanding in powers of the coupling constant without having to fix the gauge as is done in the conventional approach. The only breaking of gauge invariance is in the boundary conditions at $t=0: A_\mu(x,t)|_{t=0}=0$, but the large t behaviour is independent of the boundary conditions.

What happens at higher orders? Diagrams must be regularized by dimensional regularization or lattice regularization (i.e., by a gauge invariant procedure). In principle, it is known that for a renormalizable theory dynamic correlation functions are finite only after a renormalization of the matrix $M(x,y)$ in Eq. (2.15). This phenomenon which is wellknown in the theory of the second-order phase transitions[4], does not modify the static (equal-time) correlation functions at large times. We have therefore two possibilities:

1) Add the counterterms in $M(x,y)$ in order to have finite results at all times.

2) Take the limit $t\to\infty$ before sending the cutoff to infinity or the space dimensions D to the physical one. The first alternative would be the best, if we want to find explicitly the convergence rate of the Langevin equation (e.g., to compare with Monte Carlo procedures), while the second alternative is the simplest if we are interested only in equilibrium properties.

In the next section we will argue that also in non-Abelian theories we obtain the correct results which correspond to the effect of the Faddeev-Popov ghost.

VI. General Considerations

The general theorems on the approaching to equilibrium of the solution of the Langevin (or the Fokker-Planck) equation can be applied to our case, therefore there is no doubt that the large-time behaviour of the correlation functions is the correct one. The main problem is to show that for gauge invariant quantities equilibrium is approached uniformly in g and in t so that the Taylor expansion in g and the limit $t \to \infty$ can be freely exchanged. As is seen in the previous sections, this is not true for quantities which are not gauge invariant. Therefore, it is better to present the argument in detail (although it is rather similar to that in Sec. IV).

It is convenient to introduce the quantities $A^T_\mu(x,t)$ $\alpha(x,t)$ and $\phi^T(x,t)$ defined by

$$\left.\begin{aligned} &\partial_\mu A^T_\mu(x,t) = 0, \\ &-ieA^T_\mu(x,t) = \exp[-ie\alpha(x,t)](\partial_\mu - ieA_\mu)\exp[ie\alpha(x,t)], \\ &\phi^T(x,t) = \exp[-ie\alpha(x,t)]\phi(x,t). \end{aligned}\right\} \tag{6.1}$$

In perturbation theory in e these equations fix uniquely $\alpha(x,t)$, $A^T_\mu(x,t)$ and $\phi^T(x,t)$ in terms of $A_\mu(x,t)$ and $\phi(x,t)$[1]. $A^T_\mu(x,t)$ and $\phi^T(x,t)$ are gauge invariant quantities, i.e., they do not change under a gauge transformation for $A_\mu(x,t)$. All gauge invariant quantities can be written in terms of $A^T_\mu(x,t)$ and $\phi^T(x,t)$.

The gauge invariance of the Langevin equation or of the associated Fokker-Planck equation implies that the evolution of $A^T_\mu(x,t)$ and $\phi^T(x,t)$ is independent of the evolution of $\alpha(x,t)$. Therefore we can write

$$\left.\begin{aligned} \dot{A}^T_\mu(x,t) &= F_1(A^T_\mu, \phi^T, \eta^T), \\ \dot{\phi}^T(x,t) &= F_2(A^T_\mu, \phi^T, \eta^T). \end{aligned}\right\} \tag{6.2}$$

We are not interested in the detailed form of these equations or of the associated Fokker-Planck equations. From Eq. (4.2) we have already seen that the approach to equilibrium is controlled by the smallest positive eigenvalues λ_1 of an operator H which acts on gauge invariant quantities, so that the slow approach to equilibrium for gauge non-invariant quantities like the distribution of $\alpha(x,t)$ has no effect on the evolution of gauge invariant quantities. Thus, apart from possible ultraviolet or infrared divergences[1], equilibrium is approached in the interacting theory at roughly the same rate as in the free theory. Divergences should be eliminated by introducing the needed cutoffs and counterterms in order to obtain finite results.

These arguments show that no term linear in t appears in the expectation value of gauge invariant quantities, and that one finds automatically the correct results which corresponds to the contribution of the Faddeev-Popov ghost. Of course, this

1) More precisely, the uniqueness of $\alpha(x,t)$ holds only for fixed $A_\mu(x,t)$ and sufficiently small e. The existence of $A_\mu(x,t)$ such that there are many solutions for $\alpha(x,t)$ is just the Gribov ambiguity.

2) In order to make the argument complete, one should first consider the case of finite lattice and impose the appropriate boundary conditions in such a way that the spectrum of H becomes discrete and that the corrections to the equilibrium distributions are exponentially small.

contribution would show up as a finite remainder of incomplete cancellation of terms which are dependent on t, if we do perturbative expansions directly for $A_\mu(x,t)$.

VII. Conclusions

The method we have proposed here is not very useful for practical computation. The number of diagrams is much higher than that in the conventional approach and the algebraic operation is longer. Some additional difficulties arise in the computation of the S matrix for charged fields. Indeed, the Green functions of charged fields are zero at equilibrium, so that the LSZ formalism cannot be used. We have not investigated the possibility of using the gauge invariant path-dependent operators of Mandelstam in the LSZ formalism. In principle, all physical measurable informations, like cross sections, can be extracted from the Green functions of gauge invariant quantities, although this operation is very complicated in practice. It is important, however, to know that, at least in principle, we can avoid the use of the Faddeev-Popov trick, whose correctness has been questioned beyond perturbation theory.

We note that the equality between the Green functions of a field theory and the equal-time stochastic correlation functions of the Langevin equation can be the starting point of a reformulation of field theory using a different language. In this paper, we show that this formulation may be useful to construct a perturbative expansion for systems for which a nonlinear transformation is needed in the conventional approach to obtain the correct results. It is possible that the same technique an be of a wider application. Indeed, the Langevin equation (or its discretized version, the Monte Carlo procedure) is a really constructive approach to field theory in the sense that it can be used as a practical starting point for computer simulations.

We are very grateful to Prof. Hao Bailin for many discussions and suggestions. One of us (G. Parisi) is happy to thank the Institute of Theoretical Physics, Academia Sinica for the warm hospitality extended to him.

References

[1] For a review, see, e.g. Abers, E. & Lee, B. W., *Phys. Rep.*, **9C** (1973), 1; Itzykson, C. & Zuber, J. B., *Introduction to Quantum Field Theory*, McGraw-Hill, New York, 1980.
[2] Gribov, V. N., *Nucl. Phys.*, **B139** (1978), 1.
[3] For a review, see, e.g. Drouffe J. M. & Itzykson, C., *Phys. Rep.*, **38C** (1978), 133.
[4] See, e.g. Graham, R., *Springer Tracts in Modern Physics*, **66** (1973), 1; Hohenberg P. C. & Halperin, B., *Rev. Mod. Phys.*, **49** (1977), 435; Fox, R. F., *Phys. Rep.*, **48C** (1978), 179; Zhou Guangzhao, Su Zhaobin, Hao Bailin & Yu Lu, *Acta Physica Sinica*, **29** (1980) (to be published); *Phy. Rev.*, **B** (1980), (to be published).
[5] Creutz, M., Jacobs L. & Rebbi, C., *Phys. Rev. Lett.*, **42** (1979), 1390; Wilson, K., *Proceedings of Cargese Summer School*, 1979, Plenum Press (to be published).
[6] Parisi G. & Sourlas, N. *ibid.*, **43** (1979), 744.
[7] De Dominicis. C., *Lett. Nuovo Cimento*, **12** (1975), 567.
[8] Zumino, B., *J. Math. Phys.*, **1** (1960), 1.

Perturbation Theory

Nuclear Physics B214 (1983) 392–404

EQUIVALENCE OF STOCHASTIC AND CANONICAL QUANTIZATION IN PERTURBATION THEORY

E. FLORATOS

I.T.P., University of Bern, Switzerland

J. ILIOPOULOS

Laboratoire de Physique Théorique de l'Ecole Normale Supérieure, Paris, France*

Received 2 November 1982

It is shown that the stochastic quantization method introduced by Parisi and Wu reproduces, order by order, the ordinary perturbation expansion. The proof is valid for any field theory, including gauge theories, provided one considers gauge-invariant quantities.

1. Introduction

The relation between euclidean quantum field theory and stochastic differential equations has attracted considerable attention recently*. It offers a new way to quantization which, sometimes, presents conceptual advantages. For example, by considering a stochastic equation in real space, one can show that a supersymmetry naturally arises [2]. In a more traditional approach, one may introduce a fifth dimension in the form of a new "time" t and consider a stochastic equation in this variable [3]. We shall study some aspects of this method here.

Let $\mathcal{L}(A(x))$ be the lagrangian density describing the dynamics of a physical system. $\mathcal{L}$ is a function of a set of fields in d-dimensional euclidean space, which we denote collectively by $A(x)$, and of their first derivatives. $S[A]$ is the corresponding action functional. The Green functions of the theory are the vacuum expectation values of products of fields and are given by

$$\langle A(x_1)\dots A(x_n)\rangle_0 = \frac{\int \mathcal{D}[A]\mathrm{e}^{-S[A]}A(x_1)\dots A(x_n)}{\int \mathcal{D}[A]\mathrm{e}^{-S[A]}}\,. \tag{1.1}$$

* Laboratoire Propre du Centre National de la Recherche Scientifique associé à l'Ecole Normale Supérieure et à l'Université de Paris Sud. Postal address: 24, rue Lhomond, 75231 Paris Cedex 05-France.

* For a recent introduction to stochastic differential equations see ref. [1a]. The use of stochastic methods in quantum theory is discussed in [1b].

In the stochastic approach we introduce an additional dimension and we consider fields which depend on $d+1$ variables $A_\eta(x,t)$. The evolution in this extra time t is assumed to be governed by a Langevin equation:

$$\frac{\partial A_\eta(x,t)}{\partial t} = -\frac{\delta S[A_\eta]}{\delta A_\eta(x,t)} + \eta(x,t), \tag{1.2}$$

with the initial condition $A_\eta(x,0)=0$. In fact this condition can be replaced by a more general one $A_\eta(x,0)=C(x)$, with $C(x)$ an arbitrary function. In appendix C we show that the large time limits are independent of the choice of C. $\eta(x,t)$ is a white, gaussian noise. Correlation functions are now introduced as averages over η:

$$\langle A_\eta(x_1,t_1)\ldots A_\eta(x_n,t_n)\rangle_\eta = \int \mathcal{D}[\eta]\mathrm{e}^{-\frac{1}{4}\int\eta^2(x,t)\mathrm{d}x\mathrm{d}t} A_\eta(x_1,t_1)\ldots A_\eta(x_n,t_n). \tag{1.3}$$

A convenient way to rewrite (1.3) is to introduce the probability density $P[A,\tau]$ which is defined by

$$P[A,\tau] = \int \mathcal{D}[\eta]\mathrm{e}^{-\frac{1}{4}\int\eta^2(x,t)\mathrm{d}x\mathrm{d}t}\prod_y \delta\big(A(y)-A_\eta(y,\tau)\big). \tag{1.4}$$

In terms of P we can write the equal-time correlation function as

$$\langle A_\eta(x_1,\tau)\ldots A_\eta(x_n,\tau)\rangle_\eta = \int \mathcal{D}[A]A(x_1)\ldots A(x_n)P[A,\tau]. \tag{1.5}$$

The probability density P satisfies the Fokker-Planck equation:

$$\frac{\partial}{\partial\tau}P[A,\tau] = \int \mathrm{d}x \frac{\delta}{\delta A(x)}\left\{\frac{\delta}{\delta A(x)} + \frac{\delta S}{\delta A(x)}\right\}P[A,\tau], \tag{1.6}$$

with the initial condition $P[A,0]=\prod_y \delta(A(y))$. On the r.h.s. of (1.6) we assume that a suitable regularization has been used.

The stochastic approach to quantization is contained in the following property [3]:

$$\lim_{\tau\to\infty}\langle A_\eta(x_1,\tau)\ldots A_\eta(x_n,\tau)\rangle_\eta = \langle A(x_1)\ldots A(x_n)\rangle_0, \tag{1.7}$$

or, equivalently,

$$\underset{\tau\to\infty}{\mathrm{w.lim}}\, P[A,\tau] = P^{\mathrm{eq}}[A] = \frac{\mathrm{e}^{-S[A]}}{\int\mathcal{D}[A]\mathrm{e}^{-S[A]}}, \tag{1.8}$$

i.e. at large times the probability distribution reaches the equilibrium one given by

(1.8). Here the limit is supposed to be taken "weakly", as expressed in (1.7); in other words $P[A, \tau]$ will always be applied to a string of fields. The probability density $P[A, \tau]$ will be used only as a short-hand notation for the Green functions of the theory.

The property (1.7) or (1.8) has been shown to hold for several systems [4]*. In ref. [3] it was conjectured that it was true even for gauge theories, provided one looked at gauge-invariant quantities. More precisely, it was argued that the limit (1.7) does not exist for arbitrary gauge non-invariant Green functions, but it exists and gives the correct result for gauge-invariant quantities. This is very interesting because it provides us with a new quantization method which does not require any gauge-fixing term and is therefore free from gauge ambiguities [6]. In this paper we shall prove this statement order by order in perturbation theory. In sect. 2 we present our proof for the simple case of theories without gauge invariance. This proof is extended to gauge theories in sect. 3. Some simple computations and remarks are included in three appendices.

2. Field theories without gauge invariance

We consider in this section the case of a local, polynomial field theory defined in a d-dimensional euclidean space whose classical action has no local gauge invariance. The important property of such a theory is that one can always introduce a suitable regulator such that the regularized probability distribution $e^{-S[A]}/\int\mathcal{D}[A]e^{-S[A]}$ is well defined order by order in perturbation theory. We shall prove the property (1.7) for such a theory. In order to simplify the notation we shall present the case of a single, self-interacting scalar field $\varphi(x)$, but the generalization is straightforward. We write

$$\mathcal{L} = \mathcal{L}_0 + g\mathcal{L}_I, \qquad \mathcal{L}_0 = -\tfrac{1}{2}(\partial_\mu\varphi)^2 - \tfrac{1}{2}m^2\varphi^2. \tag{2.1}$$

In perturbation theory Green functions are computed as power series in g. Similarly, we expand the probability density $P[\varphi, \tau]$:

$$P[\varphi, \tau] = \sum_{k=0}^{\infty} g^k P_k[\varphi, \tau]. \tag{2.2}$$

The n-point equal time correlation function in the kth order of perturbation theory is given by

$$\langle\varphi_\eta(x_1, \tau)\dots\varphi_\eta(x_n, \tau)\rangle_\eta = g^k\int\mathcal{D}[\varphi]\varphi(x_1)\dots\varphi(x_n)P_k[\varphi, \tau], \tag{2.3}$$

* The property (1.8) has been shown in perturbation for a φ^4 theory in ref. [5].

and is computed as the sum of the corresponding regularized Feynman diagrams. This is the meaning we attach to $P_k[\varphi, \tau]$ and all our arguments should be interpreted in this way.

The Fokker-Planck equation (1.6) becomes

$$\frac{\partial}{\partial\tau} P_k[\varphi,\tau] = \int \mathrm{d}x \frac{\delta}{\delta\varphi(x)} \left\{ \frac{\delta}{\delta\varphi(x)} + \frac{\delta S_0}{\delta\varphi(x)} \right\} P_k[\varphi,\tau]$$

$$+ \int \mathrm{d}x \frac{\delta}{\delta\varphi(x)} \frac{\delta S_\mathrm{I}}{\delta\varphi(x)} P_{k-1}[\varphi,\tau], \tag{2.4}$$

with the initial conditions

$$P_0[\varphi,0] = \prod_y \delta(\varphi(y)) \tag{2.4a}$$

$$P_k[\varphi,0] = 0, \qquad k = 1,2,\dots . \tag{2.4b}$$

The proof is inductive. We assume that, for $1 \leqslant l \leqslant k-1$, $P_l[\varphi, \tau]$ is uniformly bounded and satisfies

$$\underset{\tau\to\infty}{\mathrm{w.lim}}\, P_l[\varphi,\tau] = P_l^{\mathrm{eq}}[\varphi]. \tag{2.5}$$

The reason why we do not start the induction from $l = 0$ is that $P_0[\varphi, \tau]$ is not bounded at $\tau \to 0$ because it satisfies the initial condition (2.4a). However, as we shall see, this is harmless and, in fact, $P_0[\varphi, \tau]$ does satisfy (2.5). In (2.5) w.lim denotes again the limit in the weak sense, as explained in sect. 1 and $P_l^{\mathrm{eq}}[\varphi]$ is the coefficient of g^l in a perturbation expansion of the equilibrium density (1.8). $P_l^{\mathrm{eq}}[\varphi]$ is again a short-hand notation for the lth order Green functions of the ordinary regularized perturbation theory. It is easy to verify – and we shall do so presently – that this inductive hypothesis is satisfied for $P_1[\varphi, \tau]$. We must prove it for $l = k$. The proof is based on the following lemma.

Lemma:

$$\underset{\tau\to\infty}{\mathrm{w.lim}} \frac{\partial}{\partial\tau} P_k[\varphi,\tau] = 0. \tag{2.6}$$

Proof: We first transform eq. (2.4) in integral form:

$$P_k[\varphi,\tau] = \int \mathcal{D}[\varphi'] \int_0^\tau \mathrm{d}\tau' \mathcal{D}_0[\varphi,\varphi';\tau-\tau'] \int \mathrm{d}x \frac{\delta}{\delta\varphi'(x)} \frac{\delta S_\mathrm{I}}{\delta\varphi'(x)} P_{k-1}[\varphi',\tau'], \tag{2.7}$$

where $\mathcal{D}_0[\varphi, \varphi'; \tau - \tau']$ is the Green functional of the free Fokker-Planck equation:

$$\frac{\partial}{\partial\tau}\mathcal{D}_0[\varphi, \varphi'; \tau - \tau'] = \int \mathrm{d}x \frac{\delta}{\delta\varphi(x)}\left\{\frac{\delta}{\delta\varphi(x)} + \frac{\delta S_0}{\delta\varphi(x)}\right\}\mathcal{D}_0[\varphi, \varphi'; \tau - \tau'] + \delta(\tau - \tau')\prod_x \delta(\varphi(x) - \varphi'(x)), \tag{2.8}$$

with the boundary condition

$$\mathcal{D}_0[\varphi, \varphi'; \tau - \tau']|_{\tau = \tau'_+} = \prod_x \delta(\varphi(x) - \varphi'(x)). \tag{2.9}$$

By solving eq. (2.8) we can construct $\mathcal{D}_0[\varphi, \varphi'; \tau - \tau']$ explicitly. Similarly, we can construct $P_0[\varphi, \tau]$ by solving eq. (2.4) for $g = 0$. Both solutions are presented in appendix A. The results are

$$P_0[\varphi, \tau] = N_0^{-1}\exp\left\{-\tfrac{1}{2}\int \mathrm{d}k\, \varphi(k)\frac{k^2 + m^2}{1 - \mathrm{e}^{-2\tau(k^2+m^2)}}\varphi(-k)\right\}, \tag{2.10}$$

$$\mathcal{D}_0[\boldsymbol{\varphi}, \varphi'; \tau] = \theta(\tau)\Delta_0^{-1}\exp\left\{-\tfrac{1}{2}\int \mathrm{d}k\, \boldsymbol{\varphi}(k) D(\tau) \boldsymbol{\varphi}(-k)\right\}, \tag{2.11}$$

where N_0 and Δ_0 are normalization factors such that

$$\int \mathcal{D}[\varphi] P_0[\varphi, \tau] = 1, \tag{2.10a}$$

$$\int \mathcal{D}[\varphi] \mathcal{D}_0[\varphi, \varphi'; \tau] = \theta(\tau). \tag{2.11a}$$

In (2.11) $\boldsymbol{\varphi}(k)$ denotes $(\varphi(k), \varphi'(k))$ and $D(\tau)$ is the following 2×2 matrix:

$$D_{11}(\tau) = \frac{k^2 + m^2}{1 - \mathrm{e}^{-2\tau(k^2+m^2)}}, \tag{2.12a}$$

$$D_{12}(\tau) = D_{21}(\tau) = -D_{11}(\tau)\mathrm{e}^{-\tau(k^2+m^2)}, \tag{2.12b}$$

$$D_{22}(\tau) = D_{11}(\tau)\mathrm{e}^{-2\tau(k^2+m^2)}. \tag{2.12c}$$

We can easily verify that, in the limit $\tau \to \infty$, P_0 and $\mathcal{D}_0$ satisfy the relations

$$P_0[\varphi, \tau] \underset{\tau \to \infty}{\to} P_0^{\mathrm{eq}}[\varphi] = N_0^{\mathrm{eq}^{-1}} \exp\left\{ -\tfrac{1}{2} \int \mathrm{d}k\, \varphi(k)(k^2+m^2)\varphi(-k) \right\}, \tag{2.13}$$

$$\frac{\partial P_0[\varphi, \tau]}{\partial \tau} \underset{\tau \to \infty}{\to} \mathrm{O}(\mathrm{e}^{-m^2\tau}), \tag{2.14}$$

$$\mathcal{D}_0[\varphi, \varphi'; \tau] \underset{\tau \to \infty}{\to} P_0^{\mathrm{eq}}[\varphi], \tag{2.15}$$

$$\frac{\partial \mathcal{D}_0}{\partial \tau} \underset{\tau \to \infty}{\to} \mathrm{O}(\mathrm{e}^{-m^2\tau}), \tag{2.16}$$

$$\frac{\delta \mathcal{D}_0}{\delta \varphi'(x)} \underset{\tau \to \infty}{\to} \mathrm{O}(\mathrm{e}^{-m^2\tau}). \tag{2.17}$$

We now look at $P_k[\varphi, \tau]$ given by eq. (2.7). By partial functional integration we obtain

$$P_k[\varphi, \tau] = -\int \mathcal{D}[\varphi'] \int \mathrm{d}x \frac{\delta S_{\mathrm{I}}}{\delta \varphi'(x)} \int_0^\tau \mathrm{d}\tau' \frac{\delta \mathcal{D}_0[\varphi, \varphi'; \tau - \tau']}{\delta \varphi'(x)} P_{k-1}[\varphi', \tau']. \tag{2.18}$$

We are also interested in the derivative $\partial P_k / \partial \tau$:

$$\frac{\partial P_k}{\partial \tau} = -\int \mathcal{D}[\varphi'] \int \mathrm{d}x \frac{\delta S_{\mathrm{I}}}{\delta \varphi'(x)} \frac{\partial}{\partial \tau} \int_0^\tau \mathrm{d}\tau' \frac{\delta \mathcal{D}_0[\varphi, \varphi'; \tau - \tau']}{\delta \varphi'(x)} P_{k-1}[\varphi', \tau']. \tag{2.19}$$

The exchange of the orders of differentiation and integration over φ' and τ is harmless because, as we have already emphasized, (2.18) or (2.19) represent the Green functions of the theory which are sufficiently regularized even in the presence of local vertex insertions. We must first consider eq. (2.18) for $k = 1$, in order to verify the first step of the inductive hypothesis. The reason is that $P_0[\varphi, \tau]$ is not finite for $\tau \to 0$ because it satisfies the initial condition (2.4a). However this δ-function is multiplied by $\delta S_{\mathrm{I}}/\delta\varphi'$, which is a monomial in φ'. Therefore, using the explicit expressions of P_0 and $\mathcal{D}_0$ given by (2.10) and (2.11), it is straightforward to verify that $P_1[\varphi, \tau]$ exists, is bounded uniformly in φ for all values of τ, has a well-defined limit as $\tau \to \infty$, and vanishes at $\tau = 0$, as required by (2.4b). We must still prove that $P_1[\varphi,\infty]=P_1^{\rm eq}[\varphi]$ but we prefer to complete, first, the proof of the lemma (2.6) for arbitrary k. This is simple. $P_{k-1}[\varphi',\tau']$ is bounded and $\delta\mathcal{D}_0/\delta\varphi'$ is integrable for all φ. The functional integration over φ' in (2.18) is always possible because P_{k-1} contains the gaussian measure of the free theory. Therefore, $P_k[\varphi,\tau]$, as given by (2.18), is bounded for all φ and has a limit $P_k[\varphi,\infty]$. A similar argument, applied to eq. (2.19), shows that $\partial P_k/\partial\tau$ also exists for all τ and all φ and has a well-defined limit when $\tau\to\infty$. It follows that $\partial P_k/\partial\tau\to 0$. This completes the proof of the lemma.

Armed with this result we proceed to the proof of our theorem, eq. (2.5), for $l=k\geqslant 1$. We consider the Fokker-Planck equation (2.4). In the limit $\tau\to\infty$ the left-hand side vanishes. The second term of the right-hand side contains $P_{k-1}[\varphi,\tau]$ which, according to the inductive hypothesis (2.5), goes to $P_{k-1}^{\rm eq}[\varphi]$. This is true even for $k=1$, which is not included in the inductive hypothesis, but it has been proven explicitly in (2.13). Therefore $P_k[\varphi,\tau]$ approaches, in the weak sense, a limiting distribution $P_k[\varphi,\infty]$ which satisfies

$$\int \mathrm{d}x \frac{\delta}{\delta\varphi(x)}\left\{\frac{\delta}{\delta\varphi(x)}+\frac{\delta S_0}{\delta\varphi(x)}\right\}P_k[\varphi,\infty]=-\int \mathrm{d}x\frac{\delta}{\delta\varphi(x)}\frac{\delta S_{\rm I}}{\delta\varphi(x)}P_{k-1}^{\rm eq}[\varphi]. \tag{2.20}$$

In a diagrammatic expansion we can consider that (2.20) defines the distribution $P_k[\varphi,\infty]$ in terms of $P_{k-1}^{\rm eq}[\varphi]$ which generates the Feynman diagrams of the ordinary perturbation series. In fact, (2.20) is identical to the Schwinger-Dyson equation [7] (see appendix B) and is therefore satisfied when $P_k[\varphi,\infty]$ is replaced by $P_k^{\rm eq}[\varphi]$. On the other hand the Schwinger-Dyson equation has a unique solution in regularized but unrenormalized perturbation theory and the same is true for the renormalized series up to renormalization group transformations. It follows that $P_k[\varphi,\infty]$ is identical to $P_k^{\rm eq}[\varphi]$, which completes our proof.

Before going to the more complicated case of the next section we want to add some remarks concerning massless field theories. Our argument goes through as before, except that now the estimations (2.14), (2.16) and (2.17) change. It is easy to verify that all three quantities still vanish in the limit $\tau\to\infty$ but the approach is an inverse power of τ and not an exponential. None of the previous conclusions is affected. In particular $\partial P_k/\partial\tau$ still goes to zero as τ goes to infinity but now as an inverse power of τ and not as an exponential. Needless to say that an important implicit assumption in our proof was that the equilibrium distribution $P_k^{\rm eq}[\varphi]$, or equivalently the Green functions of the ordinary perturbation theory, exist for all k. This is not true for massless superrenormalizable theories or for gauge theories without a gauge fixing term.

3. Gauge theories

In theories possessing a local invariance at the classical level the proof of the previous section fails because, already for the free field theory, there is no limiting distribution $P_0[A,\infty]$. One expects the limit to exist only for gauge-invariant quantities. Let $F[A]$ be a functional of the gauge potential $A^a_\mu(x)$. The average of F is defined by

$$\langle F\rangle_\eta = \int \mathcal{D}[A]F[A]P[A,\tau], \tag{3.1}$$

where P is again the probability distribution. It satisfies again the Fokker-Planck equation, eq. (1.6), with $S[A]$ the gauge-invariant classical action. The claim is that, although $P[A,\tau]$ has no limit when $\tau\to\infty$, the average (3.1) does possess such a limit for gauge-invariant functionals $F[A]$ which, furthermore, equals the result obtained by the usual method. We shall prove this statement, order by order in perturbation theory, following the method of ref. [8].

Under an infinitesimal gauge transformation with parameters $\theta^a(x)$ the fields $A^a_\mu(x)$ transform as

$$\delta A^a_\mu(x) = D^{ab}_\mu\theta^b(x) \equiv \delta^{ab}\partial_\mu\theta^b(x) - f^{abc}\theta^b(x)A^c_\mu(x), \tag{3.2}$$

and the functional $F[A]$ as

$$\delta F[A] = -\int \mathrm{d}x\,\theta^a(x)D^{ab}_\mu\frac{\delta}{\delta A^b_\mu(x)}F[A]. \tag{3.3}$$

A gauge-invariant functional satisfies

$$D^{ab}_\mu\frac{\delta}{\delta A^b_\mu(x)}F[A] = 0. \tag{3.4}$$

The crucial observation [8] is that, if one wants to compute the average of such a gauge-invariant functional, one may replace the stochastic equation (1.2) with the following one:

$$\frac{\partial A^a_\mu(x,t)}{\partial t} = -\frac{\delta S[A]}{\delta A^a_\mu(x,t)} - D^{ab}_\mu\mathcal{V}^b + \eta(x,t), \tag{3.5}$$

where, in order to simplify the notation, we have omitted the subscript η in the stochastic fields. D^{ab}_μ is given by (3.2) and $\mathcal{V}^a = \mathcal{V}^a[A,x]$ is an arbitrary gauge-non-invariant functional of A. Indeed, using (3.5), we can compute the time evolution of a gauge-invariant functional $F[A]$ which satisfies (3.4):

$$\frac{\partial F}{\partial t}=\int\frac{\partial A_\mu^a(x,t)}{\partial t}\frac{\delta F}{\delta A_\mu^a(x,t)}\mathrm{d}x, \tag{3.6}$$

and we can see that, by virtue of (3.4), the term proportional to $\mathcal{V}^a$ does not contribute. The average of F can be computed using a modified probability distribution $P_{\mathcal{V}}[A,\tau]$ which satisfies the following Fokker-Planck equation:

$$\frac{\partial P_{\mathcal{V}}}{\partial \tau}=\int\mathrm{d}x\frac{\delta}{\delta A_\mu^a(x)}\left\{\frac{\delta}{\delta A_\mu^a(x)}+\frac{\delta S}{\delta A_\mu^a(x)}+D_\mu^{ab}\mathcal{V}^b\right\}P_{\mathcal{V}}[A,\tau]. \tag{3.7}$$

We see again that, because of (3.4), the average value of a gauge-invariant quantity F does not depend on the choice of $\mathcal{V}^a$ and is the same irrespective of whether one uses the distribution $P[A,\tau]$ or the modified one $P_{\mathcal{V}}[A,\tau]$. Of course, this is not true for gauge-non-invariant quantities.

The advantage of this approach is now obvious. If $\mathcal{V}^a$ is gauge non-invariant the distribution $P_{\mathcal{V}}[A,\tau]$ does possess a well-defined limit when $\tau\to\infty$, at least in perturbation theory, and our previous proof can be adjusted to cover this case. In fact, the required changes in our formulation of the previous section are essentially notational. We shall present here the special case $\mathcal{V}^a(x)=(1/\alpha)\partial_\mu A_\mu^a(x)$ with α an arbitrary constant. Because S contains terms proportional to both g and g^2, the Fokker-Planck equation (2.3) will connect $P_{\mathcal{V},k}$ with $P_{\mathcal{V},k-1}$ and $P_{\mathcal{V},k-2}$ for $k\geqslant 2$. $P_{\mathcal{V},1}$ will be only connected to $P_{\mathcal{V},0}$. We again make the inductive hypothesis (2.5) which we shall verify for $P_{\mathcal{V},0}$. The analogs of eqs. (2.10), (2.11) are

$$P_{\mathcal{V},0}[A,\tau]=N_0^{-1}\exp\left\{-\tfrac{1}{2}\int\mathrm{d}k\,A_\mu^a(k)\left(\Delta_{\mu\nu}^{ab}\right)^{-1}A_\nu^b(-k)\right\}, \tag{3.8}$$

with

$$\Delta_{\mu\nu}^{ab}=\delta^{ab}\frac{1}{k^2}\left[\left(\delta_{\mu\nu}-\frac{k_\mu k_\nu}{k^2}\right)\left(1-\mathrm{e}^{-2k^2\tau}\right)+\alpha\frac{k_\mu k_\nu}{k^2}\left(1-\mathrm{e}^{-2k^2\tau/\alpha}\right)\right]. \tag{3.9}$$

As expected, for $\alpha=1$ we obtain the "Feynman propagator" for finite τ. For $\alpha\to\infty$ we recover the propagator given in ref. [3]. We see that

$$\underset{\tau\to\infty}{\mathrm{w.lim}}\,P_{\mathcal{V},0}[A,\tau]=P_{\mathcal{V},0}^{\mathrm{eq}}[A], \tag{3.10}$$

which verifies the inductive hypothesis. Similarly we find for $\mathcal{D}_0[A,A';\tau]$:

$$\mathcal{D}_0=\theta(\tau)\Delta_0^{-1}\exp\left\{-\tfrac{1}{2}\int\mathrm{d}k\,A_\mu^a(k)D_{\mu\nu}^{ab}A_\nu^b(-k)\right\}, \tag{3.11}$$

$$D_{\mu\nu}^{ab}=\delta^{ab}\left\{\left(\delta_{\mu\nu}-\frac{k_\mu k_\nu}{k^2}\right)D(\tau,m=0)+\frac{1}{\alpha}\frac{k_\mu k_\nu}{k^2}D\left(\frac{\tau}{\alpha},m=0\right)\right\}, \tag{3.12}$$

where again A^a_μ denotes $(A^a_\mu, A^{a\prime}_\mu)$ and the 2×2 matrix D is the scalar matrix given in eqs. (2.12) with $m=0$.

From this point the proof goes as before with the estimations which correspond to a massless theory. From the explicit construction of $P_{\mathcal{V},0}$ we first prove the theorem for $P_{\mathcal{V},1}$; then, using both, we proceed to $P_{\mathcal{V},2}$, etc. The conclusion is that, in the limit $\tau\to\infty$, we obtain a regularized perturbation series which, however, is not the one we would derive from the usual theory with a gauge fixing term proportional to $(\partial_\mu A_\mu)^2$ [8]. This is obvious from the absence of explicit Faddeev-Popov ghost terms whose contribution is replaced by the more complicated structure of the vertices. There is a choice of $\mathcal{V}^a$ which precisely reproduces the Faddeev-Popov perturbation theory, but in the stochastic language, it is a non-local functional of A [8].

We wish to acknowledge very helpful discussions with Drs. H. Epstein, V. Glaser, N. Sourlas and R. Stora.

Appendix A

In this appendix we calculate the quantities $P_0[\varphi,\tau]$ and $\mathcal{D}_0[\varphi,\varphi';\tau]$. By definition $P_0[\varphi,\tau]$ is

$$P_0[\varphi,\tau]=\int\mathcal{D}[\eta]\mathrm{e}^{-\frac{1}{4}\int\eta^2(x,t)\mathrm{d}x\mathrm{d}t}\prod_y\delta\big(\varphi(y)-\varphi^0_\eta(y,\tau)\big), \tag{A.1}$$

where φ^0_η satisfies the free Langevin equation,

$$\frac{\partial\varphi^0_\eta}{\partial t}=(\partial^2-m^2)\varphi^0_\eta+\eta(x,t). \tag{A.2}$$

The solution of this equation with initial condition $\varphi^0_\eta(x,0)=0$ is the following:

$$\varphi^0_\eta(x,\tau)=\int\mathrm{d}y\int_0^\tau\mathrm{d}t\,G(x-y;\tau-t)\eta(y,t), \tag{A.3}$$

where the Green function $G(x-y;\tau-t)$ is given by

$$G(x-y;\tau-t)=\frac{1}{(2\pi)^d}\int\mathrm{d}k\,\mathrm{e}^{ik(x-y)-(\tau-t)(k^2+m^2)}. \tag{A.4}$$

We shall also need the correlation function D,

$$\begin{aligned}D(x,y;\tau,\tau')&\equiv\langle\varphi(x,\tau)\varphi(y,\tau')\rangle_\eta\\&=2\int\mathrm{d}z\int_0^{\min(\tau,\tau')}\mathrm{d}t\,G(x-z;\tau-t)G(y-z;\tau'-t).\end{aligned} \tag{A.5}$$

To find $P_0[\varphi, \tau]$, we write eq. (A.1) in the form

$$P_0[\varphi, \tau] = \int \mathcal{D}[\eta]\mathcal{D}[\xi] e^{-\frac{1}{4}\int dx\, dt\, \eta^2(x,t)} e^{i\int dx\, \xi(x)[\varphi(x) - \varphi_\eta^0(x,\tau)]}. \tag{A.6}$$

Then we introduce the expression (A.3) for φ_η^0 and we do the gaussian, η and ξ integrations. The result is

$$P_0[\varphi, \tau] = N_0^{-1} \exp\left\{ -\tfrac{1}{2} \int dx\, dx' \varphi(x) D^{-1}(x, x'; \tau, \tau) \varphi(x') \right\}, \tag{A.7}$$

where D^{-1} is defined as

$$\int dz\, D(x, z; \tau, \tau) D^{-1}(z, x'; \tau, \tau) = \delta(x - x'). \tag{A.8}$$

From (A.5) and (A.4) we find

$$D(x, y; \tau, \tau) = \frac{1}{(2\pi)^{d/2}} \int dk\, e^{ik(x-y)} \frac{1 - e^{-2\tau(k^2 + m^2)}}{k^2 + m^2}, \tag{A.9}$$

and so in momentum space (A.7) reduces to the expression (2.10).

For the functional propagator $\mathcal{D}_0[\varphi, \varphi'; \tau]$ we use its definition

$$\mathcal{D}_0[\varphi, \varphi'; \tau - \tau'] \equiv P_0[\varphi, \varphi'; \tau, \tau'] / P_0[\varphi', \tau'] \tag{A.10}$$

as the conditional probability density, where

$$P_0[\varphi, \varphi'; \tau, \tau'] = \int \mathcal{D}[\eta] e^{-\frac{1}{4}\int \eta^2 dx\, dt} \prod_y \delta\big(\varphi(y) - \varphi_\eta^0(y, \tau)\big) \delta\big(\varphi'(y) - \varphi_\eta^0(y, \tau')\big). \tag{A.11}$$

We follow the same method as for $P_0[\varphi, \tau]$ and forming the ratio (A.10) we find the expression (2.11).

Appendix B

In sect. 2, in the main proof of the uniqueness of $P^{\mathrm{eq}}[\varphi]$, we used the statement that eq. (2.20)

$$\int dx \frac{\delta}{\delta\varphi(x)} \left\{ \frac{\delta}{\delta\varphi(x)} + \frac{\delta S_0}{\delta\varphi(x)} \right\} P_k[\varphi, \infty] = -\int dx \frac{\delta}{\delta\varphi(x)} \frac{\delta S_{\mathrm{I}}}{\delta\varphi(x)} P_{k-1}^{\mathrm{eq}}[\varphi] \tag{B.1}$$

is equivalent to the Schwinger-Dyson equations. We demonstrate this statement in this appendix. Let us define

$$P_k[\varphi,\infty]-P_k^{\mathrm{eq}}[\varphi]=\mathrm{e}^{-S_0}G_k[\varphi]. \tag{B.2}$$

Since $P_k^{\mathrm{eq}}[\varphi]$ satisfies (B.1), it follows that G_k satisfies

$$\int \mathrm{d}x\frac{\delta}{\delta\varphi(x)}\left\{\mathrm{e}^{-S_0}\frac{\delta G_k[\varphi]}{\delta\varphi(x)}\right\}=0. \tag{B.3}$$

Eq. (B.3) implies that G_k, as defined in (B.2), is independent of φ. A simple way to see it is to multiply eq. (B.3) by G_k and integrate over φ. We obtain

$$\begin{aligned}0&=\int\mathcal{D}[\varphi]\int\mathrm{d}x\,G_k[\varphi]\frac{\delta}{\delta\varphi(x)}\left\{\mathrm{e}^{-S_0}\frac{\delta G_k[\varphi]}{\delta\varphi(x)}\right\}\\&=-\int\mathcal{D}[\varphi]\int\mathrm{d}x\,\mathrm{e}^{-S_0}\left\{\frac{\delta G_k[\varphi]}{\delta\varphi(x)}\right\}^2.\end{aligned} \tag{B.4}$$

The partial integration in (B.4) is legitimate because both $P_k[\varphi,\infty]$ and $P_k^{\mathrm{eq}}[\varphi]$ contain the gaussian factor e^{-S_0}. The integrand of (B.4) is positive, therefore G_k is a constant. It follows that $P_k[\varphi,\infty]$ satisfies the Schwinger-Dyson equation, or, equivalently, the equations of motion. The arbitrariness in the determination of $P_k[\varphi,\infty]$ is therefore a multiplet of the free equilibrium distribution e^{-S_0}. This arbitrariness is also present in the Schwinger-Dyson equations. It can be fixed by using either an initial condition $P_k^{\mathrm{eq}}[\varphi=0]=0$ for $k\neq 0$ or the normalization condition $\int P_k^{\mathrm{eq}}[\varphi]\mathcal{D}[\varphi]=0$ for $k\neq 0$. This same condition is also satisfied by $P_k[\varphi,\tau]$ given by eq. (2.7). We conclude that $G_k=0$ and $P_k[\varphi,\infty]=P_k^{\mathrm{eq}}[\varphi]$.

Appendix C

In this appendix we show that the asymptotic behaviour of $\mathcal{D}_0[\varphi,\varphi';\tau]$ and $P_0[\varphi,\tau]$ for large times τ is not changed if we choose a different initial condition for the solution of the free Langevin equation. Let us consider the Langevin equation

$$\frac{\partial\varphi_\eta^0}{\partial\tau}=(\partial^2-m^2)\varphi_\eta^0+\eta(x,\tau), \tag{C.1}$$

where

$$\varphi_\eta^0(x,0)=C(x). \tag{C.2}$$

Then the solution of (C.1) is

$$\varphi_\eta^0(x,\tau) = \int \mathrm{d}y\, G(x-y;\tau) C(y) + \int_0^\tau \mathrm{d}t \int \mathrm{d}y\, G(x-y;\tau-t)\eta(y,t), \quad \text{(C.3)}$$

where G is given in (A.4). We repeat the calculation for $P_0[\varphi,\tau]$, but now $\varphi_\eta^0(x,\tau)$ of (A.3) is replaced by that of (C.3). We can easily see that expression (2.10) which gives $P_0[\varphi,\tau]$ for the case of the previous initial condition $C=0$ is simply replaced by

$$P_0[\varphi,\tau] = N_0^{-1}\exp\left\{ -\tfrac{1}{2}\int \mathrm{d}k \left[\varphi(k) - \mathrm{e}^{-\tau(k^2+m^2)} C(k)\right] \frac{k^2+m^2}{1-\mathrm{e}^{-2\tau(k^2+m^2)}} \times \left[\varphi(-k) - \mathrm{e}^{-\tau(k^2+m^2)} C(-k)\right]\right\}, \quad \text{(C.4)}$$

that is, φ is replaced by

$$\varphi(x) \to \varphi(x) - \int \mathrm{d}y\, G(x-y;\tau) C(y). \quad \text{(C.5)}$$

The same is true for the evaluation of $\mathcal{D}_0[\varphi,\varphi';\tau]$ with the initial condition (C.2). We see again that for any $C(x)$

$$P_0[\varphi,\tau] \underset{\tau\to\infty}{\to} P_0^{\mathrm{eq}}[\varphi]. \quad \text{(C.6)}$$

The inductive argument of the main theorem as well as the lemma still remain valid.

References

[1] (a) N.G. van Kampen, Stochastic processes in physics and chemistry (North-Holland, Amsterdam, 1981)
(b) J. Glimm and A. Jaffe, Quantum physics (Springer-Verlag, New York, 1981)
[2] G. Parisi and N. Sourlas, Phys. Rev. Lett. 43 (1979) 744; Nucl. Phys. B206 (1982) 321;
S. Cecotti and L. Girardello, to be published
[3] G. Parisi and Y.S. Wu, Sci. Sin. 24 (1981) 483
[4] W.D. Wick, Comm. Math. Phys. 81 (1981) 361 and references therein
R.F. Fox, Phys. Reports 48 (1978) 179
[5] C. De Dominicis, Nuovo Cim. Lett. 12 (1975) 567
[6] V.N. Gribov, Nucl. Phys. B139 (1978) 1;
I. Singer, Comm. Math. Phys. 60 (1978) 7
[7] G. Marchesini, Nucl. Phys. B191 (1981) 214
[8] D. Zwanziger, Nucl. Phys. B192 (1981) 259;
L. Baulieu and D. Zwanziger, Nucl. Phys. B193 (1981) 163

Z. Phys. C - Particles and Fields 18, 129-134 (1983)

Zeitschrift für Physik C Particles and Fields

Perturbation Theory from Stochastic Quantization of Scalar Fields

W. Grimus

CERN, CH-1211 Genf 23, Switzerland

H. Hüffel*

Institut für Theoretische Physik, Universität Wien, A-1090 Wien, Austria

Received 4 November 1982

Abstract. By using a diagrammatical technique it is shown that in scalar theories the stochastic quantization method of Parisi and Wu gives the usual perturbation series in Feynman diagrams.

1. Introduction

Parisi and Wu [1] introduced a stochastic quantization method which is based on the Langevin equation [2] of non-equilibrium statistical mechanics. They applied their method to the quantization of scalar and gauge field theories with the main aim of constructing a new perturbation theory for gauge theories without introducing gauge fixing. Therefore for non-Abelian gauge theories their method is free of the Gribov ambiguity [3].

In this paper we deal with scalar theories and show diagrammatically that the perturbation theory of Parisi and Wu gives the usual Feynman diagrams when calculating Green functions. Our paper is organized as follows: In Sect. II we recall the basic features of the stochastic quantization method and how to obtain "stochastic diagrams" from an iterative solution of the Langevin equation. In Sect. III we show how to perform the time integrations of stochastic diagrams by introducing time ordering of the vertices and in Sect. IV we prove the equivalence of sums of stochastic diagrams and Feynman diagrams in the case of self-interacting scalar fields. According to a remark in [1] this has been shown in [4] but we find it hard to see what has been proved in this connection.

* Supported in part by a fellowship of the French Government

2. Langevin Equation and Stochastic Diagrams

The starting point of the discussion of stochastic diagrams is the introduction of a fictitious time for the fields whose time evolution is given by the Langevin equation

$$\frac{\partial \Phi(x,t)}{\partial t} = -\frac{\delta S}{\delta \Phi(x,t)} + \eta(x,t) \tag{2.1a}$$

$$S = \int d^D x \left\{ \frac{1}{2}(\partial \Phi)^2 + \frac{1}{2} m^2 \Phi^2 + \frac{1}{3!} g \Phi^3 + \frac{1}{4!} \lambda \Phi^4 \right\}. \tag{2.1b}$$

We have written down (2.1) for a real self-interacting scalar field. S is the Euclidean action in D space-time dimensions. The fictitious time t should not be confused with the physical time contained in x. η is a random source with Gaussian distribution

$$\langle \eta(x,t)\, \eta(x',t') \rangle_\eta = 2\delta(x-x')\, \delta(t-t') \tag{2.2a}$$

$$\langle \eta(x_1,t_1) \dots \eta(x_{2n+1},t_{2n+1}) \rangle_\eta = 0 \tag{2.2b}$$

$$\langle \eta(x_1,t_1) \dots \eta(x_{2n},t_{2n}) \rangle_\eta = \sum_{\substack{\text{possible} \\ \text{pair comb.}}} \prod_{\text{pairs}} \langle \eta(x_i,t_i)\, \eta(x_j,t_j) \rangle_\eta . \tag{2.2c}$$

Explicitly we can write the random average over η

$$\langle \dots \rangle_\eta = \frac{\int [d\eta] (\dots) \exp(-\frac{1}{4}\int d^D x\, dt\, \eta^2(x,t))}{\int [d\eta] \exp(-\frac{1}{4}\int d^D x\, dt\, \eta^2(x,t))} . \tag{2.3}$$

Now the crucial point is that in the limit $t \to \infty$ equilibrium is reached and the random average of correlation functions of $\phi(x,t)$ tends to the corresponding Green functions of the quantum field theory with action (2.1b)

$$\lim_{t\to\infty} \langle \Phi(x_1,t) \dots \Phi(x_L,t) \rangle_\eta = \langle \phi(x_1) \dots \Phi(x_L) \rangle . \tag{2.4}$$

This has been shown in [1] for scalar theories by using the Fokker-Planck equation [2]. For gauge theories the situation is more complicated [5]. In contrast to [1] we prove (2.4) by a diagrammatical technique in the case of scalar fields.

The Langevin equation (2.1a) can be transformed into an integral equation in momentum space

$$\Phi(k,t)=\int_0^t d\tau\, G(k,t-\tau)\Big\{\eta(k,\tau) \tag{2.5a}$$

$$-\frac{1}{2!}g\int\frac{d^Dp}{(2\pi)^D}\,\Phi(p,\tau)\,\Phi(k-p,\tau)$$

$$-\frac{1}{3!}\lambda\int\frac{d^Dp}{(2\pi)^D}\int\frac{d^Dq}{(2\pi)^D}\,\Phi(p,\tau)\,\Phi(q,\tau)\,\Phi(k-p-q,\tau)\Big\}$$

$$G(k,t-\tau)=\exp(-(t-\tau)(k^2+m^2)) \tag{2.5b}$$

where the boundary condition $\phi(x,0)=0$ has been used. However, any trace of a specific boundary condition should die out with $t\to\infty$ due to the equilibrium property of the system. In momentum space (2.2a) reads

$$\langle\eta(k,t)\,\eta(k',t')\rangle_\eta=2(2\pi)^D\,\delta(k+k')\,\delta(t-t'). \tag{2.6}$$

Solving (2.5) by iteration one arrives at a power series expansion of Φ in the coupling constants, which can be written diagrammatically

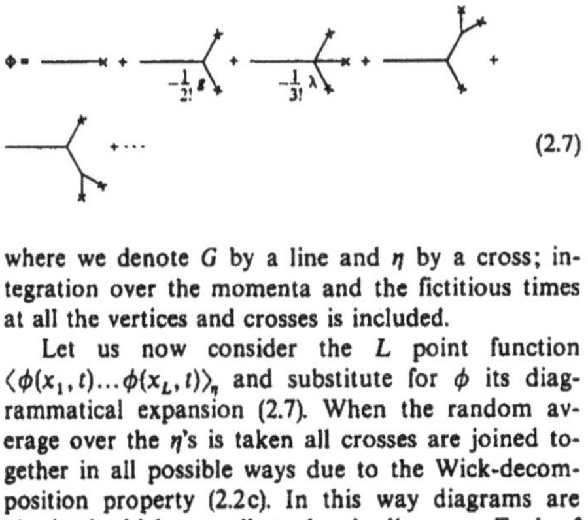

where we denote G by a line and η by a cross; integration over the momenta and the fictitious times at all the vertices and crosses is included.

Let us now consider the L point function $\langle\phi(x_1,t)\ldots\phi(x_L,t)\rangle_\eta$ and substitute for ϕ its diagrammatical expansion (2.7). When the random average over the η's is taken all crosses are joined together in all possible ways due to the Wick-decomposition property (2.2c). In this way diagrams are obtained which we call stochastic diagrams. Each of them has the form of an ordinary Feynman diagram of the theory described by the action S apart from crosses on the lines where two η's have been joined together (Fig. 1). Conversely, to every Feynman diagram there exists a number of stochastic diagrams with the same topology [6]. Actually we will show in Sect. IV that the sum of all stochastic diagrams to a given Feynman diagram yields just this Feynman diagram. Obviously cutting a stochastic diagram of

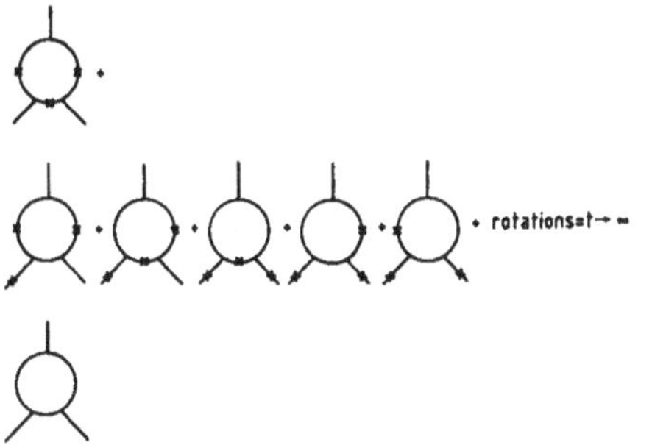

Fig. 1. All stochastic diagrams belonging to a given Feynman diagram. This is an example of what is proved in Sect. IV

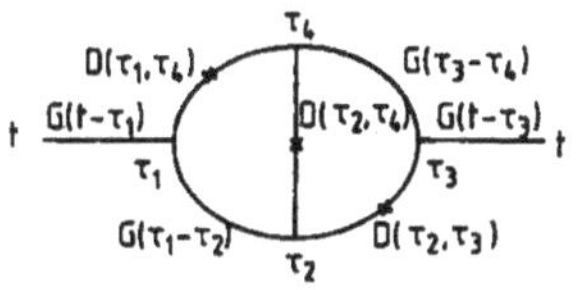

Fig. 2. A stochastic diagram and the times and propagators associated to it. In each of the G's the first time must be greater than the second one because of the time integration in the expansion (2.7). Momenta have been omitted

an L point function at all crosses gives L connected trees contained in (2.7).

The momentum δ function in (2.6) guarantees momentum conservation at a cross and eliminates one momentum integration, so momentum integration is reduced just to the usual one of Feynman diagrams. Also the time integrations on both sides of a cross can be performed and lead to the propagator

$$D(k,\tau,\tau')=2\int_0^{\min(\tau,\tau')} d\tau''\,G(k,\tau-\tau'')\,G(k,\tau'-\tau'')$$

$$=\frac{1}{k^2+m^2}(\exp(-|\tau-\tau'|(k^2+m^2))$$

$$-\exp(-(\tau+\tau')(k^2+m^2))). \tag{2.8}$$

τ, τ' are the times of the neighbour vertices. This means that in a stochastic diagram each line represents the Green function G and each crossed line the propagator D respectively (Fig. 2).

3. Integrations Over the Fictitious Times

In this section we will show how to perform the time integration of stochastic diagrams by introducing time ordering of the vertices. Let us consider a stochastic diagram belonging to some Feynman diagram with L external lines and N vertices. Each vertex carries a time τ_i which has to be integrated over between 0 and the time of some next vertex according to (2.7). Due to the absolute values of time differences in the D propagators it is however convenient to introduce fixed time orderings of the vertices. Given the Feynman graph we will allow all time orderings of vertices which are compatible with at least one of the corresponding stochastic diagrams (e.g. in Fig. 2 τ_2 or τ_4 cannot be greater than all the other τ's, because one could not go then from vertex 2(4) to an external leg via an increasing sequence of times at the vertices for any distribution of the crosses). In general, a given allowed time ordering belongs to several stochastic diagrams. In performing the time integrations we will only keep the largest terms thus leaving out the second term in D and always dropping terms, which come from the lower boundaries of the integrations. We will see that the integration over the largest terms compensates exactly the exponential t dependence of the G's on the external lines and all smaller terms will go to zero exponentially in t.

Let us concentrate now on a fixed time ordering of the N vertices which we can choose to be

$$0<\tau_1<\tau_2<\ldots<\tau_N<t. \tag{3.1}$$

We denote the set of momenta of the i-th vertex by V_i and leave out the momentum denominators of the D's for the moment. For convenience of notation we substitute $p^2+m^2 \to p^2$ for all momenta. Thus masses are trivially contained in our discussion. Now we integrate over τ_1. Because it is the smallest time all τ_1 exponents in G's and D's have positive signs (except those we have already dropped from D, which do not concern us anymore) and we obtain

$$\int_0^{\tau_2} d\tau_1 \exp(\tau_1 \sum_{V_1} p^2) = \frac{\exp(\tau_2 \sum_{V_1} p^2)-1}{\sum_{V_1} p^2}. \tag{3.2}$$

We drop the 1 in (3.2) and go to the τ_2 integration. Now there are two possibilities:

a) vertex 2 is not a neighbour vertex of the first one, which means that all neighbour vertices of vertex 2 have times larger than τ_2. In this case again all τ_2 exponents have positive signs and

$$\int_0^{\tau_3} d\tau_2 \exp(\tau_2(\sum_{V_1} p^2 + \sum_{V_2} p^2)) = \frac{\exp(\tau_3 \sum_{V_1 \cup V_2} p^2)-1}{\sum_{V_1 \cup V_2} p^2}. \tag{3.3}$$

The 1 will be dropped as before.

b) vertex 2 is a neighbour of vertex 1. In this case special care has to be taken for the lines which connect vertices 1 and 2 as the corresponding G's and C's depend on both τ_1 and τ_2; they are of the form $\exp(-(\tau_2-\tau_1)p^2)$ before τ_1 integration. From this one sees immediately that all τ_2 exponents coming from momenta, which connect vertices 1 and 2, are cancelled by τ_1 integration. On the other hand all other neighbour vertices of 2 will give positive τ_2 exponents. Explicitly we have

$$\int_0^{\tau_3} d\tau_2 \exp(\tau_2(\sum_{V_1} p^2 - \sum_{V_1 \cap V_2} p^2 + \sum_{V_2 - V_1 \cap V_2} p^2)) = \frac{\exp(\tau_3 \sum_{W_2} p^2)-1}{\sum_{W_2} p^2} \tag{3.4a}$$

$$W_2 = V_1 \cup V_2 - V_1 \cap V_2 \tag{3.4b}$$

(3.4) also includes (3.3), so it is the general result of the second integration. Using the same arguments as in (3.4) one obtains for the k-th integration

$$\int_0^{\tau_{k+1}} d\tau_k \exp(\tau_k(\sum_{W_{k-1}} p^2 - \sum_{\substack{\cup(V_i \cap V_k)\\ 1 \leqq i<k}} p^2 + \sum_{\substack{V_k - \cup(V_i \cap V_k)\\ 1 \leqq i<k}} p^2)) = \frac{\exp(\tau_{k+1} \sum_{W_k} p^2)-1}{\sum_{W_k} p^2} \tag{3.5a}$$

$$\begin{aligned} W_k &= (W_{k-1} - \bigcup_{1 \leqq i<k} (V_i \cap V_k)) \cup (V_k - \bigcup_{1 \leqq i<k} (V_i \cap V_k)) \\ &= \bigcup_{i=1}^{k} V_k - \bigcup_{1 \leqq i<j \leqq k} (V_i \cap V_j) \\ W_1 &\equiv V_1, \quad \tau_{N+1} \equiv t \end{aligned} \tag{3.5b}$$

(3.5b) can easily be shown by induction.

Now we would like to make some remarks to (3.5).

a) The procedure is independent of the number of momenta leading to a vertex.

b) The contribution from the N-th time integration is given by

$$\frac{\exp(t \sum_{\substack{\text{external}\\ \text{lines}}} p^2)-1}{\sum_{\substack{\text{external}\\ \text{lines}}} p^2} \tag{3.6}$$

so that the t exponents of the G's and D's corresponding to external lines are exactly cancelled. We see at this point that the neglect of non-leading con-

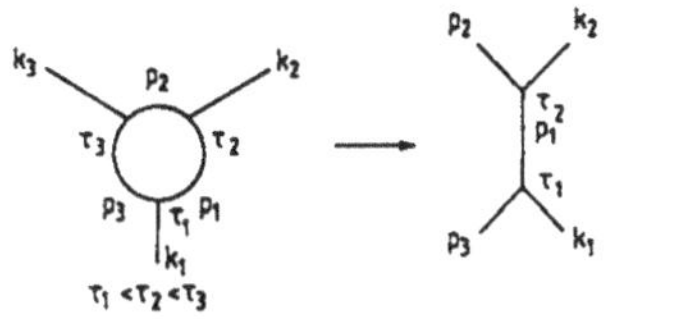

Fig. 3. This figure illustrates remark d) in Sect. III. The second diagram has been obtained from the first one by removing the vertex with the largest time τ_3 and the external line k_3

tributions is justified due to their exponential fall off in t.

c) From all time integrations we finally get for

$$\prod_{k=1}^{N} \frac{1}{\sum\limits_{W_k} p^2}. \tag{3.7}$$

d) Note that leaving out the last time integration in (3.7) we get exactly the expression of the time integration of a diagram obtained from the original one by dropping the τ_N vertex with its external line(s). In the example of Fig. 3 for instance we get

$$\frac{1}{k_1^2+p_1^2+p_3^2}\,\frac{1}{k_1^2+k_2^2+p_2^2+p_3^2}\,\frac{1}{k_1^2+k_1^2+k_3^2}$$
$$\to \frac{1}{k_1^2+p_1^2+p_3^2}\,\frac{1}{k_1^2+k_2^2+p_2^2+p_3^2}. \tag{3.8}$$

e) Given a time ordered stochastic diagram we find therefore

$$\prod_{k=1}^{N} \frac{1}{\sum\limits_{W_k} p^2} \prod_{\text{crosses}} \frac{1}{p^2} \tag{3.9}$$

where the second product of momenta arises simply from the so far neglected denominators of the D's.

f) To obtain the full contribution of a stochastic diagram one has to sum (3.9) over all allowed time orderings.

4. Equivalence Proof

In this section we prove by induction on the number of vertices N of a given Feynman diagram that the sum over all stochastic diagrams with the same topology gives exactly the Feynman diagram in the case of the real scalar field theory (2.1 b).

For $N=1$ we only have the 3 point and the 4 point vertex function in lowest order. According to the simplest application of the time integration rule (3.7) and (3.9) we get

$$-\frac{1}{2!}g$$

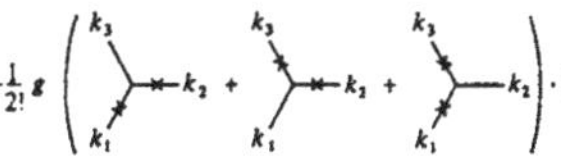

(4.1 a)

$$= -g\frac{1}{k_1^2+k_2^2+k_3^2}\left(\frac{1}{k_1^2k_2^2}+\frac{1}{k_2^2k_3^2}+\frac{1}{k_3^2k_1^2}\right)=\frac{-g}{k_1^2k_2^2k_3^2}$$

$$-\frac{1}{3!}\lambda$$

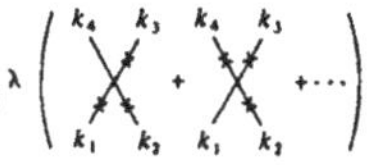

$$= -\lambda\frac{1}{k_1^2+k_2^2+k_3^2+k_4^2}\left(\frac{1}{k_1^2k_2^2k_3^2}+\frac{1}{k_2^2k_3^2k_4^2}+\dots\right)$$
$$=\frac{-\lambda}{k_1^2k_2^2k_3^2k_4^2}. \tag{4.1 b}$$

2! (3!) is the number of possibilities of joining the η's together to obtain the desired Feynman diagram and for the crosses we have inserted the propagators stemming from the D's according to (3.9). We have exactly obtained the corresponding Feynman diagrams. Because of the expansion (2.7) all Green functions we consider are non-truncated.

Now we assume that we have proved the equivalence to Feynman diagrams for all numbers of vertices smaller than N with any number of external lines.

Let us consider a Feynman diagram F with N vertices and L external lines and a stochastic diagram S_F with the same topology. In F and S_F momentum integration is not contained. It is easy to see that S_F has the number of crosses

$$Z=\frac{L+N_3+2N_4}{2} \tag{4.2}$$

where $N_{3(4)}$ is the number of 3(4) vertices of F and $N=N_3+N_4$.

Now we introduce time ordering of the vertices compatible with S_F. The largest time has to be at a vertex with at least one external line (see (2.7) and Sect. III). For an external line there are five topological possibilities shown in Fig. 4. All lines from vertex N leading into the blobs of Fig. 4 are inner lines. We call the largest time τ_N. With k's we denote external momenta and with p's inner momenta. Integration over τ_N yields

$$\frac{1}{\sum\limits_{i=1}^{L} k_i^2} \equiv K. \tag{4.3}$$

We discuss now the different cases of Fig. 4:

Case (a). Because τ_N is the largest time there cannot be a cross on k_1. Removing the momentum k_1 and

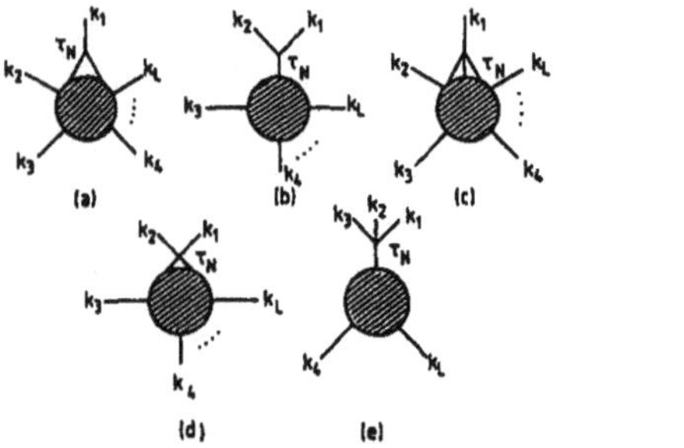

Fig. 4a–e. The topological possibilities for an external line in a self-interacting scalar field theory with action (2.1 b)

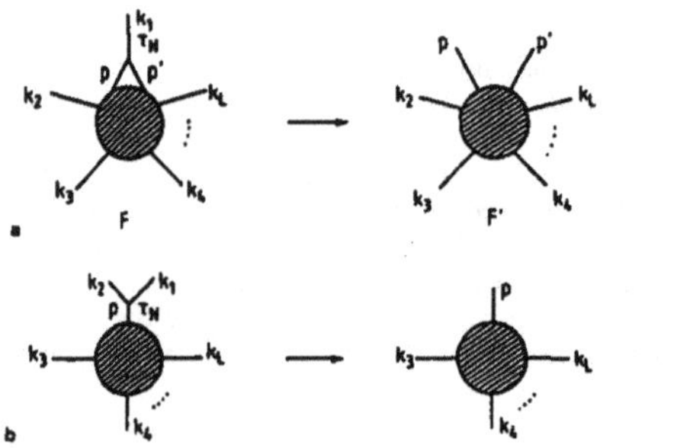

Fig. 5a and b. Removing the vertex with the largest time τ_N and its external lines from F one obtains F'

$-1/2!\ g$ from F we obtain a Feynman diagram F' (Fig. 5a). Doing the same with S_F we get $\tilde{S}_F$. The factor 1/2! comes from the first iteration of $\phi(k_1)$ and is exactly needed for a loop with two identical bosons. The change in the number of crosses going from topology F to F' is $\Delta Z=0$ (4.2). Because there was no cross on k_1 this means that $\tilde{S}_F$ has the right number of crosses to be a stochastic diagram with topology F' and because of remark d) in Sect. III the time integration of S_F with the given time ordering is just (time integration of $\tilde{S}_F) \times K$. Now summing over all time ordered stochastic diagrams S_F keeping τ_N as largest time one obtains (sum over all stochastic diagrams of $F') \times K$. According to our induction assumption the sum over all stochastic diagrams of F' gives F', because F' has one vertex less than F. So we obtain for the sum over all time ordered stochastic diagrams of F with τ_N as largest time

$$-\frac{1}{2!} g KF' = k_1^2 K \left(-\frac{1}{2!} g\right) \frac{1}{k_1^2} F' = \frac{k_1^2}{\sum\limits_{i=1}^{L} k_i^2} F. \tag{4.4}$$

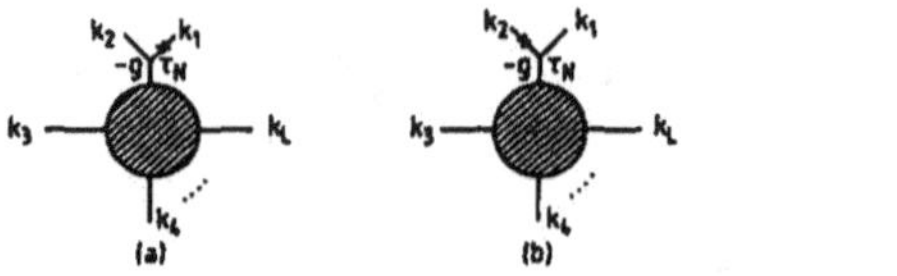

Fig. 6a and b. The two possibilities for crosses on the external lines of the vertex with τ_N of Fig. 4b

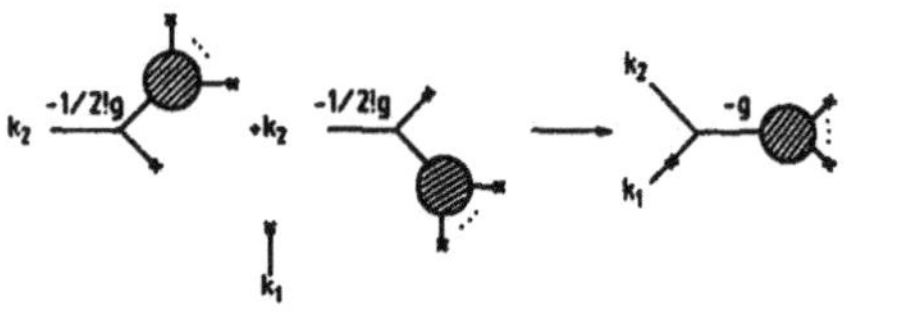

Fig. 7. This figure shows how the 1/2! at the vertex with time τ_N in Fig. 6a is cancelled by two identical terms in the expansion of $\phi(k_2)$

Case (b). There must be now either a cross on k_1 or k_2, so we consider the two stochastic diagrams of Fig. 6. Because the iterative expansion (2.7) of $\phi(k_2)$ in Fig. 6a for example contains two identical pieces we have only $-g$ at the vertex τ_N (see Fig. 7) in accordance with the Feynman rules.

Now we remove k_1, k_2 and $-g$ from F and obtain F' (Fig. 5b) and $\tilde{S}_F$. The change in the number of crosses going from F to F' is $\Delta Z = -1$. Again $\tilde{S}_F$ is a correct stochastic diagram with topology F'. Doing the same summation as in case (a) we get

$$-g\left(\frac{1}{k_1^2}+\frac{1}{k_2^2}\right) KF' = \frac{k_1^2+k_2^2}{\sum\limits_{i=1}^{L} k_i^2} F \tag{4.5}$$

$1/k_1^2$ and $1/k_2^2$ come from the D propagators at the places of the crosses in Fig. 6.

Analogously case (c) yields the same as (a), case (d) the same as (b) and case (e) gives

$$\frac{k_1^2+k_2^2+k_3^2}{\sum\limits_{i=1}^{L} k_i^2} F. \tag{4.6}$$

Thus summing over all possible places for the largest time τ_N we get a contribution $k_i^2 KF$ from each external leg with momentum k_i and all the contributions sum up to give F.

We want to stress that the correct momentum integration of the Feynman diagram has already been obtained by the iterative expansion of ϕ (2.7) and the random average over the η's (2.6).

We have shown now the equivalence in the case of one real scalar field. It is straightforward to generalize our considerations to many scalar fields and complex scalar fields. Then every field has its own random source, also the charge conjugate of a complex field [1]. The time integrations are the same as in Sect. III and the only point, where one has to be careful in Sect. IV, is to get the right combinatorial factors of the Feynman diagrams, but this can easily be seen from the iterative expansion of the fields.

Acknowledgements. H.H. is very grateful for the hospitality at the Laboratoire de Physique Théorique de l'Ecole Normale Supérieure in Paris and thanks especially J. Iliopoulos for stimulating discussions. Both authors want to thank E. Etim for valuable discussions and J.S. Bell for a critical reading of the manuscript.

References

1. G. Parisi, Wu Yongshi: Sci. Sin. **24**, 483 (1981)
2. See e.g. A.H. Jazwinski: Stochastic processes and filtering theory. In: Mathematics in science and engineering. Bellmann, R. (ed.), Vol. 64. New York: Academic Press 1970
3. V.N. Gribov: Nucl. Phys. **B 139**, 1 (1978)
4. C. De Dominicis: Lett. Nuovo Cimento **12**, 567 (1975)
5. D. Zwanziger: Nucl. Phys. **B 192**, 259 (1981); D. Zwanziger: Phys. Lett. **114 B**, 337 (1982); L. Baulieu, D. Zwanziger: Nucl. Phys. **B 193**, 163 (1981)
6. A. Niemi, L.C.R. Wijewardhana: Ann. Phys. **140**, 247 (1982). Though their method of stochastic quantization is quite different from that of Parisi and Wu, the diagrammatic representation is the same in both methods, but the interpretation is different

Nuclear Physics B260 (1985) 545-568

STOCHASTIC DIAGRAMS AND FEYNMAN DIAGRAMS

H. HÜFFEL and P.V. LANDSHOFF[1]

CERN, Geneva, Switzerland

Received 20 March 1985

We study the relationship between ordinary perturbation theory and perturbation theory obtained from stochastic quantization. We give a simple proof that, except in gauge theories, the several stochastic diagrams of a given topology are together equivalent to the corresponding Feynman diagram. Our analysis is presented in Minkowski space, but most of it may readily be adapted to euclidean space. The field propagator may be a non-diagonal matrix, such as is the case in real-time thermal field theory.

We present a new version of the Langevin equation which directly reproduces the usual axial-gauge perturbation theory. Otherwise, we find that for gauge theories the relationship between ordinary and stochastic perturbation theory is not simple, and we present a recursive method of reconstructing Feynman diagrams from stochastic diagrams, without the need explicitly to introduce ghost fields. We consider both the original Parisi-Wu version of the Langevin equation, and Zwanziger's modified version with its stochastic gauge-fixing term.

1. Introduction

In Parisi and Wu's stochastic approach to the quantization of field theory [1], one introduces an additional coordinate t (a fictitious time), and a white-noise random source $\eta(x,t)$. A classical field $\phi(x,t)$ is defined as the solution of the Langevin equation

$$\frac{\partial}{\partial t}\phi(x,t)=i\frac{\delta S}{\delta\phi(x,t)}+\eta(x,t) \tag{1.1}$$

subject to a suitable boundary condition. One then calculates stochastic correlation functions by taking stochastic averages

$$\langle\phi(x_1,t)\phi(x_2,t)\cdots\phi(x_N,t)\rangle \tag{1.2}$$

and finds that in the equilibrium limit these approach the quantum Green functions

$$\langle 0|\mathrm{T}\phi(x_1)\phi(x_2)\cdots\phi(x_N)|0\rangle\,. \tag{1.3}$$

[1] On leave of absence from DAMTP, University of Cambridge.

The equilibrium limit usually requires $t \to +\infty$, but with a suitable boundary condition it may already be achieved at finite t (see sect. 2).

In this paper, ϕ may represent a whole set of fields, including fermionic fields (see sect. 4) and matrix fields. We work in Minkowski space, which is why [2] the action term in the Langevin equation (1.1) appears with the factor i. However, most of our equations may readily be adapted to euclidean space. If we write

$$i\frac{\delta S}{\delta\phi} = iD\left(i\frac{\partial}{\partial x}\right)\phi(x,t) + i\frac{\delta S^{\mathrm{INT}}}{\delta\phi} \tag{1.4}$$

where the first term corresponds to the free-field part of the action, Feynman's $i\varepsilon$ prescription includes in D an infinitesimal imaginary part which is positive and, at least in each term of perturbation theory, it ensures convergence to equilibrium. In the case of zero-temperature field theory, this imaginary part is just a multiple of the unit matrix; at finite temperature [3], it is a specific non-diagonal infinitesimal matrix whose eigenvalues are positive.

The stochastic average (1.2) may be calculated as a perturbation expansion [1], whose terms correspond to diagrams which we call stochastic diagrams. In the equilibrium limit several stochastic diagrams together correspond to a given Feynman diagram of the perturbation expansion of the quantum Green function (1.3). Various authors [4–7] have explored by various methods the relationship between the two types of diagram.

In sect. 2 we give a new, simple analysis of this relationship. We work with a Fourier transform with respect to the fictitious time t, which allows us rather easily to take the equilibrium limit early in the calculation. We require that D be invertible (its inverse is the Feynman propagator) and we show that in perturbation theory the limit of the stochastic average is indeed the quantum Green function.

A particular interest of the stochastic approach lies in the quantization of gauge theories. Parisi and Wu [1] have proposed that it is not necessary explicitly to include either gauge-fixing terms or ghosts in the Langevin equation. However, then D is not invertible and our analysis of sect. 2 is not immediately applicable. In sect. 3 we present a new version of the Langevin equation for which the work of sect. 2 is valid; this equation yields in the equilibrium limit the usual perturbation theory in axial gauge.

As we describe in sect. 4, a more general form of the Langevin equation includes an additional "kernel" operator K:

$$\frac{\partial}{\partial t}\phi(x,t) = i\int \mathrm{d}y\, K(x-y)\frac{\delta S}{\delta\phi(y,t)} + \zeta(x,t). \tag{1.5}$$

If ϕ represents a set of fields, K will generally be a matrix operator. The white-noise function ζ satisfies

$$\langle \zeta(x,t) \rangle = 0,$$

$$\langle \zeta(x,t)\zeta(y,t') \rangle = [K(x-y) + K(y-x)]\delta(t-t'), \tag{1.6}$$

with appropriate relations for stochastic averages of more than two ζ's. We show that when D is invertible, and K fulfils certain conditions given in sect. 4, the analysis of sect. 2 may easily be modified and in the equilibrium limit the stochastic average (1.2) again approaches the quantum Green function (1.3), independent of K. Introducing a kernel allows one easily to handle fermions [8,9]. The corresponding components of ϕ must be anticommuting Grassmann fields, but this detail will not affect our work in any essential way. In Minkowski space, there are complications because the Dirac matrix-operator D contains complex γ matrices, so that convergence to equilibrium is not ensured by the $i\varepsilon$ prescription. Introducing the same kernel as is used for euclidean space [8] overcomes this problem. Another interesting choice for the kernel is the free-field propagator itself; in this case each stochastic diagram becomes equal to the corresponding Feynman diagram times a constant fraction. In any case one needs a kernel in order to match the dimensions of the fictitious time t in the boson and fermion Langevin equations.

Our demonstration that one may introduce a kernel is not valid when D is not invertible. Nevertheless, our axial-gauge Langevin equation of sect. 3 may be obtained formally from the Parisi-Wu form of the Langevin equation, by introducing a particular kernel. A different kernel, similarly applied, gives an equation which in the equilibrium limit yields the usual perturbation theory in the Landau gauge. However, ghost contributions are absent, so that this procedure is not valid for a non-abelian gauge theory. In general, then, a kernel may not be used in non-abelian gauge theory. We have no fundamental understanding of why it happens to work in the particular case of our axial-gauge equation.

In sect. 5 we discuss Zwanziger's approach [10,11] in which a t-dependent gauge transformation is applied to the Parisi-Wu form of the Langevin equation. This is chosen so as to yield a new D that is invertible. However, in the non-abelian case it also introduces a new interaction vertex into the theory. This vertex is not derivable from an action and we show that in consequence the relation between stochastic diagrams and Feynman diagrams is very complicated. We give a recursive method of constructing Feynman diagrams in this theory, which allows the calculation of gauge invariant quantities without the explicit inclusion of ghosts. By means of a simple example, we show that it is again not generally valid in the Zwanziger approach to introduce a kernel.

In sect. 6, we discuss the direct use of the Parisi-Wu form of the Langevin equation without explicitly introducing gauge fixing or ghosts. We show that it is essential to take the limit $\varepsilon \to 0^+$ before going to the equilibrium limit. The boundary condition applied to the solution of the Langevin equation now has an important effect. There is again a complicated relationship between stochastic and ordinary Feynman diagrams.

We should perhaps emphasize that all our work in this paper is within the framework of perturbation theory.

2. Stochastic quantization and Feynman diagrams

In this section we develop a perturbation theory for calculating the equilibrium limit of the stochastic average (1.2). We show that, provided D in (1.4) is invertible, we retrieve the ordinary Feynman-diagram expansion of the matrix element (1.3).

For simplicity of presentation, we restrict ourselves to real scalar fields [so that D is symmetric and $D(p)=D(-p)$] and suppose that the interaction is trilinear and totally symmetric, with no derivatives:

$$\mathcal{L}^{\mathrm{INT}}=\frac{1}{3!}\lambda_{ijk}\phi_i\phi_j\phi_k\,. \tag{2.1}$$

These details are not essential to our analysis (fermions are discussed in sect. 4).

We work with the Fourier transform of the field:

$$\phi(p,\omega)=\int \mathrm{d}^n x\,\mathrm{d}t\,\mathrm{e}^{ip\cdot x+i\omega t}\phi(x,t)\,. \tag{2.2}$$

The Fourier transform with respect to t was previously used [7] in a superfield formulation of stochastic quantization. We show its usefulness when applied directly to the basic Langevin equation (1.1). We have

$$-i(\omega I+D(p))\phi(p,\omega)=\eta(p,\omega)+\tfrac{1}{2}i\lambda\phi^2(p,\omega)\,, \tag{2.3}$$

where I is the unit matrix and

$$\begin{aligned}\left[\lambda\phi^2(p,\omega)\right]_i=\frac{\lambda_{ijk}}{(2\pi)^{n+1}}\int \mathrm{d}^n p_1\,\mathrm{d}^n p_2\,\mathrm{d}\omega_1\,\mathrm{d}\omega_2\,\delta^{(n)}(p_1+p_2-p)\\ \times\delta(\omega_1+\omega_2-\omega)\phi_j(p_1,\omega_1)\phi_k(p_2,\omega_2)\,.\end{aligned} \tag{2.4}$$

It is important to notice that by taking the Fourier transform with respect to t, we have suppressed that part of $\phi(x,t)$ which depends on whatever initial condition we impose on the solution of the Langevin equation. Because D implicitly contains Feynman's $i\varepsilon$, this part of the solution is transient: it dies away exponentially like $\mathrm{e}^{-\varepsilon t}$ in the equilibrium limit. Because it explodes exponentially for large negative t, its Fourier transform does not exist and our procedure does not take account of it. As we are interested only in the equilibrium limit, this removal represents a considerable simplification. A similar procedure is familiar in the analysis of alternating-current electric circuits, where the usual expression $V/Z(\omega)$ for the

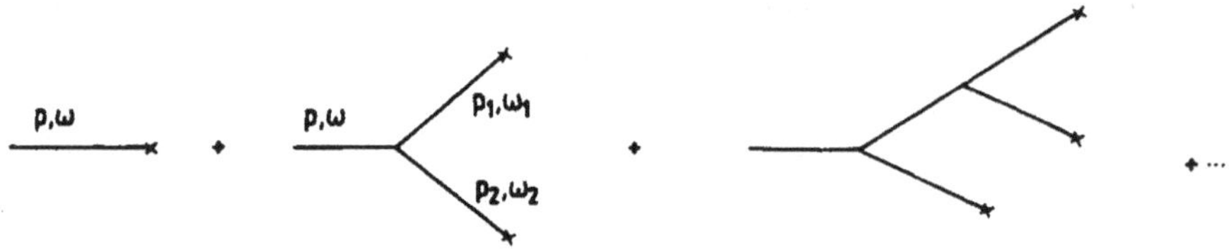

Fig. 1. Diagrams for the perturbation expansion of the stochastic field.

electric current in terms of the complex impedance is valid only in the steady-state limit, when the transients associated with the precise way in which the circuit is switched on have died away. Suppressing the transients amounts, at least formally, to imposing the initial condition [12]

$$\phi(x,t)\to 0, \qquad \text{as} \qquad t\to -\infty, \tag{2.5}$$

so that the stochastic process relaxes to equilibrium already at finite fictitious time t. We shall indeed find that when D is invertible our procedure yields stochastic averages that are independent of t at finite t. When D is not invertible, as in the case of the Parisi-Wu form of the gauge-field Langevin equation, special care is needed; we discuss this in sect. 6. In this section we suppose that D is invertible, and then there is no subtlety.

Introducing the stochastic Green function

$$G(p,\omega)=[\omega I+D(p)]^{-1}, \tag{2.6}$$

we may solve (2.3) iteratively and represent the resulting perturbation expansion for $\phi(p,\omega)$ graphically: see fig. 1. With each line of a graph we associate a pair of variables (p,ω) and a stochastic Green function $iG(p,\omega)$. There is an integration over all the (p,ω) other than those on the left-most line. Each vertex represents the interaction $\frac{1}{2}i\lambda/(2\pi)^{n+1}$; because of the δ-functions in (2.4) the total (p,ω) flowing from left to right at each vertex is conserved. Each cross represents a factor η, whose arguments are the values of (p,ω) on the line attached to it.

When, as in this section, K is the unit operator, the Fourier transforms of the relations (1.6) are

$$\langle\eta(p,\omega)\rangle=0,$$
$$\langle\eta(p,\omega)\eta(p',\omega')\rangle=2(2\pi)^{n+1}\delta^{(n)}(p+p')\delta(\omega+\omega'). \tag{2.7}$$

To calculate the stochastic average

$$\langle\phi(p_1\omega_1)\phi(p_2\omega_2)\cdots\phi(p_N\omega_N)\rangle, \tag{2.8}$$

we first draw all possible stochastic diagrams, obtained by inserting the expansion of fig. 1 for each ϕ. A typical example is drawn in fig. 2. It is convenient to draw the diagrams such that the lines $(p_1,\omega_1),(p_2,\omega_2),\ldots,(p_N,\omega_N)$ all enter from the left.

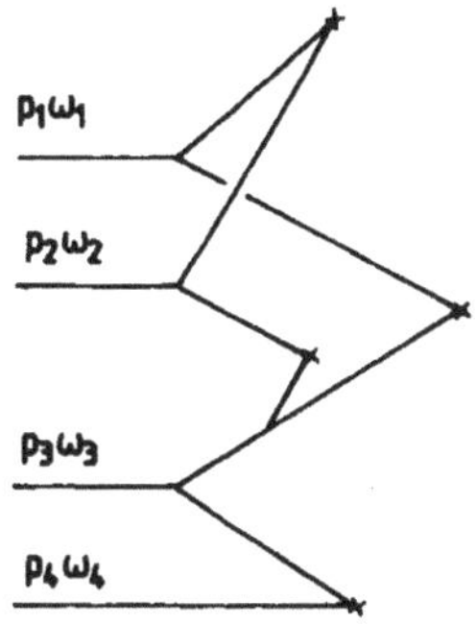

Fig. 2. Diagrams for the perturbation expansion of a stochastic average (2.8).

The rules for calculating these diagrams are the same as for fig. 1, except that because of (2.7) a cross now denotes $2\cdot(2\pi)^{n+1}$, with the pair of ω variables entering it from the left adding to zero, and likewise the pair of p variables. Also, if one draws each stochastic diagram of a given topology only once, the factor $\frac{1}{2}$ in the interaction vertex should be omitted. This takes account of the different ways in which the crosses in the expansion of ϕ can be paired in (2.8). (There are complications in the case of diagrams, such as bubble diagrams, that have internal symmetry. We do not go into these complications: they are just the same as in ordinary perturbation theory.)

Because of these rules, each diagram contains the δ-function

$$\delta(\omega_1+\omega_2+\cdots+\omega_N), \tag{2.9}$$

and

$$V_N=\langle\phi(p_1t)\phi(p_2t)\cdots\phi(p_Nt)\rangle, \tag{2.10}$$

which is obtained by applying to (2.8) the integration

$$\frac{1}{(2\pi)^N}\int \mathrm{d}\omega_1\,\mathrm{d}\omega_2\cdots\mathrm{d}\omega_N\,\mathrm{e}^{-i(\omega_1+\omega_2+\cdots+\omega_N)t}, \tag{2.11}$$

is independent of t. From what we have said, it should therefore be equal to the matrix element

$$\langle 0|\mathrm{T}\phi(p_1)\phi(p_2)\cdots\phi(p_N)|0\rangle \tag{2.12}$$

[by which we mean just the Fourier transform of the matrix element (1.3)]. Because our Fourier transform procedure suppresses the transients, the equilibrium limit has in effect already been taken.

Because of (2.11) and the δ-function (2.9), we must now integrate over *all* the ω variables, including those on the external lines of the stochastic diagram. The integrand is a product of poles, so that the ω integrations just give a sum of residues. However, if the matrix $D(p)$ is not diagonal, this is not completely straightforward.

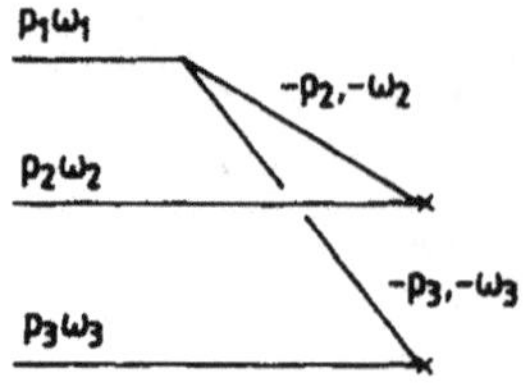

Fig. 3. One of the stochastic diagrams in the calculation of the three-point function in lowest order.

As a simple example, consider the lowest-order contribution to the three-point function [$N=3$ in (2.10)]. There are three stochastic diagrams, fig. 3 and two similar diagrams. Fig. 3 contributes

$$V_3^{(1)} = (2\pi)^n \delta^{(n)}(p_1+p_2+p_3)\frac{1}{(2\pi)^2}\int \mathrm{d}\omega_1\,\mathrm{d}\omega_2\,\mathrm{d}\omega_3\,\delta(\omega_1+\omega_2+\omega_3)U_3^{(1)}, \tag{2.13}$$

where

$$U_3^{(1)} = \begin{bmatrix} G(p_1,\omega_1) \\ 2G(p_2,\omega_2)G(-p_2,-\omega_2) \\ 2G(p_3,\omega_3)G(-p_3,-\omega_3) \end{bmatrix}\lambda. \tag{2.14}$$

We have used a generalized matrix notation, in which the ith row of the square bracket corresponds to the ith line on the left of the diagram. From (2.6) it is trivial to derive the identity

$$D(p)G(p,\omega) = I - \omega G(p,\omega). \tag{2.15}$$

Hence

$$\begin{bmatrix} D(p_1) \\ I \\ I \end{bmatrix} U_3^{(1)} + \begin{bmatrix} I \\ D(p_2) \\ I \end{bmatrix} U_3^{(1)} + \begin{bmatrix} I \\ I \\ D(p_3) \end{bmatrix} U_3^{(1)}$$

$$= \begin{bmatrix} I \\ 2G(p_2,\omega_2)G(-p_2,-\omega_2) \\ 2G(p_3,\omega_3)G(-p_3,-\omega_3) \end{bmatrix}\lambda + \begin{bmatrix} G(p_1,\omega_1) \\ 2G(-p_2,-\omega_2) \\ 2G(p_3,\omega_3)G(-p_3,-\omega_3) \end{bmatrix}\lambda$$

$$+ \begin{bmatrix} G(p_1,\omega_1) \\ 2G(p_2,\omega_2)G(-p_2,-\omega_2) \\ 2G(-p_3,-\omega_3) \end{bmatrix}\lambda - (\omega_1+\omega_2+\omega_3)U_3^{(1)}. \tag{2.16}$$

When the integration (2.13) is applied to this equation, the last term on the right-hand side vanishes because of the δ-function. We use the δ-function to perform the ω_1 integration in the other terms on the right-hand side. Then by closing the contour of the ω_2 integration with an infinite semicircle in the lower half complex plane, and remembering the $i\varepsilon$ prescription included in G, we see that the second term also vanishes. Similarly, the ω_3 integration makes the third term vanish. In the first term the ω_2 and ω_3 integrals are both of the form

$$\int \mathrm{d}\omega\, 2G(p,\omega)G(-p,-\omega). \tag{2.17}$$

Because we have said that we are assuming that $D(p) = D(-p)$, the matrices $G(p,\omega)$ and $G(-p,-\omega)$ commute and so the integral is straightforward: it is equal to $-2\pi i D^{-1}(p)$. (It is not immediately obvious that things work out the same way for fermion fields, but nevertheless it is true, as we indicate in sect. 4.)

If we now add to $V_3^{(1)}$ the two other diagrams, obtained by permuting the left-hand lines in fig. 3, we therefore have

$$\begin{bmatrix} D(p_1) \\ I \\ I \end{bmatrix} V_3 + \cdots = -(2\pi)^n \delta^{(n)}(p_1+p_2+p_3) \begin{bmatrix} I \\ D^{-1}(p_2) \\ D^{-1}(p_3) \end{bmatrix} \lambda + \cdots, \tag{2.18}$$

where the terms not written explicitly are obtained by cyclic permutation. If we now diagonalize each of the matrices $D(p)$, which is possible by means of an orthogonal transformation because we have said they are symmetric, and use the fact that none of their eigenvalues vanish if they are invertible, we obtain finally (at least for almost all values of the momenta p)

$$V_3 = -(2\pi)^n \delta^{(n)}(p_1+p_2+p_3) \begin{bmatrix} D^{-1}(p_1) \\ D^{-1}(p_2) \\ D^{-1}(p_3) \end{bmatrix} \lambda, \tag{2.19}$$

which is the correct three-point function in lowest order.

Notice the dual role of Feynman's $i\varepsilon$ prescription in this work. Not only does it ensure convergence towards the equilibrium limit; it also makes sure that we end up with the correctly-defined propagators.

We now derive the corresponding result for general diagrams. Our proof is by induction. It is along the lines described in ref. [4], but it is simpler because of our ω-space technique. We assume first that (2.10) and (2.12) are equal up to some order in the coupling λ, and show that this implies that they are equal also up to the next order. This induction argument begins with the lowest-order result (2.19) above.

A general stochastic diagram gives a contribution which may be written in the form

$$V_N^{(\alpha)} = \int d\omega_1 \, d\omega_2 \cdots d\omega_N \, \delta(\omega_1 + \omega_2 + \cdots + \omega_N) \begin{bmatrix} G(p_1\omega_1) \\ G(p_2\omega_2) \\ \vdots \\ G(p_N\omega_N) \end{bmatrix} \tilde{V}_N^{(\alpha)}. \tag{2.20}$$

From the identity (2.15), applied in the same way as in (2.16) above,

$$\sum_{r=1}^{N} \begin{bmatrix} I \\ I \\ \vdots \\ I \\ D(p_r) \\ I \\ \vdots \\ I \end{bmatrix} V_N^{(\alpha)} = \int d\omega_1 \cdots d\omega_N \, \delta(\omega_1 + \cdots + \omega_N) \sum_{r=1}^{N} \begin{bmatrix} G(p_1\omega_1) \\ \vdots \\ G(p_{r-1}\omega_{r-1}) \\ I \\ G(p_{r+1}\omega_{r+1}) \\ \vdots \\ G(p_N\omega_N) \end{bmatrix} \tilde{V}_N^{(\alpha)}. \tag{2.21}$$

Consider the rth term on the right-hand side. If the (p_r, ω_r) line connects directly to a cross (as is the case for the bottom line in fig. 2), the term vanishes. The argument is the same as for the second and third terms on the right-hand side of (2.16). There are two other cases shown in fig. 4. In each of these, replacing $G(p_r, \omega_r)$ by I is equivalent to removing the corresponding external line from the diagram. For fig. 4a if we also remove the vertex within the dotted box, that is the factor $i\lambda/(2\pi)^{n+1}$, and possibly also an integration over a loop momentum, we obtain just the integral corresponding to the residual stochastic diagram. Although this has one more incoming line than we started with, it has one less vertex. Hence, if we sum over all stochastic diagrams α of the given order, we can apply our induction hypothesis. In

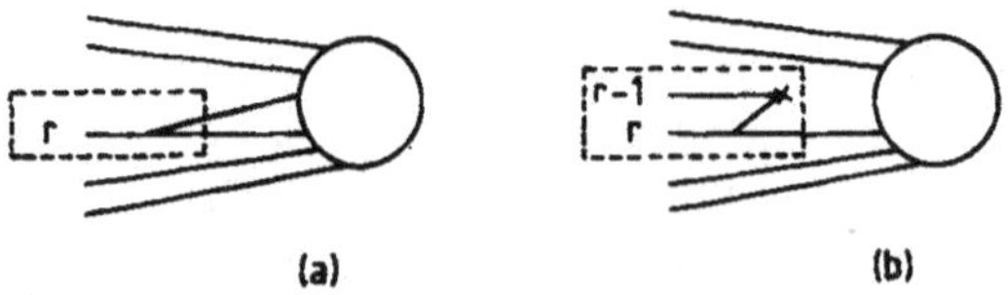

Fig. 4. Two different ways in which external lines couple into stochastic diagrams.

the case of fig. 4b, we have to remove the structure within the dotted box by also explicitly performing the ω_{r-1} integration, but the conclusion is similar.

What we have proved, then, is that if V_N is the sum of all stochastic diagrams of a given order in the coupling, and F_N is the corresponding sum of Feynman diagrams,

$$\sum_{r=1}^{N} \begin{bmatrix} I \\ \vdots \\ I \\ D(p_r) \\ I \\ \vdots \\ I \end{bmatrix} (V_N - F_N) = 0. \tag{2.22}$$

As in the discussion following (2.18), this implies that $V_N = F_N$.

We note that actually a slightly stronger result may be proved in the same way: the sum of all stochastic diagrams of a given topology is equal to the Feynman diagram of the same topology.

3. Axial gauge

For a non-abelian gauge theory, with lagrangian

$$\mathcal{L} = -\tfrac{1}{4} F^a_{\mu\nu} F^{a\mu\nu}, \tag{3.1}$$

the Langevin equation (1.1) reads

$$\frac{\partial}{\partial t} A^a_\mu(x,t) = -i\Delta^{ab\nu} F^b_{\mu\nu}(x,t) + \eta^a_\mu(x,t), \tag{3.2}$$

where Δ is the covariant derivative

$$\Delta^{ab}_\nu = \partial_\nu \delta^{ab} - g f^{abc} A^c_\nu, \tag{3.3}$$

and the white-noise function obeys

$$\langle \eta^a_\mu(x,t)\eta^b_\nu(y,t')\rangle = -2\delta^{ab} g_{\mu\nu} \delta^{(n)}(x-y)\delta(t-t'). \tag{3.4}$$

In momentum space (3.2) takes the form

$$\frac{\partial}{\partial t} A^a_\mu(p,t) = i\left[(p^2 + i\varepsilon) g_{\mu\nu} - p_\mu p_\nu\right] A^{a\nu}(p,t) + I^a_\mu(p,t) + \eta^a_\mu(p,t), \tag{3.5}$$

where I^a_μ denotes the three-gluon and four-gluon interaction terms. We have explicitly written Feynman's $i\varepsilon$. Note that in (3.2), (3.4) and (3.5) we have introduced a "kernel" $K = -1$ [see (1.5) and (1.6)] in order that the $i\varepsilon$ gives convergence to the equilibrium limit. Less trivial kernels are discussed later in this paper.

The work of sect. 2 is not applicable to this Langevin equation, because the coefficient of $A^{a\nu}$ on the right-hand side is not invertible when $\varepsilon \to 0$. As is well known, this is connected with the fact that the field equations contain unphysical degrees of freedom. The Langevin equation (3.5) is supposed [1] to yield the correct results only for gauge-invariant quantities, but if we attempt an analysis such as that associated with fig. 4, described in sect. 2, we encounter non-gauge-invariant subamplitudes. We have not been able to set up a consistent induction procedure to deal with this. Ideally, one should try to build up a formulation that involves only gauge-invariant quantities, such as Wilson loops [13], but this seems to be very difficult.

The best we can do in this direction is to write down a Langevin equation which in the equilibrium limit yields Feynman perturbation theory in axial gauge. We define

$$N^{\mu\nu}(p) = g^{\mu\nu} - \frac{p^\mu n^\nu}{n \cdot p}. \tag{3.6}$$

This has the properties

$$N^{\mu\rho}(p) N_\rho{}^\nu(p) = N^{\mu\nu}(p), \tag{3.7}$$

$$N^{\mu\rho}(p) N^\nu{}_\rho(p) = g^{\mu\nu} - \frac{p^\mu n^\nu + n^\mu p^\nu}{n \cdot p} + n^2 \frac{p^\mu p^\nu}{(n \cdot p)^2}. \tag{3.8}$$

Our Langevin equation is

$$\frac{\partial}{\partial t} A^a_\mu = i(p^2 + i\varepsilon) A^a_\mu + N_{\mu\nu}(N^{\rho\nu} I^a_\rho + \eta^{a\nu}), \tag{3.9}$$

where I^a is the same interaction as in (3.5) and η^a is also the same white-noise function, obeying (3.4).

It is straightforward to work through the analysis of sect. 2 using this Langevin equation. Because (3.8) is the numerator of the usual axial-gauge propagator, the stochastic perturbation theory based on (3.9) reproduces the usual axial-gauge perturbation theory in the equilibrium limit. We remark that we do not intend to discuss here problems connected with the zeros of the denominator in (3.6); whatever is the appropriate prescription [14] in the usual theory, whether it be principal value or otherwise, must be applied in (3.6).

4. Introduction of a kernel

Under certain conditions, the Langevin equation may be modified by the introduction of a "kernel" operator K: see (1.5) and (1.6). If we impose certain conditions on K, it is straightforward to adapt the analysis of sect. 2. (These conditions may perhaps be unnecessary, but they are needed if we want to use the methods of sect. 2).

In the Fourier-transformed Langevin equation (2.3), $D(p)$ will be replaced by $K(p)D(p)$ and the interaction term also is multiplied by $K(p)$. We must retain the usual Feynman $i\varepsilon$ prescription in D, in order that the propagators obtained in the equilibrium limit be correctly defined. But we must also ensure that $\operatorname{Im} KD > 0$; otherwise there will be no equilibrium limit. Because the interaction is multiplied by $K(p)$, a vertex in a stochastic diagram now corresponds to $iK(p)\lambda/(2\pi)^{n+1}$, where p is the momentum flowing into the vertex from the left. Instead of (2.7) we require the Fourier transform of (1.6):

$$\langle \zeta(p,\omega)\zeta(p',\omega')\rangle = \left(K(p)+K(p')\right)(2\pi)^{n+1}\delta^{(n)}(p+p')\delta(\omega+\omega'), \tag{4.1}$$

and then a cross in a stochastic diagram will denote $(K(p)+K(-p))(2\pi)^{n+1}$.

$D(p)$ is replaced by $K(p)D(p)$ in the Green function $G(p,\omega)$ of (2.6) and in the identity (2.15) which it satisfies. When we calculate the three-point function V_3, instead of (2.17) we encounter the integral

$$\int d\omega G(p,\omega)(K(p)+K(-p))G(-p,-\omega). \tag{4.2}$$

In order to be able readily to evaluate this, and obtain a recognisable result, we now impose on $K(p)$ the condition that it commutes with $D(p)$:

$$[K(p),D(p)] = 0. \tag{4.3}$$

Because we are still assuming that $D(-p)=D(p)$, then $K(-p)$ will also commute with $D(p)$, and the three factors in the integrand of (4.2) commute with each other. Hence the integral is straightforward and equal to $-2\pi i D^{-1}(p)$, the same as (2.17). $D(p)$ is replaced by $K(p)D(p)$ also in the analogues of (2.18) and (2.22), so we must assume that KD is symmetric in order to be able to diagonalize it. If it is also invertible, so that it does not have zero eigenvalues, we again conclude that $V_N = F_N$, independent of K.

An interesting choice for K which satisfies the conditions we have imposed is

$$K(p) = iD^{-1}(p). \tag{4.4}$$

In this case the ω integrations in the stochastic diagrams decouple from the momentum integrations: each stochastic diagram is equal to the corresponding Feynman diagram times a constant fraction.

Formally, our axial-gauge Langevin equation (3.9) may be obtained by applying to the original Parisi-Wu equation (3.5) the kernel $N^{\mu\rho}N^{\nu}{}_{\rho}$ of (3.8), or in matrix notation

$$K = NN^{\mathrm{T}}. \tag{4.5}$$

We identify $\zeta = N\eta$; because η satisfies (3.4), ζ will satisfy (4.1). The stochastic Green function is

$$G(p,\omega) = \frac{1}{\omega + NN^{\mathrm{T}}D} = \frac{1}{\omega + Np^2}, \tag{4.6}$$

where, from (3.5),

$$D^{\mu\nu} = p^2 g^{\mu\nu} - p^{\mu}p^{\nu}. \tag{4.7}$$

Because of the factor N in ζ and because the interaction is multiplied by K, in stochastic diagrams G always appears multiplied on the right by N. But because of the identity (3.7),

$$GN = \frac{1}{\omega + p^2}N. \tag{4.8}$$

That is, we may use $G = (\omega + p^2)^{-1}$ instead of (4.6). But this is just the stochastic Green function corresponding to our modified Langevin equation (3.9).

We have not paid attention to the $i\varepsilon$ prescription in this "derivation" because, of course, KD is not invertible, so that it does not satisfy the condition we imposed in our general discussion above. Thus, although we showed in sect. 3 that (3.9) does lead to the correct axial-gauge perturbation series, we are not able to derive it from the original Parisi-Wu equation (3.5). It is easy to see that such a derivation is not generally possible. If in eqs. (4.5) to (4.8) we use for N not (3.6) but rather $N = D/p^2$, the corresponding equation (3.9) would then reproduce perturbation theory in the Landau gauge. However, it would not include any ghost terms, and so would be wrong in non-abelian gauge theory.

In order to include fermions in the formalism, it is necessary to introduce a kernel in the corresponding Langevin equations. If for no other reason, this is because otherwise the fictitious time t in the fermion equations would have a different dimension from that in the boson equations. More seriously, the Langevin equation for the fermion fields has

$$D(p) = \gamma \cdot p - m + i\varepsilon. \tag{4.9}$$

At least one of the γ matrices is inevitably complex, which means that the $i\varepsilon$ is not sufficient to give convergence to equilibrium*. Furthermore, in our analysis we required $K(p)D(p)$ to be a symmetric matrix. These difficulties are most easily overcome by introducing the kernel

$$K(p) = D(-p). \tag{4.10}$$

This choice is also made for a slightly different reason in work [8] on stochastic quantization of fermion fields in euclidean space. [An alternative would be again the choice (4.4).]

Our general analysis is now valid, except for some technical points. First, to end up with Fermi statistics we must make the fermion field and its stochastic white-noise source anticommuting Grassmann functions. Secondly, when taking Fourier transforms it is convenient to treat ψ and $\bar{\psi}$ as independent, that is we use the definition (2.2) for each of them. In consequence, the momentum of the $\bar{\psi}$ fields is $-p$ rather than p, and its inverse propagator is

$$\bar{D}(p) = D(-p), \tag{4.11}$$

and, of course, we choose $\bar{K}(p) = K(-p)$. Then $\bar{G}(p,\omega) = G(-p,\omega) = G(p,\omega)$. Instead of (4.1) we impose

$$\langle\zeta(p,\omega)\zeta(p',\omega')\rangle = \langle\bar{\zeta}(p,\omega)\bar{\zeta}(p',\omega')\rangle = 0,$$

$$\langle\zeta(p,\omega)\bar{\zeta}(p',\omega')\rangle = 2D(p)(2\pi)^{n+1}\delta^{(n)}(p+p')\delta(\omega+\omega'). \tag{4.12}$$

Everything is then straightforward. For example, if we evaluate the three-point function which couples ψ and $\bar{\psi}$ to the gauge field, we encounter instead of (4.2) the integral

$$\int \mathrm{d}\omega\, G(p,\omega) 2D(p)\bar{G}(-p,-\omega) = \frac{-2\pi i}{\gamma\cdot p - m}. \tag{4.13}$$

Finally in this section, we remark that if one chooses to handle a non-abelian gauge theory by including in the action both a gauge-fixing term and a ghost field, and then writing a Langevin equation for each field, it is straightforward to see from our analysis that the correct matrix elements will be obtained in the equilibrium limit.

* One of us is grateful to F. Englert for a discussion on this.

5. Stochastic gauge fixing

Although stochastic quantization based on the Parisi-Wu form (3.2) of the Langevin equation is believed to be consistent without any gauge fixing, it is often useful to impose a generalized gauge fixing [10, 11]. In the case of the Parisi-Wu equation, the ω integrations remain finite for gauge-invariant sums of diagrams in the equilibrium limit, but not for individual stochastic diagrams. With appropriate gauge fixing, they remain finite for all diagrams.

The basic idea [10] is to transform the field $A_\mu^a(x,t)$ into a new field so that in the Langevin equation for this new field the matrix D of (4.7) is replaced by an invertible matrix. The transformation is required to leave unchanged all gauge-invariant quantities, and so must be a gauge transformation which, in order to be useful, depends on t. The Langevin equation (3.2) then acquires [10] on its right-hand side an additional term

$$\Delta_\mu^{ab}\chi^b, \tag{5.1}$$

where χ is an arbitrary function. With the simple choice

$$\chi^b(x,t) = -i\partial^\nu A_\nu^b(x,t) \tag{5.2}$$

the new Langevin equation becomes, when Fourier transformed,

$$-i(\omega+p^2+i\varepsilon)A_\lambda^a(p,\omega) = I_\lambda^a(p,\omega) + J_\lambda^a(p,\omega) + \eta_\lambda^a(p,\omega). \tag{5.3}$$

Here I^a denotes the usual interaction terms and J^a is a new interaction derived from (5.1) and (5.2):

$$J_\lambda^a(p,\omega) = \frac{gf^{abc}}{2(2\pi)^{n+1}}\int d^nq\, d^nr\, d\omega_1\, d\omega_2\, \delta^{(n)}(p+q+r)$$
$$\times\delta(\omega_1+\omega_2-\omega)A^{b\mu}(-q,\omega_1)A^{c\nu}(-r,\omega_2)v_{\lambda\mu\nu}^J(p,q,r) \tag{5.4}$$

with

$$v_{\lambda\mu\nu}^J(p,q,r) = g_{\lambda\mu}r_\nu - g_{\lambda\nu}q_\mu\,. \tag{5.5}$$

This new interaction is a three-gluon interaction: see fig. 5. Unlike the usual three-gluon interaction contained in I^a, for which

$$v_{\lambda\mu\nu}^I(p,q,r) = g_{\lambda\mu}(p-q)_\nu + \text{cyclic perms}, \tag{5.6}$$

it is a directed vertex because it is symmetric under the interchange of two lines only.

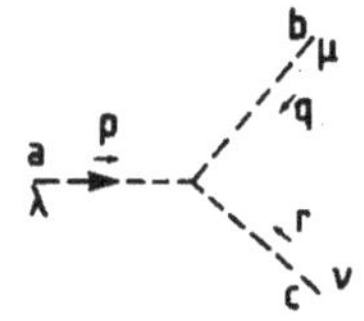

Fig. 5. The new vertex v^J.

We have indicated this by an arrow in the figure: in our stochastic diagrams the arrow will always point into the vertex from the left.

The directedness complicates the analysis of this new interaction and makes the relation between stochastic diagrams and Feynman diagrams non-trivial. This complication is presumably because the new interaction is not derivable from a lagrangian [10, 15].

Consider first the contribution from the new vertex to V_3 in lowest order. It takes a different form in each of the three stochastic diagrams (fig. 3 and two similar diagrams). The three terms sum to

$$-i(2\pi)^n\delta^{(n)}(p_1+p_2+p_3)g(p_1^2+i\varepsilon)^{-1}(p_2^2+i\varepsilon)^{-1}(p_3^2+i\varepsilon)^{-1}$$

$$\times f^{abc}v^K_{\lambda\mu\nu}(p_1,p_2,p_3)$$

with

$$v^K_{\lambda\mu\nu}(p_1,p_2,p_3)=(p_1^2+p_2^2+p_3^2)^{-1}\big[(p_1^2-p_2^2)g_{\lambda\mu}p_{3\nu}+\text{cyclic perms}\big]. \tag{5.7}$$

This is an effective, undirected vertex, which we shall denote in diagrams by a black spot.

Stochastic diagrams of a given topology that contain only vertices of the ordinary type v^I sum to the corresponding Feynman diagram; this follows directly from the work of sect. 2. Because of our choice (5.2) for χ, this Feynman diagram is calculated in Feynman gauge. However, for stochastic diagrams that contain one or more vertices of the new type v^J the situation is much more complicated.

If we sum all such diagrams, and use them to calculate a gauge-invariant quantity, we expect [10] that they should just reproduce the usual ghost contributions. The methods of sect. 2 may still be used to express the sum of all the stochastic diagrams of a given topology in terms of residual Feynman-like diagrams having one less vertex. This is much simpler than calculating all the stochastic diagrams individually. However, because the residual diagrams are not gauge-invariant, even when summed, we are unable to set up a simple induction procedure like that of sect. 2 which will

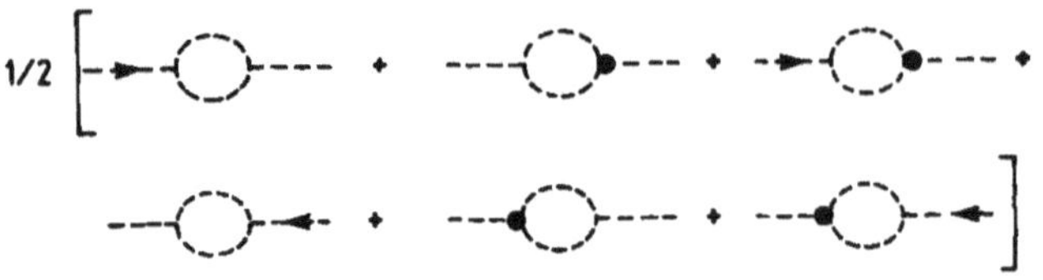

Fig. 6. Feynman-like diagrams encountered in the calculation of the two-point function in order g^2 from the Langevin equation (5.3). The black spots denote the effective vertex v^K.

prove that finally the ghost contributions, and only the ghost contributions, will emerge.

As an example, consider the order-g^2 stochastic diagrams for the two-point function V_2. The details are in the appendix. The stochastic diagrams that do not contain a vertex of the new type indeed sum to the ordinary Feynman graphs. With the help of the recursive methods of sect. 2, we find that the stochastic diagrams having either or both of their vertices of the new type v^J sum to the Feynman-like diagrams of fig. 6. If we write down the integrals corresponding to these diagrams, but do not perform the integration over the loop momentum, we recognize a term which is just what is normally associated with a ghost loop. However, the remaining contribution is unrecognizable. We expect that this unrecognizable term should cancel if we use V_2 to calculate some gauge-invariant quantity.

We have checked this in two examples. The first is quark-quark scattering in order g^4, and the second is

$$\langle 0|F^a_{\mu\nu}(x)F^a_{\rho\sigma}(x)|0\rangle \tag{5.8}$$

in order g^2. In the first example, we suppose that we are using dimensional regularization, so that the quark self-energy and vertex corrections are gauge-invariant. Then the remaining diagrams must together be gauge invariant: they are shown in fig. 7. In order to keep the calculation simple, we do not apply stochastic quantization to the external quark lines: we merely use them to project out a gauge-invariant quantity from the gluon fields:

$$u_{ri}(p_1)u_{r'i'}(p_2)\bar{u}_{sj}(p_3)\bar{u}_{s'j'}(p_4)\Big\{\langle 0|A^a_\lambda(p)A^{a'}_{\lambda'}(p')|0\rangle\gamma^\lambda_{rs}\gamma^{\lambda'}_{r's'}t^a_{ij}t^{a'}_{i'j'}$$
$$-\frac{g}{(2\pi)^n}\int d^nq\,\langle 0|A^a_\mu(q)A^b_\nu(r)A^c_{\lambda'}(p')|0\rangle$$
$$\times\frac{(\gamma^\mu\gamma\cdot(p_1-q)\gamma^\nu)_{rs}}{q^2-2p_1\cdot q}\gamma^{\lambda'}_{r's'}t^a_{il}t^b_{lj}t^c_{i'j'}+\cdots\Big\}, \tag{5.9}$$

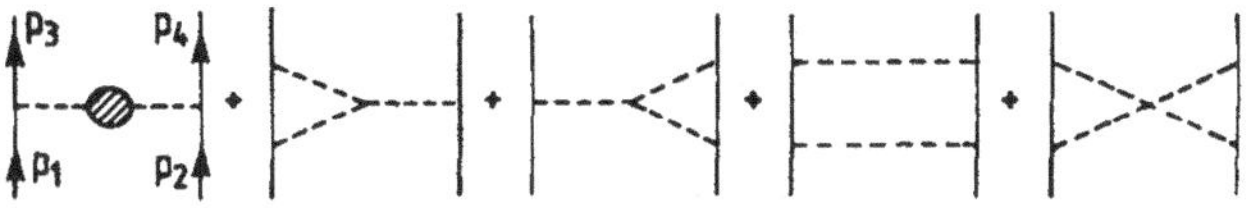

Fig. 7. Contributions to quark-quark scattering in order g^4.

where $p = p_1 - p_3$, $p' = p_2 - p_4$ and $p' + q + r = 0$. According to what we have said, the contributions from the ordinary vertices v^I are the normal ones. The anomalous contributions to the terms of fig. 7 are those in the gluon propagator (fig. 6) in the first term and a new vertex contribution to the second and third terms. Without doing any loop integration, we find that the unrecognizable integrands just cancel, leaving the wanted ghost term behind.

In the case of $\langle FF \rangle$ in (5.8) the result is similar, but the details are different. The unwanted integrands do not cancel, but it is easy to see that they integrate to zero in any regularization scheme. (We find that the terms disagree in detail with those of ref. [6], though the conclusion is the same.)

It is tempting to modify the Langevin equation after applying stochastic gauge fixing by including an additional kernel operator K. With the choice

$$K = \frac{i}{p^2 + i\varepsilon}, \tag{5.10}$$

things apparently become very much simpler, in particular the effective vertex v^K of (5.7) then vanishes and so only the first and fourth terms of fig. 6 survive. However, we cannot use the proof of sect. 4 to show that the sum of stochastic diagrams is independent of K. If we repeat the calculation of $\langle FF \rangle$, we do find that the answer is the same. But this is not true for quark-quark scattering: the unrecognizable terms no longer cancel, even after integration. We conclude that it is not generally valid to introduce a kernel after the Langevin equation has been modified by stochastic gauge fixing.

Finally in this section, we mention that there is a choice of χ which seems to yield the axial gauge. Having added (5.1) to the right-hand side of the Parisi-Wu form (3.5) of the Langevin equation, we choose χ by imposing

$$n \cdot A(p,t) = n \cdot \dot{A}(p,t) = 0. \tag{5.11}$$

This gives

$$\chi = ip \cdot A - n \cdot (I + \eta)/n \cdot p.$$

The result is a Langevin equation rather like (3.9), except that there are additional interaction terms which make things very much more complicated. Presumably, they cancel when gauge-invariant quantities are calculated.

Our main conclusion in this section is that the relationship between the stochastic approach to non-abelian gauge theory and the ordinary perturbation expansion is not at all simple. That, of course, is its interest.

6. The Parisi-Wu approach to gauge theory

We have emphasized the role of Feynman's $i\varepsilon$ in our analysis. Although the coefficient of $A^{a\nu}$ in (3.5) is not invertible when $\varepsilon \to 0$, there is no problem before this limit is taken. Then

$$\left[D^{ab}_{\mu\nu}(p)\right]^{-1} = \delta^{ab}\left[\frac{g^{\mu\nu} + p^{\mu}p^{\nu}/i\varepsilon}{p^2 + i\varepsilon}\right], \tag{6.1}$$

and the vertices of the theory are solely those of the conventional type. The work of sect. 2 is then valid, so that the sum of the stochastic diagrams gives ordinary Feynman diagrams with the propagator (6.1). We have repeated the calculations of $\langle FF\rangle$ and quark-quark scattering in this formalism. When $\varepsilon \to 0$, the negative powers of ε cancel in these gauge-invariant quantities and leave behind a term which is just what is normally associated with a ghost loop. However, it is too small by a factor of two.

This discovery is not new [16]. By first setting $\varepsilon \neq 0$, we have effectively given the gauge field a (complex) mass, and it is known that the zero-mass limit of a massive "Yang-Mills" theory has just this disease.

In the present context, the problem arises because we have taken the equilibrium limit of the stochastic process before $\varepsilon \to 0^+$. If we take these limits in the reverse order, the result is different and apparently then correct.

We explore this. Introduce the transverse and longitudinal projection operators

$$T^{\mu\nu}(p) = g^{\mu\nu} - p^{\mu}p^{\nu}/p^2, \qquad L^{\mu\nu}(p) = p^{\mu}p^{\nu}/p^2. \tag{6.2}$$

With D as in (6.1), the stochastic Green function (2.6) is

$$G^{ab}_{\mu\nu}(p,\omega) = \delta^{ab}\left[\frac{T_{\mu\nu}(p)}{\omega + p^2 + i\varepsilon} + \frac{L_{\mu\nu}(p)}{\omega + i\varepsilon}\right], \tag{6.3}$$

and its Fourier transform is

$$G^{ab}_{\mu\nu}(p,t) = -i\theta(t)\delta^{ab}\left[T_{\mu\nu}(p)e^{i(p^2+i\varepsilon)t} + L_{\mu\nu}(p)e^{-\varepsilon t}\right]. \tag{6.4}$$

To zeroth order this gives

$$\begin{aligned} A^{a}_{\mu}(p,t) &= \frac{i}{2\pi}\int d\omega\, e^{-i\omega t}G^{ab}_{\mu\nu}(p,\omega)\eta^{b\nu}(p,\omega) \\ &= i\int dt'\, G^{ab}_{\mu\nu}(p,t-t')\eta^{b\nu}(p,t'). \end{aligned} \tag{6.5}$$

The θ-function in (6.4) gives an effective upper limit $t'=t$ in this last integration, but there is no lower limit. This is because, as we explained in sect. 2, the Fourier transform method of solving the Langevin equation gives a particular solution, which vanishes as $t\to-\infty$. Consequently, our stochastic correlation functions at finite t are already the equilibrium Green functions. If we let $\varepsilon\to 0^+$ in our correlation functions, this is after we have gone to the equilibrium limit, which gives the wrong result when it is applied to the Parisi-Wu form of the Langevin equation.

If, as instructed by Parisi and Wu [1], we impose the boundary condition that the stochastic field vanish at some finite t, which we may as well take to be $t=0$, when $\varepsilon=0$ the part of the solution which depends on the initial conditions is no longer transient: we cannot neglect it when t now becomes large. This complication may be traced back to the occurrence of the pole at $\omega=0$ in the longitudinal part of G: see (6.3). Such a pole arises because D is not invertible when $\varepsilon=0$.

We impose our initial condition on $A(p,t)$ by including an additional factor $\theta(t')$ under the second integral in (6.5). So effectively $\eta(p,t')$ is multiplied by $\theta(t')$, which means that its Fourier transform $\eta(p,\omega)$ must be replaced with

$$\frac{i}{2\pi}\int \mathrm{d}\Omega\,\frac{1}{\Omega+i0}\eta(p,\omega-\Omega)\,. \tag{6.6}$$

The effect on stochastic diagrams is that the sum of the ω's on the two lines attached to a cross no longer vanishes. This means that we no longer have the overall δ-function (2.9), and the stochastic average (2.10) now varies with t. It is convenient to modify the stochastic diagrams by including an additional line attached to each cross, emerging from it to the right. This line carries the sum Ω of the two ω's entering the cross from the left and has a "propagator"

$$\frac{i}{2\pi}\,\frac{1}{\Omega+i0}\,. \tag{6.7}$$

As the simplest possible example, consider the modified stochastic diagram of fig. 8, corresponding to the two-point function v_2 to zeroth order in the interaction. Its contribution to the stochastic average

$$\langle A^a_\mu(p_1,\omega_1)A^b_\nu(p_2,\omega_2)\rangle \tag{6.8}$$

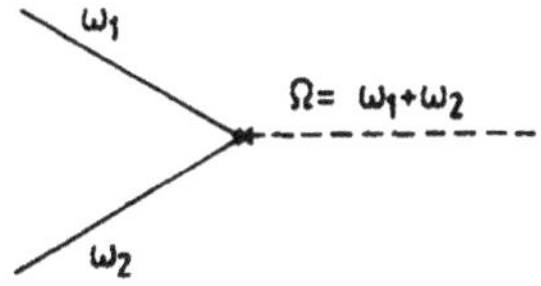

Fig. 8. A generalized stochastic diagram, which takes account of the initial condition on the stochastic field.

is

$$2(2\pi)^{n+1}\delta^{(n)}(p_1+p_2)G^{ac}_{\mu\rho}(p_1,\omega_1)G^{bcp}_{\nu}(p_2,\omega_2)\frac{i}{2\pi}\frac{1}{\omega_1+\omega_2+i0}. \qquad (6.9)$$

We go over to t-space by applying the Fourier integration (2.11) (with $N=2$). Because we are interested in $t>0$, we complete the contours of the ω_1 and ω_2 integrations with infinite semicircles in the lower half planes, and then evaluate the integrals by taking the residues at the pole of the integrand. This gives

$$\langle A^a_\mu(p_1,t)A^b_\nu(p_2,t)\rangle = -(2\pi)^n\delta^{(n)}(p_1+p_2)\delta^{ab}$$
$$\times i\left[\frac{T_{\mu\nu}(p_1)}{p_1^2+i\varepsilon}\left(1-e^{2it(p_1^2+i\varepsilon)}\right)+\frac{L_{\mu\nu}(p_1)}{i\varepsilon}\left(1-e^{-2\varepsilon t}\right)\right]. \qquad (6.10)$$

If, incorrectly, we now take the limit $t\to\infty$, the exponentials disappear. They are the part of the solution which depends on our initial condition. If, however, we let $\varepsilon\to 0^+$ before letting t become large, although the first exponential again becomes suppressed because it oscillates infinitely rapidly, the coefficient of $L_{\mu\nu}$ in the square bracket becomes $-2it$. According to Parisi and Wu [1], when one calculates a gauge-invariant expression such powers of t cancel, and leave behind terms that in the usual quantization procedure are associated with ghosts.

It is fairly easy to check this in simple examples, such as we described in sect. 3. In calculating stochastic diagrams more complicated than fig. 8, it is helpful again to use an identity like (2.21). However, because we no longer have the $\delta(\omega_1+\omega_2+\cdots\omega_N)$, the right-hand side has the additional term $-i(\partial/\partial t)V_N^{(\alpha)}$. The resulting differential equation is easy to solve, but we have not been able to derive from it any general results that may be stated in a concise form.

We have to conclude that, in the Parisi-Wu approach, the connection between stochastic diagrams and Feynman diagrams is at least as complicated as in the Zwanziger approach. In this sense, both quantization procedures are very different from the canonical one.

We are grateful to several members of TH Division at CERN for patient discussions, in particular D. Amati and P. Damgaard.

Appendix

In this appendix we show how the new interaction (5.4) contributes, in gauge-invariant quantities, terms normally associated with a ghost field.

If we use the recursive method of sect. 2 to sum the appropriate stochastic diagrams, we find that in $O(g^2)$ the contributions to

$$\langle 0|\mathrm{T}A_\lambda^a(p)A_{\lambda'}^b(p')|0\rangle \tag{A.1}$$

from terms involving the new vertex are those shown in fig. 6. There the vertex with the arrow is the new vertex v^J of (5.5), and the black spot denotes the effective vertex v^K of (5.7). Four of the diagrams include also an ordinary vertex (5.6). The diagrams are Feynman-like: they imply the usual integration over the internal loop momentum. An elementary calculation of these diagrams gives

$$-\delta^{(n)}(p+p')g^2 f^{acd}f^{bcd}\int \mathrm{d}^n q\,\mathrm{d}^n r\,\delta^{(n)}(p+q+r)\frac{1}{p^2q^2r^2}$$

$$\times\left[\frac{q_\lambda r_{\lambda'}}{p^2}-g_{\lambda\lambda'}\frac{2p^2-q^2-r^2}{p^2+q^2+r^2}\right]+\cdots, \tag{A.2}$$

where the usual $i\varepsilon$ prescription is to be applied to the denominators. The terms which we have not written explicitly are proportional to $p_\lambda p_{\lambda'}$; these will not contribute to the gauge-invariant quantities which we calculate below. The first term in the square bracket corresponds just to the usual ghost term. We show that the other term does not contribute to the two gauge-invariant quantities discussed in sect. 5.

In the case of quark-quark scattering, we must insert (A.2) into the first term of (5.9). We must also use the effective vertex v^K to calculate the second term of (5.9), together with another term in which the roles of the two quarks are reversed. With some elementary Dirac and colour algebra, we find that these two terms yield an integral which exactly cancels the contribution from the unwanted term in (A.2). (The integrands cancel; we do not have to do the loop integration.)

In the case of the gauge-invariant expression $\langle FF\rangle$ of (5.8), things work out differently. We interpret (5.8) as

$$\lim_{y\to x}\langle 0|\mathrm{T}F_{\mu\nu}^a(x)F_{\rho\sigma}^{a'}(y)|0\rangle. \tag{A.3}$$

The contribution of the two-point function to this is

$$-\frac{1}{(2\pi)^{2n}}\int \mathrm{d}^n p\,\mathrm{d}^n p'\,\mathrm{e}^{-i(p+p')\cdot x}\Big\{p_\mu p'_\rho\langle 0|\mathrm{T}A_\nu^a(p)A_\sigma^{a'}(p')|0\rangle$$

$$-(\mu\leftrightarrow\nu)-(\rho\leftrightarrow\sigma)+(\mu\leftrightarrow\nu,\rho\leftrightarrow\sigma)\Big\}. \tag{A.4}$$

Fig. 9. A contribution to $\langle FF \rangle$ of (A.3).

If we insert (A.2) into this expression, the unwanted term in the square bracket of (A.2) contributes

$$\frac{-1}{(2\pi)^{2n}} g^2 f^{abc} f^{a'bc} \int \mathrm{d}^n p\, \mathrm{d}^n q\, \mathrm{d}^n r\, \delta^{(n)}(p+q+r)$$
$$\times \left\{ g_{\nu\sigma} p_\mu p_\rho \frac{q^2+r^2-2p^2}{p^2 q^2 r^2 (p^2+q^2+r^2)} + \cdots \right\}, \qquad \text{(A.5)}$$

where the unwritten terms in the curly bracket denote the same permutations as in (A.4).

We obtain (A.4) by using the terms linear in A in both $F^a_{\mu\nu}$ and $F^{a'}_{\rho\sigma}$. If instead we use the quadratic term in one of these two operators, we obtain in order g^2 terms such as depicted in fig. 9. When the vertex here is the effective vertex v^K, we obtain an integral like (A.5), but with the curly bracket replaced with

$$-p_\mu v^K_{\nu\rho\sigma}(p,q,r) + p_\nu v^K_{\mu\rho\sigma}(p,q,r)$$
$$+ p_\rho v^K_{\sigma\mu\nu}(-p,-q,-r) - p_\sigma v^K_{\rho\mu\nu}(-p,-q,-r). \qquad \text{(A.6)}$$

In order to see that these terms, when integrated, just cancel (A.5) one needs two identities:

$$\int \mathrm{d}^n q\, \mathrm{d}^n r\, \delta^{(n)}(p+q+r) \frac{q^2-r^2}{q^2 r^2 (p^2+q^2+r^2)} = 0,$$

$$\int \mathrm{d}^n p\, \mathrm{d}^n q\, \mathrm{d}^n r\, \delta^{(n)}(p+q+r) p_\lambda q_{\lambda'} \frac{p^2-q^2}{p^2 q^2 r^2 (p^2+q^2+r^2)} = 0. \qquad \text{(A.7)}$$

Each of these identities follows trivially from the symmetry properties of the integrand, and each would be valid also if we were using some regularization scheme different from dimensional regularization.

References

[1] G. Parisi and Wu Yongshi, Sci. Sin. 24 (1981) 483
[2] H. Hüffel and H. Rumpf, Phys. Lett. 148B (1984) 104; Univ. of Vienna preprint UWThPh-1984-30 (to appear in Z. Phys. C.);
E. Gozzi, Phys. Lett. 150B (1985) 119
[3] H. Umezawa, H. Matsumoto and M. Tachiki, Thermo field dynamics and condensed states, (North-Holland, Amsterdam, 1982);
A.W. Niemi and G.W. Semenoff, Ann. of Phys. 152 (1984) 105; Nucl. Phys. B230 [FS 10] (1984) 181
[4] W. Grimus and H. Hüffel, Z. Phys. C18 (1983) 129
[5] M. Namiki, I. Ohba, K. Okano and Y. Yamanaka, Prog. Theor. Phys. 69 (1983) 1580;
H. Nakazato, M. Namiki, I. Ohba and K. Okano, Prog. Theor. Phys. 70 (1983) 298;
C. De Dominicis, Lett. Nuovo Cim. 12 (1975) 567
[6] H. Nakagoshi, M. Namiki, I. Ohba and K. Okano, Prog. Theor. Phys. 70 (1983) 326
[7] E. Egorian and S. Kalitzin, Phys. Lett. 129B (1983) 320
[8] J.D. Breit, S. Gupta and A. Zaks, Nucl. Phys. B233 (1984) 61;
P.H. Damgaard and K. Tsokos, Nucl. Phys. B235 (1984) 75
[9] T. Fukai, H. Nakazato, I. Ohba, K. Okano and Y. Yamanaka, Prog. Theor. Phys. 69 (1983) 361
[10] D. Zwanziger, Nucl. Phys. B192 (1981) 259
[11] L. Baulieu and D. Zwanziger, Nucl. Phys. B193 (1981) 163;
E. Floratos and J. Iliopoulos, Nucl. Phys. B214 (1983) 392;
E. Seiler, Acta Phys. Austr., Suppl. 26 (1984) 259
[12] Y. Nakano, Prog. Theor. Phys. 69 (1983) 361;
E. Gozzi, Phys. Rev. D28 (1983) 1922
[13] G. Aldazabal, E. Dagotto, A. Gonzales-Arroyo and N. Parga, Phys. Lett. 125B (1983) 305;
G. Marchesini, Nucl. Phys. B239 (1984) 135
[14] W. Kummer, Acta Phys. Austr. 41 (1975) 315;
S. Caracciolo, G. Curci and P. Menotti, Phys. Lett. 113B (1982) 311
[15] E.G. Floratos, J. Iliopoulos and D. Zwanziger, Univ. Bern preprint BUTP-83/14 (1983);
A. Burnel, Phys. Rev. D29 (1984) 2344; D30 (1984) 2244
[16] H. van Dam and M. Veltman, Nucl. Phys. B22 (1970) 397

Gauge Fields

Nuclear Physics B192 (1981) 259–269

COVARIANT QUANTIZATION OF GAUGE FIELDS WITHOUT GRIBOV AMBIGUITY

Daniel ZWANZIGER[1]

Department of Physics, New York University, New York, New York 10003, USA

Received 30 March 1981

Recently Parisi and Wu proposed a method of quantizing gauge fields whereby euclidean expectation values are obtained by relaxation to equilibrium of a stochastic process depending on an artificial fifth time parameter. In the present work the equilibrium distribution is determined directly, without reference to the artificial time, by a stationary condition which is an eigenfunction equation in the euclidean Hilbert space. The solution has a perturbative expansion which appears renormalizable by naive power counting. Because of gauge freedom, a free dimensionless gauge parameter appears in the theory although no gauge condition such as $\partial \cdot A = 0$ is imposed.

Parisi and Wu [1] proposed that the euclidean expectation value $\langle \Phi \rangle$ of a functional $\Phi[A]$ of the euclidean gauge field $A^a_\mu(x)$ be defined as the limit to which the mean relaxes,

$$\langle \Phi \rangle = \lim_{t\to\infty} \langle \Phi \rangle_t \equiv \lim_{t\to\infty} \int \mathrm{d}A \Phi[A] P[A, t] , \tag{1}$$

where the probability distribution $P[A, t]$, depending on an artificial fifth time parameter, is determined by a stochastic process, namely

$$\dot{P}[A, t] = \int \mathrm{d}^4x \frac{\delta}{\delta A^a_\mu(x)} \left\{ \frac{\delta}{\delta A^a_\mu(x)} + \frac{\delta S[A]}{\delta A^a_\mu(x)} \right\} P[A, t] , \tag{2}$$

with

$$S[A] = \tfrac{1}{4} \int \sum_{\mu,\nu,a} (F^a_{\mu\nu})^2 \, \mathrm{d}^4x , \tag{3}$$

$F^a_{\mu\nu} = \partial_\mu A^a_\nu - \partial_\nu A^a_\mu + g c^{abc} A^b_\mu A^c_\nu$, for some initial data $P[A, 0] = P_0[A] \geq 0$, $\int \mathrm{d}A P_0[A] = 1$. If the Boltzmann distribution e^{-S} is normalizable, $P[A, t]$ relaxes to $\lim_{t\to\infty} P[A, t] = \mathrm{e}^{-S} / \int \mathrm{d}A \, \mathrm{e}^{-S}$, so eq. (1) reduces to the standard expression, for any P_0. In gauge field theories e^{-S} is not normalizable because S is constant along gauge orbits so $\int \mathrm{d}A \, \mathrm{e}^{-S}$ has the divergence of the infinite volume of the gauge group. Nevertheless the stochastic process (2) is supposed to have the virtue that the mean $\langle \Phi \rangle_t$ relaxes to a value $\langle \Phi \rangle$ which is universal, in the sense that it is independent of P_0, even though $P[A, t]$ has no limit as a normalized probability distribution.

[1] Research supported in part by National Science Foundation grant no. PHY78-21503.

To exploit analogies with stochastic mechanics we expand $A^a_\mu(x) = \sum_i q_i \psi^a_{i,\mu}(x)$ in a complete orthonormal set $\psi^a_{i,\mu}(x)$, and use the discrete variables q_i, so eq. (2) reads

$$\dot{P}(q, t) = \sum_i \frac{\partial}{\partial q_i}\left[\frac{\partial}{\partial q_i} - f_i(q)\right] P(q, t), \tag{4}$$

where $f_i(q) = -\partial S(q)/\partial q_i$. This is Smoluchowski's equation* with drift force f_i, in units such that the diffusion constant and friction coefficient are unity. With

$$L \equiv -\sum_i \left[\frac{\partial}{\partial q_i} + f_i(q)\right] \frac{\partial}{\partial q_i}, \tag{5a}$$

$$L^* \equiv -\sum_i \frac{\partial}{\partial q_i}\left[\frac{\partial}{\partial q_i} - f_i(q)\right], \tag{5b}$$

we have

$$P(q, t) = e^{-L^* t} P_0(q), \tag{6}$$

so

$$\langle \Phi \rangle_t = \int dq \Phi(q) [e^{-L^* t} P_0(q)], \tag{7}$$

$$\langle \Phi \rangle_t = \int dq [e^{-Lt} \Phi(q)] P_0(q). \tag{8}$$

As $\langle \Phi \rangle_t$ is supposed to relax to $\langle \Phi \rangle$ independent of P_0, we may choose $P_0(q) = \delta(q - q_1)$, q_1 arbitrary. Hence with

$$\Phi(q, t) \equiv e^{-Lt} \Phi(q), \tag{9}$$

we obtain

$$\langle \Phi \rangle = \lim_{t \to \infty} \Phi(q, t). \tag{10}$$

We thus arrive at the interesting conclusion that if $\Phi(q, t)$ is the solution of

$$\dot{\Phi}(q, t) = -L\Phi(q, t) \tag{11}$$

with initial data

$$\Phi(q, 0) = \Phi(q), \tag{12}$$

then $\Phi(q, t)$ relaxes to a constant** which defines the mean value $\langle \Phi \rangle$. Reverting to continuum notation, we note that if $\Phi[A]$ is gauge invariant, then both the initial condition $\Phi[A, 0] = \Phi[A]$ and the equation of motion

$$\dot{\Phi}[A, t] = -L\Phi[A, t], \tag{13a}$$

* The Fokker–Planck equation in velocity space is similar.

** This conclusion is supported by the strong maximum principle for parabolic differential equations, at least in a finite number of dimensions. See ref. [2].

$$L = -\int d^4x \left[\frac{\delta}{\delta A_\mu^a(x)} - \frac{\delta S}{\delta A_\mu^a(x)}\right] \frac{\delta}{\delta A_\mu^a(x)} \tag{13b}$$

are gauge invariant. This equation in discrete form could be used in computer calculations in close analogy to computer calculations in lattice gauge theories whereby the euclidean functional integral is evaluated by a Monte Carlo relaxation procedure. At present, however, we shall eliminate the artificial time and obtain a perturbative expansion.

As it stands, eq. (13) does not lend itself to a perturbative expansion because it provides no restoring force in the longitudinal modes, so the propagator contains [1] the apparently unrenormalizable term $tk_\mu k_\nu/k^2$ instead of $k_\mu k_\nu/(k^2)^2$. Such terms should cancel, in principle, if $\Phi[A]$ is gauge invariant. However, they would destroy renormalizability of individual graphs unless gauge invariance is systematically exploited to make each graph renormalizable. The generator of local gauge transformations is

$$G^a(x) = -D_\mu^{ab} \frac{\delta}{\delta A_\mu^b(x)} = -[\partial_\mu \delta^{ac} + gc^{abc} A_\mu^b]\delta/\delta A_\mu^c(x) \tag{14}$$

satisfying

$$[G^a(x), G^b(y)] = gc^{abc}\delta^4(x-y)G^c(x)\,. \tag{15}$$

The operator L, eq. (13b), satisfies the strong gauge invariance condition

$$[G^a(x), L] = 0\,, \tag{16}$$

and gauge invariance of the functional Φ is the condition

$$G^a(x)\Phi[A] = 0\,. \tag{17}$$

Thus when $\Phi[A]$ is gauge invariant, the solution to eq. (13) may be written

$$\Phi[A, t] = e^{-Lt}\Phi[A] = e^{-L_v t}\Phi[A]\,, \tag{18}$$

where

$$L_v = L + \int d^4x\, v^a(x) G^a(x) \tag{19}$$

with $v^a(x)$ arbitrary. If L_v is to describe a manifestly Lorentz invariant, renormalizable theory, and if $v^a(x)$ is local in $A(x)$, the only possibility is $v^a = -\alpha^{-1}\partial\cdot A^a$ because all other expressions have higher dimension. Here α is a dimensionless gauge parameter with $\alpha = 1$ being the Feynman gauge and $\alpha \to 0$ the Landau gauge. Henceforth we consider only L_v of the form

$$L_\alpha \equiv L + \alpha^{-1} \int d^4x\, \partial\cdot A^a D_\mu^{ab} \frac{\delta}{\delta A_\mu^b(x)}\,, \tag{20}$$

$$L_\alpha = -\int d^4x \left[\frac{\delta}{\delta A_\mu^a(x)} + f_{\alpha,\mu}^a(x)\right]\frac{\delta}{\delta A_\mu^a(x)}, \tag{21a}$$

$$L_\alpha^* = -\int d^4x \frac{\delta}{\delta A_\mu^a(x)}\left[\frac{\delta}{\delta A_\mu^a(x)} - f_{\alpha,\mu}^a(x)\right], \tag{21b}$$

where the drift force depends on α:

$$f_{\alpha,\mu}^a(x) = -\frac{\delta S}{\delta A_\mu^a(x)} - \alpha^{-1}\frac{\delta S_G}{\delta A_\mu^a(x)} + \alpha^{-1} C_\mu^a(x). \tag{22}$$

Here S is given by eq. (3), S_G is the traditional gauge-breaking action

$$S_G = \tfrac{1}{2}\int d^4x \sum_a (\partial \cdot A^a)^2 \tag{23}$$

and

$$C_\mu^a(x) = g c^{abc} A_\mu^b\, \partial \cdot A^c \tag{24}$$

replaces the Faddeev–Popov ghost interaction.

In terms of the discrete variables introduced in eq. (4), one has

$$L_\alpha = -\sum_i \left[\frac{\partial}{\partial q_i} + f_{\alpha,i}(q)\right]\frac{\partial}{\partial q_i}, \tag{25a}$$

$$L_\alpha^* = -\sum_i \frac{\partial}{\partial q_i}\left[\frac{\partial}{\partial q_i} - f_{\alpha,i}(q)\right], \tag{25b}$$

with drift force

$$f_{\alpha,i}(q) = -\frac{\partial S}{\partial q_i} - \alpha^{-1}\frac{\partial S_G}{\partial q_i} + \alpha^{-1} C_i. \tag{26}$$

If $\Phi[A]$ is gauge invariant, eqs. (8) and (18) give

$$\langle \Phi \rangle_t = \int dq [e^{-L_\alpha t}\Phi(q)] P_0(q), \tag{27}$$

$$\langle \Phi \rangle_t = \int dq \Phi(q) [e^{-L_\alpha^* t} P_0(q)]. \tag{28}$$

Thus, if $P_\alpha(q, t)$ is the solution of

$$\dot P_\alpha(q, t) = -L_\alpha^* P_\alpha(q, t) \tag{29}$$

with initial data

$$P_\alpha(q, 0) = P_0(q), \tag{30}$$

for gauge invariant $\Phi(q)$ we have

$$\langle\Phi\rangle_t = \int dq\Phi(q)P_\alpha(q,t)\,. \tag{31}$$

The gauge-fixing term $\alpha^{-1}S_G$ provides the desired restoring force along gauge orbits and we thus expect that for finite, positive α, $P_\alpha(q,t)$ relaxes to an α-dependent equilibrium distribution

$$P_{\alpha,E}(q) = \lim_{t\to\infty} P_\alpha(q,t)\,, \tag{32}$$

with the desired euclidean expectation value $\langle\Phi\rangle = \lim_{t\to\infty}\langle\Phi\rangle_t$,

$$\langle\Phi\rangle = \int dq\Phi(q)P_{\alpha,E}(q)\,. \tag{33}$$

The remaining term $\alpha^{-1}C_i$ in the drift force, eq. (26), does not destroy this expectation. For one may verify from eq. (24) that $\partial C_i/\partial q_i = 0$, $\partial C_j/\partial q_i - \partial C_i/\partial q_j \neq 0$, so C_i is a pure circulation, with no flux lines of $C_i(q)$ ending. Nor do they go to infinity with q because of the antisymmetry of c^{abc} in eq. (24). Thus, at equilibrium C_i produces a stationary circulating current, and $P_E(q)$ represents a generalization of the Boltzmann distribution to non-conservative drift forces.

From eq. (29) it follows that the equilibrium distribution satisfies the stationary condition

$$L_\alpha^* P_{\alpha,E}[A] = 0\,, \tag{34}$$

where L_α^* is given in eq. (21b), and we have reverted to continuum notation. One may easily verify that this equation is not satisfied by the Faddeev–Popov distribution [3]

$$P_{\alpha,FP}[A] = N\exp[-S-\alpha^{-1}S_G]\,\mathrm{Det}\,(\partial^\mu D_\mu/\partial^2)\,. \tag{35}$$

This shows that some euclidean Green functions calculated according to $P_{\alpha,E}$ and $P_{\alpha,FP}$ are different. For the physical content of the two theories to be the same it is sufficient that euclidean Green functions (i.e. euclidean expectation values) of gauge-invariant quantities be the same. As the Faddeev–Popov distribution P_{FP} is subject to the Gribov ambiguity [6], presumably only a comparison of the perturbative expansion of these quantities is meaningful.

We now develop the perturbative expansion of the equilibrium distribution P_E. With $g=0$, the operator L_α, eq. (21), becomes

$$L_\alpha^{(0)} = -\int d^4x\left[\frac{\delta}{\delta A_\mu^a(x)} - \frac{\delta S_\alpha^{(0)}}{\delta A_\mu^a(x)}\right]\frac{\delta}{\delta A_\mu^a(x)} \tag{36}$$

$$S_\alpha^{(0)} = \int d^4x \sum_a\left[\tfrac{1}{4}\sum_{\mu,\nu}(\partial_\mu A_\nu^a - \partial_\nu A_\mu^a)^2 + \tfrac{1}{2}\alpha^{-1}(\partial\cdot A^a)^2\right], \tag{37}$$

so

$$P_\alpha^{(0)}[A] = \frac{e^{-S_\alpha^{(0)}[A]}}{\int dA\, e^{-S_\alpha^{(0)}[A]}} \tag{38}$$

is the normalized solution to $L_\alpha^{(0)*}P_{\alpha,E}^{(0)} = 0$. We introduce an inner product on polynomial functions by

$$(\Phi_1, \Phi_2) \equiv \frac{\int dA\, \Phi_1^*[A]\Phi_2[A]\, e^{-S_\alpha^{(0)}[A]}}{\int dA\, e^{-S_\alpha^{(0)}[A]}} \tag{39}$$

[a more precise definition is given below, eq. (56)] and complete the space of polynomials in the corresponding norm to a euclidean Hilbert space $\mathscr{H}$. We factorize $P_\alpha^{(0)}$ out of $P_{\alpha,E}$ and define $\Phi_{\alpha,E}[A]$ by

$$P_{\alpha,E}[A] = P_\alpha^{(0)}[A]\Phi_{\alpha,E}[A], \tag{40}$$

so the expectation value (33) reads

$$\langle \Phi \rangle = (\Phi_{\alpha,E}, \Phi). \tag{41}$$

With this inner product, hermitian adjoints are

$$A_\mu^a(x)^\dagger = A_\mu^a(x), \tag{42a}$$

$$\left[\frac{\delta}{\delta A_\mu^a(x)}\right]^\dagger = -\frac{\delta}{\delta A_\mu^a(x)} + \frac{\delta S_\alpha^{(0)}}{\delta A_\mu^a(x)}, \tag{42b}$$

so the stationary condition $L_\alpha^* P_{\alpha,E} = 0$ reads, by eq. (21),

$$L_\alpha^\dagger \Phi_{\alpha,E} = 0, \tag{43}$$

where

$$L_\alpha^\dagger = \int d^4x \left[\frac{\delta}{\delta A_\mu^a(x)}\right]^\dagger \left\{\frac{\delta}{\delta A_\mu^a(x)} - f_{\mathrm{int},\mu}^a(A)\right\}, \tag{44}$$

$$f_{\mathrm{int},\mu}^a(A) = -\frac{\delta S_{\mathrm{int}}[A]}{\delta A_\mu^a(x)} + \alpha^{-1} C_\mu^a(x), \tag{45}$$

and S_{int} is the cubic and quartic parts of the Yang–Mills action. The introduction of the euclidean Hilbert space $\mathscr{H}$ promotes the stationary condition (34), which is an elliptic partial differential equation, into the eigenfunction equation (43) which determines $\Phi_{\alpha,E}$. The artificial fifth time parameter and its attendant parabolic partial differential equation, are eliminated.

It would be of interest to attempt a non-perturbative variational calculation by minimizing

$$(L^\dagger \Phi, L^\dagger \Phi)/(\Phi, \Phi). \tag{46}$$

At present however we seek the perturbative expansion of Φ_{E}. (We suppress the index α.) With $g=0$ the operator $L^{\dagger}$ becomes

$$L^{(0)}=L^{(0)\dagger}=\int \mathrm{d}^4x\left[\frac{\delta}{\delta A_{\mu}^{a}(x)}\right]^{\dagger}\frac{\delta}{\delta A_{\mu}^{a}(x)}, \tag{47}$$

which is self-adjoint and positive. The Fourier transforms

$$A_{\mu}^{a}(k)=(2\pi)^{-2}\int \mathrm{d}^4x\,\mathrm{e}^{-ik\cdot x}A_{\mu}^{a}(x), \tag{48a}$$

$$\frac{\delta}{\delta A_{\mu}^{a}(k)}=(2\pi)^{-2}\int \mathrm{d}^4x\,\mathrm{e}^{ik\cdot x}\frac{\delta}{\delta A_{\mu}^{a}(x)}, \tag{48b}$$

satisfy

$$A_{\mu}^{a}(k)^{\dagger}=A_{\mu}^{a}(-k), \tag{49a}$$

$$\left[\frac{\delta}{\delta A_{\mu}^{a}(k)}\right]^{\dagger}=-\frac{\delta}{\delta A_{\mu}^{a}(-k)}+k^2A_{\mu}^{\mathrm{T}a}(k)+\alpha^{-1}k^2A_{\mu}^{\mathrm{L}a}(k), \tag{49b}$$

where T and L refer to transverse and longitudinal parts. Observe that

$$a_{\mu}^{a}(k)\equiv\frac{1}{\sqrt{k^2}}\frac{\delta}{\delta A_{\mu}^{\mathrm{T}a}(k)}+\sqrt{\frac{\alpha}{k^2}}\frac{\delta}{\delta A_{\mu}^{\mathrm{L}a}(k)} \tag{50}$$

enjoys the defining property of an annihilation operator

$$[a_{\mu}^{a}(k),a_{\nu}^{b}(k')^{\dagger}]=\delta_{\mu\nu}\delta^{ab}\delta^4(k-k'), \tag{51}$$

with vacuum state

$$\Phi_0=1, \qquad a_{\mu}^{a}(k)\Phi_0=0. \tag{52}$$

The operator $L^{(0)}$ is simply

$$L^{(0)}=\int \mathrm{d}^4k[k^2a^{\mathrm{T}}(k)^{\dagger}a^{\mathrm{T}}(k)+\alpha^{-1}k^2a^{\mathrm{L}}(k)^{\dagger}a^{\mathrm{L}}(k)], \tag{53}$$

where summation on Lorentz and color indices is implied. This expression provides an immediate euclidean particle interpretation of the states in the spectrum of $L^{(0)}$, which form a complete basis in $\mathscr{H}$. The inversion formulae

$$A(k)=\frac{1}{\sqrt{k^2}}[a^{\mathrm{T}}(k)^{\dagger}+a^{\mathrm{T}}(-k)]+\sqrt{\frac{\alpha}{k^2}}[a^{\mathrm{L}}(k)^{\dagger}+a^{\mathrm{L}}(-k)], \tag{54a}$$

$$\frac{\delta}{\delta A(k)}=\sqrt{k^2}a^{\mathrm{T}}(k)+\sqrt{\alpha^{-1}k^2}a^{\mathrm{L}}(k), \tag{54b}$$

allow the generic operator to be expressed in terms of $a(k)$ and $a(k)^{\dagger}$. Similarly the generic functional polynomial in A may be expressed in terms of the states

$$\Phi=\left[c^{(0)}+\int d^4k c^{(1)}(k)a^{\dagger}(k)+\tfrac{1}{2}\int d^4k_1\, d^4k_2 c^{(2)}(k_1,k_2)a^{\dagger}(k_1)a^{\dagger}(k_2)\right]\Phi_0+\cdots \tag{55}$$

with inner product

$$(\Phi_1,\Phi_2)=c_1^{(0)*}c_2^{(0)}+\int d^4k c_1^{(1)*}(k)c_2^{(1)}(k)$$
$$+\tfrac{1}{2}\int d^4k_1\, d^4k_2 c_1^{(2)*}(k_1,k_2)c_2^{(2)}(k_1,k_2)+\cdots. \tag{56}$$

As an example, let

$$\Phi[A]=\tfrac{1}{2}\int d^4k_1\, d^4k_2 c^{ab}_{\mu\nu}(k_1,k_2)A^a_\mu(k_1)A^b_\nu(k_2),$$

$$\Phi[A]=\tfrac{1}{2}\int d^4k c^{ab}_{\mu\nu}(k,-k)\left[\frac{1}{k^2}P^{\mathrm{T}}_{\mu\nu}(k)+\frac{\alpha}{k^2}P^{\mathrm{L}}_{\mu\nu}\right]\delta^{ab}\Phi_0$$
$$+\tfrac{1}{2}\int d^4k_1\, d^4k_2 c^{ab}_{\mu\nu}(k_1,k_2)(k_1^2)^{-1/2}(k_2^2)^{-1/2}$$
$$\times[a^{\mathrm{T}a}_\mu(k_1)^\dagger+\alpha^{1/2}a^{\mathrm{L}a}_\mu(k_2)^\dagger][a^{\mathrm{T}b}_\nu(k_2)^\dagger+\alpha^{1/2}a^{\mathrm{L}b}_\nu(k_2)^\dagger]\Phi_0,$$

where $P^{\mathrm{L}}_{\mu\nu}(k)=k_\mu k_\mu/k^2$, $P^{\mathrm{T}}_{\mu\nu}(k)=\delta_{\mu\nu}-P^{\mathrm{L}}_{\mu\nu}(k)$. In zeroth order $\Phi_{\mathrm{E}}=\Phi_0=1$ and the zeroth order euclidean expectation value (41) gives in this case

$$\langle\Phi\rangle^{(0)}=(\Phi_0,\Phi)=\tfrac{1}{2}\int d^4k c^{ab}_{\mu\nu}(k,-k)\left[\frac{1}{k^2}P^{\mathrm{T}}_{\mu\nu}(k)+\frac{\alpha}{k^2}P^{\mathrm{L}}_{\mu\nu}(k)\right],$$

whereby the familiar free two-point Green function

$$D^{ab}_{\mu\nu}(k)=\frac{1}{k^2}[P^{\mathrm{T}}_{\mu\nu}(k)+\alpha P^{\mathrm{L}}_{\mu\nu}(k)]\delta^{ab}$$

has made its appearance.

To solve the eigenfunction equation $L^\dagger\Phi_{\mathrm{E}}=0$ perturbatively, we write

$$L^\dagger=L^{(0)}+V,\qquad V=-\int d^4x\left[\frac{\delta}{\delta A^a_\mu(x)}\right]^\dagger f^a_{\mathrm{int},\mu}(x), \tag{57}$$

and pose

$$\Phi_{\mathrm{E}}=\sum_{n=0}^{\infty}\Phi^{(n)} \tag{58}$$

with $\Phi^{(0)}=\Phi_0=1$ and obtain

$$L^{(0)}\Phi_n=-V\Phi_{n-1}. \tag{59}$$

A valuable consistency check on the present approach is that $L^{\dagger}\Phi_{\mathrm{E}}=0$ has no solution unless 0 is an eigenvalue of $L^{\dagger}$. Perturbatively, eq. (59) has no solution, consistent with $L^{(0)}\Phi_0=0$ and $L^{(0)}=L^{(0)\dagger}$ unless $\lambda_n=(\Phi_0,\ V\Phi_{n-1})=(V^{\dagger}\Phi_0,\ \Phi_{n-1})$, which is the nth order shift in eigenvalue, vanishes. Since $V^{\dagger}=-\int \mathrm{d}^4x f^a_{\mathrm{int},\mu}(x)\delta/\delta A^a(x)$ and $\Phi_0=1$, this is true and the consistency test is passed. The component of $\Phi^{(n)}$ parallel to Φ_0 is determined by the condition $\langle 1\rangle=(\Phi_{\mathrm{E}},\ \Phi_0)=1$, so with $\Phi^{(0)}=\Phi_0=1$, $\Phi^{(n)}$ must be orthogonal to Φ_0, $(\Phi_0,\ \Phi^{(n)})=1$ for $n>0$. The desired perturbative expansion of Φ_{E} follows:

$$\Phi_{\mathrm{E}}=\Phi_0-\frac{P}{L^{(0)}}V\Phi_0+\frac{P}{L^{(0)}}V\frac{P}{L^{(0)}}V\Phi_0-\cdots\,, \tag{60}$$

where P is the projector onto the space orthogonal to Φ_0.

To obtain a diagrammatic representation of this perturbation series, express V in terms of creation and annihilation operators:

$$V=-\int \mathrm{d}^4k\sqrt{k^2}[a^{\mathrm{T}a}_{\mu}(-k)^{\dagger}+\alpha^{-1/2}a^{\mathrm{L}a}_{\mu}(-k)^{\dagger}]f^a_{\mathrm{int},\mu}(k)\,, \tag{61a}$$

$$f^a_{\mathrm{int},\mu}(k)=-\frac{\delta S_{\mathrm{int}}[A]}{\delta A^a_{\mu}(k)}+\alpha^{-1}C^a_{\mu}(k)\,, \tag{61b}$$

$$C^a_{\mu}(k)=\int \mathrm{d}^4k_1\,\mathrm{d}^4k_2\delta^4(k_1+k_2-k)g'c^{abc}A^b_{\mu}(k_1)ik_2\cdot A^c(k_2)\,, \tag{62}$$

where $g'=g/(2\pi)^2$, and $A(k)$ is expressed in terms of $a(k)$ and $a^{\dagger}(k)$ in eq. (54). The diagrammatic representation has the old fashioned non-covariant form which distinguishes creation operators, which correspond to lines leaving a vertex toward the left, from annihilation operators, which correspond to lines entering a vertex from the right. Typical vertex parts are shown in fig. 1. Transverse and longitudinal gluons are represented by wavy and straight lines respectively, and vertices coming from the first and second terms in eq. (61b) are distinguished by labels S and C. (There are also 4-vertex parts from S.) Because of the $k\cdot A(k)$ factor in eq. (62), every C-vertex contains at least one longitudinal gluon. A typical intermediate or final state

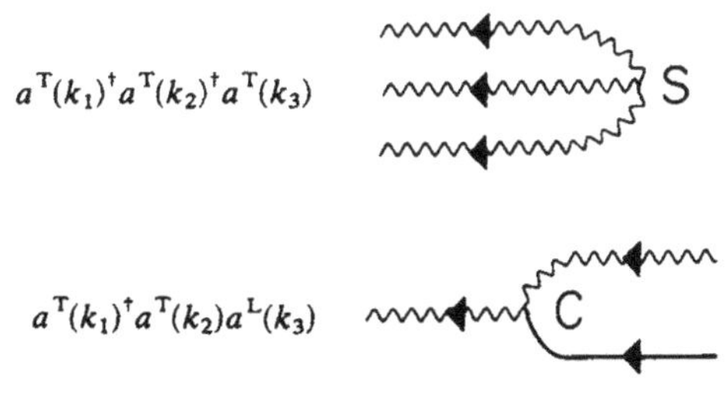

Fig. 1. Typical vertex parts.

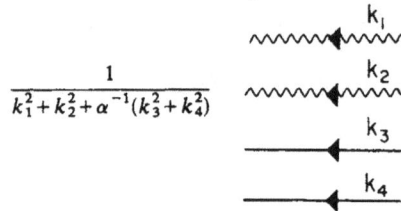

$$\frac{1}{k_1^2+k_2^2+\alpha^{-1}(k_3^2+k_4^2)}$$

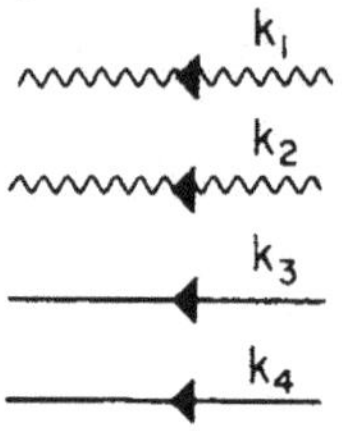

Fig. 2. Typical intermediate or final state.

corresponding to P/L_0, whose value is determined by eq. (53) is represented in fig. 2. Note that the vacuum state is on the far right in eq. (60) and that, according to eq. (61) every vertex has at least one creation operator, so the vacuum state never reappears and consequently all vacuum-vacuum graphs are absent from this series. All external lines appear at the far left. A particular term in the series is represented by the graph in fig. 3.

If the non-conservative force $\alpha^{-1}C$ were absent, then the series would sum to $e^{-S_{\text{int}}}/(\Phi_0, e^{-S_{\text{int}}})$, as one may easily verify. This is the sum of all ghostless diagrams in the Faddeev–Popov series. As the Landau gauge is approached, $\alpha \to 0$, longitudinal gluons become ghosts in the sense that external lines become transverse, but longitudinal gluons survive in intermediate states, as one may also verify. This occurs because energy denominators with longitudinal gluons are suppressed by a factor α, like the one in fig. 2,

$$\frac{1}{k_1^2+k_2^2+\alpha^{-1}(k_3^2+k_4^2)} \to \frac{\alpha}{k_3^2+k_4^2},$$

but compensating negative powers of α occur in the vertices. However, proof of the equivalence of the two approaches remains incomplete.

In practice, the perturbation series (60) is not particularly convenient as it stands because of the multiplication of terms arising from seperate creation and annihilation parts. However, the exact equation $L^{\dagger}\Phi_E[A]=0$, which determines the equilibrium distribution $\Phi_E[A]$ as an eigenfunction of the euclidean Hilbert space $\mathcal{H}$, is not without interest from a constructive or non-perturbative point of view. For unlike the Faddeev–Popov distribution, its derivation has required no gauge condition such as

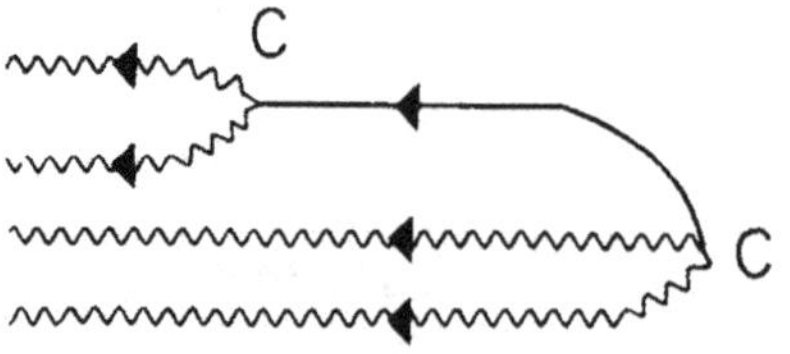

Fig. 3. Graph representing a term contained in $(P/L^{(0)})V(P/L^{(0)})V\Phi_0$.

$\partial \cdot A = 0$ which is supposed to intersect each gauge orbit only once, but in fact does so more than once [4].

Much of the work reported here was done while the author was a visitor at the Laboratoire de Physique Théorique de l'Ecole Normale Supérieure. He is grateful for the hospitality accorded him there. He recalls with pleasure stimulating conversations with Edouard Brezin, Jean-Loup Gervais, Giorgio Velo, Gianfausto Dell'Antonio and John Lowenstein at the inception and during the course of this work.

References

[1] G. Parisi and Wu Yong-Shi, Sci. Sin. 24 (1981) 483
[2] A. Friedman, Partial differential equations of parabolic type, (Prentice Hall, 1964) theorem 18, ch. 2 and theorem 2, ch. 6
[3] L. Faddeev and V. Popov, Phys. Lett. 25B (1967) 29
[4] V.N. Gribov, Nucl. Phys. B139 (1978) 1

Nuclear Physics B193 (1981) 163–172

EQUIVALENCE OF STOCHASTIC QUANTIZATION AND THE FADDEEV–POPOV ANSATZ

L. BAULIEU[1,2]

Columbia University and Harvard University, USA

Daniel ZWANZIGER[3]

New York University and Brookhaven National Laboratory, USA*

Received 17 July 1981

We prove the equivalence of stochastic quantization to the Faddeev–Popov ansatz for covariant and axial gauges. A principal ingredient of the proof is a theorem which asserts that, for a certain large class of stochastic processes, the time-dependent distribution relaxes to an equilibrium distribution.

1. Introduction

We briefly review the method of stochastic quantization [1, 2] in this section, then prove its equivalence to the Faddeev–Popov ansatz in sect. 2, and conclude with some general remarks in sect. 3.

Let $A = A^a_\mu(x)$ be a euclidean gauge field, where a is a color index, $\mu = 1, \dots, 4$ is a Lorentz index, and x is a euclidean 4-vector**. It will frequently be convenient to replace $A^a_\mu(x)$ by discrete variables q_i, $i = 1, \dots, \infty$ according to $A^a_\mu(x) = \sum_i q_i \psi^a_{i\mu}(x)$, where the $\psi^a_{i\mu}(x)$ are a complete orthonormal set. Thus, an arbitrary functional $\phi[A]$, for example

$$\phi[A] = A^a_\mu(x) A^b_\nu(y)\,, \qquad \text{or} \qquad \phi[A] = \mathrm{tr}\,\mathrm{P}\exp i\oint \lambda^a A^a_\mu(x)\,\mathrm{d}x^\mu\,,$$

will be written $\phi[A] = \phi(q)$. The stochastic method*** introduces an artificial fifth time parameter t and promotes $\phi(q)$ to a stochastic variable $\phi(q, t)$ governed by the diffusion equation with a drift force $f(q)$:

$$\dot{\phi}(q, t) = -L_q \phi(q, t)\,, \tag{1}$$

$$L_q = -[\Delta + f \cdot \nabla]\,, \tag{2}$$

[1] Research supported in part by the National Science Foundation under grant no. PHY77-22864.
[2] On leave from Ecole Normal Superieure, 24 rue Lhomond 75005, Paris.
[3] Research supported in part by the National Science Foundation under grant no. PHY78-21503 and the US Department of Energy under contract no. DE-AC02-76CH00016.
* Permanent address.
** Arbitrary dimension D may replace 4 throughout.
*** We follow the approach developed in ref. [2].

supplemented by the initial condition

$$\phi(q,0)=\phi(q)\,. \tag{3}$$

Here

$$\dot{\phi}\equiv\partial\phi/\partial t\,,\qquad \Delta\equiv\sum_i \partial^2/\partial q_i^2\,,\qquad f\cdot\nabla\equiv\sum_i f_i(q)\partial/\partial q_i\,,$$

and the drift force $f_i(q)$ is derived from the classical action

$$f_i(q)=-\frac{\partial S(q)}{\partial q_i}\,, \tag{4}$$

which for a gauge theory is the Yang–Mills action

$$S(q)=S_{\rm cl}[A]=\tfrac{1}{4}\int\sum_{\mu,\nu,a}[F^a_{\mu\nu}(x)]^2\,{\rm d}x^4\,, \tag{5}$$

where $F^a_{\mu\nu}=\partial_\mu A^a_\nu-\partial_\nu A^a_\mu+f^{abc}A^b_\mu A^c_\nu$. Here and in the following the coupling constant g has been absorbed in the structure constant of the gauge group f^{abc}. Stochastic quantization is completed by the ansatz for calculating euclidean expections values,

$$\langle\phi\rangle=\lim_{t\to\infty}\phi(q,t)\,, \tag{6}$$

which is equivalent to the proposal of Parisi and Wu [1]. Clearly, for this formula to make sense, the limit must exist and be independent of q. We shall show that in a non-gauge theory it reduces to the standard functional integral formula

$$\langle\phi\rangle=\frac{\int {\rm d}q\phi(q)\,{\rm e}^{-S(q)}}{\int {\rm d}q\;{\rm e}^{-S(q)}}\,, \tag{7}$$

and that for a gauge theory it agrees with the Faddeev–Popov ansatz for gauge-invariant ϕ.

As a trivial example of how this can be, let q_i be the discrete Fourier components of A in a box quantization, let $S[A]$ be the free-field action $S[A]=\frac{1}{2}\sum_i(k_i^2+m^2)q_i^2$, and take $\phi(q)$ to be the 2-point function $\phi(q)=q_iq_j$. One may easily verify that the corresponding stochastic equation,

$$\dot{\phi}(q,t)=\sum_i[\partial^2/\partial q_i^2-(k_i^2+m^2)q_i\partial/\partial q_i]\phi(q,t)\,,$$

with initial condition $\phi(q,0)=q_iq_j$ has the solution

$$\phi(q,t)=\frac{\delta_{ij}}{(k_i^2+m^2)}[1-{\rm e}^{-2(k_i^2+m^2)t}]+q_iq_j\,{\rm e}^{-(k_i^2+k_j^2+2m^2)t}\,.$$

One sees that $\lim_{t\to\infty}\phi(q,t)=\delta_{ij}/(k_i^2+m^2)$, so the stochastic prescription does give the familiar free-field propagator $\langle q_i q_j\rangle$.

Consider now the case of a gauge theory. The infinitesimal gauge transformation $\delta A_\mu^a = D_\mu^{ab} w^b$, where $D_\mu^{ac}=\partial_\mu \delta^{ac}+f^{abc}A_\mu^b$, induces the change in ϕ given by

$$\delta\phi[A]=-\int d^4x w^a D_\mu^{ab}\frac{\delta}{\delta A^b(x)}\phi[A],$$

so gauge-invariant functions are characterized by

$$G^a(x)\phi[A]=0\,, \tag{8}$$

where $G^a(x)=-D_\mu^{ab}\delta/\delta A_\mu^b(x)$, satisfies the commutation relations of the local gauge group, $[G^a(x), G^b(y)]=\delta^4(x-y)f^{abc}G^c(x)$. With L, eq. (2), written in functional notation

$$L=-\int d^4x\left[\frac{\delta}{\delta A_\mu^a(x)}-\frac{\delta S_{cl}[A]}{\delta A^a(x)}\right]\frac{\delta}{\delta A_\mu^a(x)}, \tag{9}$$

one easily verifies that L is strongly gauge invariant, namely $[G^a(x), L]=0$. Consequently for gauge-invariant ϕ, the formal solution to eqs. (1) and (3), namely, $\phi[A,t]=\exp(-Lt)\phi[A]$, may also be written

$$\phi[A,t]=\exp(-L_v t)\phi[A]\,, \tag{10}$$

where

$$L_v=L+\int v^a(x)G^a(x)\,d^4x=L+\int d^4x(D_\mu v)^a(x)\frac{\delta}{\delta A_\mu^a(x)}, \tag{11}$$

where $v^a(x)=v^a(x,A)$ is an arbitrary functional of A which, in the full enjoyment of gauge freedom, we may specify at our pleasure. Reverting to discrete notation, we see that L_v is again of the diffusion form,

$$L_v=-(\Delta+f_v\cdot\nabla)\,, \tag{12}$$

with a drift force $f_v(q)$ that depends on v determined from

$$\sum_i f_{v,i}(q)\psi_{i\mu}^a(x)=-\frac{\delta S_{cl}[A]}{\delta A_\mu^a(x)}-(D_\mu v)^a(x,A)\,. \tag{13}$$

Let $P(q_2,q_1,t)$ be the kernel of the time-translation operator (10),

$$\phi(q_1,t)=\int\phi(q_2)P(q_2,q_1,t)\,dq_2\,. \tag{14}$$

By eq. (1), it satisfies

$$\dot P(q_2,q_1,t)=-L_{v,q_1}P(q_2,q_1,t)\,, \tag{15}$$

which is sometimes called the backwards diffusion equation. From the composition law,

$$\int P(q_3, q_2, t_3 - t_2)P(q_2, q_1, t_2 - t_1)\,\mathrm{d}q_2 = P(q_3, q_1, t_3 - t_1)\,,$$

the forwards equation follows:

$$\dot{P}(q_2, q_1, t) = -L^*_{v,q_2}P(q_2, q_1, t)\,, \tag{16}$$

where

$$L^*_{v,q}u = -[\Delta u - \nabla \cdot (f_v u)]\,, \tag{17}$$

and one has

$$P(q_2, q_1, 0) = \delta(q_2 - q_1)\,. \tag{18}$$

The form of the last three equations allows one to interpret $P(q_2, q_1, t)$ as the probability that a brownian motion particle subject to a drift force f_v will arrive at q_2 at time t if it was originally at q_1.

The relevant theorem on stochastic processes has been proven only for a finite number of degrees of freedom, and we suppose that our equations may be regularized by keeping only a finite number N of modes q_i, $i = 1, \ldots, N$. In this case the following theorem* applies:

Given a drift force $f(q)$, if there exists a smooth positive function, call it $P_\mathrm{E}(\boldsymbol{q})$, which satisfies

$$L^*P_\mathrm{E}(q) = 0\,, \qquad P_\mathrm{E}(q) \geqslant 0\,, \qquad \int P_\mathrm{E}(q)\,\mathrm{d}q = 1\,, \tag{19}$$

where $L^*P_\mathrm{E}(q) = -\Delta P_\mathrm{E}(q) + \nabla \cdot (f(q)P_\mathrm{E}(q))$, then $P_\mathrm{E}(q)$ is unique and, moreover, $P(q, q_1, t)$ relaxes to it:

$$\lim_{t\to\infty} P(q, q_1, t) = P_\mathrm{E}(q)\,. \tag{20}$$

We call $P_\mathrm{E}(q)$ the equilibrium distribution. Thus, if there is a distribution satisfying (19) then, according to eq. (14) the stochastic prescription (6) reduces to

$$\langle \phi \rangle = \int \phi(q) P_\mathrm{E}(q)\,\mathrm{d}q\,. \tag{21}$$

* More precisely: Let $L^*P_\mathrm{E}(q) = 0$, where $P_\mathrm{E} \geqslant 0$ is smooth and $\int P_\mathrm{E}(q)\,\mathrm{d}q = 1$. Let K be any compact set. Then

$$\sup_{q_1 \in K} \int |P(q, q_1, t) - P(q)|\,\mathrm{d}q \to 0\,, \qquad \text{as } t \to \infty.$$

The existence of $P(q, q_1, t)$ is assumed [3].

In a non-gauge theory ($v=0$) one has $f(q)=-\nabla S(q)$, and it is easy to verify that $P_{\rm E}(q)={\rm e}^{-S(q)}/\int {\rm d}q\,{\rm e}^{-S(q)}$ (suitably regularized) satisfies conditions (19) and so, in virtue of the theorem, the stochastic prescription (6) coincides with the standard formula (7).

2. Proof of equivalence

The important point for a gauge theory is that the above theorem holds even if $f(q)$ is not derivable from a potential, and consequently formula (21) may be used provided that for some $v^a(q)$ there exists a $P_{\rm E}(q)$ satisfying conditions (19) [with $L=L_v$ and $f(q)=f_v(q)$ given by eqs. (11) or (13)]. Equivalence to the Faddeev–Popov ansatz follows because the equation

$$L_v^* P_{\rm E}(q)=0\,, \tag{22a}$$

or

$$L^* P_{\rm E}[A]=-\int {\rm d}^4x D_\mu^{ac}\frac{\delta}{\delta A_\mu^c(x)}\{v^a(x,A)P_{\rm E}[A]\}\,, \tag{22b}$$

does, in fact, hold for

$$v^a(x,A)=-\int \frac{\partial G^{ab}}{\partial y^\nu}(x,y;A)J_\nu^b(y,A)\,{\rm d}^4y\,, \tag{23}$$

with $P_{\rm E}[A]$ the Faddeev–Popov distribution

$$P_{\rm E}[A]=P_{\rm FP}[A]=N\det(\partial\cdot D)\exp\left[-S_{\rm cl}-(2\alpha)^{-1}\int(\partial\cdot A)^2\,{\rm d}^4x\right]. \tag{24}$$

Here $G^{ab}(x,y;A)$ is the Green function or ghost field propagator defined by

$$-\partial\cdot D^{(x)ac}G^{cb}(x,y;A)=-D\cdot\partial^{(y)bc}G^{ac}(x,y;A)=\delta^4(x-y)\delta^{ab}\,, \tag{25}$$

and the effective induced current $J_\nu^b(y,A)$ is given by*

$$J_\nu^b(y,A)=D_\mu^{bc}F_{\mu\nu}^c(y)+\alpha^{-1}\partial_\nu\partial\cdot A^b(y)+f^{bcd}\partial_\nu^{(z)}G^{cd}(y,z;A)|_{z=y}\,. \tag{26}$$

It is possible to verify eq. (22) directly using properties of determinants. However, the verification is simpler if v and $P_{\rm FP}$ are expressed in terms of the usual ghost and antighost fields C and $\bar{C}$. With the gauge-fixing action given by

$$S_{\rm GF}=\int[(2\alpha)^{-1}(\partial\cdot A)^2+\partial_\mu\bar{C}D_\mu C]\,{\rm d}^4x\,, \tag{27}$$

and

$$S_{\rm tot}=S_{\rm cl}+S_{\rm GF}\,, \tag{28}$$

* There is a charge renormalization divergence due to the coincidence of arguments in $f^{bcd}\partial_\nu^{(z)}G^{cd}(y,z)|_{z=y}$. It is to be regularized by retaining only a finite number of modes in the ghost field, as for the A field.

one has

$$P_{\mathrm{FP}} = N \int \mathrm{d}C\, \mathrm{d}\bar{C} \exp(-S_{\mathrm{tot}}) , \tag{29}$$

and it is easy to verify using Wick's theorem that

$$v^a(x, A) = P_{\mathrm{FP}}^{-1} N \int \mathrm{d}C\, \mathrm{d}\bar{C} C^a(x) \partial_\nu \bar{C}^b(y) \frac{\delta S_{\mathrm{tot}}}{\delta A_\nu^b(y)} \exp(-S_{\mathrm{tot}})\, \mathrm{d}^4 y \tag{30}$$

agrees with eq. (23), where

$$-\frac{\delta S_{\mathrm{tot}}}{\delta A_\mu^a(x)} = D_\lambda F_{\lambda\mu}^a(x) + \alpha^{-1} \partial_\mu \partial \cdot A^a(x) + f^{abc} C^b(x) \partial_\mu \bar{C}^c(x) . \tag{31}$$

We now verify eq. (22). From eq. (9) one has

$$L^* = -\int \mathrm{d}^4 x \frac{\delta}{\delta A_\mu^a(x)} \left[\frac{\delta}{\delta A_\mu^a(x)} + \frac{\delta S_{\mathrm{cl}}}{\delta A_\mu^a(x)} \right] \tag{32}$$

and so, by eqs. (28) and (30), the left-hand side of eq. (22b) gives

$$L^* P = \int \mathrm{d}^4 x \frac{\delta}{\delta A_\mu^a(x)} N \int \mathrm{d}C\, \mathrm{d}\bar{C} \frac{\delta S_{\mathrm{GF}}}{\delta A_\mu^a(x)} \exp(-S_{\mathrm{tot}}) . \tag{33}$$

On the other hand, by eq. (30), the right-hand side of eq. (22b) may be written

$$R = \int \mathrm{d}^4 x\, \mathrm{d}^4 y D_\mu^{ac} \frac{\delta}{\delta A_\mu^c(x)} \frac{\delta}{\delta A_\nu^b(y)} N \int \mathrm{d}C\, \mathrm{d}\bar{C} C^a(x) \partial_\nu \bar{C}^b(y) \exp(-S_{\mathrm{tot}}) .$$

Upon commuting $D_\mu^{ac} \delta / \delta A_\mu^c(x)$ with $\delta / \delta A_\nu^b(y)$ one obtains

$$R = R_1 - \int \mathrm{d}^4 x \frac{\delta}{\delta A_\nu^c(x)} N \int \mathrm{d}C\, \mathrm{d}\bar{C} f^{abc} C^a(x) \partial_\nu \bar{C}^b(x) \exp(-S_{\mathrm{tot}}) , \tag{34}$$

where

$$R_1 = \int \mathrm{d}^4 y \frac{\delta}{\delta A_\nu^b(y)} W_\nu^b(y) , \tag{35}$$

$$W_\nu^b(y) = \int \mathrm{d}^4 x D_\mu^{ac} \frac{\delta}{\delta A_\mu^c(x)} N \int \mathrm{d}C\, \mathrm{d}\bar{C} C^a(x) \partial_\nu \bar{C}^b(y) \exp(-S_{\mathrm{tot}}) ,$$

or

$$W_\nu^b(y) = -\int \mathrm{d}^4 x N \int \mathrm{d}C\, \mathrm{d}\bar{C} C^a(x) \partial_\nu \bar{C}^b(y) D_\mu^{ac} \frac{\delta S_{\mathrm{GF}}}{\delta A_\mu^c(x)} \exp(-S_{\mathrm{tot}}) ,$$

because $D_\mu^{ac} \delta S_{\mathrm{cl}} / \delta A_\mu^c(x) = 0$, and so

$$W_\nu^b(y) = -\int \mathrm{d}^4 x N \int \mathrm{d}C\, \mathrm{d}\bar{C} D_\mu^{ac} C^c(x) \partial_\nu \bar{C}^b(y) \times [\alpha^{-1} \partial_\mu \partial \cdot A^a(x) + f^{ade} C^d(x) \partial_\mu \bar{C}^e(x)] \exp(-S_{\mathrm{tot}}) , \tag{36}$$

by eq. (31). The last term in this expression integrates to zero. To see this, note that

$$D_\mu^{ac}C^c(x)f^{ade}C^d(x)=\tfrac{1}{2}D_\mu^{ea}[f^{acd}C^c(x)C^d(x)],$$

where the anticommutativity of the ghost fields has been used, and thus the contribution of the second term in eq. (36), call it $W^{(2)}$, reads

$$W_\nu^{(2)b}(y)=N\int dC\,d\bar{C}\,d^4x\tfrac{1}{2}f^{acd}C^c(x)C^d(x)(D\cdot\partial\bar{C})^a(x)\partial_\nu\bar{C}^b(y)\exp(-S_{\rm tot}), \tag{37}$$

$$W_\nu^{(2)b}(y)=N\int dC\,d\bar{C}\,d^4x\tfrac{1}{2}f^{acd}\left[C^cC^d\frac{\delta}{\delta C^a}\right](x)\,\partial_\nu\bar{C}^b(y)\exp(-S_{\rm tot}), \tag{38}$$

by eq. (27). This vanishes on partial integration with respect to C:

$$W_\nu^{(2)b}(y)=0, \tag{39}$$

and hence

$$W_\nu^b(y)=N\int dC\,d\bar{C}\,d^4x(\partial\cdot DC)^a(x)\partial_\nu\bar{C}^b(y)\alpha^{-1}\partial\cdot A^a(x)\exp(-S_{\rm tot}). \tag{40}$$

By eq. (27) we find

$$W_\nu^b(y)=N\int dC\,d\bar{C}\,d^4x\frac{\delta}{\delta\bar{C}^a(x)}[\exp(-S_{\rm tot})]\partial_\nu\bar{C}^b(y)\alpha^{-1}\partial\cdot A^a(x),$$

and so, by partial integration with respect to $\bar{C}$,

$$W_\nu^b(y)=-\alpha^{-1}\partial_\nu\partial\cdot A^b(y)N\int dC\,d\bar{C}\exp(-S_{\rm tot}). \tag{41}$$

From eqs. (35) and (34) one obtains

$$R=-\int d^4x\frac{\delta}{\delta A_\mu^a(x)}N\int dC\,d\bar{C}\,[\alpha^{-1}\partial_\mu\partial\cdot A^a+f^{abc}C^b(x)\partial_\mu\bar{C}^c(x)]$$
$$\times\exp(-S_{\rm tot}), \tag{42}$$

which, by eq. (31) agrees with eq. (33). Q.E.D.

3. Remarks

(i) The Landau gauge, $\alpha\to0$, yields $P_{\rm FP}$ proportional to $\delta(\partial\cdot A)$ which violates the smoothness hypothesis of the theorem. Our proof holds for $\alpha>0$, and $\alpha\to0$ is a singular limit.

(ii) We regard $\det(\partial\cdot D)$, which appears in the Faddeev–Popov ansatz as a formal perturbation series and leave open the verification of the positivity hypothesis in (19) beyond zeroth order.

(iii) For values of A such that the operator $\partial\cdot D$ has a vanishing eigenvalue, the corresponding Green function, which appears in v, eqs. (23) and (26), is singular, and consequently the drift force f_v is also. Again, f_v is considered to be a formal power series in the coupling constant.

(iv) As a by-product of our proof it follows that the limit which appears in the stochastic ansatz, namely

$$\langle\phi\rangle = \lim_{t\to\infty} \phi[A, t],$$

with $\phi[A, 0] = \phi[A]$, has the required properties for gauge-invariant $\phi[A]$. Namely, this limit, given by eq. (21), exists and is independent of A.

(v) The proof given above may be applied step by step to the class of axial gauges represented by

$$P_\eta[A] = N\int dC\, d\bar{C} \exp(-S_{\text{tot}}), \tag{43a}$$

$$S_{\text{tot}} = S_{\text{cl}} + \int [(2\alpha)^{-1}(\partial\cdot A)^2 + \bar{C}\eta\cdot DC]\, d^4x, \tag{43b}$$

where η is a fixed 4-vector, for

$$v^a(x, A) = -P_\eta^{-1}N\int dC\, d\bar{C}\, d^4y \frac{\delta}{\delta A_\nu^b(y)} C^a(x)\eta_\nu\bar{C}^b(y) \exp(-S_{\text{tot}}). \tag{44}$$

Again, the limit $\alpha\to 0$ corresponding to P_η proportional to $\delta(\eta\cdot A)$ is singular.

(vi) The above proof may also be extended to the more general case, first considered by Curci and Ferrari [4], which includes a 4-ghost coupling:

$$P_{\text{CF}} = N\int dC\, d\bar{C} \exp(-S_{\text{tot}}), \tag{45a}$$

$$S_{\text{tot}} = S_{\text{cl}} + \int \left[(2\alpha)^{-1}B^2 + \partial_\mu\bar{C}D_\mu C + \tfrac{1}{2}\lambda \sum_a (f^{abc}\bar{C}^bC^c)^2\right] d^4x, \tag{45b}$$

$$B^a \equiv \partial\cdot A^a + \lambda f^{abc}\bar{C}^bC^c. \tag{45c}$$

Here λ is a parameter which governs the hermiticity of the ghost-gauge boson interaction [5]. For $\lambda = 0$ the Faddeev-Popov distribution is recovered. The action (45b) represents the most general renormalizable, Lorentz and BRS and anti-BRS invariant expression one can construct from A, C and $\bar{C}$ [5]. It is remarkable that in the more general case $\lambda \neq 0$, $v^a(x, A)$ is given by the identical expression (30) (with $P_{\text{FP}} \to P_{\text{CF}}$). However, because of the quartic ghost coupling in S_{tot}, it is no longer possible to integrate out the ghost fields to obtain a more explicit expression for v analogous to eq. (23). Details of the proof are presented in the appendix.

(vii) We have seen that familiar "gauges" of various types can be accommodated in the stochastic framework. Here they are not characterized by a gauge *condition* $f^a(x, A) = 0$, such as $\partial\cdot A^a(x) = 0$ or $\eta\cdot A^a(x) = 0$. Indeed this would give a probability distribution proportional to $\delta(f)$ which violates the regularity hypothesis of the theorem and can only be reached as a singular limit. Instead, in the stochastic approach, "gauges" are characterized by a local gauge *transformation* with generator $v^a(x, A)$, eqs. (10) and (11). Because gauge conditions of the type $f^a(x, A)=0$ are subject to the Gribov ambiguity [6], gauges characterized instead by a regular generator $v^a(x, A)$ may be of some advantage in a non-perturbative treatment*.

(viii) Extension of the proof to include matter fields and spontaneous symmetry breaking does not appear to present any obstacles.

One of us (D.Z.) is grateful to Prof. Varadhan for clarifying discussions.

Appendix

We indicate the minor modifications of the proof given in sect. 2 which are required for the Curci–Ferrari gauge, eq. (45). Eq. (30) holds, but in eq. (31) $\partial \cdot A$ is replaced by B, given in eq. (45c). Eqs. (32) to (35) are unchanged, but in eq. (36), $\partial \cdot A$ is replaced by B, which gives

$$W^b_\nu(y) = -N\int dC\, d\bar{C}\, d^4x D^{ac}_\mu C^c(x)\partial_\nu \bar{C}^b(y)$$
$$\times[\alpha^{-1}\partial_\mu B^a(x) + f^{ade}C^d(x)\partial_\mu \bar{C}^e(x)]\exp(-S_{\text{tot}})\,. \tag{A.1}$$

Eq. (37) is unchanged, but eq. (38) is modified because of the term proportional to λ in

$$\frac{\delta S_{\text{GF}}}{\delta C^a} = -(D\cdot\partial\bar{C})^a + \lambda f^{abc}[\alpha^{-1}B^b + f^{bde}\bar{C}^d C^e]\bar{C}^c\,, \tag{A.2}$$

so eq. (39) is replaced by

$$W^{(2)b}_\nu(y) = N\int dC\, d\bar{C}\, d^4x \tfrac{1}{2} f^{acd}C^c(x)C^d(x)$$
$$\times\lambda f^{aef}[\alpha^{-1}B^e + f^{egh}\bar{C}^g C^h](x)\bar{C}^f(x)\partial_\nu\bar{C}^b(y)\exp(-S_{\text{tot}})\,. \tag{A.3}$$

The second term in the square bracket may be shown to give a vanishing contribution by using the Jacobi identity twice and so

$$W^{(2)b}_\nu(y) = N\int dC\, d\bar{C}\, d^4x \tfrac{1}{2} f^{acd}C^c(x)C^d(x)$$
$$\times\lambda\alpha^{-1} f^{aef}B^e(x)\bar{C}^f(x)\partial_\nu\bar{C}^b(y)\exp(-S_{\text{tot}})\,. \tag{A.4}$$

* Among these, the choice $v^a(x) = -\alpha^{-1}\partial\cdot A^a(x)$, considered in ref. [2], is the only one which is Lorentz invariant, renormalizable (by naive power counting) and local in A. It is not subject to the caveat (iii).

The contribution of the first term in the square brackets in eq. (A.1), call it $W^{(1)}$, reads

$$W_\nu^{(1)b}(y)=N\int dC\,d\bar{C}\,d^4x(\partial\cdot DC)^a(x)\partial_\nu\bar{C}^b(y)\alpha^{-1}B^a(x)\exp(-S_{\rm tot})\,. \qquad (A.5)$$

Use of the identity

$$\frac{\delta S_{\rm GF}}{\delta\bar{C}^a}=-(\partial\cdot DC)^a+\lambda f^{abc}C^b[\alpha^{-1}B^c+f^{cde}\bar{C}^dC^e] \qquad (A.6)$$

gives

$$W_\nu^{(1)b}(y)=N\int dC\,d\bar{C}\,d^4x\left\{-\frac{\delta S_{\rm GF}}{\delta\bar{C}^a}+\lambda f^{acd}C^c[\alpha^{-1}B^d+f^{def}\bar{C}^eC^f]\right\}(x)$$
$$\times\alpha^{-1}B^a(x)\partial_\nu\bar{C}^b(y)\exp(-S_{\rm tot})\,. \qquad (A.7)$$

The term containing $f^{acd}B^dB^a$ vanishes and the contribution from the last term in the square brackets may be shown to equal $-W^{(2)}$. Hence from $W=W^{(1)}+W^{(2)}$ we have

$$W_\nu^b(y)=N\int dC\,d\bar{C}\,d^4x\frac{\delta\exp(-S_{\rm tot})}{\delta\bar{C}^a(x)}\alpha^{-1}B^a(x)\partial_\nu\bar{C}^b(y)\,. \qquad (A.8)$$

Eqs. (41) and (42) follow, provided $\partial\cdot A$ is replaced by B, which again agrees with eq. (33). Q.E.D.

References

[1] G. Parisi and Wu, Y.-S. Sci. Sin. 24 (1981) 483
[2] D. Zwanziger, Nucl. Phys. B192 (1981) 259
[3] S.R.S. Varadhan, Diffusion problems and partial differential equations (published for Tata Institute of Fundamental Research by Springer-Verlag, 1980), p. 251
[4] G. Curci and R. Ferrari, Nuovo Cim. 32A (1976) 151
[5] L. Baulieu and J. Thierry-Mieg, Columbia preprint CUTP 196
[6] V.N. Gribov, Nucl. Phys. B139 (1978) 1

Progress of Theoretical Physics, Vol. 69, No. 5, May 1983

Stochastic Quantization of Non-Abelian Gauge Field

—— *Unitarity Problem and Faddeev-Popov Ghost Effects* ——

Mikio NAMIKI, Ichiro OHBA, Keisuke OKANO
and Yoshiya YAMANAKA

Department of Physics, Waseda University, Tokyo 160

(Received November 8, 1982)

The stochastic quantization method is applied to the non-Abelian gauge field up to the second order perturbation. It is shown that the stochastic quantization method automatically produces the same correct results as given by the well-known Faddeev-Popov trick but never requires to introduce artificially any ghost field. The gauge fixing problem in this method is also examined in detail. Finally preliminary discussions are given as to whether the physical state condition is automatically satisfied.

§ 1. Introduction and summary

Recently Parisi and Wu[1] have proposed a new quantization method, that is, the stochastic quantization different from the conventional ones, the canonical and path-integral methods. They have described the stochastic quantization by means of the Langevin equation governing a hypothetical stochastic process of the Wiener type with respect to an additional and fictitious time. The stochastic process is so designed as to give the ordinary field theory at the stationary limit. Our main interests in their theory are the following: The first interest is in their emphasis that the stochastic quantization method enables us to quantize gauge fields without gauge fixing. The second is in their conjecture that the stochastic quantization method, when applied to the non-Abelian gauge field, will automatically lead us to the same correct results as given by the Faddeev-Popov trick in the conventional field theory without resort to artificial input of any ghost field. Zwanziger et al[2] has discussed the two problems on the basis of the equivalent Fokker-Planck equation but not directly on the Langevin equation. In this approach the Fokker-Planck operator is so modified with additional terms as to keep expectation values of gauge invariant quantities unchanged. Their conclusion is that the stationary distribution given by the modified Fokker-Planck equation is equivalent to the well-known results brought about by the Faddeev-Popov ghost field. However, one may point out that their rather heavy modification of the Fokker-Planck operator is nothing other than a sort of artificial input of the gauge fixing and the ghost field, or at least, may say that his approach steps out of the original line given by Parisi and Wu. The stochastic quantization is attractive to us for its simplicity in the principle.

In this paper we return to the original Langevin equation to discuss the above two problems through straightforward perturbation calculations up to the second order. The main purposes of this paper are: First we show that Parisi and Wu's conjecture is certainly justified, that is to say, that the stochastic quantization method automatically yiélds the Faddeev-Popov ghost effects without resort to artificial input of any ghost field or artificial modification of the Fokker-Planck operator. Second we show that the stochastic quantization method implicitly introduces a sort of gauge fixing mechanism to give a gauge parameter in the field theoretical propagator, even though the method is free from the usual gauge fixing to reduce the degree of freedom of fields by modifying the original field Lagrangian.

In § 2 we first outline the stochastic quantization of field and then apply it to the Abelian gauge field together with discussions on the gauge fixing problem. Section 3 is devoted to the central part of this paper in which we deduce the above-mentioned conclusion in the case of the non-Abelian gauge field. In § 4 we make preliminary discussions as to whether the physical state condition can be realized automatically in the stochastic quantization method. Concluding remarks are given in § 5.

§ 2. Abelian gauge field and gauge fixing problem

In the Euclidean field theory we have the well-known path-integral formula

$$\Delta(x-y)=\frac{\int d\phi\phi(x)\phi(y)e^{-S[\phi]}}{\int d\phi e^{-S[\phi]}} \tag{2·1}$$

for the propagator of a neutral scalar field $\phi(x)$, where x and y are 4-dimensional Euclidean coordinates and $S[\phi]$ stands for the action integral. The stochastic quantization method gives us a simple prescription to derive $\Delta(x-y)$ based on the Langevin equation

$$\frac{\partial}{\partial t}\phi(x,t)=-\frac{\delta S[\phi]}{\delta\phi}\bigg|_{\phi=\phi(x,t)}+\eta(x,t) \tag{2·2}$$

which describes a hypothetical stochastic process of a random field $\phi(x,t)$ with respect to fictitious time t (different from the real time). In (2·2) $\eta(x,t)$ is a Gaussian random source field characterized by the statistical properties

$$\langle\eta(x,t)\rangle_\eta=0\,,\quad \langle\eta(x,t)\eta(x',t')\rangle_\eta=2\delta^4(x-x')\delta(t-t'), \tag{2·3}$$

where $\langle f(\eta)\rangle_\eta$ means the average over η given by

$$\langle f(\eta)\rangle_\eta=\frac{\int d\eta f(\eta)\exp[-\frac{1}{4}\int d^4x dt\{\eta(x,t)\}^2]}{\int d\eta\exp[-\frac{1}{4}\int d^4x dt\{\eta(x,t\}^2]}. \tag{2·4}$$

Since (2·2) gives us its solution $\phi(x,t)$ as a functional of η (and also initial distribution), we can obtain the two point correlation function by the averaging procedure

$$D(x,t|y,t')=\langle\phi(x,t)\phi(y,t')\rangle_\eta. \tag{2·5}$$

Recalling the fact that the Langevin equation (2·2) together with (2·3) or (2·4) leads us to the stationary distribution $\exp\{-S[\phi]\}$ as t goes to infinity, we can easily show

$$\Delta(x-y)=\lim_{t\to\infty}D(x,t|y,t). \tag{2·6}$$

This is the central prescription of the stochastic quantization.

Now let us apply the stochastic quantization method to the Abelian gauge field $A_\mu(x)$ whose action integral is given by

$$S_0[A_\mu]=\frac{1}{2}\int d^4x A_\mu(x)(-\Box\delta_{\mu\nu}+\partial_\mu\partial_\nu)A_\nu(x). \tag{2·7}$$

In terms of Fourier transforms*) $A_\mu(k,t)$ and $\eta_\mu(k,t)$, we have the Langevin equation

$$\dot{A}_\mu(k,t)=-k^2\left(\delta_{\mu\nu}-\frac{k_\mu k_\nu}{k^2}\right)A_\nu(k,t)+\eta_\mu(k,t), \tag{2·8}$$

where $\dot{A}_\mu(k,t)=(\partial/\partial t)A_\mu(k,t)$ and

$$\langle\eta_\mu(k,t)\rangle_\eta=0,\quad \langle\eta_\mu(k,t)\eta_\nu(k',t')\rangle_\eta=2\delta_{\mu\nu}\delta^4(k+k')\delta(t-t'). \tag{2·9}$$

Solving (2·8) under a special initial condition $A_\mu(k,0)=0$, we can easily get

$$A_\mu(k,t)=\int_0^\infty G_{\mu\nu}(k:t-t')\eta_\nu(k,t')dt', \tag{2·10}$$

where $G_{\mu\nu}(k:t-t')$ is the Green function given by

$$G_{\mu\nu}(k:t-t')=\left\{\left(\delta_{\mu\nu}-\frac{k_\mu k_\nu}{k^2}\right)e^{-k^2|t-t'|}+\frac{k_\mu k_\nu}{k^2}\right\}\theta(t-t'), \tag{2·11}$$

which is the solution of the equation

*) Note that

$$A_\mu(k,t)=\frac{1}{(2\pi)^2}\int d^4x e^{-ik\cdot x}A_\mu(x,t)=A_\mu^*(-k,t).$$

$$\left[\delta_{\mu\kappa}\frac{\partial}{\partial t}+(k^2\delta_{\mu\kappa}-k_\mu k_\kappa)\right]G_{\kappa\nu}(k:t-t')=\delta_{\mu\nu}\delta(t-t') \tag{2·12}$$

under the condition $G_{\mu\nu}(k:+0)=\delta_{\mu\nu}$ and $G_{\mu\nu}(k:-0)=0$. The prescription of the stochastic quantization requires us to prepare the correlation function in the following way:

$$\begin{aligned}D^{(0)}_{\mu\nu}(k,t|k',t')&=\langle A_\mu(k,t)A_\nu(k',t')\rangle_\eta\\&=\int_0^\infty dt''\int_0^\infty dt'''\,G_{\mu\kappa}(k:t-t'')G_{\nu\lambda}(k':t'-t''')\langle\eta_\kappa(k,t'')\eta_\lambda(k',t''')\rangle_\eta\\&=\delta^4(k+k')D^{(0)}_{\mu\nu}(k:t,t'),\end{aligned} \tag{2·13a}$$

$$D^{(0)}_{\mu\nu}(k:t,t')=\Delta^{(0)T}_{\mu\nu}(k)[e^{-k^2|t-t'|}-e^{-k^2|t+t'|}]+2t_<\frac{k_\mu k_\nu}{k^2}, \tag{2·13b}$$

where $t_<$ stands for the smaller one of t and t', and

$$\Delta^{(0)T}_{\mu\nu}(k)=\frac{1}{k^2}\left(\delta_{\mu\nu}-\frac{k_\mu k_\nu}{k^2}\right), \tag{2·14}$$

which is nothing but the ordinary field propagator with the Landau gauge. Consequently, the transverse component of the equal time correlation function $D^{(0)}_{\mu\nu}(x,t|x',t)$ tends to $\Delta^{(0)T}_{\mu\nu}(x-x')$ as t goes to infinity, while the longitudinal component $2tk_\mu k_\nu/k^2$ will be cancelled out in gauge invariant quantities as already observed in the Abelian gauge theory.[1),3)] This is the basic idea of the stochastic quantization.

At first sight the above approach seems to be able to quantize the gauge field as if the gauge fixing procedure were not necessary. Really it is true that the stochastic quantization method never requires any modification of the Lagrangian to fix gauge, that is to say, that the method is free from the usual gauge fixing in this sense. However, we have to pay attention to the special choice of the initial condition $A_\mu(k,0)=0$, which is closely related to the gauge parameter fixing problem in the following sense. To examine this problem, it is convenient to decompose the Langevin equation (2·8) into two parts, transverse and longitudinal ones, as follows:

$$\dot{A}_\mu{}^T=-k^2A_\mu{}^T+\eta_\mu{}^T, \tag{2·15a}$$

$$\dot{A}_\mu{}^L=\eta_\mu{}^L, \tag{2·15b}$$

where $A_\mu{}^T(k,t)=(\delta_{\mu\nu}-k_\mu k_\nu/k^2)A_\nu(k,t)$ and $A_\mu{}^L(k,t)=(k_\mu k_\nu/k^2)A_\nu(k,t)$ stand for the transverse and longitudinal parts, respectively, and $\eta_\mu{}^T$ and $\eta_\mu{}^L$ for the corresponding parts of η_μ. Speaking in general, (2·15a) has the solution

$$A_\mu{}^T(k,t)=A_\mu{}^{T(0)}(k,t)+A_\mu{}^T(k,0)e^{-k^2t}, \tag{2·16}$$

where $A_\mu{}^{T(0)}(k, t)$ is the solution under the special initial condition $A_\mu{}^{T(0)}(k, 0)=0$ and $A_\mu{}^T(k, 0)$ represents an arbitary initial value of $A_\mu{}^T(k, t)$. It is then obvious that $A_\mu{}^T(k, t)$ tends to $A_\mu{}^{T(0)}(k, t)$ irrespectively of the choice of initial condition as t goes to infinity, in other words, that we can use the initial condition $A_\mu{}^T(k, 0)=0$ without lack of generality. This fact comes from the presence of the 'damping force' $(-k^2 A_\mu{}^T)$ in (2·15a). Contrary to this, (2·15b) has no 'damping force', so that the initial value never disappears even for very large t. If we put the initial condition $A_\mu{}^L(k, 0)=(k_\mu/k^2)\phi(k)$ in which $\phi(k)$ is a scalar function, then we have another solution

$$A_\mu{}^L(k, t)=\frac{k_\mu}{k^2}\phi(k)+A_\mu{}^{L(0)}(k, t), \tag{2·17}$$

where $A_\mu{}^{L(0)}(k, t)=\int_0^\infty dt''\theta(t-t'')\eta_\mu(k, t'')$ is the solution of (2·15b) under the initial condition $A_\mu{}^{L(0)}(k, 0)=0$. Corresponding to this choice of initial condition, we obtain another correlation function

$$\langle A_\mu(k, t)A_\nu(k', t')\rangle_\eta=\frac{k_\mu k_\nu'}{k^2 k'^2}\phi(k)\phi(k')+D_{\mu\nu}^{(0)}(k: t, t'). \tag{2·18}$$

It is, however, noted that the simple product $\phi(k)\phi(k')$ never yields the important factor $\delta^4(k+k')$ which is required from the translational invariance or the uniformity of the space-time. Thus $\phi(k)=0$ as far as we adhere to put a sharp functional distribution, $\phi(k)$, for the initial condition. However, we have to remark that, in the theory of stochastic processes, a sharp functional distribution is exceptional but a functional probability distribution over random field is usually set down as its initial condition. Therefore, what we must have as the correlation function is not (2·18) but its functional average over ϕ, namely

$$\overline{\langle A_\mu(k, t)A_\nu(k', t')\rangle_\eta}^\phi=\frac{k_\mu k_\nu'}{k^2 k'^2}\Phi(k, k')+D_{\mu\nu}^{(0)}(k, t|k', t'), \tag{2·19}$$

where $\Phi(k, k')=\overline{\phi(k)\phi(k')}^\phi$ stands for the functional average of $\phi(k)\phi(k')$ over ϕ with a given distribution. It is now easy to show that an adequate distribution around zero field can give us $\Phi(k, k')=-a\delta^4(k+k')$, a being a dimensionless positive parameter.*) Thus, we get the correlation function

*) From the reality condition $A_\mu{}^L(-k, t)=A_\mu{}^{L*}(k, t)$ we get $\phi(-k)=-\phi^*(k)$ and then $\Phi(k, k')=-\overline{\phi(k)\phi^*(-k')}^\phi$. If we expand $\phi(k)=\sum_i c_i u_i(k)$ in a complete orthonormal set $\{u_i(k)\}$, c_i being real variable, and average $\phi(k)\phi^*(-k')=\sum_{i,j}c_i c_j u_i(k)u_j{}^*(-k')$ with the distribution functional $W[\phi]=\exp[-(1/2a)\sum_i c_i{}^2]/\prod_i\int_{-\infty}^{+\infty}\exp[-(1/2a)c_i{}^2]dc_i$, then we get $\Phi(k, k')=-a\sum_i u_i(k)u_i{}^*(-k')=-a\delta^4(k+k')$. Note that $a>0$. A more general form compatible with the translational invariance would be $\Phi(k, k')=\gamma(k^2)\delta^4(k+k')$, but $\gamma(k^2)$ should simply be a constant unless we introduce parameters with dimension into the distribution functional.

$$\overline{\langle A_\mu(k,t)A_\nu(k',t')\rangle_\eta}^{\phi}=\alpha\frac{k_\mu k_\nu}{(k^2)^2}\delta^4(k+k')+D^{(0)}_{\mu\nu}(k,t|k',t')$$

$$\xrightarrow[t=t'\to\infty]{}\delta^4(k+k')\Big[\frac{1}{k^2}\Big\{\delta_{\mu\nu}-(1-\alpha)\frac{k_\mu k_\nu}{k^2}\Big\}+2t\frac{k_\mu k_\nu}{k^2}\Big]. \tag{2·20}$$

This equation implies that the choice of initial state distribution just corresponds to fixing of the gauge parameter. Hence we cannot assert that the stochastic quantization method is quite free from the gauge fixing. It is, however, emphasized that the stochastic quantization method can determine the gauge field propagator without resort to the usual gauge fixing procedure of introducing a gauge fixing term like $(2\alpha)^{-1}(\partial A)^2$ into the Lagrangian,*) that is to say, to reduce the degree of freedom of fields by the constraint condition.

The reason why the usual gauge fixing procedure was not necessary in the above approach can be understood from the character of the operator $[\delta_{\mu\kappa}\partial/\partial t+(k^2\delta_{\mu\kappa}-k_\mu k_\kappa)]$ in (2·8) or (2·12). Contrary to the conventional theory in which we have not $\delta_{\mu\kappa}\partial/\partial t$ but only the projection operator $(k^2\delta_{\mu\kappa}-k_\mu k_\kappa)$, this operator is not singular and hence has its inverse explicitly given by $G_{\mu\nu}(k:t-t')$. The situation will become clear if we describe (2·8) or (2·12) in terms of the Fourier transform with respect to the fictitious time. Introducing

$$\tilde{G}_{\mu\nu}(k,\omega)=\int G_{\mu\nu}(k,t)e^{-i\omega t}d\omega\,, \tag{2·21}$$

then we have

$$[(-i\omega+k^2)\delta_{\mu\kappa}-k_\mu k_\kappa]\tilde{G}_{\kappa\nu}(k,\omega)=\delta_{\mu\nu}\,. \tag{2·22}$$

Due to the presence of $-i\omega$, the matrix $\|(-i\omega+k^2)\delta_{\mu\kappa}-k_\mu k_\kappa\|$ can have its inverse explicitly written down as

$$\tilde{G}_{\mu\nu}(k,\omega)=\frac{1}{k^2-i\omega}\Big[\delta_{\mu\nu}-\frac{k_\mu k_\nu}{i\omega}\Big], \tag{2·23}$$

where $-i\omega$ behaves as if it were the gaugeon mass squared. It is easily observed that introduction of the fictitious time and the longitudinal random source field has given an additional degree of freedom to the field into the present theory. In fact, our correlation function (2·13) or its limit $\lim_{t\to\infty}D^{(0)}_{\mu\nu}(x,t|x',t)$ contains the longitudinal part besides the transverse part $\Delta^{(0)T}_{\mu\nu}(x-x')$.

Since we can use the above method for the zeroth-order propagator in the perturbation theory, we can easily guess that the same prescription of stochastic

*) In this case we have the 'damping force' $-\alpha^{-1}k^2A^L_\mu$ on the right-hand side of (2·15b) and then can obtain the usual correlation function with gauge parameter α at $t\to\infty$, irrespectively of initial conditions. As is well known, however, such a modification of the Lagrangian should be followed by introduction of Faddeev-Popov ghost field in the case of the non-Abelian gauge field.

quantization can be applied to the non-Abelian gauge field without resort to the usual gauge fixing procedure. On the other hand, it is well-known that the usual gauge fixing should be followed by artificial input of the Faddeev-Popov ghost field. For these reasons we are now led to a natural conjecture that the non-Abelian gauge field can be quantized without help of any ghost field within the framework of stochastic quantization. In the next section we justify this conjecture by showing that the stochastic quantization method automatically produces the Faddeev-Popov ghost effects without resort to artificial input of any ghost field. The natural occurrence of the Faddeev-Popov ghost effects is closely related to the above-mentioned additional degree of field brought from introduction of the fictitious time and the longitudinal random source field as mentioned above. The above averaging procedure over initial-state distribution and also the occurrence of additional degree of freedom will be more clearly described in the framework of the operator theory of stochastic quantization.[4)]

§ 3. Non-Abelian gauge field and Faddeev-Popov ghost effects

A non-Abelian gauge field in the D-dimensional Euclidean space is characterized by the action integral

$$S[A]=\frac{1}{4}\int d^D x F^a_{\mu\nu}(x)F^a_{\mu\nu}(x), \tag{3·1a}$$

$$F^a_{\mu\nu}(x)=\partial_\mu A_\nu{}^a(x)-\partial_\nu A_\mu{}^a(x)-gf^{abc}A_\mu{}^b(x)A_\nu{}^c(x), \tag{3·1b}$$

where a, b and c stand for the color indices. From (3·1) we get the Langevin equation

$$\dot{A}_\mu{}^a(k,t)=-(k^2\delta_{\mu\nu}-k_\mu k_\nu)A_\nu{}^a(k,t)+Y_\mu{}^a(k,t), \tag{3·2a}$$

where

$$\begin{aligned}Y_\mu{}^a(k,t)=\eta_\mu{}^a(k,t)&+\frac{g}{(2\pi)^{D/2}}\int d^D k_1 d^D k_2 \delta^D(k-k_1-k_2)\\&\times V^{abc}_{\mu\kappa\lambda}(k,-k_1,-k_2)A_\kappa{}^b(k_1,t)A_\lambda{}^c(k_2,t)\\&+\frac{g^2}{(2\pi)^D}\int d^D k_1 d^D k_2 d^D k_3 \delta^D(k-k_1-k_2-k_3)\\&\times W^{abcd}_{\mu\nu\kappa\lambda}A_\nu{}^b(k_1,t)A_\kappa{}^c(k_2,t)A_\lambda{}^d(k_3,t),\end{aligned} \tag{3·2b}$$

in which $V^{abc}_{\mu\kappa\lambda}(k,-k_1,-k_2)$ and $W^{abcd}_{\mu\nu\kappa\lambda}$ are, respectively, the three point and four point vertex factors listed in Table I together with the corresponding diagrams. Note that $V^{abc}_{\mu\nu\lambda}(k,-k_1,-k_2)=\frac{1}{2}V^{(0)abc}_{\mu\kappa\lambda}(k,-k_1,-k_2)$ and $W^{abcd}_{\mu\nu\kappa\lambda}=\frac{1}{6}W^{(0)abcd}_{\mu\nu\kappa\lambda}$ in terms of the conventional Feynman rule. Equation (3·2a) under the initial

Table I. Diagrams of Green function, propagator and vertices.

DIAGRAM	NOTATION	FORMULA
	$G^{ab}_{\mu\nu}(k : t-t')$	$\delta^{ab}\left\{\left(\delta_{\mu\nu}-\frac{k_\mu k_\nu}{k^2}\right)e^{-k^2\|t-t'\|}+\frac{k_\mu k_\nu}{k^2}\right\}\theta(t-t')$
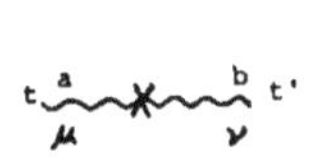	$D^{ab}_{\mu\nu}(k : t, t')$	$\delta^{ab}\left\{\frac{1}{k^2}\left(\delta_{\mu\nu}-\frac{k_\mu k_\nu}{k^2}\right)(e^{-k^2\|t-t'\|}-e^{-k^2\|t+t'\|}) + 2t_< \frac{k_\mu k_\nu}{k^2}\right\}$
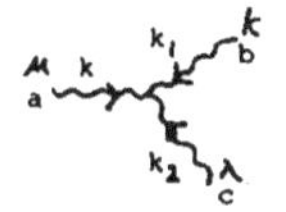	$gV^{abc}_{\mu\kappa\lambda}(k, k_1, k_2)$	$\left(-\frac{i}{2}\right)gf^{abc}$ $[(k-k_1)_\lambda\delta_{\mu\kappa}+(k_1-k_2)_\mu\delta_{\kappa\lambda}+(k_2-k)_\kappa\delta_{\mu\lambda}]$
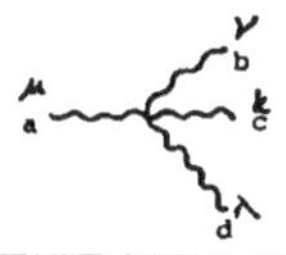	$g^2W^{abcd}_{\mu\nu\kappa\lambda}$	$-\frac{1}{6}g^2[f^{abe}f^{cde}(\delta_{\mu\kappa}\delta_{\nu\lambda}-\delta_{\mu\lambda}\delta_{\nu\kappa})$ $+f^{ace}f^{bde}(\delta_{\mu\nu}\delta_{\kappa\lambda}-\delta_{\mu\lambda}\delta_{\nu\kappa})$ $+f^{ade}f^{cbe}(\delta_{\mu\kappa}\delta_{\nu\lambda}-\delta_{\mu\nu}\delta_{\kappa\lambda})]$

condition $A_\mu{}^a(k,0)=0$ is equivalent to the integral equation

$$A_\mu{}^a(k,t)=\int_0^\infty G^{ab}_{\mu\nu}(k : t-t')Y_\nu{}^b(k,t')dt' , \tag{3·3}$$

where $G^{ab}_{\mu\nu}(k : t-t')=\delta^{ab}G_{\mu\nu}(k : t-t')$ is the zeroth order Green function discussed in detail in § 2. It may be convenient to write (3·3) in a symbolic way as

$$A=G(\eta+gVAA+g^2WAAA). \tag{3·4}$$

Solving (3·3) or (3·4) by means of the iteration method, we obtain its solution in a perturbation expansion as follows:

$$\begin{aligned}A=&G\eta+gGV(G\eta)(G\eta)+g^2GV\{GV(G\eta)(G\eta)\}(G\eta)\\&+g^2GV(G\eta)\{GV(G\eta)(G\eta)\}+g^2W(G\eta)(G\eta)(G\eta)+\cdots ,\end{aligned} \tag{3·5}$$

which is graphically represented in Fig. 1. Here, following Parisi and Wu[1)], we have represented $G_{\mu\nu}$ by a wavy line, η_μ by a cross and gV and g^2W, respectively, by a three and a four-point vertex.

Our first task is to calculate the correlation function $\langle A_\mu{}^a(k,t)A_\nu{}^b(k',t')\rangle$. As anticipated by discussion in § 2, we shall have in the correlation function those

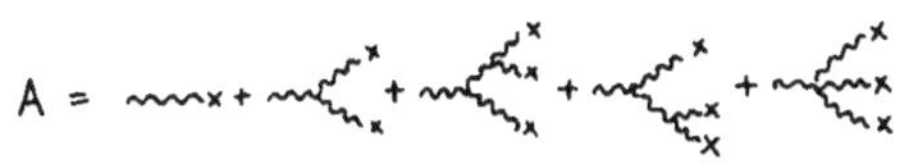

Fig. 1. Perturbative expansion of $A_\mu{}^a(k,t)$.

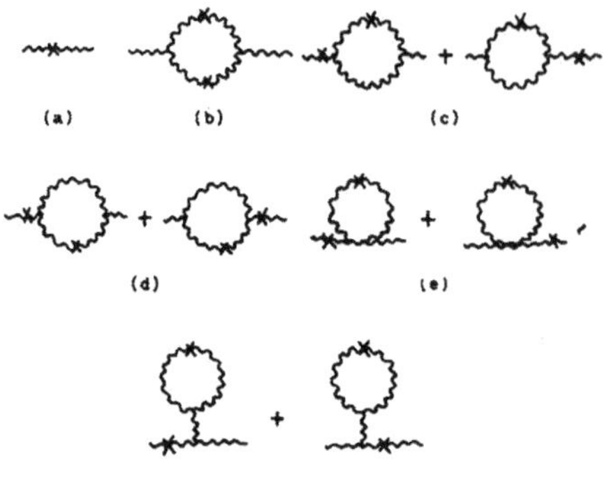

Fig. 2. Graphical representations of $D^{ab}_{\mu\nu}(k: t, t')$.

terms which diverge as t or t' goes to infinity. Those divergent terms are considered to be characteristic to gauge non-invariant quantities such as $\langle A_\mu{}^a(k, t)A_\nu{}^b(k, t)\rangle$ and to be cancelled out if we make gauge invariant or covariant quantities. Indeed we can show by straightforward calculations that this is the case for $\lim_{t\to\infty}\langle F^a_{\mu\nu}(x, t)F^a_{\kappa\lambda}(x, t)\rangle$ for example. In what follows, consequently, we keep only those terms which remain finite as t or t' goes to infinity. Using (3·5) and averaging over η by (2·9), we get the second-order pertubation terms as follows:

$$\langle A_\mu{}^a(k, t)A_\nu{}^b(k', t')\rangle=\delta^D(k+k')D^{ab}_{\mu\nu}(k: t, t') \qquad (3\cdot 6a)$$

with

$$D^{ab}_{\mu\nu}(k: t, t')=(a)+2(b)+2(c)+2(d)+3(e)+(f), \qquad (3\cdot 6b)$$

where (a)~(f) stand for the contributions from the corresponding diagrams shown in Fig. 2. Diagram (a) (wavy line with a cross) represents the free propagator (2·13), and diagrams (b), (c), (d) and (e), respectively, correspond to the following contributions:

$$2(b)=\frac{1}{(2\pi)^D}\int d^Dk_1d^Dk_2\delta^D(k-k_1-k_2)$$
$$\times gV^{acd}_{\kappa\lambda\epsilon}(k, -k_1, -k_2)gV^{bcd}_{\sigma\zeta\eta}(-k, k_1, k_2)$$
$$\times 2\int_0^\infty dt'''\int_0^\infty dt''\, G_{\mu\kappa}(k: t-t'')$$
$$\times D^{(0)}_{\lambda\zeta}(k_1: t'', t''')D^{(0)}_{\epsilon\eta}(k_2: t''', t'')G_{\nu\sigma}(k: t'-t'''), \qquad (3\cdot 7)$$

$$2\{(c)+(d)\}=\frac{1}{(2\pi)^D}\int d^Dk_1d^Dk_2\delta^D(k-k_1-k_2)$$

$$\times g V^{acd}_{\kappa\lambda\xi}(k, -k_1, -k_2) g V^{bcd}_{\sigma\zeta\eta}(-k, k_1, k_2)\Big[2\int_0^\infty dt''' \int_0^\infty dt''$$

$$\times \{D^{(0)}_{\mu\kappa}(k : t, t'')D^{(0)}_{\lambda\zeta}(k_1 : t'', t''')G_{\xi\eta}(k_2 : t'''-t'')G_{\nu\sigma}(k : t'-t''')$$

$$+G_{\mu\kappa}(k : t-t'')D^{(0)}_{\lambda\zeta}(k_1 : t'', t''')G_{\xi\eta}(k_2 : t''-t''')D^{(0)}_{\nu\sigma}(k : t', t''')\}$$

$$+\begin{pmatrix} k_1 \leftrightarrow k_2 \\ \lambda \leftrightarrow \xi \\ \eta \leftrightarrow \zeta \end{pmatrix}\Big], \tag{3·8}$$

$$3(\mathrm{e}) = \frac{1}{(2\pi)^D}\int d^D k_1 g^2 W^{abcd}_{\nu\lambda\xi\zeta}$$

$$\times\Big[3\int_0^\infty dt'' D^{(0)}_{\mu\kappa}(k : t, t'')D^{(0)}_{\lambda\zeta}(k_1 : t'', t''')G_{\sigma\nu}(k : t'-t'') + \begin{pmatrix} \mu\leftrightarrow\sigma \\ \chi\leftrightarrow\nu \end{pmatrix}\Big]. \tag{3·9}$$

Diagram (f) gives vanishing contribution due to the anti-symmetry of $V^{abc}_{\nu\kappa\lambda}$ with respect to a, b and c, namely, $V^{abc}_{\nu\kappa\lambda}\delta^{bc}=0$. Here it should be noted that we can discard longitudinal components of external lines for the following reasons: (i) After integrating over loop momenta, we can write the cross terms from longitudinal to transverse corresponding to Fig. 3(a) in the form $O^T_{\mu\kappa}(k)\boldsymbol{\Pi}_{\kappa\sigma}(k) \times O^L_{\sigma\nu}(k)$, where

$$O^T_{\mu\kappa}(k) = \left(\delta_{\mu\kappa} - \frac{k_\mu k_\kappa}{k^2}\right), \quad O^L_{\sigma\nu}(k) = \frac{k_\sigma k_\nu}{k^2}. \tag{3·10}$$

Because of the transformation property, $\Pi_{\kappa\sigma}(k)$ is a linear combination of $\delta_{\kappa\sigma}$ and $k_\kappa k_\sigma$, so that the above cross terms should vanish. (ii) The longitudinal-longitudinal terms corresponding to Fig. 3(b) can be written as $O^L_{\mu\kappa}(k)\Pi_{\kappa\sigma}(k)O^L_{\sigma\nu}(k)$. These terms are propotional to $O^L_{\mu\nu}(k)$ and hence cancelled out in gauge covariant combinations like $\lim_{t\to\infty}\langle F^a_{\mu\nu}(k, t)F^b_{\kappa\sigma}(k', t)\rangle$, as we shall explicitly see later. On the other hand, those longitudinal components which take place in internal lines will produce important contributions just corresponding to the Faddeev-Popov ghost effects in the conventional theory. Recall that those longitudinal components originate in an additional degree of freedom coming from introduction of the ficititious time as remarked at the end of the previous section. We summarize finite contributions in the limit $t=t'\to\infty$ in Table II. The contri-

(a) (b)

Fig. 3. (a) Cross term of propagator. T and L stand for the transverse and longitudinal components, respectively.

(b) Longitudinal component of propagator.

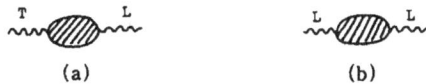

Table II. Contributions from the diagrams in Fig. 2.

				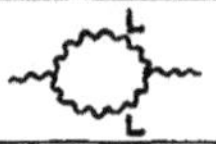
	$O^T_{\mu\kappa}(k)[VO^T(k_1)$ $\times O^T(k_2)V]^{ab}_{cd}O^T_{\nu\sigma}(k)$	$O^T_{\mu\kappa}(k)[VO^T(k_1)O^L(k_2)V]^{ab}_{cd}O^T_{\nu\sigma}(k)$	$O^T_{\mu\kappa}(k)[VO^L(k_1)O^T(k_2)V]^{ab}_{cd}O^T_{\nu\sigma}(k)$	$O^T_{\mu\kappa}(k)[VO^L(k_1)O^L(k_2)V]^{ab}_{cd}O^T_{\nu\sigma}(k)$
(b)	$\frac{1}{k^2k_1^2k_2^2(k^2+k_1^2+k_2^2)}$	$-\frac{1}{k^2k_1^2}\left\{\frac{1}{k^2(k^2+k_1^2)}+\frac{2}{(k^2+k_1^2)^2}\right\}$	$-\frac{1}{k^2k_2^2}\left\{\frac{1}{k^2(k^2-k_2^2)}+\frac{2}{(k^2+k_2^2)^2}\right\}$	$\frac{14}{(k^2)^4}$
(c)	$\frac{1}{(k^2)^2k_1^2(k^2+k_1^2+k_2^2)}$	$\frac{1}{(k^2)^2k_1^2(k^2+k_1^2)}$	$\frac{1}{(k^2)^2k_2^2(k^2+k_2^2)}$	$-\frac{3}{(k^2)^4}$
(d)	$\frac{1}{(k^2)^2k_2^2(k^2+k_1^2+k_2^2)}$	$-\frac{1}{(k^2)^2}\left\{\frac{1}{k^2(k^2+k_1^2)}+\frac{2}{(k^2+k_1^2)^2}\right\}$	$-\frac{1}{(k^2)^2}\left\{\frac{1}{k^2(k^2-k_2^2)}+\frac{2}{(k^2+k_2^2)^2}\right\}$	$-\frac{3}{(k^2)^4}$

	$O^T_{\mu\kappa}(k_1)[WO^T(k_1)]^{ab}_{cd}O^T_{\nu\sigma}(k)$	$O^T_{\mu\kappa}(k)[WO^L(k_1)]^{ab}_{cd}O^T_{\nu\sigma}(k)$
(e)	$\frac{1}{(k^2)^2k_1^2}$	$\frac{-1}{(k^2)^3}$

butions from the transverse components (Fig. 4) give

$$2[\{(b)+(c)+(d)\}_T]=-\frac{g^2}{2}\cdot\frac{1}{(2\pi)^D}\int d^Dk_1d^Dk_2\delta^D(k-k_1-k_2)$$
$$\times V^{(0)ced}_{\kappa\lambda\xi}(k,-k_1,-k_2)\Delta^{(0)T}_{\lambda\zeta}(k_1)\Delta^{(0)T}_{\xi\eta}(k_2)V^{(0)hed}_{\sigma\zeta\eta}(k,-k_1,-k_2),\tag{3·11}$$

$$3[(e)_T]^{ch}_{\kappa\sigma}=\frac{g^2}{2}\int\frac{d^Dk_1}{(2\pi)^D}W^{(0)ceeh}_{\kappa\lambda\zeta\sigma}\Delta^{(0)T}_{\lambda\zeta}(k),\tag{3·12}$$

where we have suppressed the factor $\Delta^{(0)T}_{\mu\kappa}(k)\Delta^{(0)T}_{\nu\sigma}(k)$ corresponding to the external propagators. It is evident that (3·11) and (3·12) coincide, respectively, with the gluon loop and tadpole terms in the conventional Landau gauge field theories. Then we denote the sum of (3·11) and (3·12) by $\Pi^{ch}_{\kappa\sigma}(LG)$. The

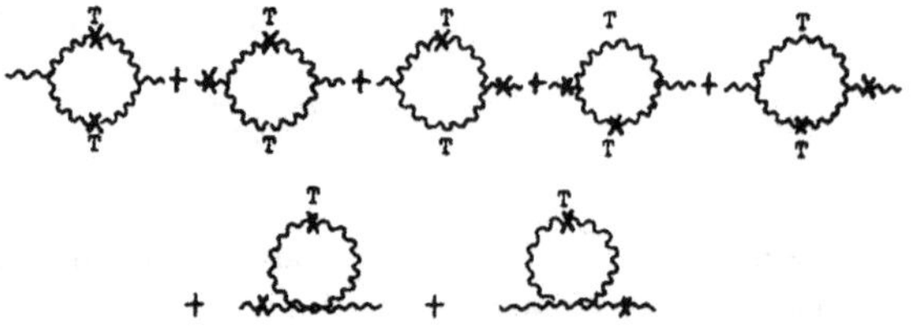

Fig. 4. Diagrams including exclusively transverse internal lines.

remaining terms (Fig. 5) become

$$2[\{(\mathrm{b})+(\mathrm{c})+(\mathrm{d})\}_{LT}]^{ch}_{\kappa\sigma}$$

$$=\frac{2g^2}{(2\pi)^D}\int d^Dk_1 d^Dk_2 \delta^D(k-k_1-k_2)$$

$$\times\Big[\frac{-1}{k_1^2}\Big(\frac{1}{k^2}+\frac{1}{k^2+k_1^2}\Big)[VO^T(k_1)O^L(k_2)V]^{ch}_{\kappa\sigma}$$

$$+\frac{-1}{k_2^2}\Big(\frac{1}{k^2}+\frac{1}{k^2+k_2^2}\Big)[VO^L(k_1)O^T(k_2)V]^{ch}_{\kappa\sigma}$$

$$+\frac{8}{(k^2)^2}[VO^L(k_1)O^L(k_2)V]^{ch}_{\kappa\sigma}\Big]$$

$$=\frac{g^2}{(2\pi)^D}\int d^Dk_1 d^Dk_2 \delta^D(k-k_1-k_2)\frac{-2}{k_1^2k_2^2}\Big[\Big\{\frac{(2k^2+k_1^2)(k^2-k_1^2)}{8k^2k_1^2(k^2+k_1^2)}$$

$$+\frac{(2k^2+k_2^2)(k^2-k_2^2)^2}{8k^2k_2^2(k^2+k_2^2)}+1\Big\}(k_{1\kappa}k_{2\sigma}+k_{1\sigma}k_{2\kappa})$$

$$+\Big\{\frac{(2k^2+k_1^2)(k^2-k_1^2)^2}{4k^2(k^2+k_1^2)}+\frac{(2k^2+k_2^2)(k^2-k_2^2)^2}{4k^2(k^2+k_2^2)}\Big\}\delta_{\kappa\sigma}\Big]. \tag{3·13}$$

$$3[(\mathrm{e})_L]^{ch}_{\kappa\sigma}=\int\frac{d^Dk_1}{(2\pi)^D}\frac{-3}{k^2}[WO^L(k_1)]^{ch}_{\kappa\sigma}$$

$$=g^2f^{cde}f^{hde}\int\frac{d^Dk_1}{(2\pi)^D}\frac{1}{k^2}\Big(\delta_{\kappa\sigma}-\frac{k_{1\kappa}k_{1\sigma}}{k_1^2}\Big)$$

$$=\frac{g^2}{(2\pi)^D}f^{cde}f^{hde}\int d^Dk_1 d^Dk_2\delta^D(k-k_1-k_2)\frac{1}{k^2}\Big\{\delta_{\kappa\sigma}+\frac{1}{2k_1^2}(k_{1\kappa}k_{2\sigma}+k_{2\kappa}k_{1\sigma})\Big\}, \tag{3·14}$$

where we have used the formulas

$$[VO^T(k_1)O^L(k_2)V]^{ch}_{\kappa\sigma}\equiv V^{cde}_{\kappa\lambda\xi}(k,-k_1,-k_2)O^T_{\lambda\zeta}(k_1)O^L_{\xi\eta}(k_2)V^{hde}_{\sigma\zeta\eta}(-k,k_1,k_2)$$

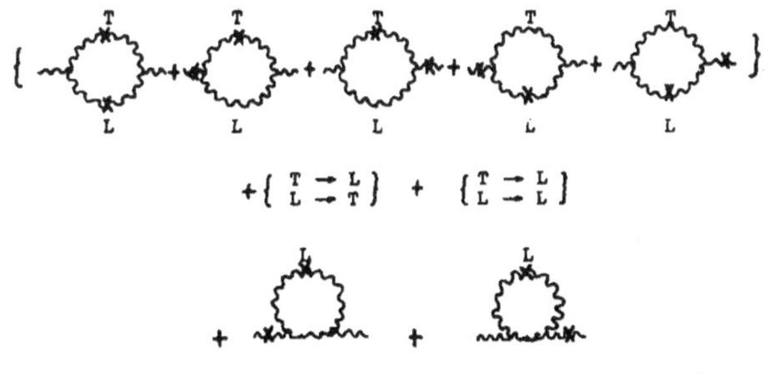

Fig. 5. Diagrams including longitudinal internal lines.

$$=f^{cde}f^{hde}\frac{1}{4}\left[\frac{1}{k_2^2}(k_2^2+2k_1k_2)^2\delta_{\kappa\sigma}\right.$$

$$\left.+\frac{1}{2k_1^2k_2^2}(k_2^2+2k_1k_2)^2(k_{1\kappa}k_{2\sigma}+k_{1\sigma}k_{2\kappa})\right],$$

$$[VO^L(k_1)O^T(k_2)V]^{ch}_{\kappa\sigma}=(k_1\leftrightarrow k_2 \text{ in } [VO^T(k_1)O^L(k_2)V]^{ch}_{\kappa\sigma}),$$

$$[VO^L(k_1)O^L(k_2)V]^{ch}_{\kappa\sigma}=-f^{cde}f^{hde}\frac{(k^2)^2}{8k_1^2k_2^2}(k_{1\kappa}k_{2\sigma}+k_{1\sigma}k_{2\kappa}),$$

$$[WO^L(k_1)]^{ch}_{\kappa\sigma}=W^{cdeh}_{\kappa\lambda\zeta\sigma}\delta_{de}O^L_{\lambda\zeta}(k_1)=-f^{cde}f^{hde}\frac{1}{3}\left(\delta_{\kappa\sigma}-\frac{k_{1\kappa}k_{1\sigma}}{k_1^2}\right), \tag{3·15}$$

and have taken the presence of the external transverse propagators (suppressed in (3·13)~ (3·15)) on both sides into account. Both (3·13) and (3·14) include those quartic divergences (for $D=4$) which are not renormalizable. It is, however, easy to see that those divergences cancell each other when (3·13) and (3·14) are summed. Therefore, we conclude that the correlation function or the propagator is free from the quartic divergence and then renormalizable, even though the zeroth order Green function (2·11) and the zeroth order correlation function (2·13) have the extra term $(k_\mu k_\nu/k^2)$ going to constant as $k\to\infty$. Summing up, we obtain

$$\lim_{t\to\infty}D^{ab}_{\mu\nu}(k:t,t)=\delta_{ab}\Delta^{(0)T}_{\mu\nu}(k)+\Delta^{(0)T}_{\mu\kappa}(k)\Pi^{ab}_{\kappa\sigma}\Delta^{(0)T}_{\kappa\sigma}(k), \tag{3·16}$$

where

$$\Pi^{ab}_{\kappa\sigma}=[\Pi(\mathrm{LG})]^{ab}_{\kappa\sigma}+[\Pi(\mathrm{FP})]^{ab}_{\kappa\sigma}+\Xi^{ab}_{\kappa\sigma}\,; \tag{3·17}$$

with

$$[\Pi(\mathrm{FP})]^{ab}_{\kappa\sigma}=g^2f^{acd}f^{bcd}\int\frac{d^Dk_1}{(2\pi)^D}\frac{k_{1\kappa}k_{2\sigma}}{k_1^2k_2^2}, \tag{3·18}$$

$$\Xi^{ab}_{\kappa\sigma}=g^2f^{acd}f^{bcd}\left[2k^2\int\frac{d^Dk_1}{(2\pi)^D}\frac{k_{1\kappa}k_{1\sigma}}{(k_1^2)^2k_2^2}+4\int\frac{d^Dk_1}{(2\pi)^D}\frac{k_{1\kappa}k_{1\sigma}}{k_2^2(k^2+k_1^2)}\right.$$

$$\left.-2k^2\int\frac{d^Dk_1}{(2\pi)^D}\frac{\delta_{\kappa\sigma}}{k_1^2k_2^2}+4k^2\int\frac{d^Dk_1}{(2\pi)^D}\frac{\delta_{\kappa\sigma}}{k_2^2(k^2+k_1^2)}\right], \tag{3·19}$$

where $k_2=k-k_1$. Here it is emphasized that we have obtained in (3·17) the Faddeev-Popov ghost loop $\Pi(\mathrm{FP})$ given by the conventional theory. In (3·17), however, we have the additional term Ξ which does not vanish even after evaluation of the loop integration by means of the dimensional regularization, for example, must be due to the gauge non-invariance of the correlation function $\langle A_\mu{}^a(k,t)A_\nu{}^b(k',t')\rangle$. Thus we are led to calculate the gauge covariant correlation function $\lim_{t=t'\to\infty}\langle F^a_{\mu\nu}(k,t)F^b_{\kappa\lambda}(k',t')\rangle$.

After elementary calculations we get

$$\langle F^a_{\mu\nu}(k,t)F^b_{\kappa\lambda}(k',t)\rangle=\delta^D(k+k')\Bigg[\Big\{k_\mu k_\kappa D^{ab}_{\nu\lambda}(k:t,t)-(\mu\leftrightarrow\nu)-(\kappa\leftrightarrow\lambda)+\begin{pmatrix}\mu\leftrightarrow\nu\\ \kappa\leftrightarrow\lambda\end{pmatrix}\Big\}$$

$$+\left\{k_\mu I^{ab}_{\nu\kappa\lambda}(-k:t)-(\mu\leftrightarrow\nu)+\begin{pmatrix}\mu\leftrightarrow\kappa\\ \nu\leftrightarrow\lambda\\ k\to-k\end{pmatrix}-\begin{bmatrix}\mu\to\lambda\\ \nu\to\kappa\\ \kappa\to\mu\\ \lambda\to\nu\\ k\to-k\end{bmatrix}\right\}$$

$$+I^{ab}_{\mu\nu\kappa\lambda}(k)\Bigg], \tag{3·20}$$

where $D^{ab}_{\mu\nu}(k:t,t')$ is already given by (3·6) and (3·16), and we have put

$$\delta^D(k+k')I^{ab}_{\nu\kappa\lambda}(-k,t)=i\frac{g}{(2\pi)^{D/2}}f^{beh}\int d^Dk_1'd^Dk_2'\delta^D(k'-k_1'-k_2')$$

$$\times\langle A_\nu{}^a(k,t)A_\kappa{}^e(k_1',t)A_\lambda{}^h(k_2',t)\rangle, \tag{3·21}$$

$$\delta^D(k+k')I^{ab}_{\mu\nu\kappa\lambda}(k,t)=\frac{g^2}{(2\pi)^D}f^{acd}f^{beh}$$

$$\times\int d^Dk_1d^Dk_2d^Dk_1'd^Dk_2'\delta^D(k-k_1-k_2)\delta^D(k-k_1'-k_2')$$

$$\times\langle A_\mu{}^c(k_1,t)A_\nu{}^d(k_2,t)A_\kappa{}^e(k_1',t)A_\lambda{}^h(k_2',t)\rangle. \tag{3·22}$$

By making use of (2·9), we can reduce the three-point function in the integrand of (3·21) to the following form:

$$\delta^D(k'-k_1'-k_2')\langle A_\nu{}^a(k,t)A_\kappa{}^e(k_1',t)A_\lambda{}^h(k_2',t)\rangle$$

$$=\delta^D(k+k')[2(\mathrm{i})+2(\mathrm{j})+2(\mathrm{k})], \tag{3·23}$$

where (i), (j) and (k) are contributions from diagrams (i), (j) and (k), respectively, in Fig. 6, and explicitly written as

$$\delta^D(k+k')I^{ab}_{\nu\kappa\lambda}(-k,t)$$

$$=2i\frac{g}{(2\pi)^D}f^{bcd}\int d^Dk_1d^Dk_2\delta^D(k-k_1-k_2)gV^{acd}_{\varepsilon\xi\eta}(k,-k_1,-k_2)$$

$$\times\int_0^\infty dt''\Big\{G_{\nu\varepsilon}(k:t-t'')D^{(0)}_{\xi\kappa}(k_1:t,t'')D^{(0)}_{\eta\lambda}(k_2:t,t'')$$

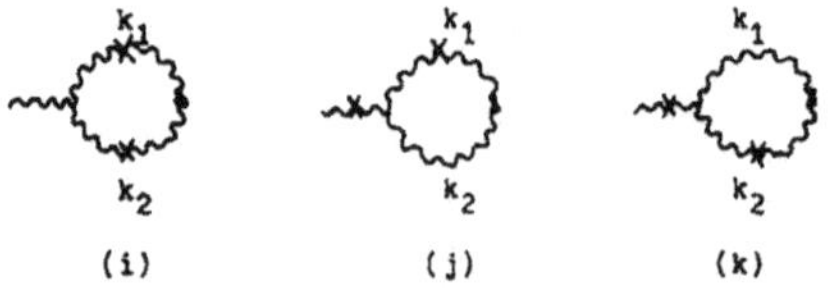

Fig. 6. Diagrams of three point function in the field strength propagator.

$$+\begin{pmatrix} k\leftrightarrow k_2 \\ \nu\leftrightarrow\zeta \\ \xi\leftrightarrow\chi \end{pmatrix}+\begin{pmatrix} k\leftrightarrow k_2 \\ \nu\leftrightarrow\eta \\ \xi\leftrightarrow\lambda \end{pmatrix}\Bigg\}. \tag{3·24}$$

Performing the time integration and taking the limit $t\to\infty$, we get contributions from the above diagrams as listed in Table III. Summing up all contributions, we obtain

$$\lim_{t=t'\to\infty}\delta^D(k+k')I^{ab}_{\nu\kappa\lambda}(-k,t)=\delta^D(k+k')[I^{ab}_{\nu\kappa\lambda}(\mathrm{LG})+I^{ab}_{\nu\kappa\lambda}(1)+I^{ab}_{\nu\kappa\lambda}(2)], \tag{3·25}$$

where

$$I^{ab}_{\mu\nu\lambda}(\mathrm{LG})=\frac{2ig^2}{(2\pi)^D}f^{bcd}\int d^Dk_1d^Dk_2\delta^D(k-k_1-k_2)V^{acd}_{\epsilon\zeta\eta}(k,-k_1,-k_2)$$
$$\times O^T_{\nu\epsilon}(k)O^T_{\kappa\zeta}(k_1)O^T_{\lambda\eta}(k_2)\frac{1}{k^2k_1^2k_2^2}, \tag{3·26}$$

$$I^{ab}_{\nu\kappa\lambda}(1)=\frac{2ig^2}{(2\pi)^D}f^{bcd}O^T_{\nu\epsilon}(k)\int d^Dk_1d^Dk_2\delta^D(k-k_1-k_2)V^{acd}_{\epsilon\zeta\eta}(k,-k_1,-k_2)$$
$$\times\Bigg[O^T_{\xi\kappa}(k_1)O^L_{\eta\lambda}(k_2)\frac{-1}{k^2k_1^2(k^2+k_1^2)}+O^L_{\xi\kappa}(k_1)O^T_{\eta\lambda}(k_2)\frac{-1}{k^2k_2^2(k^2+k_2^2)}$$
$$+O^L_{\xi\kappa}(k_1)O^L_{\eta\lambda}(k_2)\frac{4}{(k^2)^3}\Bigg], \tag{3·27}$$

Table III. Contributions from the diagrams in Fig. 7.

	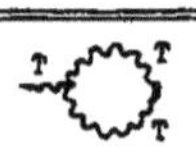							
	$O^T_{\nu\epsilon}(k)\times[VO^T(k_1)\times O^T(k_2)]_{\epsilon\kappa\lambda}$	$O^T_{\nu\epsilon}(k)\times[VO^T(k_1)\times O^L(k_2)]_{\epsilon\kappa\lambda}$	$O^T_{\nu\epsilon}(k)\times[VO^L(k_1)\times O^T(k_2)]_{\epsilon\kappa\lambda}$	$O^T_{\nu\epsilon}(k)\times[VO^L(k_1)\times O^L(k_2)]_{\epsilon\kappa\lambda}$	$O^L_{\nu\epsilon}(k)\times[VO^T(k_1)\times O^T(k_2)]_{\epsilon\kappa\lambda}$	$O^L_{\nu\epsilon}(k)\times[VO^T(k_1)\times O^L(k_2)]_{\epsilon\kappa\lambda}$	$O^L_{\nu\epsilon}(k)\times[VO^L(k_1)\times O^T(k_2)]_{\epsilon\kappa\lambda}$	$O^L_{\nu\epsilon}(k)[VO^L(k_1)\times O^L(k_2)]_{\epsilon\kappa\lambda}$
(i)	$\frac{1}{k_1^2k_2^2(k^2+k_1^2+k_2^2)}$	$\frac{-2}{k_1^2(k^2+k_1^2)^2}$	$\frac{-2}{k_2^2(k^2+k_1^2)^2}$	$\frac{8}{(k^2)^3}$	$\frac{1}{k_1^2k_2^2(k_1^2+k_2^2)}$	$\frac{-2}{(k_1^2)^3}$	$\frac{-2}{(k_2^2)^3}$	0
(j)	$\frac{1}{k^2k_1^2(k^2+k_1^2+k_2^2)}$	$\frac{-2}{k^2(k^2+k_1^2)^2}$	$\frac{1}{k^2k_2^2(k^2+k_2^2)}$	$\frac{-2}{(k^2)^3}$	$\frac{-2}{k_2^2(k_1^2+k_2^2)}$	$\frac{-2}{(k_1^2)^3}$	$\frac{8}{(k_2^2)^3}$	0
(k)	$\frac{1}{k^2k_2^2(k^2+k_1^2+k_2^2)}$	$\frac{1}{k^2k_1^2(k^2+k_1^2)}$	$\frac{-2}{k^2(k^2+k_2^2)^2}$	$\frac{-2}{(k^2)^3}$	$\frac{-2}{k_1^2(k_1^2+k_2^2)^2}$	$\frac{8}{(k_1^2)^3}$	$\frac{-2}{(k_2^2)^3}$	0

$$I^{ab}_{\nu\kappa\lambda}(2)=\frac{2ig^2}{(2\pi)^D}f^{bcd}O^L_{\nu\epsilon}(k)\int d^Dk_1d^Dk_2\delta^D(k-k_1-k_2)V^{acd}_{\epsilon\xi\eta}(k,\ -k_1,\ -k_2)$$

$$\times\left[O^T_{\xi\kappa}(k_1)O^T_{\eta\lambda}(k_2)\frac{-1}{k_1{}^2k_2{}^2(k_1{}^2+k_2{}^2)}+O^T_{\xi\kappa}(k_1)O^L_{\eta\lambda}(k_2)\frac{4}{(k_1{}^2)^3}\right.$$

$$\left.+O^L_{\xi\kappa}(k_1)O^T_{\eta\lambda}(k_2)\frac{4}{(k_2{}^2)^3}\right]. \qquad (3\cdot28)$$

We can easily observe that $I^{ab}_{\nu\kappa\lambda}$(LG) coincides with the conventional field theoretical result in the Landau gauge. Since $I^{ab}_{\nu\kappa\lambda}(2)$ vanishes because of its antisymmetry in κ and λ and operation of $O^L_{\nu\epsilon}(k)$ from the left, we have only $I^{ab}_{\nu\kappa\lambda}(1)$. By means of the formulas

$$V^{abc}_{\epsilon\xi\eta}(k,\ -k_1,\ -k_2)O^T_{\xi\kappa}(k_1)O^L_{\eta\lambda}(k_2)$$

$$=\frac{-i}{2}f^{abc}\left[(k^2-k_1{}^2)\left(\delta_{\epsilon\kappa}-\frac{k_{1\epsilon}k_{1\kappa}}{k_1{}^2}\right)-k_\epsilon\left(\delta_{\kappa\alpha}-\frac{k_{1\kappa}k_{1\alpha}}{k_1{}^2}\right)k_\alpha\right]\frac{k_{2\lambda}}{k_2{}^2}\,.$$

$$V^{abc}_{\epsilon\xi\eta}(k,\ -k_1,\ -k_2)O^L_{\xi\kappa}(k_1)O^T_{\eta\lambda}(k_2)$$

$$=\frac{-i}{2}f^{abc}\left[(k^2-k_2{}^2)\left(\delta_{\epsilon\lambda}-\frac{k_{2\epsilon}k_{2\lambda}}{k_2{}^2}\right)-k_\epsilon\left(\delta_{\lambda\alpha}-\frac{k_{2\lambda}k_{2\alpha}}{k_2{}^2}\right)k_\alpha\right]\frac{k_{1\kappa}}{k_1{}^2}\,,$$

$$V^{abc}_{\epsilon\xi\eta}(k,\ -k_1,\ -k_2)O^L_{\xi\kappa}(k_1)O^L_{\eta\lambda}(k_2)$$

$$=\frac{-i}{2}f^{abc}[k^2k_{1\epsilon}-(k\cdot k_1)k_\epsilon]\frac{k_{1\kappa}k_{2\lambda}}{k_1{}^2k_2{}^2}\,, \qquad (3\cdot29)$$

$I^{ab}_{\nu\kappa\lambda}(1)$ can be reduced to

$$I^{ab}_{\nu\kappa\lambda}(1)=g^2f^{acd}f^{bcd}\frac{O^T_{\nu\epsilon}(k)}{k^2}\left[k_\lambda\left\{\int\frac{d^Dk_1}{(2\pi)^D}\frac{k_{1\epsilon}k_{1\kappa}}{(k_1{}^2)^2k_2{}^2}\right.\right.$$

$$+\frac{2}{k^2}\int\frac{d^Dk_1}{(2\pi)^D}\frac{k_{1\epsilon}k_{1\kappa}}{k_2{}^2(k^2+k_1{}^2)}-\delta_{\epsilon\kappa}\int\frac{d^Dk_1}{(2\pi)^D}\frac{1}{k_1{}^2k_2{}^2}$$

$$\left.\left.+2\delta_{\epsilon\kappa}\int\frac{d^Dk_1}{(2\pi)^D}\frac{1}{k_2{}^2(k^2+k_1{}^2)}\right\}+\delta_{\epsilon\kappa}\int\frac{d^Dk_1}{(2\pi)^D}\frac{(k^2-k_1{}^2)k_{1\lambda}}{k_1{}^2k_2{}^2(k^2+k_1{}^2)}\right]$$

$$-(\kappa\leftrightarrow\lambda), \qquad (3\cdot30)$$

where $k_2=k-k_1$. For the sake of later convenience, we rewrite (3·30) as

$$k_\mu I^{abc}_{\nu\kappa\lambda}(1)=\{k_\mu k_\lambda\Delta^{(0)T}_{\nu\alpha}(k)\tilde{\Xi}^{ab}_{\alpha\beta}\Delta^{(0)T}_{\kappa\beta}(k)+J^{ab}_{\mu\nu\kappa\lambda}\}-\{\kappa\leftrightarrow\lambda\}, \qquad (3\cdot31)$$

where

$$\tilde{\Xi}^{ab}_{\alpha\beta}=g^2f^{acd}f^{bcd}\left[k^2\int\frac{d^Dk_1}{(2\pi)^D}\frac{k_{1\alpha}k_{1\beta}}{(k_1{}^2)^2k_2{}^2}+2\int\frac{d^Dk_1}{(2\pi)^D}\frac{k_{1\alpha}k_{1\beta}}{k_2{}^2(k^2+k_1{}^2)}\right.$$

$$-k^2\int\frac{d^Dk_1}{(2\pi)^D}\frac{\delta_{\alpha\beta}}{k_1^2k_2^2}+2k^2\int\frac{d^Dk_1}{(2\pi)^D}\frac{\delta_{\alpha\beta}}{k_2^2(k^2+k_1^2)}\Bigg] \tag{3·32}$$

and

$$J^{ab}_{\mu\nu\kappa\lambda}=g^2f^{acd}f^{bcd}O^T_{\nu\kappa}(k)\int\frac{d^Dk_1}{(2\pi)^D}\frac{(k^2-k_1^2)k_\mu k_{1\lambda}}{k^2k_1^2k_2^2(k^2+k_1^2)}\,. \tag{3·33}$$

Our final task is to calculate $\delta(k+k')I^{ab}_{\mu\nu\kappa\lambda}(k,t)$ appearing in (3·20) and given by (3·22). Up to the second order perturbation, we can easily get

$$\begin{aligned}I^{ab}_{\mu\nu\kappa\lambda}(k,t)=&\frac{g^2}{(2\pi)^D}f^{acd}f^{bcd}\int d^Dk_1d^Dk_2\delta^D(k-k_1-k_2)\\&\times\left[\Delta_{\mu\kappa}(k_1,t)(1-e^{-2k_1^2t})+2\frac{k_{1\kappa}k_{1\sigma}}{k_1^2}t\right]\\&\times\left[\Delta_{\nu\lambda}(k_2,t)(1-e^{-2k_2^2t})+2\frac{k_{2\kappa}k_{2\sigma}}{k_2^2}t\right]\\&-(\kappa\leftrightarrow\lambda),\end{aligned} \tag{3·34}$$

corresponding to Fig. 7. It follows from (3·33) that we have the finite contribution

$$\begin{aligned}I^{ab}_{\mu\nu\kappa\lambda}(\mathrm{LG})=&g^2f^{acd}f^{bcd}\int d^Dk_1d^Dk_2\delta^D(k-k_1-k_2)\\&\times[\Delta^{(0)T}_{\mu\kappa}(k_1)\Delta^{(0)T}_{\nu\lambda}(k_2)-\Delta^{(0)T}_{\mu\lambda}(k_1)\Delta^{(0)T}_{\nu\kappa}(k_2)].\end{aligned} \tag{3·35}$$

as $t\to\infty$. This term corresponds exactly to the one in the conventional Landau gauge field theory.

Substituting (3·16), (3·31) and (3·35) into (3·20), we obtain the final result

$$\lim_{t\to\infty}\langle F^a_{\mu\nu}(k,t)F^b_{\kappa\lambda}(k',t)\rangle=\delta^D(k+k')[\Sigma^{ab}_{\mu\nu\kappa\lambda}(\mathrm{FTh})+\Sigma^{ab}_{\mu\nu\kappa\lambda}(\mathrm{Res})], \tag{3·36}$$

where

$$\begin{aligned}\Sigma^{ab}_{\mu\nu\kappa\lambda}(\mathrm{FTh})=&\Big[k_\mu k_\kappa(\delta_{ab}\Delta^{(0)T}_{\nu\lambda}(k)+\Delta^{(0)T}_{\nu\alpha}(k)\{\Pi^{ab}_{\alpha\beta}(\mathrm{LG})+\Pi^{ab}_{\alpha\beta}(\mathrm{FP})\}\Delta^{(0)T}_{\beta\lambda}(k))\\&-(\mu\leftrightarrow\nu)-(\kappa\leftrightarrow\lambda)+\begin{pmatrix}\mu\leftrightarrow\nu\\ \kappa\leftrightarrow\lambda\end{pmatrix}\Big]\end{aligned}$$

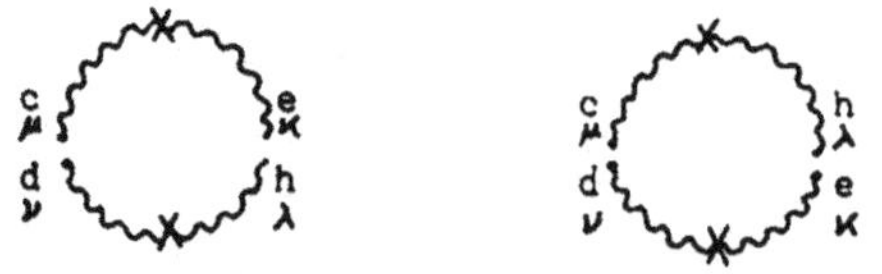

Fig. 7. Diagrams of four point function in the field strength propagator.

$$+\left[k_\mu I^{ab}_{\nu\kappa\lambda}(\mathrm{LG})-(\mu\leftrightarrow\nu)+\begin{pmatrix}\mu\leftrightarrow\kappa\\ \nu\leftrightarrow\lambda\\ k\to -k\end{pmatrix}-\begin{bmatrix}\mu\to\lambda\\ \nu\to\kappa\\ \kappa\to\mu\\ \lambda\to\nu\\ k\to -k\end{bmatrix}\right]$$

$$+I^{ab}_{\mu\nu\kappa\lambda}(\mathrm{LG}), \tag{3·37}$$

$$\Sigma^{ab}_{\mu\nu\kappa\lambda}(\mathrm{Res})=\Big[k_\mu k_\kappa \Delta^{(0)T}_{\nu\alpha}(k)(\Xi^{ab}_{\alpha\beta}-2\tilde{\Xi}^{ab}_{\alpha\beta})\Delta^{(0)T}_{\kappa\beta}(k)$$

$$-(\mu\leftrightarrow\nu)-(\kappa\leftrightarrow\lambda)+\begin{pmatrix}\mu\leftrightarrow\nu\\ \kappa\leftrightarrow\lambda\end{pmatrix}\Big]$$

$$+g^2 f^{acd}f^{bcd}\int\frac{d^Dk_1}{(2\pi)^D}\frac{(k^2-k_1{}^2)K_{\mu\nu\kappa\lambda}}{k^2k_1{}^2(k-k_1)^2(k^2+k_1{}^2)}, \tag{3·38}$$

where $K_{\mu\nu\kappa\lambda}=-\delta_{\mu\kappa}(k_\nu k_{1\lambda}+k_\lambda k_{1\nu})-(\mu\leftrightarrow\nu)-(\kappa\leftrightarrow\lambda)+(\mu\leftrightarrow\nu,\ \kappa\leftrightarrow\lambda)$. Equation (3·37) is nothing but the field strength propagator given by the conventional Landau gauge field theory with the Faddeev-Popov trick. The other contribution (3·38) vanishes by the following reasons. First, $\Xi^{ab}_{\alpha\beta}-2\tilde{\Xi}^{ab}_{\alpha\beta}=0$ as we see directly from (3·19). Second

$$\int d^Dk\Sigma^{ab}_{\mu\nu\kappa\lambda}(\mathrm{Res})\propto\int d^Dk d^Dk_1\frac{(k^2-k_1{}^2)K_{\mu\nu\kappa\lambda}}{k^2k_1{}^2(k-k_1)^2(k^2+k_1{}^2)}=0, \tag{3·39}$$

because the integrand is anti-symmetric with respect to k and k_1. Note that the k integral appears in the gauge invariant quantity $\lim_{y\to x}\langle F^a_{\mu\nu}(x,t)F^b_{\kappa\lambda}(y,t)\rangle \propto\int d^Dk\langle F^a_{\mu\nu}(k,t)F^b_{\kappa\lambda}(-k,t)\rangle$. We can therefore conclude that, up to the second order perturbation, the stochastic quantization method automatically leads us to the correct gauge field propagator including the Faddeev-Popov ghost effects without resort to any ghost field.

§ 4. Preliminary discussion on physical state condition and unitarity relation

Here we first obtain the vertex function with three lines having momenta k, k_1 and k_2, which is given by $\lim_{t\to\infty}\langle A_\mu{}^a(k,t)A_\nu{}^b(k_1,t)A_\kappa{}^c(k_2,t)\rangle$ in the stochastic quantization method. Apart from those terms which diverge as $t\to\infty$, the lowest order approximation of the three line vertex part can be obtained from (3·24) and (3·26)~(3·28) by picking up their integrands, i. e.,

$$A^{abc}_{\mu\nu\kappa}(k;k_1;k_2)=-igV^{(0)abc}_{\mu\nu\kappa}(k,-k_1,-k_2)\Big[\varepsilon_\mu{}^T(k)\varepsilon_\nu{}^T(k_1)\varepsilon_\kappa{}^T(k_2)$$

$$+\varepsilon_\mu{}^T(k)\varepsilon_\nu{}^T(k_1)\varepsilon_\kappa{}^L(k_2)\frac{-k_2{}^2}{k^2+k_1{}^2}+\varepsilon_\mu{}^T(k)\varepsilon_\nu{}^L(k_1)\varepsilon_\kappa{}^T(k_2)\frac{-k_1{}^2}{k^2+k_2{}^2}$$

$$+\varepsilon_\mu{}^T(k)\varepsilon_\nu{}^L(k_1)\varepsilon_\kappa{}^L(k_2)4\frac{k_1{}^2k_2{}^2}{(k^2)^2}+\varepsilon_\mu{}^L(k)\varepsilon_\nu{}^T(k_1)\varepsilon_\kappa{}^T(k_2)\frac{-k^2}{k_1{}^2+k_2{}^2}$$

$$+\varepsilon_\mu{}^L(k)\varepsilon_\nu{}^T(k_1)\varepsilon_\kappa{}^L(k_2)4\frac{k^2k_2{}^2}{(k_1{}^2)^2}+\varepsilon_\mu{}^L(k)\varepsilon_\nu{}^L(k_1)\varepsilon_\kappa{}^T(k_2)4\frac{k^2k_1{}^2}{(k_2{}^2)^2}\Bigg], \tag{4·1}$$

where $\varepsilon_\mu{}^T(k)$ and $\varepsilon_\mu{}^L(k)$ are, respectively, polarization vectors of transverse and longitudinal gluons with momentum k. Analytically continuating (4·1) from Euclidean to Minkowski space, we can easily see that those terms which have longitudinal gluon lines will vanish on their mass shell. Therefore, if we choose one or two of gluons as incoming lines to be physical, namely, transverse, then we are left keeping only transverse outgoing gluons on mass shell ——in other words, keeping the first term of (4·1) alone. This is nothing other than the physical state condition in the first order perturbation theory. The physical state condition derived here is a natural result of the stochastic quantization method. This fact suggests us that the physical state condition would be achieved generally and automatically in the stochastic quantization method, and consequently, that the unitarity would naturally be verified. Needless to say, the first term of (4·1) satisfies the second order unitarity relation together with the propagator obtained in the previous section.

§ 5. Concluding remarks

In this paper we have shown that the stochastic quantization of the non-Abelian gauge field can give us the correct field-theoretical propagator including the Faddeev-Popov ghost effects, up to the second order perturbation theory, in a natural way without resort to introduction of any ghost field. Also a preliminary discussion indicates that the physical state condition should be automatically satisfied in the stochastic quantization method. Furthermore, the mechanism by which the stochastic quantization method is free from the usual gauge fixing procedure but not free from the gauge parameter fixing have been found to relate to the choice of initial state distributions. As was seen through the above discussion, the stochastic quantization method is very attractive to us for its simplicity in the principle, among the conventional theories which tend to be more and more tricky by introduction of complexties such as ghost fields etc. More extensive development of the stochastic quantization method will be given in forthcoming papers.

References

1) G. Parisi and Y. S. Wu, Sci. Sinica **24** (1981), 483.
2) D. Zwanziger, Nucl. Phys. **B192** (1981), 259.
L. Baulieu and D. Zwanziger, Nucl. Phys. **B193** (1981), 163.
3) Y. Kakudo, Y. Taguchi, A. Tanaka and K. Yamamoto, Prog. Theor. Phys. **69** (1983), 1225.
4) M. Namiki and Y. Yamanaka, Prog. Theor. Phys. **69** (1983), No. 6.

Nuclear Physics B241 (1984) 221–227

A COVARIANT GHOST-FREE PERTURBATION EXPANSION FOR YANG-MILLS THEORIES*

E.G. FLORATOS

Institute for Theoretical Physics, University of Bern, Sidlerstrasse 5, CH-3012 Bern, Switzerland

J. ILIOPOULOS

Laboratoire de Physique Théorique de l'Ecole Normale Supérieure, 24 rue Lhomond, F-75231 Paris Cedex 01, France

D. ZWANZIGER**

Department of Physics, New York University, New York, New York 10003, USA

Received 8 August 1983
(Revised 24 January 1984)

Starting from the equilibrium Fokker-Planck equation, we develop a perturbation expansion for Yang-Mills theories, in a covariant gauge, which is power-counting renormalizable and free from Faddeev-Popov ghosts.

Recent investigations on the stochastic quantization method have shown that it is possible to quantize gauge theories without relying on the Faddeev-Popov construction [1–4]. In the original approach the fields are promoted to stochastic variables depending on a fifth dimension, a "time" τ. One assumes an evolution equation on τ of the Langevin type with an external gaussian noise. Correlation functions are computed as averages over the noise and the Green functions of the original quantum field theory are obtained at the limit, as $\tau \to \infty$, of the equal τ correlation functions. For finite τ the correlation functions can be calculated in perturbation theory and one of the advantages of the method, is that, for gauge theories, this perturbation expansion is well-defined even in the absence of any gauge-fixing term [2]. On the other hand, a shortcoming is that it yields a perturbation theory which is non-renormalizable by power-counting. At the limit $\tau \to \infty$ gauge invariant quantities give the results obtained by the standard methods, while gauge non-invariant

* Work supported in part by Schweizerischer Nationalfonds.
** Supported in part by National Service Foundation Grant no. PHY-8116102.

quantities diverge. In this paper we elaborate on a different perturbation expansion for the Green functions of the gauge theory, without any reference to an artificial fifth time. The expansion, which has been presented already in ref. [2], is renormalizable by power-counting and, although it depends on a covariant gauge breaking term, it does not contain any Faddeev-Popov ghost fields. Whereas in ref. [2] the perturbative expansion of the distribution density $P[A]$ was given, in the present article, the corresponding but much simpler, expansion of the generating functional of connected Green's function $W[j]$ is derived as a straightforward application of the formalism developed in ref. [4].

We start by considering the distribution density $P[A]$ through which one defines the Green functions of the theory:

$$\left\langle T\left(A^{a_1}_{\mu_1}(x_1)\ldots A^{a_n}_{\mu_n}(x_n)\right)\right\rangle_0 = \int \mathcal{D}[A]P[A]A^{a_1}_{\mu_1}(x_1)\ldots A^{a_n}_{\mu_n}(x_n). \quad (1)$$

In the Faddeev-Popov approach $P[A]$ is given by

$$P_G[A] = N \det m_{FP} \exp\left[-S_{YM} - \tfrac{1}{2}\int \mathrm{d}x\, G^2\right], \quad (2)$$

where S_{YM} is the usual, gauge invariant Yang-Mills action, G is a gauge-fixing term, $\det m_{FP}$ is the Faddeev-Popov determinant and N is a normalization factor. $P_G[A]$ depends on the choice of G, but gauge invariant quantities do not. For our purposes it is convenient to write the Fokker-Planck equation:

$$\int \mathrm{d}x \frac{\delta}{\delta A^a_\mu(x)}\left\{\frac{\delta}{\delta A^a_\mu(x)} + \frac{\delta S_{YM}}{\delta A^a_\mu(x)} - D^{ab}_\mu V^b[A]\right\}P_V[A] = 0. \quad (3)$$

V is an otherwise arbitrary gauge non-invariant function of A and D_μ is the Yang-Mills covariant derivative. One can show that physical quantities do not depend on V [2] although the distribution $P_{V=0}$ does not exist. This corresponds to the well-known fact that the usual Green functions cannot be defined unless one specifies the gauge. In what follows we shall restrict ourselves to the choice

$$V^b[A] = \frac{1}{\alpha}\partial_\mu A^b_\mu(x), \quad (4)$$

so that the gauge-breaking term in eq. (3) reads:

$$D^{ab}_\mu V^b[A] = \frac{1}{\alpha}\partial_\mu\partial_\nu A^a_\nu(x) + \frac{ig}{\alpha}f^{abc}\partial_\nu A^b_\nu(x)A^c_\mu(x). \quad (5)$$

It is important to realize that (5) cannot be written as the variation with respect to A

of some action functional because one cannot find an action $\hat{S}[A]$ such that

$$\frac{\delta \hat{S}}{\delta A_\mu^a(x)} = D_\mu^{ab} V^b, \tag{6}$$

with V^b given by eq. (4). For this it is sufficient to compute the variation of the l.h.s. of (6) with respect to $A_\nu^c(y)$ and show [2] that

$$\frac{\delta D_\mu^{ab} V^b(x)}{\delta A_\nu^c(y)} - \frac{\delta D_\nu^{cb} V^b(y)}{\delta A_\mu^a(x)} \neq 0. \tag{7}$$

In other words the gauge-breaking procedure of eq. (3) is not equivalent to the choice $G \sim \partial \cdot A$ in eq. (2). Nevertheless one can solve directly eq. (3) in perturbation theory by separating the free and the interaction parts of S_{YM}

$$S_{\mathrm{YM}} = S_0 + gS_1 + g^2 S_2$$

$$= \int \mathrm{d}x \left\{ \tfrac{1}{4}\left(\partial_\mu A_\nu^a - \partial_\nu A_\mu^a\right)^2 + g f^{abc} \partial_\mu A_\nu^a A_\mu^b A_\nu^c + \tfrac{1}{4} g^2 f^{abc} f^{ab'c'} A_\mu^b A_\nu^c A_\mu^{b'} A_\nu^{c'} \right\}. \tag{8}$$

From now on we shall assume that we have introduced a suitable regularization scheme (in practice dimensional regularization is the most convenient) which defines, in the sense of perturbation theory, all formal expressions like eq. (3).

Let us introduce the generating functional of all Green functions $Z[j_\mu^a]$ which is related to $P[A]$ by

$$Z[j] = \int \mathcal{D}[A] \mathrm{e}^{\int \mathrm{d}x A_\mu^a(x) j_\mu^a(x)} P_V[A]. \tag{9}$$

It satisfies the equation:

$$\int \mathrm{d}x \left(-j_\mu^a(x)\right) \left\{ -j_\mu^a(x) + \left. \frac{\delta S_{\mathrm{YM}}}{\delta A_\mu^a(x)} \right|_{A \to \delta/\delta j} - \left. D_\mu^{ab} V^b \right|_{A \to \delta/\delta j} \right\} Z[j] = 0. \tag{10}$$

We can solve this equation in power series in g. It is more convenient, however, to do this for the generating functional of the connected Green functions $W[j] = \ln Z[j]$ which satisfies the equation:

$$\int \mathrm{d}x \left(-j_\mu^a(x)\right) \left\{ -j_\mu^a(x) + \left. \frac{\delta S_{\mathrm{YM}}}{\delta A_\mu^a(x)} \right|_{A \to \delta W/\delta j + \delta/\delta j} - \left. D_\mu^{ab} V^b \right|_{A \to \delta W/\delta j + \delta/\delta j} \right\} \mathbf{1} = 0. \tag{11}$$

In perturbation we write

$$W[j]=\sum_{n=0}^{\infty} g^n W_n[j], \tag{12}$$

and eq. (11) give for W_0

$$\int \mathrm{d}x\left(-j_\mu^a(x)\right)\left\{-j_\mu^a(x)+\left.\frac{\delta S_0}{\delta A_\mu^a(x)}\right|_{A\to\delta W_0/\delta j}-\frac{1}{\alpha}\partial_\mu\,\partial_\nu A_\nu^a(x)|_{A\to\delta W_0/\delta j}\right\}=0, \tag{13}$$

with the familiar result:

$$W_0[j]=\tfrac{1}{2}\int \mathrm{d}k j_\mu^a(k)\frac{1}{k^2}\left[g_{\mu\nu}-(1-\alpha)\frac{k_\mu k_\nu}{k^2}\right]j_\nu^a(-k). \tag{14}$$

For the higher orders we obtain the equation:

$$-\int \mathrm{d}k j_\mu^a(k)\Delta_{0\mu\nu}^{-1}(k)\frac{\delta W_n}{\delta j_\nu^a(k)}$$

$$=\int \mathrm{d}k j_\mu^a(k)\left\{K_{1,\mu}^a\left(\frac{\delta W}{\delta j}+\frac{\delta}{\delta j}\right)_{n-1,\text{order}}+K_{2,\mu}^a\left(\frac{\delta W}{\delta j}+\frac{\delta}{\delta j}\right)_{n-2,\text{order}}\right\}, \tag{15}$$

where we have defined

$$K_{1,\mu}^a(A)=\frac{\delta S_1}{\delta A_\mu^a}-\frac{1}{\alpha}f^{abc}\partial_\nu A_\nu^b A_\mu^c, \tag{16}$$

$$K_{2,\mu}^a(A)=\frac{\delta S_2}{\delta A_\mu^a}, \tag{17}$$

$$\Delta_{0,\mu\nu}^{-1}=k^2\left[g_{\mu\nu}-\left(1-\frac{1}{\alpha}\right)\frac{k_\mu k_\nu}{k^2}\right]. \tag{18}$$

The important thing is that the r.h.s. of eq. (15) contains W only up to terms of order W_{n-1} and W_{n-2} respectively and, thus, this equation can be solved by iteration. We can solve for W_n directly, but it is easier to observe that the propagator (18) simplifies if one uses the projections into transverse and longitudinal sources. If

$T_{\mu\nu}(k) = g_{\mu\nu} - k_\mu k_\nu/k^2$ and $L_{\mu\nu} = k_\mu k_\nu/k^2$ we write $j_\mu^T = T_{\mu\nu} j_\nu$ and $j_\mu^L = L_{\mu\nu} j_\nu$. Eq. (15) becomes

$$-\int \mathrm{d}k\, k^2 \left[j_\nu^{aT}(k) \frac{\delta W_n}{\delta j_\nu^{aT}(k)} + \frac{1}{\alpha} j_\nu^{aL}(k) \frac{\delta W_n}{\delta j_\nu^{aL}(k)} \right] = Q_n[j], \tag{19}$$

where Q_n represents the known right-hand side. At a given order in n, Q_n is a polynomial in $j(k)$ and each term can again be written in terms of j^T and j^L. It follows that $W_n[j]$ will be a polynomial of the same degree and eq. (19) will determine the coefficients as functions of the momenta. If we write W_n, suppressing the internal symmetry indices, as

$$W_n[j] = \sum_{p=2}^{n+2} \int \mathrm{d}k_1 \ldots \mathrm{d}k_p\, \delta(k_1 + \cdots + k_p) R^{i_1 \ldots i_p}_{\mu_1 \ldots \mu_p}(k_1, \ldots, k_p)\, j^{i_1}_{\mu_1}(k_1) \ldots j^{i_p}_{\mu_p}(k_p), \tag{20}$$

where i_l $(l = 1 \ldots p) = T$ or L, the coefficient functions R are symmetric with respect to separate interchange of any pair of external longitudinal or transverse lines. A similar expression holds for $Q_n[j]$ with coefficients $Q^{i_1 \ldots i_p}_{\mu_1 \ldots \mu_p}$ which are known functions of the momenta. It follows that R is given by an expression of the form:

$$R^{i_1 \ldots i_q i_{q+1} \ldots i_p}_{\mu_1 \ldots \mu_q \mu_{q+1} \ldots \mu_p}(k_1, \ldots, k_p) = -\frac{Q^{i_1 \ldots i_q i_{q+1} \ldots i_p}_{\mu_1 \ldots \mu_q \mu_{q+1} \ldots \mu_p}}{k_1^2 + \cdots + k_q^2 + \alpha^{-1}\left(k_{q+1}^2 + \cdots + k_p^2\right)}, \tag{21}$$

where we have assumed that i_l $(l = 1, \ldots, q) = T$ and i_l $(l = q+1, \ldots, p) = L$.

As an example, let us compute explicitly the first non-trivial term, namely W_1. It contains two terms: the first, which we call W_1^{old}, is the usual three-point function determined by the interaction term S_1. The second, W_1^{new}, is given by the gauge-breaking term DV. The equation for W_1^{new} reads:

$$\begin{aligned}
&\int \mathrm{d}k\, k^2 \left[j_\mu^{aT}(k) \frac{\delta W_1^{\text{new}}}{\delta j_\mu^{aT}(k)} + \frac{1}{\alpha} j_\mu^{aL}(k) \frac{\delta W_1^{\text{new}}}{\delta j_\mu^{aL}(k)} \right] \\
&\qquad = -\frac{i}{(2\pi)^2} f^{abc} \int \mathrm{d}k_1\, \mathrm{d}k_2\, \mathrm{d}k_3\, \delta(k_1 + k_2 + k_3) \frac{k_{3\rho}}{k_3^2} j_\rho^{cL}(k_3) \\
&\qquad \times \left\{ \frac{1}{k_2^2} \left[j_\mu^{aT}(k_1) j_\mu^{bT}(k_2) + \alpha j_\mu^{aL}(k_1) j_\mu^{bL}(k_2) \right] \right. \\
&\qquad \left. + \left(\frac{1}{k_2^2} - \frac{\alpha}{k_1^2} \right) j_\mu^{aL}(k_1) j_\mu^{bT}(k_2) \right\}.
\end{aligned} \tag{22}$$

W_1^{new} may be written as:

$$W_1^{\text{new}} = f^{abc} \int \mathrm{d}k_1\, \mathrm{d}k_2 \mathrm{d}k_3\, \delta(k_1 + k_2 + k_3)$$
$$\times \{ j_\mu^{aL}(k_1) j_\nu^{bL}(k_2) j_\rho^{cL}(k_3) R_{\mu\nu\rho}^{LLL}(k_1, k_2, k_3)$$
$$+ j_\mu^{aT}(k_1) j_\nu^{bL}(k_2) j_\rho^{cL}(k_3) R_{\mu\nu\rho}^{TLL}(k_1, k_2, k_3)$$
$$+ j_\mu^{aT}(k_1) j_\nu^{bT}(k_2) j_\rho^{cL}(k_3) R_{\mu\nu\rho}^{TTL}(k_1, k_2, k_3) \}. \tag{23}$$

Notice that the r.h.s. of eq. (22) tells us that there is no need to introduce an R^{TTT} term. Since the longitudinal and transverse sources are functionally independent, we obtain from eq. (22)

$$R_{\mu\nu\rho}^{LLL}(k_1, k_2, k_3) = -\frac{i\alpha}{(2\pi)^2} \frac{1}{k_1^2 + k_2^2 + k_3^2} \left[\frac{k_{3\rho} g_{\mu\nu}}{k_3^2 k_2^2} \right]_{\text{sym}}, \tag{23a}$$

$$R_{\mu\nu\rho}^{TLL}(k_1, k_2, k_3) = -\frac{i\alpha}{(2\pi)^2} \frac{1}{\alpha k_1^2 + k_2^2 + k_3^2} \left[\frac{k_{3\rho} g_{\mu\nu}}{k_3^2} \left(\frac{\alpha}{k_2^2} - \frac{1}{k_1^2} \right) \right]_{\text{sym}}, \tag{23b}$$

$$R_{\mu\nu\rho}^{TTL}(k_1, k_2, k_3) = -\frac{i\alpha}{(2\pi)^2} \frac{1}{\alpha(k_1^2 + k_2^2) + k_3^2} \left[\frac{k_{3\rho} g_{\mu\nu}}{k_3^2 k_2^2} \right]_{\text{sym}}. \tag{23c}$$

The subscripts "sym" in eqs. (23) mean that the corresponding expressions have to be symmetrized with respect to the exchange of any pair of lines of the same type, T or L. Since W_1 is already antisymmetric with respect to the exchange of internal symmetry indices, it follows that (23a) must be antisymmetrized with respect to the exchange of any pair of (μ, ν, ρ) and (k_1, k_2, k_3), (23b) with respect to (ν, ρ) and (k_2, k_3) and (23c) with respect to (μ, ν) and (k_1, k_2). Some remarks are in order: W_1^{new} vanishes when $\alpha \to 0$. However W_1^{new} will, in general, contribute to closed loops even as $\alpha \to 0$. The peculiar structure of the denominators in (23) is due to the eigenvalues of the operator in the l.h.s. of eq. (15). When diagonalized in the $L - T$ basis it has eigenvalues k^2 and $\alpha^{-1}k^2$. The appearance of these denominators, rather than the usual propagators, as well as the absence of the R^{TTT} term, reflects the fact that this term is not derivable from a lagrangian. In fact, if we Legendre transform $W[j]$ in order to obtain an "effective action", the resulting term will be non-local because the denominators in (23) cannot be removed by multiplying the external lines with the inverse propagators.

We can compute any desired order in this perturbation expansion. A diagrammatic representation is given in ref. [2]. All terms are obtained by combining vertices and propagators, but the usual rules are modified because one must take into account the special form of the denominators in (21) and (23). These rules are reminiscent of old-fashioned perturbation theory where in the denominators one has the sum over the energies of the intermediate states. This interpretation is accurate if one goes back to the stochastic Langevin equation and computes the finite τ correlation functions. As announced, this series is renormalizable by power-counting, makes no reference to Faddeev-Popov ghosts and, one can show, that it is equivalent to the usual perturbation for gauge invariant quantities.

An advantage of the present method, based on the diffusion equation (3), as opposed to the Faddeev-Popov formula (2), is that the exact solution, $P_V(A)$, is manifestly positive (for the euclidean theory), if it exists at all. Consequently, once it is suitably discretized and regularized, euclidean expectation values may be evaluated numerically by highly efficient and non-perturbative Monte Carlo methods. In fact such numerical calculations have been done recently for a lattice analog of eq. (3) [5]. However to interpret the numerical results, comparison should be made with analytic calculations based on the perturbative renormalization group. In particular the β function is needed to low order. To calculate it, the perturbative expansion developed here is the appropriate vehicle.

References

[1] G. Parisi and Y.S. Wu, Sci. Sinc. 24 (1981) 483
[2] D. Zwanziger, Nucl. Phys. B192 (1981) 259
[3] L. Baulieu and D. Zwanziger, Nucl. Phys. B193 (1981) 163
[4] E.G. Floratos and J. Iliopoulos, Nucl. Phys. B214 (1983) 392
[5] E. Seiler, I.O. Stamatescu and D. Zwanziger, Nucl. Phys. B239 (1984) 177

Fermions

1600

Progress of Theoretical Physics, Vol. 69, No. 5, May 1983

Stochastic Quantization Method of Fermion Fields

Tomoki FUKAI, Hiromichi NAKAZATO, Ichiro OHBA,
Keisuke OKANO and Yoshiya YAMANAKA

Department of Physics, Waseda University, Tokyo 160

(Received November 15, 1982)

Stochastic quantization of fermion fields is formulated, where the probability distribution P and the expectation value in P can be defined properly. The equivalence between the stochastic quantization and the path integral in Euclidean field theory is shown in free case. The general treatment of interacting case is also discussed.

§ 1. Introduction

Recently, Parisi and Wu[1] proposed a stochastic method of quantizing boson fields. In this method, an extra fifth time t is introduced and the evolution of classical boson fields with respect to the time t is described by the Langevin equation. Then, it can be proved that, in the t-infinity limit, correlation functions in the stochastic process are identical with the corresponding ones which are given in the usual path integral form of Euclidean field theories. It is essential for this equivalence between the stochastic process and the path integral that the probability distribution has the unique equilibrium when t goes to infinity. The approach to the equilibrium is realized by the damping factor coming from the drift force proportional to Euclidean Klein-Gordon operator $(-\Box+m^2)$.

It is the purpose of this paper to investigate whether and how the stochastic quantization can be extended to fermion. At the first sight, some difficulties may arise in its extension:

(i) There is no classical analogue because of the anti-commuting nature of fermion fields.

(ii) What is the origin of the damping factor in fermionic case where the classical equation of motion is described by a first differential equation (Dirac equation)?

The point (i) leads us to use Grassmann numbers, which seem to be incompatible with the concept of probability.*) However, we can safely construct the probabilistic interpretation even in this case. As for the point (ii), the mass term of fermion field gives the damping factor, when the Langevin equations are set up properly.

---

*) The alternative way to represent the anti-communing nature is seen in Ref. 2).

We set up the Langevin equations and, solving the free equations, discuss the damping factor in § 2. In § 3 we first define the probability distribution in a consistent way and derive the Fokker-Planck equation from the Langevin equation. In order to prove the equivalence between our method and the path integral quantization, the spectrum of the Fokker-Planck operator is discussed in free and interacting cases, respectively. Section 4 is devoted to concluding remarks. The notation of Euclidean field theory and the properties of Grassmann number are summarized in Appendices A and B, respectively. In Appendix C, a simplified model with one degree of freedom is solved.

§ 2. Langevin equation

We consider a system of a Dirac fermion $\psi(x)$ in four-dimensional Euclidean space. The action S is given by

$$S=S_0-gS_{\rm int}\,, \tag{2·1}$$

where

$$S_0=-\int d^4x\bar{\psi}(x)(i\not\partial-m)\psi(x) \tag{2·2}$$

is the free part and $S_{\rm int}$ represents a fermion self-interaction part. Here $\psi(x)$ and $\bar{\psi}(x)$ are Grassmann numbers. The notation here is explained in Appendix A.

Let us proceed to formulate the stochastic quantization of fermion fields. Suppose that fermion fields $\psi(x)$ and $\bar{\psi}(x)$ depend on the fictitious time t and are random fields. By complete analogy with bosonic cases, we assume that the time evolution of $\psi(x,t)$ and $\bar{\psi}(x,t)$ is described by the following Langevin equations:[3)]

$$\begin{aligned}\alpha\frac{\partial}{\partial t}\psi_\eta(x,t)&=-\gamma\frac{\delta S}{\delta\bar{\psi}_\eta(x,t)}+\xi_\eta(x,t)\\&=\gamma(i\not\partial-m)_{\eta\zeta}\psi_\zeta(x,t)-g\gamma\frac{\delta S_{\rm int}}{\delta\bar{\psi}_\eta(x,t)}+\xi_\eta(x,t),\\\alpha\frac{\partial}{\partial t}\bar{\psi}_\eta(x,t)&=\gamma\frac{\delta S}{\delta\psi_\eta(x,t)}+\bar{\xi}_\eta(x,t)\\&=-\gamma\bar{\psi}_\zeta(x,t)(-i\overleftarrow{\not\partial}-m)_{\zeta\eta}+g\gamma\frac{\delta S_{\rm int}}{\delta\psi_\eta(x,t)}+\bar{\xi}_\eta(x,t),\end{aligned} \tag{2·3}$$ *)

where Grassmann random variables $\xi_\eta(x,t)$ and $\bar{\xi}_\eta(x,t)$ satisfy the following

*) Notice that "left-derivative" is used throughout this paper. See Appendix B.

correlations:

$$\langle\xi_\eta(x,t)\bar{\xi}_\zeta(x',t')\rangle=-\langle\bar{\xi}_\zeta(x',t')\xi_\eta(x,t)\rangle$$
$$=2\beta^{-1}\delta_{\eta\zeta}\delta^4(x-x')\delta(t-t'),$$
$$\langle\xi_\eta(x,t)\rangle=\langle\bar{\xi}_\eta(x,t)\rangle=0,$$
$$\langle\xi_\eta(x,t)\xi_\zeta(x',t')\rangle=\langle\bar{\xi}_\eta(x,t)\bar{\xi}_\zeta(x',t')\rangle=0,$$
$$\langle\xi_\eta\xi_\zeta\xi_\sigma\rangle=\langle\xi_\eta\xi_\zeta\bar{\xi}_\sigma\rangle=\langle\xi_\eta\bar{\xi}_\zeta\bar{\xi}_\sigma\rangle=\langle\bar{\xi}_\eta\bar{\xi}_\zeta\bar{\xi}_\sigma\rangle=0,$$
$$\langle\xi_\eta\bar{\xi}_\zeta\xi_\sigma\bar{\xi}_\rho\rangle=\langle\xi_\eta\bar{\xi}_\zeta\rangle\langle\xi_\sigma\bar{\xi}_\rho\rangle-\langle\xi_\eta\bar{\xi}_\rho\rangle\langle\xi_\sigma\bar{\xi}_\zeta\rangle,\quad \text{etc.}\tag{2·4}$$

The definition of these expectation values $\langle\cdots\rangle$ will be clear in § 3 after the physical meaning has been given to the ***probability distribution of Grassmann fields.*** Positive parameters α, β and γ in Eqs. (2·3) and (2·4) are introduced to adjust dimensions.

Because of the anti-commuting nature of Grassmann fields, we may change signs of drift forces in Eqs. (2·3) even if we take the convention of "left-derivatives". The Langevin equations (2·3) are correct choices. See the discussion below (2·8).

Now we will solve Eqs. (2·3) and calculate correlation functions in free case ($g=0$). It is convenient to introduce the free Green's functions $G(x,t)$ and $\bar{G}(x,t)$ which satisfy

$$\alpha\frac{\partial}{\partial t}G_{\zeta\eta}(x,t)-\gamma(i\nabla\!\!\!\!/-m)_{\zeta\sigma}G_{\sigma\eta}(x,t)=\delta_{\zeta\eta}\delta^4(x)\delta(t),$$
$$\alpha\frac{\partial}{\partial t}\bar{G}_{\zeta\eta}(x,t)-\gamma\bar{G}_{\zeta\sigma}(x,t)(-i\overleftarrow{\nabla\!\!\!\!/}-m)_{\sigma\eta}=\delta_{\zeta\eta}\delta^4(x)\delta(t)\tag{2·5}$$

and

$$G(x,t)=\bar{G}(x,t)=0,\qquad t<0.$$

Random fields $\psi^0(x,t)$ and $\bar{\psi}^0(x,t)$ are written as

$$\psi_\eta{}^0(x,t)=\int_0^t dt'd^4x'G_{\eta\zeta}(x-x',t-t')\xi_\zeta(x',t'),$$
$$\bar{\psi}_\eta{}^0(x,t)=\int_0^t dt'd^4x'\bar{\xi}_\zeta(x',t')\bar{G}_{\zeta\eta}(x-x',t-t'),\tag{2·6}$$

if we take

$$\psi_\eta{}^0(x,0)=\bar{\psi}_\eta{}^0(x,0)=0$$

as the initial conditions. Equations (2·5) are easily solved to give the explicit expressions for G and $\bar{G}$,

$$\left.\begin{aligned} G(x,t)&=\frac{\theta(t)}{\alpha}\int\frac{d^4p}{(2\pi)^4}\exp\left(-\frac{\gamma}{\alpha}(\not p+m)t+ipx\right),\\ \bar{G}(x,t)&=\frac{\theta(t)}{\alpha}\int\frac{d^4p}{(2\pi)^4}\exp\left(-\frac{\gamma}{\alpha}(\not p+m)t-ipx\right). \end{aligned}\right\} \tag{2·7}$$

The matrix $(\not p+m)$ with two degenerate eigenvalues $\pm i\sqrt{p^2}+m$ is diagonalized by a unitary matrix U_p:

$$(\not p+m)=U_p^{-1}\begin{pmatrix} i\sqrt{p^2}+m & & & 0\\ & i\sqrt{p^2}+m & & \\ & & -i\sqrt{p^2}+m & \\ 0 & & & -i\sqrt{p^2}+m\end{pmatrix}U_p\,. \tag{2·8}$$

Substituting Eq. (2·8) into Eqs. (2·7), it turns out that G and $\bar{G}$ decrease as $\exp((-\gamma/\alpha)mt)$ with an oscillating factor. Compare this t-dependence with that in bosonic case $\exp(-(p^2+m^2)t)$ without oscillation; the latter depends on four momentum p while the former not. Instead of Eqs. (2·3), alternative choice of signs of drift forces leads to diverging G and $\bar{G}$.

Using Eqs. (2·4), (2·6) and (2·7), we easily obtain correlation functions, e.g.,

$$\begin{aligned} &\langle\psi_\xi{}^0(x,t)\bar{\psi}_\eta{}^0(x',t')\rangle\\ &=\int_0^t dt_1\int_0^{t'}dt_2 d^4x_1 d^4x_2 G_{\xi\sigma}(x-x_1,t-t_1)\bar{G}_{\rho\eta}(x'-x_2,t'-t_2)\\ &\quad\times\langle\xi_\sigma(x_1,t_1)\bar{\xi}_\rho(x_2,t_2)\rangle\\ &=\frac{1}{\alpha\beta\gamma}\int\frac{d^4p}{(2\pi)^4}\exp(ip(x-x'))\\ &\quad\times\left(\frac{\exp\left(-\frac{\gamma}{\alpha}(\not p+m)|t-t'|\right)-\exp\left(-\frac{\gamma}{\alpha}(\not p+m)(t+t')\right)}{\not p+m}\right)_{\xi\eta}. \end{aligned} \tag{2·9}$$

Then taking the limit $t=t'\to\infty$, we get

$$\lim_{t=t'\to\infty}\langle\psi^0(x,t)\bar{\psi}^0(x',t')\rangle=\frac{1}{x}\int\frac{d^4p}{(2\pi)^4}\exp(ip(x-x'))\frac{1}{\not p+m} \tag{2·10}$$

which coincides with free fermion propagator in Euclidean field theory when $x=\alpha\beta\gamma$ is set equal to unity.

For $g\neq 0$, the formal solutions of Eqs. (2·3) can be expanded by use of G and $\bar{G}$ as

$$\psi_\eta(x,t)=\int_0^t dt'd^4x'G_{\eta\xi}(x-x',t-t')\left(\xi_\xi(x',t')-g\gamma\frac{\delta S_{\rm int}}{\delta\bar{\psi}_\xi(x',t')}\right),$$

$$\bar{\psi}_\eta(x,t)=\int_0^t dt' d^4x' \left(\bar{\xi}_\xi(x',t')+g\gamma \frac{\delta S_{\rm int}}{\delta\psi_\xi(x',t')}\right)\bar{G}_{\xi\eta}(x-x',t-t'). \quad (2\cdot11)$$

Perturbative expansion can be performed by iteration in Eqs. (2·11) just as in bosonic case.

In the above discussion, we have shown that the correlation function $\langle\psi^0(x,t)\bar{\psi}^0(x',t')\rangle$ becomes the free propagator in the limit $t=t'\to\infty$ because of the damping factor. However, it must be noted that the damping factor does not necessarily assure the existence and the uniqueness of the equilibrium state. In order to prove the complete equivalence between path integral quantization and stochastic one, we must check whether the well-defined probability distribution, whose time evolution is governed by the Fokker-Planck equation, approaches the equilibrium state in the limit $t\to\infty$. This is the crucial point which we discuss in the next section.

§3. Probability distribution and its behaviour at t infinity in Fokker-Planck formalism

As we have already mentioned, there is no classical analogue of ψ and $\bar{\psi}$ in fermionic case, and we have to introduce Grassmann numbers to represent fermion fields. It seems difficult to construct a conventional c-number probability distribution which we must use to describe the time evolution of stochastic processes in the Fokker-Planck equation. However, only if we extend a probability distribution P to be a function of Grassmann numbers, we can find a possible way to obtain correlation functions consistently. So P itself does not necessarily have a physical meaning, but expectation values in P become meaningful.

Let us consider P and expectation values as follows. For simplicity, we take up the case of one degree of freedom ϕ and $\bar{\phi}$ instead of fields ψ and $\bar{\psi}$.

(I) $P(\phi,\bar{\phi},t)$ is a function of Grassmann numbers ϕ, $\bar{\phi}$ and the fictitious time t, and it can be expanded in powers of ϕ and $\bar{\phi}$:

$$P=C_0+C_1\phi+C_2\bar{\phi}+C_3\bar{\phi}\phi\,. \quad (3\cdot1)$$

(II) The expectation value at time t is defined by the following relation:

$$\langle f(\phi,\bar{\phi})\rangle\equiv\int f(\phi,\bar{\phi})P(\phi,\bar{\phi},t)d\phi d\bar{\phi}\,. \quad (3\cdot2)$$

For example, using Eq. (3·1),

$$\langle 1\rangle=\int P(\phi,\bar{\phi})d\phi d\bar{\phi}=-C_3=1\,,\quad \text{(normalization condition)}$$

$$\langle\phi\rangle=\int\phi P(\phi,\bar{\phi})d\phi d\bar{\phi}=-C_2\,,$$

$$\langle\bar{\phi}\rangle=\int\bar{\phi}P(\phi,\bar{\phi})d\phi d\bar{\phi}=C_1\,,$$

$$\langle\bar{\phi}\phi\rangle=\int\bar{\phi}\phi P(\phi,\bar{\phi})d\phi d\bar{\phi}=-C_0\,,\tag{3·3}$$

where integral over Grassmann variables is normalized as $\int\phi\bar{\phi}d\phi d\bar{\phi}=1$. (See Appendix B.) It must be assumed about the nature of the coefficients $\{C_i\}$ that C_0 and C_3 are commuting numbers while C_1 and C_2 anti-commuting ones, in order that the expectation values in fermionic case are properly defined in Eq. (3·2). Then P commutes with any Grassmann number. Extension to the case of field is straightforward.

Next we discuss the time evolution of P only based on the Langevin equation and statistical properties of random variables ξ and $\bar{\xi}$ assumed in the preceding section.[4] To this end, let us define Ξ_η and $\bar{\Xi}_\eta$ from Grassmann random variables ξ_η and $\bar{\xi}_\eta$ over infinitesimal regions Δt and Δx,

$$\Xi_\eta(x,t)=\int_{\Delta x,\Delta t}\xi_\eta(x,t)dxdt\simeq\xi_\eta(x,t)\Delta x\Delta t\,,$$

$$\bar{\Xi}_\eta(x,t)=\int_{\Delta x,\Delta t}\bar{\xi}_\eta(x,t)dxdt\simeq\bar{\xi}_\eta(x,t)\Delta x\Delta t\,,\tag{3·4}$$

whose correlations are given by means of Eq. (2·4):

$$\langle\Xi_\xi(x,t)\bar{\Xi}_\eta(x,t)\rangle=2\beta^{-1}\delta_{\xi\eta}\Delta x\Delta t\,,$$

$$\langle\Xi_\xi\Xi_\eta\rangle=\langle\bar{\Xi}_\xi\bar{\Xi}_\eta\rangle=0\,,\quad\text{etc}\,.\tag{3·5}$$

These correlations can be realized by the following Gaussian distribution of Grassmann variables:

$$\Psi_{\Xi\bar{\Xi}}(X,\bar{X})=\frac{\exp\left(-\frac{1}{2\beta^{-1}\Delta x\Delta t}\sum_b\bar{X}_bX_b\right)}{\int\exp\left(-\frac{1}{2\beta^{-1}\Delta x\Delta t}\sum_b\bar{X}_bX_b\right)\prod_b dX_bd\bar{X}_b}\tag{3·6}$$

Now the expression of the transition probability $T[\psi'',\bar{\psi}'',t+\Delta t|\psi',\bar{\psi}',t]$ at an infinitesimal interval $[t,t+\Delta t]$ can be derived from the Langevin equation and the distribution $\Psi_{\Xi\bar{\Xi}}$. That is, by substituting the following expressions for X_b and $\bar{X}_b$ into Eq. (3·6),

$$X_b\simeq\alpha(\psi''-\psi')\Delta x+\gamma\frac{\delta S}{\delta\bar{\psi}'}\Delta x\Delta t\,,$$

$$\bar{X}_b\simeq\alpha(\bar{\psi}''-\bar{\psi}')\Delta x-\gamma\frac{\delta S}{\delta\psi'}\Delta x\Delta t\,,$$

$$\prod_b dX_b d\bar{X}_b \propto \mathcal{D}\psi' \mathcal{D}\bar{\psi}' . \qquad (3\cdot7)^{*)}$$

T can be written as

$$T[\psi'', \bar{\psi}'', t+\Delta t|\psi', \bar{\psi}', t]$$

$$= N \exp\left(-\frac{1}{2\beta^{-1}\Delta x\Delta t}\sum_x\left(a(\bar{\psi}''-\bar{\psi}')\Delta x - \gamma\frac{\delta S}{\delta\psi'}\Delta x\Delta t\right)\right.$$

$$\left.\times\left(a(\psi''-\psi')\Delta x + \gamma\frac{\delta S}{\delta\bar{\psi}'}\Delta x\Delta t\right)\right), \qquad (3\cdot8)$$

where N is a normalization constant. This explicit expression leads to the fact that T satisfies the Kolmogoroff-Chapman equation in our Markoffian process with Gaussian random variables. Therefore, a probability distribution at t_f, $P[\psi_f, \bar{\psi}_f, t_f]$, is connected with one at t_i, $P[\psi_i, \bar{\psi}_i, t_i]$, by the transition probability at finite time interval, $T[\psi_f, \bar{\psi}_f, t_f|\psi_i, \bar{\psi}_i, t_i]$:

$$P[\psi_f, \bar{\psi}_f, t_f] = \int T[\psi_f, \bar{\psi}_f, t_f|\psi_i, \bar{\psi}_i, t_i]\mathcal{D}\psi_i\mathcal{D}\bar{\psi}_i P[\psi_i, \bar{\psi}_i, t_i], \qquad (3\cdot9)$$

$$T[\psi_f, \bar{\psi}_f, t_f|\psi_i, \bar{\psi}_i, t_i] = \int_{\substack{\psi(t_i)=\psi_i, \bar{\psi}(t_i)=\bar{\psi}_i \\ \psi(t_f)=\psi_f, \bar{\psi}(t_f)=\bar{\psi}_f}} \mathcal{D}\psi(t)\mathcal{D}\bar{\psi}(t)\exp\left(-\frac{1}{2\beta^{-1}}\int \Lambda dt\right), \qquad (3\cdot10)$$

where

$$\Lambda[\psi, \bar{\psi}] = \int d^4x\left(a\frac{\partial}{\partial t}\bar{\psi}(x,t) - \gamma\frac{\delta S}{\delta\psi(x,t)}\right)\left(a\frac{\partial}{\partial t}\psi(x,t) + \gamma\frac{\delta S}{\delta\bar{\psi}(x,t)}\right). \qquad (3\cdot11)$$

Since T is expressed in the path integral form, we can easily write down the equation of motion for P, namely, the Fokker-Planck equation,

$$2\beta^{-1}\frac{\partial}{\partial t}P = \mathcal{H}P , \qquad (3\cdot12)$$

just like the Schrödinger equation in quantum mechanics. Here $\mathcal{H}$ is the "Hamiltonian" operator which could be derived from the "Lagrangian" Λ by complete analogy with the usual method of the Legendre transformation, where "momenta" are replaced with differential operators. However, there remain some uncertainties about the ordering of the operators in $\mathcal{H}$; one coming from taking continuous limit of path integral like in bosonic case[5] and, in addition to this, another due to the anti-commuting nature of fermion fields.

First we discuss the latter uncertainty, repeating to derive the Fokker-Planck equation by use of the exlicit form of P. Let us take the total time derivative of P along the classical path:

---

*) This expression holds only in an integral.

$$\frac{dP}{dt}=\lim_{\Delta t\to 0}\frac{P[\psi+\dot{\psi}\Delta t,\ \bar{\psi}+\dot{\bar{\psi}}\Delta t,\ t+\Delta t]-P[\psi,\ \bar{\psi},\ t]}{\Delta t}$$

$$=\frac{\partial P}{\partial t}+\int\dot{\bar{\psi}}_\eta(x)\frac{\delta P}{\delta\bar{\psi}_\eta(x)}d^4x+\int\dot{\psi}_\eta(x)\frac{\delta P}{\delta\psi_\eta(x)}d^4x\,. \qquad (3\cdot 13)^{*)}$$

The terms $\delta P/\delta\psi$, $\delta P/\delta\bar{\psi}$ and dP/dt are written as

$$\frac{\delta P}{\delta\psi_\eta(x)}=-\frac{1}{2\beta^{-1}}\pi_\eta(x)P\,,$$

$$\frac{\delta P}{\delta\bar{\psi}_\eta(x)}=-\frac{1}{2\beta^{-1}}\bar{\pi}_\eta(x)P\,, \qquad (3\cdot 14\text{a})$$

$$\frac{dP}{dt}=-\frac{1}{2\beta^{-1}}\Lambda P\,, \qquad (3\cdot 14\text{b})$$

where

$$\pi_\eta(x)=\frac{\delta\Lambda}{\delta\dot{\psi}_\eta(x)}=-\alpha\left(a\dot{\bar{\psi}}_\eta(x)-\gamma\frac{\delta S}{\delta\psi_\eta(x)}\right),$$

$$\bar{\pi}_\eta(x)=\frac{\delta\Lambda}{\delta\dot{\bar{\psi}}_\eta(x)}=\alpha\left(a\dot{\psi}_\eta(x)+\gamma\frac{\delta S}{\delta\bar{\psi}_\eta(x)}\right). \qquad (3\cdot 15)$$

Notice that Eqs. (3·14a, b) mean that momenta π and $\bar{\pi}$ conjugate to ψ and $\bar{\psi}$ can be represented by functional derivatives,

$$\pi_\eta(x)=-2\beta^{-1}\frac{\delta}{\delta\psi_\eta(x)}\,,$$

$$\bar{\pi}_\eta(x)=-2\beta^{-1}\frac{\delta}{\delta\bar{\psi}_\eta(x)}\,. \qquad (3\cdot 16)$$

Substituting Eqs. (3·14) into Eq. (3·13), we obtain

$$\mathcal{H}=\int d^4x\{\dot{\psi}_\eta(x)\pi_\eta(x)+\dot{\bar{\psi}}_\eta(x)\bar{\pi}_\eta(x)\}-\Lambda[\psi,\ \bar{\psi},\ \dot{\psi},\ \dot{\bar{\psi}}]. \qquad (3\cdot 17)$$

The orderings of operators in the above curly parentheses come from a convention of "left-derivative".

We return to the uncertainty coming from taking a continuous limit of path integral. The orderings of operators must be so arranged to ensure the conservation of probability $(d/dt)\int P\mathcal{D}\psi\mathcal{D}\bar{\psi}=0$; differential operators π and $\bar{\pi}$ should be placed on the left of other operators of ψ and $\bar{\psi}$ because, then, by use of Eq. (3·12) one can easily show

*) See Eq. (B·2) in Appendix B.

$$\frac{d}{dt}\int P\mathscr{D}\psi\mathscr{D}\bar{\psi}=\int\left(\frac{\delta}{\delta\psi}(\cdots P)+\frac{\delta}{\delta\bar{\psi}}(\cdots P)\right)\mathscr{D}\psi\mathscr{D}\bar{\psi}=0\,.$$

(See Eq. (B·3) in Appendix B.) This determines the explicit form of the "Hamiltonian" uniquely,

$$\begin{aligned}\mathscr{F}&=\int d^4x(-\pi_\eta(x)\dot{\psi}_\eta(x)-\bar{\pi}_\eta(x)\dot{\bar{\psi}}_\eta(x))-\Lambda[\psi,\bar{\psi},\dot{\psi},\dot{\bar{\psi}}]\\&=\int d^4x\left(\frac{1}{\alpha^2}\bar{\pi}_\eta(x)\pi_\eta(x)+\frac{\gamma}{\alpha}\left[\pi_\eta(x)\frac{\delta S}{\delta\bar{\psi}_\eta(x)}-\bar{\pi}_\eta(x)\frac{\delta S}{\delta\psi_\eta(x)}\right]\right)\\&=-\frac{1}{\alpha^2\beta^2}\int d^4x\left(\frac{\delta^2}{\delta\psi_\eta(x)\delta\bar{\psi}_\eta(x)}\right.\\&\qquad\left.+\frac{\chi}{2}\left(\frac{\delta}{\delta\psi_\eta(x)}\left(\frac{\delta S}{\delta\bar{\psi}_\eta(x)}\right)-\frac{\delta}{\delta\bar{\psi}_\eta(x)}\left(\frac{\delta S}{\delta\psi_\eta(x)}\right)\right)\right),\end{aligned}\tag{3·18}$$

where on the first line of the above expression, the anti-commuting character has been used.

Next, in the Fokker-Planck equation (3·12), we investigate whether a unique equilibrium distribution in the limit $t\to\infty$ exists or not and what the equilibrium is, if it exists. For this purpose, it is sufficient to solve the eigenvalue problem of $\mathscr{F}$,

$$\mathscr{F}u_k=-\lambda_k u_k\,.\tag{3·19}$$ *)

Expanding the probability distribution P in terms of the eigenfunctions u_k's, we can show the existence and the uniqueness of the equilibrium distribution only when the lowest eigenvalue λ_0 is equal to zero, nondegenerate and discrete.

First we take up the free case and then discuss the interacting case.

3.1. *Free case*

Fortunately, we can find a similarity transformation by which the "Hamiltonian" is diagonalized. In what follows, let us show how to solve this problem explicitly.

The free "Hamiltonian" $\mathscr{F}_0$ is given by

$$\mathscr{F}_0=\sum_n f_{0,n}\,,$$

$$f_{0,n}=-\frac{4}{\alpha^2\beta^2}\left(\frac{\partial^2}{\partial\psi_{n,\eta}\partial\bar{\psi}_{n,\eta}}+\frac{\chi}{2}(M_n)_{\zeta\eta}\left(\frac{\partial}{\partial\psi_{n,\zeta}}\psi_{n,\eta}+\frac{\partial}{\partial\bar{\psi}_{n,\eta}}\bar{\psi}_{n,\zeta}\right)\right),$$

*) We label λ_k such that $\mathrm{Re}\,\lambda_0\leqq\mathrm{Re}\,\lambda_1\leqq\cdots$. Generally speaking, since we cannot directly use a similar transformation analogous to bosonic case, eigenvalue problem cannot be set up in fermion theory except for in free case.

$$(M_n)_{\xi\eta}=(\not{p}_n+m)_{\xi\eta}\,. \tag{3·20}$$

Here we have used the box normalization,

$$\psi_\eta(x)=\frac{1}{\sqrt{V}}\sum_n \psi_{n,\eta}\exp(ip_n x),$$

$$\bar{\psi}_\eta(x)=\frac{1}{\sqrt{V}}\sum_n \bar{\psi}_{n,\eta}\exp(-ip_n x), \tag{3·21}$$

where V is a four-dimensional Euclidean volume and subscripts n represent discretized four momentum p. It is convenient to perform the following unitary transformation:

$$\frac{\partial}{\partial\psi'_{n,\eta}}=(U_n^{-1})_{\sigma\eta}\frac{\partial}{\partial\psi_{n,\sigma}}, \qquad \psi'_{n,\xi}=(U_n)_{\xi\rho}\psi_{n,\rho},$$

$$\frac{\partial}{\partial\bar{\psi}'_{n,\xi}}=(U_n)_{\xi\rho}\frac{\partial}{\partial\bar{\psi}_{n,\rho}}, \qquad \bar{\psi}'_{n,\eta}=(U_n^{-1})_{\sigma\eta}\bar{\psi}_{n,\sigma}. \tag{3·22}$$

Here U_n, by which M_n can be diagonalized, has been defined in Eq. (2·8). After this transformation, $f_{0,n}$ takes a diagonalized form with respect to spin indices,

$$\left.\begin{aligned} f_{0,n}&=-\frac{4}{\alpha^2\beta^2}\left(\frac{\partial^2}{\partial\psi'_{n,\eta}\partial\bar{\psi}'_{n,\eta}}-\frac{x}{2}(\mathcal{M}_n)_{\xi\eta}\left(\psi'_{n,\xi}\frac{\partial}{\partial\psi'_{n,\eta}}+\bar{\psi}'_{n,\eta}\frac{\partial}{\partial\bar{\psi}'_{n,\xi}}\right)+4xm\right),\\ (\mathcal{M}_n)_{\xi\eta}&=\begin{pmatrix}\omega_{n,1} & 0\\ 0 & \omega_{n,2}\end{pmatrix}_{\xi\eta}, \quad \omega_{n,1}=i\sqrt{p_n{}^2}+m\,, \quad \omega_{n,2}=-i\sqrt{p_n{}^2}+m\,. \end{aligned}\right\} \tag{3·23}$$

Now if we solve the simplified eigenvalue problem,

$$f_{0,n}u^0_{n,k}=-\lambda_{n,k}u^0_{n,k}\,, \tag{3·24}$$

the probability distribution P is given by

$$P_0=\prod_n P_{0,n}\,,$$

$$P_{0,n}=\sum_k C_{n,k}\exp\left(-\frac{1}{2\beta^{-1}}\lambda_{n,k}t\right)u^0_{n,k}\,. \tag{3·25}$$

This eigenvalue equation (3·24) can be diagonalized by the straightforward extension of the simplified model discussed in Appendix C to our spinor case as follows. We introduce the following abstract representations:

$$\psi'_{n,\xi}\to \mathbf{a}^\dagger_{n,\xi}\,, \qquad \frac{\partial}{\partial\psi'_{m,\eta}}\to \mathbf{a}_{m,\eta}\,,$$

$$\bar{\psi}'_{n,\xi}\to \mathbf{b}^\dagger_{n,\xi}\,, \qquad \frac{\partial}{\partial\bar{\psi}'_{m,\eta}}\to \mathbf{b}_{m,\eta}\,,$$

$$\{\mathbf{a}_{n,\zeta}, \mathbf{a}^{\dagger}_{m,\eta}\} = \delta_{nm}\delta_{\zeta\eta}, \quad \{\mathbf{b}_{n,\zeta}, \mathbf{b}^{\dagger}_{m,\eta}\} = \delta_{nm}\delta_{\zeta\eta},$$

$$\mathbf{a}_{n,\zeta}|0\rangle = \mathbf{b}_{n,\zeta}|0\rangle = 0. \tag{3·26}$$

(See Appendix C.) By a similarity transformation $\mathbf{O}_n = \exp(\mathbf{a}_{n,\zeta}\mathbf{b}_{n,\eta}(1/\chi(\mathcal{M}_n^{-1})_{\zeta\eta}))$, $\mathbf{f}_{0,n}$ is transformed into a diagonalized one,

$$\begin{aligned}\tilde{\mathbf{f}}_{0,n} &= \mathbf{O}_n \mathbf{f}_{0,n} \mathbf{O}_n^{-1} \\ &= -\frac{4}{\alpha^2\beta^2}\left(-\frac{\chi}{2}(\mathcal{M}_n)_{\zeta\eta}(\mathbf{a}^{\dagger}_{n,\zeta}\mathbf{a}_{n,\eta} + \mathbf{b}^{\dagger}_{n,\zeta}\mathbf{b}_{n,\eta}) + 4\chi m\right).\end{aligned} \tag{3·27}$$

Then the whole set of eigenvalues $\lambda_{n,k}$ and eigenstates $|\tilde{u}_{n,k}\rangle$ are as follows:

$$\lambda_{n,0} = 0, \qquad |\tilde{u}_{n,0}\rangle = \prod_{\zeta=1}^{4} \mathbf{a}^{\dagger}_{n,\zeta} \prod_{\eta=1}^{4} \mathbf{b}^{\dagger}_{n,\eta}|0\rangle,$$

$$\begin{aligned}\lambda_{n,1\sim4} &= \frac{4}{\alpha^2\beta^2}\left(4\chi m - \frac{3}{2}\chi\omega_{n,1} - \frac{4}{2}\chi\omega_{n,2}\right) \\ &= \frac{4}{\alpha^2\beta^2}\frac{1}{2}\chi\omega_{n,1}, \qquad |\tilde{u}_{n,1}\rangle = \prod_{\zeta\neq1}^{4} \mathbf{a}^{\dagger}_{n,\zeta} \prod_{\eta=1}^{4} \mathbf{b}^{\dagger}_{n,\eta}|0\rangle,\end{aligned}$$

$$|\tilde{u}_{n,2}\rangle = \prod_{\zeta\neq2}^{4} \mathbf{a}^{\dagger}_{n,\zeta} \prod_{\eta=1}^{4} \mathbf{b}^{\dagger}_{n,\eta}|0\rangle,$$

$$|\tilde{u}_{n,3}\rangle = \prod_{\zeta=1}^{4} \mathbf{a}^{\dagger}_{n,\zeta} \prod_{\eta\neq1}^{4} \mathbf{b}^{\dagger}_{n,\eta}|0\rangle,$$

$$|\tilde{u}_{n,4}\rangle = \prod_{\zeta=1}^{4} \mathbf{a}^{\dagger}_{n,\zeta} \prod_{\eta\neq2}^{4} \mathbf{b}^{\dagger}_{n,\eta}|0\rangle,$$

$$\vdots$$

$$\lambda_{n,2^8-1} = \frac{4}{\alpha^2\beta^2}(4\chi m), \qquad |\tilde{u}_{n,2^8-1}\rangle = |0\rangle, \tag{3·28}$$

where $\omega_{n,i}$ ($i=1, 2$) is given in Eqs. (3·23). This tells us:

(1) The existence of the unique equilibrium is assured because the lowest eigenvalue $\lambda_{n,0}$ is zero, nondegenerate and discrete.

(2) The state which $|P_{0,n}\rangle_t$ reaches as t goes to infinity is obtained from the explicit form of the ground state $|\tilde{u}_{n,0}\rangle$ and by the inverse transformation $\mathbf{O}_n^{-1}$:

$$|P_{0,n}\rangle_t \xrightarrow[t\to\infty]{} \frac{1}{\chi^4}\det(\mathcal{M}_n^{-1})\exp(\chi\mathbf{a}^{\dagger}_{n,\zeta}(\mathcal{M}_n)_{\zeta\eta}\mathbf{b}^{\dagger}_{n,\eta})|0\rangle. \tag{3·29}$$

Coming back from the abstract representation, Eq. (3·29) means that

$$P_{0,n}(\psi_n', \bar{\psi}_n', t) \xrightarrow[t\to\infty]{} \frac{1}{\chi^4}\det(\mathcal{M}_n^{-1})\exp(-\chi\bar{\psi}'_{n,\eta}(\mathcal{M}_n)_{\eta\zeta}\psi'_{n,\zeta}). \tag{3·30}$$

Therefore, recalling the definitions (3·21), (3·22) and (3·25), the total distribution $P_0[\psi, \bar{\psi}, t]$ approaches the state apart from the irrelevant normalization factor,

$$P_0[\psi, \bar{\psi}, t] \xrightarrow[t\to\infty]{} \exp(-xS_0), \tag{3·31}$$

where S_0 is the free action given in Eq. (2·2). The expectation value $\langle F[\psi, \bar{\psi}]\rangle_t$,

$$\langle F[\psi, \bar{\psi}]\rangle_t = \int F[\psi, \bar{\psi}] P_0[\psi, \bar{\psi}, t] \mathcal{D}\psi \mathcal{D}\bar{\psi}$$

becomes

$$\lim_{t\to\infty}\langle F[\psi, \bar{\psi}]\rangle_t = \frac{\int F[\psi, \bar{\psi}]\exp(-xS_0[\psi, \bar{\psi}])\mathcal{D}\psi\mathcal{D}\bar{\psi}}{\int \exp(-xS_0[\psi, \bar{\psi}])\mathcal{D}\psi\mathcal{D}\bar{\psi}}. \tag{3·32}$$

Thus we have proved that our stochastic quantization for free Dirac field is equivalent to that of path integral, only if the parameter x is set equal to unity as it should be in Eq. (2·10).

3.2. *Interacting case*

When we discuss the interacting case in the Fokker-Planck formalism, we must solve the "generalized eigenvalue problem",

$$(\mathcal{F} + \lambda_k)^\nu u_{k,\nu} = 0\,, \quad (\nu \geq 1, \text{ integer}) \tag{3·33}$$

to expand the probability distribution P as in Eq. (3·25) and to investigate its behaviour at t-infinity limit. In general, it seems difficult to solve the above equation completely just as in the interacting bosonic case. However we can show that $P_{st} = \exp(-xS)$ is a stationary solution,

$$\mathcal{F} P_{st} = 0\,, \tag{3·34}$$

which means that P_{st} is an eigenfunction belonging to $\lambda = 0$ with $\nu = 1$ in Eq. (3·33). As we have already mentioned, the crucial point for the existence and the uniqueness of the equilibrium state is that the lowest eigenvalue λ_0 is equal to zero, nondegenerate and discrete. Along with a similar line of thought in free case, the discreteness seems to be guaranteed from the finite degrees of freedom in fermi statistics. So far we have not succeeded in proving that λ_0 is zero, and nondegenerate. If this holds, P_{st} becomes the unique equilibrium distribution which gives the corresponding correlation functions in Euclidean field theory. In fact, we have verified that the explicit perturbative calculations of the correlation functions in such case as $(\bar{\psi}\psi)^2$ become identical with those in Euclidean field theory as t goes to infinity.

§ 4. Concluding remarks

In this paper, we formulate the stochastic quantization method of fermion fields; the probability distribution P and the expectation value in P can be defined properly in spite of using Grassmann random variables. Then we have proved the equivalence between our stochastic quantization and path integral in free case completely. The general treatment of interacting case has also been discussed.

As for a massless fermion where the damping factor like $\exp(-(\gamma/a)mt)$ does not exist, we should pay special attention to it. However, it is needless to say that there is a simple way to define it as a massless limit of massive theory.

Our formalism can be extended to the coupled system of bosons and fermions without any difficulty. Our method will be able to be applied to the fermion on a lattice and may be useful for the purpose to simplify its numerical calculations.

Finally, we will comment on the operator formalism for stochastic quantization which has been invented to reduce the hard work to calculate the correlation functions in the Langevin formalism for boson fields.[6] The operator formalism for fermion field as well is in preparation.

Acknowledgements

The authors would like to express their sincere thanks to Professor M. Namiki for his continuous encouragements.

Appendix A

In this Appendix, we fix the notation of Euclidean field theories used in the present paper. The action of Dirac fermion system is given in Minkowski space as follows:

$$S=\int d^4x\mathcal{L}(x),$$

$$\mathcal{L}(x)=\bar{\psi}(x)(i\nabla\!\!\!\!/-m)\psi(x)+g\mathcal{L}_{\rm int}(x), \qquad (\text{A}\cdot 1)$$

where we have used Bjorken-Drell notations. The Euclidean action S_E is defined from the above S by performing Wick rotation, $x_0\to -ix_4$:

$$S_E=\int d^4x_E[-\mathcal{L}(x_0=-ix_4, \boldsymbol{x})], \quad d^4x_E=d^3\boldsymbol{x}dx_4\,. \qquad (\text{A}\cdot 2)$$

Then we have the following expression for S_E with the definition $\gamma_4=i\gamma_0$,

$$S_E=S_{E,0}-gS_{E,\rm int}\,,$$

$$S_{E,0}=-\int d^4x_E\bar{\psi}(x_E)(i\partial\!\!\!/_E-m)\psi(x_E),$$

$$S_{E,\text{int}}=-\int d^4x_E\mathcal{L}_{\text{int}}(x_E), \qquad (\text{A}\cdot3)$$

where $\partial\!\!\!/_E=\gamma_4\partial_4+\boldsymbol{\gamma}\cdot\boldsymbol{\nabla}$ and $\{\gamma_\mu,\gamma_\nu\}=-2\delta_{\mu\nu}$. Throughout the text, we simply omit the index E from the coordinates and the action.

Appendix B

In this Appendix, we summarize the properties of Grassmann numbers ψ_i and $\bar{\psi}_i$. We follow the convention given in Ref. 7). We define the integration over Grassmann numbers as follows:

$$\int d\psi_i=0\,,\quad \int\psi_i d\psi_i=i\,,\quad \int d\bar{\psi}_i=0 \quad\text{and}\quad \int\bar{\psi}_i d\bar{\psi}_i=i\,. \qquad (\text{B}\cdot1)$$

Let us follow the convention of the "left-derivatives" with respect to both ψ_i and $\bar{\psi}_i$, for example,

$$\frac{\partial}{\partial\psi_2}(\psi_1\psi_2\bar{\psi}_1\bar{\psi}_2)=-\frac{\partial}{\partial\psi_2}(\psi_2\psi_1\bar{\psi}_1\bar{\psi}_2)=-\psi_1\bar{\psi}_1\bar{\psi}_2\,,$$

$$\frac{\partial}{\partial\bar{\psi}_2}(\psi_1\psi_2\bar{\psi}_1\bar{\psi}_2)=-\frac{\partial}{\partial\bar{\psi}_2}(\bar{\psi}_2\psi_1\psi_2\bar{\psi}_1)=-\psi_1\psi_2\bar{\psi}_1\,.$$

Likewise, we can define the "functional left-derivative" of a functional $\psi(x)$ and $\bar{\psi}(x)$, $F[\psi,\bar{\psi}]$ as follows:

$$F[\psi+\varepsilon\chi,\bar{\psi}+\bar{\varepsilon}\bar{\chi}]-F[\psi,\bar{\psi}]\equiv\varepsilon\int\chi\frac{\delta F}{\delta\psi}dx+\bar{\varepsilon}\int\bar{\chi}\frac{\delta F}{\delta\bar{\psi}}dx\,, \qquad (\text{B}\cdot2)$$

where ε and $\bar{\varepsilon}$ are infinitesimal c-number and $\chi(x)$ and $\bar{\chi}(x)$ are arbitrary functions which are Grassmann numbers themselves. Notice that the above orderings of χ and $\delta F/\delta\psi$, and $\bar{\chi}$ and $\delta F/\delta\bar{\psi}$ are consistent with the convention of "left-derivative". We will show a simple example $F[\psi,\bar{\psi}]=\int dxA(x)\psi(x)\bar{\psi}(x)$ where $A(x)$ is a c-number function:

$$F[\psi+\varepsilon\chi,\bar{\psi}+\bar{\varepsilon}\bar{\chi}]-F[\psi,\bar{\psi}]$$
$$=\varepsilon\int dx\chi(x)A(x)\bar{\psi}(x)+(-1)\bar{\varepsilon}\int dx\bar{\chi}(x)A(x)\psi(x),$$

while in the convention of "left-derivative",

$$\frac{\delta F}{\delta\psi(x)}=A(x)\bar{\psi}(x) \quad\text{and}\quad \frac{\delta F}{\delta\bar{\psi}(x)}=-A(x)\psi(x).$$

Finally, we have important formulas for an arbitrary function of ψ_i and $\bar{\psi}_i$, $f(\psi, \bar{\psi})$,

$$\int \frac{\partial}{\partial \psi_i} f(\psi, \bar{\psi}) d\psi_i = \int \frac{\partial}{\partial \bar{\psi}_i} f(\psi, \psi) d\bar{\psi}_i = 0\,, \quad \text{(for fixed } i\text{)} \tag{B·3}$$

because the integration can be replaced with the derivation and the second derivative of $f(\psi, \bar{\psi})$ with respect to the same Grassmann number is equal to zero.

Appendix C

In order to solve the eigenvalue equation (3·24), let us consider a simplified model with ϕ and $\bar{\phi}$ (one degree of freedom),

$$\frac{\partial}{\partial t} P(\phi, \bar{\phi}, t) = hP(\phi, \bar{\phi}, t),$$

$$h = -\frac{\partial^2}{\partial\phi\partial\bar{\phi}} - \frac{1}{2}\left(\frac{\partial}{\partial\phi}\phi + \frac{\partial}{\partial\bar{\phi}}\bar{\phi}\right), \tag{C·1}$$

where P is expanded as in Eq. (3·1). Instead of dealing with the above differential equation of Grassmann numbers, it is convenient to convert this into an algebraic one by introducing abstract representations. To get this representation, we replace ϕ, $\bar{\phi}$, $\partial/\partial\phi$ and $\partial/\partial\bar{\phi}$ with "creation" and "annihilation" operators,

$$\phi \to \mathbf{a}^\dagger\,, \quad \frac{\partial}{\partial\phi} \to \mathbf{a}\,, \quad \bar{\phi} \to \mathbf{b}^\dagger\,, \quad \frac{\partial}{\partial\bar{\phi}} \to \mathbf{b}\,, \tag{C·2}$$

$$\{\mathbf{a}, \mathbf{a}^\dagger\} = 1\,, \qquad \{\mathbf{b}, \mathbf{b}^\dagger\} = 1\,, \tag{C·3}$$

and define the "vacua" $|0\rangle$ and $\langle 0|$,

$$\mathbf{a}|0\rangle = \mathbf{b}|0\rangle = 0\,, \qquad \langle 0|\mathbf{a}^\dagger = \langle 0|\mathbf{b}^\dagger = 0\,. \tag{C·4}$$

Then any distribution P can be represented by the "ket" $|P\rangle$:

$$|P\rangle = (C_0 + C_1\mathbf{a}^\dagger + C_2\mathbf{b}^\dagger + C_3\mathbf{b}^\dagger\mathbf{a}^\dagger)|0\rangle\,, \tag{C·5}$$

and the Fokker-Planck equation becomes in the operator form,

$$\frac{d}{dt}|P\rangle = \mathbf{h}|P\rangle\,,$$

$$\mathbf{h} = -\mathbf{ab} + \frac{1}{2}(\mathbf{a}^\dagger\mathbf{a} + \mathbf{b}^\dagger\mathbf{b}) - 1\,. \tag{C·6}$$

Now we will diagonalize the eigenvalue equation,

$$\mathbf{h}|u_k\rangle = -\lambda_k|u_k\rangle\,, \tag{C·7}$$

by the similarity transformation $\mathbf{O}=\exp(\mathbf{ab})$. With this, operator $\mathbf{Q}$ and the state $|P\rangle$ are transformed into

$$\tilde{\mathbf{Q}}=\mathbf{OQO}^{-1}\,, \qquad |\tilde{P}\rangle=\mathbf{O}|P\rangle\,. \tag{C·8}$$

Transformed operators are

$$\tilde{\mathbf{a}}=\mathbf{a}\,,\quad \tilde{\mathbf{b}}=\mathbf{b}\,,\quad \tilde{\mathbf{a}}^\dagger=\mathbf{a}^\dagger-\mathbf{b}\,,\quad \tilde{\mathbf{b}}^\dagger=\mathbf{b}^\dagger+\mathbf{a}\,, \tag{C·9}$$

$$\tilde{\mathbf{h}}=\frac{1}{2}(\mathbf{a}^\dagger\mathbf{a}+\mathbf{b}^\dagger\mathbf{b})-1\,. \tag{C·10}$$

It is clear that the whole set of eigenvalues and eigenvectors for $\mathbf{h}$ is as follows:

$$\begin{aligned} &\lambda_0=0\,, && |\tilde{u}_0\rangle=\mathbf{b}^\dagger\mathbf{a}^\dagger|0\rangle\,,\\ &\lambda_1=\frac{1}{2}\,, && |\tilde{u}_1\rangle=\mathbf{a}^\dagger|0\rangle\,,\\ &\lambda_2=\frac{1}{2}\,, && |\tilde{u}_2\rangle=\mathbf{b}^\dagger|0\rangle\,,\\ &\lambda_3=1\,, && |\tilde{u}_3\rangle=|0\rangle\,. \end{aligned} \tag{C·11}$$

Therefore, under the arbitrary initial condition,

$$\begin{aligned} &|\tilde{P}\rangle_{t=0}=(d_0+d_1\mathbf{a}^\dagger+d_2\mathbf{b}^\dagger+d_3\mathbf{b}^\dagger\mathbf{a}^\dagger)|0\rangle\,,\\ &d_3=-1 \text{ (normalization condition)}, \end{aligned} \tag{C·12}$$

the general solution of the Fokker-Planck equation is expressed by

$$|\tilde{P}\rangle_t=(e^{-t}d_0+e^{-(1/2)t}d_1\mathbf{a}^\dagger+e^{-(1/2)t}d_2\mathbf{b}^\dagger+d_3\mathbf{b}^\dagger\mathbf{a}^\dagger)|0\rangle\,, \tag{C·13}$$

and approaches a definite state $d_3\mathbf{b}^\dagger\mathbf{a}^\dagger|0\rangle$ as t goes to infinity,

$$|\tilde{P}\rangle_t \xrightarrow[t\to\infty]{} d_3\mathbf{b}^\dagger\mathbf{a}^\dagger|0\rangle\,. \tag{C·14}$$

The original state $|P\rangle_t$ also reaches a unique state,

$$|P\rangle_t \xrightarrow[t\to\infty]{} e^{-\mathbf{b}^\dagger\mathbf{a}^\dagger}|0\rangle\,, \tag{C·15}$$

where the relation $d_3=-1$ is used.

Thus we have proved that starting from an arbitrary condition, the distribution approaches the equilibrium one $\exp(-\bar{\phi}\phi)$ in the t-infinity limit. The crucial point is that the lowest eigenvalue λ_0 for this "Hamiltonian" h is equal to zero, nondegenerate and discrete.

References

1) G. Parisi and Wu Yongshi, Scientia Sinica **24** (1981), 483.
2) J. R. Klauder, Ann. of Phys. **11** (1960), 123.
3) Y. Kakudo, Y. Taguchi, A. Tanaka and K. Yamamoto, Osaka Univ. Preprint OS-GE-82-39 (1982).
4) N. Saito and M. Namiki, Prog. Theor. Phys. **16** (1956), 71.
5) F. Langouche, D. Rockaerts and E. Tirapebui, Nuovo Cim. **53B** (1979), 135; also see the references cited therein.
6) M. Namiki and Y. Yamanaka, Prog. Theor. Phys. **69** (1983), No. 6.
7) Y. Ohnuki, KEK Lecture Note, KEK-77-11 (1977).
Y. Ohnuki and T. Kashiwa, Prog. Theor. Phys. **60** (1978), 548.

Nuclear Physics B235[FS11] (1984) 75–92

STOCHASTIC QUANTIZATION WITH FERMIONS

P.H. DAMGAARD

NORDITA, Blegdamsvej 17, DK-2100, Copenhagen Ø, Denmark

K. TSOKOS

Physics Department, University of Maryland, College Park, MD 20742, USA

Received 27 June 1983
(Revised 14 September 1983)

The method of stochastic quantization of euclidean quantum field theory is extended to the case of theories with fermion fields. A modified Langevin equation for spinors is introduced, and the corresponding master equation for the stochastic probability distribution is derived. The formal solution of this generalized Fokker-Planck equation is constructed, and the convergence towards a thermodynamic equilibrium is demonstrated. We illustrate the method in a perturbative setting by one-loop calculations in a $(\bar{\psi}\psi)^2$ theory and in gauge theories.

1. Introduction

The method of stochastic quantization introduced by Parisi and Wu [1] provides an intriguing alternative to the standard methods in euclidean quantum field theory. Originally introduced mainly to circumvent the introduction of gauge fixing terms in non-abelian gauge theories (and hence provide a quantization prescription free of Gribov ambiguities [2] and, in a perturbative context, free of ghost fields), it has by now been recognized to have applications far beyond its original scope. The areas in which stochastic quantization ideas now play a prominent role range from applications in Monte-Carlo simulations of lattice gauge theories [3] and constructions of the quenched momentum prescription in $N = \infty$ master field [4] to non-perturbative numerical methods for solving ordinary quantum field theories [5]. The method is quite likely to find further applications in the future.

The main purpose of the present paper is to show that the method of stochastic quantization can be generalized to include the quantization of fermion fields as well. That this is a non-trivial issue will become apparent when we set up the appropriate formalism; convergence towards equilibrium will only be guaranteed if one uses a carefully chosen generalization of the Langevin equation. The root of this problem lies in the fact that there exists no classical analogue of fermionic fields. In the operator language this manifests itself in the appearance of operators that are not positive definite. Not surprisingly, the solution to this problem lies in properly choosing a bosonized version of the Langevin equation. This will be discussed in detail in sect. 2 of this paper.

Stochastic quantization in the presence of fermion fields has previously been considered from two different points of view. The first is the so-called pseudofermion method approach to lattice gauge theories by Fucito, Marinari, Parisi and Rebbi [3]. This approach, which is applicable only in theories where the fermion fields appear as bilinears, consists in the formal rewriting of the functional integral containing Grassmann variables as a functional integral over bosonic fields. For a gauge theory it is based on the identity

$$\det[\not{D} + m] = \det[\gamma_5(\not{D} + m)\gamma_5] = \det[-\not{D} + m], \tag{1.1}$$

which leads to

$$\begin{aligned} \det[\not{D} + m] &= \{\det[\not{D} + m]\det[-\not{D} + m]\}^{1/2} \\ &= \{\det[-\not{D}^2 + m^2]\}^{1/2}, \end{aligned} \tag{1.2}$$

so that one can replace the operator $(\not{D} + m)$ by the positive definite operator $(-\not{D}^2 + m^2)$. In the upgrading of Monte-Carlo sweeps in lattice gauge theories one encounters ratios of the determinants above. The idea of the authors of ref. [3] is to calculate such determinants by a Monte-Carlo routine on a corresponding bosonic functional. However, such a method does not produce a viable procedure in the continuum; it is simply a way of removing anticommuting variables in Monte-Carlo routines.

The second approach, which is much closer to the method we shall outline in this paper, has recently been suggested by Fukai, Nakazato, Ohba, Okano and Yamanaka [6]. These authors base their stochastic quantization on Green functions that are not diagonal in their spinor indices. Upon diagonalization the Green functions turn out to be convergent only in the case of *massive* fermions. This lack of convergence can, of course be traced directly back to lack of positivity of the operator $(\not{D} + m)$ as discussed above. With their method one can therefore only quantize massless fermions by the introduction of fictitious fermion mass terms during the quantization itself. This is clearly an unwanted situation, particularly if one is interested in the chiral properties of the theory.

As we shall see, it is possible to solve this problem of chiral invariance by a certain choice of the generalized Langevin equation. With this particular choice, convergence of the Green functions will always be assured.

The outline of the paper is as follows. In sect. 2 we very briefly review the formalism of stochastic quantization of spin-0 fields, and we then attempt to generalize this scheme in the most straightforward way to spin-$\frac{1}{2}$ fields. The resulting Langevin equations turn out to be those discussed by the authors of ref. [6], and hence are beset by the same problems. Instead, a set of generalized Langevin equations is introduced, which leads to well-behaved and convergent Green functions. These Green functions are diagonal in spinor indices. The corresponding generalized Fokker-Planck equation is derived in sect. 3, and we show that in perturbation theory, the equilibrium configurations obtained agree with the standard vacuum expectation values of the underlying quantum field theory. In sect. 4 we demonstrate explicitly how this is achieved in a perturbative context by calculating, up to one-loop order, the connected 2- and 4-point functions of a $(\bar{\psi}\psi)^2$-type theory in 2 space-time dimensions. The coupling to gauge fields is discussed in sect. 5 and illustrated in perturbation theory by a calculation of the vacuum polarization in QED to one-loop order. Finally, sect. 6 contains a brief summary and conclusions.

2. Setting up the Langevin equation

Let us first briefly review the idea of stochastic quantization for scalar fields, as proposed by Parisi and Wu [1]. Consider the euclidean action $S[\phi]$ of some spin-0 field $\phi(x)$ in D-dimensional spacetime. Now imagine this system embedded in a $(D+1)$-dimensional heat reservoir, the extra dimension being supplied by a fictitious, time direction t*. The field ϕ can now take values in this $(D+1)$-dimensional space: $\phi=\phi(x,t)$. As $t\to\infty$ a thermodynamic equilibrium will be established. The path to this equilibrium can be chosen to follow the path dictated by the Langevin equation

$$\frac{\partial}{\partial t}\phi(x,t)=-\frac{\delta S}{\delta\phi}+\eta(x,t), \tag{2.1}$$

where $\eta(x,t)$ is a random noise field with the corresponding stochastic expectation values:

$$\langle F[\eta]\rangle_\eta=\frac{\int d\eta\, F(\eta)e^{-\frac{1}{4}\int d^Dx\, dt\, \eta^2(x,t)}}{\int d\eta\, e^{-\frac{1}{4}\int d^Dx\, dt\, \eta^2(x,t)}}. \tag{2.2}$$

In particular,

$$\langle\eta(x,t)\rangle_\eta=0,\qquad \langle\eta(x,t)\eta(x',t')\rangle_\eta=2\delta^D(x-x')\delta(t-t'), \tag{2.3}$$

and similarly for higher n-point functions.

Note that through the Langevin equation (2.1) the field has implicitly become a functional of the stochastic noise field η which we will indicate by writing the solution as $\phi^{[\eta]}(x,t)$.

* Here, and in the following, the word "time" will always refer to the fictitious time direction.

It is a fundamental property of stochastic quantization that as $t \to \infty$, and the thermodynamic equilibrium is reached, the stochastic expectation values will coincide with the corresponding *vacuum* expectation values of the underlying euclidean quantum field theory. Hence,

$$\lim_{t \to \infty} \langle F[\phi^{[\eta]}(x,t)] \rangle_\eta = \langle F[\phi(x,t)] \rangle, \tag{2.4}$$

for any functional $F[\phi]$. We will return to this fundamental identity in sect. 3.

We wish to generalize this method to an action $S[\psi, \bar{\psi}]$ which is a functional of spin-$\frac{1}{2}$ fields. The Langevin equations in this case are:

$$\frac{\partial}{\partial t}\psi(x,t) = -\frac{\delta S}{\delta \bar{\psi}} + \eta(x,t),$$

and*

$$\frac{\partial}{\partial t}\bar{\psi}(x,t) = \frac{\delta S}{\delta \psi} + \bar{\eta}(x,t). \tag{2.5}$$

By analogy with the bosonic case one would define stochastic expectation values

$$\langle \eta \rangle_\eta = \langle \bar{\eta} \rangle_\eta = 0, \qquad \langle \eta_\alpha(x,t) \bar{\eta}_\beta(x',t') \rangle_\eta = 2\delta^D(x-x')\delta(t-t')\delta_{\alpha\beta}, \tag{2.6}$$

and similarly for higher n-point functions, taking into account the anticommuting nature of the noise fields.

One way to solve the set of Langevin equations (2.5) is to write them as two coupled integral equations, and then solve them by iteration. This expansion will formally generate the standard perturbation of ordinary Feynman diagrams, once all stochastic fields have been contracted away. However the choice (2.5) will not generate the proper equilibrium configurations as $t \to \infty$ in the general case. We only have to consider the free fermionic theory to see this.

So consider now the free euclidean action

$$S[\psi, \bar{\psi}] = -i \int \mathrm{d}^D x\, \bar{\psi}(\not{\partial} + im)\psi. \tag{2.7}$$

We use euclidean gamma matrices normalized as $\{\gamma^\mu, \gamma^\nu\} = -2\delta^{\mu\nu}$.

The Langevin equation (2.7) then becomes

$$\frac{\partial}{\partial t}\psi(x,t) = (i\not{\partial} - m)\psi(x,t) + \eta(x,t), \tag{2.8}$$

* In the following the conjugate equation will not be written down explicitly.

which is easily solved by the introduction of the Green function

$$\left(\frac{\partial}{\partial t} - i\partial\!\!\!/ + m\right) g(x, t) = \delta^D(x)\delta(t).$$

Note that although the left-hand side of eq. (2.12) is of course diagonal in the spinor indices, $g(x, t)$ itself is not. Formally

$$g(x, t) = \theta(t)\int \frac{d^Dp}{(2\pi)^D} e^{-(p\!\!\!/+m)t+ipx}, \tag{2.9}$$

so that the solution for $\psi(x, t)$ can be written

$$\psi(x, t) = \int_0^{\infty} dt' \int d^Dx\, g(x-x', t-t')\eta(x', t'). \tag{2.10}$$

Superficially it looks as if $g(x, t) \to 0$ as $t \to \infty$ as is required for convergence towards equilibrium. However, as mentioned, $g(x, t)$ is not diagonal. Upon diagonalization one finds $g(x, t) \sim \exp(-mt)$, so that $g(x, t) \to 0$ as $t \to \infty$, *only* if $m \neq 0$ [6].

What is the solution of this problem? First, it is clear (and was in fact already noted in the original paper of Parisi and Wu [1]) that the Langevin equation (2.1) is only one very particular choice out of a large class of relaxation equations. One way to generalize (2.1) is to include a kernel $V(x, y)$:

$$\frac{\partial}{\partial t}\phi(x, t) = -\int d^Dy\, V(x, y)\frac{\delta S}{\delta\phi(y, t)} + \eta(x, t), \tag{2.11}$$

provided one simultaneously changes the stochastic expectation values to

$$\langle\eta(x, t)\eta(x', t')\rangle_\eta = 2V(x, x')\delta(t-t'). \tag{2.12}$$

The special choice $V(x, y) = \delta^D(x-y)$ leads to the Langevin equation (2.1). (This can in fact be even further generalized by the introduction of a "memory kernel", which is non-local in time [7]. We will not need such an extension here.)

In (2.11) the integrand on the RHS has to be positive in order to ensure that integrals are properly damped in the $t \to \infty$ limit. For bosonic theories, where $\delta S/\delta\phi$ is positive, this requires that V also be positive. In fermionic theories, however, $\delta S/\delta\phi$ can be negative. This means that V has to be chosen such that the negative eigenvalues of $\delta S/\delta\phi$ correspond to negative eigenvalues of V, resulting in a positive integrand. One such choice is the kernel

$$K(x, y) = (i\partial\!\!\!/_x + m)\delta(x-y) = i\delta\!\!\!/'(x-y) + m\delta(x-y), \tag{2.13}$$

where we have defined

$$\not{\delta}'(x-y)=\gamma_\mu\frac{\partial}{\partial x_\mu}\delta^D(x-y).$$

As in the bosonic case we introduce a random noise field $\theta(x,t)$; here it is fermionic and is determined by its stochastic expectation values

$$\langle\theta(x,t)\rangle_\theta=0=\langle\bar{\theta}(x,t)\rangle_\theta,$$
$$\langle\theta_\alpha(x,t)\bar{\theta}_\beta(x',t')\rangle_\theta=2[i\not{\delta}'(x-x')+m\delta(x-x')]_{\alpha\beta}\delta(t-t'). \tag{2.14}$$

The generalized Langevin equation

$$\frac{\partial}{\partial t}\psi(x,t)=-\int d^Dy K(x,y)\frac{\delta S}{\delta\bar{\psi}(y,t)}+\theta(x,t), \tag{2.15}$$

becomes simply

$$\frac{\partial}{\partial t}\psi(x,t)=(\partial^2-m^2)\psi(x,t)+\theta(x,t), \tag{2.16}$$

which is just like the bosonic Langevin equation: all knowledge of the fermionic nature of the theory is contained in the stochastic expectation values of the θ's.

For an arbitrary functional of the θ's we have:

$$\langle F[\theta,\bar{\theta}]\rangle_\theta=\frac{\int d\theta\, d\bar{\theta}\, F[\theta,\bar{\theta}]\exp\{-\frac{1}{2}\int d^Dx\, dt\, \bar{\theta}(i\not{\partial}+m)^{-1}\theta\}}{\int d\theta\, d\bar{\theta}\exp\{-\frac{1}{2}\int d^Dx\, dt\, \bar{\theta}(i\not{\partial}+m)^{-1}\theta\}}. \tag{2.17}$$

In particular, for the $2n$-point functions

$$\langle\theta(x_1,t_1)\ldots\bar{\theta}(x_n,t_n)\rangle_\theta=\sum_{\text{perm}}\varepsilon_p\prod_{\text{pairs}}\langle\theta(x_i,t_i)\bar{\theta}(x_j,t_j)\rangle_\theta. \tag{2.18}$$

Eq. (2.16) is easily solved by the introduction of the corresponding diagonal Green function

$$\left\{\frac{\partial}{\partial t}-\partial^2+m^2\right\}G(x,t)=\delta^D(x)\delta(t),$$
$$G(x,t)=0,\qquad \text{for } t<0, \tag{2.19}$$

which leads to

$$G(x,t)=\theta(t)\int\frac{d^Dp}{(2\pi)^D}e^{-(p^2+m^2)t+ipx}. \tag{2.20}$$

(The corresponding Green function $G(x, t)$ for the conjugate Langevin equation is simply $\bar{G}(x, t) = G(x, t)^*$.)

For the free theory we therefore have the solution $\psi^{[\theta]}(x, t)$ as

$$\psi(x, t) = \int_0^t dt' \int d^D x' G(x - x')\theta(t - t'). \tag{2.21}$$

Note that this converges properly as $t \to \infty$, even in the massless case. In the interacting case we can write an integral equation for $\psi^{[\theta]}(x, t)$

$$\psi^{[\theta]}(x, t) = \int_0^\infty dt' \int d^D x' G(x - x', t - t') \left(\theta(x', t') - \frac{\delta S_{\text{int}}[\psi, \bar{\psi}]}{\delta \bar{\psi}^{[\theta]}(x', t')} \right), \tag{2.22}$$

which, when coupled with its conjugate equation, uniquely determines $\psi^{[\theta]}(x, t)$.

3. The generalized Fokker-Planck equation

Having proposed a proper choice of the Langevin equations to use in the fermionic case, we now have to demonstrate that as $t \to \infty$ the stochastic expectation values defined by eq. (2.17) actually reduce to regular vacuum expectation values, i.e.

$$\lim_{t \to \infty} \langle F[\psi, \bar{\psi}] \rangle_\theta = \langle F[\psi, \bar{\psi}] \rangle. \tag{3.1}$$

This can be shown by setting up the master equation corresponding to the Langevin equation (2.15); this equation is known in statistical mechanics as the Fokker-Planck equation.

Before proceeding to the case at hand, let us first present a standard heuristic proof of the corresponding Fokker-Planck equation in the case of scalar fields. Having done this we can see what steps need to be modified due to the introduction of Grassmann variables. So consider again a scalar euclidean action, and let us for simplicity use the stochastic expectation values defined by eq. (2.3).

We can formally define a probability distribution

$$P[\phi, t] = \frac{\int d\eta\, \delta(\phi(x) - \phi^{[\eta]}(x, t)) e^{-\frac{1}{2}\int d^D x\, dt\, \eta^2(x, t)}}{\int d\eta\, e^{-\frac{1}{2}\int d^D x\, dt\, \eta^2(x, t)}} \tag{3.2}$$

where the δ-function is defined in the usual functional sense as the product over each space-time point. This definition of $P[\phi, t]$ implies according to eq. (2.2) that stochastic expectation values can be written

$$\langle F[\phi^{[\eta]}(x, t)] \rangle_\eta = \int d\phi\, F[\phi(x)]\, P[\phi, t]. \tag{3.3}$$

We wish to show that as $t \to \infty$,

$$P[\phi, t] \to \frac{e^{-S[\phi]}}{\int d\phi e^{-S[\phi]}}, \tag{3.4}$$

so that

$$\langle F[\phi^{[\eta]}(x, t)] \rangle_\eta \to \frac{\int d\phi F[\phi] e^{-S[\phi]}}{\int d\phi e^{-S[\phi]}}. \tag{3.5}$$

Using the boundary condition $\phi^{[\eta]}(x, t=0) = 0$ it follows from eq. (3.2) that we can view $P[\phi, t]$ as the probability functional of an evolution from $\phi(x) = 0$ at $t = 0$ to $\phi(x)$ at time t. We can indicate this explicitly by writing $P[\phi, 0; t, 0]$.

Consider now,

$$\langle F[\phi^{[\eta]}(x, t+\Delta t)] \rangle_\eta = \int d\phi F[\phi] P[\phi, 0; t+\Delta t, 0]. \tag{3.6}$$

Here

$$P[\phi, 0; t+\Delta t, 0] = \int d\tilde{\phi} P[\phi, \tilde{\phi}; t+\Delta t, t] P[\tilde{\phi}, 0; t, 0], \tag{3.7}$$

due to the stochastic nature of the noise field. Eq. (3.7) expresses the crucial markovian property of the probability distribution, which is needed in order to derive the Fokker-Planck equation. We have

$$\begin{aligned}\langle F[\phi^{[\eta]}(x, t+\Delta t)] \rangle_\eta &= \int d\phi d\tilde{\phi} P[\phi, \tilde{\phi}; t+\Delta t, \ ] P[\tilde{\phi}, 0; t, 0] \\ &\quad \times \left(F[\phi] + \int dx \Delta\phi(x) \frac{\delta F}{\delta \phi} \right. \\ &\quad \left. + \tfrac{1}{2} \int dx dx' \Delta\phi(x) \Delta\phi(x') \frac{\delta^2 F}{\delta \phi^2} + \cdots \right), \end{aligned} \tag{3.8}$$

which implies that

$$\begin{aligned} &\int d\phi F[\phi] (P[\phi, 0; t+\Delta t, 0] - P[\phi, 0; t, 0]) \\ &= \int d\phi P[\phi, 0; t, 0] \left(\int dx \langle \Delta\phi \rangle_\eta \frac{\delta F}{\delta \phi} \right. \\ &\quad \left. + \tfrac{1}{2} \int dx dx' \langle \Delta\phi(x) \Delta\phi(x') \rangle_\eta \frac{\delta^2 F}{\delta \phi^2} + \cdots \right). \end{aligned} \tag{3.9}$$

Note that up to order $(\Delta t)^2$ the stochastic expectation values in eq. (3.9) are easily found from the regular Langevin equation:

$$\langle \Delta\phi \rangle_\eta = -\frac{\delta S}{\delta\phi}\Delta t,$$

$$\langle \Delta\phi(x_1)\Delta\phi(x_2)\rangle_\eta = 2\delta^D(x_1 - x_2)\,\Delta t + \mathrm{O}\big((\Delta t)^2\big). \tag{3.10}$$

After two functional integrations by parts, and using the fact that the resulting equation must be valid for any arbitrary functional $F[\phi]$, we find as $\Delta t \to 0$:

$$\frac{\partial}{\partial t}P[\phi,t] = \int \mathrm{d}x\left(\frac{\delta}{\delta\phi}\left(P[\phi,t]\frac{\delta S}{\delta\phi}\right) + \frac{\delta^2}{\delta\phi^2}P[\phi,t]\right). \tag{3.11}$$

This is the Fokker-Planck equation for scalar fields. Note that the normalized equilibrium distribution is precisely:

$$P[\phi] = \frac{\mathrm{e}^{-S[\phi]}}{\int \mathrm{d}\phi\,\mathrm{e}^{-S[\phi]}}, \tag{3.12}$$

as is needed in order to obtain the fundamental property expressed by eq. (3.5).

It is easy to generalize this result to the case of modified Langevin equations as given by (2.11). Now we have

$$\langle \Delta\phi \rangle_\eta = -\int \mathrm{d}y\, V(x,y)\frac{\delta S}{\delta\phi(y)}\Delta t,$$

$$\langle \Delta\phi(x)\,\Delta\phi(y)\rangle_\eta = 2V(x,y)\,\Delta t + \mathrm{O}\big((\Delta t)^2\big), \tag{3.13}$$

which immediately leads to the generalized Fokker-Planck equation

$$\frac{\partial}{\partial t}P[\phi,t] = \int \mathrm{d}x\,\mathrm{d}y\,V(x,y)\left(\frac{\delta}{\delta\phi}\left[P\frac{\delta S}{\delta\phi}\right] + \frac{\delta^2 P}{\delta\phi^2}\right). \tag{3.14}$$

As can be seen, the normalized equilibrium probability distribution is again given by eq. (3.12).

The fermionic case can be treated in a completely analogous manner. Introduce a distribution functional by

$$P[\psi,\bar{\psi},t] = \frac{\int \mathrm{d}\theta\,\mathrm{d}\bar{\theta}\,\delta(\psi - \psi^{[\theta]})\delta(\bar{\psi} - \bar{\psi}^{[\theta]})\exp\left\{-\frac{1}{2}\int \mathrm{d}^D x\,\mathrm{d}t\,\bar{\theta}(i\partial\!\!\!/ + m)^{-1}\theta\right\}}{\int \mathrm{d}\theta\,\mathrm{d}\bar{\theta}\exp\left\{-\frac{1}{2}\int \mathrm{d}^D x\,\mathrm{d}t\,\bar{\theta}(i\partial\!\!\!/ + m)^{-1}\theta\right\}}, \tag{3.15}$$

so that the stochastic θ expectation values can be written

$$\langle F[\psi^{[\eta]}(x,t),\bar{\psi}^{[\eta]}(y,t)]\rangle_\theta = \int \mathrm{d}\psi\,\mathrm{d}\bar{\psi}\,F[\psi(x),\bar{\psi}(y)]\,P[\psi,\bar{\psi},t]. \quad (3.16)$$

Treating ψ and $\bar{\psi}$ as independent variables, and going through the same steps as in the bosonic case, we find the fermionic Fokker-Planck equation

$$\frac{\partial}{\partial t}P[\psi,\bar{\psi},t] = \int \mathrm{d}x\,\mathrm{d}y\,K(x,y)\left[\frac{\delta}{\delta\bar{\psi}}\left(\frac{\delta S}{\delta\psi}P\right) - \frac{\delta}{\delta\psi}\left(\frac{\delta S}{\delta\bar{\psi}}P\right) + 2\frac{\delta^2 P}{\delta\bar{\psi}\,\delta\psi}\right], \quad (3.17)$$

with the kernel $K(x, y)$ as defined in eq. (2.13). Alternatively one can start directly from the defining eq. (3.15), differentiate with respect to t and substitute the Langevin equation (2.15). The result is given again by (3.17).

To see if this Fokker-Planck equation will drive the system correctly towards the equilibrium distribution, we expand P in eigenstates Ψ_n of H, the operator appearing on the RHS of (3.17):

$$P[\psi,\bar{\psi}] = \sum_n \Psi_n \mathrm{e}^{-\lambda_n t}, \qquad H\Psi_n = -\lambda_n\Psi_n. \quad (3.18)$$

If we can show that the lowest eigenvalue of the system described by (3.18) is zero and that all others have a positive real part then in the limit $t\to\infty$, $P[\psi,\bar{\psi}]\to\Psi_0$ the "ground" state wave functional. It is easy to see that $\Psi_0 = \exp(-S)$ is indeed a solution of (3.18) with $\lambda = 0$ so that the Fokker-Planck equation drives the system to the correct equilibrium distribution. Thus the problem reduces to considering the non-zero eigenvalues of H. Note that in the bosonic case the "hamiltonian" H has positive eigenvalues *before* introducing the kernel V, thus V must also be positive. This is not true in the fermionic case however: the "hamiltonian" without the kernel is not even hermitian and its eigenvalues are therefore not constrained to be real. Thus the kernel that has to be introduced need not be positive. This was apparent from our discussion of the Langevin equations (2.15). With our choice of K, eq. (2.13), it can be shown, for example by employing the coherent representation of H, that the eigenvalues of H are positive, if S is the free fermionic action. In the interacting case we may write these eigenvalues as $\lambda_n = |\lambda_n^0| + g\mu_n + O(g^2)$ where g is a coupling constant, $|\lambda^0|$ is the (positive) free eigenvalue and μ_n is arbitrary. It follows that as long as we restrict ourselves to weak coupling ($g \ll 1$) the $t\to\infty$ limit of (3.18) is again $P = \exp(-S)/\int \mathrm{d}\psi\,\mathrm{d}\bar{\psi}\exp(-S)$ as required, and therefore

$$\langle F[\psi^{[\theta]},\bar{\psi}^{[\theta]}]\rangle_\theta \to \langle F[\psi,\bar{\psi}]\rangle. \quad (3.19)$$

The advantage of our choice of kernel, eq. (2.13), is that it reproduces the usual Feynman perturbative expansion in a straightforward way; this will be illustrated in the next two sections.

To treat the case of strong coupling, we restrict attention to theories that are bilinear in the fermion fields i.e. $S = \int d^Dx\, \bar{\psi} M \psi$, and take $K(x, y) = M(x, y)^{-1}$. It can again be checked that the resulting hamiltonian has positive eigenvalues ensuring the correct $t \to \infty$ limit.

We end this section by pointing out that just as in the bosonic case, we have

$$\langle F[\psi, \bar{\psi}] \rangle = \lim_{T\to\infty} \left(\frac{1}{T} \int_0^T dt\, F[\psi^{[\theta]}, \bar{\psi}^{[\theta]}] \right), \tag{3.20}$$

as a consequence of the ergodicity of the system. This property will be useful in numerical applications of stochastic quantization.

4. Perturbation theory

To illustrate how this method works within the context of standard perturbation theory, consider first a purely fermionic theory. For definiteness, choose

$$S_{\text{int}}[\psi, \bar{\psi}] = -g \int d^Dx\, \bar{\psi}\psi\bar{\psi}\psi. \tag{4.1}$$

One can, for simplicity, think of this as the interaction of two types of fermions, each carrying some internal quantum number that distinguishes them. Further, although we will keep an arbitrary dimensionality D, one can always imagine doing the calculation in 2 dimensions, in which case this theory is renormalizable.

Substituting the interaction term (4.1) into the exact integral equation (2.22) we obtain a highly non-linear equation for the solution $\psi^{[\theta]}(x, t)$ of the Langevin equation. However, treating g as a small parameter we can solve the equation by iteration. This is illustrated in fig. 1. Green functions are indicated by lines ("stochastic propagators"); dashed lines correspond to the bosonic Green functions $G(x, t)$ whereas fully drawn lines are the *fermionic* Green functions corresponding to

$$\Gamma(x, t) = \int d^Dy\, K(x, y) G(y, t). \tag{4.2}$$

Uncontracted noise fields $\theta(x, t)$ and $\bar{\theta}(x, t)$ are indicated by crosses and boxes, respectively.

Let us consider the 2-point function $\langle \psi\bar{\psi} \rangle$. To lowest order this is simply given by

$$\langle \psi(x, t)\bar{\psi}(x', t') \rangle$$
$$= \int_0^\infty d\tau_1\, d\tau_2 \int d^Dx_1\, d^Dx_2\, G(x - x_1, t - \tau_1) G(x' - x_2, t' - \tau_2) \langle \theta\bar{\theta} \rangle_\theta. \tag{4.3}$$

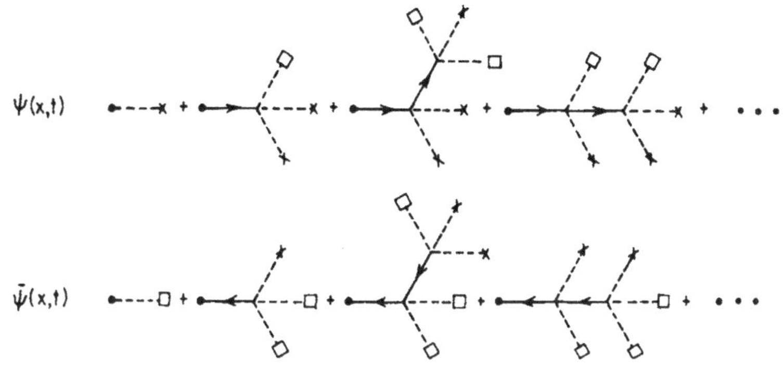

Fig. 1. The perturbative expansion of the fermion fields; crosses and boxes represent η and $\bar{\eta}$ fields respectively; solid lines indicate Γ and broken lines G.

Using the stochastic expectation values (2.14) and the momentum-space representation (2.20) of the Green functions, this immediately leads to the momentum space propagator

$$\Delta(p;t,t') = \langle \psi(p,t)\bar{\psi}(-p,t')\rangle_{\theta}$$
$$= \frac{\not{p}-m}{p^2+m^2} e^{-(p^2+m^2)(t+t')}\left[e^{-2(p^2+m^2)\bar{t}} - 1\right], \tag{4.4}$$

where $\bar{t} = \min\{t_1, t_2\}$. For practical purposes the "time ordered" propagator $t < t'$ is useful:

$$\Delta(p;t,t') = \frac{\not{p}-m}{p^2+m^2} e^{-(p^2+m^2)(t+t')}\left[e^{-2(p^2+m^2)\bar{t}} - 1\right]. \tag{4.5}$$

Note that for equal times this simply reduces to the standard euclidean fermion propagator:

$$\Delta(p;t,t') \to \frac{1}{\not{p}+m}, \qquad \text{as} \qquad t = t' \to \infty, \tag{4.6}$$

as required.

All of the standard perturbation theory results can be generated in this way, using the expansion represented in fig. 1 and contracting all stochastic noise fields by means of the rules (2.14) and (2.18). As an illustration consider the one-loop corrections to the 2-point function, (4.3). These are given by the diagrams of fig. 2.

Fig. 2. The one-loop corrections to the 2-point function in a $(\bar{\psi}\psi)^2$ theory.

The two graphs are identical and equal to

$$(\mathrm{a}) = \int_0^\infty \mathrm{d}\tau \int \frac{\mathrm{d}^D p}{(2\pi)^D} \Gamma(k; t_1, \tau) \mathrm{Tr}(\Delta(p; \tau, \tau)) \Delta(k; t_2, \tau) . \tag{4.7}$$

In the limit $t_1 = t_2 \to \infty$ we obtain

$$(\mathrm{a}) + (\mathrm{b}) = g \int \frac{\mathrm{d}^D p}{(2\pi)^D} \frac{1}{\not{k} + m} \mathrm{Tr}\left(\frac{1}{\not{p} + m} \right) \frac{1}{\not{k} + m} , \tag{4.8}$$

the usual Feynman result.

The computation of the 4-point function is equally simple. Fig. 3a shows the stochastic graphs contributing to $\Gamma^{(4)}$ and fig. 3b the corresponding graphs of ordinary perturbation theory. To zeroth order we have

$$\begin{aligned} \Gamma^{(4)} &= \int_0^\infty \mathrm{d}\tau\, \Gamma(p_1, t_1 - \tau) \Delta(p_2, t_2 - \tau) \Delta(p_4, t_4 - \tau) \Delta(p_3, t_3 - \tau) + \text{permutations} \\ &= (\not{p}_1 - m) \frac{(\not{p}_2 - m)}{p_2^2 + m^2} \frac{(\not{p}_4 - m)}{p_4^2 + m^2} \frac{(\not{p}_3 - m)}{p_3^2 + m^2} \frac{1}{\Sigma p_i^2 + 4m^2} + \text{permutations}, \end{aligned} \tag{4.9}$$

as $t \to \infty$. Adding the 4 permutations we clearly get what corresponds to the single Feynman diagram. It is not hard to see how this works to higher orders in

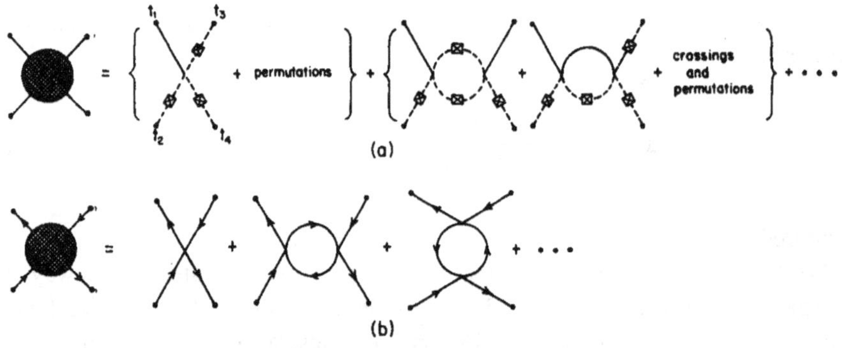

Fig. 3. The 4-point function in $(\bar{\psi}\psi)^2$ theory to one-loop order.

perturbation theory. For example the one-loop 4-point function is given by ("s-channel")

$$\Gamma^{(4)} = \int \frac{d^D q}{(2\pi)^D} \int d\tau_1 \int d\tau_2 (\Gamma(p_1; t_1, \tau_1)\Delta(p_4, t_4, \tau_1)$$

$$\times \mathrm{Tr}\{\Delta(q; \tau_1, \tau_2)\Delta(k; \tau_2, \tau_1)\}\Gamma(p_3; t_4, \tau_2)\Delta(p_2; t_2, \tau_2)$$

$$+\Gamma(p_1; t_1, \tau_1)\Delta(p_4; t_4, \tau_1)$$

$$\times \mathrm{Tr}\{\Gamma(q; \tau_1, \tau_2)\Delta(k; \tau_2, \tau_1)\}\Delta(p_3; t_3, \tau_2)\Delta(p_2; t_2, \tau_2))$$

$$+ \text{terms with } q \leftrightarrow k. \tag{4.10}$$

Here $k = p_1 + p_4 - q$.

It is a straightforward calculation to show that in the limit $t_1 = t_2 = t_3 = t_4 \to \infty$ one indeed recovers the usual Feynman result.

5. Coupling to gauge fields

The formalism discussed in the previous sections is only slightly modified by the coupling of fermions to other fields. Of most interest, of course, is the coupling to gauge fields. In this section we shall briefly discuss the appropriate stochastic relaxation equations and illustrate how the method works in perturbation theory by calculating explicitly the one-loop vacuum polarization graph of QED.

Consider the euclidean action for a set of fermion fields coupled to gauge fields in a local SU(N) invariant way,

$$S[A, \psi, \bar{\psi}] = \int d^D x \left(\tfrac{1}{4} F^a_{\lambda\nu} F^{\lambda\nu}_a - \bar{\psi}(i\not{D} - m)\psi\right), \tag{5.1}$$

with the standard notation:

$$F^a_{\lambda\nu} = \partial_\lambda A^a_\nu - \partial_\nu A^a_\lambda - g f^{abc} A^b_\lambda A^c_\nu, \qquad \not{D} = \not{\partial} - ig\not{A}.$$

Choosing for the A^a_λ field the standard bosonic Langevin eq. (2.1) and for the ψ fields the Langevin eq. (2.19) leads to

$$\frac{\partial}{\partial t} A^a_\lambda(x, t) = D_\nu F^a_{\lambda\nu}(x, t) + J^a_\lambda(x, t) + \eta^a_\lambda(x, t), \tag{5.2a}$$

$$\frac{\partial}{\partial t}\psi(x, t) = (\partial^2 - m^2)\psi(x, t) + g\int d^D y K(x, y)\not{A}(y, t)\psi(y, t) + \theta(x, t), \tag{5.2b}$$

where $J^a_\lambda = \frac{1}{2} g\bar{\psi}(x, t)\gamma_\lambda \lambda^a \psi(x, t)$

These (exact) coupled differential equations can again be solved iteratively. To do that, one can introduce the Green function $G^{ab}_{\lambda\nu}(x,t)$ for the vector potential $A^a_\lambda(x,t)$:

$$G^{ab}_{\lambda\nu}(x,t)=\theta(t)\delta^{ab}\int\frac{\mathrm{d}^Dp}{(2\pi)^D}\left(\left[\delta_{\lambda\nu}-\frac{p_\lambda p_\nu}{p^2}\right]\mathrm{e}^{-p^2t+ipx}+\frac{p_\lambda p_\nu}{p^2}\mathrm{e}^{ipx}\right),$$

which is the solution of

$$\left(\delta_{\lambda\alpha}\left(\frac{\partial}{\partial t}-\partial^2\right)+\partial_\lambda\partial_\alpha\right)G^{ab}_{\alpha\nu}=\delta^{ab}\delta_{\lambda\nu}\delta^D(x)\delta(t). \qquad (5.4)$$

But first a brief comment about gauge invariance. In the action (5.1) we did not include any gauge fixing term, and we did not discuss the possible introduction of ghost field terms. As long as one calculates gauge invariant quantities neither term is needed [1]. But in fact, one has far more freedom: it is completely consistent to add a gauge fixing term to the lagrangian, still without the introduction of ghost fields. This issue has already been discussed at length in the literature [8]. Here we simply note that in a perturbative context it is actually far more convenient to operate with gauge fixing. Choosing the Feynman gauge we find, instead of eq. (5.3), the more tractable Green function

$$G^{ab}_{\lambda\nu}(x,t)=\delta^{ab}\delta_{\lambda\nu}\theta(t)\int\frac{\mathrm{d}^Dp}{(2\pi)^D}\mathrm{e}^{-p^2t+ipx}, \qquad (5.5)$$

corresponding to the gauge fixed Langevin equation

$$\frac{\partial}{\partial t}A^a_\lambda(x,t)=D^{ab}_\nu F^b_{\lambda\nu}(x,t)+\partial_\lambda\partial_\nu A^a_\nu-gf^{abc}A^b_\lambda(x,t)\partial_\nu A^c_\nu(x,t)$$
$$+J^a_\lambda(x,t)+\eta^a_\lambda(x,t). \qquad (5.6)$$

The time-ordered 2-point function corresponding to (5.4) reads

$$D^{ab}_{\lambda\nu}(p;t,t')=\langle A^a_\lambda(p,t)A^b_\nu(-p,t')\rangle_\eta$$
$$=\delta^{ab}\delta_{\lambda\nu}\frac{1}{p^2}\mathrm{e}^{-p^2(t-t')}\left[1-\mathrm{e}^{-2p^2t'}\right], \qquad (5.7)$$

in momentum space. We have used the stochastic expectation values

$$\langle\eta^a_\lambda(x,t)\eta^b_\nu(x',t')\rangle_\eta=2\delta^{ab}\delta_{\lambda\nu}\delta^D(x-x')\delta(t-t'). \qquad (5.8)$$

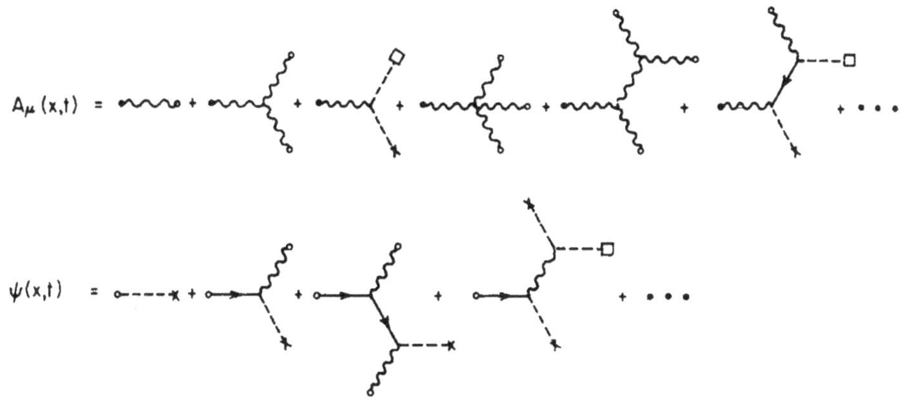

Fig. 4. Perturbative expansion of fields in a gauge theory: open circles denote noise fields $\eta_\mu(x,t)$.

The iterative solution of the Langevin equations can be illustrated diagrammatically as shown in fig. 4, the η's being indicated by open circles. Using this expansion we can calculate what corresponds to the single one-loop QED vacuum polarization diagram in ordinary field theory language. The stochastic diagrams for this quantity are shown in fig. 5.

With the notation indicated in the figure and in the previous sections we have

$$
\begin{aligned}
(\mathrm{a}) = \int \frac{\mathrm{d}^D q}{(2\pi)^D} \int_0^\infty \mathrm{d}\tau_1\,\mathrm{d}\tau_2\,\mathrm{Tr}\big(\Delta(q_1;\tau_1,\tau_2)\gamma_\alpha\Delta(q_2;\tau_1,\tau_2)\gamma_\beta\big) \\
\times G_{\lambda\alpha}(p,t_1-\tau_1)G_{\beta\nu}(p,t_2-\tau_2),
\end{aligned}
\tag{5.9a}
$$

$$
\begin{aligned}
(\mathrm{b}) = \int \frac{\mathrm{d}^D q}{(2\pi)^D} \int_0^\infty \mathrm{d}\tau_1\,\mathrm{d}\tau_2\,\mathrm{Tr}\big(\Gamma(q_1;\tau_1-\tau_2)\gamma_\alpha\Delta(q_2;\tau_1,\tau_2)\gamma_\beta\big) \\
\times D_{\lambda\alpha}(p;t_1,\tau_1)G_{\beta\nu}(p,t_2-\tau_2),
\end{aligned}
\tag{5.9b}
$$

and similarly for (c).

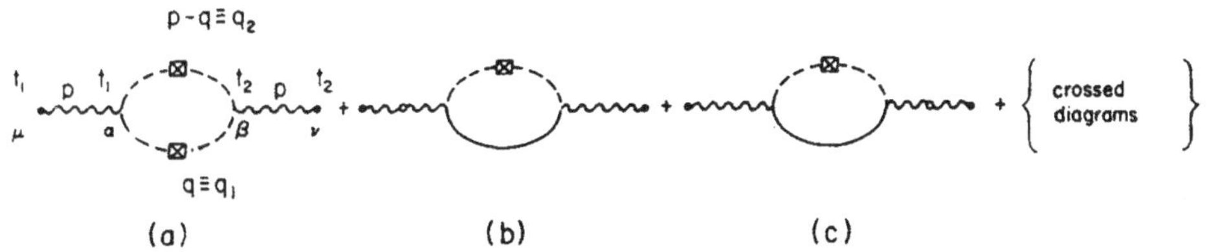

Fig. 5. The one-loop vacuum polarization in QED.

Adding the crossed diagrams and taking the limit $t_1 = t_2 \to \infty$ we are left with

$$(\mathrm{a}) = \int \frac{\mathrm{d}^D q}{(2\pi)^D} \tilde{\Pi}_{\lambda\nu}(p,q) \frac{1}{p^2} \frac{1}{p^2 + q_1^2 + q_2^2 + 2m^2} \frac{1}{q_1^2 + m^2} \frac{1}{q_2^2 + m^2}, \tag{5.10a}$$

$$(\mathrm{b}) = \tfrac{1}{2} \int \frac{\mathrm{d}^D q}{(2\pi)^D} \tilde{\Pi}_{\lambda\nu}(p,q) \left(\frac{1}{p^2}\right)^2 \frac{1}{p^2 + q_1^2 + q_2^2 + 2m^2} \left(\frac{1}{q_1^2 + m^2} + \frac{1}{q_2^2 + m^2}\right), \tag{5.10b}$$

$$(\mathrm{c}) = (\mathrm{b}), \tag{5.10c}$$

where $\tilde{\Pi}_{\lambda\nu}(p,q) = \mathrm{Tr}((\not{q} - m)\gamma_\lambda(\not{q}_2 - m)\gamma_\nu)$. Finally adding (5.10a), (5.10b), and (5.10c) we are left with the standard polarization tensor of QED.

6. Summary and conclusions

As we have shown in this paper, the method of stochastic quantization can be extended to cases where the action contains fermionic degrees of freedom. Setting up a proper Langevin equation for the case of fermions forces one to introduce a kernel $K(x, y)$ whose function is to ensure convergence as $t \to \infty$. With this kernel, convergence is assured both on a perturbative and a non-perturbative level. We illustrated in a perturbative context how convergence towards the underlying quantum field theory emerges from the corresponding stochastic diagrams. Convergence towards regular vacuum expectation values follows from the master equation for the probability distribution, the Fokker-Planck equation.

The construction for stochastic quantization of theories involving fermion fields may have several immediate applications. With a combined stochastic quantization scheme available for both fermions and bosons the road is open for a consistent stochastic quantization procedure or supersymmetric theories. At a practical computational level, the method may be useful in numerical simulations of lattice gauge theories with fermions, here using a discretized version of the Langevin equation. Also the numerical method of ref. [5] which is based on a variational principle related to the Langevin equation, may now be applied directly to full 4-dimensional gauge theories such as QCD.

We would like to thank Bill Caswell, Jeff Greensite, Ken Macrae and Poul Olesen for helpful discussions.

Note added

After completion of this work we received a preprint by J.D. Breit, S. Gupta and A. Zaks (IAS, Princeton preprint March 1983) which contains ideas similar to ours.

References

[1] G. Parisi and Y. Wu, Scient. Sin. 24 (1981) 483
[2] V.N. Gribov, Nucl. Phys. B139 (1978) 1
[3] F. Fucito, E. Marinari, G. Parisi and C. Rebbi, Nucl. Phys. B180[FS2] (1981) 369;
G. Parisi, Nucl. Phys. B180[FS2] (1981) 378;
F. Fucito and E. Marinari, Nucl. Phys. B190[FS3] (1981) 31
[4] J. Greensite and M.B. Halpern, Nucl. Phys. B211 (1983) 343;
J. Alfaro and B. Sakita, Phys. Lett. 121B (1983) 339
[5] P.A. Amundsen and P.H. Damgaard, Phys. Rev. D29 (1984), 323
[6] T. Fukai, H. Nakazato, I. Ohba, K. Okano and Y. Yamanaka, Waseda University preprint WU-HEP-82-7 (1982)
[7] R.F. Fox, Phys. Reports 48 (1978) 179
[8] D. Zwanziger, Nucl. Phys. B192 (1981) 259;
L. Baulieu and D. Zwanziger, Nucl. Phys. B193 (1981) 163;
E. Floratos and J. Iliopoulos, Nucl. Phys. B214 (1983) 392

Gravity

PHYSICAL REVIEW D VOLUME 33, NUMBER 4 15 FEBRUARY 1986

Stochastic quantization of Einstein gravity

Helmut Rumpf
Institut für Theoretische Physik, Universität Wien, A-1090 Vienna, Austria
(Received 29 July 1985; revised manuscript received 7 October 1985)

We determine a one-parameter family of covariant Langevin equations for the metric tensor of general relativity corresponding to DeWitt's one-parameter family of supermetrics. The stochastic source term in these equations can be expressed in terms of a Gaussian white noise upon the introduction of a stochastic tetrad field. The only physically acceptable resolution of a mathematical ambiguity in the ansatz for the source term is the adoption of Ito's calculus. By taking the formal equilibrium limit of the stochastic metric a one-parameter family of covariant path-integral measures for general relativity is obtained. There is a unique parameter value, distinguished by any one of the following three properties: (i) the metric is harmonic with respect to the supermetric, (ii) the path-integral measure is that of DeWitt, (iii) the supermetric governs the linearized Einstein dynamics. Moreover the Feynman propagator corresponding to this parameter is causal. Finally we show that a consistent stochastic perturbation theory gives rise to a new type of diagram containing "stochastic vertices."

I. INTRODUCTION

The stochastic quantization method of Parisi and Wu[1] provides an interesting alternative to the standard quantization of gauge theories, as no gauge fixing and associated Faddeev-Popov ghosts are required. It appears worthwhile to apply this scheme also to the gravitational field, although, pessimistically, one might expect that all the well-known difficulties of the standard (i.e., Hamiltonian and path integral) approaches to quantum gravity will show up again, though, maybe, in a different guise. From the optimist's point of view it is precisely this transformation of an old problem into a new guise that might offer some new insight, however. We feel that stochastic quantization does indeed live up to this expectation in the case of the gravitational field: We hope that our main result, contained in Sec. V, is of interest also to those who have reservations about the stochastic approach.

The main obstacle to applying the Parisi-Wu method in its original form to the metric tensor field of general relativity is the fact that the Euclidean Einstein-Hilbert action is not bounded from below. This necessitates a modification of the standard Euclidean path integral[2] and hence also of the stochastic formalism in which this path integral is expected to define the equilibrium distribution of a stochastic relaxation process. Several *ad hoc* modifications of stochastic quantization have been proposed to overcome this difficulty.[3–5] In our opinion the physically most promising one is to generalize stochastic quantization so as to make it applicable to fields in physical space-time with Lorentzian metric signature.[6,7] The linearized gravitational field has already been treated in this manner.[8] In this paper we propose a nonperturbative stochastic quantization scheme for the metric tensor in the full nonlinear Einstein theory. This scheme differs from another one proposed some time ago.[9] The perturbation theory implied by our scheme is found to be different from that employed in Ref. 5.

In Sec. II we show that the principle of general covariance with respect to field redefinitions determines a one-parameter family of Langevin equations for the metric tensor field of general relativity. The parameter is the same as in DeWitt's metric on the space of four-metrics. In Sec. III the stochastic source term of the Langevin equation is expressed explicitly in terms of a stochastic vielbein functional and a Gaussian white noise. A so-called Ito-Stratonovich ambiguity in this expression is resolved by requiring that the stochastic metric be independent of the choice of the vielbein functional, while maintaining the covariance of the Langevin equation with respect to general coordinate transformations. In Sec. IV we associate an "analytically continued" Fokker-Planck equation with every complex Langevin equation and derive from its equilibrium limit a covariant path-integral measure, again depending on one parameter, for general relativity. In Sec. V the existence of a preferred parameter is pointed out, distinguished by (i) harmonicity of the metric with respect to the DeWitt metric, (ii) coincidence of the measure with the DeWitt measure, (iii) the special role it plays in the linearized Einstein dynamics and causality of the corresponding propagator. The linearization corresponding to this parameter is not the standard linearization, however, since it has been shown[8] that in the latter the propagator of stochastic quantization is noncausal. In Sec. VI we sketch the perturbation theory implied by our covariant quantization scheme and show that it gives rise to a new type of diagram containing "stochastic vertices." A final section is reserved for concluding remarks.

II. COVARIANT LANGEVIN EQUATIONS

Our main guideline for the formulation of a Langevin equation for the metric tensor field $g_{\alpha\beta}(x)$ will be the notion of general covariance with respect to field redefinitions in field configuration space. The usefulness of this concept in quantum field theory has been stressed recently.[10,11] For conciseness we shall adopt the notation of DeWitt (see, e.g., Ref. 12) and represent a general stochastic field by $\Phi^A(s)$ where s is the fictitious evolution parameter and $A=(\alpha,\beta,\ldots;x)$ comprises component indices as well as the space-time coordinates. By covariance, then, the general form of the Lorentzian generalization[6] of the Parisi-Wu ansatz[1] for the Langevin equation for Φ^A must be

$$\dot{\Phi}^A \equiv \frac{\partial \Phi^A}{\partial s} = iG^{AA'}\frac{\delta S[\Phi]}{\delta \Phi^{A'}(s)} + \xi^A(s) . \tag{2.1}$$

Here $G^{AA'}$ is the inverse of a metric tensor functional $G_{AA'}[\Phi]$ in field configuration space, and the summation over A' includes a space-time integration. The stochastic source term $\xi^A(s)$ will be defined here only implicitly by requiring that all its correlations vanish with the exception of

$$\langle \xi^A(s)\xi^{A'}(s')\rangle = 2\langle G^{AA'}[\Phi(s)]\rangle \delta(s-s') . \tag{2.2}$$

A more explicit definition of ξ^A will be given in the next section.

Note that (2.2) implies

$$\langle \xi^A(s)\xi^{A'}(s')\rangle = 2G^{AA'}\delta(s-s') \tag{2.3}$$

only if $G_{AA'}$ is independent of Φ (and hence of ξ). For nongravitational fields there exists a field coordinate system (defining the "natural field variables") for which $G_{AA'}$ is indeed independent of Φ. However, for the gravitational field

$$\Phi^A \equiv g_{\alpha\beta}(x) , \tag{2.4}$$

$G_{AA'}$ must depend on Φ for reasons of ordinary general covariance (except we adopt a bimetric theory). [The index positions in (2.4) should cause no confusion, as usually the covariant metric is taken as the gravitational field variable, which, when considered as a field coordinate, must carry an upper index.] Indeed, ordinary covariance requires that the actions of general coordinate transformations on $g_{\alpha\beta}(x)$ be isometries with respect to the field metric $G_{AA'}$. The most general *local* metric $G_{AA'}$ that has this property is known[12] to be

$$\begin{aligned} G_{AA'} &= G^{\alpha\beta,\alpha'\beta'}(x,x') \\ &= \frac{C}{2}\,|g|^{1/2}(g^{\alpha\alpha'}g^{\beta\beta'} + g^{\alpha\beta'}g^{\beta\alpha'} + \lambda g^{\alpha\beta}g^{\alpha'\beta'}) \\ &\quad \times \delta^{(4)}(x-x') , \end{aligned} \tag{2.5}$$

$$g = \det(g_{\alpha\beta}) , \tag{2.6}$$

$$\lambda \neq -\tfrac{1}{2} . \tag{2.7}$$

The constant C in (2.5) must be different from zero, but otherwise its choice is irrelevant: It can be seen from (2.1) and (2.2) that a positive rescaling of C, $C \to \gamma C$, $\gamma > 0$, corresponds merely to a rescaling of the parameter s, $s \to \gamma s$. Even a change of the sign of C does not affect the equilibrium limit as will be seen in Sec. IV. We are thus left with a one-parameter (λ) family of field metrics. For later reference we note the determinant and the inverse of the metric (2.5):

$$G \equiv \det(G_{AA'}) = \prod_x C^{10}(-1-2\lambda) , \tag{2.8}$$

$$\begin{aligned} G^{AA'} &\equiv G_{\alpha\beta,\alpha'\beta'}(x,x') \\ &= (2C\,|g|)^{-1/2}\left[g_{\alpha\alpha'}g_{\beta\beta'} + g_{\alpha\beta'}g_{\beta\alpha'} - \frac{\lambda}{2\lambda+1}g_{\alpha\beta}g_{\alpha'\beta'} \right] \\ &\quad \times \delta^{(4)}(x-x') . \end{aligned} \tag{2.9}$$

In particular the determinant G is constant in the usual parametrization of geometries, a property that holds only if the number of space-time dimensions is four.[12] Moreover we see from (2.8) that the signature of $G_{AA'}$ depends on λ.

Let us now insert the Einstein-Hilbert action

$$S[g] = \frac{1}{2\kappa}\int d^4x\,|g|^{1/2}R , \tag{2.10}$$

$$\frac{\delta S}{\delta g_{\alpha\beta}} = -\frac{1}{2\kappa}|g|^{1/2}(R^{\alpha\beta} - \tfrac{1}{2}g^{\alpha\beta}R) , \tag{2.11}$$

into the Langevin equation (2.1). As we want to retain the canonical dimension L^2 of s, we choose

$$C = 4\kappa . \tag{2.12}$$

Then we obtain

$$\dot{g}_{\alpha\beta} = -2i\left[R_{\alpha\beta} - \frac{\lambda+1}{2(2\lambda+1)}g_{\alpha\beta}R \right] + \xi_{\alpha\beta} . \tag{2.13}$$

Note that ξ does not have its usual canonical dimension L^{-3} in Eq. (2.13). By the field rescaling

$$\bar{g}_{\alpha\beta} = g_{\alpha\beta}/(2\kappa^{1/2}) \tag{2.14}$$

one recovers this canonical dimension, however, and moreover, $\tilde{G}_{AA'}$ becomes dimensionless with

$$|\tilde{G}| = 1 \quad \text{if } \lambda = 0 . \tag{2.15}$$

As is easily seen, the drift term in the Langevin equation (2.13) is proportional to the Einstein tensor only if $\lambda=0$. Also, only in this case does one obtain the standard Langevin equation for the linearized gravitational field discussed in Ref. 8 [the numerical factor in (2.12) has been chosen such that then also the parameter s is the same]. This appears to give the value $\lambda=0$ a preferred status. It will be argued in Sec. V, however, that a different value ($\lambda=-1$) has all the prospects of describing the correct physics. For the present we shall leave λ unspecified.

III. RESOLUTION OF AN ITO-STRATONOVICH DILEMMA

Definitions (2.1) and (2.2) of the stochastic processes Φ^A and ξ^A are purely formal and not useful for practical calculations. Our aim is to reduce (2.1) to a more familiar type of stochastic differential equation by giving a more explicit definition of ξ^A. This may be accomplished as follows. We introduce a field-independent "reference metric" $G^{(0)}_{MM'}$ in field configuration space and associate with it a stochastic vielbein functional $E^M{}_A[\phi]$ obeying

$$G^{(0)}_{MM'}E^M{}_A[\phi]E^{M'}{}_{A'}[\phi']=G_{AA'}[\phi] \ . \tag{3.1}$$

With the help of the vielbein we may define the anholonomic "vector components" $\xi^{(0)M}$ by

$$\xi^{(0)M}=E^M{}_A[\phi]\xi^A \tag{3.2}$$

or

$$\xi^A=E_M{}^A[\phi]\xi^{(0)M} \tag{3.3}$$

with

$$E_M{}^A E^M{}_{A'}=\delta^A{}_{A'} \ . \tag{3.4}$$

It is then natural to take (3.3) as an ansatz to satisfy (2.2) and to define $\xi^{(0)M}$ as a generalized Gaussian white noise with nonvanishing correlation

$$\langle \xi^{(0)M}(s)\xi^{(0)M'}(s')\rangle=2G^{(0)MM'}\delta(s-s') \ , \tag{3.5}$$

where $G^{(0)MM'}$ is the inverse of $G^{(0)}_{MM'}$.

When (3.3) is substituted into (2.1), the Langevin equation is recognized to be of the type studied by Ito[13] and Stratonovich.[14] It is well known (see, e.g., Ref. 15) that a stochastic differential equation of this type is not well defined because the Wiener-type process $W^M(s)$ whose increment

$$dW^M=\xi^{(0)M}ds \tag{3.6}$$

appears in the associated stochastic integral is of unbounded variation (or, put differently, $\xi^{(0)M}$ is a generalized process, its sample "functions" being distributions). The product $E_N{}^A[\Phi]dW^N(s)$ may be interpreted in infinitely many different ways:

$$E_N{}^A[\phi]dW^N(s) = \lim_{\Delta s\to 0^+}\{(1-\alpha)E_N{}^A[\Phi(s)]+\alpha E_N{}^A[\Phi(s+\Delta s)]\} \times[W^N(s+\Delta s)-W^N(s)] \ , \tag{3.7}$$

$$0\le\alpha\le 1 \ . \tag{3.8}$$

From the mathematical point of view there are two natural choices of α: $\alpha=0$, defining Ito's calculus, and $\alpha=\frac{1}{2}$, defining Stratonovich's calculus. In the following, we show that $\alpha=0$ is dictated by physical requirements. We shall make extensive use of results on covariant stochastic calculus obtained by Graham.[16]

Let us consider the Stratonovich interpretation first. Its attractive feature is that it implies that each term in (2.1) transforms as a "contravariant vector" under arbitrary coordinate transformations in field configuration space. On the other hand, for $\alpha>0$ (3.7) and (2.1) imply that $\Phi^A(s)$ depends statistically on $W^N(s+ds)-W^N(s)$, and hence in general

$$\langle F[\Phi(s)]\xi^A(s)\rangle\neq 0 \ , \tag{3.9}$$

where F is an arbitrary functional of $\Phi(s)$. Consequently there appear spurious terms in

$$\langle \xi^A(s)\xi^{A'}(s')\rangle=\langle E_M{}^A(s)E_{M'}{}^{A'}(s')\rangle \times\langle \xi^{(0)M}(s)\xi^{(0)M'}(s')\rangle+\cdots \tag{3.10}$$

that violate (2.2). Equation (2.1) with the Stratonovich interpretation for (3.3) is also not form invariant under a Φ-dependent change of the vielbein, in contrast with naive expectations. Indeed, if we perform a Φ-dependent, locally $G^{(0)}$-isometric transformation of the vielbein functional,

$$E_M{}^A[\Phi]\to\Lambda_M{}^N[\Phi]E_N{}^A[\Phi] \ , \tag{3.11}$$

$$\xi^{(0)M}\to\Lambda^M{}_N[\Phi]\xi^{(0)N} \ , \tag{3.12}$$

$$G^{(0)MN}\Lambda_M{}^P\Lambda_N{}^Q=G^{(0)PQ} \ , \tag{3.13}$$

$$\Lambda_M{}^P\Lambda^N{}_P=\delta_M{}^N \ , \tag{3.14}$$

then a spurious term

$$-\frac{\delta\Lambda^M{}_P}{\delta\Phi^B}\Lambda_N{}^P E_M{}^A E^{NB} \ , \tag{3.15}$$

$$E^{NB}:=G^{BC}E^N{}_C \ , \tag{3.16}$$

adds to the "drift vector" in (2.1). (It is even possible to transform the drift vector to zero by an appropriate Φ-dependent "rotation."[17]) Therefore a different choice of vielbein defines a different process Φ^A, which is clearly not desirable.

The physical consequences of the vielbein dependence can be read off from the Fokker-Planck (FP) equation implied by the Stratonovich interpretation of (3.3) in (2.1). Since Φ^A is a complex process, this FP equation involves partial derivatives with respect to the real and imaginary parts of Φ^A. Because we are interested in the limit $s\to\infty$ which is expected to exist only in the distributional sense,[6] we shall not write down this correct FP equation, but a modified one, obtained from the FP equation for the Euclidean analog of Φ^A by the analytic continuation

$$S_{\rm Eucl}\to -iS \ , \tag{3.17}$$

where $S_{\rm Eucl}$ is the Euclidean action. It has been conjectured by Parisi[18] that the *real* process (with *complex* probabilities) described by this modified FP equation has the same expectation values as the original *complex* process with *real* probabilities).

The modified FP equation for the Stratonovich interpretation of (3.3) and (2.1) reads

$$\frac{\partial P[\Phi,s]}{\partial s}=\left[-\frac{\delta}{\delta\Phi^A}K^A[\Phi] + \frac{\delta^2}{\delta\Phi^A\delta\Phi^B}G^{AB}[\Phi]\right]P[\Phi,s] \ , \tag{3.18}$$

$$K^A=iG^{AB}\frac{\delta S}{\delta\Phi^B}+\frac{\delta E_M{}^A}{\delta\Phi^B}E_N{}^B G^{(0)MN} \ , \tag{3.19}$$

$$\langle K^A\rangle=\langle\dot\Phi^A\rangle \ . \tag{3.20}$$

Equation (3.19) shows the explicit dependence of the probability distribution $P[\Phi,s]$ on the choice of the vielbein even in the equilibrium case $\partial P/\partial s=0$. The Stratonovich interpretation is therefore not viable in our case. The same conclusion may be reached for all other values of α in (3.7) different from zero.

If the Langevin equation (2.1) with the ansatz (3.3) and (3.5) is interpreted in the sense of Ito, the unphysical dependence on the choice of the vielbein disappears. In particular (3.10) reduces to (2.2) because of the "nonanticipating" nature of $\Phi(s)$, and hence of $E_M{}^A[\Phi(s)]$, with respect to $dW(s)$. On the other hand, in Ito's calculus $\dot\Phi^A$ is not a contravariant "vector," but transforms inhomogeneously according to

$$\dot\Phi'^A=\frac{\delta\Phi'^A}{\delta\Phi^B}\dot\Phi^B+G^{BC}\frac{\delta^2\Phi'^A}{\delta\Phi^B\delta\Phi^C} \tag{3.21}$$

(the same rule holds for K^A in Stratonovich's calculus). Therefore the Langevin equation (2.1) is *not* manifestly covariant with respect to field redefinitions, but assumes its particular form only in a special class of "coordinate" systems.

A generally covariant stochastic quantization scheme may be based on the covariantized Langevin equation

$$\dot\Phi^A-\Delta_G\Phi^A=iG^{AB}\frac{\delta S[\Phi]}{\delta\Phi^B}+\xi^A\,, \tag{3.22}$$

where Δ_G denotes the Laplace-Beltrami operator of the field metric G_{AB}. The left-hand side of (3.22) is indeed a "vector," because

$$\Delta_G\Phi^A=|G|^{-1/2}\frac{\delta}{\delta\Phi^B}(|G|^{1/2}G^{AB}) \tag{3.23}$$

transforms as

$$\Delta_{G'}\Phi'^A=\Delta_G\Phi^A+G^{BC}\frac{\delta^2\Phi'^A}{\delta\Phi^B\delta\Phi^C}\,. \tag{3.24}$$

In the present paper, however, we shall stick mainly to the "naive" Langevin equation (2.1) with the field variable (2.4), because only this version allows us to obtain also the noncovariant (in field configuration space) path-integral measures that have been proposed in the literature. Note that (2.1) is still covariant with respect to diffeomorphisms, because the latter act linearly on Φ^A and hence $\delta^2\Phi'^A/\delta\Phi^B\delta\Phi^C$ vanishes.

Finally we mention a further advantage of Ito's calculus. As indicated above, the process $\Phi^A(s)$ is "causal," $\Phi^A(s)$ and $dW^N(s')$ being uncorrelated for $s\le s'$. Therefore a discretized version of equation (2.1) can in principle be used to simulate sample "paths" of $\Phi^A(s)$ on a computer by repetitive calls of a Gaussian random number generator for the Wiener increments.[19]

IV. FORMAL EQUILIBRIUM LIMIT AND PATH-INTEGRAL MEASURE

The modified FP equation implied by the Ito version of the Langevin equation (2.1) is

$$\frac{\partial P[\Phi,s]}{\partial s}=\left[-\frac{\delta}{\delta\Phi^A}F^A[\Phi]+\frac{\delta^2}{\delta\Phi^A\delta\Phi^B}G^{AB}[\Phi]\right]P[\Phi,s]\,, \tag{4.1}$$

$$F^A=iG^{AB}\frac{\delta S}{\delta\Phi^B}\,, \tag{4.2}$$

$$\langle F^A\rangle=\langle\dot\Phi^A\rangle\,. \tag{4.3}$$

The probability distribution $P[\Phi,s]$ being a scalar density in field configuration space, it is natural to introduce the corresponding "scalar field" Q defined by

$$Q[\Phi,s]=|G|^{-1/2}P[\Phi,s]\,. \tag{4.4}$$

Equation (4.1) implies

$$\frac{\partial Q}{\partial s}=-|G|^{-1/2}\frac{\delta}{\delta\Phi^A}|G|^{1/2}\times\left[F^A-\Delta_G\Phi^A-G^{AB}\frac{\delta}{\delta\Phi^B}\right]Q\,. \tag{4.5}$$

This equation is not covariant, because the term in parentheses is not a "vector." This is not surprising, as we have seen in the last section that (2.1) is not covariant in the Ito interpretation. If we adopted (3.22) instead of (2.1), then the spurious term $-\Delta_G\Phi^A$ would disappear in the parentheses in (4.5) and we would obtain

$$\frac{\partial Q}{\partial s}=-\nabla^B\left[\frac{\delta S}{\delta\Phi^B}Q-\frac{\delta Q}{\delta\Phi^B}\right] \tag{4.6}$$

which is manifestly covariant. Here $\nabla^B=G^{BC}\nabla_C$ with ∇_C the covariant derivative with respect to Φ^C implied by the Levi-Cività connection $\Gamma^A{}_{BC}$ of the field metric G_{AB}:

$$\Gamma^A{}_{BC}=\frac{1}{2}G^{AJ}\left[\frac{\delta G_{JB}}{\delta\Phi^C}+\frac{\delta G_{JC}}{\delta\Phi^B}-\frac{\delta G_{BC}}{\delta\Phi^J}\right]. \tag{4.7}$$

With regard to the question of the equivalence of stochastic quantization with path-integral quantization it is interesting to check whether the modified FP equation admits an equilibrium solution Q_{eq} proportional to e^{iS}. The stochastic equilibrium averages will then coincide with the Schwinger averages of quantum field theory. A sufficient condition for the stationarity of Q_{eq} in (4.5) is

$$\frac{\delta Q_{eq}}{\delta\Phi^A}=\left[i\frac{\delta S}{\delta\Phi^A}-G_{AB}\Delta_G\Phi^B\right]Q_{eq}\,. \tag{4.8}$$

Evaluation of the second term on the right-hand side (RHS) of this equation gives

$$-G_{AB}\Delta_G\Phi^B\equiv-\int d^4y\,G^{\alpha\beta,\gamma\delta}(x,y)\frac{\delta G_{\gamma\delta,\alpha'\beta'}(y,x')}{\delta g_{\alpha'\beta'}(x')}=-\tfrac{9}{2}(1+\lambda)g^{\alpha\beta}(x)\delta^{(4)}(0)\,. \tag{4.9}$$

Now

$$\delta^{(4)}(0)g^{\alpha\beta}(x)=\frac{\delta}{\delta g_{\alpha\beta}(x)}\left[\sum_y\ln|g(y)|\right]. \tag{4.10}$$

Therefore

$$Q_{eq}[g]\propto\prod_x|g(x)|^\gamma e^{iS[g]}\,, \tag{4.11}$$

$$\gamma=-\tfrac{9}{2}(1+\lambda)\,. \tag{4.12}$$

Thus Q_{eq} contains in addition to e^{iS} also a nontrivial path-integral measure, and we obtain the following formal expression for the equilibrium expectation value of an arbitrary functional $F[g]$:

$$\langle F[g]\rangle \propto \int \prod_x \prod_{\alpha\le\beta} |g(x)|^{\gamma} dg_{\alpha\beta}(x) F[g] e^{iS[g]} . \quad (4.13)$$

This expression shows that the Langevin equation (2.1) with $\Phi^A = g_{\alpha\beta}$ can reproduce any covariant (with respect to diffeomorphisms) measure based on the integration variable $g_{\alpha\beta}$ with the exception of $\gamma = -\frac{9}{4}$. The values $\gamma = -\frac{5}{2}$ (Refs. 20–22) and $\gamma = 0$ (Refs. 23 and 24) have been considered most frequently. A unique noncovariant measure with weight $|g|^{-5/2}|^{(3)}g|$, resulting from a lattice discretization[25] and also from a canonical formulation[26], has also been proposed. Note that DeWitt's choice $\gamma = 0$ is obtained only for $\lambda = -1$. Some unique implications of this parameter value will be pointed out in the next section.

As is well known, the path integral (4.13) is not normalizable even in a formal sense because of the invariance of S under diffeomorphisms. By analogy with the situation in gauge theories[27,28] one is led to conjecture, however, that stochastic gauge fixing[29] will yield an equilibrium distribution that is formally identical with the Faddeev-Popov distribution of standard quantization in an appropriate gauge. An investigation in this direction has already been carried out by Sakamoto,[9] who uses an approach different from ours.

V. HARMONICITY, DeWITT MEASURE AND LINEARIZED EINSTEIN DYNAMICS

The DeWitt path-integral measure [$\gamma = 0$ in (4.13)] appears attractive because of its invariance under general coordinate transformations. It is obtained from (2.1) only if $\lambda = -1$. The reason is that only in this case

$$\Delta_G \Phi^A \equiv \nabla_C \nabla^C \Phi^A = 0 ; \quad (5.1)$$

i.e., $\lambda = -1$ is the unique parameter value for which $g_{\alpha\beta}(x)$ are harmonic "coordinates."

Of course the covariantized Langevin equation (3.22) implies via (4.6) the generally covariant equilibrium distribution

$$P_{eq}[\Phi] \propto |G[\Phi]|^{1/2} e^{iS[\Phi]} \quad (5.2)$$

which is independent of λ because of (2.8) and coincides with (4.11) in the case $\lambda = -1$.

We note in passing that besides $g_{\alpha\beta}$ also $|g|^{5/2} g_{\alpha\beta}$ is harmonic for $\lambda = -1$. In general there are two real solutions of the type $|g|^r g_{\alpha\beta}$ to the harmonic coordinate condition (5.1) except in the interval $\lambda > \frac{7}{16}$ where r becomes complex. There is a unique parameter λ for a given exponent r such that $|g|^r g_{\alpha\beta}$ is harmonic, with the exception of $r = -\frac{1}{4}$ and $-\frac{9}{4}$, where no solution exists (note that for $r = -\frac{1}{4}$, $|g|^r g_{\alpha\beta}$ are not valid coordinates because they comprise only the conformally invariant part of the metric $g_{\alpha\beta}$). Another remarkable feature of $\lambda = -1$ is that only in this case $G_{AB} d\Phi^B$ is a differential, namely,

$$\delta(-|g|^{1/2} g^{\alpha\beta}) = \int d^4x' G^{\alpha\beta,\alpha'\beta'} \delta g_{\alpha'\beta'} \quad \text{if } \lambda = -1 .$$

Therefore, and because G is constant, the integration variable $g_{\alpha\beta}$ in (4.13) may be replaced by $|g|^{1/2} g^{\alpha\beta}$, if $\lambda = -1$. The variable $g_{\alpha\beta}$ is the only one for which a relationship of this type and harmonicity hold at the same value of λ.

Up to this point in the paper our considerations have been completely independent of the special choice of action $S[\Phi]$. We now want to examine the stochastic dynamics generated by the Einstein action in the linear limit defined by

$$g_{\alpha\beta} = \eta_{\alpha\beta} + 2\kappa^{1/2}\psi_{\alpha\beta} , \quad (5.3)$$

$$S[g] = S^{(0)}[\psi] + S^{(\text{int})}[\psi] , \quad (5.4)$$

$$S^{(0)}[\psi] = \tfrac{1}{2} \int d^4x\, \psi_{\alpha\beta} V^{\alpha\beta\gamma\delta} \psi_{\gamma\delta} . \quad (5.5)$$

Here η is the Minkowski metric, and all indices will be transvected with this metric. For the parameter value $\lambda = 0$ the Langevin equation for $\psi_{\alpha\beta}(x,s)$ implied by the action $S^{(0)}$ has already been studied in an earlier paper,[8] to which we refer the reader for more details on the technical points of the discussion that follows. Our aim is to derive the Feynman propagator for a general value of λ. It is convenient to work in momentum space where the linear operator V of (5.5) is given by

$$V = k^2 (P^2 - 2P^{0'}) . \quad (5.6)$$

Here P^2 and $P^{0'}$ are members of a complete orthogonal set $\{P^2, P^1, P^0, P^{0'}\}$ of projection operators on the space of symmetric tensor fields, the superscript indicating their (massive) spin value.[30]

In the linear limit the Langevin equation (2.15) becomes

$$\dot\psi_{\alpha\beta} = i W_{\alpha\beta}{}^{\gamma\delta} \psi_{\gamma\delta} + \xi_{\alpha\beta} , \quad (5.7)$$

$$W_{\alpha\beta}{}^{\gamma\delta} \equiv G^{-1}{}_{\alpha\beta\mu\nu} V^{\mu\nu\gamma\delta} , \quad (5.8)$$

where the inverse supermetric is simply

$$G^{-1}{}_{\alpha\beta\gamma\delta} = \tfrac{1}{2}(\eta_{\alpha\gamma}\eta_{\beta\delta} + \eta_{\alpha\delta}\eta_{\beta\gamma} + \mu\eta_{\alpha\beta}\eta_{\gamma\delta}) , \quad (5.9)$$

$$\mu \equiv -\lambda/(2\lambda+1) , \quad (5.10)$$

and ξ is Gaussian with correlation

$$\langle \xi_{\alpha\beta}(k,s) \xi_{\alpha'\beta'}(k',s') \rangle = 2(2\pi)^4 G^{-1}{}_{\alpha\beta\alpha'\beta'} \delta^{(4)}(k+k') \delta(s-s') . \quad (5.11)$$

If $\psi(x,s)$ is subjected to the initial condition $\psi(x,0) = 0$ as usual (for a discussion of different initial conditions see Ref. 8), then the solution of (5.7) is given by

$$\psi(s) = \int_0^s H(s-s') \xi(s') ds' , \quad (5.12)$$

where H is the "Schrödinger kernel"

$$H(s) = e^{iWs} \quad (5.13)$$

corresponding to the "Hamiltonian" W. Equation (5.8) implies

$$W_{\alpha\beta}{}^{\gamma\delta} = V_{\alpha\beta}{}^{\gamma\delta} - \mu k^2 \eta_{\alpha\beta} T^{\gamma\delta} , \quad (5.14)$$

$$T^{\gamma\delta} := \eta^{\gamma\delta} - L^{\gamma\delta} , \quad (5.15)$$

$$L^{\gamma\delta}:=k^\gamma k^\delta/(k^2+i0)\,. \tag{5.16}$$

Since W is not a linear combination of orthogonal projections, H is most practically calculated using the relation

$$G(k,s)=\theta(s)H(k,s)\,, \tag{5.17}$$

where G is the retarded Green's function of the deterministic part of Eq. (5.7) obeying

$$\left[\frac{\partial}{\partial s}-iW\right]G(k,s)=\delta(s)\mathbf{1}\,. \tag{5.18}$$

Here $\mathbf{1}$ is the unit operator on symmetric tensor fields,

$$\mathbf{1}_{\alpha\beta}{}^{\gamma\delta}=\delta_{(\alpha}{}^{\gamma}\delta_{\beta)}{}^{\delta}\,. \tag{5.19}$$

The Green's function G can be computed by Fourier transforming (5.18) with respect to s:

$$G(k,s)=\frac{i}{2\pi}\int_{-\infty}^{\infty}d(m^2)K(k,m^2-i0)e^{im^2s}\,, \tag{5.20}$$

$$(W-m^2\mathbf{1})K(m^2)=\mathbf{1}\,. \tag{5.21}$$

Equation (5.21) defines the "propagator" corresponding to the "field equation"

$$(W-m^2\mathbf{1})\phi=0\,, \tag{5.22}$$

$$K=\frac{P^2}{k^2-m^2}-\frac{P^1}{m^2}-\frac{P^0}{m^2}-\frac{P^{0'}}{2k^2+m^2}+\frac{\lambda k^2\eta T}{[(3\lambda+2)k^2+m^2](2k^2+m^2)}+\frac{2\lambda k^4LT}{m^2[(3\lambda+2)k^2+m^2](2k^2+m^2)}\,, \tag{5.23}$$

where $(\eta T)_{\alpha\beta}{}^{\gamma\delta}=\eta_{\alpha\beta}T^{\gamma\delta}$, etc. Using complex integration in (5.20) we obtain via (5.7)

$$H=e^{ik^2s}P^2+P^1+P^0+e^{-2ik^2s}P^{0'}+\tfrac{1}{3}(e^{-i(3\lambda+2)k^2s}-e^{-2ik^2s})\eta T-\left[\frac{\lambda}{3\lambda+2}+\frac{2}{3(3\lambda+2)}e^{-i(3\lambda+2)k^2s}-\tfrac{1}{3}e^{-2ik^2s}\right]LT$$

$$=e^{ik^2s}P^2+P^1+P^0+\tfrac{1}{3}e^{-i(3\mu+2)k^2s}\eta T-\left[\frac{\mu}{3\mu+2}+\frac{2}{3(3\mu+2)}e^{-i(3\mu+2)k^2s}\right]LT\,. \tag{5.24}$$

From (5.12), (5.24), and (5.11) we may calculate the "stochastic propagator"

$$D_{\alpha\beta\alpha'\beta'}(k,s;k',s'):=\langle\psi_{\alpha\beta}(k,s)\psi_{\alpha'\beta'}(k',s')\rangle \tag{5.25}$$

$$=2(2\pi)^4\delta^{(4)}(k+k')\int_0^{\min(s,s')}d\sigma\,H_{\alpha\beta}{}^{\gamma\delta}(k,s-\sigma)H_{\alpha'\beta'}{}^{\gamma'\delta'}(k',s'-\sigma)G^{-1}{}_{\gamma\delta\gamma'\delta'}\,. \tag{5.26}$$

In this section we are not interested in its explicit form, but only in its equilibrium limit, which defines the Feynman propagator K:

$$\lim_{s\to\infty}D_{\alpha\beta,\alpha'\beta'}(k,s;k',s)=i(2\pi)^4\delta^{(4)}(k+k')K_{\alpha\beta,\alpha'\beta'}(k)\,. \tag{5.27}$$

Since $H(k)=H(-k)$, we have

$$K_{\alpha\beta,\alpha'\beta'}(k)=-2i\int_0^{\infty}d\tau\left[H_{\alpha\beta}{}^{\gamma\delta}(k,\tau)H_{\alpha'\beta'\gamma\delta}(k,\tau)+\frac{\mu}{2}H_{\alpha\beta}{}^{\gamma}{}_{\gamma}(k,\tau)H_{\alpha'\beta'}{}^{\delta}{}_{\delta}(k,\tau)\right]. \tag{5.28}$$

The explicit result is

$$K=K^{(0)}+R\,, \tag{5.29}$$

$$K^{(0)}=\frac{P^2}{k^2+i0}-\frac{1}{2}\frac{P^{0'}}{k^2-i0}-i\,\infty^2(P^1+P^0)\,, \tag{5.30}$$

$$R=\frac{1}{6}\left[\frac{1}{k^2-i0}-\frac{1}{k^2-iq0}\right]\eta\eta+\frac{1}{3}\left[\frac{1}{(3\mu+2)k^2-i0}-\frac{1}{2}\frac{1}{k^2-i0}\right](\eta L+L\eta)+\left[\frac{1}{6k^2-i0}-\frac{2}{3(3\mu+2)}\frac{1}{3(\mu+2)k^2-i0}-i\frac{\mu}{3\mu+2}\infty^2\right]LL\,, \tag{5.31}$$

$$q:=\operatorname{sgn}(3\mu+2)\,, \tag{5.32}$$

$$\infty^{2n}=2\int_0^{\infty}ds\,s^{n-1}\,. \tag{5.33}$$

One recognizes in $K^{(0)}$ the propagator of the standard linearization[8] ($\lambda=\mu=0$) and checks easily that R vanishes for $\mu=0$. Recalling that

$$P^{0'}=\tfrac{1}{3}TT\,, \tag{5.34}$$

$$P^0=LL\,, \tag{5.35}$$

we conclude from (5.29)–(5.32) that

$$K=\frac{P^2}{k^2+i0}-\frac{1}{6}\frac{\eta\eta}{k^2-iq0}-i\infty^2P^1-\left[\frac{2}{3(3\mu+2)}\frac{1}{(3\mu+2)k^2-i0}+\frac{4\mu+2}{3\mu+2}i\infty^2\right]P^0+\frac{1}{3}\frac{1}{(3\mu+2)k^2-i0}(\eta L+L\eta)\ . \tag{5.36}$$

Thus the whole propagator is causal if

$$\mu<-\tfrac{2}{3}\Longleftrightarrow -2<\lambda<-\tfrac{1}{2}\ . \tag{5.37}$$

The quadratic divergences in K are harmless as they do not contribute to gauge-invariant quantities. In fact the only part of K contributing to gauge-invariant expectation values is

$$K^{(\mathrm{inv})}_{\alpha\beta,\alpha'\beta'}(k)=\tfrac{1}{2}(\eta_{\alpha\alpha'}\eta_{\beta\beta'}+\eta_{\alpha\beta'}\eta_{\beta\alpha'}-\tfrac{2}{3}\eta_{\alpha\beta}\eta_{\alpha'\beta'})\frac{1}{k^2+i0}-\frac{1}{6}\eta_{\alpha\beta}\eta_{\alpha'\beta'}\frac{1}{k^2-iq0} \tag{5.38}$$

$$=\frac{G^{-1}{}_{\alpha\beta,\alpha'\beta'}(\lambda=-1)}{k^2+i0}\quad \text{if } \mu<-\tfrac{2}{3}\ . \tag{5.39}$$

We find it remarkable that the field metric with parameter value $\lambda=-1$ appears in the propagator. Another remarkable property is that the whole of K assumes an especially simple form if $\lambda=-1$. Making use of

$$P^2{}_{\alpha\beta\gamma\delta}=\tfrac{1}{2}(T_{\alpha\gamma}T_{\beta\delta}+T_{\alpha\delta}T_{\beta\gamma})-\tfrac{1}{3}T_{\alpha\beta}T_{\gamma\delta} \tag{5.40}$$

we obtain

$$K^{(-1)}_{\alpha\beta,\alpha'\beta'}=\tfrac{1}{2}(\eta_{\alpha\alpha'}\eta_{\beta\beta'}+\eta_{\alpha\beta'}\eta_{\beta\alpha'}-\eta_{\alpha\beta}\eta_{\alpha'\beta'}+\eta_{\alpha\alpha'}k_\beta k_{\beta'}+\eta_{\alpha\beta'}k_\beta k_{\alpha'}+\eta_{\beta\alpha'}k_\alpha k_{\beta'}+\eta_{\beta\beta'}k_\alpha k_{\alpha'})\frac{1}{k^2+i0}-i\infty^2P^1-2i\infty^2P^0\ . \tag{5.41}$$

The case $\mu=-\frac{2}{3}$ must be treated separately. Since in this case there is a double pole in the last term of (5.23), the last term of (5.24) becomes

$$(-1+\tfrac{2}{3}ik^2s)LT\ . \tag{5.42}$$

Consequently

$$R=\frac{\eta\eta}{6k^2-i0}+\left[\frac{i}{3}\infty^2-\frac{1}{6k^2-i0}\right](\eta L+L\eta)+\left[\frac{1}{6k^2-i0}-\frac{i}{3}\infty^2-\frac{2}{3}k^2\infty^4\right]LL\ ,\quad (\mu=-\tfrac{2}{3})\ . \tag{5.43}$$

Note that the first term on the RHS of (5.43) differs from what would be obtained if the limit $\mu\to-\frac{2}{3}$ were taken after the limit $s\to\infty$, and in a symmetrical manner, namely,

$$\frac{1}{6}\left[\frac{1}{k^2-i0}-P\frac{1}{k^2}\right]\eta\eta\ , \tag{5.44}$$

P denoting the principal value. This discrepancy will be seen shortly to be related to the so-called van Dam–Veltman mass discontinuity.[31]

Having evaluated the Feynman propagator implied by stochastic quantization it is natural to seek an inverse to it. This is a nontrivial problem because of the divergences present in K. In standard quantization the Feynman propagator is obtained by inverting the linear differential operator defined by the free part of the action. If a gauge invariance is present, as in the case of the linearized gravitational field, the propagator does not exist. The usual way around this difficulty is to break the gauge invariance of the action by adding a gauge-fixing term to the Lagrangian. Adding a mass term to the Lagrangian may be considered as a special case of gauge fixing. In the following we show that K may be identified with the inverse of a certain massive extension of linearized gravity in the (singular) limit as the mass goes to zero.

It is natural to consider the massive extension

$$V^{\alpha\beta\gamma\delta}\to V^{\alpha\beta\gamma\delta}-M^2G^{\alpha\beta\gamma\delta} \tag{5.45}$$

with G the inverse of the field metric (5.9). The extension (5.45) implies

$$W_{\alpha\beta}{}^{\gamma\delta}\to W_{\alpha\beta}{}^{\gamma\delta}-M^2 1_{\alpha\beta}{}^{\gamma\delta}\ , \tag{5.46}$$

$$H\to e^{-iM^2s}H\ . \tag{5.47}$$

The propagator corresponding to the massive extension is

$$K_{M^2}=\frac{P^2}{k^2-M^2+i0}-\frac{1}{6}\frac{3\mu+2}{(3\mu+2)k^2+M^2-i0}\eta\eta-\frac{P^1}{M^2-i0}-\left[\frac{2}{3(3\mu+2)}\frac{1}{(3\mu+2)k^2+M^2-i0}+\frac{4\mu+2}{3\mu+2}\frac{1}{M^2-i0}\right]P^0+\frac{1}{3}\frac{1}{(3\mu+2)k^2+M^2-i0}(\eta L+L\eta)\quad(\mu\neq-\tfrac{2}{3})\ . \tag{5.48}$$

If $\mu \neq -\frac{2}{3}$, the P^0 term in (5.48) has to be replaced by

$$\left[\frac{2k^2}{3(M^2-i0)^2}-\frac{4}{3}\frac{1}{M^2-i0}\right]P^0 . \tag{5.49}$$

Comparison of (5.48) and (5.49) with (5.31) and (5.43), respectively, shows that indeed

$$K_{M^2} \underset{M^2\to 0}{\longrightarrow} K , \tag{5.50}$$

where also the correct singularity structure of K is implied: If $i0$ is replaced by $i\epsilon$ in K_{M^2}, then (5.33) is consistent with

$$\infty^{2n} = \lim_{\epsilon\to 0+} \epsilon^{-n} . \tag{5.51}$$

This is a confirmation of the fact that K can be obtained also from the Langevin equation with real damping,

$$\dot{\psi}_{\alpha\beta} = -\epsilon\psi_{\alpha\beta} + iW_{\alpha\beta}{}^{\gamma\delta}\psi_{\gamma\delta} + \xi_{\alpha\beta} \tag{5.52}$$

if the limit $\epsilon\to 0+$ is taken after the calculation has been performed.

It is easy to evaluate $K^{-1}{}_{M^2}$ once it is noted that the operator W is Hermitian with respect to the scalar product

$$\langle\psi,\psi\rangle = \int d^4x\, \psi_{\alpha\beta}G^{\alpha\beta\gamma\delta}\psi_{\gamma\delta} \equiv \int d^4x\, \psi^T G\psi . \tag{5.53}$$

Denoting the Hermitian adjoint of W by $\tilde{W}$, we have

$$\tilde{W} = G^{-1}W^TG , \tag{5.54}$$

where T denotes the ordinary matrix transposition. Since $W = G^{-1}V$ [cf. (5.8)] and $G = G^T, V = V^T$, it follows that

$$\tilde{W} = W . \tag{5.55}$$

Therefore the stochastic propagators (5.25) and (5.26) can also be written as

$$D(k,s;k',s') = 2(2\pi)^4\delta^{(4)}(k+k') \times \int_0^{\min(s,s')} d\sigma\, e^{iW(s-\sigma)}G^{-1}(e^{iW(s'-\sigma)})^T$$

$$= 2(2\pi)^4\delta^{(4)}(k+k') \times \int_0^{\min(s,s')} d\sigma\, e^{iW(s+s'-2\sigma)}G^{-1} \tag{5.56}$$

and

$$K = -2i\int_0^\infty d\tau\, e^{2iW\tau}G^{-1} . \tag{5.57}$$

If $W \to W - M^2\mathbf{1}$, we obtain

$$K_{M^2} = -2i\int_0^\infty d\tau\, e^{2i(W-M^2)\tau}G^{-1} = (W - M^2\mathbf{1})^{-1}G^{-1} \tag{5.58}$$

$$= (V - M^2G)^{-1} . \tag{5.59}$$

Thus all the propagators K are "inverses" of the singular differential operator V in a certain sense specified by (5.59) and (5.50). With the exception of $\lambda = -2$ ($\mu = -\frac{2}{3}$) the principal value of the gauge-independent part is the same for all propagators. The "critical" value $\lambda = -2$ corresponds to the so-called Fierz-Pauli mass term[32] appearing in the Lagrangian for a massive pure spin-2 particle,

$$L_M^{(2)} = \tfrac{1}{2}\psi^T V\psi - \frac{M^2}{2}(\psi_{\alpha\beta}\psi^{\alpha\beta} - \psi^\gamma{}_\gamma\psi^\delta{}_\delta) . \tag{5.60}$$

The "anomalous" behavior of the corresponding propagator for $M^2\to 0$ is known as the van Dam–Veltman mass discontinuity.[31]

The Fierz-Pauli extension is one of the two massive extensions characterized by the property that only one physical mass value is present [in the other cases there is a second pole of $K_{M^2}(k)$ at $k^2 = -M^2/(3\mu+2)$]. The other extension is that of $\lambda = -1$. In this case we have

$$K_{M^2}^{(-1)} = \frac{1}{k^2 - M^2 + i0}[P^2 - \tfrac{1}{6}(\eta + 2L)(\eta + 2L)] - \frac{P^1}{M^2 - i0} - \frac{2P^0}{M^2 - i0} . \tag{5.61}$$

Summarizing the results of this section, we have seen that the parameter value $\lambda = -1$ is preferred both kinematically and dynamically. It is distinguished dynamically because (i) it governs the linearized Einstein equations [cf. (5.39)], (ii) the corresponding Feynman propagator is causal (though this is the case also for other parameter values), and (iii) considerable simplifications occur in K [cf. (5.41)] and K_{M^2} [cf. (5.61)]. Another manifestation of property (i) is the fact that by adding the gauge-fixing term

$$S_{gf} = \int d^4x\,(\partial_\gamma\psi_\alpha{}^\gamma - \tfrac{1}{2}\partial_\alpha\psi_\gamma{}^\gamma) \tag{5.62}$$

to the action $S^{(0)}$ one obtains

$$S^{(0)} + S_{gf} = \int d^4x\, \psi_{\alpha\beta}G^{\alpha\beta\gamma\delta}(\lambda = -1)(-\Box)\psi_{\gamma\delta} . \tag{5.63}$$

It is not possible to change λ in (5.63) by a different gauge fixing. It is amusing to observe that the usual harmonic coordinate condition involved in (5.62) serves to make the harmonicity of the metric itself, as implied by the Einstein action, manifest.

VI. PERTURBATION THEORY

The appearance of the stochastic source (3.3) in the Langevin equation (2.1) implies that in the case of the gravitational field a new type of diagram will contribute in the perturbative calculation of stochastic averages. In the following we sketch this perturbation theory briefly. It can be formulated for the stochastic perturbation $\psi_{\alpha\beta}(x,s)$ of an arbitrary deterministic and s-independent background metric $g^{\rm cl}_{\alpha\beta}(x)$. The full metric is split according to

$$g_{\alpha\beta}(x,s) = g^{\rm cl}_{\alpha\beta}(x) + 2\kappa^{1/2}\psi_{\alpha\beta}(x,s) . \tag{6.1}$$

A convenient choice of vielbein functional $E_M{}^A$ is the following. We introduce a deterministic, nonsingular "reference metric" $g^{(0)}_{ab}(x,s)$ and an associated stochastic tetrad field (in the ordinary sense) $e^a{}_\alpha(x,s)$ with

$$g^{(0)}_{ab}e^a{}_\alpha e^b{}_\beta = g_{\alpha\beta} \,. \tag{6.2}$$

Since $g_{\alpha\beta}$ is a complex process, we admit also complex $e^a{}_\alpha$. It is therefore not necessary to impose a signature restriction on $g^{(0)}_{ab}$. Then

$$E_M{}^A = |g|^{-1/4} e^a{}_\alpha e^b{}_\beta \tag{6.3}$$

yields (3.1) via (3.4) with

$$G^{(0)}_{MM'} = \frac{C}{2}|g^{(0)}|^{1/2}[g^{(0)aa'}(x)g^{(0)bb'}(x') + g^{(0)ab'}(x)g^{(0)ba'}(x') + \lambda g^{(0)ab}(x)g^{(0)a'b'}(x')]\delta^{(4)}(x-x') \,. \tag{6.4}$$

For simplicity we shall choose in the following for the reference metric

$$g^{(0)}_{ab}(x,s) = \delta_{ab} \tag{6.5}$$

and for the background metric

$$g^{\rm cl}_{\alpha\beta}(x) = \eta_{\alpha\beta} \,. \tag{6.6}$$

Moreover we make a concrete choice of orthonormal tetrad fields:

$$e^a{}_\alpha(x,s) = (g_+^{1/2})_{a\alpha} \,. \tag{6.7}$$

Here $g_+^{1/2}$ is defined by

$$g_+^{1/2} = \eta_+^{1/2}(1 + 2\kappa^{1/2}\eta\psi)^{1/2}$$
$$= \eta_+^{1/2}\left[1 + \kappa^{1/2}\eta\psi - \frac{\kappa}{2}(\eta\psi)^2 + \cdots\right] , \tag{6.8}$$

where $\eta_+^{1/2}$ is the square root of η with one positive imaginary eigenvalue.

We may now expand the drift term and the noise term in the Langevin equation for the rescaled metric $\tilde{g}_{\alpha\beta}$ defined by (2.14) in powers of $\kappa^{1/2}$:

$$|g|^{-1/2}\left[g_{\alpha\gamma}g_{\beta\delta} + \frac{\mu}{2}g_{\alpha\beta}g_{\gamma\delta}\right]\frac{\delta S}{\delta\psi_{\gamma\delta}} = W_{\alpha\beta}{}^{\gamma\delta}\psi_{\gamma\delta} + I_{\alpha\beta}(\psi,\partial\psi) \,, \tag{6.9}$$

$$|g|^{-1/4}e^a{}_\alpha e^b{}_\beta \xi^{(0)}_{ab} = 2\kappa^{1/2}[\tilde{\xi}^{(0)}_{\alpha\beta} + J_{\alpha\beta}{}^{\gamma\delta}(\psi)\tilde{\xi}^{(0)}_{\gamma\delta}] \,, \tag{6.10}$$

$$I = \sum_{n=2}^{\infty}\kappa^{(n-1)/2}I^{(n)} \,, \tag{6.11}$$

$$J = \sum_{n=1}^{\infty}\kappa^{n/2}J^{(n)} \,, \tag{6.12}$$

$$\langle\tilde{\xi}^{(0)}_{\alpha\beta}(x,s)\tilde{\xi}^{(0)}_{\alpha'\beta'}(x',s')\rangle = (\eta_{\alpha\alpha'}\eta_{\beta\beta'} + \eta_{\alpha\beta'}\eta_{\beta\alpha'} + \mu\eta_{\alpha\beta}\eta_{\alpha'\beta'})\delta^{(4)}(x-x')\delta(s-s') \,. \tag{6.13}$$

The last equation is the Fourier transform of (5.11) with respect to k and k'. The Langevin equation for ψ implied by (2.13) is

$$\dot{\psi} - iW\psi = iI(\psi,\partial\psi) + J(\psi)\tilde{\xi}^{(0)} + \tilde{\xi}^{(0)} \,. \tag{6.14}$$

This may be solved iteratively using the Schrödinger kernel H introduced in (5.13). With the initial condition $\psi(x,0)=0$ Eq. (6.14) implies

$$\psi(s) = \int_0^\infty d\sigma\, H(s-\sigma)[\, iI(\psi(\sigma),\partial\psi(\sigma)) + J(\psi(\sigma))\tilde{\xi}^{(0)}(\sigma) + \tilde{\xi}^{(0)}(\sigma)] \,. \tag{6.15}$$

The iterative solution of (6.15) yields a series of tree diagrams whose first terms are depicted in Fig. 1. The main new feature in these gravitational tree diagrams is the appearance of "stochastic vertices" corresponding to the terms $J^{(n)}$, with $n+1$ prongs, $n=1,2,\ldots$. These vertices are represented by an encircled cross in Fig. 1.

The tree series expansion implies a perturbative expansion of the stochastic average of products of ψ fields. The corresponding "stochastic diagrams" are obtained by joining all possible pairs of crosses (including the encircled ones) in the tree diagrams. (Because of the Gaussian character of $\tilde{\xi}^{(0)}$ no joinings of higher order need be considered.) A joined pair of crosses is usually denoted by just one cross, because diagrams containing single crosses do not contribute to expectation values, again because of the Gaussian character of $\tilde{\xi}^{(0)}$. We note that because we have adopted Ito's interpretation of the stochastic integral (6.15), all diagrams containing bubbles attached to stochastic vertices vanish (Fig. 2). This brings about a great simplification as compared with the rules that would follow from Stratonovich's calculus. Figure 3 shows the lowest-order nonvanishing contributions to the 1-, 2-, and 3-point functions. The crossed line denotes the stochastic propagator D introduced in (5.25).

Because of the divergences in the Feynman propagator (5.36) all stochastic diagrams will actually diverge in the limit $s\to\infty$, although the divergences will not contribute to gauge-invariant expectation values. For practical calculations it is more convenient to use the method of stochastic gauge fixing introduced by Zwanziger.[29] A complete discussion of all covariant linear stochastic gauges in linearized gravity for the case $\lambda=0$ can be found in Ref.

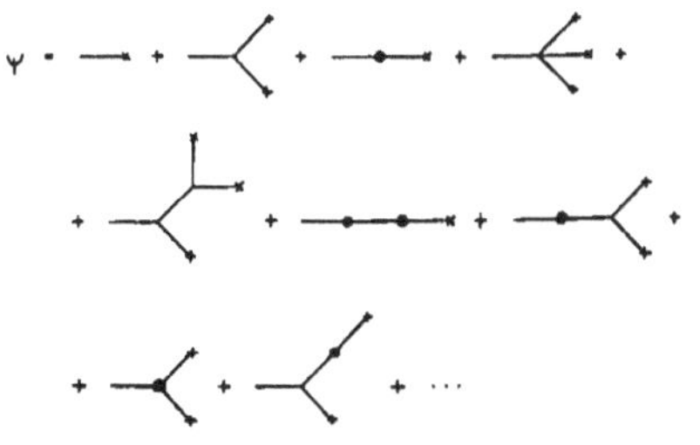

FIG. 1. Diagrammatical representation of the first terms in the tree series expansion for $\psi(s)$. The line denotes the Schrödinger kernel H, the cross denotes the stochastic source $\tilde{\xi}^{(0)}$, and the encircled cross with $n+1$ prongs denotes $\kappa^{n/2}J^{(n)}\tilde{\xi}^{(0)}$.

FIG. 2. Example of a subdiagram that implies the vanishing of a stochastic diagram.

8. In higher-order perturbation theory the method was first applied to the gravitational field by Fukai and Okano.[3] It appears, however, that their discussion is incomplete as they did not take into account the J term in (6.14) and the associated stochastic vertices.

VII. CONCLUSION

There are two different contexts in which the results of this paper may be interpreted. The point of view that was adopted throughout the paper was to accept the Langevin equation (2.1), though it turned out to be not covariant with respect to general field transformations, but only diffeomorphisms. Only in this way was it possible to obtain a whole family of nontrivial equilibrium path-integral measures. From this point of view the following interpretation of our results is possible: A characteristic feature of stochastic quantization is that it is based directly on a classical field equation, and no classical action or Hamiltonian is required. Classically equivalent field equations may give rise to inequivalent quantizations. This is what happens in the case of the vacuum Einstein field equations. As can be seen from (2.13), the one-parameter family of Langevin equations corresponds to the classical field equations

$$G_{\alpha\beta}-\frac{\mu}{2}g_{\alpha\beta}R=0\,, \tag{7.1}$$

where $G_{\alpha\beta}$ is the Einstein tensor. Although $\mu=0$ looks like a natural choice, it implies the rather odd exponent $\gamma=-\frac{9}{2}$ in the equilibrium path-integral measure contained in (4.14), and the Feynman propagator of the corresponding linearization turns out to be noncausal. We have argued in this paper that the natural choice is $\mu=-1$, in which case (7.1) becomes

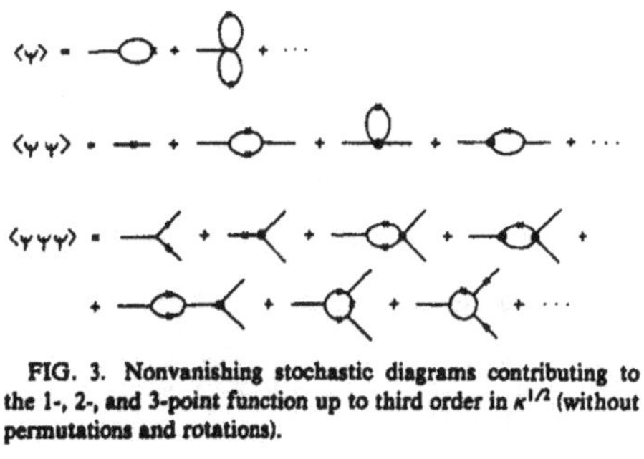

FIG. 3. Nonvanishing stochastic diagrams contributing to the 1-, 2-, and 3-point function up to third order in $\kappa^{1/2}$ (without permutations and rotations).

$$R_{\alpha\beta}=0 \tag{7.2}$$

with $R_{\alpha\beta}$ the Ricci tensor. Note the relation

$$R_{\alpha\beta}(x)=\int d^4x' G_{\alpha\beta,\alpha'\beta'}(x,x')G^{\alpha'\beta'}(x') \quad \text{for } \mu=-1 \tag{7.3}$$

in which the Ricci and Einstein tensors appear with their natural index positions.

We found that $\mu=-1$ implies $\gamma=0$ and that the corresponding field metric appears in the gauge-invariant part of the Feynman propagator. The propagator corresponding to $\mu=-1$ is causal and assumes a particularly simple form in its gauge-dependent part. Causality alone, however, allows a whole range of parameter values μ. The corresponding range of exponents γ is

$$-\tfrac{9}{4}<\gamma<9 \tag{7.4}$$

which excludes Misner's choice $\gamma=-\frac{5}{2}$.

Our results may, however, also be seen in a different context: Stochastic quantization should be generally covariant also with respect to changes of the field variable. Therefore the covariantized Langevin equation (3.22) appears to be more fundamental than (2.1). This observation does not make our calculations obsolete. Since the spurious term $-\Delta_G\Phi^A$ in (3.22) is divergent unless it vanishes, e.g.,

$$\Delta_G g_{\alpha\beta}(x)=\tfrac{9}{2}(1+\mu)\delta^{(4)}(0)\,|g\,|^{-1/2}g_{\alpha\beta}(x) \tag{7.5}$$

the Langevin equation can be solved only if the field variable is harmonic. In particular the linearization is meaningful only in harmonic "coordinates." The propagator in these coordinates may be easily inferred from the result of Sec. V. Since it is related to the latter by a bilinear transformation, the causality properties are the same. The conclusion then is that one and the same equilibrium distribution, viz., (5.2), gives rise to different stochastic perturbation theories based on different variables. In general these perturbation theories are inequivalent; in particular the causality properties of the corresponding propagators may be different.

It appears worthwhile to study the perturbative aspects of the stochastic quantizations of the gravitational field. It cannot be excluded that there is a parameter value for which the theory is finite. If nontrivial divergences are present, however, a subtraction scheme at finite s will have to be devised before meaningful calculations can be made.

Finally we remark that in the causal range $\mu<-\frac{2}{3}$ the stochastic perturbation theory may even be performed in Euclidean space-time, because then the original Parisi-Wu ansatz implies only decreasing exponentials in (5.24).

ACKNOWLEDGMENT

I am grateful to Professor John R. Klauder for valuable comments and encouragement at an early stage of this work.

[1]G. Parisi and Wu Yong-Shi, Sci. Sin. 24, 483 (1981).
[2]G. W. Gibbons, S. W. Hawking, and M. J. Perry, Nucl. Phys. B138, 141 (1978).
[3]H. Rumpf, Acta Phys. Austriaca Suppl. XXVI, 435 (1984).
[4]J. Greensite and M. B. Halpern, Nucl. Phys. B242, 167 (1984).
[5]T. Fukai and K. Okano, Prog. Theor. Phys. 73, 790 (1985).
[6]H. Hüffel and H. Rumpf, Phys. Lett. 148B, 104 (1984).
[7]E. Gozzi, Phys. Lett. 150B, 119 (1985).
[8]H. Hüffel and H. Rumpf, Z. Phys. C 29, 319 (1985).
[9]J. Sakamoto, Prog. Theor. Phys. 70, 1424 (1983).
[10]G. A. Vilkovisky, Nucl. Phys. B234, 125 (1984).
[11]B. S. DeWitt, contribution to the Fradkin Festschrift (unpublished).
[12]B. S. DeWitt, in *General Relativity*, edited by S. W. Hawking and W. Israel (Cambridge University Press, Cambridge, 1979).
[13]K. Ito, Proc. Imp. Acad. 20, 519 (1944).
[14]R. L. Stratonovich, *Conditional Markov Processes and Their Application to the Theory of Optimal Control* (Elsevier, New York, 1968).
[15]L. Arnold, *Stochastic Differential Equations* (Wiley-Interscience, New York, 1974).
[16]R. Graham, Z. Phys. B 26, 397 (1977); Phys. Lett. 109A, 209 (1985).
[17]D. Ryter and U. Deker, J. Math. Phys. 21, 2662 (1980).
[18]G. Parisi, Phys. Lett. 131B, 393 (1983).
[19]P. Zoller, Acta Phys. Austriaca Suppl. XXVI, 75 (1984).
[20]C. W. Misner, Rev. Mod. Phys. 29, 497 (1957).
[21]B. Laurent, Ark. Fys. 16, 279 (1959).
[22]J. R. Klauder, Nuovo Cimento 19, 1059 (1961).
[23]B. S. DeWitt, J. Math. Phys. 3, 1073 (1962).
[24]K. Fujikawa and O. Yasuda, Nucl. Phys. B245, 436 (1984).
[25]H. Leutwyler, Phys. Rev. 134, B1155 (1964).
[26]E. S. Fradkin and G. A. Vilkovisky, Phys. Rev. D 8, 4241 (1973).
[27]L. Baulieu and D. Zwanziger, Nucl. Phys. B193, 163 (1983).
[28]M. Horibe, A. Hosoya, and J. Sakamoto, Prog. Theor. Phys. 70, 1636 (1983).
[29]D. Zwanziger, Nucl. Phys. B192, 259 (1981).
[30]P. van Nieuwenhuizen, Nucl. Phys. B60, 478 (1973).
[31]H. van Dam and M. Veltman, Nucl. Phys. B22, 397 (1970).
[32]G. Wentzel, *Quantum Theory of Fields* (Interscience, New York, 1949).

VOLUME 43, NUMBER 11 PHYSICAL REVIEW LETTERS 10 SEPTEMBER 1979

Random Magnetic Fields, Supersymmetry, and Negative Dimensions

G. Parisi
Istituto Nazionale di Fisica Nucleare, Frascati, Italy

and

N. Sourlas
Laboratoire de Physique Théorique de l'Ecole Normale Supérieure, 75231 Paris Cédex 05, France
(Received 26 June 1979)

We prove the equivalence, near the critical point, of a D-dimensional spin system in a random external magnetic field with a $(D-2)$-dimensional spin system in the absence of a magnetic field. This is due to the hidden supersymmetry of the associated stochastic differential equation. We identify a space with one anticommuting coordinate with a space having negative dimensions -2.

The critical behavior of a spin system in a random external magnetic field (i.e., the infrared behavior of a scalar field theory in presence of a random external source) has recently been investigated.[1] By explicit computations it has been found[1,2] that the values of the most-infrared-divergent diagrams[3] in dimensions D are equal to the values of the same diagrams without magnetic fields in dimensions $D-2$. In this Letter we show that this apparently mysterious result has a simple geometrical interpretation, which stems from a hidden supersymmetry[4] of the associated stochastic equation.

Let us define the free energy F_R averaged over a Gaussian random magnetic field by the functional integrals[5]:

$$F[h]=\ln\int \mathfrak{D}_\varphi \exp\{-\int d^Dx[\mathcal{L}(x)+h(x)\varphi(x)]\},$$
$$F_R=\int \mathfrak{D}h F[h]\exp[-\tfrac{1}{2}\int d^Dx\, h^2(x)], \quad (1)$$
$$\mathcal{L}(x)=-\tfrac{1}{2}\varphi(x)\Delta\varphi(x)+V(\varphi(x)).$$

For definiteness we can consider $V(\varphi)=\frac{1}{2}m^2\varphi^2+g\varphi^4$. The perturbative expansion in g for F_R can be easily constructed either by direct inspection, or by using the replica trick.[6] The most-infrared-divergent diagrams contain the maximum number of h^2 insertions, as follows from dimensional analysis. If the other diagrams are neglected we obtain the tree approximation for $F(h)$. Using the correspondence between the tree approximation[7] and the classical nonlinear differential equation, we find that in this limit the two-point Green's function is given by

$$\langle\varphi(x)\varphi(0)\rangle_R \sim \int \mathfrak{D}h\,\varphi_h(x)\varphi_h(0)\exp[-\tfrac{1}{2}\int d^Dy\, h^2(y)], \quad (2)$$

where $\varphi_h(x)$ is the solution of the equation

$$-\Delta\varphi+V'(\varphi)+h=0. \quad (3)$$

Equation (3) can also be regarded as a differential stochastic equation, h being a stochastic Gaussian function having autocorrelation $\langle h(x)h(y)\rangle=\delta^D(x-y)$. The results of Refs. 1 and 2 imply that the Green's functions of the stochastic differential equations (3) are the same as those generated by the Lagrangian of Eq. (1) in $D-2$ dimensions. Let us see why.

Using standard manipulations,[8] we find

$$\langle\varphi(x)\varphi(0)\rangle\sim\int\mathfrak{D}\varphi\mathfrak{D}h\,\varphi(x)\varphi(0)\delta(-\Delta\varphi+V'(\varphi)+h)\det[-\Delta+V''(\varphi)]\exp[-\tfrac{1}{2}\int h^2(y)d^Dy]$$
$$\sim\int\mathfrak{D}\varphi\mathfrak{D}\omega\mathfrak{D}\psi\exp[-\int d^Dy\mathcal{L}_R(y)]\,\varphi(x)\varphi(0), \quad (4)$$
$$\mathcal{L}_R=-\tfrac{1}{2}\omega^2+\omega[-\Delta\varphi+V'(\varphi)]+\bar\psi[-\Delta+V''(\varphi)]\psi,$$

where ψ is an anticommuting scalar field[9] (a ghost field). The Lagrangian $\mathcal{L}_R$ is invariant under the supersymmetry transformations:

$$\delta\varphi=-\bar a\epsilon_\mu x_\mu\psi,\quad \delta\omega=2\bar a\epsilon_\mu\partial_\mu\psi,$$
$$\delta\psi=0,\quad \delta\bar\psi=\bar a(\epsilon_\mu x_\mu\omega+2\epsilon_\mu\partial_\mu\varphi), \quad (5)$$

$\bar a$ being an infinitesimal anticommuting number and ϵ_μ an arbitrary vector. The invariance under these supersymmetry transformations [Eq. (5)] is quite unexpected.[10] It is useful to introduce the superspace[4] characterized by a D-dimensional commuting coordinate x and by an anticommuting coordinate θ ($\theta^2=\bar\theta^2=\theta\bar\theta+\bar\theta\theta=0$) and the superfield,

$$\Phi(x,\theta)=\varphi(x)+\bar\theta\psi(x)+\bar\psi(x)\theta+\theta\bar\theta\omega(x). \quad (6)$$

Higher orders in θ are identically zero as a result of the anticommuting properties of θ. The action [Eq. (4)] can be written as $\int d^Dx d\theta \mathcal{L}_{ss}(\Phi)$, with

$$\mathcal{L}_{ss}(\Phi) = -\tfrac{1}{2}\Phi\Delta_{ss}\Phi + V(\Phi), \qquad (7)$$

where $\Delta_{ss} = \Delta + \partial^2/\partial\bar{\theta}\partial\theta$ is the Laplacian in the superspace and the integration in θ selects the term proportional to $\theta\bar{\theta}$ [e.g., $\int d\theta\, \Phi(x,\theta) = -\pi^{-1}\omega(x)$].[11]

The supersymmetry transformations [Eq. (5)] are simply rotations in superspace leaving invariant the metric $x^2 + \theta\bar{\theta}$. We argue that the superspace (x,θ) is equivalent to an ordinary $(D-2)$-dimensional space. Indeed a space with only one anticommuting coordinate θ is equivalent to an ordinary space with negative dimensions -2. (Space with negative dimensions are defined by analytic continuations[12] from positive dimensions.) This can be seen from the relation

$$\int d\theta f(\bar{\theta}\theta) = -\frac{1}{\pi}\frac{d}{dz}f(z)\Big|_{z=0} = \lim_{D\to -2}\int d^D r f(r^2) = \lim_{D\to -2} S_D \int r^{D-1} dr f(z^2), \quad S_D = 2\pi^{D/2}/\Gamma(\tfrac{1}{2}D). \qquad (8)$$

S_D is the surface of the unit sphere in D dimensions.

Let us consider a space of dimension $D-2$ and formally decompose it as the sum of a space of dimension D and of another space of dimension -2. The previous argument implies that an ordinary space of dimension $D-2$ is equivalent to the D-dimensional superspace. The precise meaning of the equivalence is the following:

$$\int d^{D-2}x F(Y_i, x, x^2) = \int d^D x d\theta\, F(Y_i, x, x^2 + \bar{\theta}\theta), \qquad (9)$$

where Y_i are some $(D-2)$-dimensional vectors. For example,

$$\int d^{D-2}x f(x^2) = \int d^D x d\theta f(x^2 + \bar{\theta}\theta) = -\pi^{-1}\int d^D x f'(x^2) = \int d^{D-2}x f(x^2). \qquad (10)$$

Equation (9) is sufficient to prove, at all orders in perturbation theory, that the Green's functions computed in the $D-2$ space are the same as those computed in the D-dimensional superspace. Indeed, the perturbative expansion for the Lagrangian (7) [which is equivalent to the stochastic Eq. (3)] can be written directly from Feynmann's rule in configuration superspace using the technique of superpropagator.[13] The final integrals have the form of Eq. (9) and the equivalence of the stochastic Eq. (3) with the $(D-2)$-dimensional field theory is therefore proved in the perturbative expansion. It has its root in the hidden supersymmetry of the system and in the geometrical equivalence of an anticommuting-variable space with a negative-dimensional space.

It may be useful to establish this equivalence rigorously beyond perturbation theory. The stochastic differential Eq. (3) may provide us with a different framework to study the properties of a field theory. It would be quite interesting to see if and how this formalism can be extended to gauge theories.

Laboratoire de Physique Théorique de l'Ecole Normale Supérieure is a laboratoire propre du Centre National de la Recherche Scientifique, associé à l'Ecole Normale Supérieure et à l'Université de Paris-Sud.

[1]Y. Imry and S.-k. Ma, Phys. Rev. Lett. 35, 1399 (1975).

[2]A. P. Young, J. Phys. C 10, L257 (1977); E. Brézin and G. Parisi, unpublished.

[3]The approximation of keeping only the most-infrared-divergent diagrams may be justified near the critical point.

[4]A supersymmetry transformation mixes commuting and anticommuting (boson and fermion) fields. For a review on supersymmetry, see for example, P. Fayet and S. Ferrara, Phys. Rep. 32C, 249 (1977).

[5]Green's functions can also be defined using a similar procedure.

[6]S. F. Edwards and P. W. Anderson, J. Phys. F 5, 965 (1975).

[7]K. Symanzik, in *Lectures in Theoretical Physics*, edited by E. Brittin, B. W. Downs, and J. Downs (Interscience, New York, 1961), Vol. III.

[8]P. C. Martin, E. D. Siggia, and H. Rose, Phys. Rev. A 8, 423 (1973); B. I. Halperin, P. C. Hohenberg, and S.-k. Ma, Phys. Rev. B 10, 139 (1974).

[9]The definition of anticommuting functional intergrations can be found in F. Berezin, *The Method of Second Quantization* (Academic, New York, 1966).

[10]This Eq. (5) is similar to the Becchi-Rouet-Stora transformation in gauge theories; C. Becchi, A. Rouet, and R. Stora, Ann. Phys. (N. Y.) 98, 287 (1976).

[11]The factor $-1/\pi$ has been introduced for later convenience.

[12]C. G. Bollini and I. J. Giambiagi, Phys. Lett. 40B, 566 (1972); K. G. Wilson and M. E. Fisher, Phys. Rev. Lett. 28, 240 (1972); G. 't Hooft and M. Veltman, Nucl. Phys. B44, 189 (1972).

[13]A. Salam and J. Strathdee, Nucl. Phys. B76, 477 (1974), and B86, 142 (1975); R. Delbourgo, Nuovo Cimento 25A, 646 (1975).

VOLUME 43, NUMBER 11 PHYSICAL REVIEW LETTERS 10 SEPTEMBER 1979

Magnetic Moment of Weak Bosons Produced in pp and $p\bar{p}$ Collisions

K. O. Mikaelian and M. A. Samuel
Physics Department, Oklahoma State University, Stillwater, Oklahoma 74074

and

D. Sahdev
Physics Department, Case Western Reserve University, Cleveland, Ohio 44106

(Received 5 June 1979)

We suggest that the reactions $pp\to W^{\pm}\gamma X$ and $p\bar{p}\to W^{\pm}\gamma X$ are good candidates for measuring the magnetic moment parameter κ in $\mu_W=(e/2M_W)(1+\kappa)$. The angular distribution of the W bosons in $p\bar{p}\to W^{\pm}\gamma X$ is particularly sensitive to this parameter. For the gauge-theory value of $\kappa=1$, we have found a peculiar zero in $d\sigma(d\bar{u}\to W^-\gamma)/d\cos\theta$ at $\cos\theta=-\frac{1}{3}$, the location of this zero depending on the quark charge through $\cos\theta=-(1+2Q_d)$. A similar zero occurs in $d\sigma(u\bar{d}\to W^+\gamma)/d\cos\theta$. We can offer no explanation for this behavior.

Expectations are high that the weak intermediate vector bosons $W^{\pm}$ and Z^0 will be discovered in the next generation of accelerators[1] (the CERN $p\bar{p}$ project, ISABELLE at Brookhaven, and the $p\bar{p}$ project at Fermilab). The theoretical predictions appear to be on firm ground: The highly successful Weinberg-Salam SU(2)⊗U(1) theory predicts the boson masses to lie below 100 GeV/c^2, and this theory coupled with the Drell-Yan picture of quark-antiquark annihilation in hadron-hadron collisions predicts[2] cross sections around 10^{-33} cm^2, which is well matched by the luminosities of the projects mentioned above.

Assuming that such a discovery will indeed take place, we ask ourselves what properties of weak bosons are likely to be measured. The mass and spin can probably be measured by looking at decay distributions,[2,3] which also serve as a signal for the production process. Another property in which we are interested is the magnetic moment μ_W of the $W^{\pm}$ bosons or, more generally, the coupling between the weak and electromagnetic fields $W_\mu^{\pm}$ and A_μ. (The interaction of $W_\mu^{\pm}$ fields with the Z_μ field is even more interesting from the gauge-theory point of view but, of course, it is harder to detect, requiring W^+W^- and $W^{\pm}Z^0$ pair production.[4])

In principle the electromagnetic coupling of weak bosons could be measured in the reaction[5] $e^+e^-\to W^+W^-$ or in photoproduction[6] $\gamma p\to W^{\pm}X$. However, it will be a long time before we achieve the necessary center-of-mass energies in these reactions. We have, therefore, concentrated on pp and $p\bar{p}$ collisions, and, since an electromagnetic field must clearly be involved, we are naturally led to the processes

$$p+p \text{ or } p+\bar{p}\to W^{\pm}+\gamma+X, \tag{1}$$

which we believe are very good candidates for measuring μ_W. We have used the commonly accepted quark-parton model to describe those processes which involve hard collisions to produce a massive W boson accompanied by a hard photon. In the course of our investigation we found a peculair angular distribution which we cannot yet explain.

In the usual manner we calculate first the basic process with quarks before embedding them in physical particles. In this case the process is

$$q_i+\bar{q}_j\to W^{\pm}+\gamma. \tag{2}$$

The Feynman diagrams and our notation are shown in Fig. 1. Of course, we are after diagram (c). To see how reaction (2) depends on μ_W, we use the standard generalization of the $W_\alpha(p+k)$-

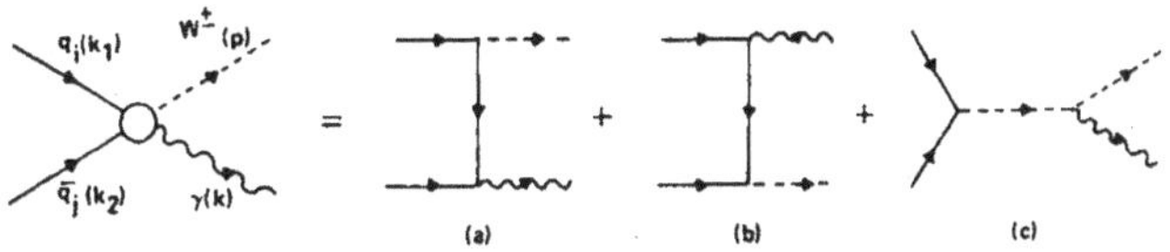

FIG. 1. Feynman diagrams for the process $q_i(k_1)\bar{q}_j(k_2)\to W^{\pm}(p)\gamma(k)$.

$\gamma_\mu(k)$-$W_\beta(p)$ vertex[7]:

$$V_{\alpha\mu\beta} = -ie[g_{\alpha\beta}(2p+k)_\mu - g_{\alpha\mu}(p+k+\kappa k)_\beta - g_{\beta\mu}(p-\kappa k)_\alpha]. \quad (3)$$

The magnetic moment is given by[7]

$$\mu_W = (e/2M_W)(1+\kappa).$$

In the standard model $\kappa = 1$ and, therefore, $\mu_W = e/M_W$.

We have calculated the differential cross section and find

$$\frac{d\sigma}{dt}(q_i\bar{q}_j \to W^-\gamma) = \frac{\alpha}{s^2}\frac{M_W^2 G_F}{\sqrt{2}} g_{ij}^2 \left\{ \left(Q_i + \frac{1}{1+t/u}\right)^2 \frac{t^2+u^2+2sM_W^2}{tu} + (\kappa-1)\left(Q_i + \frac{1}{1+t/u}\right)\frac{t-u}{t+u} + \frac{(\kappa-1)^2}{2(t+u)^2}\left[tu + (t^2+u^2)\frac{s}{4M_W^2}\right]\right\}, \quad (4)$$

where $s = (k_1+k_2)^2$, $t = (p-k_1)^2$, and $u = (p-k_2)^2$, with $s+t+u = M_W^2$, $g_{ij} = \cos\theta_C$ for $q_i\bar{q}_j = d\bar{u}$ and $s\bar{c}$, and $g_{ij} = \sin\theta_C$ for $q_i\bar{q}_j = s\bar{u}$ and $d\bar{c}$. $Q_i|e|$ is the charge of the quark q_i; we have set $Q_j = Q_i + 1$. All quark masses have been dropped. The cross section for $q_j q_i \to W^+\gamma$ is also given by Eq. (4) with t taken to be the four-momentum transferred between the antiquark and W^+. For definiteness, we will concentrate on $W^-\gamma$ production. We have checked Eq. (4) against the corresponding expression for $\gamma q_i \to W^+ q_j$ in Ref. 6.

The peculiar behavior referred to earlier can be seen in Fig. 2, where we plot the angular distribution of the W^- obtained from Eq. (4):

$$d\sigma(d\bar{u} \to W^-\gamma)/d\cos\theta = \tfrac{1}{2}(s - M_W^2)d\sigma(d\bar{u} \to W^-\gamma)/dt, \quad (5)$$

where θ is the angle between W^- and d (γ and $\bar{u}$) in the $W^-\gamma$ center-of-mass frame: $t = -\frac{1}{2}(s - M_W^2)(1-\cos\theta)$. The smooth behavior for $\kappa = -1$ or 0 is radically changed when we set $\kappa = 1$, its gauge-theory value. An unexpected zero appears now at $\cos\theta = -\frac{1}{3}$. Since the same curves apply for $s\bar{c} \to W^-\gamma$ and, except for an overall factor of $\tan^2\theta_C$, to $s\bar{u} \to W^-\gamma$ and to $d\bar{c} \to W^-\gamma$ also, we conclude that *in all cases the standard theory predicts the vanishing of $d\sigma/dt$ at a particular angle.* This zero in $d\sigma/dt$ can be traced to the coefficient $[Q_i + (1+t/u)^{-1}]^2$ in Eq. (4), which vanishes for $t^*/u^* = -(1+1/Q_i)$ implying

$$\cos\theta^* = -(1+2Q_i). \quad (6)$$

Since $Q_i = -\frac{1}{3}$, we get $\cos\theta^* = -\frac{1}{3}$. We have no physical explanation for this zero; we know of no other case where a quark cross section vanishes at a particular value of $\cos\theta$ depending on the charge of the quark.[8] Furthermore, we have shown that no other set of κ, Q_i, etc., gives a zero in the physical region.

We point out that the condition $\cos\theta^* = -(1+2Q_i)$ does *not* involve M_W^2/s, which is the only other dimensionless quantity that could possibly appear in Eq. (6). This has interesting implications for the physical processes in Eq. (1), which we now consider.

The measurable quantity that comes closest to the q-$\bar{q}$ differential cross section is $d\sigma/d\cos\theta_{c.m.}$, the angular distribution of W bosons or photons in the $W\gamma$ center-of-mass frame, where $\theta_{c.m.}$ is the angle, say between the W and the direction of one of the colliding beams, which we take to be a proton beam. Let us consider $p\bar{p}$ collisions; then

$$\frac{d\sigma}{d\cos\theta_{c.m.}}(p\bar{p} \to W^-\gamma X) = \frac{1}{3}\sum_{i=d,s}\iint dx_A\, dx_B\, P_i^p(x_A) P_{\bar{u}}^{\bar{p}}(x_B)\left(\frac{s-M_W^2}{2}\right)\frac{d\sigma}{dt}(q_i\bar{q}_u \to W^-\gamma) + \frac{1}{3}\sum_{i=d,s}\iint dx_A\, dx_B\, P_{\bar{u}}^p(x_A) P_i^{\bar{p}}(x_B)\left(\frac{s-M_W^2}{2}\right)\frac{d\sigma}{dt}(q_i\bar{q}_u \to W^-\gamma)_{u \leftrightarrow t}. \quad (7)$$

In the first term $\theta = \theta_{c.m.}$, while in the second term $\theta = \pi - \theta_{c.m.}$. In both terms $s = x_A x_B S$, where S is the square of the $p\bar{p}$ c.m. energy, $S = (p_p + p_{\bar{p}})^2$. To be specific we use $\sqrt{S} = 540$ GeV, which is the projected energy for the CERN $p\bar{p}$ machine.

Were the p and $\bar{p}$ to consist purely of q's and $\bar{q}$'s, respectively, $d\sigma/d\cos\theta_{c.m.}$ would vanish identically at $\cos\theta_{c.m.} = -\frac{1}{3}$ if $\kappa = 1$. This is because the second term in Eq. (7), the so-called sea contribution, would be absent, and in the first term (valence contribution) the zero in $d\sigma(d\bar{u}\to W^-\gamma)/dt$ is not affected by the integration over x_A and x_B. [Remember, the condition $\cos\theta^* = -(1+2Q_d)$ does not involve s over which we are integrating in Eq. (7).]

We find that the sea contribution does not change this picture very much. Though finite, $(d\sigma/d\cos\theta_{c.m.})_{\kappa=1}$ is only 0.76×10^{-39} cm^2 at $\cos\theta_{c.m.} = -\frac{1}{3}$, compared to 0.32×10^{-36} cm^2 at $\cos\theta_{c.m.} = +\frac{1}{3}$, for example. In Fig. 3 we plot Eq. (7) for $\kappa = -1$, 0, 1. The quark distributions P_i, etc., are parametrized in the same way as in Ref. 4. To make sure that the photon is observed, we have made a cut of $E_\gamma > (E_\gamma)_{min} = 30$ GeV, where E_γ is the energy of the photon is the laboratory frame (which we always take to be the pp or $p\bar{p}$ c.m. frame).

We see that $d\sigma/d\cos\theta_{c.m.}$ is highly sensitive to the value of κ. This sensitivity is increased if we increase $(E_\gamma)_{min}$, but then, of course, the rates get smaller. The converse occurs if we decrease $(E_\gamma)_{min}$. 30 GeV appears to be a reasonable compromise between sensitivity to κ and acceptable rates, though other values are certainly possible. Note that we are *not* doing radiative corrections to single-W production.

A measurement of $d\sigma/d\cos\theta_{c.m.}$ requires detailed measurement of the momenta and directions of both the $W^\pm$ and the photon to reconstruct their center-of-mass motion event by event. As an example of a much cruder experiment, we show in Fig. 4 the behavior of $d\sigma/d\cos\theta_{lab}$, where θ_{lab} is the direction of the W^- with respect to one of the colliding beams, which we again take to be a proton beam, in the laboratory frame. This quantity requires a complicated phase-space in-

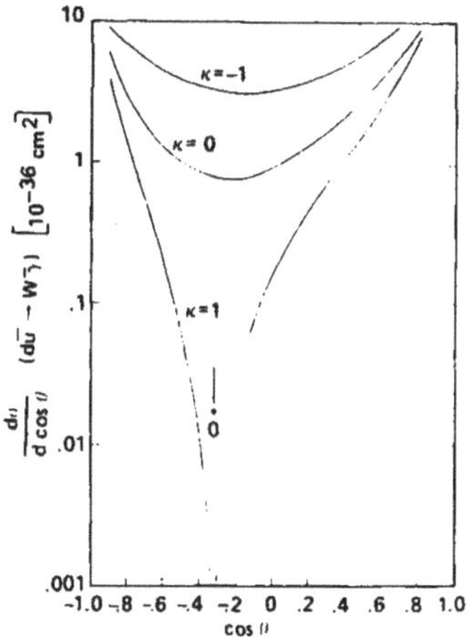

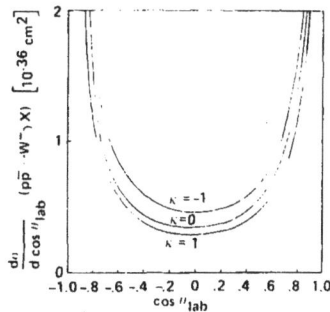

FIG. 2. The differential cross section for $d\bar{u}\to W^-\gamma$. θ is the angle between W^- and d, or between γ and $\bar{u}$, in the c.m. frame. $\sqrt{s} = 200$ GeV and $M_W = 85$ GeV/c^2.

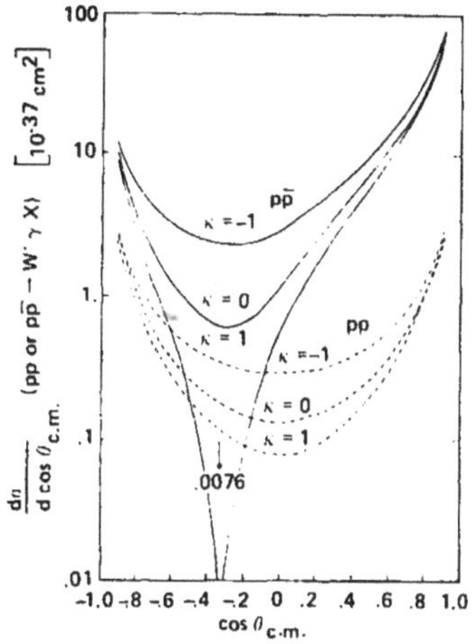

FIG. 3. The differential cross section for $pp\to W^-\gamma X$ and $p\bar{p}\to W^-\gamma X$, with a photon energy cut $E_\gamma > 30$ GeV. $\theta_{c.m.}$ is the angle between the W^- and the proton direction in the $W^-\gamma$ c.m. frame. $\sqrt{S} = 540$ GeV and $M_W = 85$ GeV/c^2.

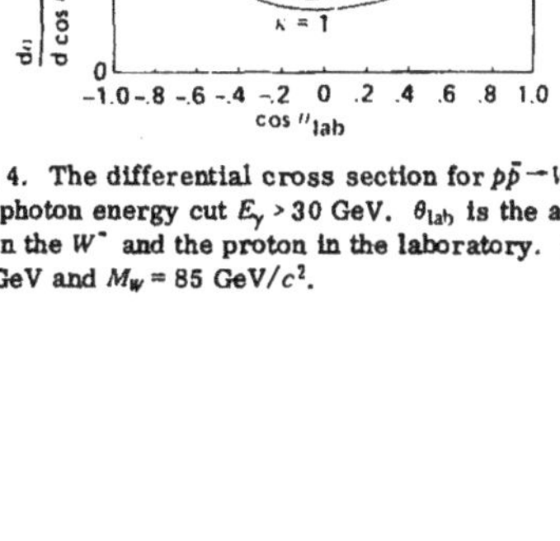

FIG. 4. The differential cross section for $p\bar{p}\to W^-\gamma X$ with a photon energy cut $E_\gamma > 30$ GeV. θ_{lab} is the angle between the W^- and the proton in the laboratory. $\sqrt{S} = 540$ GeV and $M_W = 85$ GeV/c^2.

VOLUME 43, NUMBER 11 PHYSICAL REVIEW LETTERS 10 SEPTEMBER 1979

tegration and we shall omit the details. The same $E_\gamma > 30$ GeV cut has been applied in Fig. 4.

We see that $d\sigma/d\cos\theta_{\rm lab}$ is mildly sensitive to κ, with a 30–40% difference around $\theta_{\rm lab} = 90°$ where the detectors will be. The reason for this behavior is the following: As one integrates over all variables except $\theta_{\rm lab}$, one sweeps over a fairly large range of θ, and the dip at $\cos\theta^* = -\frac{1}{3}$ gets filled in, though not completely. Obviously, one can do better by measuring a doubly differential cross section. The same remarks made earlier concerning the value of $(E_\gamma)_{\rm min}$ apply here also.

Both $d\sigma/d\cos\theta_{\rm c.m.}$ and $d\sigma/d\cos\theta_{\rm lab}$ are large near the forward-backward directions where they do not discriminate between different values of κ. This means that if one performs the third and last integral to obtain $\sigma_{\rm total}$, the results will be rather insensitive to κ. Indeed, we find[9] only a 7% difference between $\sigma_{\rm total}(\kappa = -1)$ and $\sigma_{\rm total}(\kappa = +1)$.

Finally, we consider $W^+\gamma$ production and pp vs $p\bar{p}$ collisions:

(a) Everything we said about the W^- applies unchanged to the W^+ where now the various angles θ, $\theta_{\rm c.m.}$, or $\theta_{\rm lab}$ are taken between the W^+ and the antiquark or antiproton; e.g., in $p\bar{p} \to W^+\gamma X$ for $\kappa = 1$, there is a dip when the $W^+(\gamma)$ makes an angle of about 110° (70°) with the direction of the $\bar{p}$.

(b) pp collisions differ in two ways from $p\bar{p}$ collisions. The absence of valence antiquarks implies lower rates in general, and we find that the differential cross sections in pp collisions are an order of magnitude smaller than in $p\bar{p}$ collisions over most of the phase-space region. This, however, may be compensated by the larger luminosities of pp machines. A more serious setback for pp collisions perhaps is the decrease in sensitivity to κ. For example, in the equation corresponding to Eq. (7), both terms will contribute with equal weights and if $d\sigma(d\bar{u} \to W^-\gamma)/dt$ vanishes in the first term then it does not vanish[10] in the second (note that the $q\bar{q}$ cross section is symmetric under $x_A \leftrightarrow x_B$ because it depends only on $s = x_A x_B S$ and $\cos\theta$). Hence, each one of the terms will substantially fill in the dip left by the other. For comparison we have included in Fig. 3 curves for pp collisions at the same energy $\sqrt{S}$. Our remarks are clearly substantiated by these curves. We should add, however, that it may be possible to find a totally or doubly differential cross section in pp collisions which reflects the κ dependence of the $q\bar{q}$ process, and we hope to do this in the future.[11]

We have benefited much from discussions with R. W. Brown. This research was supported in part by the U. S. Department of Energy and by the National Science Foundation.

[1] J. D. Bjorken, Stanford Linear Accelerator Center Report No. SLAC-PUB-2281 (unpublished); B. Richter, Stanford Linear Accelerator Center Report No. SLAC-PUB-2290 (unpublished).

[2] C. Quigg, Rev. Mod. Phys. 49, 297 (1977); R. F. Peierls, T. F. Trueman, and L. L. Wang, Phys. Rev. D 16, 1397 (1977).

[3] J. Kogut and J. Shigenitsu, Nucl. Phys. B129, 461 (1977); F. Halzen and D. M. Scott, Phys. Lett. 78B, 318 (1978).

[4] R. W. Brown and K. O. Mikaelian, Phys. Rev. D 19, 922 (1979); R. W. Brown, D. Sahdev, and K. O. Mikaelian, Phys. Rev. D (to be published).

[5] F. M. Renard, to be published.

[6] K. O. Mikaelian, Phys. Rev. D 17, 750 (1978).

[7] T. D. Lee and C. N. Yang, Phys. Rev. 128, 885 (1962).

[8] For $\bar{\nu}_e e \to W^-\gamma$, $Q_i = -1$ and $d\sigma/d\cos\theta$ vanishes in the forward direction.

[9] To obtain $\sigma_{\rm total}$ a finite quark mass m_q must be used in the propagators: $1/t \to 1/(t - m_q^2)$ and $1/u \to 1/(u - m_q^2)$. We have set $m_q = 300$ MeV/c^2.

[10] It could vanish if $d\sigma/d\cos\theta$ (see Fig. 2) had another zero at $\cos\theta^* = \frac{1}{3}$, or if $\theta^* = \pi - \theta^* = 90°$, which would be the case if $Q_d = -\frac{1}{2}$.

[11] The idea is to take advantage of the fact that the antiquarks in the sea are found at low x_A or x_B, and study the correlation between the energy and direction of the W bosons or the associated photons.

Reprinted from ANNALS OF PHYSICS
All Rights Reserved by Academic Press, New York and London

Vol. 145, No. 1, January 1983
Printed in Belgium

Stochastic and Parastochastic Aspects of Supersymmetric Functional Measures: A New Non-Perturbative Approach to Supersymmetry

S. CECOTTI*

Lyman Laboratory of Physics, Harvard University, Cambridge, Massachusetts 02138 and Scuola Normale Superiore, 56100 Pisa, Italy

AND

L. GIRARDELLO

Istituto di Fisica, Università di Milano, Milano, Italy and INFN, Sez. di Milano, Milano, Italy

Received May 12, 1982

We study the properties of supersymmetric models having a *local* Nicolai mapping. In these cases the Nicolai mapping can be interpreted as a stochastic differential equation, and hence we can use all the standard stochastic techniques to extract physical information from the theory. The corresponding Langevin equation does not describe, in general, a system approaching (asymptotically) a thermal equilibrium. We construct explicitly and non-perturbatively the Nicolai mapping for a large class of two dimensional models. In particular, this is the first non-perturbative proof of the existence of the mapping. The properties of the mapping agree with the expectations from general arguments. We show how the Nicolai mapping can be used to eliminate completely the fermions from the perturbative expansion, leaving a simpler set of diagrammatic rules involving only scalars. Finally, we argue that the present approach may be very powerful for studying finiteness properties of extended supersymmetric theories.

I. INTRODUCTION: A STOCHASTIC INTERPRETATION OF THE NICOLAI MAPPING

It is well known that, among all possible (quantum) field theories, the supersymmetric ones enjoy a rather particular status. For instance, they do not need normal-order procedures for the Hamiltonian operator and present very remarkable cancellations of divergences which would be allowed by naive arguments [1]. Most of the particular properties of these theories can be traced back to the fundamental fact that the Hamiltonian H (and hence the whole dynamics) enters non-trivially in the

* Address after September 1st: Scuola Normale Superiore, 56100 Pisa, Italy.

0003-4916/83/010081-19$07.50/0

super-algebra and, moreover, that $\langle H \rangle$ itself is an order parameter for the symmetry, owing to the presence of anticommutators in the corresponding (graded) algebra.

The situation is even more remarkable if the supersymmetry is extended: in this case there are *exact* (quantum) mass-formulas [2], and, as far as we know [3], the $N=4$ super-Yang–Mills is finite at the quantum level, i.e., $\beta(g) \equiv 0$ and all dimensions are canonical. This is an interesting phenomenon which has, at the moment, no explanation (but there is a plausible conjecture, due to Rossi [4], connecting this finiteness with the apparently exact self-duality of the model).

The purpose of the present paper is to study the non-perturbative structure of the quantum theory from a functional-integral point of view. In a particular class of supersymmetric models (which in dimension greater than one, tends to be identified with the extended case), we find a stochastic interpretation which renders the theory equivalent (in a sense to be specified in the following) to a *simpler* theory *without* fermions. If these methods can be extended to the relevant cases, probably they can be used to simplify the analysis of divergences.

In the functional-integral approach, the special properties of the supersymmetric theories are best stated as general "miracles" of the (supersymmetric) functional measure.

The two fundamental, not uncorrelated, miracles of the measure are:

(i) The supersymmetric measure $[dS(x_\mu, \theta_\alpha)]$ is independent of the particular superfield $S(x_\mu, \theta_\alpha)$ used, as long as the superfield respects the Grassman structure of superspace [5]. This means that the measure is unique and universal.

(ii) There exists a Nicolai mapping [6].

In a supersymmetric field theory, the Nicolai mapping [6], is a non-linear (and, generally speaking, non-local) transformation of the bosonic field configurations $\phi_i(x)$ $(i = 1,..., n)$

$$\phi_i(x) \to \xi_i(x) \tag{I.1}$$

such that:

(a) The bosonic part of the Euclidean Lagrangian[1] is given simply by

$$\mathscr{L}_B = \tfrac{1}{2} \sum_i \xi_i^2 + \text{total divergence}. \tag{I.2}$$

(b) The Jacobian of transformation (I.1), $\det\{\delta\xi/\delta\phi\}$ is given by the Matthews–Salam [7] determinant for the fermions, i.e., in formulas[2]

$$\mathscr{D}\mu[\phi] = [d\phi] \int [d\psi\, d\bar{\psi}]\, e^{-\int \mathscr{L}_F d^4x} = [d\phi] \det\left[\frac{\delta\xi}{\delta\phi}\right] = [d\xi]. \tag{I.3}$$

[1] After the elimination of the auxiliary fields.

[2] $\mathscr{L}_F = \mathscr{L} - \mathscr{L}_B$.

From Eqs. (I.2) and (I.3), it follows that the bosonic measure (after the integration over fermionic fields) becomes, in terms of the fields ξ, Gaussian with zero mean and covariance one.

EXAMPLE [8]. Supersymmetric quantum mechanics with one degree of freedom [12]. The Euclidean Lagrangian is

$$L = \frac{1}{2}\dot{x}^2 + \frac{1}{2}\left(\frac{\partial V}{\partial x}\right)^2 + \bar{\eta} i d_t \eta + i\bar{\eta}\frac{\partial^2 V}{\partial x^2}\eta \tag{I.4}$$

(where η, $\bar{\eta}$ are complex Grassman variables). The Nicolai mapping is

$$\xi(t) = \dot{x}(t) + \frac{\partial V(x)}{\partial x} \tag{I.5}$$

and we have

$$\frac{1}{2}\xi^2 = L_B + \frac{dV}{dt}, \tag{I.6}$$

$$\begin{aligned} \text{Det}\left[\frac{\delta\xi}{\delta\phi}\right] &= \text{Det}\left[d_t + \frac{\partial^2 V}{\partial\phi^2}\right] \\ &= \int [d\eta\, d\bar{\eta}] \exp\left\{-\int_0^\beta [\bar{\eta} i d_t \eta + i\bar{\eta}(\partial^2 V/\partial x^2)\eta]\, dt\right\}, \end{aligned} \tag{I.7}$$

where we suppose periodic boundary conditions in order to cancel the surface terms (see Ref. [8]).

Nicolai [6], noted that all the supersymmetric theories (in which there is no dynamical supersymmetry breaking [6, 8]) are characterized by the existence of such a mapping transforming the (bosonic) functional measure in Gaussian.[3]

At this point there is a natural question: What is the physical meaning of the Nicolai mapping (if any)? Or, to be more precise, is there an interpretation of theGaussian field $\xi(x)$ in Eq. (I.1)?

At this point we present a digression on this aspect of the mapping which we hope will clarify the main ideas behind the present paper. The answer to the previous question is easy in at least one case, that of the most famous Nicolai mapping which is known: the Parisi–Sourlas [9] theorem[4] on the critical behavior of a spin system in an external random (i.e., Gaussian) magnetic field.

[3] In reality, Nicolai [7] proved this result only in perturbation theory and in the hyperlocal limit: in this paper we give a non-perturbative proof of the existence of the Nicolai mapping in some classes of models in which it is local.

[4] In order to avoid any misunderstanding, we stress that we are interested in situations completely different with respect to the Parisi–Sourlas case. Our fermions are physical degrees of freedom which carry spin, have Dirac Lagrangian and interpolate particles (for weak coupling), not just dummy variables.

In that case, the effective Lagrangian is

$$\mathscr{L}_{\text{eff}} = -\frac{1}{2}F^2 + F\left[-\Delta\phi + \frac{\partial V}{\partial\phi}\right] + \bar{\psi}\left[-\Delta + \frac{\partial^2 V}{\partial\phi^2}\right]\psi \tag{I.8}$$

which, after the elimination of the auxiliary field F, is the same of that of Eq. (I.4), with the identifications

$$\begin{aligned} \Delta &\leftrightarrow d_t, \\ \phi &\leftrightarrow x, \\ \psi, \bar{\psi} &\leftrightarrow \eta, i\bar{\eta} \end{aligned} \tag{I.9}$$

and hence, comparing with Eq. (I.5), the Nicolai mapping is

$$\xi = -\Delta\phi + \frac{\partial V}{\partial\phi} \tag{I.10}$$

and the corresponding measure is

$$\mathscr{D}\mu[\xi] = [d\xi] \exp\left\{-\frac{1}{2}\int d^n x\, \xi^2\right\}. \tag{I.11}$$

Comparing Eqs. (I.10) and (I.11) with Ref. [9], we see that ξ is just the random magnetic field (Gaussian with zero mean and covariance one) interacting with our spin system whose dynamics is given by the Lagrangian

$$\tilde{\mathscr{L}} = -\tfrac{1}{2}\phi\,\Delta\phi + V(\phi). \tag{I.12}$$

Then, the functional transformation Eq. (I.11) is the stochastic differential equation connecting ϕ with the random magnetic field. The perturbative solution of this stochastic equation is, as is well known [10], just given by the tree diagrams corresponding to the Lagrangian, Eq. (I.12), with the maximal number of insertions of ξ fields.

By analogy, we can say that (*at least in the case at hand*) the Nicolai mapping transforms the original fields into random would-be magnetic fields, and to the transformation itself we can give an interpretation in terms of stochastic equation of motion.

Now comes the physically relevant point. The analogy is more general and deep. If we interpret Eq. (I.5) as a stochastic equation, we see that it is the most famous of all the stochastic differential equations, namely, the *Langevin equation* [11]. The Langevin equation is (see Ref. [11] and references therein)

$$\frac{dx}{dt} = -\frac{\partial V}{\partial x} + \xi. \tag{I.13}$$

Now, the important fact is that Eq. (I.13) is not just a *formal* Langevin equation, but it is the *correct* (i.e., physical) Langevin equation giving the stochastic equations of motion corresponding to the Lagrangian (I.4).

Indeed, the classical stochastic versions of Newton's equations are

$$\dot{x} = -\frac{\partial S}{\partial x} + \xi, \qquad m = 1, \tag{I.14}$$

where $S(x)$ is a particular (to be specified) solution of the Hamilton–Jacobi equation and ξ is a random Gaussian field. Now the *classical Euclidean* (reduced) *Hamilton—Jacobi* equation for the sub-manifold of phase space in which[5] $H(x, p) = E = 0$ (classical Euclidean vacuum) is

$$\frac{1}{2}\left(\frac{\partial S}{\partial x}\right)^2 - \frac{1}{2}\left(\frac{\partial V}{\partial x}\right)^2 = E \equiv 0. \tag{I.15}$$

The solutions of Eq. (I.15) are $S = \pm V + \text{const}$ which substituted in the stochastic Newton equations (I.14) reproduce the *two* Nicolai mappings of the system, namely, Eq. (I.13). Hence *the Nicolai mapping is the classical stochastic process associated with the classical Euclidean vacuum.*[6]

In more general examples, the situation is more complex and the previous sentence is not true in the strict sense used here.

Using the *Fokker–Planck* equation associated with our *Langevin* equation (see Ref. [11]) we see that, *if* $V(x) \to \infty$ as $|x| \to \infty$, we have a large time limit corresponding to a *thermic equilibrium*.

Thix means that we have

$$\langle x^n \rangle = \lim_{t\to\infty} \langle x(t)^n \rangle = \frac{\int dx\, x^n e^{-2V(x)}}{\int dx\, e^{-2V(x)}} \tag{I.16}$$

which, formally, can be interpreted as a field theory in one dimensionless (in this case just an ordinary integral). Of course, its real meaning is

$$\exp -2V(x) = \int d\eta\, d\bar{\eta}\, \Psi^*_{\text{vac}}[\eta, \bar{\eta}, x]\, \psi_{\text{vac}}[\eta, \bar{\eta}, x], \tag{I.17}$$

where $\psi_{\text{vac}}[\eta, \bar{\eta}, x]$ is the vacuum wave-functional.

[5] $H(x, p)$ is the bosonic part o the Euclidean Hamiltonian, i.e., $H(x, p) = p^2/2 - 1/2(\partial V/\partial x)^2$.

[6] Witten [12] and others [13] have suggested that instantons are relevant for dynamical supersymmetry breaking. If the model has instantons, they are just solutions of Eq. (I.15) and they just change the structure of the classical process of the vacuum. Let us note that the instantonic calculus is *exact* if we use instantons not as an ingredient for WKB approximations but for classical stochastic mechanics. (Of course, the classical vacuum has also non-instantonic components).

And hence the theory can be solved exactly if the Nicolai mapping satisfies the two conditions:

(i) It is a stochastic differential equation.

(ii) This stochastic equation describes an off-equilibrium system which asymptotically reaches a thermic equilibrium.

The condition (i) is equivalent so saying that the Nicolai mapping is *time-local*, namely, that $\xi(t)$ is equal to an expression involving only a finite number of (time) derivatives of $\phi(t)$ at the same time t. In Section II, we will show some classes of supersymmetric models in dimension greater than one with property (i) by constructing explicitly the mapping.[7]

In order to see what condition (ii) is about (from a physical point of view) let us consider what happens if $V(x) \to -\infty$ as $|x| \to +\infty$. In this case we can redefine the Nicolai mapping using the second solution of the Hamilton–Jacobi equation (i.e., $S = -V$)

$$\dot{x} = \frac{\partial V}{\partial x} + \xi \tag{I.18}$$

which corresponds to the change $t \leftrightarrow -t$ (i.e., backward stochastic process). In order to keep the Lagrangian invariant under $V \to -V$ we have to interchange $\eta \leftrightarrow \bar{\eta}$.[8] This means that the vacuum, instead of being killed by the fermionic annihilation operator, is destroyed by the fermionic creation operator (and thus has fermionic number $+1$). The (normalizable) vacuum wave function is then $\exp + V(x)$.

From this we see that the substitution *fermion* $\leftrightarrow$ *hole* corresponds, from the point of view of the stochastic process, to the substitution *forward process* $\leftrightarrow$ *backward process*.

A necessary (and, if Witten's index Δ [8, 14] is not zero, also sufficient) condition for a theory (with many degrees of freedom) to have an equilibrium limit, is that we can define the Nicolai mapping in such a way that it is a forward stochastic process for all the degrees of freedom, in which case a Boltzman distribution is obtained for $t = +\infty$. Because of the previous argument, this means that the vacuum is destroyed by all the fermionic annihilation operators. (This can be shown explicitly.)

Unfortunately, this is not possible for $d > 1$. What happens is that the Nicolai mapping, interpreted as stochastic process, is a forward process for particles with positive energy, whereas it is a backward process for negative energy particles, in agreement with the general structure of QFT.[9] The coupling of negative and positive energy particles maintains the effective "stochastic" system out of equilibrium in any

[7] In a regularized version of the model, of course.

[8] Let $\bar{\eta} = z - iw$, $\eta = z + iw$. Now $\bar{\eta} i d_t \eta = i(z\dot{z} + w\dot{w}) - id_t(zw)$ changes only by a derivative if $w \leftrightarrow -w$, whereas $\bar{\eta}\eta = (z - iw)(z + iw) = 2izw$ changes sign.

[9] Of course, one can redefine the negative energy particles as positive energy anti-particles, but this gives us only a perturbative procedure for solving the theory.

time limit, i.e., the Nicolai mapping does not give us directly the exact solution of the theory.

Despite the failure of condition (ii), the knowledge of an explicit (exact) Nicolai mapping satisfying the condition (i) helps a lot in understanding the non-perturbative structure of the theory, and in doing explicit calculations. And, in any case, it simplifies enormously the perturbative computations.

We hope that, after the previous example, the aims of the present paper would be clear to the reader. We want to study the supersymmetric models having *local* (in time, at least) Nicolai mappings (explicitly and non-perturbatively constructible).

These models possess very peculiar properties at the non-perturbative level, as a consequence of their almost stochastic structure (called parastochastic in the rest of the paper) analogous to the stochastic structure of the previous one-dimensional model.

The most interesting point has to do with divergence cancellations. The supersymmetric quantum theory is equivalent to a classical theory *but* with acausal propagation from the point of view of classical physics. *Very* naively, one would expect that purely classical equations of motion are not affected by quantum corrections, whether or not, from the point of view of classical physics, they respect causality. Of course, this is true only in part, because we have to take into account the quantum divergences in the process of constructing the "classical equations" out of the full quantum theory.[10]

The paper is organized as follows: in Section II we study the explicit construction of local Nicolai mappings in dimension greater than one, more as an existence theorem for models which present the discussed mechanism than for their intrinsic interest. The explicit construction of (local) Nicolai mappings for theories of direct physical interest will be the subject of a future paper. We limit ourselves to a cut-off version of the general $N=2$ model containing only chiral superfields (in two dimensions). We introduce a new perturbative technique (infra-diagrams) for the parastochastic reconstruction of the bosonic Green's functions out of the Nicolai mapping, which is easier than the standard ones and in which the "miraculous" cancellations are automatic and, moreover, deeply "stochastically" motivated.

II. Nicolai Mappings and Parastochasticity in Dimension Larger than One

II.1. *Models in Two Dimensions*

Consider the $N=2$ supersymmetric models in $1+1$ dimensions with (Minkowskian) Lagrangian

[10] In a renormalizable theory these divergences are controlled by the usual renormalization techniques.

$$\mathscr{L} = \partial_\mu \phi^* \partial^\mu \phi + \bar{\psi} i \not\partial \psi - \frac{\partial V}{\partial \phi^*} \frac{\partial V}{\partial \phi} + \bar{\psi} \left(\frac{1-\gamma_3}{2} \right) \bar{\psi} \frac{\partial^2 V}{\partial \phi^2}$$

$$+ \bar{\psi} \left(\frac{1+\gamma_3}{2} \right) \psi \frac{\partial^2 V}{\partial \phi^{*2}}, \tag{II.1}$$

where V is an analytic function of one variable (polynomial usually) and ψ is a Dirac spinor.

In the following we shall suppose that the model is quantized in an Euclidean box, using periodic boundary conditions for *all* the fields, and that, in order to give a definite mathematical meaning to our formal manipulations, in the functional integral we integrate only over a large, but finite, number of Fourier components, namely, on all the frequencies (ω, p) contained in a square with side Λ. This regularization respects all the symmetries of the periodic box and it is supersymmetry preserving, *if* it is done symmetrically on all the components of the supermultiplet, auxiliary fields comprehended, *before* their elimination from the Lagrangian.

In order to simplify the notation, we shall always use the continuum notation, without indicating explicitly the cut-off procedure.

Now that we have a finite number of degrees of freedom, we expect that the problem of finding the Nicolai mapping can be reconducted to the general ansatz for the quantum mechanical case with n degrees of freedom.[11] For a mechanical system with superpotential $W(x_1,..., x_n)$ the general ansatz is

$$\xi_i = \dot{x}_i - \frac{\partial W}{\partial x_i}. \tag{II.2}$$

In the case of Eq. (II.1) we define an effective superpotential $W[\phi_x, \phi_x^*]$:

$$W[\phi_x, \phi_x^*] = \int dx_1 \left\{ V(\phi_x) + V(\phi_x^*) + \frac{i}{2} \phi_x^* \overleftrightarrow{\partial}_1 \phi_x \right\}. \tag{II.3}$$

Then the Nicolai Gaussian fields have the same form of the one-dimensional case[12]

$$\xi_x = \dot{\phi}_x - \frac{\delta W}{\delta \phi_x^*} = \dot{\phi}_x - \frac{\partial V}{\partial \phi_x^*} - \frac{i}{2} \{\phi_z \partial_1 \delta_{zx} - \delta_{xz} \partial_1 \phi_z\}, \tag{II.4}$$

$$\xi_x^* = \dot{\phi}_x^* - \frac{\delta W}{\delta \phi_x} = \dot{\phi}_x^* - \frac{\partial V}{\partial \phi_x} + \frac{i}{2} \{\phi_z^* \partial_1 \delta_{zx} - \delta_{xz} \partial_1 \phi_z^*\}. \tag{II.5}$$

[11] Continuous repeated indices are integrated over x denotes the space coordinate.

[12] Technically, this construction is unnecessarily complicated; trivial properties of the γ-matrices are enough. However, from a general point of view the presented construction is preferable. The construction is based on the classical action principle *via* the Hamilton–Jacobi equation, as explained in Section I (W is a solution). The action principle gives the only general ansatz for the local Nicolai mapping, generating all the inequivalent (from a stochastic point of view) mappings. This ansatz is likely to be generalizable to higher dimension.

If V is a polynomial, we have

$$\int dx_1\, \xi_x^* \xi_x = L_B + \frac{dW}{dt} + \int dx_1\, \partial_1 \text{ (some polynomial in } \phi, \phi^*), \tag{II.6}$$

where L_B is the Euclidean Lagrangian

$$L_B = \int dx_1 \left\{ \partial_\mu \phi^* \partial_\mu \phi + \frac{\partial V}{\partial \phi^*} \frac{\partial V}{\partial \phi} \right\}. \tag{II.7}$$

The last term in the rhs of Eq. (II.6) vanishes by virtue of the periodic boundary conditions.

Denote with A the operator

$$A = A_{yx} = \tfrac{1}{2}\delta_{yz}\partial_1 \delta_{zx} - \tfrac{1}{2}\delta_{xz}\partial_1 \delta_{zy} = \partial_1 \delta_{yx}. \tag{II.8}$$

Then the functional Jacobian of transformations (II.4) and (II.5) is

$$J_{xy} = \begin{pmatrix} \dfrac{\delta\xi_x}{\delta\phi_y} & \dfrac{\delta\xi_x}{\delta\phi_y^*} \\ \dfrac{\delta\xi_x^*}{\delta\phi_y} & \dfrac{\delta\xi_x^*}{\delta\phi_y^*} \end{pmatrix} = \begin{pmatrix} d_t \delta_{yx} - iA_{yx} & \dfrac{\partial^2 V}{\partial \phi^2}\delta_{yx} \\ \dfrac{\partial^2 V}{\partial \phi^{*2}}\delta_{yx} & d_t \delta_{yx} + iA_{yx} \end{pmatrix}, \tag{II.9}$$

namely,

$$J = d_t - i\sigma_3 A + \sigma_1 \operatorname{Re}\frac{\partial^2 V}{\partial \phi^2} - \sigma_1 \operatorname{Im}\frac{\partial^2 V}{\partial \phi^2}. \tag{II.10}$$

If we choose the following representation for the two-dimensional Dirac matrices,

$$\gamma_0 = \sigma_1, \qquad \gamma_1 = \sigma_2, \qquad \gamma_3 = \sigma_3, \tag{II.11}$$

we have

$$J = \gamma^0 \left\{ \gamma^0 d_t + \gamma^1 \partial_1 + \operatorname{Re}\frac{\partial^2 V}{\partial \phi^2} + i\gamma_3 \operatorname{Im}\frac{\partial^2 V}{\partial \phi^2} \right\} = \gamma^0 D[\phi, \phi^*], \tag{II.12}$$

where $D[\phi, \phi^*]$ is defined by the equation

$$\mathscr{L}_F = \bar{\psi} i D[\phi, \phi^*]\, \psi \tag{II.13}$$

and $\mathscr{L}_F$ is the fermionic part of the Euclidean version of the Lagrangian in Eq. (II.1).

Now, because our fermions ψ are Dirac, we have

$$\int [d\psi\, d\bar{\psi}] \exp\left(- \int \mathscr{L}_F\, d^2x \right) = \det(D[\phi, \phi^*]) = \det(J). \tag{II.14}$$

By virtue of our regularization, Eq. (II.14) is the determinant of a finite dimensional matrix and hence is unambiguously defined. If we had not regularized the theory, Eq. (II.14) would need a regularization and the two sides of the equation would not be guaranteed to be equal. This means that we must use an explicitly supersymmetric regularization in order to construct the regularized Nicolai mapping (see subsection II.2).

Equation (II.14) proves that our ansatz, Eqs. (II.4) and (II.5), is a Nicolai mapping. This is the first non-perturbative construction of a Nicolai mapping in supersymmetric field theory and constitutes a proof of Nicolai's conjectures [6] based on perturbative arguments. Let us remark that, in agreement with the general arguments of Ref. [8], a non-perturbative Nicolai mapping exists *only* in an Euclidean box with *periodic* boundary conditions. In the infinite volume limit one does not expect to find an exact Nicolai mapping because the process of transforming the functional measure in Gaussian and that of decomposing the functional measure in its ergodic components (different vacua) do not, in general, commute (see Ref. [8]).

The equations

$$\dot{\phi} = \frac{\delta W}{\delta \phi^*} + \xi, \qquad \dot{\phi}^* = \frac{\delta W}{\delta \phi} + \xi^* \tag{II.15}$$

have the general structure of the *Langevin* equation. However, as anticipated in the introduction, they do not have the correct causal behavior for a transport phenomenon. This can be most easily seen in the free case

$$W_{\text{free}} = \int dx_1 \left\{ \frac{i}{2} \phi^* \overleftrightarrow{\partial}_1 \phi + \frac{1}{2} m\phi^2 + \frac{1}{2} m\phi^{*2} \right\}, \tag{II.16}$$

$$\begin{aligned} \dot{\phi}|_{\text{free}} &= i\,\partial_1 \phi + m\phi^* + \xi, \\ \dot{\phi}^*|_{\text{free}} &= -i\,\partial_1 \phi^* + m\phi + \xi^*. \end{aligned} \tag{II.17}$$

If the Langevin equations (II.17) would have a thermic equilibrium limit, we would have a vacuum wave-functional

$$\sim \exp(-W_{\text{free}}[\phi, \phi^*]), \tag{II.18}$$

namely,

$$\sim \exp\left[-\frac{1}{2} \sum_n \omega_n (\phi^*_{+,n} \phi_{+,n} - \phi^*_{-,n} \phi_{-,n})\right], \tag{II.19}$$

where $\omega_n = (\vec{p}_n^{\,2} + m^2)^{1/2}$ and $\phi_{+,n}$ (resp. $\phi_{-,n}$) are the corresponding positive (resp.

negative) frequency Fourier coefficients. Equation (II.19) is wrong only because of the minus sign inside the parenthesis that makes it non-normalizable.

According to our discussion in the introduction, this corresponds to the fact that the real vacuum is not destroyed by all the fermionic annihilation operators, but only by those corresponding to positive energy states.

Of course, we can redefine the creation operators for negative energy states to be annihilation operators for positive energy anti-particles. Then the real vacuum is destroyed by all the fermionic annihilation operators and there exists a causal stochastic process. The stochastic process that we obtain so is

$$\dot{\phi} = -(\sqrt{-\partial_1^2 + m^2}\phi^*) + \xi, \qquad \dot{\phi}^* = -(\sqrt{-\partial_1^2 + m^2}\phi) + \xi^2 \tag{II.20}$$

whose canonical partition function (in the limit $t \to +\infty$) is the correct wave-functional of the vacuum. Equation (II.20) is an alternative Nicolai mapping.

Let us note that the drift coefficients of the Langevin equtions, Eqs. (II.15) and (II.20), satisfy the (Euclidean) Hamilton–Jacobi equation for the (classical) zero-energy sub-manifold, namely (for the Lagrangian in Eq. (II.1)),

$$\frac{\delta W}{\delta \phi_x^*}\frac{'\delta W}{\delta \phi_x} - \int dx_1 \left\{ |\partial_1 \phi|^2 + \left| \frac{\partial V}{\partial \phi} \right|^2 \right\} \equiv 0. \tag{II.21}$$

With more than one degree of freedom, there are solutions of Eq. (II.19) which *are not the classical stochastic process of the vacuum*, as we have seen in the example of the free theory. Thus the problem is to single out the classical process of the vacuum from the set of all the solutions of Eq. (II.21) (*if such a process exists*). Only in one case we have an effective algorithm for selecting the process out of Eq. (II.21), namely, when Eq. (II.21) is completely separable, in which case we have a complete set of conserved (classical) quantities in involution, the classical analogous of a complete set of commuting conserved charges (Heisenberg representation), wich can be used to determine the vacuum process as the solution of Eq. (II.21) with the "quantum numbers" of the vacuum.

However, we emphasize that, in general, there is no classical stochastic process for the (quantum) vacuum,[13] and hence we must be content with the *parastochastic* structure, which is very relevant and interesting for itself.

For instance, it can be used as a simplifying change of variables in the (Euclidean) functional integral. In the perturbative approach this corresponds to using the *classical* stochastic perturbation theory. This will be the argument of subsection (II.2).

The reason of the utility of the Nicolai mapping is that the generator of the bosonic Green's functions, $Z\{J(x)\}$ for the *classical (para-) stochastic process* of Eq. (II.15) is the same of that obtained from the full supersymmetric quantum theory, indeed

[13] Because the functional Jacobian is not guaranteed to match the fermionic determinant by the classical action principle; there are many cases, however, in which they match.

$$Z\{J(x)\} = \int [d\xi\, d\xi^*\, d\phi\, d\phi^*] \exp\left[i \int d^2x(J\phi^* + J^*\phi)\right] \delta\left\{\dot{\phi} - \frac{\delta W}{\delta\phi^*} - \xi\right\}$$

$$\times \delta\left\{\dot{\phi}^* - \frac{\delta W}{\delta\phi} - \xi^*\right\} \operatorname{Det}\frac{\delta(\xi,\xi^*)}{\delta(\phi,\phi^*)} \exp\left[-\int \xi^*\xi\, d^2x\right]$$

$$= \int [d\phi\, d\phi^*\, d\psi\, d\bar{\psi}] \exp\left\{-\int \mathscr{L}(\phi,\phi^*,\psi,\bar{\psi}) + iJ\phi^* + iJ^*\phi]\, d^2x\right\}. \qquad \text{(II.22)}$$

$$\operatorname{Det}\frac{\delta(\xi,\xi^*)}{\delta(\phi,\phi^*)} = \int [d\psi\, d\bar{\psi}] \exp\left\{-\int \mathscr{L}_F(\phi,\phi^*,\psi,\bar{\psi})\, d^2x\right\}. \qquad \text{(II.23)}$$

We conclude this subsection with the observation that a straightforward generalization to four dimensions of the ansatz Eq. (II.13) for the *relevant solution* of the action principle works only for the (massive) free theory. This fact agrees with the general ideas presented in the introduction. Indeed, it is well known that the free theory is the only $N = 2$ supersymmetric theory (with only chiral superfields) which is renormalizable. If the only divergences are those encountered in constructing the Langevin equations, they are unlikely to be worse than logarithmic (*if* there exists a Nicolai mapping, local both in space and time, of the form given by the ansatz Eq. (II.13)). Hence, naively, the existence of a Nicolai mapping of this form in that class of models would be inconsistent. Let us note that the conjecture that the relevant W is not renormalized is verified in the case of Eq. (II.3).

II.2. *New Perturbative Techniques: Infradiagrams*

As we have said, we can use the knowledge of an explicit (regularized) local Nicolai mapping for constructing a new perturbative scheme along the lines of the classic stochastic perturbation theory.

Having integrated over the fermions (in order to cancel the functional Jacobian) our perturbation scheme contains only the bosonic fields (scalars in our examples) and hence the usual cancellations of divergences between bosonic and fermionic loops are automatic in the formalism. This means a better power-counting convergence of the diagrams (when these ideas are used in dimension larger than two), connected with the fact that all the vertices involved in this new scheme have a considerably lower order with respect to those entering the standard Feynman rules (with *both* fermionic and bosonic propagators).

This perturbative scheme allows the computation of all the Green's functions containing only bosonic fields if (and only if) the supersymmetry is not dynamically broken. The rest of the physical content of the theory can be reconstructed using supersymmetry and/or unitarity (see the following).

Because the present idea is just the opposite of that behind the superdiagrammatic approach, i.e., instead of using supersymmetry for putting together bosons and fermions in the same superpropagator, we use it for *eliminating* the fermions from the computation. We call this approach infradiagrammatic.

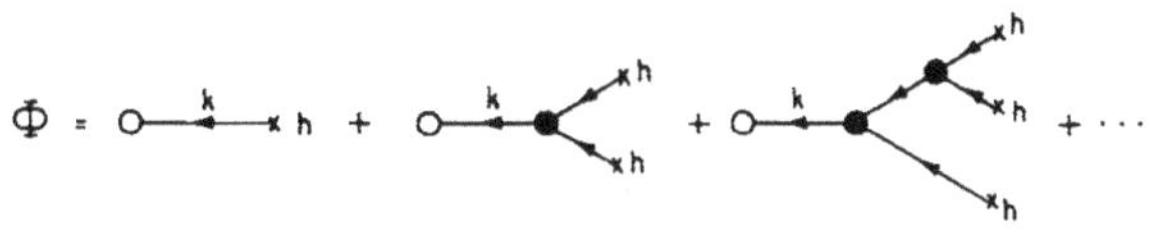

FIG. 1. Diagrammatic expansion for the perturbative solution of the (para-)stochastic Eq. (II.21).

In this paper, we limit ourselves to the examples of subsection (II.1). Let us assume that in Eqs. (II.4) and (II.5), V has the form

$$V(\Phi)=\frac{1}{2}m\Phi^2+\frac{g}{n+1}\Phi^{n+1}$$

denoting the Φ vector (Re Φ, Im Φ); Eqs. (II.4) and (II.5) become

$$\dot{\Phi}+i\sigma_2\partial_1\Phi-m\sigma_3\Phi=H+g\sigma_3\Phi^{[n]}, \tag{II.24}$$

where $\Phi^{[n]}=(\mathrm{Re}\,\Phi^n, \mathrm{Im}\,\Phi^n)$ and H is a Gaussian field with covariance 1.

As it is well known (see, for example, Refs. [11, 15]), the diagrammatic expansion of Eq. (II.24) is given by all the tree diagrams (classical theory) with insertions of H fields, as (for the case $n=3$) is shown in Fig. 1.

In Fig. 1 an empty circle O denotes the Φ field, a black circle a vertex and a cross × an insertion of the Gaussian field H.

The propagator $K(\omega, p)$ (recall that ω and p assume discrete values because of our periodic boundary conditions) is

$$K(\omega,p)=(-i\omega+\sigma_2p-m\sigma_3)^{-1}=\frac{i\omega+\sigma_2p-m\sigma_3}{(\omega^2+p^2+m^2)} \tag{II.25}$$

(ω, p are counted positively if they enter the propagator from the side of the field ϕ).

For the example in Fig. 1 the explicit form of the vertex (Fig. 2) is

$$A_{\alpha\beta\gamma}=\begin{pmatrix}\sigma^3_{\beta\gamma}\\ \sigma^1_{\beta\gamma}\end{pmatrix}_\alpha, \qquad \alpha,\beta,\gamma=1,2. \tag{II.26}$$

Using the methods of Ref. [11], it is easy to reconstruc the (bosonic) Green's functions from the perturbative solution of the Nicolai mapping in Fig. 1. For example, for the two-point function, using the fact that $\langle H\rangle=0$ and $\langle H(x)\,H(y)\rangle=\delta(x-y)$, we have the diagrammatic expansion in Fig. 3.

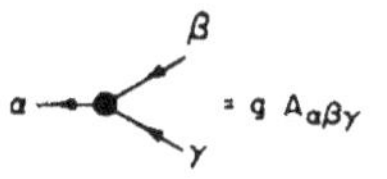

FIG. 2. Vertex for the two-dimensional Wess–Zumino model.

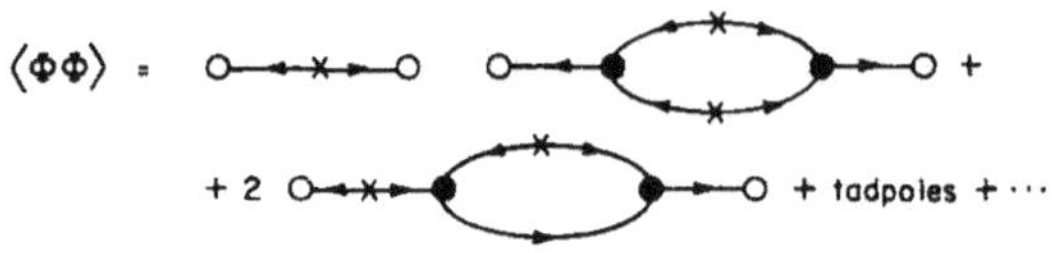

FIG. 3. Diagrammatic expansion for the two-point function in the Wess–Zumino model. The cross means that two h-field insertions are averaged to 1, according to the Gaussian property.

In the free case, we have

$$\langle \phi_\alpha(\omega, p)\, \phi_\beta(\omega', p')\rangle|_{\text{free}} = \langle ((-i\omega + \sigma_2 p - m\sigma_3)^{-1}_{\alpha\rho} H_\rho(\omega, p)(-i\omega' + \sigma_2 p' - m\sigma_3)^{-1}_{\beta\sigma} H_\sigma(\omega', p')\rangle; \quad \text{(II.27)}$$

using

$$\langle H_\alpha(\omega, p)\, H_\beta(\omega', p')\rangle = \delta_{\alpha\beta}\delta_{\omega,-\omega'}\delta_{p,-p'}, \quad \text{(II.28)}$$

we have

$$\begin{aligned}\langle \phi_\alpha(\omega, p)\, \phi_\beta(\omega', p')\rangle|_{\text{free}} &= (-i\omega + \sigma_2 p - m\sigma_3)^{-1}_{\alpha\beta} (i\omega - \sigma_2 p - m\sigma_3)^{-1}_{\beta\rho} \delta_{\omega,-\omega'}\delta_{p,-p'} \\ &= (\omega^2 + p^2 + m^2)^{-1} \delta_{\alpha\beta}\delta_{\omega,-\omega'}\delta_{p,-p'} \end{aligned} \quad \text{(II.29)}$$

which, of course, is the usual Euclidean propagator.

Now we see why we had to regularize the Nicolai mapping; we had to do so in order that the regularization (and renormalization) of the Green's functions in Fig. 3 be consistent with that of the Nicolai mapping, in such a way that the super-Ward identities be true for the renormalized Green's functions. The expansion in Fig. 3 for the Green's functions has to be renormalized according to the standard methods.

The reader can see by himself that the "super-miraculous" cancellations now are automatic.

Now we discuss briefly how the infradiagrammatic technique automatically takes care of the fermions. Obviously, the infradiagrams do not satisfy the usual cutting rules guaranteeing unitarity for Feynman diagrams. This corresponds to the fact that the unitarity equations *are not* saturated by the bosonic particles alone and hence our infradiagrams must have some discontinuity corresponding to fermions in the intermediate state, *even* if the rules themselves do not contain explicitly fermions. The general situation is represented in Fig. 4, where --- means scalar and →— means fermionic propagators.

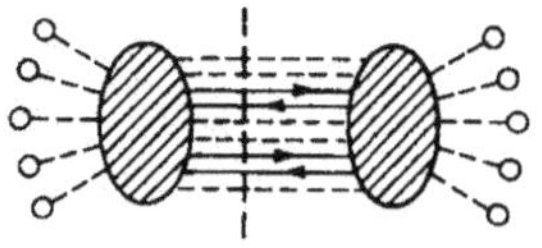

FIG. 4. A general contribution to the discontinuity of a bosonic Green's function. Full lines represent fermions.

This implies that the infradiagrams contain, in principle, all information for fermionic scattering through unitarity (of course, an easier procedure is to relate the fermionic Green's functions to the bosonic ones using supersymmetry).

Note the possibility that both scalar and fermionic propagators' discontinuities are represented by the *same* infrapropagator

$$K(\omega, p) = \frac{i\omega + \sigma_2 p - m\sigma_3}{(\omega^2 + p^2 + m^2)} \tag{II.30}$$

(which, by the way, is not more complicated than the standard fermionic propagator), is due to the fact that, because of supersymmetry, the fermionic and scalar discontinuities are at the same value of p^2.[14]

That the fermions are correctly treated by the infradiagrammatic method is guaranteed by the general arguments of subsection (II.1) which we had used in order to construct the Nicolai mapping, Equations (II.22) and (II.23) contain in closed form all the combinatorics required to prove the equivalence of the infradiagrams with the standard Feynman diagrams. A diagrammatic analysis of Eqs. (II.22) and (II.23) can be found, in a different context, in Refs. [11, 15].

In this paper we are more interested in the non-perturbative structure of the theory from an abstract point of view than in obtaining an improved perturbative scheme out of that. Hence, we do not proceed here to a systematic diagrammatic analysis of the unitarity relations, but simply show, in a very easy case, how the fermions are kept into account.

Consider the one-dimensional system with (Euclidean) Lagrangian

$$L = \tfrac{1}{2}\dot{x}^2 + \tfrac{1}{2}(mx + gx^3)^2 + \bar{\eta} i d_t \eta + i\bar{\eta}(m + 3gx^2)\,\eta \tag{II.31}$$

which has $\Delta = 1$.

The contribution of order g to the (Euclidean) two-point function F.T. $\{\langle x(t)\, x(0)\rangle\}$ is given by the diagram in Fig. 5 which has the value

$$(\text{Fig. } 5) = -\frac{6mg}{(\omega^2 + m^2)^2}\int \frac{dh}{2\pi}(h^2 + m^2)^{-1}. \tag{II.32}$$

To the same order, we have two infradiagrams in Fig. 6; keeping into account the combinatoric factors, we have

$$\begin{aligned}(\text{Fig. } 6) &= -\frac{3g}{-i\omega + m}\int \frac{dh}{2\pi}(h^2 + m^2)^{-1}\frac{1}{\omega^2 + m^2} \\ &\quad - \frac{3g}{\omega^2 + m^2}\int \frac{dh}{2\pi}(h^2 + m^2)^{-1}\frac{1}{i\omega + m} \\ &= -\frac{6mg}{(\omega^2 + m^2)^2}\int \frac{dh}{2\pi}(h^2 + m^2)^{-1}\end{aligned} \tag{II.33}$$

[14] Note that the infradiagrammatic technique, as the Nicolai mapping itself, has a meaning only if $\Delta \neq 0$, and hence supersymmetry is unbroken.

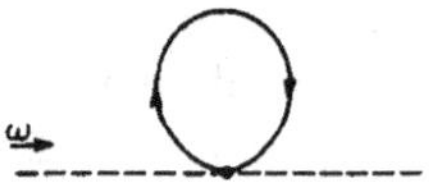

FIG. 5. Order g contribution to the coordinate two-point Green's function $T\langle x(t)\, x(0)\rangle$ in a supersymmetric version of the g^2x^6 anharmonic quantum oscillator.

which is equal to the computation in which the fermions are explicitly used, Eq. (II.32).

III. CONCLUSIONS

In this paper we have discussed how the existence of a time-local Nicolai mapping in a supersymmetric theory implies an underlying structure of classical stochastic differential equations. The presence of this structure seems more probable in the case of extended supersymmetry, in which case we have Dirac fermions which give us a determinant of a local operator rather that the square root obtained from integration over Majorana fermions.

In this paper, we have addressed ourselves to a simple class of models in two dimensions, just to show how the parastochastic ideas work. Nevertheless our result are a first non-perturbative proof of the existence of a Nicolai mapping and, moreover, they confirm some of the general ideal of Ref. [8]. In a future paper, we shall consider models more relevant from a physical point of view.

In general, the stochastic equations do not admit a thermal equilibrium limit. This means that one cannot read the exact solution of the model directly out of the analytic expression of a local Nicolai mapping. However, the (non-perturbative) knowledge of a local Nicolai mapping can help very much in understanding the deep properties of a given supersymmetric theory.

We think that the main advantage of the parastochastic structure consists in the possibility of constructing general non-perturbative arguments about the dynamics; nevertheless this structure allows a perturbative scheme which seems very interesting. In such a scheme, the fermionic fields are completely eliminated by exact analytic integration over them. The corresponding rules are very simple and can be used in actual computations.

However, the pure fact that such rules, involving only bosons, exist—if the theory possesses enough structure—is by far more interesting (and useful) than the technical advantage for perturbation theory which we can obtain from them.

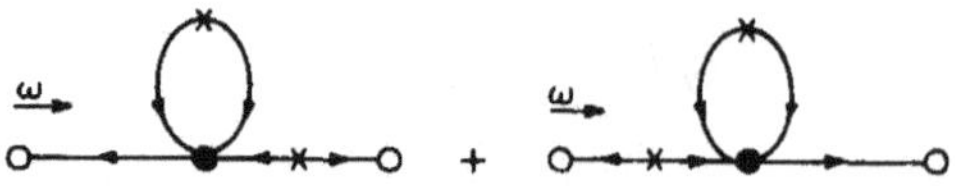

FIG. 6. The infradiagrams corresponding to the diagram of Fig. 5.

This state of affairs suggests that this approach can be very useful for understanding the "miraculous" cancellations of divergences in a supersymmetric field theory. And, indeed, this was the original motivation for the search of a hidden classic (stochastic) physics behind a quantum supersymmetric theory.

Our results go in the right direction, but, of course, there remains a lot of work to do before obtaining a (para-)stochastic interpretation of the conjectured finiteness of four-dimensional supersymmetric field theories.

APPENDIX: Some Remarks on the Witten Index in One and More Dimensions

In general, because of mathematical simplicity, one uses one-dimensional supersymmetry as an example of general ideas on the breaking. However, one cannot transport acritically one-dimensional results to higher dimension because there is a difference of paramount importance between the two cases. The point is trivial but can be confusing for non-experts, so we briefly discuss it.

The Witten index is defined as [8]

$$\begin{aligned}\Delta &= (\text{Number of bosonic zero-energy states})\\ &\quad - (\text{Number of fermionic zero-energy states})\\ &= \int_{\text{box}} [d\phi]_p \exp -S(\phi).\end{aligned} \tag{A.1}$$

In Ref. [8] we gave a practical recipe for computing Δ: first one has to concentrate the functional measure over the classical zero-action configurations. If the quadratic fluctuations around these classical vacua are non-degenerate, each vacuum contributes with 1 to the functional integral in Eq. (A.1). This is due to a theorem [5] that states that $\int [d\phi] \exp - \int \bar{\phi} A \phi = 1$ for every non-degenerate, supersymmetry preserving operator A. Hence, in absence of degeneration

$$\Delta = \text{number of classical vacua}. \tag{A.2}$$

Now, Eq. (A.2) *is not true in less than four dimensions* (but *it is* true if the symmetry is extended). This is due to the fact that, because of the ambiguity in defining the fermionic grading of a state, the Gaussian functional measure is not universal but has two different orientations, namely,

$$\int [d\phi] \exp \left\{ - \int \bar{\phi} A \phi \right\} = (-)^F \tag{A.3}$$

$F =$ (Morse index of the quadratic form A) = (number of negative eigenvalues). For instance, with one degree of freedom, Eq. (A.3) is

$$(1 - e^{-\beta\omega})/|1 - e^{-\beta\omega}| = \pm 1 \tag{A.4}$$

and if $\omega < 0$ the vacuum containts one fermion and hence Eq. (A.3) gives the correct fermionic grading of the vacuum (see Ref. [8]). This can happen because the concept of fermionic number if purely conventional in less than four dimensions.[15] Then Δ can vanish just because the number of fermionic and bosonic classical vacua are equal [12]. *This cannot happen in four dimensions* (or $N = 2$ in $d = 2$). Indeed, from Eq. (A.3) we see that the fermionic grading of a classical vacuum is given by minus to the number of negative eigenvalues of the mass-matrix (in a functional sense) for the fermions around that vacuum (i.e., $(-)^{F_i}$, where F_i is the Morse index for the ith (*non-degenerate*) critical point of the superpotential: $\partial V/\partial x_i = 0$ [14]). In four dimension $(-)^{F_i}$ is independent of i (and hence +), because F_i for two different i differs by $2x$ (integer) because, by PCT, the sign of the mass for a fermion and its PCT conjugate are always the same. Moreover, PCT requires complex scalars in the chiral multiplet. This means there cannot be compensation between fermionic and bosonic classical vacua and Eq. (A.2) holds. *Hence, the mechanism that breaks supersymmetry in one dimension* [12] *definitely cannot break in four.* This is connected with our discussion of the acausality of the Nicolai–Langevin equation and the corresponding fermionic structure of the vacuum. During the completion of this work, similar results were derived by G. Parisi and N. Sourlas [16]. We thank them for communicating their results before publication.

Acknowledgment

We are very grateful to Dr. C. Gomez for useful discussions.

References

1. For a review on supersymmetry, see P. Fayet and S. Ferrara, *Phys. Rep. C* **32** (1977), 249. For a review of the divergence problem see M. Dugg, Ultraviolet divergences in extended supergravity, CERN preprint, 1982.
2. E. Witten and D. Olive, *Phys. Lett. B* **78** (1978), 97.
3. M. T. Grisaru and W. Siegel, Supergraphity (II), preprint CALT-68-892, 1982, and references therein.
4. P. Rossi, *Phys. Lett. B* **99** (1981), 229.
5. H. Nicolai, *Phys. Lett. B* **101** (1981), 396.
6. H. Nicolai, *Phys. Lett. B* **89** (1980), 341; H. Nicolai, *Nucl. Phys. B* **176** (1980), 419; H. Nicolai, Supersymmetry without anticommuting variables, *in* "Unification of the Fundamental Particle Interactions" (S. Ferrara, J. Ellis, and P. van Nieuwenhuizen, Eds.), Plenum, New York, 1980.

[15] In four dimensions $(-)^F = e^{2\pi i J_z}$; in tree dimensions this formula is not true for massless particles.

7. T. Matthews and A. Salam, *Nuovo Cimento* **12** (1954), 563; **2** (1955), 120.
8. S. Cecotti and L. Girardello, *Phys. Lett. B* **110** (1982), 39.
9. G. Parisi and N. Sourlas, *Phys. Rev. Lett.* **43** (1979), 744. Similar ideas in a somewhat different context can be found in Ref. [15].
10. See Refs. [11, 15] and references therein.
11. G. Parisi and Wu Yongshi, *Sci. Sin.* **24** (1981), 483.
12. E. Witten, *Nucl. Phys. B.* **188** (1981), 513.
13. P. Salomonson and J. W. van Holten, *Nucl. Phys. B* **196** (1982), 509.
14. E. Witten, Constraints on supersymmetry breaking, Princeton preprint, 1982.
15. A. Niemi, L.C.R. Wijewardhana, preprint CPT #966, 1982; B. McClain, A. Niemi, and C. Taylor, preprint CTP #967, 1982, and references therein.
16. G. Parisi and N. Sourlas, LPTENS preprint 82/6, (1982), to appear in *Nuclear Phys. B.*

PHYSICAL REVIEW D VOLUME 28, NUMBER 8 15 OCTOBER 1983

Functional-integral approach to Parisi-Wu stochastic quantization: Scalar theory

E. Gozzi
Physics Department, City College of C.U.N.Y., New York, New York 10031
(Received 23 June 1983)

The 5th-time stochastic-quantization approach to field theory, recently proposed by Parisi and Wu, is put in a path-integral form. The procedure of taking the limit $\tau\to\infty$ is analyzed and based on new grounds through the introduction of the vacuum-vacuum generating functional. Different aspects of the interplay between forward and backward Fokker-Planck dynamics are studied in detail in connection with the supersymmetry recently discovered in Gaussian stochastic processes.

INTRODUCTION

Recently, Parisi and Wu[1] proposed a new and interesting method for the quantization of physical systems. The idea was to introduce a 5th time τ and postulate a stochastic Langevin dynamics for the system. Those authors[1] showed that, at least at the perturbative level, the usual quantum theory is reproduced by the equilibrium limit $\tau\to\infty$ of that dynamics. The first advantage of this method is the possibility to quantize gauge theories without fixing the gauge, and much work[1,2] has already been done in this direction. Another recent application is the use of the Langevin equation for the computer simulation of lattice-field-theory models[3]: simulation that should have better properties than the usual Monte Carlo one. The third application,[4] and we hope not the last, is a better and deeper understanding of the so-called quenched reduced models.

In view of all these connections and hoping for more to come, we give in this paper a functional-integral reformulation of this new method of quantization. The hope is to be able to use all the techniques developed in recent years in path integration and bring out the rich content, still undiscovered, in this approach of Parisi and Wu.

The paper is organized as follows: in Sec. I we briefly review the work of Parisi and Wu. In Sec. II, starting from the Langevin equation, we derive the corresponding generating functional. In Sec. III we impose on the Langevin equation to describe a stationary process, and the procedure of taking the limit $\tau\to\infty$ is put on a new basis through the introduction of the vacuum-vacuum generating functional. In Sec. IV we make contact with the recently discovered hidden supersymmetry in stochastic Gaussian processes, and analyze in detail the nice interplay of forward and backward Fokker-Planck dynamics present in the supersymmetric form of the generating functional.

In this work we limit ourselves to scalar theories without any internal symmetry.

I. REVIEW OF PARISI-WU STOCHASTIC QUANTIZATION

We all know that the "quantum" correlation functions for a Euclidean system, described by an action $S[\phi]$, are given by

$$\langle 0|T\phi(x_1)\phi(x_2)\cdots\phi(x_l)|0\rangle=\frac{\int\mathscr{D}\phi[\phi(x_1)\cdots\phi(x_l)]e^{-S[\phi]}}{\int\mathscr{D}\phi e^{-S[\phi]}}\,. \tag{1}$$

Parisi and Wu[1] proposed the following alternative method to get the quantum averages:

(i) Introduce a 5th time τ, in addition to the usual four-space-time x^μ, and postulate the following Langevin equation for the dynamics of the field ϕ in this extra time τ:

$$\frac{\partial\phi(x,\tau)}{\partial\tau}=-\frac{\delta S[\phi]}{\delta\phi}+\eta(x,\tau)\,. \tag{2}$$

η is a Gaussian random variable,

$$\begin{aligned}&\langle\eta(x,\tau)\rangle_\eta=0\,,\\&\langle\eta(x,\tau)\eta(x'\tau')\rangle_\eta=2\delta(x-x')\delta(\tau-\tau')\,,\\&\langle\eta\cdots\eta\rangle_\eta=0\,.\end{aligned} \tag{3}$$

The angular brackets denote connected average with respect to the random variable η.

(ii) Evaluate the stochastic average of fields ϕ_η satisfying Eq. (2), that means

$$\langle\phi_\eta(x_1\tau_1)\phi_\eta(x_2\tau_2)\cdots\phi_\eta(x_l\tau_l)\rangle_\eta\,. \tag{4}$$

(iii) Put $\tau_1=\tau_2=\cdots=\tau_l$ in (4) and take the limit $\tau_1\to\infty$.

It is possible to prove,[1] at least perturbatively, that

$$\lim_{\tau_1\to\infty}\langle\phi_\eta(x_1\tau_1)\phi_\eta(x_2\tau_1)\cdots\phi_\eta(x_l\tau_1)\rangle_\eta=\frac{\int\mathscr{D}\phi[\phi(x_1)\cdots\phi(x_l)]e^{-S[\phi]}}{\int\mathscr{D}\phi e^{-S[\phi]}}\,. \tag{5}$$

To understand this relation we have to introduce the notion of probability $P(\phi,\tau)$, that is, the probability of having the system in the configuration ϕ at time τ. There exists for $P(\phi,\tau)$ an equation that describes its evolution in the time τ. It is called the Fokker-Planck (FP) equation and it has been derived many times in the literature[5]:

$$\frac{\partial P}{\partial \tau}=\frac{\partial^2 P}{\partial \phi^2}+\frac{\partial}{\partial \phi}\left[P\frac{\partial S}{\partial \phi}\right]. \tag{6}$$

It is possible to recast this equation in a Schrödinger-type form:

$$\frac{\partial \Psi}{\partial \tau}=-2\hat{H}^{\,FP}\Psi\,, \tag{7}$$

where

$$\Psi \equiv P(\phi,\tau)e^{S[\phi]/2}$$

and

$$\hat{H}^{\,FP}\equiv -\frac{1}{2}\frac{\delta^2}{\delta\phi^2}+\frac{1}{8}\left[\frac{\partial S}{\partial \phi}\right]^2-\frac{1}{4}\frac{\partial^2 S}{\partial \phi^2}\,.$$

Because of this form, we call $\hat{H}^{\,FP}$ the Fokker-Planck Hamiltonian. It is a positive semi-definite operator $\hat{H}^{\,FP}\Psi_n=E_n\Psi_n$, $E_n\geq 0$ whose ground state $E_0=0$ is $\Psi_0=e^{-S[\phi]/2}$. The solution of (7) is

$$\Psi(\phi,\tau)=\sum_{\eta} c_n\Psi_n e^{-2E_n\tau}\,,$$

where c_n are normalizing constants. The probability can be written as

$$P(\phi,\tau)=e^{-S[\phi]/2}\sum_n c_n\Psi_n e^{-2E_n\tau}$$

In the limit $\tau\to\infty$ the only term that does not disappear in this expression is Ψ_0, so we have

$$\lim_{\tau\to\infty} P(\phi,\tau)=c_0e^{-S(\phi)/2}e^{-S(\phi)/2}=c_0e^{-S(\phi)}\,. \tag{8}$$

This is the reason why (5) holds.

II. FUNCTIONAL-INTEGRAL APPROACH TO STOCHASTIC QUANTIZATION

In this section we would like to reformulate the Parisi-Wu method in a path-integral form.

We want to build a generating functional (that we will call $Z^{FP}[J]$ for the Fokker-Planck generating functional) from which the correlations (4) can be derived in the usual fashion:

$$\langle \phi_\eta(x_1\tau_1)\cdots\phi_\eta(x_l\tau_l)\rangle_\eta=\left.\frac{\delta^l Z^{FP}[J]}{\delta J(x_1\tau_0)\cdots\delta J(x_l\tau_l)}\right|_{J=0}.$$

This can be easily done retracing steps (i), (ii), and (iii) of Sec. I. $Z^{FP}[J]$ becomes

$$Z^{FP}[J]=\mathcal{N}\int \tilde{\mathscr{D}}\phi\tilde{\mathscr{D}}\eta P(\phi(0))\delta(\phi-\phi_\eta)\exp\left[-\int_0^\tau J\phi\, d\tau'\right]\exp\left[-\int_0^\tau \frac{\eta^2}{4}d\tau'\right]. \tag{9}$$

ϕ_η that appears in (9) is the solution of the Langevin equation (2), solved with some initial probability $P(\phi(0))$; $\mathcal{N}$ is a normalizing constant and $\tilde{\mathscr{D}}\phi=\lim_{N\to\infty}\prod_{i=0}^{N}\mathscr{D}\phi_{\tau_i}$ where ϕ_{τ_i} are the field configurations at the time τ_i, having sliced the interval 0 to τ in N infinitesimal parts ϵ with $\tau_i=i\epsilon$. This measure is a product of the usual four-dimensional path-integral measures. The $\delta(\phi-\phi_\eta)$ in (9) is a "formal" expression that we can write as

$$\delta(\phi-\phi_\eta)=\delta\left[\dot{\phi}+\frac{\partial S}{\partial \phi}-\eta\right]\left|\left|\frac{\delta\eta}{\delta\phi}\right|\right|, \tag{10}$$

where $||\delta\eta/\delta\phi||$ is the Jacobian of the transformation $\eta\to\phi$, that is,

$$\left|\left|\frac{\delta\eta}{\delta\phi}\right|\right|=\det\left[\left[\partial_\tau+\frac{\partial^2 S}{\partial\phi(\tau)\delta\phi(\tau')}\right]\delta(\tau-\tau')\right]. \tag{11}$$

With well-known manipulations we can write this as

$$\left|\left|\frac{\delta\eta}{\delta\phi}\right|\right|=\exp\left[\mathrm{tr}\ln\left[\partial_\tau+\frac{\partial^2 S}{\partial\phi(\tau)\delta\phi(\tau')}\right]\delta(\tau-\tau')\right]$$
$$=\exp\left[\mathrm{tr}\ln\partial_\tau\left[\delta(\tau-\tau')+\partial_{\tau'}{}^{-1}\frac{\partial^2 S}{\partial\phi(\tau)\delta\phi(\tau')}\right]\right],$$

where $(\partial_{\tau'})^{-1}$ is just to indicate the Green's function $G(\tau-\tau')$ that satisfies

$$\partial_\tau G(\tau-\tau')=\delta(\tau-\tau')\,. \tag{12}$$

The solutions are $G(\tau-\tau')=\theta(\tau-\tau')$ if we choose propagation forward in time, or $G(\tau-\tau')=-\theta(\tau'-\tau)$ for propagation backward in time. It is also possible to choose $G(\tau-\tau')=\frac{1}{2}[\theta(\tau-\tau')-\theta(\tau'-\tau)]$ but we will concentrate on the first two. In the first case, propagation forward in time (that is, the one chosen by the Parisi and Wu), we get

$$\left|\left|\frac{\delta\eta}{\delta\phi}\right|\right|=\exp\left\{\mathrm{tr}\left[\ln\partial_\tau+\ln\left[\delta(\tau-\tau')+\theta(\tau-\tau')\frac{\partial^2 S}{\partial\phi(\tau)\delta\phi(\tau')}\right]\right]\right\}$$
$$=\exp(\mathrm{tr}\ln\partial_\tau)\exp\left[\mathrm{tr}\ln\left[\delta(\tau-\tau')+\theta(\tau-\tau')\frac{\delta^2 S}{\delta\phi(\tau)\delta\phi(\tau')}\right]\right]. \tag{13}$$

The term $\exp(\mathrm{tr}\ln\partial_\tau)$ can be dropped, as it cancels with the same term in the denominator of (9), once we normalize $\hat{Z}[J]=Z[J]/Z[0]$. So in (13) we are left with

$$\left\|\frac{\delta\eta}{\delta\phi}\right\|=\exp\left[\mathrm{tr}\ln\left[\delta(\tau-\tau')+\theta(\tau-\tau')\frac{\delta^2 S}{\delta\phi(\tau)\delta\phi(\tau')}\right]\right]$$

Doing the usual expansion for the ln, we obtain

$$\left\|\frac{\delta\eta}{\delta\phi}\right\|=\exp\left[\mathrm{tr}\left[\theta(\tau-\tau')\frac{\delta^2 S}{\delta\phi(\tau)\delta\phi(\tau')}+\theta(\tau-\tau')\theta(\tau'-\tau)\frac{\partial^2 S}{\delta\phi(\tau)\delta\phi(\tau')}\frac{\partial^2 S}{\delta\phi(\tau')\delta\phi(\tau'')}+\cdots\right]\right]$$

$$=\exp\left[\int d\tau\,\theta(0)\frac{\delta^2 S}{\delta\phi^2(\tau)}+\int d\tau' d\tau\,\theta(\tau-\tau')\theta(\tau'-\tau)\frac{\partial^2 S}{\partial\phi(\tau)\delta\phi(\tau')}\frac{\delta^2 S}{\delta\phi(\tau')\delta\phi(\tau)}+\cdots\right].$$

The second term in this expression is zero because $\theta(\tau-\tau')\theta(\tau'-\tau)=0$ and the same for all the subsequent terms. The only one left is the first term and choosing $\theta(0)=\frac{1}{2}$ (Ref. 6) we get

$$\left\|\frac{\delta\eta}{\delta\phi}\right\|=\exp\left[\frac{1}{2}\int_0^{\tau} d\tau'\frac{\partial^2 S}{\partial\phi^2}\right]. \tag{14}$$

Inserting (14) and (10) back into (9) and performing the η integration, we have

$$Z^{\mathrm{FP}}[J]=\int \mathscr{D}\phi P(\phi(0))\exp\left\{-\int_0^{\tau}\left[\frac{1}{4}\left[\dot\phi+\frac{\partial S}{\partial\phi}\right]^2-\frac{1}{2}\frac{\partial^2 S}{\partial\phi^2}\right]d\tau'-\int_0^{\tau}J\phi\,d\tau'\right\}. \tag{15}$$

If we want also to specify that we are interested only in the correlations at the same 5th time τ_1, we have just to choose $J(x,\tau')$ of the form $J(x,\tau')=\tilde{J}(x)\delta(\tau'-\tau_1)$, $\tau_1<\tau$.

Expression (15) then becomes

$$Z^{\mathrm{FP}}[J]=\int \mathscr{D}\phi P(\phi(0))\exp\left\{-\int_0^{\tau}\left[\frac{1}{4}\left[\dot\phi+\frac{\partial S}{\partial\phi}\right]^2-\frac{1}{2}\frac{\partial^2 S}{\partial\phi^2}\right]d\tau'-\tilde{J}(x\tau_1)\phi(x\tau_1)\right\}.$$

In all this we have to remember, of course, that once we send $\tau_1\to\infty$ we have also to extend the interval of integration from $\int_0^{\tau}$ to $\int_0^{\infty}$. In (15) we are neglecting the normalizing constant $\mathscr{N}$ and all the usual four-space integration. It is easy, anyhow, to reinstate them when necessary. The Lagrangian in the exponent of (15), which we call the FP Lagrangian, does not seem to have any relation to the Hamiltonian in (7). It is easy, anyhow, to see the connection: let us first, in the action of (15), perform the integration of the term $\int_0^{\tau}\frac{1}{2}\dot\phi(\partial S/\partial\phi)d\tau'=\frac{1}{2}[S(\phi(\tau))-S(\phi(0))]$ (Ref. 7) and second, let us rescale the time $\tau'\to\tau'/2$, so that we get

$$Z^{\mathrm{FP}}[J]=\int \mathscr{D}\phi(0)P(\phi(0))e^{+S(\phi(0))/2}\mathscr{D}(\phi(2\tau))e^{-S(\phi(2\tau))/2}\mathscr{D}''\phi\exp\left[-\int_0^{2\tau}\mathscr{L}^{\mathrm{FP}}d\tau'-\int_0^{2\tau}J\phi\,d\tau'\right], \tag{16}$$

where

$$\mathscr{D}''\phi=\lim_{N\to\infty}\prod_{i=1}^{N-1}\mathscr{D}\phi_{\tau_i}$$

and

$$\int_0^{2\tau}\mathscr{L}^{\mathrm{FP}}d\tau'=\int_0^{2\tau}\left[\dot\phi^2/2+\frac{1}{8}\left[\frac{\partial S}{\partial\phi}\right]^2-\frac{1}{4}\frac{\partial^2 S}{\partial\phi^2}\right]d\tau'.$$

Now it is clear why we called $\mathscr{L}^{\mathrm{FP}}$ the Fokker-Planck Lagrangian. It is a sort of "Euclidean" Lagrangian for the Hamiltonian $\hat{H}^{\mathrm{FP}}$ in (7), and for this reason we also call Z^{FP} the generating functional of the Fokker-Planck dynamics.

If in (13) we had made the choice $G(\tau-\tau')=-\theta(\tau'-\tau)$, then the action in (16) would have been

$$\int_0^{2\tau}\mathscr{L}^{\mathrm{FP}}d\tau'=\int_0^{2\tau}\left[\frac{1}{2}\dot\phi^2+\frac{1}{8}\left[\frac{\partial S}{\partial\phi}\right]^2+\frac{1}{4}\left[\frac{\partial^2 S}{\partial\phi^2}\right]\right]d\tau'.$$

The only difference from (16) is in the sign of the third term. The corresponding Hamiltonian has been known in the literature for a long time as the Kolmogoroff-Fokker-Planck backward Hamiltonian:

$$\hat{H}^{\mathrm{FP}}_{\mathrm{backward}}=-\frac{1}{2}\frac{\partial^2}{\partial\phi^2}+\frac{1}{8}\left[\frac{\partial S}{\partial\phi}\right]^2+\frac{1}{4}\frac{\partial^2 S}{\partial\phi^2}. \tag{17a}$$

Going back to the derivation of (15) we want to stress what has been done: We have integrated away the η and replaced the role it plays in the Langevin equation with a sort of "effective" action $\mathscr{L}^{\mathrm{FP}}$. This Lagrangian might look very complicated but it contains only the field ϕ as a dynamical variable (Fokker-Planck dynamics). In the Langevin equation the dynamical variables were both ϕ and η on the same ground and they were interlocked in a dynamics (Langevin dynamics) that only apparently looked simpler. With the generating functional (15) we can, of course, develop perturbation theory using the Feynman rules dictated by $\mathscr{L}^{FP}$. This perturbation theory is the parallel of the one[1] that has been developed starting from the Langevin equation. Differently from that we do not, anyhow, have to integrate over the η at the level of Feynman graphs, as this is already done at the level of Z^{FP}. The number of graphs is very large in both approaches: In our case the high number comes from the extra vertices contained in $\frac{1}{8}(\partial S/\partial\phi)^2-\frac{1}{4}\partial^2 S/\partial\phi^2$.

Before concluding this section we want to make a remark concerning the Jacobian (11). All the steps from (9) to (16), that we have done to derive the generating function Z^{FP}, are possible only if the Jacobian is not identically zero. If this happens the Z^{FP} itself is zero. The same Langevin equation, starting point of the stochastic quantization, loses all its meaning. In fact, $\|\delta\eta/\delta\phi\|=0$ means that there is no field associated, through (2), to a particular η. In technical language this can be expressed by saying that the winding number of the transformation $\eta\rightarrow\phi$ is zero. This number called Δ has been studied in great detail in Ref. 9 in connection with supersymmetry and it is known as the Witten index. Our conclusion is that, in case $\Delta=0$, the stochastic quantization does not hold any more. In this case, anyhow, the same traditional method of quantization [given by (1)] does not hold. In fact, it has been shown in Ref. 9 that $\Delta=0$ implies non-normalizability for $e^{-S/2}$; that means the "quantum" probability e^{-S} cannot be used in (1) any more.

III. VACUUM-VACUUM GENERATING FUNCTIONAL

Of the random process (3), we have used, up to now, only the property that it is Gaussian. The action (15) that we have obtained is a consequence of this. Besides this property there is another very interesting one: the stochastic process (3) is stationary. By stationary[10] we mean a process whose "momenta" $c(\tau_1\tau_2\cdots\tau_l)$,

$$c(\tau_1\cdots\tau_l)\equiv\langle\eta(\tau_1)\cdots\eta(\tau_l)\rangle_\eta ,$$

are functions only of the differences $(\tau_i-\tau_j)$. The process (3) has exactly this feature. A question that arises naturally is if also the correlation functions

$$\langle\phi_\eta(x_1\tau_1)\cdots\phi_\eta(x_l\tau_l)\rangle_\eta \tag{17b}$$

manifest this property. The answer is generally no. In fact the averages that we perform are not only in η, but also on the initial configuration $\phi(0)$ for which we give the $P(\phi(0))$. It is the form of this $P(\phi(0))$ that determines if (17b) is a function only of $(\tau_i-\tau_j)$. The choice of Parisi and Wu was $P(\phi(0))=\delta(\phi(0)-\phi_1)$ (with ϕ_1 a definite configuration) and the perturbative calculation done by them (for details see Ref. 1) showed that their choice of $P(\phi(0))$ does not make (17b) stationary. To find out which is the right one, let us start supposing (17b) is stationary:

$$\langle\phi_\eta(x_1\tau_1)\cdots\phi_\eta(x_l\tau_l)\rangle_{\eta,P(\phi(0))} =f(\tau_1-\tau_2,\tau_2-\tau_3,\ldots,\tau_l-\tau_{l-1}) , \tag{18}$$

where we use the notation $\langle\ \ \rangle_{\eta,P(\phi(0))}$ to remind us of the average over both η and $\phi(0)$. From (18) we see that, if we rescale all the τ_i of a quantity T, nothing changes on the right-hand side, so

$$\langle\phi_\eta(x_1\tau_1)\cdots\phi_\eta(x_l\tau_l)\rangle_{\eta,P(\phi(0))} =\langle\phi_\eta(x_1,\tau_1+T)\cdots\phi_\eta(x_l,\tau_l+T)\rangle_{\eta,P(\phi(0))} .$$

Let us first put on both sides $\tau_1=\tau_2\cdots=\tau_l$,

$$\langle\phi_\eta(x_1\tau_1)\phi_\eta(x_2\tau_1)\cdots\phi_\eta(x_l,\tau_1)\rangle_{\eta,P}=\langle\phi_\eta(x_1,\tau_1+T)\phi_\eta(x_2,\tau_1+T)\cdots\phi_\eta(x_l,\tau_1+T)\rangle_{\eta,P}$$

and second let us take the limit of $T\rightarrow\infty$,

$$\lim_{T\rightarrow\infty}\langle\phi_\eta(x_1\tau_1)\phi_\eta(x_2\tau_1)\cdots\phi_\eta(x_l\tau_1)\rangle_{\eta,P}=\lim_{T\rightarrow\infty}\langle\phi_\eta(x_1,\tau_1+T)\phi_\eta(x_2,\tau_1+T)\cdots\phi_\eta(x_l,\tau_1+T)\rangle_{\eta,P} .$$

The left-hand side does not depend on T, while the right-hand side of (5) is the "quantum" correlation function [see (5)], so we have

$$\langle\phi_\eta(x_1\tau_1)\phi_\eta(x_2\tau_1)\cdots\phi_\eta(x_l\tau_1)\rangle_{\eta,P} =\frac{\int\mathscr{D}\phi[\phi(x_1)\cdots\phi(x_l)]e^{-S(\phi)}}{\int\mathscr{D}\phi e^{-S(\phi)}} . \tag{19}$$

As the left-hand side is stationary, we can rescale all the fields backward of τ_1; that means

$$\langle\phi_\eta(x_10)\phi_\eta(x_20)\cdots\phi_\eta(x_l0)\rangle_{\eta,P} =\langle\phi_\eta(x_1\tau_1)\phi_\eta(x_2\tau_1)\cdots\phi_\eta(x_l\tau_1)\rangle_{\eta,P}$$

and using (19) we conclude

$$\langle\phi_\eta(x_10)\phi_\eta(x_20)\cdots\phi_\eta(x_l0)\rangle_{\eta,P} =\frac{\int\mathscr{D}\phi[\phi(x_1)\cdots\phi(x_l)]e^{-S(\phi)}}{\int\mathscr{D}\phi e^{-S(\phi)}} . \tag{20}$$

From this expression we can explicitly derive the form of $P(\phi(0))$: the left-hand side of (20) is at $\tau=0$, so we do not have any random effect caused by η (η has not been switched on yet), the only average is with respect to $P(\phi(0))$; that means the left-hand side of (20) is

$$\int\mathscr{D}\phi(0)P(\phi(0))[\phi(x_1,0)\cdots\phi(x_l,0)] .$$

Comparing with its right-hand side we get

$$P(\phi(0))=\frac{e^{-S(\phi(0))}}{\int\mathscr{D}\phi(0)e^{-S(\phi(0))}} . \tag{21}$$

From (19) we can derive a second conclusion: for that particular form of $P(\phi(0))$, for which the correlation functions are stationary, we do not need to take the limit $\tau_1\rightarrow\infty$. At every finite τ_1, we have that the stochastic average is already the quantum one. The physical meaning of this is very clear: From the beginning we put the system in the equilibrium distribution $P(\phi)=e^{-S(\phi)}$ and the presence of the Langevin dynamics $\dot\phi=-\partial S/\partial\phi+\eta$ does not modify this. On the contrary, in the case of the choice[1] $P(\phi(0))=\delta(\phi(0)-\phi_1)$ we started with every field in configuration ϕ_1 and then the Langevin dynamics was able to spread them to the equilibrium form at time $\tau_1\to\infty$.

Inserting (21) back into (16), the new generating functional looks like

$$Z^{FP}_{\substack{\text{vacuum}\\\text{vacuum}}}[J]=\int \mathcal{D}\phi(0)e^{-S(\phi(0))/2}\mathcal{D}\phi(\tau)e^{-S(\phi(\tau))/2}\mathcal{D}''\phi\exp\left[-\int_0^\tau \mathcal{L}^{FP}d\tau'\right]. \tag{22}$$

We like to call this the "*vacuum-vacuum generating functional.*" The reason is clear if we remember that the ground-state (vacuum) of $\hat H^{FP}$ is $\Psi_0=e^{-S(\phi)/2}$.

Another way (less transparent) to get stationary correlation functions is to start from (15) and take the limit of integration from $\int_0^\tau$ to $\int_{-\infty}^\tau$. What happens is that the Fokker-Planck dynamics builds up a probability between $-\infty$ and 0 equal to $e^{-S(\phi)}$; that means equal to the one we inserted at $\tau=0$ by (21).

Before concluding, a word of caution is needed: Stochastic quantization does not compel us to choose (21). Any normalizable form of $P(\phi(0))$ is acceptable: the result, the limit of $\tau_1\to\infty$, is independent of $P(\phi(0))$.

The particular choice (21) has the advantage that it avoids step (iii) of the Parisi-Wu prescription.

Somehow this $Z^{FP}_{\substack{\text{vacuum}\\\text{vacuum}}}$ is another method for representing the traditional quantum generating functional $Z=\int \mathcal{D}\phi e^{-S(\phi)}$, and in Ref. 11 its connection to the new functional method recently proposed by De Alfaro, Fubini, and Furlan[12] has been shown, using nonperturbative techniques.

IV. HIDDEN SUPERSYMMETRY

A. General notation

In this section we want to study another form for Z^{FP} The expression for the Jacobian that we derived in (14) is not the only manner in which to write it. Another way to do so is by using anticommuting variables $\psi,\bar\psi$:

$$\left\|\frac{\delta\eta}{\delta\phi}\right\| =\int \mathcal{D}\bar\psi\mathcal{D}\psi\exp\left\{-\int_0^\tau\left[\bar\psi\left[\frac{\partial}{\partial\tau'}+\frac{\partial^2 S}{\partial\phi^2}\right]\psi\right]d\tau'\right\}. \tag{23}$$

With this form for the Jacobian and rescaling the time $\tau'\to\tau'/2$ the generating functional Z^{FP} becomes

$$Z^{FP}_{SS}=\int \mathcal{D}\phi\mathcal{D}\bar\psi P(\phi(0))\exp\left\{-\int_0^{2\tau}\left[\frac12\left[\dot\phi+\frac12\frac{\partial S}{\partial\phi}\right]^2+\bar\psi\left[\partial_{\tau'}+\frac12\frac{\partial^2 S}{\partial\phi^2}\right]\psi\right]d\tau'-\int_0^{2\tau}J\phi\, d\tau'\right\}. \tag{24}$$

The reason for the notation SS is that, with a proper choice of $P(\phi(0))$ and of boundary conditions for $\psi,\bar\psi$, this system reveals a hidden supersymmetry recently discussed in Ref. 13. Let us choose $P(\phi(0))=\delta(\phi(0)-\phi(2\tau))$. Then Z^{FP}_{SS} can be written as

$$Z^{FP}_{SS}=\int {}'\mathcal{D}\phi'\mathcal{D}\bar\psi'\mathcal{D}\psi\exp\left\{-\int_0^{2\tau}\left[\frac12\dot\phi^2+\frac18\left[\frac{\partial S}{\partial\phi}\right]^2+\bar\psi\left[\partial_{\tau'}+\frac12\frac{\partial^2 S}{\partial\phi^2}\right]\psi\right]d\tau'\right\} \tag{25}$$

with

$${}'\mathcal{D}\phi=\lim_{N\to\infty}\prod_{i=1}^N\mathcal{D}\phi_{\tau_i}\,.$$

The Lagrangian that appears in Z^{FP}_{SS} is

$$\mathcal{L}^{FP}_{SS}=\tfrac12\dot\phi^2+\frac18\left[\frac{\partial S}{\partial\phi}\right]^2+\bar\psi\left[\partial_\tau+\frac12\frac{\partial^2 S}{\partial\phi^2}\right]\psi\,. \tag{26}$$

The corresponding "Euclidean" Hamiltonian

$$H^{FP}_{SS}=-\frac12\frac{\delta^2}{\delta\phi^2}+\frac18\left[\frac{\partial S}{\partial\phi}\right]^2+\frac14[\bar\psi,\psi]\frac{\partial^2 S}{\partial\phi^2} \tag{27}$$

is well known in the literature and it has been studied in great detail in Ref. 14. It manifests a sort of *nonrelativistic supersymmetry* whose conserved charges are

$$Q_\psi=\left[\pi_\phi+\frac12\frac{\partial S}{\partial\phi}\right]\psi\,,\qquad \pi_\phi=\frac{\delta}{\delta\phi}\,, \tag{28}$$
$$\bar Q_{\bar\psi}=\bar\psi\left[\pi_\phi+\frac12\frac{\partial S}{\partial\phi}\right].$$

The symmetry transformations generated by them are

$$\delta\phi=+\epsilon_\psi\psi+\epsilon_{\bar\psi}\bar\psi\,,$$
$$\delta\psi=\epsilon_{\bar\psi}\left[\pi_\phi+\frac12\frac{\partial S}{\partial\phi}\right], \tag{29}$$
$$\delta\bar\psi=\epsilon_\psi\left[\pi_\phi-\frac12\frac{\partial S}{\partial\phi}\right],$$

where ϵ_ψ and $\epsilon_{\bar\psi}$ are infinitesimal anticommuting parameters. $\hat H^{FP}_{SS}$ itself can be written, in perfect supersymmetric fashion, as

$$\hat H^{FP}_{SS}=\tfrac12\{\bar Q_{\bar\psi},Q_\psi\}\,.$$

We can even bring the notation a step further with the use of superfields.[15] Let us *first* rewrite (24) with an auxiliary field ω (that in statistical mechanics is known as a response field)

$$Z_{SS}^{FP}[J]=\int {}'\mathscr{D}\phi\,{}'\mathscr{D}\omega\,{}'\mathscr{D}\bar{\psi}\,{}'\mathscr{D}\psi \exp\left\{+\int_0^{2\tau}\left[\omega^2+\omega\left[\dot{\phi}+\frac{1}{2}\frac{\partial S}{\partial\phi}\right]-\bar{\psi}\left[\frac{\partial}{\partial\tau}+\frac{1}{2}\frac{\partial^2 S}{\partial\phi^2}\right]\psi\right]d\tau'\right\}. \tag{30}$$

Second let us introduce the superfield Φ,

$$\Phi=\phi(x)+\bar{\theta}\psi+\theta\bar{\psi}+\theta\bar{\theta}\omega ,$$

where $\theta,\bar{\theta}$ are elements of a Grassmann algebra. Then the action in (30) can be written, in a very compact form, as

$$A[\Phi]\equiv\int [\tfrac{1}{2}(D_\theta\Phi)(D_{\bar{\theta}}\Phi)-S(\Phi)]d\tau\, d\theta\, d\bar{\theta} . \tag{31}$$

S is the usual action of the system (1) from which we started, but whose argument in (31) is the superfield Φ and $D_\theta=\partial_\theta-\bar{\theta}\partial_\tau$ is the so-called covariant derivative.[14] As we can see S plays the role of a sort of "potential" and in supersymmetric jargon its proper name is superpotential. The "space" over which the Lagrangian in (31) is integrated is $\tau,\theta,\bar{\theta}$, and it is the superspace[14] of our system. The symmetry (29), under which $\mathscr{L}_{SS}^{FP}$ is invariant, can be seen as a transformation on the fields $\phi,\psi,\bar{\psi}$ induced by the following "supertranslation" in superspace:

$$\delta\theta=\epsilon_\psi ,$$

$$\delta\bar{\theta}=\epsilon_{\bar{\psi}} ,$$

$$\delta\tau=-(\bar{\theta}\epsilon_\psi-\epsilon_{\bar{\psi}}\theta) .$$

B. Forward and backward Fokker-Planck dynamics

The manner in which we wrote $||\delta\eta/\delta\phi||$ in (23) deserves a deeper analysis. Being $||\delta\eta/\delta\phi||$ nothing else than a determinant (11), we can think of evaluating it as the product of its eigenvalues, that means

$$\left\|\frac{\delta\eta}{\delta\phi}\right\| =\int \mathscr{D}\bar{\psi}\mathscr{D}\psi \exp\left[-\int_0^{2\tau}\bar{\psi}\left[\frac{\partial}{\partial\tau'}+\frac{1}{2}\frac{\partial^2 S}{\partial\phi^2}\right]\psi d\tau'\right] =\prod_{n=-\infty}^{+\infty}\alpha_n , \tag{32}$$

where α_n are the eigenvalues of the equation

$$\left[\partial_\tau+\frac{1}{2}\frac{\partial^2 S}{\partial\phi^2}\right]\Psi_n=\alpha_n\Psi_n . \tag{33}$$

The solutions of (33) are

$$\Psi_n=K\exp\left[\int_0^\tau d\tau'\left[\alpha_n-\frac{1}{2}\frac{\partial^2 S}{\partial\phi^2}\right]\right]$$

where K is a normalizing constant) (from now on we will call the extreme of integration τ and not 2τ, for convenience).

If we impose, following Ref. 16, antiperiodic boundary conditions,

$$\Psi_n(\tau)=-\Psi_n(0) ,$$

we have

$$\alpha_n=\frac{i(2n+1)\pi}{\tau}+\frac{1}{\tau}\int_0^\tau d\tau'\frac{1}{2}\frac{\partial^2 S}{\partial\phi^2} .$$

Unfortunately we cannot make this choice in our case. Our system is supersymmetric and the choice of periodic boundary conditions on the bosonic variable induces the same boundary conditions on the $\psi,\bar{\psi}$. If we had chosen antiperiodic ones, we would have broken supersymmetry explicitly (for more details see Ref. 9).

Taking $\Psi_n(\tau)=\Psi_n(0)$ we have

$$\alpha_n=\frac{2ni\pi}{\tau}+\frac{1}{\tau}\int_0^\tau d\tau'\frac{1}{2}\frac{\partial^2 S}{\partial\phi^2} \tag{34}$$

and, substituting this in (32), we get

$$\begin{aligned}\prod_{n=-\infty}^{+\infty}\alpha_n&=\prod_{n=-\infty}^{+\infty}\left[\frac{i2n\pi}{\tau}+\frac{1}{\tau}\int_0^\tau\frac{1}{2}\frac{\partial^2 S}{\partial\phi^2}d\tau'\right]\\&=\left[\prod_{n=-\infty}^{+\infty}\left[\frac{i2n\pi}{\tau}\right]\right]\left[\frac{1}{\tau}\int_0^\tau\frac{1}{2}\frac{\partial^2 S}{\partial\phi^2}\right]\prod_{n=-\infty}^{+\infty}{}'\left[1+\frac{\int_0^\tau\frac{1}{2}2(\partial^2 S/\partial\phi^2)d\tau'}{2\pi ni}\right]\\&=c'\left[\frac{1}{\tau}\int_0^\tau\frac{1}{2}\frac{\partial^2 S}{\partial\phi^2}d\tau'\right]\prod_{n=-\infty}^{\infty}{}'\left[1+\frac{\int_0^\tau\frac{1}{2}(\partial^2 S/\partial\phi^2)d\tau'}{2\pi ni}\right]\end{aligned}$$

(Π' is to indicate that we exclude the term $n=0$ from the product)

$$\begin{aligned}&=c'\sinh\tfrac{1}{4}\int_0^\tau d\tau'\frac{\partial^2 S}{\partial\phi^2}\\&=\frac{c'}{2}\left[\exp\left[\tfrac{1}{4}\int_0^\tau\frac{\partial^2 S}{\partial\phi^2}d\tau'\right]-\exp\left[-\tfrac{1}{4}\int_0^\tau\frac{\partial^2 S}{\partial\phi^2}d\tau'\right]\right]\end{aligned} \tag{35}$$

(c' is a constant). Inserting this expression into (25), we get

$$Z_{SS}^{FP}=\frac{c'}{2}\left[\int{}'\mathcal{D}\phi\exp\left\{-\int_0^\tau\left[\dot\phi^2/2+\frac{1}{8}\left[\frac{\partial S}{\partial\phi}\right]^2-\frac{1}{4}\frac{\partial^2 S}{\partial\phi^2}\right]d\tau'\right\}\right.$$
$$\left.-\int{}'\mathcal{D}\phi\exp\left\{-\int_0^\tau\left[\dot\phi^2/2+\frac{1}{8}\left[\frac{\partial S}{\partial\phi}\right]^2+\frac{1}{4}\frac{\partial^2 S}{\partial\phi^2}\right]d\tau'\right\}\right] \tag{36}$$

We see that the first term on the right-hand side is the usual Z^{FP} studied in Sec. II, while the second term is the generating functional corresponding to the backward Kolmogoroff-Fokker-Planck Lagrangian presented in (17).

We can indicate this in a compact way as

$$Z_{SS}^{FP}[J]=\frac{c'}{2}(Z_{\text{forward}}^{FP}[J]-Z_{\text{backward}}^{FP}[J]) , \tag{37}$$

where the notation is self-explanatory.

If we had chosen antiperiodic boundary conditions for ψ and $\bar\psi$, we would have gotten

$$Z_{SS}^{FP}[J]=\frac{c}{2}(Z_{\text{forward}}^{FP}[J]+Z_{\text{backward}}^{FP}[J]) .$$

In this case the right-hand side, besides the notation, would not have been supersymmetric.

This nice interplay of forward and backward FP dynamics is also evident at the level at $\hat H_{SS}^{FP}$ (27). If we in fact represent ψ and $\bar\psi$ as 2×2 matrices (see Ref. 14),

$$\psi=\begin{bmatrix}0&1\\0&0\end{bmatrix},\quad \bar\psi=\begin{bmatrix}0&0\\1&0\end{bmatrix},$$

$$\psi^2=\bar\psi^2=0,\quad \{\psi,\bar\psi\}=1 ,$$

$\hat H_{SS}^{FP}$ can be written as

$$\hat H_{SS}^{FP}=-\frac{1}{2}\frac{\partial^2}{\partial\phi^2}+\frac{1}{8}\left[\frac{\partial S}{\partial\phi}\right]^2-\frac{1}{4}\sigma_3\left[\frac{\partial^2 S}{\partial\phi^2}\right]$$
$$=\begin{bmatrix}\hat H_{\text{forward}}^{FP}&\\&\hat H_{\text{backward}}^{FP}\end{bmatrix}, \tag{38}$$

where $\hat H_{\text{forward}}^{FP}$ is the expression in (7) while $\hat H_{\text{backward}}^{HP}$ is the one in (17a). Both of these Hamiltonians are positive semi-definite: in fact, $\hat H_{\text{forward}}^{HP}=\frac{1}{2}QQ^\dagger$ and $\hat H_{\text{backward}}^{FP}=\frac{1}{2}Q^\dagger Q$ with $Q=\partial/\partial\phi-1/2\,\partial S/\partial\phi$. As we said in Sec. I, the ground state of $\hat H_{\text{forward}}^{FP}$ is at $E_0=0$ and is $\Psi_0^{\text{forward}}=e^{-S/2}$. Also $\hat H_{\text{backward}}^{FP}$ has a state at $E_0=0$ and is $\Psi_0^{\text{backward}}=e^{+S/2}$, but we have to be cautious about this state. If we assume, in fact, that the $\Psi_0^{\text{forward}}=e^{-S/2}$ is normalizable [and we have to assume that for the traditional quantization (1) to hold], then $\Psi_0^{\text{backward}}=e^{+S/2}$ is not normalizable and cannot be accepted as part of the spectrum of $\hat H_{\text{backward}}^{FP}$. This means that there is no physical state for $\hat H_{\text{backward}}^{FP}$ at $E_0=0$: All its states are at $E_n>0$.

These conditions can also be expressed using the language of supersymmetry: there is only one ground state for $\hat H_{SS}^{FP}$, that is

$$\begin{bmatrix}e^{-S/2}\\0\end{bmatrix};$$

the other one,

$$\begin{bmatrix}0\\e^{+S/2}\end{bmatrix},$$

cannot be accepted. This is equivalent to saying that supersymmetry is unbroken (see Ref. 14).

We can thus conclude that the request that the traditional quantization holds implies for the hidden supersymmetry of Z_{SS}^{FP} be unbroken.

The presence of both dynamics in Z_{SS}^{FP} is very amusing but it may bother the careful reader who knows that the prescription of Parisi and Wu[1] is to choose the forward one. It should be remembered, anyhow, that in the stochastic quantization we have to take the limit of integration τ in (36) to infinity. In this limit only the Z_{forward}^{FP} is left in (37). In fact,

$$\lim_{\tau\to\infty}Z_{SS}^{FP}[J]=\lim_{\tau\to\infty}(Z_{\text{forward}}^{FP}-Z_{\text{backward}}^{FP})$$
$$=\lim_{\tau\to\infty}(\operatorname{tr}e^{-\hat H_{\text{forward}}^{FP}\tau}-\operatorname{tr}e^{-\hat H_{\text{backward}}^{FP}\tau})$$
$$=\lim_{\tau\to\infty}\left[\sum_n e^{-E_{\text{forward}}^n\tau}-\sum_n e^{-E_{\text{backward}}^n\tau}\right]$$
$$=\lim_{\tau\to\infty}Z_{\text{forward}}^{FP}[J] . \tag{39}$$

(All the E_{backward}^n are positive so the last term goes to zero.) (We have neglected the constant c' because we can get rid of it by properly normalizing Z_{SS}^{FP}.)

The presence of this hidden supersymmetry can be further exploited deriving, for example, the corresponding Ward identities. This has been done in Ref. 13, but it does not throw any new light on the problem. These Ward identities only express the fact that correlations involving the ϕ and $\psi,\bar\psi$ fields can be reexpressed as correlations involving only ϕ fields. This is clear already in (37) where we succeed in integrating the $\psi,\bar\psi$ away, leaving only ϕ fields.

We want to derive some different identities here that also stem from the hidden supersymmetry.

Let us define the following generating functional:

$$Z_\alpha^{[SS]}=\int e^{-(1+\alpha)S^{[SS]}}\,{}'\mathcal{D}\phi\,{}'\mathcal{D}\bar\psi\,{}'\mathcal{D}\psi ,$$

where $S^{[SS]}$ is a supersymmetric action and α is a parameter. It has been shown in Ref. 17 that, if supersymmetry is unbroken, then

$$\frac{\partial Z_\alpha^{[SS]}}{\partial\alpha}=0 ,$$

that means $Z_\alpha^{[SS]}$ is independent of α. Let us write this down for our Z_{SS}^{FP} in (24):

$$Z^{FP}_{(\alpha)SS}=\int {}'\mathcal{D}\phi\,{}'\mathcal{D}\bar{\psi}\,{}'\mathcal{D}\psi \exp\left[-(1+\alpha)S_B^{FP}-(1+\alpha)\int_0^\tau \bar{\psi}\left[\partial_{\tau'}+\frac{1}{2}\frac{\partial^2 S}{\partial\phi^2}\right]\psi\, d\tau'\right], \tag{40}$$

where S_B^{FP} is the bosonic part of the FP action (26),

$$S_B^{FP}=\int_0^\tau\left[\dot{\phi}^2/2+\frac{1}{8}\frac{\partial S}{\partial\phi}\right]d\tau' .$$

If we make the following rescaling in the fermionic part of the action $\psi\to\sqrt{1+\alpha}\,\psi\equiv\psi'$, $\bar{\psi}\to\sqrt{1+\alpha}\,\bar{\psi}\equiv\bar{\psi}'$ we get

$$Z^{FP}_{(\alpha)SS}=\int \exp\left[-(1+\alpha)S_B^{FP}-\int_0^\tau \bar{\psi}'\left[\partial_{\tau'}+\frac{1}{2}\frac{\partial^2 S}{\partial\phi^2}\right]\psi'\right]\frac{{}'\mathcal{D}\phi\,{}'\mathcal{D}\bar{\psi}'\,{}'\mathcal{D}\psi'}{1+\alpha} .$$

Performing the integration in $\psi',\bar{\psi}'$, we obtain

$$Z^{FP}_{(\alpha)SS}[J]=\frac{\int {}'\mathcal{D}\phi\, e^{-(1+\alpha)S_B^{FP}+S_2}-\int {}'\mathcal{D}\phi\, e^{-(1+\alpha S_B^{FP}-S_2}}{2(1+\alpha)}$$

with

$$S_2=\frac{1}{4}\int_0^\tau \frac{\partial^2 S}{\partial\phi^2}d\tau' .$$

This Z_α^{FP} is independent of α, so we can set its derivative equal to zero. What we obtain is the following relation:

$$\int (1+S_B^{FP})\exp\left[-\int_0^\tau \mathcal{L}^{FP}_{\text{forward}}d\tau'\right]{}'\mathcal{D}\phi$$

$$=\int {}'\mathcal{D}\phi(1+S_B^{FP})\exp\left[-\int_0^\tau \mathcal{L}^{FP}_{\text{backward}}d\tau'\right].$$

This identity has never been derived before for stochastic processes and it expresses a sort of "time symmetry" between forward and backward Fokker-Planck dynamics. This time symmetry is a reflection of the supersymmetry of the sytem.

CONCLUSION

In this paper we have put the basis of a functional-integral approach to stochastic quantization. We have done this, not to merely develop once again perturbation theory, but with the goal of having a new tool to study the rich nonperturbative content of field theory. The traditional generating functional has proved, in the last 20 years, to be a very powerful instrument. We hope that our $Z^{FP}[J]$ can at least complement this.

Note added. After this work was completed, I was informed of past and recent works on the functional approach to stochastic problems: F. Langouche *et al.*, Physica 95A, 252 (1979); Y. Nakano, University of Alberta report, 1982 (unpublished); M. Namiki *et al.*, Prog. Theor. Phys. (to be published); C. M. Bender *et al.*, Nucl. Phys. B219, 61 (1983).

ACKNOWLEDGMENTS

A preliminary report of this work was mentioned by Professor B. Sakita at the "Symposium on High Energy Physics," Tokyo, Sept. 1982. I thank him for many discussions and my friend V. Sarzi for convincing me to write up these results. This work was supported in part by the National Science Foundation Grant No. 82-15364 and CUNY-PSC-BHE Faculty Research award.

[1]G. Parisi and Wu-yong-shi, Sci. Sin. 24, 484 (1981).
[2]D. Zwanziger, Nucl. Phys. B192, 259 (1981); B193, 163 (1981); G. Marchesini, *ibid.* B191, 214 (1981).
[3]G. Parisi, Nucl. Phys. B180, 378 (1981); B205, 337 (1982).
[4]J. Alfaro and B. Sakita, Phys. Lett. 121B, 339 (1983); J. Greensite and M. B. Halpern, Nucl. Phys. B211, 343 (1983); A. Guha and S. C. Lee, Phys. Rev. D 27, 2412 (1983).
[5]M. C. Wang and G. E. Uhlenbeck, Rev. Mod. Phys. 17, 323 (1945).
[6]This choice corresponds to a midpoint prescription for the path integral in (9).
[7]This integration can be done in the usual way, even inside the path integral if we choose the midpoint prescription (see Ref. 8 and 6).
[8]L. S. Schulman, *Techniques and Applications of Path Integration* (Wiley, New York, 1980), p. 29.
[9]S. Ceccotti and L. Girardello, Phys. Lett. 110B, 39 (1982).
[10]Yu. A. Rozanov, *Stationary Random Processes* (Holden-Day, San Francisco, 1967).
[11]E.Gozzi, Phys. Lett. B (to be published).
[12]V. De Alfaro, S. Fubini, and G. Furlan, Phys. Lett. 105B,
[13]G. Parisi and N. Sourlas, Nucl. Phys. B206, 321 (1982); S.

Ceccotti and L. Girardello, Harvard Report No. 82/4 HTUP, 1982 (unpublished).
[14]E. Witten, Nucl. Phys. B188, 513 (1981); P. Salomonson and J. W. Van Holten, *ibid.* B196, 509 (1982).
[15]A. Salam and J. Strathdee, Fortschr. Phys. 25, 58 (1977).
[16]R. F. Dashen, B. Hasslacher, and A. Neveu, Phys. Rev. D 12, 2443 (1975) (Appendix).
[17]H. Nicolai, Nucl. Phys. B176, 419 (1980). 462 (1981).

Volume 129B, number 5 PHYSICS LETTERS 29 September 1983

A SUPERFIELD FORMULATION OF STOCHASTIC QUANTIZATION WITH FICTITIOUS TIME

E. EGORIAN
Erevan Institute of Physics, Theoretical Physics Department, Markarian Street 2, POB 375036, Erevan, USSR

and

S. KALITZIN
Institute of Nuclear Research and Nuclear Energy, Boulevard Lenin 72, Sofia 1184, Bulgaria

Received 11 February 1983

We prove the equivalence between the D-dimensional scalar quantum field theory and the corresponding $D+1$ (with fictitious time) stochastically quantized field theory. The superdiagram technique is used within perturbation theory. The independence of the Green's functions on the fictitious time is shown.

1. Introduction. The method of stochastic quantization was proposed in ref. [1] for the D-dimensional scalar field theory. The method is based on the Langevin equation with an auxiliary (fictitious) time t:

$$\partial\varphi(x,t)/\partial t + \delta S[\varphi]/\delta\varphi(x,t) = \eta(x,t)\,, \qquad (1.1)$$

where $\eta(x,t)$ is a random source obeying the gaussian correlation law:

$$\langle\eta(x,t)\eta(x',t')\rangle = 2\delta(x-x')\delta(t-t')\,. \qquad (1.2)$$

It turns out that the averaged functionals of the scalar field theory with action $S[\varphi]$ obey the following equality

$$\langle\varphi(x_1)\ldots\varphi(x_n)\rangle = \lim_{t\to\infty}\langle\varphi_\eta(x_1,t)\ldots\varphi_\eta(x_n,t)\rangle_\eta\,. \qquad (1.3)$$

On the rhs of (1.3) the averaging is with respect to the gaussian noise $\eta(x,t)$ and φ_η are solutions of (1.1). This theorem can be proved by means of the Fokker–Planck equation [1–3] or using the perturbative solution of (1.1) [4].

In the present paper we apply a different approach toward understanding (1.3) for nonstationary solutions of (1.1). We found that in the scalar case the stochastically quantized theory in D dimensions can be presented as a $(D+1)$-dimensional superfield theory and the latter is reducible to the usual scalar quantum field theory. It turns out also that the $t\to\infty$ limit in (1.3) is irrelevant due to the t independence of (2.2) (see below).

2. A superfield formulation of the Langevin formalism. The generating functional is introduced as usual:

$$Z(h) = \int \mathcal{D}\eta \exp\left(\int d^Dx\, dt(-\tfrac{1}{4}\eta^2 + h\varphi_\eta)\right)\,. \qquad (2.1)$$

Then clearly

$$\langle\varphi_\eta(x_1,t)\ldots\varphi_\eta(x_n,t)\rangle_\eta = [\delta/\delta h(x_1,t)]\ldots[\delta/\delta h(x_n,t)]Z(h)|_{h=0}\,. \qquad (2.2)$$

Let us substitute $\eta\to\varphi_\eta$ in (2.1) by the standard decomposition of the unit:

$$1 = \int\delta(\eta - \dot\varphi - \delta S/\delta\varphi)\,\mathrm{Det}(\partial_t + \delta^2S/\delta\varphi^2)\,\mathcal{D}\varphi$$

$$\equiv \int\delta(\varphi_\eta - \varphi(x,t))\,\mathcal{D}\varphi\,.$$

(Here we assume that $S[\varphi]$ is such that the Langevin equation with some proper initial and boundary conditions has only one solution.) The rising of the determinant in the last expression in the exponent is performed by two anticommuting scalars ψ and $\bar{\psi}$. Introducing also the auxiliary field $A(x, t)$, we obtain:

$$Z(h) = \int \mathcal{D}\varphi \mathcal{D}\psi \mathcal{D}\bar{\psi} \mathcal{D}A$$

$$\times \exp\left(\int d^D x\, dt [A^2 - A(\dot{\varphi} + \delta S/\delta\varphi) + \bar{\psi}(\partial_t + \delta^2 S/\delta\varphi^2)\psi + h\varphi]\right). \tag{2.3}$$

Using the superfields

$$\phi(x, t, \theta, \bar{\theta}) = \varphi(x, t) + \bar{\theta}\psi(x, t) + \bar{\psi}(x, t)\theta + \bar{\theta}\theta A(x, t),$$

$$H(x, t, \theta, \bar{\theta}) = \bar{\theta}\theta h(x, t), \tag{2.4}$$

eq. (2.3) can be written as:

$$Z(h) = \int \mathcal{D}\phi \exp\left(-\int S[\phi]\, dt\, d\theta\, d\bar{\theta} - \int d^D x\, dt\, d\theta\, d\bar{\theta} \{\phi[\partial^2/\partial\theta\partial\bar{\theta} + \theta(\partial/\partial\theta)\partial_t - \tfrac{1}{2}\partial_t]\phi + H\phi\}\right). \tag{2.5}$$

The term $-\frac{1}{2}\phi\partial_t\phi$ is, of course, a total t derivative but, nevertheless we shall keep it in (2.5) in order to have a symmetric operator whose inverse is just the free propagator.

We notice also that the translational invariance of (2.5) leads to the t independence of (2.2). Hence we can use in (2.2) the following source:

$$H(x, t, \theta, \bar{\theta}) = \bar{\theta}\theta\delta(t)h(x). \tag{2.6}$$

We now discuss the reduction of (2.5) to the D-dimensional scalar field theory. So, we are going to prove that $Z(h)$ equals the generating functional of the "usual" theory:

$$Z(h) = \tilde{Z}(h) \equiv \int \mathcal{D}\varphi \exp\left(-S[\varphi] - \int d^D x\, h(x)\varphi(x)\right). \tag{2.7}$$

The evidence is provided within perturbation theory. The free (super) propagator in (2.5) is:

$$(\Box^2 - \partial_t^2)^{-1}[\Box - 2\partial^2/\partial\theta\partial\bar{\theta} + 2\theta(\partial/\partial\theta)\partial_t - \partial_t]. \tag{2.8}$$

Going to momentum representation for x and t, we have:

$$\langle\phi(p\omega\theta_1)\phi(p'\omega'\theta_2)\rangle = [-2 - (p^2 - i\omega)(\bar{\theta}_1 - \bar{\theta}_2)\theta_1 - (p^2 + i\omega)(\bar{\theta}_2 - \bar{\theta}_1)\theta_2](p^4 + \omega^2)^{-1} \times \delta(\omega + \omega')\delta^4(p + p'). \tag{2.9}$$

Hence the coefficient before h^2 in the $Z(h)$ decomposition is:

$$\delta^2 Z/\delta h(p)\delta h(p')|_{h=0} = \int \bar{\theta}_1\theta_1\bar{\theta}_2\theta_2 \frac{d\omega}{2\pi} d^2\theta_1 d^2\theta_2 \langle\phi(p\omega\theta_1)\phi(p'\omega'\theta_2)\rangle = p^{-2}\delta^4(p + p') = \delta^2\tilde{Z}/\delta h(p)\delta h(p')|_{h=0}.$$

So (2.7) is valid in this order. To prove it for a general diagram, let us disjoin any vertex, say ϕ^3 (see fig. 1). The expression corresponding to this diagram is:

$$\prod_{i=1}^{3} \frac{2 + (p_i^2 - i\omega_i)(\bar{\theta}_i - \bar{\theta})\theta_i + (p_i^2 + i\omega_i)(\bar{\theta} - \bar{\theta}_i)\theta}{p_i^2 + \omega_i^2} \times \delta\left(\sum_i \omega_i\right) F \prod_{i=1}^{3} \frac{d\omega_i}{2\pi} d^2\theta \prod_{i=1}^{3} d^2\theta_i, \tag{2.10}$$

where F corresponds to the remaining part of the diagram and may include terms of the type $\bar{\theta}\theta\delta(t) \times h(x)$ (external sources). Clearly F does not depend on θ and θ and does not possess poles on ω_i.

Integrating (2.10) over θ and $\bar{\theta}$, we obtain:

$$\frac{4\Sigma_i p_i^2 + 2\Pi_{i\neq j}(p_i^2 - i\omega_i)(p_j^2 + i\omega_j)\theta_i\bar{\theta}_j}{\Pi_i(p_i^4 + \omega_i^2)} \delta\left(\sum_i \omega_i\right) F \prod_{i=1}^{3} \frac{d\omega_i}{2\pi}. \tag{2.11}$$

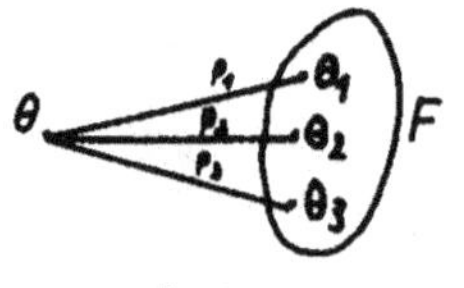

Fig. 1.

Volume 129B, number 5 PHYSICS LETTERS 29 September 1983

The second term in (2.11) equals zero after ω_i integration. This fact can be verified by using

$$2\pi\delta\left(\sum_i \omega_i\right) = \int_{-\infty}^{+\infty} \exp\left[i\tau\left(\sum_i \omega_i\right)\right] d\tau, \qquad (2.12)$$

and properly choosing for the positive and the negative values of τ the contours of the ω_i integrations (see appendix A). The first term in (2.11) yields

$$\left(\prod_i p_i^2\right)^{-1} \widetilde{F},$$

where $\widetilde{F}$ is the contribution from the remaining part of the diagram after all ω and θ integrations (see appendix B).

Thus the propagators connecting our arbitrary chosen vertex to the remainder of the diagram become exactly $1/p_i^2$ after ω and θ integrations. So, performing all these integrations, we obtain inductively the D-dimensional scalar theory.

We are grateful to Dr. S.G. Mathinian for useful discussions, and one of us (S. Kalitzin) would like to thank the Erevan Institute of Physics for the hospitality and support.

Appendices

A. We prove the equality:

$$I_1 \equiv \int \frac{(p_1^2 - i\omega_1)(p_2^2 + i\omega_2)}{\Pi_i (p_i^4 + \omega_i^2)} \delta\left(\sum_i \omega_i\right) \times F(p_i \omega_i) \prod_j \frac{d\omega_j}{2\pi}, \qquad (A.1)$$

where F is analytical on ω_i.

Let us substitute (2.12) in (A.1). Then for $t > 0$, we choose the contour of ω_i integration in the upper half-plane (for each i, $i = 1,2,3$). However, the term $p_1^2 - i\omega_1$ in the numerator cancels the ω_1 pole and so the ω_1 integration yields zero. For $t < 0$, we choose the contours in the lower half-planes and then the ω_2 integral equals zero for the same reason.

B. To prove the equality

$$I_2 \equiv \int \frac{4\Sigma_i p_i^2}{\Pi_i (p_i^4 + \omega_i^2)} \delta\left(\sum_i \omega_i\right) F(p_i \omega_i) = \left(\prod_i p_i^2\right)^{-1} \widetilde{F}(p_i), \qquad (B.1)$$

where $\widetilde{F}$ is specified below, we again use (2.12). Then I_2 can be split as $I_2 = I_2^+ + I_2^-$, where in I_2^+ $[I_2^-]$ the τ integration is performed in the interval $(0, +\infty)$ $[(0, -\infty)]$. Choosing the integration contours for ω_i as above, we obtain:

$$I_2^+ = \left(\prod_i p_i^2\right)^{-1} \tfrac{1}{2} F(p_i, i p_i^2),$$

$$I_2^- = \left(\prod_i p_i^2\right)^{-1} \tfrac{1}{2} F(p_i, -i p_i^2).$$

Then for $I_2 = I_2^+ + I_2^-$ we derive (B.1).

Note added

The former analysis holds equally well for arbitrary field theories described by an action of the general form:

$$S = \int d^D x\, \phi A \phi + S_{\text{int}}(\phi), \qquad (N.1)$$

where A is a hermitian invertible differential operator. The superpropagator now is as (2.8):

$$(A^2 - \partial_t^2)^{-1} [A - 2\partial^2/\partial\theta\partial\bar{\theta} + 2\theta(\partial/\partial\theta)\partial_t - \partial_t]. \qquad (N.2)$$

So, going to momentum representation and performing diagonalization of A, we can repeat the whole evidence proving that the stochastically quantized theory yields the usual Green's functions of (N.1)

Theories not entering this category are particularly gauge theories (the kinetic operator A is not invertible) which, however, can be treated in the same fashion after the standard gauge fixing procedure. Super theories are also treatable in this way irrespectively of the presence of explicit Grassman-coordinate dependence in A (via the supercovariant derivatives), A is purely differential in the component formalism.

References

[1] G. Parisi and Y.-S. Wu, Scientia Sinica 24 (1981) 483.
[2] T. Hori, TMUP-HEL-8207.
[3] E. Floratos and J. Illiopoulos, LPT ENS 82/31.
[4] W. Grimus and H Hüffel, CERN preprint TH 3446.

STOCHASTIC QUANTIZATION, SUPERSYMMETRY AND THE NICOLAI MAP

P.H. DAMGAARD
NORDITA, Blegdamsvej 17, DK-2100, Copenhagen Ø, Denmark

and

K. TSOKOS
Department of Physics, University of Maryland, College Park, MD 20742, U.S.A.

ABSTRACT. We present a formulation of the method of stochastic quantization of Parisi and Wu that reveals its intimate connection with supersymmetry. The crucial ingredient of this analysis is the Nicolai map. By using supersymmetric Ward identities, we derive relations between two Fokker–Planck-type Hamiltonians which arise naturally in this formalism.

1. INTRODUCTION

In field theory we are interested in computing correlation functions $G_n(x_1 \ldots x_n)$ given by

$$G_n(x_1 \ldots x_n) = \int \mathrm{d}[\phi]\, \phi(x_1) \ldots \phi(x_n)\, \mathrm{e}^{-S}, \tag{1}$$

where ϕ is a generic field and S the action. Stochastic quantization, as introduced by Parisi and Wu [1], offers an alternative method for the computation of $G_n(x_1 \ldots x_n)$. The method is essentially the following: one appends to ordinary four-dimensional spacetime a fifth coordinate t and all fields in the theory are assumed to acquire a dependence on this variable that is given by the solutions of the Langevin equation:

$$\frac{\partial}{\partial t} \phi(x, t) = -\frac{\delta S}{\delta \phi} + \eta(x, t) \tag{2}$$

where $\eta(x, t)$ is a random noise field with Gaussian distribution. The n-point functions of the Euclidean field theory are reproduced, if we evaluate the $t \to \infty$ limit of the quantity

$$\langle \phi(x, t) \ldots \phi(y, t) \rangle_\eta = \int \mathrm{d}[\eta]\, \phi(x, t) \ldots \phi(y, t)\, \mathrm{e}^{-1/4 \int \mathrm{d}^D x\, \mathrm{d}t\, \eta^2}. \tag{3}$$

This method of computation was originally introduced in order to avoid problems with gauge fixing in the perturbative definition of gauge theories, but has since proved to be of more general

Letters in Mathematical Physics 8 (1984) 535–540. 0377–9017/84/0086–0535 $00.90.

use. It is a technique that is particularly suited for numerical work, and it is now extensively used in Monte-Carlo simulations of lattice gauge theories [2]. The method of Parisi and Wu has recently been extended to include fermions as well [3] so that supersymmetric theories can now also be treated in this way [4].

2. SUPERSYMMETRY AND THE NICOLAI MAP

It is interesting to observe, that before the limit $t \to \infty$ is taken, Equation (3) defines a theory in its own right. Specifically, we can view Equation (3) as defining a field theory in a $(D+1)$-dimensional space with coordinates (x, t) and a partition function Z given by (we insert a factor $1 = \int d[\phi]\,\delta(\phi - \phi^{[\eta]})$)

$$Z = \int d[\phi]\; d[\eta] \exp(-\tfrac{1}{4}\int d^D x\, dt\, \eta^2)\,\delta(\phi - \phi^{[\eta]}) \tag{4}$$

just like the elegant construction of Parisi and Sourlas in D dimensions [5]. Here the functional data-function has been introduced in order to make sure that only fields $\phi^{[\eta]}$ that are solutions of the Langevin equation contribute to the functional integral. The superscript on the field shows explicitly that ϕ has become a functional of the noise field η.

The form of the partition function, Equation (4), is very simple; if the range of the field η is minus to plus infinity* we can evaluate $Z = 1$, which implies that the ground-state energy of the system vanishes. This is a property of theories with unbroken supersymmetry. One way of uncovering this symmetry [7] is by a use of Nicolai's map [8]. The Nicolai map is an invertible, nonlocal transformation of the bosonic fields such that (i) the Jacobian of this transformation cancels the determinant obtained by performing the integral over fermions and (ii) the new bosonic variables are free Gaussian variables**.

This last property of the Nicolai map enables us to interpret the Nicolai variable as the random source field of stochastic quantization. Thus, we are guaranteed obtaining a supersymmetric theory by using Nicolai's map in the reverse order, i.e., by starting from the partition function, Equation (4), and eliminating the noise field in favor of the original fields in the theory.

A simple illustration of this construction is provided by the Langevin equation for the space coordinate x of a classical particle

$$\frac{dx}{dt} = -\frac{\delta S}{\delta x} + \eta. \tag{5}$$

The change of variables we seek is $\eta \to x$ and, therefore, the Jacobian is given by (we write $V = \delta S/\delta x$)

$$\det\left(\frac{\delta\eta(t)}{\delta x(t')}\right) = \det((\partial_t + V')\delta(t - t')). \tag{6}$$

*The fact that the range of the Nicolai variable may not be minus to plus infinity is, in fact, a signal of supersymmetry breaking [6].

**The Nicolai map can be thought of as a special case of what in the mathematical literature is known as the Morse–Palais lemma. We are indebted to Ken Macrae for discussions on this point.

After a series of standard manipulations the partition function becomes

$$
\begin{aligned}
Z &= \int dx\, d[\eta]\; \delta(\partial_t x + V - \eta)\, \det(\partial_t + V') \exp\left(-\tfrac{1}{4}\int dt\, \eta^2\right) \\
&= \int dx\, \det(\partial_t + V') \exp\left(-\tfrac{1}{4}\int dt (\partial_t x + V)^2\right) \\
&= \int dx\, d\psi\, d\bar{\psi}\, \exp\left(-\tfrac{1}{4}\int dt (\partial_t x + V)^2 + \int dt\, \bar{\psi}(\partial_t + V')\psi\right)
\end{aligned}
\tag{7}
$$

The theory described by (7) is recognized as Witten's example of supersymmetric quantum mechanics [9]. To make that connection clearer, we evaluate the determinant in (7) explicitly. It is formally infinite so in order to extract a finite part, we have to regularize it. A convenient method of doing that is provided by the technique of zeta-function regularization [10]. We are considering the operator $N = \partial_t + V'$ so in order to construct the generalized zeta function we need the eigenvalues of N

$$
N\psi_n = \nu_n \psi_n \rightarrow \psi_n(t) = \exp\left(\int_0^t d\tau(\nu_n - V')\right)\psi_n(0). \tag{8}
$$

To find ν_n we enclose the system in a box of size T with *periodic* boundary conditions, i.e., that $\psi_n\ (t = 0) = \psi_n\ (t = T)$, which leads to

$$
\nu_N = \frac{2\pi i n}{T} + \frac{1}{T}\int_0^T d\tau\, V', \quad |n| = 0, 1, 2, \ldots . \tag{9}
$$

Now we form the zeta-function for the operator N (note that it is the absolute value of the eigenvalues which enters the sum because N is a first-order operator):

$$
\zeta_N(s) = \sum |\nu_n|^{-s}. \tag{10}
$$

Then,

$$
\log \det N = -\frac{d}{ds}\zeta_N(s)|_{s=0}. \tag{11}
$$

We can now compute

$$
\begin{aligned}
\frac{d}{ds}\zeta_N(s)|_{s=0} &= \zeta'_R(0, a) + \zeta'_R(0, -a) + \\
&+ \log\left|\frac{T}{2\pi i}\right|(\zeta_R(0, a) + \zeta_R(0, -a)) - \log\left|\frac{T}{2\pi i a}\right|,
\end{aligned}
\tag{12}
$$

where $a = 1/(2\pi i)\int_0^T d\tau\, V'$ and $\zeta_R(s, a)$ is the generalized Riemann zeta function defined by

$\zeta_R(s, a) = \Sigma_0^\infty (n + a)^{-s}$. We now use the relations [11]

$$\zeta'_R(0, a) = \log \Gamma(a) - \tfrac{1}{2}\log 2\pi; \qquad \zeta_R(0, a) = \tfrac{1}{2} - a, \tag{13}$$

$$\Gamma(a)\Gamma(-a) = -\frac{1}{a}\,\frac{\pi}{\sin(\pi a)}$$

to find

$$\det N = -2i \sinh\left(\tfrac{1}{2}\int_0^T d\tau\, V'\right). \tag{14}$$

3. THE WARD IDENTITIES

The Hamiltonian corresponding to the Lagrangian in (7) is (after a rescaling of the t coordinate)

$$H_\pm = -\frac{1}{2}\frac{\delta^2}{\delta x^2} + \frac{1}{8}\left(\frac{\delta S}{\delta x}\right)^2 \pm \frac{1}{4}\frac{\delta^2 S}{\delta x^2}. \tag{15}$$

It is interesting to observe that the Hamiltonian entering the Fokker–Planck equation in the usual formulation of stochastic quantization [1] is H_-; we see that, because of supersymmetry, we end up having to consider two related Fokker–Planck Hamiltonians, H_+. The relation between the two Hamiltonians is actually more interesting than what might at first be imagined: to see exactly what this relation is, we will need to examine the supersymmetry of the Lagrangian in (7) more carefully. To this end, we write this Lagrangian as

$$L = \tfrac{1}{2}\dot{x}^2 - \tfrac{1}{2}D^2 - \tfrac{1}{2}DV - \bar{\psi}(\partial_t + \tfrac{1}{2}V')\psi, \tag{16}$$

where D is an auxiliary field. This Lagrangian is invariant under the supersymmetry transformations

$$\begin{aligned} \delta x &= \bar{\epsilon}\psi - \bar{\psi}\epsilon, \qquad \delta D = \bar{\epsilon}\dot{\psi} + \dot{\bar{\psi}}\epsilon, \\ \delta\psi &= (\dot{x} + D)\epsilon, \qquad \delta\bar{\psi} = \bar{\epsilon}(\dot{x} - D). \end{aligned} \tag{17}$$

It follows from (17) that the commutator of two supersymmetry transformations is proportional to d/dt, as it should be.

We now introduce a set $J = \{J_x, \dot{J}_D, \zeta, \bar{\zeta}\}$ of sources, which from a scalar supermultiplet under (17); their transformation properties are given by

$$\begin{aligned} \delta J_x &= \bar{\epsilon}\dot{\zeta} + \dot{\bar{\zeta}}\epsilon, \qquad \delta J_D = \bar{\epsilon}\zeta - \bar{\zeta}\epsilon, \\ \delta\zeta &= \epsilon J_x + \epsilon\dot{J}_D, \qquad \delta\bar{\zeta} = -\bar{\epsilon}J_x + \bar{\epsilon}\dot{J}_D. \end{aligned} \tag{18}$$

Thus,

$$L \to L + xJ_x + DJ_D + \bar{\zeta}\psi + \bar{\psi}\zeta \quad \text{and} \quad \delta L = 0. \tag{19}$$

This invariance of the Lagrangian leads to

$$\begin{aligned}\delta Z = 0 &= \frac{\delta Z}{\delta J_x}\delta J_x + \frac{\delta Z}{\delta J_D}\delta J_D - \frac{\delta Z}{\delta \zeta}\delta\zeta + \delta\bar{\zeta}\frac{\delta Z}{\delta\bar{\zeta}} \\ &= \frac{\delta Z}{\delta J_x}(\bar{\epsilon}\dot{\zeta} + \dot{\bar{\zeta}}\epsilon) + \frac{\delta Z}{\delta J_D}(\bar{\epsilon}\zeta - \bar{\zeta}\epsilon) - \\ &\quad - \frac{\delta Z}{\delta\zeta}(\epsilon\dot{J}_x + \epsilon J_D) + (-\bar{\epsilon}\dot{J}_x + \bar{\epsilon}J_D)\frac{\delta Z}{\delta\bar{\zeta}}.\end{aligned} \tag{20}$$

Equation (20) contains all the Ward identities of the theory. We obtain relations between Green functions of the theory by differentiating (20) with respect to the sources and then setting them equal to zero. One such example is

$$\langle \psi\bar{\psi}\rangle - \frac{\mathrm{d}}{\mathrm{d}t}\langle x^2\rangle - \langle xV\rangle = 0. \tag{21}$$

We can now obtain information about $H_\pm$ by integrating out the fermions and recalling (14) which implies that Z can be written as

$$Z = (Z^+ - Z^-)\exp(-\bar{\zeta}(\partial_t + V')^{-1}\xi), \tag{22}$$

where

$$Z^\pm = \int \mathrm{d}x\,\mathrm{d}D\,\mathrm{e}^{-S^\pm} \tag{23}$$

and $S^\pm$ are the actions corresponding to the two Fokker–Planck Hamiltonians $H^\pm$. The relation between these two actions is now clear: let $F_Z[x, V(x)]$ stand for combinations of Green functions which obey Equation (20), i.e., $F_Z[x, V(x)] = 0$. Equations (22) and (23) them imply that

$$F_{Z^+}[x, V(x)] = F_{Z^-}[x, V(x)], \tag{24}$$

i.e., expectation values computed with H_+ and H_- are related.

This is a symmetry between two classical Hamiltonians (well known in the literature of statistical mechanics) which is, however, demanded by supersymmetry. It would be interesting to see what the effect of this symmetry is on the classical dynamics of systems obeying the Langevin Equation (2).

4. SUMMARY AND CONCLUSIONS

By using Nicolai's map to uncover the supersymmetry hidden in stochastic processes, we obtained relations between the two Fokker–Planck Hamiltonians which arose naturally in this treatment of the formalism of stochastic quantization. These relations were the direct consequence of the supersymmetric Ward identities of the theory and illustrate the statement that 'classical Brownian motion is equivalent to a macroscopic supersymmetry'.

ACKNOWLEDGEMENT

We wish to thank Ken Macrae and Paul Mansfield for many discussions.

REFERENCES

1. Parisi, G. and Wu, Y., *Sci. Sinica* **24**, 483 (1981).
2. Fucito, F., Marinari, E., Parisi, G., and Rebbi, C., *Nucl. Phys.* **B180**, 369 (1981);
   Parisi, G., *Nucl. Phys.* **B180**, 378 (1981);
   Fucito, F. and Marinari, E., *Nucl. Phys.* **B190**, 31 (1981).
3. Damgaard, P.H. and Tsokos, K., *Nucl. Phys.* **B235** [FS11], 75 (1984).
4. Breit, J.D., Gupta, S., and Zaks, A., *Nucl. Phys.* **B233**, 61 (1984).
5. Parisi, G. and Sourlas, N., *Phys. Rev. Lett.* **43**, 744, 1364 (1979).
7. Parisi, G. and Sourlas, N., *Nucl. Phys.* **B206**, 321 (1983);
   Cecotti, S. and Girardello, L., *Ann. Phys.* **145**, 81 (1983);
   Gozzi, E., City College preprint 83/16.
8. Nicolai, H., *Phys. Lett.* **89B**, 341 (1980); *Nucl. Phys.* **B176**, 419 (1980).
9. Witten, E., *Nucl. Phys.* **B188**, 513 (1981);
   Bender, C., Cooper, F., and Freedman, F., *Nucl. Phys.* **B219**, 61 (1983).
10. Hawking, S., *Comm. Math. Phys.* **55**, 133 (1977).
11. Whittaker, E.T. and Watson, G.N., *A Course in Modern Analysis*, Cambridge University Press, 1967.
6. Cecotti, S. and Girardello, L., *Phys. Lett.* **110B**, 509 (1982).

(Received August 25, 1984).

Volume 139B, number 3 PHYSICS LETTERS 10 May 1984

QUANTIZATION BY STOCHASTIC RELAXATION PROCESSES AND SUPERSYMMETRY

R. KIRSCHNER
Sektion Physik, Karl-Marx-Universität Leipzig, Leipzig, GDR

Received 24 January 1984

We show the supersymmetry mechanism responsible for the quantization by stochastic relaxation processes and for the effective cancellation of the additional time dimension against the two Grassmann dimensions. We give a non-perturbative proof of the validity of this quantization procedure.

1. Since the paper by Parisi and Wu [1] there is increasing interest in stochastic quantization.

The convergence of the stochastic relaxation process in the large time limit to the corresponding quantum theory can be shown using the Fokker–Planck equation [2–5]. In the framework of perturbation theory the validity of stochastic quantization has been shown analyzing the stochastic graphs generated by the Langevin equation [6–8].

Besides the quantization by stochastic relaxation processes there is a similar quantization procedure by stationary stochastic processes [9]. Proofs of the validity of this quantization procedure going beyond perturbation theory have been given in refs. [10,11].

In this note we consider the supersymmetry structure of stochastic relaxation processes starting from the superfield formulation of ref. [8]. We present a non-perturbative proof of the quantization by stochastic relaxation processes, which follows the scheme proposed by Cardy [10] for the case of stationary stochastic quantization.

2. Consider a system with the action

$$S = \int d^n x\, L(\varphi(x)), \qquad (1)$$

and couple it to a random force $\eta(x, t)$, which depends on the additional time t and obeys a gaussian distribution. The behaviour of the system is described by the Langevin equation for $t > 0$ with initial conditions at $t = 0$. In order to extend the range in time to the full axis we define that for $t < 0$ a process takes place which is obtained by time reflection from the process at $t > 0$.

We are interested in correlation functions involving the field φ only at a given time t and consider the corresponding generating functional $Z(j; t)$. We use the formulation of ref. [8] to write $Z(j; t)$ as

$$Z(j; t) = \int d\Phi \exp\Big(-\int_{-\infty}^{\infty} dt' \times \int d^n x\, d\theta\, d\bar\theta (\mathcal{L}(\Phi) + \Phi J_t)\Big). \qquad (2)$$

The superfield Φ contains φ as its lowest component.

$$\Phi(x, t, \theta, \bar\theta) = \varphi(x, t) + \bar\theta\psi(x, t) + \bar\psi(x, t)\,\theta + \bar\theta\theta\omega(x, t). \qquad (3)$$

θ and $\bar\theta$ are Grassmann numbers. The current J_t has the special form

$$J_t(x, t', \theta, \bar\theta) = j(x)\,\delta(t - t')\,\delta(\bar\theta)\,\delta(\theta). \qquad (4)$$

The superfield lagrangian is given by

$$\mathcal{L}(\Phi) = \mathcal{L}_K + iL(\Phi),$$

$$\mathcal{L}_K = \Phi[\partial_\theta \partial_{\bar\theta} - \tfrac{1}{2} i\, \mathrm{sgn}(t)(\partial_t \theta \partial_\theta - \partial_\theta \theta \partial_t)]\, \Phi. \qquad (5)$$

3. The theory, eq. (5), is invariant under a supersymmetry transformation of the type of supersym-

Volume 139B, number 3 PHYSICS LETTERS 10 May 1984

metric quantum mechanics. This property of stochastic relaxation systems has been considered in refs. [4, 12]. It can be obtained by analyzing the quadratic form of derivatives in $\mathcal{L}_K$, eq. (5). The line element in superspace corresponding to this quadratic form turns out to be

$$ds^2 = dx^2 + [dt + i\,\mathrm{sgn}(t)\,\theta\, d\bar{\theta}]^2 . \qquad (6)$$

The theory is invariant under transformations leaving ds^2 invariant. Alternatively, we can study the variations of the lagrangian under the transformation

$$t \to t' = t + i(\bar{\theta}\zeta\lambda^- - \bar{\zeta}\theta\lambda^+) ,$$

$$\theta \to \theta' = \theta + \zeta , \quad \bar{\theta} \to \bar{\theta}' = \bar{\theta} + \bar{\zeta} , \quad x_\mu \to x'_\mu = x_\mu . \qquad (7)$$

ζ and $\bar{\zeta}$ are Grassmann numbers, $\lambda^\pm$ are ordinary numbers. The line element is invariant and the variations of the lagrangian vanish (up to total derivatives) for

$$\lambda^+ = \mathrm{sgn}(t) , \quad \lambda^+ = 0 . \qquad (8)$$

For positive time and with these parameters the generators of the superspace transformation eq. (7) obey the usual commutation relations of supersymmetric quantum mechanics up to a rescaling of t.

Choosing

$$\lambda^- = -\lambda^+ = \tfrac{1}{2}\,\mathrm{sgn}(t) , \qquad (9)$$

we find that the lagrangian is not invariant. But in this case variations can be absorbed by a redefinition of ψ and $\bar{\psi}$.

$$\psi \to \tilde{\psi} = \psi + (\partial_t + \partial^2 L/\partial\varphi^2)^{-1}$$

$$\times\, [\zeta\partial_t\,\omega(x, t)\,\mathrm{sgn}(t)] . \qquad (10)$$

This change of variables leaves $Z(j; t)$ invariant, because of the special form of the current term in eqs. (2), (4). The only invariant of the transformation eq. (7) with the parameters eq. (9) is

$$\tau = t - \tfrac{1}{2} i\,\bar{\theta}\theta\,\mathrm{sgn}(t) . \qquad (11)$$

We call this symmetry eqs. (7), (9) reduction supersymmetry in view of the role it plays in the following.

The reduction supersymmetry is just the supertransformation invariance obtained in ref. [7] analyzing the set of stochastic graphs. To show this explicitly, we perform a time translation of the type

$$t \to t' = t + ia\,\bar{\theta}\theta , \qquad (12)$$

with all other coordinates unchanged. The lagrangian is invariant to the transformation generated by eq. (12) up to total derivatives. Choosing $a = \frac{1}{2}$, we obtain that in the new coordinates the reduction supersymmetry and the invariant coincide with the corresponding expressions in ref. [7].

$$\tau \to \tau' = t' - i\bar{\theta}\theta\,\epsilon(t') , \quad \epsilon(t') = \tfrac{1}{2}[1 + \mathrm{sgn}(t')] . \qquad (13)$$

4. We consider the more general generating functional $Z_\lambda(j; t)$ with $\mathcal{L}(\Phi)$ in eq. (2) replaced by

$$\mathcal{L}_\lambda(\Phi) = \{\lambda - (1 - \lambda)\tfrac{1}{2} i\delta(\bar{\theta})\,\delta(\theta)$$

$$\times\, [\delta(t - t') + \delta(t + t')]\}\; iL(\Phi) + \mathcal{L}_K . \qquad (14)$$

At $\lambda = 1$ it coincides with the generating functional $Z(j; t)$, eq. (2). We introduce a cut-off T in time, impose periodic boundary conditions and let t tend to T. In analogy to ref. [10] it can be shown that for $\lambda \to 0$ all field components of $\Phi(x, t, \theta, \bar{\theta})$ up to the component of $\varphi(x, t)$ constant in time can be integrated out. We arrive at the generating functional of the quantized theory with the action eq. (1). In the limit $T \to \infty$ all Green functions become independent of λ and hence coincide for $\lambda \to 1$ and $\lambda \to 0$. To prove this, we calculate

$$\frac{\partial}{\partial\lambda} \ln Z_\lambda(j; T) = -i \int d^n x\, dt'\, d\theta\, d\bar{\theta} \langle L(\Phi)\rangle_{j,\lambda}$$

$$\times\, \{1 + i\delta(\bar{\theta})\,\delta(\theta)\,\tfrac{1}{2}[\delta(t' + T) + \delta(t' - T)]\}$$

$$= \quad d^n x \langle L(\Phi)\rangle_{j,\lambda}|_{t=\theta=\bar{\theta}=0} . \qquad (15)$$

Because of the reduction supersymmetry of $Z_\lambda(j; T)$ the integrand depends on $t, \theta, \bar{\theta}$ only via the invariant τ, eq. (11). This leads to the second relation in eq. (15). In the limit $T \to \infty$ the last expression becomes independent of j, because the source acts at infinite time distances from the instant $t = 0$, and this proves our assertion.

The condition for the latter statement being true is just the requirement, that in the large time limit all equal-time correlation functions become independent of the initial conditions at $t = 0$.

5. We have shown the non-perturbative equivalence of the supersymmetric theory eq. (2), obtained as a reformulation of the stochastic relaxation system, to the ordinary quantized theory with the action eq. (1). The additional time dimension has been cancelled by two Grassmann dimensions. The equivalence is understood in the sense that the arguments of the superfield Green functions are restricted to the n-dimensional submanifold $t = T$ with vanishing Grassmann variables and that the large time limit is performed. We have constructed the proof in close analogy to the one presented by Cardy [10] for the case of stationary stochastic processes. In the latter case the rotation invariance plays the role of the reduction supersymmetry in superspace, discovered by Parisi and Sourlas [9], and an expression of the type $x^2 + i\bar{\theta}\theta$ plays the role of the invariant. In contrast to our case, there the two Grassmann dimensions cancel two dimensions out of x_μ.

From our discussion the underlying supersymmetry mechanism of the quantization by stochastic relaxation processes becomes clear. Our proof of the validity of this quantization procedure generalizes the treatments of refs. [6–8] to a non-perturbative level and represents an alternative to the known proofs using the Fokker–Planck equation [2–5].

The author is grateful to Professor J. Ranft for discussions.

References

[1] G. Parisi and Y. Wu, Scientia Sinica 24 (1981) 483.
[2] L. Baulieu and D. Zwanziger, Nucl. Phys. B193 (1981) 163.
[3] J.D. Breit, S. Gupta and A. Zaks, preprint Institute for Advanced Study (Princeton, 1983).
[4] C. Bender, F. Cooper and B. Freedman, Nucl. Phys. B219 (1983) 61.
[5] E. Floratos and J. Iliopoulos, Nucl. Phys. B214 (1983) 392.
[6] W. Grimus and H. Hüffel, preprint CERN Ref. TH 3449 (1982).
[7] H. Nakazato, M. Namiki, I. Ohba and K. Okano, Prog. Theor. Phys. 70 (1983) 298.
[8] E.S. Egorian and S. Kalitsin, Phys. Lett. 129B (1983) 320.
[9] G. Parisi and N. Sourlas, Phys. Rev. Lett. 43 (1979) 744; Nucl. Phys. B206 (1982) 321.
[10] J.L. Cardy, Phys. Lett. 125B (1983) 470.
[11] A. Klein and J. Fernando Perez, Phys. Lett. 125B (1983) 473.
[12] P.H. Damgaard and K. Tsokos, University of Maryland, preprint MD-TP-219 (1983).

Canonical Stochastic Quantization

Prog. Theor. Phys. Vol. 73, No. 5, May 1985, Progress Letters

Canonical Stochastic Quantization

Shijong RYANG, Takesi SAITO* and Kazuyasu SHIGEMOTO**

Department of Physics, Osaka University, Toyonaka 560

**Department of Physics, Kyoto Prefectural University of Medicine, Kyoto 603*

***Department of Liberal Arts, Tezukayama University, Tezukayama 4, Nara 631*

(Received January 9, 1985)

A "canonical" stochastic quantization is proposed by introducing the Langevin equation with momentum $\pi(x,t)$. The Gibbs (canonical) average method is obtained as a consequence of this canonical stochastic quantization.

A few years ago Parisi and Wu[1] recalled a stochastic quantization scheme by means of the Langevin equation of non-equilibrium statistical mechanics, while it has a long-time history.[2] The method can be now applied to the quantization of scalar, Dirac and gauge fields in Euclidean space-time,[3] and also in Minkowski space.[4] The outline of the method is as follows: For a scalar field ϕ we introduce a fictious fifth time t and consider the field ϕ as a function of four Euclidean variables x and of t: $\phi=\phi(x,t)$. The field ϕ is coupled with a heat reservior at temperature $\beta^{-1}=h$ and will reach the equilibrium distribution for large time t. The evolution equation of the field is the Langevin equation of the type

$$\frac{\partial\phi(x,t)}{\partial t}=-\gamma\frac{\delta S[\phi]}{\delta\phi(x,t)}+\eta(x,t), \tag{1}$$

where $S[\phi]$ is the classical action for ϕ, γ the friction coefficient and $\eta(x,t)$ is a Gaussian random noise characterized by the stochastic average

$$\begin{aligned}&\langle\eta(x,t)\rangle=0,\\&\langle\eta(x,t)\eta(x',t')\rangle=2\gamma\beta^{-1}\delta^4(x-x')\delta(t-t'). \end{aligned}\tag{2}$$

One can show then in many ways that

$$\lim_{t\to\infty}\langle\phi(x_1,t)\cdots\phi(x_n,t)\rangle=\int[d\phi]\phi(x_1)\cdots\phi(x_n)e^{-\beta S[\phi]}\Big/\int[d\phi]e^{-\beta S[\phi]}. \tag{3}$$

Namely the equal time non-equilibrium correlation function tends to the equilibrium one for large time, i.e., the N-point Green function in the Euclidean field theory.

In this paper we propose a "canonical" stochastic quantization by introducing the Langevin equation with momentum $\pi(x,t)$

$$\left.\begin{aligned}\frac{\partial\phi(x,t)}{\partial t}&=\frac{\delta H[\phi,\pi]}{\delta\pi(x,t)},\\ \frac{\partial\pi(x,t)}{\partial t}&=-\frac{\delta H[\phi,\pi]}{\delta\phi(x,t)}\\&\quad-\gamma\frac{\delta H[\phi,\pi]}{\delta\pi(x,t)}+\eta(x,t),\end{aligned}\right\} \tag{4}$$

where the total Hamiltonian $H[\phi,\pi]$ is defined by

$$\begin{aligned}H[\phi,\pi]&=\int d^4x\left\{\frac{1}{2}\pi^2(x,t)+\mathcal{L}(\phi(x,t))\right\}\\&=\int d^4x\left\{\frac{1}{2}\pi^2(x,t)\right\}+S[\phi],\end{aligned} \tag{5}$$

i.e., the conventional Lagrangian $\mathcal{L}(\phi)$ here is regarded as a potential energy term, γ the friction coefficient, $\eta(x,t)$ the Gaussian random variable characterized by Eq. (2). We can use Eq. (4) to compute the large time behaviors of $\phi(x,t)$ and $\pi(x,t)$ and consequently the equilibrium distribution and their correlation functions. We shall show that this equilibrium is realized by the Gibbs (canonical) average, i.e.,

$$\lim_{t\to\infty}\langle\phi(x_1,t)\cdots\phi(x_n,t)\rangle=\frac{\int[d\pi][d\phi]\phi(x_1)\cdots\phi(x_n)e^{-\beta H[\phi,\pi]}}{\int[d\pi][d\phi]e^{-\beta H[\phi,\pi]}}. \tag{6}$$

Because of the Liouville theorem the volume element $[d\pi][d\phi]$ and thus the right-hand side expression are independent of t. Since the dependence of the exponent in (6) on the momentum is purely Gaussian, the integration over π can be performed elementarily, leading to the conventional form (3). Advantage of the quantization by means of the Gibbs average (6) has been extensively stressed by Alfaro, Fubini and Furlan.[5] In this way the Gibbs (canonical) average method is obtained as a consequence of our canonical stochastic quantization.

We begin with the Lagrangian

$$\mathcal{L}(\phi)=\frac{1}{2}(\partial_\mu\phi)^2+\frac{1}{2}m^2\phi^2+\frac{1}{3}g\phi^3. \qquad (7)$$

The Langevin equation (4) is

$$\left.\begin{aligned}\dot\phi&=\pi,\\ \dot\pi&=(\partial^2-m^2)\phi-g\phi^2-\gamma\pi+\eta.\end{aligned}\right\} \qquad (8)$$

Elliminating the momentum π, this is reduced to[*)]

$$\ddot\phi=(\partial^2-m^2)\phi-g\phi^2-\gamma\dot\phi+\eta, \qquad (9)$$

or in momentum space

$$\left(\frac{\partial^2}{\partial t^2}+\gamma\frac{\partial}{\partial t}+\omega^2\right)\phi(k,t)$$

$$=\eta(k,t)-g\int\phi(k',t)\phi(k-k',t)dk', \qquad (10)$$

where $\omega^2=k^2+m^2$. Let us first study the free case $g=0$. The solution of Eq. (10) can be obtained by the standard method of variation of constants, and is written as

$$\phi(k,t)=\int_0^t dt'\Delta(k,t-t')\eta(k,t'), \qquad (11)$$

where

$$\Delta(k,t-t')=\frac{1}{\gamma_1}\{e^{\lambda_1(t-t')}-e^{\lambda_2(t-t')}\} \qquad (12)$$

and

$$\lambda_1=\frac{-\gamma+\gamma_1}{2},\quad \lambda_2=\frac{-\gamma-\gamma_1}{2},\quad \gamma_1=\sqrt{\gamma^2-4\omega^2}. \qquad (13)$$

Note that real parts of λ_1 and λ_2 take non-positive values. Solution (11) corresponds to initial values $\phi(k,0)=\dot\phi(k,0)=0$, but in our case the equilibrium value is independent of the initial conditions, because any initial values decrease exponentially with t. The corresponding momentum is then

$$\pi(k,t)=\int_0^t dt'\dot\Delta(k,t-t')\eta(k,t'). \qquad (14)$$

Since η is the Gaussian stochastic process, both ϕ and π are also the Gaussian processes. Gaussian processes are uniquely determined if their expectation values and two-point correlation functions are given. The expectation values are

$$\langle\phi(k,t)\rangle=\langle\pi(k,t)\rangle=0. \qquad (15)$$

The two-point correlation functions

$$\langle\phi(k,t)\phi(k',t)\rangle\equiv\delta^4(k+k')D(k,t), \qquad (16)$$

$$\langle\pi(k,t)\pi(k',t)\rangle\equiv\delta^4(k+k')G(k,t) \qquad (17)$$

and

$$\langle\phi(k,t)\pi(k',t)\rangle\equiv\delta^4(k+k')H(k,t) \qquad (18)$$

can be easily computed. We find

$$D(k,t)=\frac{\beta^{-1}}{\omega^2}\left\{1-e^{-\gamma t}\left(\frac{2\gamma^2}{\gamma_1^2}\sinh^2\frac{\gamma_1 t}{2}+\frac{\gamma}{\gamma_1}\sinh\gamma_1 t+1\right)\right\}, \qquad (19)$$

$$G(k,t)=\beta^{-1}\left\{1-e^{-\gamma t}\left(\frac{2\gamma^2}{\gamma_1^2}\sinh^2\frac{\gamma_1 t}{2}-\frac{\gamma}{\gamma_1}\sinh\gamma_1 t+1\right)\right\}, \qquad (20)$$

$$H(k,t)=4\gamma\beta^{-1}e^{-\gamma t}\frac{1}{\gamma_1^2}\sinh^2\frac{\gamma_1 t}{2}. \qquad (21)$$

When $t\to\infty$, we obtain the equilibrium results including the conventional propagator

$$D(k,t)\to\beta^{-1}\frac{1}{k^2+m^2}, \qquad (22)$$

$$G(k,t)\to\beta^{-1}, \qquad (23)$$

$$H(k,t)\to 0. \qquad (24)$$

Hence the probability distribution that stochastic processes $\{\phi(k,t)\}$ and $\{\pi(k,t)\}$ take values of $\{\phi(k)\}$ and $\{\pi(k)\}$ at t, respectively, is given by

$$P[\phi,\pi,t]=Z^{-1}(t)\exp\left[-\int d^4k\left\{\frac{D(k,t)|\pi(k)|^2+2H(k,t)\phi(k)\pi(-k)+G(k,t)|\phi(k)|^2}{2(DG-H^2)}\right\}\right], \qquad (25)$$

where $Z^{-1}(t)$ is the normalization factor. One can see that at large time this probability distribution tends to the equilibrium one, i.e., the Gibbs distribution in the free case

$$P[\phi,\pi,t]\to\rho[\phi,\pi]$$

$$=Z_0^{-1}\exp\left[-\beta\int d^4k\frac{1}{2}\{|\pi(k)|^2+(k^2+m^2)|\phi(k)|^2\}\right]$$

$$=Z_0^{-1}e^{-\beta H_0[\phi,\pi]}, \qquad (26)$$

where

[*)] We cannot neglect $\ddot\phi$ compared with $\dot\phi$ in Eq. (11) even in a region near equilibrium, because they are exponentially damping as the time and hence of the same order.

$$Z_0=\int[d\pi][d\phi]e^{-\beta H_0[\phi,\pi]}. \quad (27)$$

For $g\neq0$, we can write in the Euclidean space

$$\phi(x,t)=\int_0^t dt'\int d^4x'\Delta(x-x',t-t')$$
$$\times[\eta(x',t')+g\phi^2(x',t')]. \quad (28)$$

Therefore, the correlation functions can be calculated by iterating Eq. (28) in the same way as Parisi and Wu.

The Fokker-Planck equation corresponding to the Langevin equation (4) is given by, following the standard procedure of non-equilibrium statistical mechanics

$$\dot{P}[\phi,\pi,t]=\int d^4x\left\{-\frac{\delta}{\delta\phi}\frac{\delta H}{\delta\pi}+\frac{\delta}{\delta\pi}\frac{\delta H}{\delta\phi}\right.$$
$$\left.+\gamma\frac{\delta}{\delta\pi}\left(\frac{\delta H}{\delta\pi}+\beta^{-1}\frac{\delta}{\delta\pi}\right)\right\}P[\phi,\pi,t]. \quad (29)$$

For the free case ($g=0$) in Eq. (8), the solution of Eq. (29) can be easily found to be just Eq. (25). For $g\neq0$, we are hard to find the corresponding exact solution, but we can see an equilibrium solution which is independent of t, i.e.,

$$P_{eq}[\phi,\pi]=Z^{-1}e^{-\beta H[\phi,\pi]}. \quad (30)$$

This is nothing but the Gibbs distribution in the interacting case.

Next we consider the $U(1)$ gauge field. The Langevin equation (4) is, in momentum space,

$$\ddot{A}_\mu(k,t)+\gamma\dot{A}_\mu(k,t)$$
$$+(k^2\delta_{\mu\nu}-k_\mu k_\nu)A_\nu(k,t)=\eta_\mu(k,t). \quad (31)$$

This equation is divided into the transverse (T) and longitudinal (L) parts, i.e.,

$$\left(\frac{\partial^2}{\partial t^2}+\gamma\frac{\partial}{\partial t}+k^2\right)A_\mu{}^T(k,t)=\eta_\mu{}^T(k,t), \quad (32)$$

$$\left(\frac{\partial^2}{\partial t^2}+\gamma\frac{\partial}{\partial t}\right)A_\mu{}^L(k,t)=\eta_\mu{}^L(k,t). \quad (33)$$

Solutions of Eqs. (32) and (33) are

$$A_\mu{}^T(k,t)=\int_0^t dt'\Delta^T(k,t-t')\eta_\mu{}^T(k,t'), \quad (34)$$

$$A_\mu{}^L(k,t)=A_\mu{}^{L(0)}(k,t)$$
$$+\frac{1}{\gamma}\int_0^t dt'(1-e^{-\gamma(t-t')})\eta_\mu{}^L(k,t'), \quad (35)$$

where

$$\Delta^T(k,t-t')=\frac{1}{\gamma_1}\{e^{\lambda_1(t-t')}-e^{\lambda_2(t-t')}\},$$

$$A_\mu{}^{L(0)}(k,0)=A_\mu{}^L(k,t)$$
$$+\frac{\pi_\mu{}^L(k,0)}{\gamma}-\frac{\pi_\mu{}^L(k,0)}{\gamma}e^{-\gamma t}$$

and

$$\lambda_1=\frac{-\gamma+\gamma_1}{2},\quad \lambda_2=\frac{-\gamma-\gamma_1}{2},\quad \gamma_1=\sqrt{\gamma^2-4k^2}.$$

The corresponding momenta are

$$\pi_\mu{}^T(k,t)=\int_0^t dt'\dot{\Delta}^T(k,t-t')\eta_\mu{}^T(k,t'), \quad (36)$$

$$\pi_\mu{}^L(k,t)=\pi_\mu{}^L(k,0)e^{-\gamma t}$$
$$+\int_0^t dt'e^{-\gamma(t-t')}\eta_\mu{}^L(k,t'). \quad (37)$$

The transverse solutions correspond to the initial conditions $A_\mu{}^T(k,0)=\dot{A}_\mu{}^T(k,0)=0$, but in the same way as the scalar case the equilibrium values are independent of the initial conditions. On the other hand, the longitudinal part $A_\mu{}^L(k,t)$ depends on the initial conditions, just as the Parisi-Wu case.

The two-point correlation functions for the transverse parts are then given by

$$\langle A_\mu{}^T(k,t)A_\nu{}^T(k',t)\rangle$$
$$=\delta^4(k+k')\left(\delta_{\mu\nu}-\frac{k_\mu k_\nu}{k^2}\right)D^T(k,t), \quad (38)$$

$$D^T(k,t)$$
$$=\beta^{-1}\frac{1}{k^2}\left\{1-e^{-\gamma t}\left(\frac{2\gamma^2}{\gamma_1{}^2}\sinh^2\frac{\gamma_1 t}{2}+\frac{\gamma}{\gamma_1}\sinh\gamma_1 t+1\right)\right\}$$
$$\xrightarrow{t\to\infty}\beta^{-1}\frac{1}{k^2},$$

$$\langle\pi_\mu{}^T(k,t)\pi_\nu{}^T(k',t)\rangle$$
$$=\delta^4(k+k')\left(\delta_{\mu\nu}-\frac{k_\mu k_\nu}{k^2}\right)G^T(k,t),$$

$$G^T(k,t)$$
$$=\beta^{-1}\left\{1-e^{-\gamma t}\left(\frac{2\gamma^2}{\gamma_1{}^2}\sinh^2\frac{\gamma_1 t}{2}-\frac{\gamma}{\gamma_1}\sinh\gamma_1 t+1\right)\right\}$$
$$\xrightarrow{t\to\infty}\beta^{-1}. \quad (39)$$

The two-point correlation functions for the longitudinal parts are

$$\langle A_\mu{}^L(k,t)A_\nu{}^L(k',t)\rangle$$
$$=\delta^4(k+k')\frac{k_\mu k_\nu}{k^2}D^L(k,t)$$
$$+\langle A_\mu{}^{L(0)}(k,t)A_\nu{}^{L(0)}(k',t)\rangle,$$

$$D^L(k,t)$$
$$=\beta^{-1}\frac{2}{\gamma}\left\{t-\frac{2}{\gamma}(1-e^{-\gamma t})+\frac{1}{2\gamma}(1-e^{-2\gamma t})\right\}$$

$$\xrightarrow{t\to\infty}\beta^{-1}\frac{2}{\gamma}t\,, \tag{40}$$

$$\langle\pi_\mu{}^L(k,t)\pi_\nu{}^L(k',t)\rangle$$
$$=\delta^4(k+k')\frac{k_\mu k_\nu}{k^2}G^L(k,t)$$
$$+\langle\pi_\mu{}^L(k,0)\pi_\nu{}^L(k',0)\rangle e^{-2\gamma t}\,,$$

$$G^L(k,t)=\beta^{-1}(1-e^{-2\gamma t})\xrightarrow{t\to\infty}\beta^{-1} \tag{41}$$

apart from terms depending on initial values. In the same way as Parisi and Wu, Eq. (38) tends to just the usual Feynman propagator in the Landau gauge, while the longitudinal part (40) is divergent as the time. The term depending on initial values $\langle A_\mu{}^{L(0)}(k,t)A_\nu{}^{L(0)}(k',t)\rangle$ in Eq. (40) yields the gauge parameter α, i.e.,

$$\lim_{t\to\infty}\langle A_\mu{}^{L(0)}(k,t)A_\nu{}^{L(0)}(k',t)\rangle$$
$$=\delta^4(k+k')\frac{k_\mu k_\nu}{k^2}\alpha\,, \tag{42}$$

if we follow the procedure of functional averages over $A_\mu{}^L(k,0)$ and $\pi_\mu{}^L(k,0)$ proposed by Namiki et al.[6] One can see that the divergent part (40) is cancelled out in gauge invariant quantities such as $\lim_{t\to\infty}\langle F_{\mu\nu}(k,t)F_{\lambda\rho}(k',t)\rangle$.

In conclusion for the $U(1)$ gauge field we obtain the same result as in the Parisi-Wu case. We are then led to a natural conjecture that non-Abelian gauge fields can also be quantized within our framework, including the Faddeev-Popov ghost effects but not introducing explicitly such ghost fields, just as shown by Namiki et al. in the conventional stochastic quantization.[6] This and also a role of the momentum will be considered in a separated paper.

To sum up, the Gibbs (canonical) average method is obtained as a consequence of our canonical stochastic quantization. It should be noted that the reason of approaching to equilibrium of solutions of our Langevin equation is essentially due to the friction term $\gamma\delta H/\delta\pi$, whereas in the Parisi-Wu case it is due to $\gamma\delta S/\delta\phi$. The equilibrium state, however, does not depend on this parameter γ, and is controlled by the Gibbs distribution.

---

1) G. Parisi and Wu Yongshi, Sci. Sin. **24** (1981), 483.
2) N. Saito and M. Namiki, Prog. Theor. Phys. **16** (1956), 71.
D. Kershaw, Phys. Rev. **136B** (1964), 1850.
E. Nelson, Phys. Rev. **150** (1966), 1079.
A. Kracklauer, Phys. Rev. **D10** (1974), 1358.
K. Yasue, J. Math. Phys. **19** (1978), 1892.
A. Aharony, Y. Imry and S. K. Ma, Phys. Rev. Lett. **37** (1976), 1364.
A. Young, J. of Phys. **C10** (1977), 1257.
3) J. D. Breit, S. Gupta and A. Zaks, Nucl. Phys. **B233** (1984), 61 and references cited therein.
See also M. Namiki, I. Ohba, K. Okano, Y. Yamanaka, Prog. Theor. Phys. **69** (1983), 1580.
T. Fukai, H. Nakazato, I. Ohba, K. Okano and Y. Yamanaka, Prog. Theor. Phys. **69** (1983), 1600.
M. Namiki and Y. Yamanaka, Prog. Theor. Phys. **69** (1983), 1764.
H. Nakazato, M. Namiki, I. Ohba and K. Okano, Prog. Theor. Phys. **70** (1983), 298.
J. Sakamoto, Prog. Theor. Phys. **71** (1984), 881; Shimane University Preprint (1983).
Y. Nakano, Prog. Theor. Phys. **69** (1983), 361.
4) H. Hüffel and H. Rumpf, Phys. Lett. **148B** (1984), 104.
5) V. de Alfaro, S. Fubini and G. Furlan, Phys. Lett. **105B** (1981), 462.
6) M. Namiki, I. Ohba, K. Okano, Y. Yamanaka, Prog. Theor. Phys. **69** (1983), 1580.

Volume 156B, number 1,2 PHYSICS LETTERS 13 June 1985

STOCHASTIC QUANTIZATION IN PHASE SPACE

A.M. HOROWITZ

Department of Physics, University of Cincinnati, Cincinnati, OH 45221, USA

Received 14 January 1985

Hamiltonian stochastic equations are proposed for the quantization of field theories. Comparisons are made with the Langevin equation. Possible advantages for numerical simulations are discussed based on the observation that the hamiltonian approach interpolates between the Langevin and molecular dynamics approaches.

In recent years we have seen the development of formulations of quantum field theory based on differential equations with random sources [1,2]. The equations considered thus far are in the case of a scalar field the following:

$$\delta S_D/\delta\phi(x) = \eta(x)\,, \tag{1}$$

$$\gamma\partial\phi(x,t)/\partial t = -\delta S_D/\delta\phi(x,t) + \eta(x,t)\,, \tag{2}$$

where S_D is the euclidean action in D dimensions, η is a random field with gaussian measure, and γ is a constant. Now with regard to (1) it has been shown [1] that the probability distribution of ϕ on a $(D-2)$-dimensional euclidean subspace infinitely far from the boundaries is just $\exp(-S_{D-2}[\phi])$. Similarly, for (2) (the Langevin equation), the distribution on a plane given by t = constant relaxes to $\exp(-S_D[\phi])$ as this plane becomes infinitely far from that containing the boundary values.

In the language of partial differential equations, (1) and (2) are elliptic and parabolic equations respectively [*1]. In this paper we show that hyperbolic stochastic equations can also be used as a basis for quantum field theory. We will first develop the formalism and then discuss why such equations are of additional interest.

We begin with the simplest hyperbolic stochastic equation that we can write for a scalar field, which is

$$\partial^2\phi(x,t)/\partial t^2 + \gamma\partial\phi(x,t)/\partial t = -\delta S_D/\delta\phi(x,t) + \eta(x,t)$$
$$= \Box\phi(x,t) + V'(\phi) + \eta(x,t)\,. \tag{3}$$

The dissipative term proportional to γ is necessary in accordance with the fluctuation–dissipation theorem: an equilibrium distribution results from a balance between fluctuations on the one hand which tend to move ϕ away from zero, and dissipation on the other which tends to move it towards zero. As an aside, we note that the damping in (1) comes about for geometrical reasons. Like the Langevin equation (2), stochastic equations of the type (3) are not new in physics. They have been studied in connection with random systems having finite numbers of degrees of freedom, e.g. particles undergoing brownian motion or quantum mechanical dynamics. (See refs. [3] and [4] and references therein.)

Eq. (3) can be generalized in the following way. Consider an arbitrary hamiltonian (which we write for a scalar field but the formalism is general)

$$H[\phi,\pi] = \int \mathrm{d}^D x\, \mathcal{H}(\phi, \partial_\mu\phi, \pi)\,. \tag{4}$$

Then the stochastic equations for $\phi(x,t)$ and $\pi(x,t)$ are

$$\dot\pi = -\delta H/\delta\phi - \gamma\dot\phi + \eta\,, \quad \dot\phi = \delta H/\delta\pi\,. \tag{5a, b}$$

The statistical evolution of π and ϕ is a Markov process: the probability $W[\phi_2\pi_2; t_2|\phi_1\pi_1; t_1]$ of

[*1] Because ϕ might not be differentiable these equations must be interpreted carefully. See ref. [3] for the more rigorous notation. We will be careful when ambiguity threatens.

Volume 156B, number 1,2 PHYSICS LETTERS 13 June 1985

finding the configuration $\{\phi_2(x), \pi_2(x)\}$ at time t_2 depends only on the boundary values $\{\phi_1(x), \pi_1(x)\}$ at time t_1. It follows that W has the convolution property

$$W[\phi_2\pi_2; t_2|\phi_1\pi_1; t_1]$$

$$= \int D\phi\, D\pi\, W[\phi_2\pi_2; t_2|\phi\pi; t]\, W[\phi\pi; t|\phi_1\pi_1; t_1] \qquad (6)$$

for any t between t_1 and t_2. Also it follows from (5) and the probability distribution for $\xi_i(x) \equiv \int_{t_i}^{t_i+\Delta t} \eta(x, t)\, dt$

$$P[\xi_i] \propto \exp\left(-\frac{1}{4\kappa}\int d^Dx\, \xi_i^2(x)\right)$$

that the infinitesimal transition probability is

$$W[\phi_{i+1}\pi_{i+1}; t_i + \Delta t|\phi_i\pi_i; t_i]$$
$$\propto \delta[\phi_{i+1} - \phi_i - (\delta H/\delta\pi_i)\Delta t]$$
$$\times \exp\left(-\frac{1}{4\kappa}\int d^Dx[\pi_{i+1} - \pi_i + (\gamma\delta H/\delta\pi_i + \delta H/\delta\phi_i)\Delta t]^2\right). \qquad (7)$$

(With the operator ordering chosen in (7), i.e. with the time slice chosen on which to evaluate $\delta H/\delta\phi$ and $\delta H/\delta\pi$, the jacobian required in (7) between $\{\Delta\pi + (\gamma\delta H/\delta\pi + \delta H/\delta\phi)\Delta t, \Delta\phi - (\delta H/\delta\pi)\Delta t\}$ and $\{\phi, \pi\}$ is unity). Using (6) and (7) along with standard manipulations one finds the following equation for W [‡2]:

$$\frac{\partial W}{\partial t} + \int d^Dx\left[\frac{\delta W}{\delta\phi}\frac{\delta H}{\delta\pi} - \frac{\delta W}{\delta\pi}\frac{\delta H}{\delta\phi}\right]$$

$$= \int d^Dx\left[\kappa\frac{\delta^2 W}{\delta\pi^2} + \gamma\frac{\delta}{\delta\pi}\left(\frac{W\delta H}{\delta\pi}\right)\right]. \qquad (8)$$

This is recognized as a generalization of Liouville's theorem. The left-hand side is the total time derivative of W while the first and second terms on the right-hand side come from the fluctuations and dissipation respectively. One also notes from (8) that $\int D\phi\, D\pi\, W$ is constant, as required for the interpretation of W as a probability.

As is proved below the equilibrium solution to (8), that is the solution in the limit $t \to \infty$, is

$$W = c\exp[-(\gamma/\kappa)H], \quad c = \text{constant}. \qquad (9)$$

This annihilates the right- and left-hand sides separately. Thus the fluctuation and dissipation terms on the RHS are balanced identically. That (9) is in fact the equilibrium solution can be seen in the following way. Let us define the operator L by rewriting (8) as

$$\partial W/\partial t = LW. \qquad (10)$$

Let $u_n[\phi, \pi]$ and λ_n denote the nth eigenfunctional and eigenvalue of L subject to the boundary condition $u_n = 0$ for ϕ or $\pi \to \pm\infty$. It is not hard to show that the eigenvalues of the adjoint $L^\dagger$ are the same λ_n, however the eigenfunctionals, $v_n[\phi, \pi]$, are different. The u_n and v_n form a biorthogonal set of functionals, i.e., their scalar product is $(u_n, v_m) = \delta_{nm}$, and they are complete: any functional $F[\phi, \pi]$ can be expanded as $\Sigma_n(v_n, F)u_n$. Hence the formal solution to (10) is

$$W(t) = \sum_n \exp(-\lambda_n t)(v_n, W(0))u_n. \qquad (11)$$

Now the point is that $u_0 = \exp[-(\gamma/\kappa)H]$ corresponding to eigenvalue $\lambda_0 = 0$ is also the ground state of L. We expect this to be true since u_0 has no zeros. However, since the λ_n may be complex, th reader might question the meaning of the statement that λ_0 is the lowest eigenvalue. Hence I shall give an independent argument that there are no eigenvalues with negative real part. The argument is based on the fact that $\int D\phi\, D\pi\, u_n = 0$ for $n \neq 0$, which can be seen when L is expressed in the form $(\delta/\delta\phi)L_1 + (\delta/\delta\pi)L_2$. So if there was an eigenvalue, λ_{-1}, with negative real part, we see from (11) that W eventually ceases to be a positive functional. That W defined by (8) is indeed positive follows from the equivalence of (8) with (6) and (7), the latter implying that W can be expressed as a path integral with positive measure.

We now have, in view of the above, that

$$W[\phi, \pi; t] = c\exp\{-(\gamma/\kappa)H[\phi, \pi]\} + O(\exp(-\lambda_1 t)), \qquad (12)$$

which relaxes to the desired equilibrium distribution provided Re $\lambda_1 > 0$. For a free scalar field with

$$H_0 = \frac{1}{2}\int d^Dx[\pi^2 + (\partial_\mu\phi)^2 + m^2\phi^2]$$

[‡2] See ref. [4] or any textbook derivation of the Fokker–Planck equation.

the λ_n can be found exactly since we are just dealing here with a set of uncoupled harmonic oscillators with frequencies $\omega_k = (k^2 + m^2)^{1/2}$. Here (12) takes the form

$$W = c\exp[-(\gamma/\kappa)H] + \int d^Dk\, C_k[\phi,\pi]\exp(-\lambda_k t) + \dots,$$

where

$$\lambda_k = \tfrac{1}{2}\gamma - [(\tfrac{1}{2}\gamma)^2 - \omega_k^2]^{1/2}. \quad (13)$$

The minimum of these, $\frac{1}{2}\gamma - [(\frac{1}{2}\gamma)^2 - m^2]^{1/2}$, is greater than zero as required (provided $m > 0$). Now for an interacting scalar field, say with a ϕ^4 interaction, $H > H_0$ and so one expects the eigenvalues to increase. Thus the approach to equilibrium should be faster.

It is interesting to compare the rate of approach to equilibrium of the free scalar field in the hyperbolic formalism with that in the parabolic formalism. For the latter, the eigenvalues corresponding to (13) are $\lambda_k^L = \omega_k^2/\gamma$. So for modes with $\omega_k < \gamma/\sqrt{2}$, the hyperbolic equation approaches equilibrium faster while for those with $\omega_k > \gamma/\sqrt{2}$ it is slower.

Like the Langevin equation, the hyperbolic equation can also be used to generate the usual perturbative expansion in powers of the coupling constant. I have checked this for the two-point function in the scalar $g\phi^3$ model to $O(g^2)$. The calculation differs only in details from that found in ref. [2] using the Langevin equation, although the time integrals are more tedious in the hyperbolic case. Presumably one can prove equivalence with the standard expansion to all orders as has been done in ref. [5] for the Langevin equation. In the case of gauge theories, the hyperbolic formalism shares with the parabolic one the nice feature that there is no need for gauge fixing [2].

What has been shown so far is that solutions to hamiltonian stochastic field equations in the large-t limit comprise a Gibbs ensemble. Consequently we have that

$$\lim_{t\to\infty}\left[\int D\eta\, \phi_\eta(x_1,t)\dots\phi_\eta(x_m,t) \times \exp\left(-\frac{1}{4\kappa}\int d^Dx\, dt\, \eta^2(x,t)\right)\right] \times \left[\int D\eta \exp\left(-\frac{1}{4\kappa}\int d^Dx\, dt\, \eta^2(x,t)\right)\right]^{-1}$$
$$= \left(\int D\phi\, D\pi\, \phi(x_1)\dots\phi(x_m)\exp[-(\gamma/\kappa)H]\right) \times \left(\int D\phi\, D\pi \exp[-(\gamma/\kappa)H]\right)^{-1}, \quad (14)$$

where $\phi_\eta(x,t)$ is a solution of the stochastic Hamilton equations.

The advantages for field theory of the Gibbs functional integral over the usual one have been discussed fully in the paper by de Alfaro et al. [6]. The main point is that the presence of the canonical momenta insures invariance of the measure under field transformations. In practice this means that the integration over π produces the functional determinants in the usual functional integral. For example, if we are interested in a chiral model in D dimensions, we write down a chiral invariant action in $(D+1)$ dimensions

$$S = \frac{1}{2}\int d^Dx\, dt\, G^{ab}(\phi) \times [(\partial\phi^a/\partial t)\partial\phi^b/\partial t - \partial_\mu\phi^a\partial_\mu\phi^b],$$

from which follows the hamiltonian

$$H = \frac{1}{2}\int d^Dx(\pi_a G^{-1}_{ab}\pi_b + G^{ab}\partial_\mu\phi^a\partial_\mu\phi^b),$$

and the functional integration over π produces the requisite $\det^{1/2}G$ in the measure for ϕ. Now a nice thing about the stochastic equations (5) derived from this hamiltonian is that they can be written solely in terms of G and $\partial G/\partial\phi$. There is no appearance of G^{-1} or $\det G$ as there is in the Langevin equation for this model. In numerical simulations this is a significant advantage when the dimension of G is large.

We now discuss gauge theories beginning with the following trivial observation: in any method of calculation, if one uses the gauge fields A_μ themselves then there are no ghosts. Of course this is impossible in the usual functional integral approach: one must choose a gauge (ghost-free or not), which essentially amounts to a change of variables. In contrast, in the Langevin or stochastic hamiltonian approach it is possible to use the A_μ themselves. The work of de Alfaro et al. implies that if one does change variables in the stochastic hamiltonian approach then the effect of the Faddeev–Popov determinant is correctly accounted for through the canonical momenta. A change of variables in the Langevin equation would require the ad hoc addition of a term coming from the determinant.

We now discuss why the hyperbolic stochastic equations might be advantageous for numerical simulations of lattice field theories. The main point is that when γ and κ are taken to zero we are still left with equations of motion whose solutions exhibit thermodynamic behavior according to the ergodic hypothesis. (We assume of course that the equations of motion are sufficiently nonlinear). These are just the equations of the microcanonical ensemble which have recently been put to good use in numerical simulations of lattice gauge theories [7]. On the other hand, taking γ and κ to be small in the Langevin equation is equivalent to taking the discretized time intervals to be large which leads to large systematic errors proportional to $1/\gamma$ [8]. For large γ and κ the hyperbolic equation is equivalent to the Langevin equation with the same parameters. This is known to converge slowly: the computer time is proportional to γ [8]. The pure microcanonical approach is not without its problems either. For example, there is a tendency for solutions to get caught in metastable states. All things considered, perhaps the hyperbolic equation with small γ and κ is the optimal compromise between the Langevin and microcanonical approaches.

I thank Peter Suranyi and Don Sinclair for helpful discussions. This work was supported by the US Department of Energy under Contract No. DE-AC02-76ER02978.

References

[1] G. Parisi and N. Sourlas, Phys. Rev. Lett. 43 (1979) 744.
[2] G. Parisi and Wu Yongshi, Sci. Sin. 24 (1981) 483.
[3] C. Dewitt-Morette and K.D. Elworthy, ed., Phys. Rep. 77 (1981) 121.
[4] S. Chandrasekhar, Rev. Mod. Phys. 15 (1943) 1.
[5] W. Grimus and H. Huffel, Z. Phys. C18 (1983) 129.
[6] V. de Alfaro, S. Fubini and G. Furlan, Phys. Lett. 105B (1981) 462.
[7] D. Callaway and A. Rahman, Phys. Rev. Lett. 49 (1982) 613;
J. Polonyi and H.W. Wyld, Phys. Rev. Lett. 51 (1983) 2257.
[8] G. Parisi, Nucl. Phys. B180 (1981) 378.

<u>Note added</u>. I give here some more concrete remarks concerning the application of hyperbolic stochastic equations to numerical simulations and its relation to the Langevin and microcanonical equations. In a simulation one must of course use a discrete fictitious time parameter. The simplest discretization of the hyperbolic equation, (3), for the scalar field, can be read off from the transfer matrix (7). Now we are interested in corrections to the equilibrium distribution (9) when Δt is small but nonzero. One finds through $O(\Delta t)$ that the effective distribution is

$$P_{eq}\{\phi\ \pi\} = \exp\{-\frac{\beta}{2}(1-\gamma\Delta t/2)\sum_x \pi_x^2 - \beta S + \frac{\beta}{2}\Delta t \sum_x \pi_x \frac{\delta S}{\delta \phi_x}\} \quad .$$

It is easy to verify by direct Gaussian integration that P_{eq} is in fact an eigenfunction of the transfer matrix (7) to this order. We see now that correlation functions of ϕ or better yet of $\phi - \Delta t \pi/2$ will have systematic errors of $O(\Delta t^2)$. This result is to be compared with that from the Euler discretization of the Langevin equation. Here systematic errors are $O(\Delta t)$, (see e.g. Ukawa, A. and Fukugita, M., Phys. Rev. Lett. 55(1985)1854). Consequently the hyperbolic equation should be much more efficient for numerical simulations than the Langevin equation as was observed in simulations of the XY model near its critical point. The optimal value of γ on a 15^2 lattice was found to be .4, independent of β (at least near β_c). More details can be found in a separate publication.

Now some remarks concerning the relation of the hyperbolic equation to the equations of the microcanonical ensemble (ME) obtained from it by putting $\gamma=0$. An important point is that infinitesimal γ ($\gamma \approx 0$) is not just a small perturbation of the ME: the system is still in the canonical ensemble, but only weakly coupled to the heat bath. Consider as an extreme example, the fluctuations in the total energy. In the ME this is of course zero. But in the stochastic system it must be given by its canonical value independent of γ. When $\gamma \approx 0$ the total energy is approximately conserved over short time scales which implies that there must be long drifts and thus long time correlations if the correct canonical value is to be produced in the average over infinite time. The same is true for any quantity having different fluctuations in the two ensembles, e.g. the potential energy (field theory action) and correlation functions of larger separation.

Stochastic Regularization

Nuclear Physics B233 (1984) 61–87

STOCHASTIC QUANTIZATION AND REGULARIZATION

J.D. BREIT, S. GUPTA AND A. ZAKS

The Institute for Advanced Study, Princeton, New Jersey 08540, USA

Received 21 March 1983
(Revised 24 August 1983)

In this paper the method of stochastic quantization introduced by Parisi and Wu is extended to field theories that include fermions and are supersymmetric. A new non-perturbative regulator based on stochastic quantization is introduced. This regulator preserves all the symmetries of the lagrangian, including gauge, chiral, and supersymmetries, at the expense of introducing non-locality.

1. Introduction

Quantum field theory can be formulated in two different ways: the canonical and path integral formalisms. Each of these formulations has its own merits and drawbacks, and use is made of each according to the problem under consideration. Even though the two formulations are equivalent, some problems may appear simpler and their resolutions more transparent if presented within the framework of one or the other. Recently Parisi and Wu [1] introduced a new quantization procedure: stochastic quantization. In their work Parisi and Wu demonstrated that the stochastic formulation can be used to quantize various systems and pointed out some of the advantages of this new approach. They showed that the use of stochastic quantization obviates the need for a gauge-fixing term in perturbation theory and discussed the equivalence of the stochastic formulation with monte carlo simulations.

In this paper an attempt is made to investigate further the properties of this new formalism and to apply it to some problems in quantum field theory. The most intriguing possibility that the stochastic formulation offers is a new, presumably non-perturbative, regularization scheme.

To make this paper as self-contained as possible we dedicate sect. 2 to a brief review of the stochastic quantization procedure and some of its properties. Sect. 3 deals with stochastic quantization of free and interacting Fermi fields and sect. 4 with the formulation of supersymmetric quantum field theories within the framework of stochastic quantization. In sects. 5 and 6 we define and investigate a regulator defined in terms of the stochastic formalism, and in sect. 7 we discuss the role of classical solutions and initial conditions. Sect. 8 contains our conclusions.

2. Stochastic quantization

Stochastic quantization is formulated in terms of the Langevin equation

$$\frac{\partial}{\partial t}\varphi_i(x,t) = -\frac{\delta}{\delta\varphi_i(x,t)}S[\varphi] + \eta_i(x,t), \tag{2.1}$$

in which t is a fictitious time variable (the Langevin time) and η is a random variable with a gaussian distribution:

$$\langle\eta_i(x,t)\eta_j(x',t')\rangle = \delta_{ij}\delta^d(x-x')\delta(t-t'). \tag{2.2}$$

The connection with quantum field theory is made by considering the equal Langevin time averages of the solutions of eq. (2.1):

$$\left\langle F\left[\varphi_\eta(t)\right]\right\rangle_\eta \equiv \int \prod_{i,t,x} \mathrm{d}\eta_i(x,t) F\left[\varphi_\eta(t)\right] \mathrm{e}^{-(1/2)\int_0^\infty \eta_i(x,t)\eta_i(x,t)\,\mathrm{d}t\,\mathrm{d}x}. \tag{2.3}$$

These averages can be calculated directly in terms of the distribution of the φ_i, at the time t induced by the markovian process defined in eqs. (2.1) and (2.2):

$$\left\langle F\left[\varphi_\eta(t)\right]\right\rangle = \int \prod_{i,x} \mathrm{d}\varphi_i(x) F\left[\varphi_i(x)\right] P\left[\varphi_i(x);t\right]. \tag{2.4}$$

The probability distribution $P[\varphi;t]$ is defined by the Fokker-Planck equation

$$\frac{\partial P(\varphi,t)}{\partial t} = \sum_i \int \mathrm{d}^d x \left[\frac{\delta^2}{\delta\varphi_i(x)^2} + \frac{\delta}{\delta\varphi_i(x)}\left(\frac{\delta S}{\delta\varphi_i(x)}\right)\right] P(\varphi,t), \tag{2.5}$$

and by the initial condition $P[\varphi;0]$. If we define $P[\varphi;t] \equiv \exp[-\tfrac{1}{2}S(\varphi)]\Psi[\varphi;t]$, we find the equation

$$-\frac{\partial}{\partial t}\Psi = H_{\mathrm{FP}}(S)\Psi, \tag{2.6}$$

where $H_{\mathrm{FP}} = \Sigma_i \int \mathrm{d}^d x\, a_i^\dagger(x) a_i(x)$ and $a_i(x)$ is given by

$$a_i(x) = -i\left(\frac{\delta}{\delta\varphi_i(x)} + \frac{1}{2}\frac{\delta S}{\delta\varphi_i(x)}\right). \tag{2.7}$$

Since the Fokker-Planck hamiltonian H_{FP} is formally self-adjoint, one can expand the solution of eq. (2.6) in the complete set of its eigenstates $H_{FP}\Psi_n = \lambda_n \Psi_n$:

$$\Psi[\varphi; t] = \sum_{n=0}^{\infty} a_n e^{-\lambda_n t} \psi_n[\varphi]. \tag{2.8}$$

The positivity of H_{FP} ensures that $\lambda_n \geqslant 0$. It is easy to see that the functional

$$\Psi_0[\varphi] = e^{-(1/2)S[\varphi]} \tag{2.9}$$

is an eigenstate of H_{FP} with $\lambda_n = 0$. Using eqs. (2.4) and (2.9) and assuming that $\lambda_1 > \lambda_0$ (the existence of a "mass gap"), we find

$$\lim_{t\to\infty} \left\langle F\left[\varphi_\eta(t)\right]\right\rangle = \lim_{t\to\infty} \int \prod \mathrm{d}\varphi\, F[\varphi] P[\varphi; t]$$

$$= \int \prod \mathrm{d}\varphi\, F[\varphi] e^{-S[\varphi]}, \tag{2.10}$$

which establishes the relation of the stochastic averages (2.3) to the euclidean field theory.

The validity of eq. (2.10) depends, then, on the positivity of the Fokker-Planck hamiltonian and on the existence of a mass gap in its spectrum. An additional requirement is that the initial condition on $\psi[\varphi; t]$ not be orthogonal to the ground state wave functional of H_{FP}.

There are cases where the use of the Langevin equation (2.1) is impossible because the Fokker-Planck hamiltonian is not positive. For such cases we use a generalized version of eqs. (2.1) and (2.2). The generalized stochastic quantization procedure is based upon the Langevin equation

$$\frac{\partial \varphi_i(x,t)}{\partial t} = -\int K_{ij}(x,y) \frac{\delta S}{\delta \varphi_j(y,t)} \mathrm{d}^d y + \eta_i(x,t), \tag{2.11}$$

and the noise correlation

$$\langle \eta_i(x,t) \eta_j(y,t') \rangle = 2K_{ij}(x,y)\delta(t-t'), \tag{2.12}$$

where K is a kernel independent of the fields φ.

Using the same methods as in ref. [2], one finds that the probability distribution induced by eqs. (2.11) and (2.12) satisfies the Fokker-Planck equation (2.6) with the modified hamiltonian

$$\hat{H}_{FP}(S) = \sum_{ij} \int \mathrm{d}^d x\, \mathrm{d}^d y\, a_i^\dagger(x) K_{ij}(x,y) a_j(y), \tag{2.13}$$

where a and $a^\dagger$ are the same as in eq. (2.7). If $\hat{H}$ is positive and if K is invertible we find again that the lowest eigenvalue $\lambda_0 = 0$ and its eigenfunction is given by

$$a\Psi_0[\varphi] = 0 \Rightarrow \Psi_0[\varphi] = e^{-(1/2)S[\varphi]}. \tag{2.14}$$

So the equilibrium ($t \to \infty$) distribution of the system defined by eqs. (2.11) and (2.12) is again that of the euclidean field theory associated with S. The question of what K to use is a matter of convenience as long as H_{FP} is positive. If H_{FP} is not positive, however, one is forced to search for a K that ensures that $\hat{H}_{FP}$ is positive.

In practical applications of stochastic quantization one cannot solve equation (2.1) exactly, and a perturbation expansion must be used. It is easy to verify, as was done in ref. [3], that the perturbative solution of the stochastic equations in the limit $t \to \infty$ is equivalent term by term to the conventional expansion. This equivalence holds if and only if the bilinear part of the action is positive.

3. Stochastic quantization of Fermi fields

In order to be able to use the stochastic quantization method for realistic models it is necessary to learn how to apply it to Dirac fermions. Let us first try the straightforward approach to the case of a free Dirac fermion:

$$\frac{\partial}{\partial t}\psi(x,t) = -\frac{\partial S}{\partial \bar{\psi}(x,t)} + \eta(x,t), \qquad \frac{\partial}{\partial t}\bar{\psi}(x,t) = \frac{\partial S}{\partial \psi(x,t)} + \bar{\eta}(x,t), \tag{3.1}$$

$$\langle \eta_\alpha(x,t)\bar{\eta}_\beta(x',t')\rangle = 2\delta_{\alpha\beta}\delta^d(x-x')\delta(t-t'), \qquad \langle\eta\eta\rangle = \langle\bar{\eta}\bar{\eta}\rangle = 0, \tag{3.2}$$

where

$$S = \int d^dx\, \bar{\psi}(x,t)(i\gamma\cdot\partial - im)\psi(x,t).$$

Unfortunately eqs. (3.1) and (3.2) cannot produce the desired result. Consider the solution to eq. (3.1) in momentum space

$$\psi(p,t) = \int_0^t d\tau\, e^{-(t-\tau)(\gamma p - im)}\eta(p,\tau), \tag{3.3}$$

$$\langle\psi(p,t)\bar{\psi}(p;t')\rangle_\eta = e^{-(t+t')(\gamma\cdot p - im)}\frac{e^{2t'(\gamma\cdot p - im)} - 1}{\gamma\cdot p - im}. \tag{3.4}$$

In the limit $t = t' \to \infty$ we do not recover the desired limiting form $(\gamma\cdot p - im)^{-1}$ because the Dirac equation has negative eigenvalues and hence the exponentials are not damped. Fukai et al. [8] have considered this problem and proposed solving it by taking $\bar{\psi} \to i\bar{\psi}$ in the action (3.2). Then in eq. (3.4) we would have exponentials of the form $e^{-(t+t')(m+i\gamma\cdot p)}$ and the expression is damped. This procedure is useless, however, for theories in which fermions and bosons interact. In these theories one immediately discovers an inconsistency in the Langevin equations. In the fermion equations t has the dimensions of length while in the boson equations it has the dimensions of length^2.

To solve these problems we use eqs. (2.11) and (2.12) instead of (2.1) and (2.2), taking the kernel K to be $i\gamma\cdot\partial + im$ for the fermionic degrees of freedom and 1 for the bosonic degrees of freedom. To demonstrate how this formalism works we outline the calculation of the vacuum polarization in massless QED to one loop order.

The Langevin equations are:

$$\dot{\psi} = \partial^2\psi - ie\gamma\cdot\partial\gamma\cdot A\psi + \eta_\psi, \tag{3.5}$$

$$\dot{\bar{\psi}} = \partial^2\bar{\psi} - ie\gamma\cdot\partial(\bar{\psi}\gamma\cdot A)^{\mathrm{T}} + \bar{\eta}_\psi, \tag{3.6}$$

$$\dot{A}_\mu = \partial^2 A_\mu - \partial_\mu\partial\cdot A + e\bar{\psi}\gamma_\mu\psi + \eta_{A\mu}, \tag{3.7}$$

where

$$\langle\eta_\psi(x,t)\bar{\eta}_\psi(x',t')\rangle = 2i\gamma\cdot\partial\delta^d(x-x')\delta(t-t'), \tag{3.8}$$

$$\langle\eta_{A\mu}(x,t)\eta_{A\nu}(x',t')\rangle = 2\delta_{\mu\nu}\delta^d(x-x')\delta(t-t'). \tag{3.9}$$

Eqs. (3.5)–(3.7) can be easily solved and the calculation of the correlation $\langle A_\mu A_\nu\rangle$ is straightforward but tedious; the details are given in the appendix. In the limit that $t = t' \to \infty$ we recover the usual result for the one-loop vacuum polarization.

4. Stochastic supersymmetry

In this section we derive the Langevin equations for supersymmetric vector and chiral fields and show that the superfield propagators reduce to their usual forms in the limit that the Langevin time $t \to \infty$. We also show, by examining the Langevin equations for the component fields, that the method given in sect. 3 for handling Fermi fields is consistent with the supersymmetric formalism. Throughout this section we adopt the notation of ref. [4]; in particular, we work in Minkowski space. Since for large Langevin time t our expressions are properly damped only in euclidean space, the limit $t \to \infty$ is taken by first continuing to euclidean space, then letting $t \to \infty$, and then continuing back to Minkowski space.

4.1. FREE VECTOR SUPERFIELDS

We start with the action

$$S=\tfrac{1}{2}\int \mathrm{d}^4x\,\mathrm{d}^2\vartheta\,\mathrm{d}^2\bar{\vartheta}\,V\left[-\partial^2P_{\mathrm{T}}+m^2\right]V. \tag{4.1}$$

where

$$P_{\mathrm{T}}=\frac{-1}{8\partial^2}D\bar{D}^2D,$$

$$D_\alpha=\frac{\partial}{\partial\vartheta^\alpha}+i\sigma^m_{\alpha\dot{\alpha}}\bar{\vartheta}^{\dot{\alpha}}\partial_m,$$

$$\bar{D}_{\dot{\alpha}}=\frac{-\partial}{\partial\bar{\vartheta}^{\dot{\alpha}}}-i\vartheta^\alpha\sigma^m_{\alpha\dot{\alpha}}\partial_m,$$

$$D^2=D^\alpha D_\alpha,\qquad \bar{D}^2=\bar{D}_{\dot{\alpha}}\bar{D}^{\dot{\alpha}},$$

$$\partial^2=-\frac{\partial^2}{\partial t^2}+\nabla^2,$$

$$\begin{aligned}V=V^\dagger=&\,C+i\vartheta\chi-i\bar{\vartheta}\bar{\chi}+\tfrac{1}{2}i\vartheta^2(M+iN)-\tfrac{1}{2}i\bar{\vartheta}^2(M-iN)-\vartheta\sigma\bar{\vartheta}\cdot v\\&+i\vartheta^2\bar{\vartheta}\left(\bar{\lambda}+\tfrac{1}{2}i\bar{\sigma}\cdot\partial\chi\right)-i\bar{\vartheta}^2\vartheta\left(\lambda+\tfrac{1}{2}i\sigma\cdot\partial\bar{\chi}\right)+\tfrac{1}{2}\vartheta^2\bar{\vartheta}^2\left(D+\tfrac{1}{2}i\partial^2C\right).\end{aligned} \tag{4.2}$$

The Langevin equation for V is

$$\dot{V}=\left(\partial^2P_{\mathrm{T}}-m^2\right)V+\eta, \tag{4.3}$$

where η is a superfield and

$$\langle\eta\eta\rangle=2\delta(x-x')\delta(\vartheta-\vartheta')\delta(\bar{\vartheta}-\bar{\vartheta}')\delta(t-t').$$

The solution to eq. (4.3) for $V(t=0)=0$ is

$$\begin{aligned}V(t)&=\int_0^t\mathrm{d}\tau\,\mathrm{e}^{(\partial^2P_{\mathrm{T}}-m^2)(t-\tau)}\eta(\tau)\\&=\int_0^t\mathrm{d}\tau\left[\mathrm{e}^{(\partial^2-m^2)(t-\tau)}P_{\mathrm{T}}+\mathrm{e}^{-m^2(t-\tau)}(1-P_{\mathrm{T}})\right]\eta(\tau)\\&\equiv\int\frac{\mathrm{d}^4k}{(2\pi)^4}\int\mathrm{d}^4y\,\mathrm{e}^{ik(x-y)}\int_0^t\mathrm{d}\tau\left[\mathrm{e}^{(k^2-m^2)(t-\tau)}P_{\mathrm{T}}+\mathrm{e}^{-m^2(t-\tau)}(1-P_{\mathrm{T}})\right]\eta(y,\tau).\end{aligned} \tag{4.4}$$

So the propagator

$$\langle V(t_1)V(t_2)\rangle = \int_0^{t_1} d\tau_1 \left[P_{T1} e^{(\partial^2 - m^2)(t_1-\tau_1)} + (1 - P_{T1}) e^{-m^2(t_1-\tau_1)} \right]$$

$$\times \int_0^{t_2} d\tau_2 \left[P_{T2} e^{(\partial^2 - m^2)(t_2-\tau_2)} + (1 - P_{T2}) e^{-m^2(t_2-\tau_2)} \right]$$

$$\times \langle \eta(\tau_1, x_1, \vartheta_1, \bar{\vartheta}_1) \eta(\tau_2, x_2, \vartheta_2, \bar{\vartheta}_2) \rangle$$

$$= 2\int_0^{t_{\min}} d\tau \left[e^{(\partial^2 - m^2)(t_1-\tau)} P_{T1} + (1 - P_{T1}) e^{-m^2(t_1-\tau)} \right]$$

$$\times \left[e^{(\partial^2 - m^2)(t_2-\tau)} P_{T2} + (1 - P_{T2}) e^{-m^2(t_2-\tau)} \right]$$

$$\times \delta(x_1 - x_2)\delta(\vartheta_1 - \vartheta_2)\delta(\bar{\vartheta}_1 - \bar{\vartheta}_2), \tag{4.5}$$

where

$$P_{T1} = \frac{-1}{8\partial^2} D_1 \bar{D}_1^2 D_1,$$

$$D_{1\alpha} = \frac{\partial}{\partial \vartheta_1^\alpha} + i\sigma^m_{\alpha\dot{\alpha}} \bar{\vartheta}_1^{\dot{\alpha}} \partial^1_m .$$

Now

$$0 = P_{T1}(1 - P_{T2})(\vartheta_1 - \vartheta_2)^2 (\bar{\vartheta}_1 - \bar{\vartheta}_2)^2 \delta(x_1 - x_2)$$

$$= P_{T2}(1 - P_{T1})(\vartheta_1 - \vartheta_2)^2 (\bar{\vartheta}_1 - \bar{\vartheta}_2)^2 \delta(x_1 - x_2),$$

$$P_{T1} P_{T2} (\vartheta_1 - \vartheta_2)^2 (\bar{\vartheta}_1 - \bar{\vartheta}_2)^2 \delta(x_1 - x_2)$$

$$= \frac{-1}{2\partial^2} e^{i(\vartheta_2 \sigma^n \bar{\vartheta}_1 - \vartheta_1 \sigma^n \bar{\vartheta}_2)\partial^1_n} \left[4 - \partial^2 (\vartheta_1 - \vartheta_2)^2 (\bar{\vartheta}_1 - \bar{\vartheta}_2)^2 \right] \delta(x_1 - x_2),$$

$$(1 - P_{T1})(1 - P_{T2})(\vartheta_1 - \vartheta_2)^2 (\bar{\vartheta}_1 - \bar{\vartheta}_2)^2 \delta(x_1 - x_2)$$

$$= \frac{1}{2\partial^2} e^{i(\vartheta_2 \sigma^n \bar{\vartheta}_1 - \vartheta_1 \sigma^n \bar{\vartheta}_2)\partial^1_n} \left[4 + \partial^2 (\vartheta_1 - \vartheta_2)^2 (\bar{\vartheta}_1 - \bar{\vartheta}_2)^2 \right] \delta(x_1 - x_2).$$

In the limit $t_1 \to t_2 \to \infty$ the integrals in eq. (4.5) become $(\partial^2 - m^2)^{-1}$ and $-1/m^2$, and we obtain the usual vector superfield propagator without a gauge-fixing term.

The Langevin equations for the component fields are easily found in terms of the following expansion for η,

$$\eta(x,t,\vartheta,\bar{\vartheta}) = \eta_0 + \vartheta^\alpha\eta_{1\alpha} + \bar{\vartheta}_{\dot{\alpha}}\eta_2^{\dot{\alpha}} + \vartheta^2\eta_3 + \bar{\vartheta}^2\eta_4 + \vartheta\sigma^n\bar{\vartheta}\eta_{5n}$$
$$+ \bar{\vartheta}^2\vartheta^\alpha\eta_{6\alpha} + \vartheta^2\bar{\vartheta}_{\dot{\alpha}}\eta_7^{\dot{\alpha}} + \vartheta^2\bar{\vartheta}^2\eta_8, \tag{4.6}$$

where $\eta_i = \eta_i(x,t)$ only. Letting $m = 0$ we find the following equations for the component fields:

$$\dot{C} = -D + \eta_0, \tag{4.7}$$

$$\dot{\chi}_\alpha = -i\sigma^m_{\alpha\dot{\alpha}}\partial_m\bar{\lambda}^{\dot{\alpha}} - i\eta_{1\alpha}, \tag{4.8}$$

$$\dot{\bar{\chi}}_{\dot{\alpha}} = i\partial_m\lambda^\alpha\sigma^m_{\alpha\dot{\alpha}} + i\eta_{2\dot{\alpha}}, \tag{4.9}$$

$$\dot{M} + i\dot{N} = -2i\eta_3, \tag{4.10}$$

$$\dot{M} - i\dot{N} = 2i\eta_4, \tag{4.11}$$

$$\dot{v}_n = \partial^2 v_n - \partial_n\partial\cdot v - \eta_{5n}, \tag{4.12}$$

$$\dot{\bar{\lambda}}^{\dot{\alpha}} + \tfrac{1}{2}i\bar{\sigma}^{m\dot{\alpha}\alpha}\partial_m\dot{\chi}_\alpha = \tfrac{1}{2}\partial^2\bar{\lambda}^{\dot{\alpha}} - i\eta_7^{\dot{\alpha}}, \tag{4.13}$$

$$\dot{\lambda}_\alpha + \tfrac{1}{2}i\sigma^m_{\alpha\dot{\alpha}}\partial_m\dot{\bar{\chi}}^{\dot{\alpha}} = \tfrac{1}{2}\partial^2\lambda_\alpha + i\eta_{6\alpha}, \tag{4.14}$$

$$\dot{D} + \tfrac{1}{2}\partial^2\dot{C} = \tfrac{1}{2}\partial^2 D + 2\eta_8. \tag{4.15}$$

Since $\langle\eta(1)\eta(2)\rangle \propto (\vartheta_1 - \vartheta_2)^2(\bar{\vartheta}_1 - \bar{\vartheta}_2)^2$, the correlations $\langle\eta_i\eta_j\rangle$ and hence the component propagators are easily found. For example, combining eqs. (4.9) and (4.14) and eqs. (4.8) and (4.13), we find

$$\lambda_\alpha(t) = \int_0^t d\tau\, e^{\partial^2(t-\tau)}\left(i\eta_{6\alpha} + \tfrac{1}{2}\sigma^m_{\alpha\dot{\alpha}}\partial_m\eta_2^{\dot{\alpha}}\right), \tag{4.16}$$

$$\bar{\lambda}^{\dot{\alpha}}(t) = \int_0^t d\tau\, e^{\partial^2(t-\tau)}\left(-i\eta_7^{\dot{\alpha}} - \tfrac{1}{2}\bar{\sigma}^{m\dot{\alpha}\alpha}\partial_m\eta_{1\alpha}\right). \tag{4.17}$$

Letting $i\eta_6 + \frac{1}{2}\sigma\cdot\partial\eta_2 = \eta_\lambda$ and $-i\eta_7 - \frac{1}{2}\bar{\sigma}\cdot\partial\eta_i = \eta_{\bar{\lambda}}$, we find that $\langle\eta_\lambda\eta_{\bar{\lambda}}\rangle = -i\sigma\cdot\partial\delta(x-x')$. Hence the supersymmetric formalism leads us naturally to the rule for handling fermions given in sect. 3.

4.2. FREE CHIRAL SUPERFIELDS

We start with the action

$$S=\int d^4x\, d^2\vartheta\, d^2\bar{\vartheta}\left[\varphi^{\dagger}\varphi-\tfrac{1}{8}m\left(\varphi\frac{D^2}{\partial^2}\varphi+\varphi^{\dagger}\frac{\bar{D}^2}{\partial^2}\varphi^{\dagger}\right)\right]. \tag{4.18}$$

This action must be varied in such a way that the chirality condition $\bar{D}\varphi=0$ is preserved. We accomplish this by writing $\varphi=-\frac{1}{4}\bar{D}^2U$ and $\varphi^{\dagger}=-\frac{1}{4}D^2U$, where U is a real superfield. To motivate this choice note that if we write U as

$$U=f+\vartheta\varphi+\bar{\vartheta}\bar{\varphi}+\vartheta^2n^*+\bar{\vartheta}^2\eta+\vartheta\sigma\bar{\vartheta}\cdot\nu+\vartheta^2\bar{\vartheta}\bar{\lambda}+\bar{\vartheta}^2\vartheta\lambda+\vartheta^2\bar{\vartheta}^2d,$$

then

$$\bar{D}^2U=-\vartheta^2\partial^2f-\vartheta^2\partial^2(\bar{\vartheta}\bar{\varphi})+2i\vartheta\sigma\cdot\partial\bar{\varphi}-4n-\vartheta^2\bar{\vartheta}^2\partial^2n$$
$$-4i\vartheta\sigma\bar{\vartheta}\cdot\partial n-4\vartheta\lambda+2i\vartheta^2\partial\lambda\cdot\sigma\bar{\vartheta}-4\vartheta^2d-2i\vartheta^2\partial\cdot\nu.$$

So letting

$$A\equiv n,\qquad \psi\equiv\lambda-\tfrac{1}{2}i\sigma\cdot\partial\bar{\varphi},\qquad F\equiv\tfrac{1}{4}\partial^2f+\tfrac{1}{2}\partial\cdot\nu+d,$$

we have the usual expansion for a chiral superfield:

$$\varphi=A+\vartheta\psi+\vartheta^2F+i\vartheta\sigma\bar{\vartheta}\cdot\partial A+\tfrac{1}{4}\vartheta^2\bar{\vartheta}^2\partial^2A-\tfrac{1}{2}i\vartheta^2\partial\psi\cdot\sigma\bar{\vartheta}.$$

We then vary U in the action

$$S=\tfrac{1}{16}\int\left[D^2U\bar{D}^2U-\tfrac{1}{8}m\left(\bar{D}^2U\frac{D^2}{\partial^2}\bar{D}^2U+D^2U\frac{\bar{D}^2}{\partial^2}D^2U\right)\right] \tag{4.19}$$

and find, after solving the Langevin equation for U:

$$U(t)=\int_0^t d\tau\left[\left(e^{\partial^2(t-\tau)}\cosh m\sqrt{\partial^2}(t-\tau)\right)(1-P_T)\right.$$
$$\left.+\left(e^{\partial^2(t-\tau)}\sinh m\sqrt{\partial^2}(t-\tau)\right)(P_++P_-)+P_T\right]\eta(\tau), \tag{4.20}$$

where

$$P_+=\frac{D^2}{4\sqrt{\partial^2}},\qquad P_-=\frac{\bar{D}^2}{4\sqrt{\partial^2}}.$$

Note that in euclidean momentum space $\partial^2\to-p^2$ and $\sqrt{\partial^2}\to ip$; so the integrals in eq. (4.20) are always damped.

Multiplying eq. (4.20) by $-\frac{1}{4}\bar{D}^2$ and $-\frac{1}{4}D^2$ we obtain φ and $\varphi^\dagger$:

$$\varphi = -\tfrac{1}{4}\int_0^t \mathrm{d}\tau\Big[\Big(\mathrm{e}^{\partial^2(t-\tau)}\cosh m\sqrt{\partial^2}\,(t-\tau)\Big)\bar{D}^2$$

$$+\Big(\mathrm{e}^{\partial^2(t-\tau)}\sinh m\sqrt{\partial^2}\,(t-\tau)\Big)P_-D^2\Big]\eta(\tau), \qquad (4.21)$$

$$\varphi^\dagger = -\tfrac{1}{4}\int_0^t \mathrm{d}\tau\Big[\Big(\mathrm{e}^{\partial^2(t-\tau)}\cosh m\sqrt{\partial^2}\,(t-\tau)\Big)D^2$$

$$+\Big(\mathrm{e}^{\partial^2(t-\tau)}\sinh m\sqrt{\partial^2}\,(t-\tau)\Big)P_+\bar{D}^2\Big]\eta(\tau). \qquad (4.22)$$

From these equations for φ and $\varphi^\dagger$ the propagators can easily be found. For example, for $t_1 > t_2$,

$$\langle\varphi(x_1, t_1, \vartheta_1, \bar{\vartheta}_1)\varphi^\dagger(x_2, t_2, \vartheta_2, \bar{\vartheta}_2)\rangle$$

$$= \Bigg[\frac{1}{2(\partial^2 + m\sqrt{\partial^2})}\Big(\mathrm{e}^{(\partial^2 + m\sqrt{\partial^2})(t_1 - t_2)} - \mathrm{e}^{(\partial^2 + m\sqrt{\partial^2})(t_1 + t_2)}\Big)$$

$$+\frac{1}{2(\partial^2 - m\sqrt{\partial^2})}\Big(\mathrm{e}^{(\partial^2 - m\sqrt{\partial^2})(t_1 - t_2)} - \mathrm{e}^{(\partial^2 - m\sqrt{\partial^2})(t_1 + t_2)}\Big)\Bigg]$$

$$\times \mathrm{e}^{i(\vartheta_1\sigma\cdot\bar{\vartheta}_1 + \vartheta_2\sigma^n\bar{\vartheta}_2 - 2\vartheta_1\sigma^n\bar{\vartheta}_2)\partial_n^1}\delta(x_1 - x_2), \qquad (4.23)$$

which in the limit $t_1 \to t_2 \to \infty$ reduces to the usual $\varphi\varphi^\dagger$ superfield propagator. Similarly

$$\langle\varphi(x_1, t_1, \vartheta_1, \bar{\vartheta}_1)\varphi(x_2, t_2, \vartheta_2, \bar{\vartheta}_2)\rangle$$

$$= -\Bigg[\frac{\sqrt{\partial^2}}{2(\partial^2 + m\sqrt{\partial^2})}\Big(\mathrm{e}^{(\partial^2 + m\sqrt{\partial^2})(t_1 - t_2)} - \mathrm{e}^{(\partial^2 - m\sqrt{\partial^2})(t_1 - t_2)}\Big)$$

$$-\frac{\sqrt{\partial^2}}{2(\partial^2 - m\sqrt{\partial^2})}\Big(\mathrm{e}^{(\partial^2 - m\sqrt{\partial^2})(t_1 - t_2)} - \mathrm{e}^{(\partial^2 - m\sqrt{\partial^2})(t_1 + t_2)}\Big)\Bigg]$$

$$\times \mathrm{e}^{i\vartheta_1\sigma^n(\bar{\vartheta}_1 - \bar{\vartheta}_2)\partial_n^1}\delta(\vartheta_1 - \vartheta_2)\delta(x_1 - x_2). \qquad (4.24)$$

4.3. INTERACTIONS

The extension of stochastic quantization to interacting superfields is now straightforward and quite similar to the quantization of interacting scalar fields given in ref. [1]. We illustrate the method for chiral superfields with a $g\varphi^3$ interaction. The action is

$$S=\int d^4x\, d^2\vartheta\, d^2\bar{\vartheta}\left[\varphi^{\dagger}\varphi+\tfrac{1}{12}g\left(\varphi^2\frac{D^2}{\partial^2}\varphi+\varphi^{\dagger 2}\frac{\bar{D}^2}{\partial^2}\varphi^{\dagger}\right)\right]. \tag{4.25}$$

We again let $\varphi=-\frac{1}{4}\bar{D}^2U$, $\varphi^{\dagger}=-\frac{1}{4}D^2U$, and find

$$\dot{U}=\frac{\delta S}{\delta U}+\eta$$

$$=(1-P_{\mathrm{T}})\partial^2U-\tfrac{1}{16}g\left[(\bar{D}^2U)^2+(D^2U)^2\right]+\eta,$$

and hence

$$\dot{\varphi}=\partial^2\varphi-\tfrac{1}{4}\bar{D}^2\left[\eta-g(\varphi^{\dagger})^2\right], \tag{4.26}$$

$$\dot{\varphi}^{\dagger}=\partial^2\varphi^{\dagger}-\tfrac{1}{4}D^2\left[\eta-g\varphi^2\right]. \tag{4.27}$$

Note that the ϑ component of eq. (4.26) is

$$\dot{\psi}=\partial^2\psi-(\bar{D}^2\eta)_{\vartheta}+2ig\sigma\cdot\partial(A^*\bar{\psi}),$$

which reproduces the Langevin equation for interacting fermions given in sect. 3.

With the boundary condition $\varphi(t=0)=0$, eqs. (4.26) and (4.27) can be written as, in euclidean space,

$$\varphi(x,t,\vartheta,\bar{\vartheta})=-\tfrac{1}{4}\int_0^t d\tau\, d^4k\, d^4y\, e^{-k^2(t-\tau)}e^{ik(x-y)}$$
$$\times\bar{D}^2\left[\eta(y,\tau,\vartheta,\bar{\vartheta})-g\varphi^{\dagger}(y,\tau,\vartheta,\bar{\vartheta})\varphi^{\dagger}(y,\tau,\vartheta,\bar{\vartheta})\right], \tag{4.28}$$

$$\varphi^{\dagger}(x,t,\vartheta,\bar{\vartheta})=-\tfrac{1}{4}\int_0^t d\tau\, d^4k\, d^4y\, e^{-k^2(t-\tau)}e^{ik(x-y)}$$
$$\times D^2\left[\eta(y,\tau,\vartheta,\bar{\vartheta})-g\varphi(y,\tau,\vartheta,\bar{\vartheta})\varphi(y,\tau,\vartheta,\bar{\vartheta})\right]. \tag{4.29}$$

Any n-point function can be found to any order in g by iterating these equations and using the fact that $\langle\eta\eta'\rangle=2\delta(z-z')$, where $z=(x,t,\vartheta,\bar{\vartheta})$. For example, $\langle\varphi\varphi^{\dagger}\rangle$ to order g^2 has the expansion shown diagrammatically in fig. 1. This

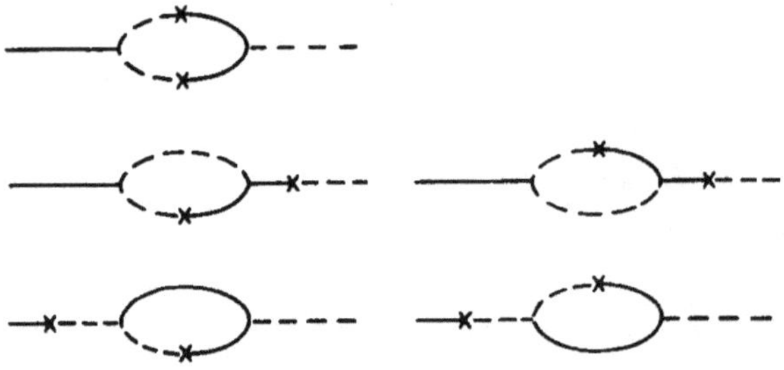

Fig. 1. Order g^2 contribution to the $\langle \varphi\varphi^\dagger \rangle$ propagator; the crosses represent $\langle \eta\eta \rangle$ correlations, the dashed lines $-\frac{1}{4}GD^2$, and the solid lines $-\frac{1}{4}G\bar{D}^2$, where G is the Green function for the free boson Langevin equation.

expansion is just that given in ref. [1] for φ^3 theory with the addition of D^2 and $\bar{D}^2$ at the vertices and superfield propagators instead of ordinary Feynman propagators. The dependence on D^2 and $\bar{D}^2$ can all be pulled onto an external leg and is identical for all five graphs. Summing the contributions from each of the graphs, we have, in the limit $t \to t' \to \infty$,

$$\langle \varphi(z)\varphi^\dagger(z') \rangle = 16g^2 \int \mathrm{d}^4k\, \mathrm{d}^4q \frac{\mathrm{e}^{ik(x-x')}}{k^2q^2(k-q)^2} \mathrm{e}^{-k(\vartheta\sigma\bar{\vartheta}+\vartheta'\sigma\bar{\vartheta}'-2\vartheta\sigma\bar{\vartheta}')} .$$

This logarithmically divergent integral contributes to the $O(g^2)$ wave function renormalization and is identical to that which would be obtained by standard superfield methods [4].

5. Stochastic regularization

Regularization schemes for quantum field theories are not very abundant, and many of the existing ones suffer from various problems that make their application to many quantum field theoretical models problematic if not impossible. The most important requirement for any regularization scheme is that it respect all the relevant symmetries of the theory. Another desirable property of a regulator is that it be possible to implement it in a non-perturbative manner. It often happens that these two requirements seem impossible to reconcile. The problem of finding a suitable regulator is not an academic one. Models like chiral gauge theories and supersymmetries are among those that suffer from the lack of a suitable regularization schemes. In light of all this it seems that a regulator capable of combining some of the properties not shared by the existing schemes would be of considerable value. The stochastic formulation of quantum field theories offers the opportunity to formulate a regularization scheme that seems not to suffer from some of the problems of the existing ones. This new regulator (henceforth referred to as stochastic regularization–SR), though not free from problems, has enough pleasing features to justify serious investigation of its properties.

5.1. DEFINITION OF THE SR SCHEME

The starting point of the SR scheme is the Langevin equation and the associated correlation functions of the random noise, eqs. (2.1) and (2.2). The Langevin equation was originally used to describe processes such as brownian motion [2]. It was soon observed [5] that a strictly markovian process, i.e. with $\langle\eta(t)\eta(t')\rangle = \delta(t-t')$, leads to inconsistencies; the averages of the solutions of (2.1) are nowhere differentiable. This problem has been solved by Doob, Ito and Stratonovich [6]. Their solution consists of giving up the strict markovian nature of the stochastic process by replacing (2.2) with

$$\langle\eta_i(t)\eta_j(t')\rangle = 2\delta_{ij}\alpha(t-t'),$$
$$\langle\eta_1\ldots\eta_n\rangle_c = 0, \tag{5.1}$$

where $\alpha(t-t')$ is some smooth function concentrated around zero. As it turns out, this is sufficient to regularize the perturbation expansions of most interacting local quantum field theories. The idea is to pick some one-parameter family of functions $\alpha_\Lambda(t)$ with a δ function limit as $\Lambda\to\infty$, recovering in the limit the unregulated theory.

Let us demonstrate how the SR works for scalar field theory. We choose $S(\varphi)$ to be

$$S[\varphi] = \int \mathrm{d}t\,\mathrm{d}^4x\left[\varphi(xt)(-\partial^2+m^2)\varphi(xt) + gV(\varphi)\right]. \tag{5.2}$$

The perturbation expansion is generated by solving eq. (2.1) iteratively [1] i.e. by iterating the formal solution

$$\varphi(xt) = \int \mathrm{d}\tau\,\mathrm{d}^4y\,G(x-y;t-\tau)\left[g\frac{\partial V(\varphi)}{\partial\varphi}(y\tau) + \eta(y\tau)\right], \tag{5.3}$$

where

$$G(xt) = \vartheta(t)\int \mathrm{d}^4k\,\mathrm{e}^{-t(k^2+m^2)+ikx}. \tag{5.4}$$

The Green function chosen here indicates the boundary conditions $P[\varphi,0] = \prod_x\delta(\varphi(x))$. If, for simplicity, we take $\partial V/\partial\varphi$ to be a monomial in φ, say φ^3, the solution to (5.3) has the simple graphical representation [1] shown in fig. 2. A line in

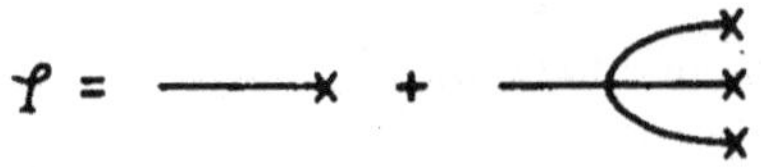

Fig. 2. Graphical representation of the iterative solution of the Langevin equation for φ^4 theory.

this representation stands for $G(k; t-\tau)$ and a cross for $\eta(k,\tau)$. Fig. 2 represents

$$\varphi(k,t)=\int_0^t \mathrm{d}\tau\left[G(k,t-\tau)\left(\eta(k,\tau)+g\int_0^\tau \prod_{i=1}^{3} \mathrm{d}\tau_i G(q_i,\tau-\tau_i)\eta(q_i,\tau_i)\right.\right.$$
$$\left.\left.\times\delta\left(k-\sum q_i\right)\mathrm{d}^4 q_i+\cdots\right)\right]. \tag{5.5}$$

Any graph in the expansion for a n-point function

$$\langle\varphi(k_1 t_1)\ldots\varphi(k_n t_n)\rangle_\eta, \tag{5.6}$$

can be obtained from the graphs of the usual perturbation expansion by placing crosses on the various lines in such a way that cutting all crossed lines will break the graph into tree graphs. The graphs for second-order two-point functions of the $g\varphi^4$ theory are depicted in fig. 3. In this expansion an uncrossed line stands for $G(k,t-t')$ and a crossed line stands for $D(k,t-t')$, where

$$D(k,t_1-t_2)=\int_0^{t_1}\mathrm{d}\tau_1\int_0^{t_2}\mathrm{d}\tau_2\, G(k,t_1-\tau_1)G(k,t_2-\tau_2)\alpha(\tau_1-\tau_2). \tag{5.7}$$

Let us choose the following Fourier decomposition for α

$$\alpha(\tau_1-\tau_2)=\int_{-\infty}^{\infty}\frac{\mathrm{d}\nu}{2\pi}\rho(\nu)e^{i\nu(\tau_1-\tau_2)} \tag{5.8}$$

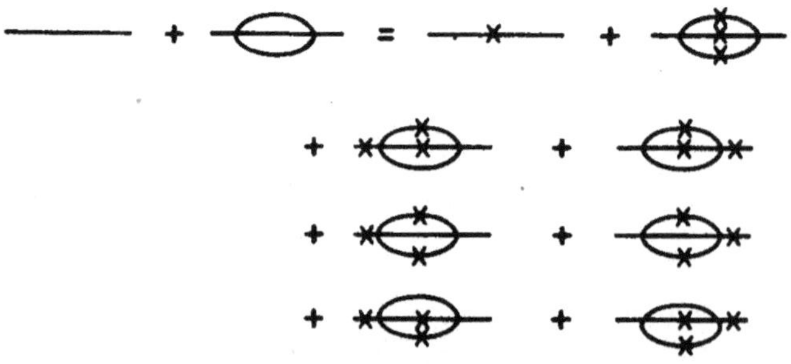

Fig. 3. Expansion up to second order of the φ^4 two-point function in stochastic perturbation theory.

and insert it into (5.7), obtaining

$$D(k, t_1 - t_2) = \int_{-\infty}^{\infty} \frac{d\nu}{2\pi} \frac{\rho(\nu)}{\omega^2(k) + \nu^2} e^{i\nu(t_1 - t_2)} + O(e^{-(t_1 + t_2)\omega}),$$

$$\omega(k) = k^2 + m^2. \tag{5.9}$$

Since each loop has at least one crossed propagator, an appropriate choice of $\rho(\nu)$ with the property

$$\rho_\Lambda(\nu) \underset{\Lambda \to \infty}{\to} 1, \tag{5.10}$$

renders all graphs finite, while (5.10) ensures that the unregulated perturbation expansion is recovered in the $\Lambda \to \infty$ limit.

The SR scheme can also be used in a straightforward manner to regularize the IR divergences of a massless theory. If we take $\rho(\nu)$ to be

$$\rho_{\Lambda, \mu}(\nu) = \rho_\Lambda(\nu) \vartheta(\nu^2 - \mu^2), \tag{5.11}$$

where ρ_Λ is any choice of ρ that regularizes the UV divergences, the resulting perturbation expansion is IR finite as well. The new scale μ serves as an IR regulator and the unregulated theory is recovered in the $\mu \to 0$ limit. To see how this choice of ρ takes care of the IR infinities let us calculate a contribution to the one-loop correction for the n-point function in the $\lambda\varphi^3$ theory with all external momenta set to zero. The most IR divergent contribution would be the one with a single internal line and all external lines, but one, crossed (fig. 4). This diagram is given by

$$\int_0^{t_1} d\tau_1 \int_0^{\tau_1} d\tau_2 \ldots \int_0^{\tau_{n-1}} d\tau_n \, D(k, \tau_1 - \tau_n) \prod_{i=1}^{n-1} G(k, \tau_i - \tau_{i+1})$$

$$= \int_0^{t_1} d\tau_1 \ldots \int_0^{\tau_{n-1}} d\tau_n \, D(k, \tau_1 - \tau_n) e^{-k^2(\tau_1 - \tau_n)}, \tag{5.12}$$

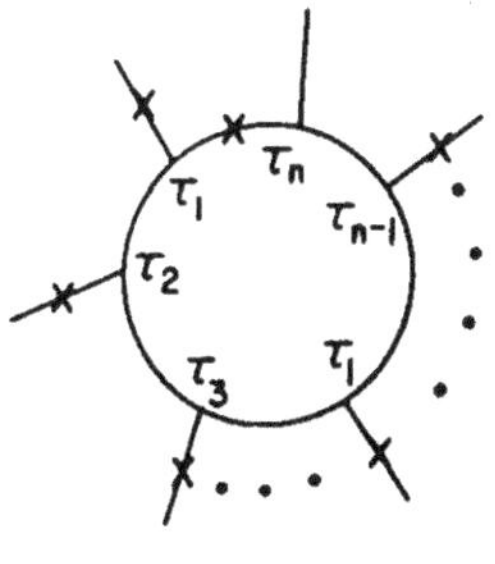

Fig. 4. Most IR divergent contribution to the n-point function in φ^3 theory.

and at $k=0$ we have

$$\int_0^{t_1} d\tau_1 \ldots \int_0^{\tau_{n-1}} d\tau_n \int_{-\infty}^{\infty} \frac{d\nu}{\nu^2} \rho_{\Lambda,\mu}(\nu) e^{i\nu(\tau_1-\tau_n)}, \tag{5.13}$$

which is finite, and the IR divergence is regulated.

5.2. TADPOLES, BUBBLES AND CONSTRAINTS ON ρ

The choice of ρ is dictated by the requirement that all the diagrams of the perturbation expansion be finite. The most severe constraint is imposed by tadpole-like graphs, where a single propagator forms a closed loop. In such a case we have integrals of the form

$$\int D(k,0)\, d^4k, \tag{5.14}$$

where $D(k,0)$ is given by $\int_{-\infty}^{\infty} \rho(\nu)(\omega^2+\nu^2)^{-1} d\nu$. Assuming that the integral can be done by contour integration, we have

$$D(k,0) = \frac{\pi}{\omega}\rho(i\omega) + 2\pi i \sum_n \frac{1}{\omega^2+\nu_n^2} \operatorname{Res} \rho(\nu_n), \tag{5.15}$$

where ν_n are the poles of ρ in the upper (lower) half plane. The k integral of the sum on the r.h.s of eq. (5.15) diverges logarithmically unless there are cancellations among the various terms. A simple analysis shows that a necessary condition for the occurrence of cancellations is $\alpha(0)=0$ or equivalently $\int d\nu\, \rho(\nu)=0$. Before going on to attempt the implementation of such a correlation function there is one point to keep in mind. The correlation functions $\langle \eta(xt)\eta(x't')\rangle$ are given by

$$2\delta(x-x')\alpha(t-t') = 2\delta(x-x')\int_{-\infty}^{\infty} \rho(\nu) e^{i\nu(t-t')} d\nu, \tag{5.16}$$

and if we wish to interpret eq. (2.1) as a gaussian stochastic process, the correlation functions must be given by a probability distribution. This imposes quite severe constraints on ρ. To have a probability distribution, the inverse of ρ must be a positive definite operator. As it turns out, a necessary condition for positive probability is $\alpha(0)>0$ which is in a marked contrast with the condition for the cancellation of divergences in (5.15). To see what is the meaning of this, recall that the objects one considers are the averages of the solutions of the Langevin equation (2.1). Since all the dynamical information about the system is encoded in those solutions, the existence of a well-defined measure (probability distribution) for η will enable us, at least in principle, to obtain the complete quantum theoretical solution of the theory by performing the η averages. If, on the other hand, all that is required is the (stochastically) regularized perturbation expansion around the $\varphi = 0$ vacuum, one can expand around $\eta = 0$ and do the η averages using the rules (5.1) as an operational prescription. This means (if finite perturbation expansion may serve as a criterion) that the stochastic regulator can be implemented non-perturbatively only for theories with no bubbles and tadpoles in their perturbation expansion.

There is, however, a way out of this impasse. Recall that the root of the problem was the fact that the requirement $\langle \eta(t)^2 \rangle = 0$ clashes with the interpretation of the correlation functions as moments of a probability distribution. To overcome this problem, let us define the random "force"

$$\eta'(xt) = \int_{-\infty}^{\infty} \mathrm{d}\tau \beta(t-\tau)\eta(x\tau), \tag{5.17}$$

where $\eta(xt)$ has a simple gaussian distribution

$$P[\eta] = \exp\left[-\int_{-\infty}^{\infty} \eta^2(xt)\,\mathrm{d}^4x\,\mathrm{d}t\right], \tag{5.18}$$

and use the Langevin equation with η' as the noise,

$$\frac{\partial \varphi}{\partial t} = -\frac{\delta S}{\delta \varphi} + \eta'. \tag{5.19}$$

The averages of η' are given by

$$\langle \eta'(xt)\eta'(x't')\rangle_\eta = 2\delta^4(x-x')\int_{-\infty}^{\infty} \mathrm{d}\tau \beta(t-\tau)\beta(t'-\tau), \tag{5.20}$$

and the requirement $\alpha(0) = 0$ reads

$$\int_{-\infty}^{\infty} \beta(t-\tau)^2 \mathrm{d}\tau = 0, \tag{5.21}$$

which can be easily achieved if β is allowed to be complex. The choice of β must be made in such a way that $\alpha(t-t')$ remains real. Such a choice guarantees the reality of the coefficients of the perturbation expansion for the n-point functions ($\langle \varphi_1 \ldots \varphi_n \rangle_\eta$. The following generic form of β is consistent with all the constraints:

$$\beta(t) = \int [\rho_1(\nu) + \rho_2(\nu)] \mathrm{e}^{i\nu t} \mathrm{d}\nu,$$
$$\rho_1(-\nu) = \rho_1(\nu), \qquad \rho_2(-\nu) = -\rho_2(\nu). \tag{5.22}$$

Using (5.22) to calculate α, we get

$$\alpha(t-t') = \int \left[\rho_1(\nu)^2 - \rho_2(\nu)^2\right] \mathrm{e}^{i\nu(t-t')} \mathrm{d}\nu, \tag{5.23}$$

and the condition $\alpha(0)=0$ reads $\int[\rho_1^2-\rho_2^2]\mathrm{d}\nu=0$ which can be easily satisfied. From (5.23) it is easy to see that α is indeed real. Finally, the condition $\alpha(t-t')\to\delta(t-t')$ as $\Lambda\to\infty$ reads $\rho_1^2-\rho_2^2\to 1/2\pi$. A suitable choice is

$$2\pi\rho_1=\frac{\Lambda^2}{\nu^2+\Lambda^2},\qquad 2\pi\rho_2=\frac{\Lambda\nu}{\nu^2+\Lambda^2},\tag{5.24}$$

which fulfills all requirements. The essence of this formulation of the stochastic process is the fact that the random "force" does not have a probability distribution. That this is so can be seen from the fact that β is non-invertible operator, so that different η's with different probabilities will give the same η'. It also means that not all "histories" in η' will be realized in the process.

As $\Lambda\to\infty$ the measure of η' not generated by the process will approach zero and eq. (5.19) will approach a normal gaussian process. The regulator defined in this section is a natural extension of the stochastic formulation of quantum field theory. It defines a finite theory in a nonperturbative manner. Since the regulator does not change the Langevin equation and does not interfere with the symmetries of the measure, it preserves the symmetries in the regulated theory. This last property will be investigated in sect. 6.

6. Symmetries and conservation laws

A major consideration in choosing a regulator for a given theory is that it respect the symmetries of the theory. The two most important regulators for particle physics, lattice and dimensional regularization, are far from perfect in this respect. The SR respects internal symmetries and local gauge symmetries in a natural way. It also conserves chiral gauge symmetries and supersymmetries, which are properties not shared by other regulators. Gauge invariance for vectorlike and chiral symmetries is a consequence of the gauge covariance of the Langevin equations

$$\frac{\partial}{\partial t}A_\mu=D_\nu F_{\mu\nu}+J_\mu+\eta_\mu,\tag{6.1a}$$

$$\frac{\partial}{\partial t}\psi=-i\gamma\cdot\partial(\gamma\cdot D\psi)+\eta_\psi,\tag{6.1b}$$

$$\frac{\partial}{\partial t}\bar\psi=-\bar\psi\gamma\cdot\overleftarrow{D}i\gamma\cdot\overleftarrow{\partial}+\bar\eta_\psi.\tag{6.1c}$$

Under the transformations

$$A_\mu^g=U(x)A_\mu U^\dagger(x)+\frac{1}{i}U(x)\partial_\mu U^\dagger(x),$$

$$\psi^g=U(x)\psi,\qquad \bar\psi^g=\bar\psi U^\dagger(x),\tag{6.2}$$

eqs. (6.1a)–(6.1c) transform to

$$\frac{\partial}{\partial t}A_\mu^g = D_\nu(A^g)F_{\mu\nu}^g + J_\mu^g + U\eta_\mu U^\dagger, \tag{6.3a}$$

$$\frac{\partial}{\partial t}\psi^g = U(-i\gamma\cdot\partial)U^\dagger\gamma\cdot D(A^g)\psi^g + U\eta^\psi, \tag{6.3b}$$

$$\frac{\partial}{\partial t}\bar\psi^g = \bar\psi^g\gamma\cdot\bar D(A^g)U(-i\gamma\cdot\bar\partial)U^\dagger + \bar\eta_\psi U^\dagger. \tag{6.3c}$$

If A_μ, ψ, and $\bar\psi$ are solutions of eqs. (6.1) with functional dependence $[\eta_\mu, \eta_\psi, \bar\eta_\psi; K]$ then A^g, ψ^g and $\bar\psi^g$ are given by the same solutions with functional dependence $[U\eta_\mu U^\dagger, U\eta_\psi, \bar\eta_\psi U^\dagger, UKU^\dagger]$. ($K$ is in our case $i\gamma\cdot\partial$, its role in the stochastic quantization scheme is defined in sects. 2 and 3.) This form of the gauge transformed solutions is correct only for gauge invariant initial conditions. If the initial conditions are not gauge invariant, they will change under a gauge transformation, and hence the functional form of A, ψ, and $\bar\psi$ will change too. This change will not, however, affect the calculation of gauge invariant quantities in the large-t limit, since they are independent of the initial conditions. Using the solutions for the gauge transformed quantities,

$$\begin{aligned}\left\langle F\left[A_\mu^g, \psi^g, \bar\psi^g\right]\right\rangle_{\eta_\mu, \eta_\psi, \bar\eta_\psi, K} &= \int\prod d\eta_\mu d\eta_\psi d\bar\eta_\psi\, F\left[A^g, \psi^g, \bar\psi^g\right] \\ &\quad\times\exp\Big\{-\int\Big[\mathrm{tr}\left[\eta_\mu(xt)\alpha^{-1}(t-t')\eta_\mu(xt')\right] \\ &\quad+\bar\eta_\psi(xt)\alpha^{-1}(t-t')K^{-1}(xy)\eta_\psi(yt')\Big]\Big\},\end{aligned} \tag{6.4}$$

changing variables in the integral to

$$\eta'_\mu = U\eta_\mu U^\dagger, \qquad \eta'_\psi = U\eta_\psi, \qquad \bar\eta'_\psi = \bar\eta_\psi U^\dagger, \tag{6.5}$$

and using the fact that the measure is invariant under the transformations (6.5), we have

$$\left\langle F\left[A_\mu^g, \psi^g, \bar\psi^g\right](K)\right\rangle_{\eta_\mu, \eta_\psi, \bar\eta_\psi, K} = \left\langle F\left[A_\mu, \psi, \bar\psi\right](K')\right\rangle_{\eta'_\mu, \eta'_\psi, \bar\eta'_\psi, K'}. \tag{6.6}$$

In sect. 2 we showed that the large-t limit is independent of the form of K as long as K is not singular. But $K' = UKU^\dagger$, so if K is not singular and gives a positive definite Fokker-Planck hamiltonian so does K'. Under these circumstances eq. (6.6) implies that

$$\langle F[A^g, \psi^g, \bar\psi^g]\rangle = \langle F[A, \psi, \bar\psi]\rangle, \tag{6.7}$$

as we have claimed. Arguments of exactly the same sort can be made for chiral gauge symmetries provided that the measure $\prod d\eta_\psi d\bar{\eta}_\psi$ is invariant [7]. The reason that the SR scheme conserves local symmetries is quite simple. The regulator does not change the Langevin equation, and as long as the gauge transformations are independent of the Langevin time they do not change the probability distribution of the random noise.

In local field theories symmetries of the action manifest themselves in conservation laws. For local gauge theories one can derive Ward-Takahashi and Slavnov-Taylor identities. while global symmetries generate conservation laws that can be derived in a simple way by using Noether's theorem. Unfortunately an attempt to derive a Noether theorem for a stochastically regulated theory does not lead to simple results. For example, let us consider an $O(n)$ scalar field theory. The symmetry of the lagrangian leads to

$$\frac{\partial L}{\partial(\partial_\mu \varphi_i)} \delta(\partial_\mu \varphi_i) + \frac{\partial L}{\partial \varphi_i} \delta\varphi_i = 0, \tag{6.8}$$

and the equation of motion

$$\dot{\varphi}_i = -\partial_\mu \frac{\partial L}{\partial(\partial_\mu \varphi_i)} + \frac{\partial L}{\partial \varphi_i} + \eta_i = 0. \tag{6.9}$$

Combining eqs. (6.8) and (6.9) and using $J_\mu = [\partial L/\partial(\partial_\mu \varphi_i)]\, \delta\varphi_i$, one finds

$$\partial_\mu J_\mu^a = (\eta_i - \dot{\varphi}_i) T_{ij}^a \varphi_j, \tag{6.10}$$

where $\delta\varphi_i = \varepsilon^a T_{ij}^a \varphi_j$ and the T^a are the generators of the $O(n)$ symmetry.

Of course, one does not expect $\partial_\mu J_\mu^a[\eta] = 0$ to hold for any given η: rather one expects

$$\lim_{t\to\infty} \left(\langle \partial_\mu J_\mu^a F[\varphi] \rangle - \langle \delta^a F[\varphi] \rangle \right) = 0. \tag{6.11}$$

where $\delta^a F(\varphi)$ is the variation of F under the symmetry. To prove eq. (6.11) we multiply equation (6.10) by $F(\varphi)$ and take the η average

$$\langle \partial_\mu J_\mu^a F[\varphi] \rangle = \langle (\eta_i - \dot{\varphi}_i) T_{ij}^a \varphi_j F[\varphi] \rangle. \tag{6.12}$$

Now the Fokker-Planck equation leads to

$$\lim_{t\to\infty} \left(\frac{\delta}{\delta\varphi_i} + \frac{\delta S}{\delta\varphi_i} \right) P(\varphi, t) = 0, \tag{6.13}$$

so that

$$\lim_{t\to\infty} \int \prod d\varphi\, F(\varphi)\left(\frac{\delta}{\delta\varphi_i} + \frac{\delta S}{\delta\varphi_i}\right) P(\varphi, t) = 0. \tag{6.14}$$

Integrating by parts, we find

$$\lim_{t\to\infty} \left\langle -\frac{\delta F}{\delta\varphi} + F(\varphi)\frac{\delta S}{\delta\varphi} \right\rangle_\eta = 0. \tag{6.15}$$

Using the fact that $\langle \delta F/\delta\varphi_i \rangle = \langle \eta_i F \rangle$, eq. (6.15) and the Langevin equation, we find

$$\lim_{t\to\infty} \langle \dot{\varphi}_i(x) F[\varphi] \rangle = 0, \tag{6.16}$$

and hence

$$\lim_{t\to\infty} \left(\langle \partial_\mu J^a_\mu F[\varphi] \rangle - \langle \eta_i T^a_{ij} \varphi_j F[\varphi] \rangle \right) = 0, \tag{6.17}$$

which is the desired result.

The derivation of eqs. (6.8)–(6.17) was done for the unregularized theory. If we try to apply a similar line of reasoning for the regularized system, eq. (6.13) and consequently the rest of the derivation fails. In fact, if we consider stochastically regularized free field theory with, say, a U(1) symmetry and calculate the current-current correlation function

$$\Pi_{\mu\nu}(x-y, t) = \langle J_\mu(xt) J_\nu(yt) \rangle,$$

we find that even in the large-t limit $\partial_\mu \Pi_{\mu\nu}$ is not zero, as required by the U(1) symmetry. To understand the reason for this failure let us derive the Fokker-Planck hamiltonian for the regularized free field theory. We assume, as in the unregularized case, that the equal time averages are given by a time dependent probability distribution $P(\varphi, t)$. For any functional F

$$\langle F[\varphi(t)] \rangle_\eta = \int \prod d\varphi\, F(\varphi) P(\varphi, t). \tag{6.18}$$

Taking the time derivative of (6.18) and using (1.1), we get (the index i stands for both space-time and internal symmetry indices)

$$\left\langle \frac{\delta F}{\delta\varphi_i(t)} \dot{\varphi}_i(t) \right\rangle_\eta = \left\langle -\frac{\delta F}{\delta\varphi_i(t)} \left(\frac{\delta S}{\delta\varphi_i(t)} - \eta_i(t) \right) \right\rangle$$

$$= \int \prod d\varphi\, F(\varphi) \dot{P}(\varphi, t). \tag{6.19}$$

Using eq. (6.18) and standard manipulations, we get

$$\int \prod \mathrm{d}\varphi\, F(\varphi)\dot{P}(\varphi, t) = \int \prod \mathrm{d}\varphi\, F(\varphi)\frac{\delta}{\delta\varphi_i}\left(\frac{\delta S}{\delta\varphi_i}P(\varphi, t)\right) + \left\langle \eta_i(t)\frac{\delta F}{\delta\varphi_i(t)}\right\rangle_\eta . \tag{6.20}$$

The last term in eq. (6.20) can be written as

$$\left\langle \eta_i(t)\frac{\delta F}{\delta\varphi_i(t)}\right\rangle_\eta = \int \mathrm{d}t'\,\alpha(t, t')\left\langle \frac{\delta}{\delta\eta_i(t')}\frac{\delta F}{\delta\varphi_i(t)}\right\rangle$$

$$= \int_0^\infty \mathrm{d}t'\,\alpha(t, t')\left\langle \frac{\delta^2 F}{\delta\varphi_i(t)\delta\varphi_j(t)}\frac{\delta\varphi_j(t)}{\delta\eta_i(t')}\right\rangle . \tag{6.21}$$

In free field theory

$$\frac{\delta\varphi_i(xt)}{\delta\eta_j(yt')} = \delta_{ij}\Theta(t-t')G(x-y, t-t') . \tag{6.22}$$

Combining (6.20)–(6.22), we find

$$\dot{P}(\varphi, t) = \int \mathrm{d}^4x\,\frac{\delta}{\delta\varphi_i(x)}\left[\left(\frac{\delta S}{\delta\varphi_i(x)} + \int \mathrm{d}^4y\, M(x-y, t)\frac{\delta}{\delta\varphi_i(y)}\right)P(\varphi, t)\right], \tag{6.23}$$

where

$$M(x-y, t) = \int_0^t \alpha(t, t')G(x-y, t-t')\,\mathrm{d}t' . \tag{6.24}$$

In the limit $t \to \infty$ we get $M(x-y, t) \to H(x-y)$, where

$$H(x-y) = \int \mathrm{d}^4k\, h(k)\mathrm{e}^{ik(x-y)},$$

$$h(k) = \int \mathrm{d}\nu \frac{\rho_\Lambda(\nu)}{k^2 + m^2 - i\nu}, \tag{6.25}$$

and we have taken

$$S = \tfrac{1}{2}\int \varphi_i(-\partial^2 + m^2)\varphi_i\,\mathrm{d}^4x\,\mathrm{d}t .$$

For a standard choice of ρ_Λ,

$$\rho_\Lambda = \frac{\Lambda^2}{\nu^2 + \Lambda^2},$$

we get

$$\lim_{t\to\infty} P(\varphi, t) = \exp\left[-\int \varphi_i(-\partial^2 + m^2)\left(1 + \frac{2}{\Lambda}(-\partial^2 + m^2) + \frac{1}{\Lambda^2}(-\partial^2 + m^2)^2\right)\varphi_i\right]. \tag{6.26}$$

Now it is easy to understand why the symmetry does not imply the expected conservation laws. The stochastic regulator introduces higher derivatives (without breaking the symmetries), and the conserved currents are not the usual ones. Even though we do not know how to derive an equation like (6.26) for the interacting theory, it is clear from the structure of perturbation theory that a similar phenomenon occurs.

Since the SR does not lead to the naive conservation laws and Ward identities, its applicability to perturbative calculation is problematic. Nonetheless it seems to us that it still may be possible to make use of this scheme in perturbation theory, provided a consistent subtraction scheme, capable of subtracting the "bad" non-local terms, can be formulated. The starting point for developing such a scheme should be equation (6.12).

7. The role of initial conditions

Throughout the previous sections of this paper we have referred only briefly to the significance of the initial conditions in the stochastic formulation of quantum field theory. If one considers the exact solution in sect. 2 of this paper, the large-t limit of the n-point functions is independent of the initial conditions. This statement must be somewhat qualified. As discussed in refs. [1,2] and in sect. 2, the equal time n-point functions can be calculated if one solves for the probability distribution $P(\varphi; t)$, eq. (2.2). The probability distribution is given by the Fokker-Planck equation

$$-\frac{\partial}{\partial t}\Psi = H_{\text{FP}}\Psi, \tag{7.1}$$

where

$$P(\varphi; t) = e^{-1/2S(\varphi)}\Psi(\varphi; t). \tag{7.2}$$

The Fokker-Planck hamiltonian is a self-adjoint operator; so its eigenfunctions form a complete set of states (in the case of field theory those are of course functionals). Consequently, given an initial condition $P(\varphi; 0)$, the probability distribution at a time t is given by

$$P(\varphi; t) = \sum_{0}^{\infty} a_n e^{-\lambda_n t} e^{-1/2S(\varphi)} \Psi_n(\varphi), \tag{7.3a}$$

$$a_n = \langle \Psi_n | \Psi(0) \rangle, \qquad \Psi(0) = e^{-(1/2)S(\varphi)} P(\varphi; 0). \tag{7.3b}$$

So if H_{FP} has a mass gap and it's ground state is not degenerate, the only requirement is that $\Psi(0)$ have non-vanishing overlap with the ground state wave function. In practical applications, like Monte Carlo simulations, it would be profitable to start with a state having as large an overlap with the ground state as possible. The case when the Fokker-Planck hamiltonian does not have a mass gap is probably not suitable for stochastic quantization even though the off-shell stochastic perturbation expansion is equivalent to the usual one.

The case of a model with spontaneously broken symmetry deserves special attention. To understand the behavior of such theories it is useful to recall the physical picture inherent to the Langevin equation. The change in time of the dynamical degree of freedom x_i is governed by two forces: the dissipative force $\partial V / \partial x_i$ and the random force η_i

$$\dot{x}_i = -\frac{\partial V}{\partial x_i} + \eta_i. \tag{7.4}$$

In the absence of the random force the variable x_i would "decay" into the first local minimum of V reachable from it's starting point by continuous descent. The influence of the random force can be now understood in the following way. Small fluctuations of η, meaning $|\eta| \ll |\partial V / \partial x|$, will cause gaussian distribution of the x_i around the local minimum, while large fluctuations will cause the system to "visit" the various minima of V. In the absence of the dissipative force the variables undergo a random walk and a limiting distribution for x_i is not reached (or rather the distribution for all initial conditions will depend on t for all t). The frequency at which the various points in configuration space are visited depends on the relative probabilities of the different η's necessary for reaching them.

To analyze the case of a spontaneously broken symmetry let us examine the theory of a complex scalar field

$$S(\varphi) = \int d^4x\, dt \left[|\partial_\mu \varphi|^2 + V(|\varphi|) \right],$$

$$V(|\varphi|) = \tfrac{1}{4}\lambda \left(|\varphi|^2 - \mu^2 \right)^2.$$

In terms of the polar variables $\varphi(xt) = \rho(xt)e^{i\vartheta(xt)}$ the Langevin equation is

$$\rho\dot{\vartheta} = \rho\partial^2\vartheta - \eta_1\sin\vartheta + \eta_2\cos\vartheta, \tag{7.5a}$$

$$\dot{\rho} = \partial^2\rho - \rho(\partial_\mu\vartheta)^2 + \frac{\partial V}{\partial\rho} + \eta_1\cos\vartheta + \eta_2\sin\vartheta. \tag{7.5b}$$

For large enough coupling we can replace ρ in (7.5a) by the expectation value ρ_0 and the solution of (7.5a) will be

$$\rho_0\vartheta(xt) = \int d^4y\,d\tau\,G(x-y, t-\tau)\left[-\eta_1(y\tau)\sin\vartheta(y\tau) + \eta_2(y\tau)\cos\vartheta(y\tau)\right]. \tag{7.6}$$

Using (7.6) to calculate the propagator for ϑ to lowest order, one gets the massless free propagator. This result indicates that in the large-t limit the system relaxes into the broken phase. This analysis will hold provided the initial condition is not $\rho(0) = 0$. If we start with $\rho = 0$, by symmetry considerations the η's that would push the system in either direction occur with equal probabilities. In that case the replacement of ρ by the expectation value would not be justified. Once the correct initial conditions are selected, the random force cannot restore the symmetry because the measure of the η's that could do that, namely η's that are space independent, is zero.

While the preceding discussion applies to the exact solution of the problem, the picture changes considerably for the perturbative approach. In perturbation theory one takes into account only small fluctuations in η so that it never can drive the system out of a local minimum of the potential (action). By choosing an initial condition φ_c such that $\delta V/\delta\varphi|_{\varphi_c} = 0$, one can expand around such a minimum. This amounts to quantizing the theory around a classical solution. One may expect to run into the usual difficulties with zero modes encountered in the usual treatment. The correct treatment of this problem within the stochastic framework merits a separate investigation, and we will not attempt it here.

8. Conclusions

In this paper we have investigated some of the properties of the stochastic quantization formalism. This formalism has been extended to theories with interacting Bose and Fermi fields and to supersymmetry. We have also used the stochastic quantization procedure to define a non-perturbative regulator for quantum field theories. This stochastic regulator conserves global and local symmetries of the theory. Unfortunately the conserved currents and Ward identities are not the standard ones, which are recovered only in the limit that the cutoff is removed. So this regulator cannot be used in perturbation theory until a subtraction procedure is found. Finally we have discussed briefly the relevance of the initial conditions used to solve the Langevin equation. For perturbative solutions, the initial conditions determine around which minimum of the action one quantizes, while for exact solutions, except for broken symmetries, the initial conditions seem to be irrelevant.

This work was supported by the Department of Energy under grant number DE-AC02-76ER02220.

Appendix

In this appendix we calculate the one-loop vacuum polarization in the stochastic formulation. The diagrammatic expansion is similar to that given in ref. [1] for a φ^3 theory. The Green functions and propagators are given by:

$$G_{\mu\nu}(q,t)=\left[\left(\delta_{\mu\nu}-\frac{q_\mu q_\nu}{q^2}\right)e^{q^2 t}+\frac{q_\mu q_\nu}{q^2}\right]\Theta(t), \tag{A1}$$

$$D_{\mu\nu}(q,t,t')=\left(\delta_{\mu\nu}-\frac{q_\mu q_\nu}{q^2}\right)\frac{1}{q^2}e^{-q^2|t-t'|}+\frac{2q_\mu q_\nu}{q^2}\min(t,t'), \tag{A2}$$

$$G_\psi(q,t,t')=G_{\bar\psi}(q,t,t')=e^{-(q^2+m^2)(t-t')}\Theta(t-t'), \tag{A3}$$

$$D_{\psi\bar\psi}(q,t,t')=\frac{-iq\cdot\gamma+m}{q^2+m^2}e^{-(q^2+m^2)|t-t'|}=-D_{\bar\psi\psi}(q,t,t'). \tag{A4}$$

The vertices are determined by the following rule. Follow the external lines into the graph. Whenever a photon branches into two fermions associate a factor of $e\gamma_\mu$ with the vertex. When a ψ branches into A_μ and ψ associate $(-\partial\cdot\gamma+m)\gamma_\mu$ with the vertex and similarly for $\bar\psi\to A_\mu$ and $\bar\psi$. The γ matrix structure of all the graphs is then $\mathrm{tr}[\gamma_\mu(-iq\cdot\gamma+m)\gamma_\nu(-i(k-q)\cdot\gamma+m)]$, where k is the external momentum and q is the loop momentum. For $t=t'\to\infty$ we have the unamputated vacuum polarization

$$\Pi_{\mu\nu}(k)=\left[T_{\mu\rho}T_{\sigma\nu}+cL_{\mu\nu\rho\sigma}\right]\int\frac{d^4q}{(2\pi)^4}\mathrm{tr}\left[\gamma_\rho(-iq\cdot\gamma+m)\gamma_\sigma(-i(k-q)\cdot\gamma+m)\right]$$
$$\times\left\{\frac{1}{k^2\left[k^2+q^2+(k-q)^2+2m^2\right]}\right.$$
$$\left.\times\left[\frac{1}{\left[(k-q)^2+m^2\right]\left[q^2+m^2\right]}+\frac{1}{k^2}\left(\frac{1}{q^2+m^2}+\frac{1}{(q-k)^2+m^2}\right)\right]\right\}, \tag{A5}$$

where $T_{\mu\nu}(k)=\delta_{\mu\nu}-k_\mu k_\nu/k^2$ and $L_{\mu\nu\rho\sigma}(k)=k_\mu k_\nu k_\rho k_\sigma$. If the loop integral is regulated in such a way that it remains transverse, we have simply

$$\Pi_{\mu\nu}(k)=T_{\mu\nu}(k)\int\frac{\mathrm{d}^4q}{(2\pi)^4}\frac{\mathrm{tr}\left[\gamma_\mu(-iq\cdot\gamma+m)\gamma_\mu(-i(k-q)\cdot\gamma+m)\right]}{(q^2+m^2)\left[(k-q)^2+m^2\right](k^2)^2}\,.$$

which is the usual result.

References

[1] G. Parisi and Yong Shi Wu, Sci. Sin. 24 (1981) 483
[2] R.F. Fox, Phys. Reports 48C (1976) 179;
J. Alfaro and B. Sakita, CCNY preprint HEP 82/16 (1982)
[3] W. Grimus and H. Hüffel, CERN preprint TH-3346 (1982);
E. Floratos and J. Iliopoulos, LPTENS preprint 82/31 (1982)
[4] J. Wess and J. Bagger, Supersymmetry and supergravity lectures at Princeton University (1982)
[5] N. Wiener, J. Math Phys. 2 (1923) 132; Acta Math. 55 (1930) 117
[6] J. L. Doob, Ann. of Math 43 (1944) 351;
K. Ito, Proc. Imp. Acad. Tokyo 20 (1944) 519; Mem. Amer. Math Soc. 4 (1951);
R. L. Stratonovich, SIAM J. of Control 4 (1966) 362
[7] K. Fujikawa, Phys. Rev. D21, 2848 (1980)
[8] T. Fukai, H. Nakazato, I. Ohba, K. Okano and Y. Yamanaka, Prog. Theor. Phy. 69 (1983) 1600

Nuclear Physics B251[FS13] (1985) 633–654

STOCHASTIC REGULARIZATION OF SCALAR ELECTRODYNAMICS*

Zvi BERN

Lawrence Berkeley Laboratory, University of California, Berkeley, California 94720, USA

Received 19 October 1984

A regularization scheme, first proposed by Breit, Gupta and Zaks and based upon the Langevin equation of Parisi and Wu, is used to regularize scalar electrodynamics. This scheme is shown to preserve the masslessness of the photon and the tensor structure of the photon vacuum polarization at the one-loop level. The scalar wave function renormalization, Z_2, is shown to be equal to the one-photon vertex renormalization, Z_1, to all orders of the stochastically regularized theory.

1. Introduction

Several years ago Parisi and Wu [1] introduced stochastic quantization. While equivalent to the more standard methods of quantization, their procedure offers some interesting insights into quantum field theory. For example, they showed that gauge theories could be constructed without the need for gauge fixing so that the Gribov ambiguity could be circumvented. They also realized that the ideas inherent in the Langevin equation are strongly connected to Monte Carlo computer simulations.

Breit, Gupta and Zaks [2] have proposed the possibility of making use of the ideas inherent in stochastic quantization to regularize field theories. However, the conclusion reached by these authors was that the applicability of this stochastic regularization to perturbation calculations is problematic. The claim was, that although the symmetries of the theory are preserved, the naive conservation laws are not preserved, so that stochastic regularization may not be a satisfactory scheme. However, the relevance of this fact to regularization and renormalization is not clear. For example, the method of higher covariant derivatives (Pauli-Villars) ruins the conservation of the naive Noether currents, but is certainly a good regularization scheme for gauge theories [3]. Another objection [4] that has been raised to the stochastic

* This work was supported by the Director, Office of Energy Research, Office of High Energy Physics and Nuclear Physics, Division of High Energy Physics of the US Department of Energy under contract DE-AC03-76SF00098.

regularization scheme is that the identity

$$\left\langle \phi(x_1)\frac{\delta S[\phi]}{\delta\phi(x_2)}\right\rangle = \delta^4(x_1-x_2), \tag{1.1}$$

where $S[\phi]$ is the action, is modified by the loop corrections in quadratically divergent theories, such as $\lambda\phi^4$ theory. Gauge theories are a different matter because, at least in the "gluon channel", they are only *superficially* quadratically divergent, if the regularization scheme preserves the gauge invariance. If in the stochastic regularization scheme such quadratic divergences do not cancel, the scheme would fail anyway. If these quadratic divergences do cancel, identity (1.1) in the "gluon channel" is, in fact, preserved. This paper will show that at the one-loop level the quadratic divergences in the photon propagator of scalar electrodynamics do indeed cancel in the stochastic regularization scheme.

As a simple example of a gauge theory, this paper discusses the stochastic regularization of scalar electrodynamics. The infinite part of the photon self-energy is calculated to one-loop order using the stochastic regularizer and the infinite part of the photon vacuum polarization tensor is shown automatically to come out transverse, as it should. The photon does not acquire a mass at the one-loop level, because at zero external momentum the photon vacuum polarization is shown to vanish. By a diagrammatic calculation it is shown that the Ward identity that equates the scalar wave function renormalization, Z_2 to the one-photon vertex renormalization, Z_1, holds to all orders of the stochastically regularized theory.

This paper is divided into five main sections. Sect. 2 contains a brief overview of the ideas inherent in stochastic quantization that are needed in order to understand the regularization scheme. Sect. 3 gives an example of how the stochastic regularizer works for the case of a scalar theory, while sect. 4 contains the explicit one-loop calculation for scalar electrodynamics. Sect. 5 contains the proof of the Ward identity to all orders of perturbation theory. In sect. 6 the conclusions and comments are given.

2. Overview of stochastic quantization

Stochastic quantization is based upon some well-known ideas in nonequilibrium statistical mechanics [5]. For simplicity, at first, the stochastic quantization of a single scalar field, ϕ, with action, $S[\phi]$, will be considered. The usual starting point of stochastic quantization [1] is the Langevin equation

$$\frac{\partial\phi(x,t^5)}{\partial t^5} = -\frac{\delta S[\phi]}{\delta\phi(x,t^5)} + \eta(x,t^5), \tag{2.1}$$

in which t^5 is a fictitious fifth-time variable, not to be confused with physical time and x represents the four physical space-time dimensions. Here, η is a five-dimensional random field with gaussian probability distribution

$$\langle F[\phi(\eta)]\rangle_\eta \equiv \frac{\int D\eta\, F[\phi(\eta)] \exp\left(-\frac{1}{4}\int \eta^2(x,t^5)\, \mathrm{d}^4x\, \mathrm{d}t^5\right)}{\int D\eta \exp\left(-\frac{1}{4}\int \eta^2(x,t^5)\, \mathrm{d}^4x\, \mathrm{d}t^5\right)}. \tag{2.2}$$

By evaluating the generating functional, $\langle \exp(\int J\eta\, \mathrm{d}^4x\, \mathrm{d}t^5)\rangle_\eta$, all the n-point η correlation functions can easily be calculated. After a simple calculation the two-point correlation is found to be

$$\langle \eta(x,t^5)\eta(x',t^{5\prime})\rangle_\eta = 2\delta^4(x-x')\delta(t^5-t^{5\prime}), \tag{2.3}$$

while all other *connected* η correlations vanish.

The connection to the standard formulation of quantum field theory is arrived at by evaluating the equal fifth-time expectation values. That is, it is possible to prove that

$$\lim_{t^5\to\infty} \langle \phi(x_1,t^5)\phi(x_2,t^5)\ldots\phi(x_n,t^5)\rangle_\eta = \frac{\int D\phi\, \phi(x_1)\phi(x_2)\ldots\phi(x_n)\mathrm{e}^{-S[\phi]}}{\int D\phi\, \mathrm{e}^{-S[\phi]}}, \tag{2.4}$$

where $S[\phi]$ is the four-dimensional action. Note that on the right-hand side of the equation the field, ϕ, is a function of the four physical space-time dimensions, while on the left-hand side of the equation, ϕ is a function of the five-dimensional extended space. By starting the Langevin system at $t_0^5 = -\infty$, the system is equilibrated for any finite fifth-time, so there is no need to take the limit of infinite fifth-time to make the correspondence to the standard formulation of field theory.

There are quite a few proofs in the literature of the equivalence of stochastic quantization to the standard procedures of quantization. One way to make the connection is by defining the Fokker-Planck probability [6], which describes the probability density of finding the field ϕ at a given value under the Langevin dynamics. By deriving an evolution equation for the Fokker-Planck probability, it is possible to show that for essentially arbitrary initial conditions, at equilibrium, the Fokker-Planck probability reduces to the probability density of the ordinary formulation. There are also proofs based on the various perturbative expansions of stochastic quantization [7]. Another rather elegant proof makes use of a hidden supersymmetry [8].

The Langevin equation can be used to perturbatively solve quantum field theories. In general, the lagrangian will consist of a kinetic term plus an interaction potential. Thus, the Langevin equation is

$$\frac{\partial\phi(x,t^5)}{\partial t^5}+(-\partial^2+m^2)\phi(x,t^5)=-V'(\phi(x,t^5))+\eta(x,t^5), \tag{2.5}$$

where $V'(\phi)$ is the derivative of the potential with respect to the field ϕ. One way to handle this equation is with the method of Green functions:

$$\frac{\partial G(x-x',t^5-t^{5\prime})}{\partial t^5}+(-\partial^2+m^2)G(x-x',t^5-t^{5\prime})=\delta^4(x-x')\delta(t^5-t^{5\prime}). \tag{2.6}$$

The causal Green function in coordinate space is

$$G(x-x',t^5-t^{5\prime})=\theta(t^5-t^{5\prime})\int\frac{\mathrm{d}^4p}{(2\pi)^4}\mathrm{e}^{-ip(x-x')}\mathrm{e}^{-(p^2+m^2)(t^5-t^{5\prime})}. \tag{2.7}$$

The Green function can be used to rewrite the differential equation as an integral equation

$$\phi(x,t^5)=\int\mathrm{d}^4x'\int_{-\infty}^{\infty}\mathrm{d}t^{5\prime}G(x-x',t^5-t^{5\prime})\left[\eta(x',t^{5\prime})-V'(\phi(x',t^{5\prime}))\right], \tag{2.8}$$

that contains the initial condition that the field vanishes at $t_0^5=-\infty$, as well as the causality requirement. To simplify matters, a compact notation is introduced:

$$G_{x1}\equiv G(x-x_1,t^5-t_1^5),\qquad \eta_1\equiv\eta(x_1,t_1^5),\qquad \int_1\equiv\int\mathrm{d}^4x_1\int_{-\infty}^{\infty}\mathrm{d}t_1^5. \tag{2.9}$$

By iteration the integral equation (2.8) can be solved as a perturbative series:

$$\phi(x,t^5)=\int_1 G_{x1}\eta_1-\int_1 G_{x1}V'\left(\int_2 G_{12}\eta_2-\int_2 G_{12}V'\left(\int_3 G_{23}\eta_3-\cdots\right)\right). \tag{2.10}$$

An explicit example of how the Langevin equation can be used to generate a perturbation series is the massive scalar ϕ^4 theory. To the first order in the coupling constant the field is taken from eq. (2.10) to be

$$\phi(x,t^5)=\int_1 G_{x1}\eta_1-\frac{\lambda}{3!}\int_1 G_{x1}\left[\int_2 G_{12}\eta_2\right]^3+\cdots. \tag{2.11}$$

Fig. 1. Perturbative Langevin expansion of ϕ in $\lambda\phi^4/4!$ theory.

The tree diagrams corresponding to the perturbation series are given in fig. 1. Each line corresponds to a Green function, while the crosses at the ends of the diagrams represent the noise term, η. The vertex factors are the same as for ordinary Feynman diagrams, up to a possible combinatoric factor.

The loop diagrams come about by piecing together the tree diagrams (fig. 2). For example, the two-point correlation function is

$$\langle\phi(x,t^5)\phi(x',t^{5\prime})\rangle_\eta = \left\langle \int_1\int_2 G_{x1}G_{x'2}\eta_1\eta_2\right\rangle_\eta$$
$$-\frac{\lambda}{3!}\left\langle \int_1\int_2 [G_{x1}G_{x'2}+G_{x'1}G_{x2}]\eta_1\left[\int_3 G_{23}\eta_3\right]^3\right\rangle_\eta + O(\lambda^2). \tag{2.12}$$

From eq. (2.2), the n-point η correlation functions are sums of products of delta functions. The delta functions can be thought of as glue that holds the tree diagrams together to form the n-point ϕ correlations. As will be discussed in the next section, stochastic regularization consists of smearing the delta function glue in fifth-time.

The zeroth-order contribution is given by

$$\langle\phi(x,t^5)\phi(x,t^{5\prime})\rangle_\eta^{(0)} \equiv D(x-x',t^5-t^{5\prime})$$
$$= 2\int_1 G_{x1}G_{x'1} \tag{2.13}$$

$$= \int \frac{d^4p}{(2\pi)^4} e^{-ip(x-x')} \frac{e^{-(p^2+m^2)|t^5-t^{5\prime}|}}{p^2+m^2}. \tag{2.14}$$

Fig. 2. Expansion of two-point function in $\lambda\phi^4/4!$ theory.

Therefore, in momentum space, the zeroth-order free propagator is given by

$$D_{12}(p)=\frac{e^{-(p^2+m^2)|t_1^5-t_2^5|}}{p^2+m^2}, \tag{2.15}$$

where the subscript on $D_{12}(p)$ refers only to the fifth-time coordinate. After replacing the η correlations with the appropriate delta functions and combining terms that differ only by dummy indices, the first-order contribution is given by

$$\langle\phi(x,t^5)\phi(x',t^{5\prime})\rangle_\eta^{(1)}=-2\lambda\int_1\int_2\int_3[G_{x1}G_{x'2}G_{21}G_{23}G_{23}+G_{x'1}G_{x2}G_{21}G_{23}G_{23}]. \tag{2.16}$$

By explicit evaluation, it is easy to check that for $t^5=t^{5\prime}$, the same result is obtained as by using ordinary Feynman diagrams.

3. Stochastic Regularization

Since there is an extra dimension present in the Langevin approach, the infinities can be smeared without destroying any symmetries that are present in the corresponding four-dimensional theory. The preservation of the symmetries that are present in the infinite theory is crucial to finding a satisfactory regularization scheme. A time-smeared system is known as a nonmarkovian system [5]. In general, such a system can be expected to be less divergent than its markovian counterpart. From the perturbative point of view, stochastic regularization can be thought of as preventing the loops of the correlation functions from completely closing on themselves in the fifth-time.

There are at least two choices for fifth-time smearing the Langevin system. Either the Langevin equation or the probability distribution of the random noise, η, can be smeared. By studying the first-order correction in the $\lambda\phi^4$ theory, it is possible to show that the nonmarkovian Langevin equation

$$\frac{\partial\phi(x,t^5)}{\partial t^5}=-\int dt^{5\prime}\alpha_\Lambda(t^5-t^{5\prime})\frac{\delta S[\phi]}{\delta\phi(x,t^{5\prime})}+\eta(x,t^5), \tag{3.1}$$

where α_Λ is a smearing function, can at best only remove two degrees of divergence in the perturbation theory. Quadratically divergent integrals become logarithmically divergent, and there does not exist a regularization function that does better.

The other possibility is to smear the η probability functional [2]. In this scheme, the Langevin equation is left alone, while eq. (2.2) is replaced by

$$\langle F[\phi(\eta)]\rangle_\eta \equiv \frac{\int D\eta\, F[\phi(\eta)]\exp\left(-\frac{1}{4}\int \eta(x,t^5)\alpha_\Lambda^{-1}(t^5-t^{5\prime})\eta(x,t^{5\prime})\,\mathrm{d}^4x\,\mathrm{d}t^5\,\mathrm{d}t^{5\prime}\right)}{\int D\eta \exp\left(-\frac{1}{4}\int \eta(x,t^5)\alpha_\Lambda^{-1}(t^5-t^{5\prime})\eta(x,t^{5\prime})\,\mathrm{d}^4x\,\mathrm{d}t^5\,\mathrm{d}t^{5\prime}\right)}. \tag{3.2}$$

This changes the η correlation to

$$\langle \eta(x,t^5)\eta(x',t^{5\prime})\rangle_\eta = 2\delta^4(x-x')\alpha_\Lambda(t^5-t^{5\prime}). \tag{3.3}$$

The smearing functions α_Λ and α_Λ^{-1} are functional inverses of each other, in the sense that

$$\int \mathrm{d}t^{5\prime\prime}\alpha_\Lambda(t^5-t^{5\prime\prime})\alpha_\Lambda^{-1}(t^{5\prime\prime}-t^{5\prime}) = \delta(t^5-t^{5\prime}). \tag{3.4}$$

The hope is that, because

$$\lim_{\Lambda\to\infty}\alpha_\Lambda(t^5-t^{5\prime}) = \delta(t^5-t^{5\prime}), \tag{3.5}$$

as Λ becomes infinite, the original theory is recovered.

Since the Langevin equation is unaffected by the stochastic regularization, the physical field is the same as in the unregularized case, so that

$$\langle \phi(x,t^5)\phi(x',t^{5\prime})\rangle_\eta^{(0)} = \int_1\int_2 G_{x1}G_{x'2}\langle \eta_1\eta_2\rangle_\eta. \tag{3.6}$$

In this case, however, the two-point η correlation is given by eq. (3.3). Working in physical momentum space the zeroth-order propagator is

$$D_{12}^\Lambda(p) \equiv 2\int \mathrm{d}t_3^5 \int \mathrm{d}t_4^5 G_{13}(p)G_{24}(p)\alpha_\Lambda(t_3^5-t_4^5) \tag{3.7}$$

$$= 2\int_{-\infty}^{t_1^5}\mathrm{d}t_3^5\int_{-\infty}^{t_2^5}\mathrm{d}t_4^5\,\mathrm{e}^{-(t_1^5-t_3^5)(p^2+m^2)}\mathrm{e}^{-(t_2^5+t_4^5)(p^2+m^2)}\alpha_\Lambda(t_3^5-t_4^5) \tag{3.8}$$

$$= 2\int \frac{\mathrm{d}E}{2\pi}\mathrm{e}^{-iE(t_1^5-t_2^5)}\frac{\tilde{\alpha}_\Lambda(E)}{(p^2+m^2)^2+E^2}, \tag{3.9}$$

where the Fourier transform of the smearing function, $\tilde{\alpha}_\Lambda(E)$, has been introduced. Since there is an extra power of p^2 in the denominator over the ordinary Feynman propagator, a reduction of two degrees of divergence can be obtained, if $\tilde{\alpha}_\Lambda(E)$ cuts off for large values of E. Since all loops in the perturbative expansion of an arbitrary theory contain at least one factor of $\tilde{\alpha}_\Lambda(E)$, the logarithmically divergent loops can be expected to be rendered finite.

It is a little more difficult to regularize a theory whose diagrams are quadratically divergent. For example, the first-order correction to the scalar propagator in ϕ^4 theory, is

$$\langle \phi(x_1, t_1^5)\phi(x_2, t_2^5)\rangle_\eta^{(1)}$$

$$= -\frac{\lambda}{3!}\left\langle \int_3\int_4 [G_{14}G_{23} + G_{13}G_{24}]\eta_3\left[\int_5 G_{45}\eta_5\right]^3\right\rangle_\eta$$

$$= -\frac{\lambda}{2}\int \frac{\mathrm{d}^4k}{(2\pi)^4}\mathrm{e}^{-ik(x_1-x_2)}\int \mathrm{d}t_3^5\left[D_{13}^\Lambda(k)G_{23}(k) + D_{23}^\Lambda(k)G_{13}(k)\right]$$

$$\times \int \frac{\mathrm{d}^4p}{(2\pi)^4}D_{33}^\Lambda(p). \tag{3.10}$$

It is possible to find a necessary condition on the set of functions that can be used as regularizers by studying the loop of the first-order correction [2]. In this case, the loop is decoupled from the rest of the diagram, so the loop can be studied by itself. The loop is given by

$$L = \int \frac{\mathrm{d}^4p}{(2\pi)^4}D_{33}^\Lambda(p) \tag{3.11}$$

$$= 2\int_0^\infty \mathrm{d}t_4^5\int_0^\infty \mathrm{d}t_5^5\int \frac{\mathrm{d}^4p}{(2\pi)^4}\mathrm{e}^{-(t_4^5+t_5^5)(p^2+m^2)}\alpha_\Lambda(t_4^5 - t_5^5) \tag{3.12}$$

$$\sim \frac{2\alpha_\Lambda(0)}{(4\pi)^2}\int_0^\varepsilon \mathrm{d}T^5\,\frac{1}{T^5}\,. \tag{3.13}$$

In order for the integral to be finite, a necessary condition on the regularization function is that [2]

$$\alpha_\Lambda(0) = 0\,. \tag{3.14}$$

Using the Fourier transform of the smearing function, $\tilde{\alpha}_\Lambda(E)$, condition (3.14) can be rewritten as

$$\int \frac{\mathrm{d}E}{2\pi} \tilde{\alpha}_\Lambda(E) = 0. \tag{3.15}$$

Therefore, to remove quadratic divergences, the support of $\tilde{\alpha}_\Lambda(E)$ is not positive. The generating functional in euclidean space, in general, will not be well defined as can be seen by looking at the generating functional written in terms of the Fourier transformed fields:

$$Z[J] = \frac{\int D\eta \exp\left(-\int (\mathrm{d}^4p\,\mathrm{d}E/(2\pi)^5)\left[\frac{1}{4}|\eta(p,E)|^2/\tilde{\alpha}_\Lambda(E) - J^*(p,E)\phi(p,E)\right]\right)}{\int D\eta \exp\left(-\int (\mathrm{d}^4p\,\mathrm{d}E/(2\pi)^5)\left[\frac{1}{4}|\eta(p,E)|^2/\tilde{\alpha}_\Lambda(E)\right]\right)}. \tag{3.16}$$

This action is unbounded from below, which seems to rule out the nonperturbative usefulness of the stochastic regularizer for quadratically divergent theories [9]. For logarithmically divergent theories, such as supersymmetric theories, the nonperturbative usefulness of the stochastic regularizer is not ruled out.

4. Stochastic regularization of perturbative scalar electrodynamics

The manifestly covariant gauge fixed four-dimensional action of euclidean scalar electrodynamics is

$$S[A_\sigma, \phi^\dagger, \phi] = \int \mathrm{d}^4x \left[-\tfrac{1}{2}A_\mu\left(T_{\mu\nu} + \frac{1}{\lambda}L_{\mu\nu}\right)\partial^2 A_\nu + |(\partial_\mu - ieA_\mu)\phi|^2 + m^2|\phi|^2\right]. \tag{4.1}$$

Using the standard Feynman diagrammatical techniques, the quantum corrections to the vacuum polarization in scalar electrodynamics can easily be calculated. In doing the calculation, care must be taken, because the diagrams are infinite [10]. For example, the first-order correction to the vacuum polarization in euclidean space is given by (fig. 3):

$$\Pi_{\mu\nu}(k) = -2e^2 \int \frac{\mathrm{d}^4p}{(2\pi)^4} \frac{\delta_{\mu\nu}}{p^2 + m^2} + e^2 \int \frac{\mathrm{d}^4p}{(2\pi)^4} \frac{(2p+k)_\mu (2p+k)_\nu}{[(k+p)^2 + m^2](p^2 + m^2)}. \tag{4.2}$$

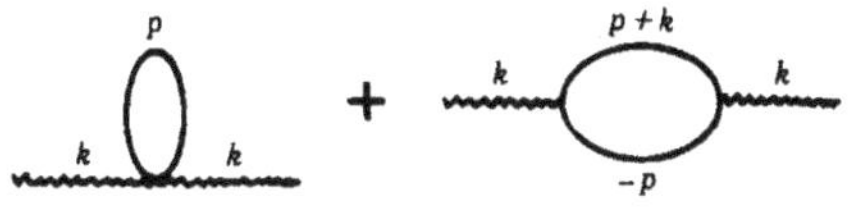

Fig. 3. One-loop correction to the photon propagator in SED using ordinary Feynman diagrams.

Using a naive momentum cutoff, Λ, on the integrals, to leading order in the cutoff, one obtains

$$\Pi_{\mu\nu}(k) \sim -\frac{e^2\Lambda^2}{16\pi^2}\delta_{\mu\nu}. \tag{4.3}$$

Thus, this naive regularizer explicitly breaks gauge invariance by giving the photon a mass.

An example of a well-known gauge invariant regularization scheme is dimensional regularization [3]. In this scheme the dimension of space-time is "analytically continued" to $4-\varepsilon$ dimensions, where the integral is finite. In this case, the photon mass correction contributions of the two diagrams just cancel to give a gauge invariant vacuum polarization.

$$\Pi_{\mu\nu}(k) = \frac{1}{3}\frac{e^2}{(4\pi)^2}\left(k_\mu k_\nu - k^2\delta_{\mu\nu}\right)\ln\frac{\Lambda^2}{m^2} + \text{regular terms}, \tag{4.4}$$

where the usual connection, $2/\varepsilon \leftrightarrow \ln\Lambda^2$, has been made and where Λ is a cutoff parameter with units of momentum.

As first discussed by Parisi and Wu [1], it is possible to formulate gauge theories without the need for gauge fixing, by using stochastic quantization. The gauge invariance manifests itself by a nonequilibrating random walk in the gauge parameter space. Since the physically interesting quantities are gauge invariant, the wandering in the gauge parameter space is essentially irrelevant. In fact, as Parisi and Wu pointed out, it is possible to rewrite the Langevin equations in terms of gauge invariant fields. Another simple way to avoid the nonequilibration of the abelian gauge field is by introducing a gauge fixing term, since the property that gauge fixing is unnecessary is unimportant for this study of regularization.

The Langevin equations of the gauge fixed scalar electrodynamics are

$$\frac{\partial\phi}{\partial t^5} = (\partial^2 - m^2)\phi - ieA_\mu\partial_\mu\phi - ie\,\partial_\mu(A_\mu\phi) - e^2A_\mu A_\mu\phi + \eta, \tag{4.5}$$

$$\frac{\partial\phi^\dagger}{\partial t^5} = (\partial^2 - m^2)\phi^\dagger + ieA_\mu\partial_\mu\phi^\dagger + ie\,\partial_\mu(A_\mu\phi^\dagger) - e^2A_\mu A_\mu\phi^\dagger + \eta^\dagger, \tag{4.6}$$

$$\frac{\partial A_\mu}{\partial t^5} = \left(T_{\mu\nu}\partial^2 + \frac{1}{\lambda}L_{\mu\nu}\partial^2\right)A_\nu - ie\phi^\dagger\left(\vec{\partial}_\mu - \overleftarrow{\partial}_\mu\right)\phi - 2e^2A_\mu\phi^\dagger\phi + \eta_\mu, \tag{4.7}$$

with unsmeared expectation values defined by

$$\langle F[A_\sigma,\phi^\dagger,\phi]\rangle_\eta \equiv \frac{\int D\eta_\mu D\eta^\dagger D\eta\, F[A_\sigma,\phi^\dagger,\phi]\exp\left(-\frac{1}{4}\int[\eta_\nu^2+2\eta^\dagger\eta]\,d^4x\,dt^5\right)}{\int D\eta_\mu D\eta^\dagger D\eta \exp\left(-\frac{1}{4}\int[\eta_\nu^2+2\eta^\dagger\eta]\,d^4x\,dt^5\right)}. \tag{4.8}$$

The causal Green function for the photon Langevin equation is

$$G_{\mu\nu}(x,t^5)=\theta(t^5)\int\frac{d^4k}{(2\pi)^4}e^{-ikx}\left[T_{\mu\nu}(k)e^{-k^2t^5}+L_{\mu\nu}(k)e^{-k^2t^5/\lambda}\right], \tag{4.9}$$

while in the unregularized theory the zeroth-order propagator is

$$D_{\mu\nu}(x,t^5)\equiv\langle A_\mu(x,t^5)A_\nu(0,0)\rangle_\eta^{(0)}$$

$$=\int\frac{d^4k}{(2\pi)^4}e^{-ikx}\left[\frac{T_{\mu\nu}(k)}{k^2}e^{-k^2t^5}+\frac{\lambda L_{\mu\nu}(k)}{k^2}e^{-k^2t^5/\lambda}\right], \tag{4.10}$$

where $T_{\mu\nu}(k)$ and $L_{\mu\nu}(k)$ are respectively the transverse and longitudinal projection operators. The two-point functions for the scalars are given in eqs. (2.7) and (2.14). As with ordinary Feynman diagrammatic calculations the simplest gauge to use is Feynman gauge, where $\lambda=1$. Henceforth, the Feynman gauge will be used exclusively.

An example of a function that satisfies the condition of eq. (3.14), and renders the loops finite, is [2]

$$\alpha_\Lambda^{(d)}(t^5-t^{5\prime})=\tfrac{1}{2}\Lambda^4|t^5-t^{5\prime}|e^{-\Lambda^2|t^5-t^{5\prime}|}, \tag{4.11}$$

The superscript refers to the fact that the Fourier transform of the above regularization function has a double pole structure. For calculational purposes it is easier to use a function whose Fourier transform has a single pole structure. Namely,

$$\alpha_\Lambda^{(s)}(t^5-t^{5\prime})=\tfrac{1}{2}\Lambda^2e^{-\Lambda^2|t^5-t^{5\prime}|}, \tag{4.12}$$

which does not satisfy the requirements of a quadratic divergence regularization function. The two functions are related by

$$\alpha_\Lambda^{(d)}(t^5-t^{5\prime})=-\Lambda^4\frac{\partial}{\partial\Lambda^2}\left[\frac{\alpha_\Lambda^{(s)}(t^5-t^{5\prime})}{\Lambda^2}\right]. \tag{4.13}$$

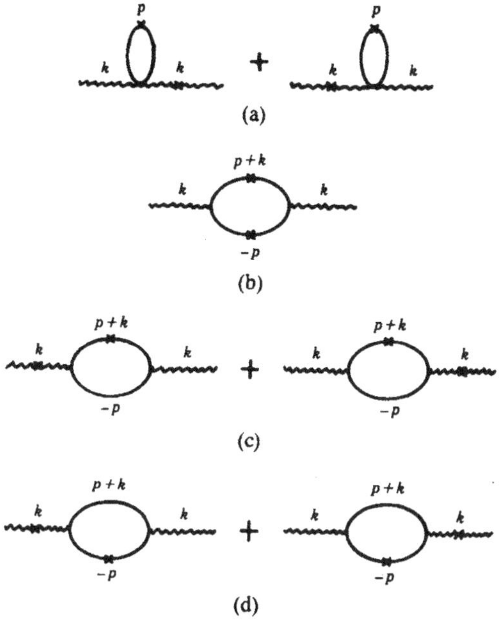

Fig. 4. One-loop correction to the photon propagator in SED using Langevin diagrams.

Therefore, $\alpha_\Lambda^{(s)}$ can be used until a divergent integral is to be evaluated, where eq. (4.13) will be used to replace $\alpha_\Lambda^{(s)}$ with $\alpha_\Lambda^{(d)}$, within the calculation.

Since this section is only concerned with the perturbative one-loop expansion of the photon propagator, the photon random noise field need not be fifth-time smeared, since only scalars appear within the loops. Using eqs. (3.8) and (4.12), the zeroth-order regularized scalar two-point function is

$$D_{12}^{(s)}(p) = e^{-|t_1^5 - t_2^5|(p^2+m^2)}\left[\frac{\Lambda^2}{(p^2+m^2)(p^2+m^2+\Lambda^2)} - \frac{\Lambda^2}{(p^2+m^2)^2-\Lambda^4}\right]$$

$$+ e^{-\Lambda^2|t_1^5 - t_2^5|}\frac{\Lambda^2}{(p^2+m^2)^2-\Lambda^4}. \tag{4.14}$$

Note that the apparent singularity at $p^2+m^2 = \pm\Lambda^2$ is fictitious.

The seven Langevin diagrams of the one-loop correction to the photon propagator in scalar electrodynamics are given in fig. 4. Since only physical expectation values are of interest, the external fifth-times are taken to be equal. Introducing the simplifying notation

$$a \equiv p^2 + m^2, \qquad b \equiv (p+k)^2 + m^2, \tag{4.15}$$

the diagrams with no external momenta in the loop (fig. 4a) are given by

$$P^{(d)1}_{\rho\sigma}(k) = -4e^2\delta_{\mu\nu}\int dt_1^5\, D^{\rho\sigma}_{01}(k) G^{\sigma\nu}_{01}(k)\int \frac{d^4p}{(2\pi)^4}\left(-\Lambda^4\frac{\partial}{\partial\Lambda^2}\right)\left\{\frac{D^{(s)}_{11}(p)}{\Lambda^2}\right\} \tag{4.16}$$

$$= -\frac{2e^2\delta_{\rho\sigma}}{k^4}\int\frac{d^4p}{(2\pi)^4}\frac{\Lambda^4}{a(a+\Lambda^2)^2}. \tag{4.17}$$

The other diagrams are significantly more complicated because of the intertwining of the external legs with the loop. In order to simplify the expressions, the vertex factors will be written as

$$V_{\mu\nu} \equiv e^2(2p+k)_\mu(2p+k)_\nu. \tag{4.18}$$

The diagram in fig. 4b is given by

$$P^{(d)2}_{\rho\sigma}(k) = \int dt_1^5\int dt_2^5\, G^{\rho\sigma}_{01}(k) G^{\sigma\nu}_{02}(k)\int\frac{d^4p}{(2\pi)^4}V_{\mu\nu}D^{(d)}_{12}(p+k)D^{(d)}_{12}(p) \tag{4.19}$$

$$\begin{aligned}= \frac{\delta_{\rho\mu}\delta_{\sigma\nu}}{k^2}&\left(-\Lambda_1^4\frac{\partial}{\partial\Lambda_1^2}\right)\left(-\Lambda_2^4\frac{\partial}{\partial\Lambda_2^2}\right)\int\frac{d^4p}{(2\pi)^4}V_{\mu\nu}\frac{1}{b^2-\Lambda_1^4}\frac{1}{a^2-\Lambda_2^4}\\
&\times\left[\frac{\Lambda_1^2\Lambda_2^2}{ab(a+b+k^2)} - \frac{\Lambda_2^2}{a(a+k^2+\Lambda_1^2)}\right.\\
&\left.\left. - \frac{\Lambda_1^2}{b(b+k^2+\Lambda_2^2)} + \frac{1}{(k^2+\Lambda_1^2+\Lambda_2^2)}\right]\right|_{\Lambda_1=\Lambda_2=\Lambda}\end{aligned} \tag{4.20}$$

$$\begin{aligned}= \frac{\delta_{\rho\mu}\delta_{\sigma\nu}}{k^4}&\left(-\Lambda_1^4\frac{\partial}{\partial\Lambda_1^2}\right)\left(-\Lambda_2^4\frac{\partial}{\partial\Lambda_2^2}\right)\int\frac{d^4p}{(2\pi)^4}V_{\mu\nu}k^2\\
&\times\Big[(\Lambda_1^2+a+b+k^2)\Lambda_2^4 + (\Lambda_2^2+a+b+k^2)\Lambda_1^4\\
&\quad+(3k^2+2b+2a)\Lambda_1^2\Lambda_2^2 + k^2(2k^2+3b+3a)(\Lambda_1^2+\Lambda_2^2)\\
&\quad+(a+b)^2(\Lambda_1^2+\Lambda_2^2) + k^4(k^2+2b+2a)\\
&\quad+k^2(a^2+3ab+b^2)+ab^2+a^2b\Big]\\
&\times\Big[ab(a+b+k^2)(\Lambda_1^2+b)(\Lambda_2^2+a)(\Lambda_1^2+a+k^2)\\
&\times(\Lambda_2^2+b+k^2)(\Lambda_1^2+\Lambda_2^2+k^2)\Big]^{-1}\Big|_{\Lambda_1=\Lambda_2=\Lambda},\end{aligned} \tag{4.21}$$

where the two regularization parameters are distinguished, in order to be able to differentiate individually each of the two regularization functions contained within the diagram. Later Λ_1 will be set equal to Λ_2. The diagrams in fig. 4c contribute a value of

$$P^{(d)3}_{\rho\sigma}(k) = 2\int dt_1^5 \int dt_2^5\, D^{\rho\sigma}_{01}(k) G^{\sigma\nu}_{02}(k) \int \frac{d^4p}{(2\pi)^4} V_{\mu\nu} D^{(d)}_{12}(p+k) G_{21}(p) \tag{4.22}$$

$$= \frac{\delta_{\rho\mu}\delta_{\sigma\nu}}{k^4} \int \frac{d^4p}{(2\pi)^4} V_{\mu\nu} \left(-\Lambda^4 \frac{\partial}{\partial\Lambda^2}\right) \left\{ \frac{1}{b^2-\Lambda^4} \left[\frac{1}{k^2+a+\Lambda^2} - \frac{\Lambda^2}{b(k^2+a+b)} \right] \right\} \tag{4.23}$$

$$= \frac{\delta_{\rho\mu}\delta_{\sigma\nu}}{k^4} \int \frac{d^4p}{(2\pi)^4} V_{\mu\nu} \left(-\Lambda^4 \frac{\partial}{\partial\Lambda^2}\right) \frac{(k^2+a+b+\Lambda^2)}{b(b+\Lambda^2)(a+b+k^2)(a+k^2+\Lambda^2)} . \tag{4.24}$$

Similarly the last two diagrams can be evaluated. The values are identical to the diagrams just calculated, as can be shown either by symmetry or by shifting the variables of integration. Therefore, the diagrams in fig. 4d contribute a value of

$$P^{(d)4}_{\rho\sigma}(k) = P^{(d)3}_{\rho\sigma}(k) . \tag{4.25}$$

In order to make the theory finite the results obtained by using $\alpha^{(s)}_\Lambda$ are taken and differentiated in order to obtain the results by using $\alpha^{(d)}_\Lambda$. For calculational purposes it is better to use the form of the vacuum polarization that contains no apparent singularities. After truncating the external photon lines the vacuum polarization of the photon is

$$\Pi^{(d)}_{\mu\nu}(k) \equiv \Pi^{(d)1}_{\mu\nu}(k) + \Pi^{(d)2}_{\mu\nu}(k) + \Pi^{(d)3}_{\mu\nu}(k) + \Pi^{(d)4}_{\mu\nu}(k), \tag{4.26}$$

where

$$\Pi^{(d)1}_{\mu\nu}(k) = -2e^2\delta_{\mu\nu} \int \frac{d^4p}{(2\pi)^4} \frac{\Lambda^4}{a(a+\Lambda^2)^2} , \tag{4.27}$$

$$\Pi^{(d)2}_{\rho\sigma}(k) = k^2 \left(-\Lambda_1^4 \frac{\partial}{\partial\Lambda_1^2}\right) \left(-\Lambda_2^4 \frac{\partial}{\partial\Lambda_2^2}\right) \int \frac{d^4p}{(2\pi)^4} V_{\mu\nu}$$
$$\times \Big[(\Lambda_1^2+a+b+k^2)\Lambda_2^4 + (\Lambda_2^2+a+b+k^2)\Lambda_1^4$$
$$+ (3k^2+2b+2a)\Lambda_1^2\Lambda_2^2 + k^2(2k^2+3b+3a)(\Lambda_1^2+\Lambda_2^2)$$
$$+ (a+b)^2(\Lambda_1^2+\Lambda_2^2) + k^4(k^2+2b+2a)$$
$$+ k^2(a^2+3ab+b^2) + ab^2 + a^2b \Big]$$

$$\times\left[ab(a+b+k^2)(\Lambda_1^2+b)(\Lambda_2^2+a)(\Lambda_1^2+a+k^2)\right.$$

$$\left.\times(\Lambda_2^2+b+k^2)(\Lambda_1^2+\Lambda_2^2+k^2)\right]^{-1}\Big|_{\Lambda_1=\Lambda_2=\Lambda}, \tag{4.28}$$

$$\Pi_{\mu\nu}^{(\mathrm{d})3}(k)=e^2\int\frac{\mathrm{d}^4p}{(2\pi)^4}\frac{\Lambda^4(2p+k)_\mu(2p+k)_\nu}{b(b+\Lambda^2)(a+b+k^2)(a+k^2+\Lambda^2)}$$

$$\times\left[\frac{a+b+k^2+\Lambda^2}{b+\Lambda^2}+\frac{a+b+k^2+\Lambda^2}{a+k^2+\Lambda^2}-1\right], \tag{4.29}$$

$$\Pi_{\mu\nu}^{(\mathrm{d})4}(k)=e^2\int\frac{\mathrm{d}^4p}{(2\pi)^4}\frac{\Lambda^4(2p+k)_\mu(2p+k)_\nu}{a(a+\Lambda^2)(a+b+k^2)(b+k^2+\Lambda^2)}$$

$$\times\left[\frac{a+b+k^2+\Lambda^2}{a+\Lambda^2}+\frac{a+b+k^2+\Lambda^2}{b+k^2+\Lambda^2}-1\right]. \tag{4.30}$$

Although these integrals may seen quite formidable, only a few of the terms will contribute to the infinite part of the vacuum polarization.

A fundamental consequence of the gauge invariance of scalar electrodynamics is that the photon does not acquire a mass by the higher-order corrections to the vacuum polarization. Setting the external momentum to zero, the exact mass correction to the photon can be found. Explicitly,

$$\Pi_{\mu\mu}^{(\mathrm{d})}(0)=\Pi_{\mu\mu}^{(\mathrm{d})1}(0)+2\Pi_{\mu\mu}^{(\mathrm{d})3}(0), \tag{4.31}$$

where

$$\Pi_{\mu\mu}^{(\mathrm{d})1}(0)=-8e^2\int\frac{\mathrm{d}^4p}{(2\pi)^4}\frac{\Lambda^4}{(p^2+m^2)(p^2+m^2+\Lambda^2)^2}, \tag{4.32}$$

$$2\Pi_{\mu\mu}^{(\mathrm{d})3}(0)=4\Lambda^4e^2\int\frac{\mathrm{d}^4p}{(2\pi)^4}\frac{p^2}{p^2+m^2}$$

$$\times\left[\frac{1}{(p^2+m^2)(p^2+m^2+\Lambda^2)^2}+\frac{2}{(p^2+m^2+\Lambda^2)^3}\right] \tag{4.33}$$

$$=4\Lambda^4e^2\int_0^1\mathrm{d}z\int_0^\infty\frac{\mathrm{d}p^2}{(4\pi)^2}\frac{6zp^4}{(p^2+m^2+z\Lambda^2)^4} \tag{4.34}$$

$$=8\Lambda^4e^2\int_0^1\mathrm{d}z\int_0^\infty\frac{\mathrm{d}p^2}{(4\pi)^2}\frac{2zp^2}{(p^2+m^2+z\Lambda^2)^3} \tag{4.35}$$

$$=8e^2\int\frac{\mathrm{d}^4p}{(2\pi)^4}\frac{\Lambda^4}{(p^2+m^2)(p^2+m^2+\Lambda^2)^2}. \tag{4.36}$$

Thus, the desired result is

$$\Pi^{(\mathrm{d})}_{\mu\mu}(0)=0, \tag{4.37}$$

and the mass correction vanishes.

A direct evaluation of the finite parts of the vacuum polarization with the stochastic regularizer is fairly involved and will not be discussed here. We have computed only the infinite part of the vacuum polarization for nonzero external momentum. The contribution to the vacuum polarization of the simplest diagrams is from eq. (4.27):

$$\Pi^{(\mathrm{d})1}_{\mu\nu}(k)=-2e^2\delta_{\mu\nu}\int\frac{\mathrm{d}^4p}{(2\pi)^4}\frac{\Lambda^4}{(p^2+m^2)(p^2+m^2+\Lambda^2)^2} \tag{4.38}$$

$$=-\frac{e^2}{8\pi^2}\delta_{\mu\nu}\left[\Lambda^2-m^2\ln\left(\frac{m^2+\Lambda^2}{m^2}\right)\right]. \tag{4.39}$$

The next contribution is given by eq. (4.28). By power counting, the integral in eq. (4.28) is finite even before differentiating with respect to Λ^2. Note that the only possible singularity as $\Lambda\to\infty$ is logarithmic. In fact, since an ultraviolet divergence in Λ^2 can only occur when there is an infrared divergence in m^2, the terms with no such divergence in m^2 can immediately be eliminated as being finite. As a further simplification, k^2 can be set to zero within the integral, without affecting the leading order in Λ^2. Also m^2 can be neglected except where it is needed to prevent an infrared divergence within the integral. After performing all these simplifications, eq. (4.28) is reduced to

$$\Pi^{(\mathrm{d})2}_{\mu\nu}(k)=\Lambda_1^4\Lambda_2^4e^2k^2\delta_{\mu\nu}$$

$$\times\frac{\partial}{\partial\Lambda_1^2}\frac{\partial}{\partial\Lambda_2^2}\int\frac{\mathrm{d}^4p}{(2\pi)^4}\frac{1}{2p^2(p^2+m^2)}\frac{\Lambda_1^2\Lambda_2^2}{(p^2+\Lambda_1^2)^2(p^2+\Lambda_2^2)^2}\Bigg|_{\Lambda_1=\Lambda_2=\Lambda}$$

$$+\text{regular terms}. \tag{4.40}$$

This integral can be evaluated with the usual Feynman parameterization to arrive at the result

$$\Pi^{(\mathrm{d})2}_{\mu\nu}(k)=\tfrac{1}{2}\delta_{\mu\nu}\frac{e^2k^2}{(4\pi)^2}\ln\frac{\Lambda^2}{m^2}+\text{regular terms}, \tag{4.41}$$

where all terms that are finite as $\Lambda\to\infty$ have not been calculated. In the remaining contributions from eqs. (4.29) and (4.30), k^2 can be neglected compared to Λ^2. As usual, this type of integral is done by first Feynman parameterization and then evaluating the momentum integrals. After neglecting all the terms that are finite as $\Lambda \to \infty$, the result is

$$\Pi^{(\mathrm{d})3}_{\mu\nu}(k) = \Pi^{(\mathrm{d})4}_{\mu\mu}(k)$$
$$= -\frac{e^2}{(4\pi)^2}\left[\delta_{\mu\nu}\left(\Lambda^2 - m^2\ln\frac{\Lambda^2}{m^2} - \tfrac{5}{12}k^2\ln\frac{\Lambda^2}{m^2}\right) + \tfrac{1}{6}k_\mu k_\nu \ln\frac{\Lambda^2}{m^2}\right] + \text{regular terms}. \tag{4.42}$$

By adding everything together, the momentum-independent pieces cancel and the infinite part of the one-loop vacuum polarization is found to be

$$\Pi^{(\mathrm{d})}_{\mu\nu} = \frac{1}{3}\frac{e^2}{(4\pi)^2}\left(k_\mu k_\nu - k^2\delta_{\mu\nu}\right)\ln\frac{\Lambda^2}{m^2} + \text{regular terms}. \tag{4.43}$$

This is precisely the correct value, as was obtained by using dimensional regularization.

As discussed by Ishikawa [4], a modification in the identity

$$\left\langle \phi(x_1)\frac{\delta S[\phi]}{\delta\phi(x_2)}\right\rangle = \delta^4(x_1 - x_2) \tag{4.44}$$

can occur in stochastically regularized quadratically divergent scalar field theories. The leading behavior of quadratically divergent loops is proportional to Λ^2, while the external legs of the Langevin diagrams may possess a Λ^{-2} dependence. The combination of these two factors can yield an extra finite nonzero contribution, in the limit that the cutoff becomes infinite.

Although it is not clear what the relevance of this fact is to regularization and renormalization, it is straightforward to show that no problem occurs at the one-loop level in the gluon channel of stochastically regularized scalar electrodynamics. In order for there to be a possibility of modifying

$$\left\langle A_\mu(x_1)\frac{\delta S[A_\sigma, \phi^\dagger, \phi]}{\delta A_\nu(x_2)}\right\rangle = \delta_{\mu\nu}\delta^4(x_1 - x_2), \tag{4.45}$$

where $S[A_\sigma, \phi^\dagger, \phi]$ is the action of scalar electrodynamics (4.1), the photon random noise field should also be fifth-time smeared. Keeping only the leading behavior of the quadratically divergent loops, explicit calculation shows that there is no modification of identity (4.45). The coefficients of the various factors of Λ^2 that occur in the one-loop evaluation of the left-hand side of eq. (4.45) can be obtained by comparison to the results for the various contributions to the photon vacuum polarization. Just as the quadratic divergences proportional to Λ^2 have cancelled in the vacuum polarization, the quadratic divergences cancel in the explicit evaluation of the left-hand side of (4.45), and no modification of the identity occurs in the stochastic regularization of scalar electrodynamics. Of course, in the charged scalar channel, the identity analogous to (4.44) would again be quadratically divergent. In the case of pure Yang-Mills or QCD with fermions, one would expect the whole phenomenon to disappear, because all quadratic divergences are spurious.

5. Diagrammatic proof of the Ward identity to all orders

By working with the standard Feynman diagrams, it is possible to prove the Ward identity [10] (in Feynman gauge)

$$\lim_{q\to 0} q^2 V_\sigma(p, p-q) = -e\frac{\partial S(p)}{\partial p_\sigma}, \tag{5.1}$$

where $V_\sigma(p, p-q)$ is the complete three-point function (fig. 5) and $S(p)$ is the complete scalar propagator (fig. 6). A regularization scheme that preserves this identity implies that $Z_1 = Z_2$. The proof using the Langevin formulation is analogous to the proof using ordinary Feynman diagrams. The main difference is that

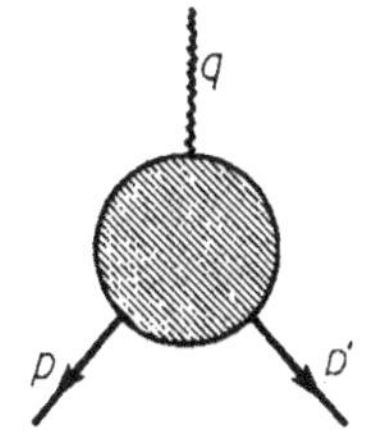

Fig. 5. Complete three-point function in scalar electrodynamics.

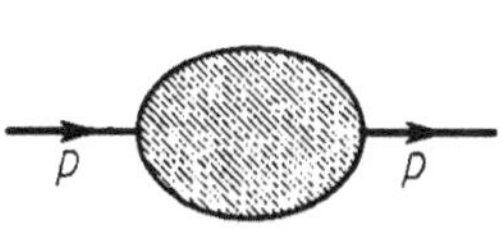

Fig. 6. Complete scalar propagator in scalar electrodynamics.

there are two types of two-point functions to consider. In the Langevin perturbative expansion, the external photon line can either be $D^{12}_{\mu\nu}(q)$ or $G^{12}_{\mu\nu}(q)$. In this section all orders of perturbation theory are being considered, so the probability density of the photon noise field must also be fifth-time smeared. In Feynman gauge, the stochastically regularized two-point functions, using $\alpha^{(d)}_{\Lambda}$, are

$$G^{12}_{\mu\nu}(q)=\delta_{\mu\nu}e^{-q^2(t_1^5-t_2^5)}\theta\left(t_1^5-t_2^5\right), \tag{5.2}$$

$$D^{12}_{\mu\nu}(q)=-\delta_{\mu\nu}\Lambda^4\frac{\partial}{\partial\Lambda^2}\left[\mathrm{e}^{-|t_1^5-t_2^5|q^2}\left(\frac{1}{q^2(q^2+\Lambda^2)}-\frac{1}{q^4-\Lambda^4}\right)+\mathrm{e}^{-\Lambda^2|t_1^5-t_2^5|}\frac{1}{q^4-\Lambda^4}\right]. \tag{5.3}$$

Since $G^{12}_{\mu\nu}(q)$ does not have a pole at $q^2=0$, vertex diagrams whose external photon line is $G^{12}_{\mu\nu}(q)$, do not contribute to the Ward identity (5.1). Vertex diagrams whose external photon line is $D^{12}_{\mu\nu}(q)$ do contribute, but in a simple way, because

$$\lim_{q\to 0} q^2 D^{12}_{\mu\nu}(q)=\delta_{\mu\nu}. \tag{5.4}$$

The proof of the Ward identity will proceed by showing that inserting a photon at zero momentum, q, at a given point in a typical Langevin diagram and then multiplying by q^2, is the same as differentiating that part of the Langevin diagram with respect to the momentum flowing through that point. Summing over all the possible ways to insert the photon into a diagram is therefore equal to summing over all the possible ways of differentiating with respect to the momentum flowing through the scalar lines. The sum over a closed scalar loop vanishes, because the loop momentum is integrated over. Thus, only the derivatives with respect to the momentum flowing through the scalar line that begins and ends externally are left. This is just the Ward identity (5.1).

All that remains to be done is to explicitly check that inserting a photon at zero momentum at a given point in a Langevin diagram is indeed equivalent to differentiating that part of the diagram with respect to the momentum flowing through the scalar. There are two types of scalar two-point functions that appear within a typical Langevin diagram and one type of vertex factor that explicitly contain momentum dependence. Since

$$\left[\underset{1}{}\overset{k}{\rule{3cm}{0.4pt}}\underset{2}{}\right]\equiv G_{12}(k)=\mathrm{e}^{-(k^2+m^2)(t_1^5-t_2^5)}\theta\left(t_1^5-t_2^5\right), \tag{5.5}$$

$$-e\frac{\partial}{\partial p_\sigma}G_{12}(k)=2ek_\sigma\left(t_1^5-t_2^5\right)G_{12}(k), \tag{5.6}$$

where $k=p-\Sigma p_i$, p is the scalar line momentum, and the p_i are the momenta of internal photons that are attached to the scalar. On the other hand,

$$\lim_{q\to 0} q^2 \Bigg[$$ 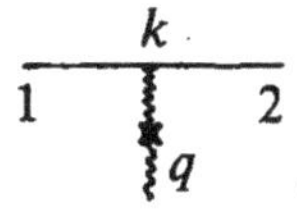 $$\Bigg] = 2ek_\sigma \int dt_3^5 G_{13}(k) G_{32}(k)$$

$$= 2ek_\sigma (t_1^5 - t_2^5) G_{12}(k) . \tag{5.7}$$

Therefore, diagrammatically,

$$-e\frac{\partial}{\partial p_\sigma}\Big[$$ 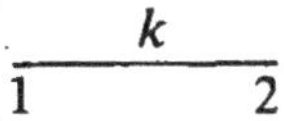 $$\Big] = \lim_{q\to 0} q^2 \Bigg[$$ 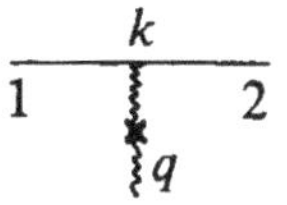 $$\Bigg] . \tag{5.8}$$

Similarly,

$$D_{12}(k) = \int dt_3^5 dt_4^5 \alpha_{34}^{\Lambda} G_{13}(k) G_{24}(k) , \tag{5.9}$$

yields

$$-e\frac{\partial}{\partial p_\sigma}\Big[$$ 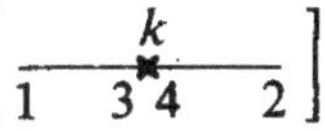 $$\Big]$$

$$= 2ek_\sigma \int dt_3^5 dt_4^5 \alpha_{34}^{\Lambda} \big[(t_1^5 - t_3^5) + (t_2^5 - t_4^5) \big] G_{13}(k) G_{24}(k) . \tag{5.10}$$

Attaching the photon in the two possible ways results in

$$\lim_{q\to 0} q^2 \Bigg[$$ $$\Bigg] = 2ek_\sigma \int dt_3^5 \int dt_4^5 \int dt_5^5 \alpha_{34}^{\Lambda} G_{13}(k) G_{25}(k) G_{54}(k)$$

$$+ 2ek_\sigma \int dt_3^5 \int dt_4^5 \int dt_5^5 \alpha_{34}^{\Lambda} G_{15}(k) G_{53}(k) G_{24}(k) \tag{5.11}$$

$$= 2ek_\sigma \int dt_3^5 \int dt_4^5 \alpha_{34}^{\Lambda} G_{13}(k) G_{23}(k)$$

$$\times \big[(t_2^5 - t_4^5) + (t_1^5 - t_3^5) \big] . \tag{5.12}$$

Thus, diagrammatically,

$$-e\frac{\partial}{\partial p_\sigma}\Big[$$ 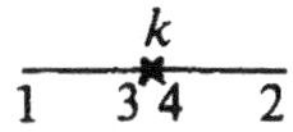 $$\Big] = \lim_{q\to 0} q^2 \Bigg[$$ 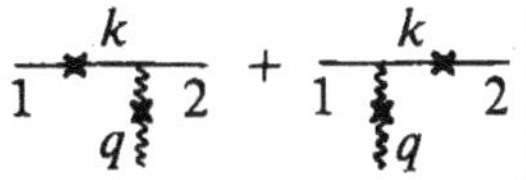 $$\Bigg] . \tag{5.13}$$

The reader may have noted that the fact that adding an external truncated photon to $D_{12}(k)$ is equivalent to differentiating with respect to the momentum flowing through the scalar is already contained in the fact that attaching an external truncated photon to $G_{12}(k)$ is equivalent to differentiating $G_{12}(k)$. However, the point was to explicitly show that the regularizer does not affect the results. Differentiating the one-photon vertex factor yields the two-photon vertex factor

$$-e\frac{\partial}{\partial p_\sigma}\left[e\left(2p-2\sum p_i-p_j\right)_\mu\right]=-2e^2\delta_{\mu\sigma}, \tag{5.14}$$

or diagrammatically,

$$-e\frac{\partial}{\partial p_\sigma}\left[\quad\right]=\lim_{q\to 0} q^2\left[\quad\right], \tag{5.15}$$

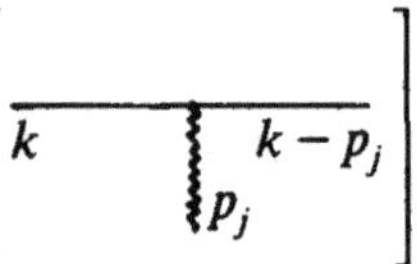

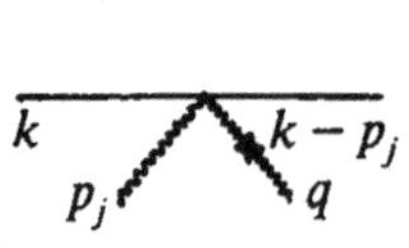

where only the vertex factor is to be differentiated on the left-hand side. Thus, it follows that the Ward identity (5.1) holds to all orders of perturbation theory.

6. Conclusions and comments

This paper showed that the stochastic regularizer does, in fact, yield the correct gauge invariant infinite part of the one-loop photon vacuum polarization. The Ward identity that equates the scalar wave function renormalization to the one-photon vertex renormalization was shown to hold to all orders of perturbation theory. Of course, it is possible that above the one-loop level, stochastic regularization breaks down, but the Ward identity would still hold. These results seem to indicate that the stochastic regularizer may be useful as a regularizer that preserves the symmetries and *relevant* identities that are present in the corresponding infinite theory. As noted previously, for logarithmically divergent theories there may be nonperturbative applications, but this requires a more detailed examination. Although this paper dealt only with scalar electrodynamics, it should be possible to extend the results presented in this paper to fermions [11] as well as to nonabelian gauge theories [12].

The author would like to thank Orlando Alvarez, Dae Sung Hwang, and Matt Visser for helpful discussions. Most of all, the author acknowledges the many helpful discussions with Professor M.B. Halpern. This work was supported by the Director, Office of Energy Research, Office of High Energy Physics and Nuclear Physics, Division of High Energy Physics of the US Department of Energy under contract DE-AC03-76SF00098.

References

[1] G. Parisi and Wu Yong-Shi, Sci. Sin. 24 (1981) 483
[2] J.D. Breit, S. Gupta and A. Zaks, Nucl. Phys. B233 (1984) 61
[3] L.D. Faddeev and A.A. Slavnov, Gauge fields: Introduction to quantum theory (Benjamin/Cummings, Reading, 1980)
[4] K. Ishikawa, Nucl. Phys. B241 (1984) 589
[5] P. Résibois and M. DeLeener, Classical kinetic theory of fluids (Wiley, New York, 1977);
N.G. van Kampen, Stochastic processes in chemistry and physics (North-Holland, Amsterdam, 1981)
[6] B. Sakita, Proc. 7th Johns Hopkins Workshop (World Scientific, 1983)
[7] W. Grimus and H. Hüffel, Z. Phys. C18 (1983) 129;
E. Egorian and S. Kalitzin, Phys. Lett. 129B (1983) 320
[8] E. Gozzi, CCNY-HEP-84/3 (1984);
R. Kirschner, Phys. Lett. 139B (1984) 180
[9] M.B. Halpern, private communication
[10] C. Itzykson and J.B. Zuber, Quantum field theory (McGraw-Hill, New York, 1980);
J.D. Bjorken and S.D. Drell, Relativistic quantum fields (McGraw-Hill, New York, 1965)
[11] P.H. Damgaard and K. Tsokos, Nucl. Phys. B235[FS11] (1984) 75;
B. Sakita, Proc. 7th Johns Hopkins Workshop (World Scientific, 1983);
J.D. Breit, S. Gupta and A. Zaks, Nucl. Phys. B233 (1984) 61;
K. Ishikawa, Nucl. Phys. B241 (1984) 589
[12] Mikio Namiki, Ichiro Ohba, Keisuke Okano and Yoshiya Yamanaka, Prog. Theor. Phys. 69 (1983) 1580

PHYSICAL REVIEW LETTERS

VOLUME 54 4 FEBRUARY 1985 NUMBER 5

Evaluation of Critical Exponents on the Basis of Stochastic Quantization

J. Alfaro[(a)]
International Centre for Theoretical Physics, Trieste, Italy

and

R. Jengo
International Centre for Theoretical Physics, Trieste, Italy, and International School for Advanced Studies, Trieste, Italy, and Istituto Nazionale di Fisica Nucleare, Sezione di Trieste, Italy

and

N. Parga[(b)]
International Centre for Theoretical Physics, Trieste, Italy
(Received 17 September 1984)

In the context of stochastic quantization of field theories we propose a method to make analytic computations of critical exponents and we evaluate them for $(\phi^2)^2_3$. It consists of regularization of the theory by use of a convenient non-Markovian process, where nonlocality in time is measured by the regularizing parameter σ. For a fixed dimensionality there is a value of σ where the theory is renormalizable and asymptotically free in the infrared, allowing a perturbative expansion around it.

PACS numbers: 03.70.+k, 05.70.Jk

Analytic computations of critical exponents for three-dimensional systems with a finite number of field components within the framework of perturbative field theory have been possible to do so far only by means of the ϵ-expansion method.[1] As is well known, the tree approximation yields mean-free values, but infrared (IR) divergences prevent us from computing loop corrections at any dimension less than four, in particular for $d=3$. A way out of this problem is to consider the case $d<4$ as an expansion around the four-dimensional theory; a second aspect of this expansion is that the β function has a zero of order $\epsilon=4-d$ and the method also provides a small parameter to make a perturbative calculation.[2]

In this note we propose an analytical method to obtain exponents in perturbative field theory directly at the physical value of the dimensionality. In particular, we shall apply it to the interesting case of $(\phi^2)^2$ in three dimensions.

The technique makes a nontrivial use of the stochastic quantization of field theory. This quantization procedure, introduced by Parisi and Wu,[3] has already been extensively used; see, e.g., Parisi *et al.*[4] It consists of introduction of an extra time dimension t and imposition of the following equation for the classical field:

$$(\partial/\partial t)\phi(x,t)=-[\delta/\delta\phi(x,t)]S+\xi(x,t), \qquad (1)$$

where S is the classical action and $\xi(x,t)$ is a Gaussian delta-correlated random force:

$$\langle\xi(x,t)\xi(x',t')\rangle=2\delta^d(x-x')\delta(t-t'). \qquad (2)$$

In this formalism, Green's functions are obtained by the taking of the infinite-time limit of the average over ξ of the corresponding product of fields $\phi(x,t)$:

$$G(x_1,x_2,\ldots,x_n)=\lim_{t\to\infty}\langle\phi(x_1,t)\phi(x_2,t)\cdots\phi(x_n,t)\rangle. \qquad (3)$$

The field theory defined in this way reproduces the usual one; therefore, it still has the usual ultraviolet (uv) divergences and first it must be regularized. An interesting property of this quantization technique is that it also suggests new regularization procedures. The idea, introduced by Breit, Gupta, and Zaks,[5] is to replace the Markovian process defined in Ref. 2 by a non-Markovian one, i.e., to replace $\delta(t-t')$ by a function g_σ such that

$$\lim_{\sigma\to 0} g_\sigma(t-t') = \delta(t-t'). \quad (4)$$

A convenient form of g_σ is[6]

$$g_\sigma(t-t') = (\sigma/2)|t-t'|^{\sigma-1}, \quad (5)$$

which has the property that uv divergences appear as poles in σ.

At this point one can wonder whether there is a value of $\sigma = \sigma^*$ such that IR divergences can be handled and the corresponding β function has a zero $O(\rho)$ with $\rho = 2(\sigma - \sigma^*)$. If such a system exists then the critical exponents of the original Markovian theory could be obtained by expansion of the non-Markovian ones. Following this plane we shall first find the value σ^* and then argue that the corresponding theory is renormalizable and asymptotically free in the IR. After that we shall compute the critical exponents to the lowest nonzero order in ρ.

Let us start by noticing that if we consider the interaction $S_{\rm int} = (\lambda_0/4!)\phi^4(x,t)$ and take $g_\sigma(t-t')$ as given in (5), then the coupling constant λ_0 has dimensions given by (μ having dimensions of mass)

$$[\lambda_0] = \mu^{2\sigma+\epsilon}, \quad (6)$$

which for $d=3$ gives a dimensionless coupling constant for $\sigma = -\frac{1}{2}$.

A power-counting analysis of the stochastic perturbative series indeed indicates that the theory is renormalizable for $\sigma = -\frac{1}{2}$. Let us recall that the usual perturbative expansion of field theory comes from stochastic quantization by solving first the Langevin equation (1) in terms of tree diagrams where one external line represents the field $\phi(t,x)$ and all the other external lines end on a stochastic source ξ, which graphically can be represented with a cross at the end. Every line corresponds to an integration over an intermediate time. A tree diagram, therefore, has one external line without cross $E_0 = 1$, a number I of internal lines $I = V - 1$, V being the number of vertices, and a number E_c of crossed external lines easily seen to be $E_c = 2V+1$. Green's functions are obtained from the average over ξ of products of ϕ's. This means contraction of the ξ's in pairs with use of Eq. (2), generating loops. We can now get diagrams with both internal and external lines with a cross coming from the contraction of two ξ's. To discuss renormalization, we focus on the one-particle irreducible parts of the Green's functions. From the integration over the intermediate times we can extract an overall integration over a time variable which we can call τ. This can be obtained for instance by use of polar coordinates in the multiple time integral. $\tau \to 0$ means that every intermediate time approaches t.

Given a diagram with L loops, m internal lines with a cross, and N intermediate times we easily obtain by inspection the overall integration:

$$\tau^{-(d/2)L}\tau^{m(\sigma-1)}\tau^{N-1}d\tau = \tau^{-D/2-1}d\tau. \quad (7)$$

Here the factor $\tau^{-(d/2)L}$ comes from the integration over momenta, $\tau^{m(\sigma-1)}$ comes from Eq. (5), and $\tau^{N-1}d\tau$ comes from the integration measure. The uv divergence comes from the behavior at $\tau \to 0$ and is expressed as a pole for $D=0$. We want to write the divergence degree D in terms of L and E_0, E_c. First, since every internal crossed line corresponds to two time variables, we have

$$N = 2m + V - 1 \quad (8)$$

and, moreover, for a Green's function with E_0, E_c external legs coming from a contraction of E_0 tree diagrams,

$$2m + E_c = 2V + E_0. \quad (9)$$

A standard counting in ϕ^4 theory gives the relation

$$V = L - 1 + (E_0 + E_c)/2, \quad (10)$$

which combined with Eq. (9) gives $m = L - 1 + E_0$; then using Eqs. (8) and (7) we obtain

$$D/2 = (d/2 - 2 - \sigma)L + 3 + \sigma - (\tfrac{3}{2} + \sigma)E_0 - E_c/2. \quad (11)$$

The theory is renormalizable whenever D is independent of L, i.e., $\sigma^* = d/2 - 2$. In such a case we can reabsorb the divergences by redefining the parameters of the Langevin equation (1) which we, therefore, write in general as

$$Z_t(\partial/\partial t)\phi_R = (Z_\phi \Box - m_R^2 - \delta m^2)\phi_R + (\lambda_R/3!)\mu^{2\sigma+\epsilon}Z_V\phi_R^3 + Z_\xi \xi, \quad (12)$$

where we also introduced μ, the momentum scale of the renormalization procedure. For $d=4$ we obtain the usual field theory $\sigma^* = 0$. For $d=3$, $\sigma^* = -\frac{1}{2}$. In this case, from Eq. (11) we get a logarithmic divergence for $E_0 = 1$, $E_c = 3$ to be reabsorbed by Z_V. For $E_0 = 1$, $E_c = 1$, we have a quadratic divergence reabsorbed by δm^2 (actually this will appear as m^2 times a pole at $D=0$) and two logarithmic ones defining Z_t and Z_ϕ.[7] Contrary to the $d=4$ case, here, there is no divergence for $E_0 = 2$, $E_c = 0$ since $D = 1$ does not correspond to a pole and therefore $Z_\xi = 1$, as it is expected since the divergences appear as local terms in time and space in the Green's functions (i.e., they are proportional to δ functions on their derivatives). But for $\sigma \neq 0$ the correlation $g_\sigma(t-t')$ is not local in time and

VOLUME 54, NUMBER 5 PHYSICAL REVIEW LETTERS 4 FEBRUARY 1985

could not reabsorb divergences.

Finally, from Eq. (12) we compute the wave-function renormalization Z. Let us rescale the time $t = \alpha \bar{t}$ and the noise $\xi = \alpha^{(\sigma-1)/2}\bar{\xi}$ such that

$$\langle \bar{\xi}(t)\bar{\xi}(t')\rangle = \sigma|\bar{t}-\bar{t}'|^{\sigma-1}\delta^d(x-x').$$

With the choice of $\alpha = Z_t/Z_\phi$ this amounts to a field rescaling $\phi = Z^{1/2}\phi_R$, where $Z^{1/2} = Z_t^{-(\sigma-1)/2} \times Z_\phi^{(\sigma+1)/2}Z_\xi^{-1}$. For $d=3$ and $\sigma = -\frac{1}{2}$ we have

$$Z^{1/2} = Z_t^{3/4}Z_\phi^{1/4}. \tag{13}$$

Of course, the usual coupling-constant renormalization constant Z_1 also involves Z_V times powers of Z_t and Z_ϕ. At the one-loop order, however, $Z_1 = Z_V$.

We can define the infinite-time limit of Green's functions also for $\sigma \neq 0$. The existence of it being ensured by a mass gap,[8] we first take $t \to \infty$ and then approach the critical point. Furthermore, we can write for these Green's functions the standard renormalization-group equation. Since uv divergences appear as poles at $\rho = 0$, we can calculate the renormalization constants by minimal subtraction of these poles. We find at the lowest order for $d=3$,

$$\begin{aligned} Z_1 &= 1 + [\lambda_R 3/(2\sqrt{\pi})^3]/\rho, \\ Z_\phi &= 1 - [\lambda_R^2 R_\phi/2(4\pi)^3]/\rho, \\ Z_t &= 1 - [\lambda_R^2 R_t/2(4\pi)^3]/\rho, \end{aligned} \tag{14}$$

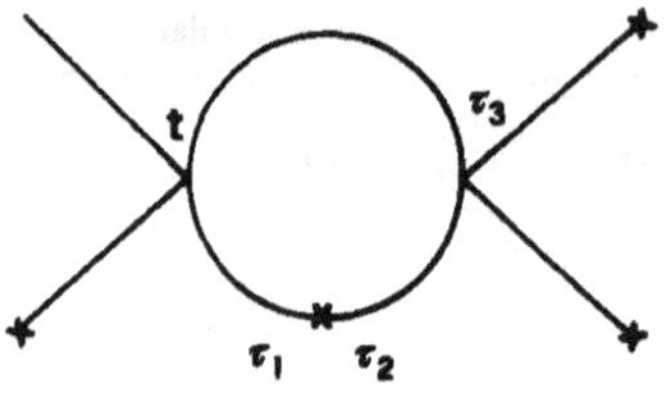

FIG. 1. The one-loop graph contributing to the β function.

where the renormalized coupling constant λ_R is given in terms of the bare one λ_0 by $\lambda_0 = \lambda_R Z_1 Z^{-2}\mu^\rho$. The residues R_ϕ and R_t are given in terms of integrals over a finite hypercube in four dimensions, which have been computed numerically and give $R_\phi = 0.196 \pm 0.003$ and $R_t = 0.264 \pm 0.004$.

As an example, we show how to evaluate the one-loop diagram which contributes to the β function (Fig. 1). In computing its divergent part we can take zero momentum. As we take into account Eq. (5), it is given by (apart from a numerical factor)

$$I = \sigma \int \frac{d^3k}{(2\pi)^3} \int_0^t d\tau_1 \int_0^t d\tau_3 \int_0^{\tau_3} d\tau_2 |\tau_1 - \tau_2|^{\sigma-1} \exp\{-k^2[(t-\tau_3)+(\tau_3-\tau_2)+(t-\tau_1)]\}. \tag{15}$$

To compute the integral over τ_1, τ_2, and τ_3 we divide it in all possible time orderings. Since t is the largest time and τ_3 is always greater than τ_2 we have only three regions: (a) $t \geq \tau_3 \geq \tau_1 \geq \tau_2$, (b) $t \geq \tau_3 \geq \tau_2 \geq \tau_1$, and (c) $t \geq \tau_1 \geq \tau_3 \geq \tau_2$, but actually the first two are equal for symmetry reasons. Considering case (a) we change variables: $\tau_1 = t(1-\alpha_1)$, $\tau_2 = t(1-\alpha_1\alpha_2)$, and $\tau_3 = t(1-\alpha_1\alpha_2\alpha_3)$, obtaining after momentum integration

$$I_a = \sigma t^{\sigma+1/2} \frac{\pi^{3/2}}{(2\pi)^3} \int_0^1 d\alpha_1\, d\alpha_2\, \alpha_1^{\sigma-1/2}\alpha_2(1-\alpha_2)^{\sigma-1}(1+\alpha_2)^{-3/2}, \tag{16}$$

and the integration over α_1 yields the pole at $\sigma = -\frac{1}{2}$. The last integral is also immediate but requires an analytical continuation in σ. After this is done we obtain, at the pole, a positive value for this integral: $I_a = [1/(2\sqrt{\pi})^3]/\rho$. I_c can be computed in a similar way, and after we consider the proper combinatorial factors we obtain Z_1 as in Eq. (14). The evaluation of Z_ϕ and Z_t is more lengthy but it involves the same difficulties.

We have also evaluated the renormalization constant Z_{ϕ^2} of the operator ϕ^2 which can be inserted into the Green's functions as a standard device to compute the dependence on the temperature of physical quantities near the critical point. We find at the lowest order

$$Z_{\phi^2} = 1 + [\lambda_R/(2\sqrt{\pi})^3]/\rho. \tag{17}$$

From the renormalization constants we can compute the renormalization-group function and the anomalous dimensions. At the lowest order we obtain

$$\beta = \mu[\partial\lambda_R/\partial\mu]_{\lambda_0} = -\rho\lambda_R + \lambda_R^2 3(2\sqrt{\pi})^{-3}, \tag{18}$$

given an IR stable fixed point at

$$\lambda_R^* = \rho(2\sqrt{\pi})^3/3 \tag{19}$$

which vanishes for $\sigma = -\frac{1}{2}$, justifying in principle our perturbative computation as in the ϵ expansion. Of course, we then extrapolate to $\sigma = 0$, i.e., $\rho = 1$. The critical exponents γ and η are given in terms of anomalous dimensions at the fixed point. Defining $\gamma_{\phi^2} = \mu[\partial \ln Z_{\phi^2}/\partial\mu]_{\lambda_0}$ to lowest order, we have the critical exponent $\gamma = 1 - \frac{1}{2}\gamma_{\phi^2}$ and

$$\begin{aligned} \eta &= \mu[\partial \ln Z/\partial\mu]_{\lambda_0} \\ &= \tfrac{1}{2}\mu(\partial/\partial\mu)(\ln Z_\phi + 3\ln Z_t). \end{aligned} \tag{20}$$

From above we get $\gamma = 1 + \rho/6$ and $\eta = \rho^2(R_\phi + 3R_t)/18$. The value of γ is the same as the one obtained from the ϵ expansion at the same order, i.e., $\gamma = 1 + \epsilon/6$, and in fact the factor 3^{-1} has in both cases the same combinatorial origin. At the lowest order in ρ we have $\gamma = 1.167$ and $\eta = 0.055$. On the other hand, the lowest order in ϵ gives $\eta = 0.019$. The high-temperature series yields[2,9] $\gamma = 1.250 \pm 0.003$ and $\eta = 0.04 \pm 0.01$. We see that both the ϵ- and ρ-expansion values for η compare badly with the high-temperature value which lies somehow in between. Let us mention a more general point of view. The idea is to approach the physical point $\rho = \epsilon = 1$ starting from an unphysical value in the line $\rho + \epsilon = 1$, which according to Eqs. (6) and (11) corresponds to a renormalizable theory. For instance, the ϵ expansion chooses the direction of approach where $\rho = 1$ and the ρ expansion the direction $\epsilon = 1$. One could think also of an intermediate situation. Further work in this direction is currently being pursued.

Let us stress, however, that the series expansion in ρ is expected to be asymptotic, and the evaluation of more terms and the use of resummation techniques will be necessary as it occurs in the ϵ expansion.[9,10]

The above calculations can be easily extended to the O(M) model, i.e., $S_{\rm int} = \lambda_0/4!M(\phi^2)^2$. Evaluating the corresponding combinatorial factors we obtain

$$\gamma = 1 + \rho(M+2)/(2M+16),$$

and

$$\eta = \rho^2(R_\phi + 3R_t)(3M+6)/2(M+8)^2,$$

i.e., M appears in the same way as in the ϵ expansion.

Two of the authors (J.A. and N.P.) would like to thank Professor Abdus Salam, the International Atomic Energy Agency, and UNESCO for hospitality at the International Centre for Theoretical Physics, Trieste, Italy.

(a)Permanent address: Laboratoire de Physique Theorique de l'Ecole Normale Superieure 24, rue Lhomond 75231, Paris, France.

(b)Permanent address: Centro Atomico Bariloche, 8400 Bariloche, Argentina.

[1]K. G. Wilson and M. E. Fisher, Phys. Rev. Lett. **28**, 240 (1972); K. G. Wilson and J. Kogut, Phys. Rep. **12**, 75 (1974).

[2]E. Brezin, J. V. Le Guillou, and J. Zinn-Justin, in *Phase Transitions and Critical Phenomena*, edited by C. Domb and M. S. Green (Academic, New York, 1976), Vol. 6, p. 125; D. J. Amit, *Field Theory, the Renormalization Group and Critical Phenomena* (McGraw-Hill, New York, 1978).

[3]G. Parisi and Y. Wu, Sci. Sin. **24**, 483 (1981).

[4]G. Parisi, Nucl. Phys. **B180** [FS2], 378 (1981), and **B205** [FS5], 337 (1982); L. Baulieu and D. Zwanziger, Nucl. Phys. **B193**, 163 (1981); F. G. Floratos, J. Iliopoulos, and D. Zwanziger, Nucl. Phys. **B241**, 221 (1984); J. Alfaro and B. Sakita, Phys. Lett. **121B**, 339 (1983); J. Alfaro, Phys. Rev. D **28**, 1001 (1983); G. Aldazabal, N. Parga, M. Okawa, and A. Gonzalez-Arroyo, Phys. Lett. **129B**, 80 (1983); R. Jengo and N. Parga, Phys. Lett. **134B**, 221 (1984).

[5]J. D. Breit, S. Gupta, and A. Zaks, Nucl. Phys. **B233**, 61 (1984).

[6]J. Alfaro, Nucl. Phys. **B253** (1985) 464.

[7]J. Alfaro, Laboratoire de Physique Theorique de l'Ecole Normale Superieure Report No. 84/8, 1984 (to be published).

[8]E. Floratos and J. Iliopoulos, Nucl. Phys. **B214**, 392 (1983); B. Sakita, City College of New York Report No. CCNY-HEP-83/14, 1983 (to be published).

[9]J. C. Le Guillou and J. Zinn-Justin, Phys. Rev. B **21**, 3976 (1980), and references quoted therein.

[10]J. Zinn-Justin, in *Recent Advances in Field Theory and Statistical Mechanics*, Proceedings of the Les Houches Summer School, Session XXXIX, edited by J. B. Zuber and R. Stora (North-Holland, Amsterdam, 1984).

Volume 165B, number 1,2,3 PHYSICS LETTERS 19 December 1985

CONTINUUM REGULARIZATION OF QCD

Z. BERN [1], M.B. HALPERN [1]

Lawrence Berkeley Laboratory and Department of Physics, University of California, Berkeley, CA 94720, USA

L. SADUN

Department of Physics, University of California, Berkeley, CA 94720, USA

and

C. TAUBES [2]

Department of Mathematics, University of California, Berkeley, CA 94720, USA

Received 30 July 1985

We introduce a new stochastic regularization for continuum gauge theory. The scheme is a covariant derivative regularization of the Parisi–Wu Langevin equation, or equivalently, the Schwinger–Dyson equations. The regularized formulation is manifestly Lorentz invariant, gauge-invariant, ghost-free and finite to all orders. We verify a vanishing gluon mass at one loop.

A number of gauge-invariant regularization schemes have been proposed for gauge theory. Of these, the most practical to date has been the lattice [1], on which numerical simulation of hadron physics has seen considerable success. For many reasons, including loss of the continuum topology of gauge theory on the lattice, the idea of a continuum computational scheme persists. Foremost among continuum regularization schemes is dimensional regularization [2], which remains at present a perturbative technique. Higher covariant derivatives in the action fail to regularize the theory, although it is claimed that Slavnov's hybrid scheme is viable [3]. Other proposals of note include the geometric [4] approach of Asorey and Mitter, and of Singer. In both these cases, we are unaware of detailed perturbative analysis. Finally we mention stochastic regularization by fifth-time smearing [5], which presumably regularizes gauge-invariant quantities to all orders. This is not a satisfactory nonperturbative scheme for QCD_4, since the superficial quadratic divergences force a bottomless action for the noise [6]. Additionally, fifth-time smearing is incompatible [6] with Zwanziger's gauge-fixing [7].

Our new stochastic regularization is a covariant derivative regularization of the Parisi–Wu Langevin equation [8] or, equivalently, the Schwinger–Dyson equations. It is intrinsically not an action regularization, for which, as mentioned above, higher covariant derivatives do not suffice. We do not smear in fifth-time. As a result, we retain, in distinction to ref. [5], all the technical advantages of a Markov process, including closed form equilibrium equations, which are the Schwinger–Dyson equations, and Zwanziger's gauge-fixing if desired. The resulting regularized theory is manifestly Lorentz invariant, gauge-invariant, ghost-free and ultraviolet finite to all orders. We are hopeful that the formulation will lend itself to non-

[1] This work was supported by the Director, Office of Energy Research, Office of High Energy and Nuclear Physics, Division of High Energy Physics of the US Department of Energy under contract DE-AC03-76SF00098 and the National Science Foundation under Research Grant No. PHY-81-18547.

[2] Address after July 1, 1985: Department of Mathematics, Harvard University, Cambridge, MA 02138, USA.

0370-2693/85/$ 03.30 © Elsevier Science Publishers B.V. (North-Holland Physics Publishing Division)

Volume 165B, number 1,2,3 PHYSICS LETTERS 19 December 1985

perturbative analysis and possibly numerical simulation. Further details of the results reported here will be given elsewhere [9].

As an introduction to our regularization of gauge theory below, we briefly mention its simpler scalar prototype. For a D-dimensional theory of a scalar field $\phi(x)$ with euclidean action S, we consider the regularized Langevin system

$$\dot{\phi}(x,t) = -\frac{\delta S}{\delta\phi}(x,t) + \int (\mathrm{d}y) R_{xy}(\Box)\eta(y,t), \qquad \text{(1a)}$$

$$\langle \eta(x,t)\eta(y,\tau)\rangle_\eta = 2\delta(t-\tau)\delta^D(x-y). \qquad \text{(1b)}$$

The regulator $R(\Box)$, which multiplies the noise, is a function of the laplacian $\Box_{xy} = \int(\mathrm{d}z)(\partial_\mu)_{xz}(\partial_\mu)_{zy}$, where $(\partial_\mu)_{xy} = \partial^x_\mu \delta^D(x-y)$. It is not difficult to see that for integer n sufficiently large, the simple choice $R^{(n)} = R_0^n$, where $R_0 \equiv (1-\Box/\Lambda^2)^{-1}$, will regularize the theory [9], since every closed loop contains at least one noise-regularized propagator. Curiously, as we will discuss below for gauge theory, the scheme is not an action regularization, since explicit divergences occur in the effective D-dimensional action, although not in the Green functions. For scalar theories, conventional higher derivative action regularization is easier and more familiar, but the generalization to gauge theory fails [3]. In contrast, the gauge extension of our scheme is successful, at least in part because it is not an action regularization.

For Yang–Mills in four dimensions, we therefore consider the Langevin system

$$\dot{A}^a_\mu(x,t) = -\frac{\delta S}{\delta A^a_\mu}(x,t) + D^{ab}_\mu Z^b(x,t) + \int(\mathrm{d}y)\, R^{ab}_{xy}(\Delta)\eta^b_\mu(y,t), \qquad \text{(2a)}$$

$$\langle \eta^a_\mu(x,t)\eta^b_\nu(y,\tau)\rangle_\eta = 2\delta^{ab}\delta_{\mu\nu}\delta(t-\tau)\delta^4(x-y), \qquad \text{(2b)}$$

where S is the usual euclidean Yang–Mills action and $D^{ab}_\mu \equiv \delta^{ab}\partial_\mu + gf^{abc}A^c_\mu$ is the covariant derivative. We have also chosen ‡1 to add a Zwanziger gauge-fixing term [7], which we will specify as $\alpha Z = \partial\cdot A$ for computational purposes. The regulator $R^{ab}_{xy}(\Delta)$ is a function of the covariant laplacian $\Delta^{ab}_{xy} \equiv \int(\mathrm{d}z) \times (D_\mu)^{ac}_{xz}(D_\mu)^{cb}_{zy}$, where $(D_\mu)^{ab}_{xy} \equiv D^{ab}_\mu(x)\delta^4(x-y)$. This construction ‡2 guarantees the gauge covariance of the Langevin equation in the absence of gauge-fixing, and that $R^{ba}_{yx} = R^{ab}_{xy}$. To maintain the naive large cutoff limit, we must also require that R approach unity as the cutoff Λ goes to infinity. This leaves a very large class of regulators, from which we study here only the simplest set $R^{(n)} = (1-\Delta/\Lambda^2)^{-n}$, although we mention that analytic forms such as the "heat kernel" regulator $R = \exp(\Delta/\Lambda^2)$ may be technically superior for nonperturbative analysis.

The extension to gauge theory is non-trivial, since the gauge field dependence of R introduces new vertices into the Langevin equation. This means, as in the action scheme, a competition between regularized propagators and higher-dimensional vertices. The outcome is seen in the expansion of the regulator

$$R^{(n)} = R_0^n + \sum_{k,l=1}^{n} \delta_{k+l,n+1} R_0^k (g\Gamma_1 + g^2\Gamma_2) R_0^l + \sum_{k,l,m=1}^{n} \delta_{k+l+m,n+2} R_0^k g\Gamma_1 R_0^l g\Gamma_1 R_0^m + O(g^3), \qquad \text{(3)}$$

which is easily carried out to arbitrarily high order. Here, R_0 is defined above, and we have defined the matrices

$$(\Gamma_1)_{xy} \equiv I^a[\partial^x_\mu A^a_\mu(x) + A^a_\mu(x)\partial^x_\mu]\delta^4(x-y)/\Lambda^2$$

$$(\Gamma_2)_{xy} \equiv I^a I^b A^a_\mu(x) A^b_\mu(x)\delta^4(x-y)/\Lambda^2, \qquad \text{(4)}$$

in which ∂^x_μ acts on everything to the right and $(I^a)^{bc} \equiv f^{abc}$. By increasing n, we obtain more powers of R_0 and hence more regularization for any fixed power of $\partial\cdot A$, $A\cdot\partial$ or A^2, so we must succeed for n sufficiently large. In fact, the theory is regularized to all orders when $n \geqslant 2$.

To expand the Langevin eq. (2a) perturbatively through a certain order of g, first expand the regulator through that order, as in eq. (3). Then standard iterative techniques yield the Langevin tree graphs

‡1 Although Zwanziger gauge-fixing is natural in our scheme, other gauge-fixings, such as that of ref. [10], may also be employed.

‡2 Gauge covariance requires that R^{ab}_{xy} is a parallel-transport from y to x, which includes weighted path integrations from y to x. Such objects are presumably related to our $R(\Delta)$ construction, but we have not studied their properties as regulators.

Volume 165B, number 1,2,3 PHYSICS LETTERS 19 December 1985

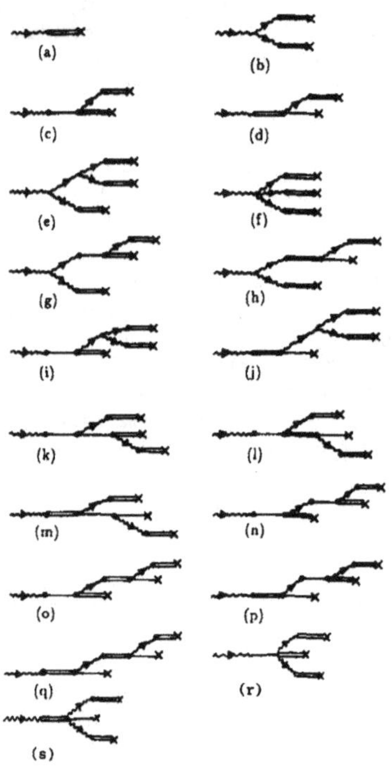

Fig. 1. Langevin tree diagrams through $O(g^2)$.

for $A^a_\mu[\eta]$, shown in fig. 1 for the case $n = 2$. A list of the propagators and vertices necessary to construct all such tree graphs are given in a set of Langevin Feynman rules in fig. 2. We call attention to the two extra vertices which arise from $R(\Delta)$, as well as the two extra regulator propagators, being single or double solid lines, which count the powers of the momentum space regulator factors $(1 + p^2/\Lambda^2)^{-1}$. There is one additional tree rule which requires an explanation: The single and double regulator factors form continuous strings of various lengths in the tree diagrams; for example, there is a string of three regulators and two strings of one regulator each in fig. (1k). The additional tree rule is that every such string contains exactly one double regulator.

The Langevin Feynman diagrams for the equal fifth-time averages are then computed as usual by averaging over the noise at large t. For example, the free gluon propagator, in terms of standard transverse and longitudinal projection operators,

$$D^{ab}_{\mu\nu}(p) = \delta^{ab}(T_{\mu\nu} + \alpha L_{\mu\nu})\Lambda^8/p^2(p^2 + \Lambda^2)^4, \tag{5}$$

is obtained by contracting two diagrams of the form (a) in fig. 1.

As a more complicated example, we have computed the one loop correction to the gluon vacuum polarization. There are forty-seven types of diagrams, all of which are finite. For brevity we show in fig. 3 only those thirteen diagrams that contribute to $\Pi^{ab}_{\mu\nu}(0)$. We note that there are two "ordinary" contributions and eleven "extra" contributions, containing regulator vertices. Contributions to $\Pi^{ab}_{\mu\nu}(0)$, in units of $g^2 f^{acd} \times f^{bcd}\delta_{\mu\nu}\Lambda^2/(4\pi)^2$, are shown with each diagram. The reader may verify that their sum is zero, so the gluon remains massless to this order. We have also computed the wave function renormalization [9]. The result agrees with dimensional regularization of the Zwanziger gauge-fixed theory, using the dictionary $\ln\Lambda \leftrightarrow (4-d)^{-1}$.

It is not difficult to survey all higher order diagrams [9]. The worst case always corresponds to subloops of the type found in diagram (m) of fig. 3, which is finite, so the theory is completely regularized for $n = 2$. In the case $n = 1$, the diagrams are similar, but double regulator lines are replaced by single regulator lines. Then diagrams (3ℓ) and (3m) are the same, which is explicitly log divergent. For general n, diagrams contain up to n-th power regulator lines, and regularization is achieved for all $n \geqslant 2$.

We turn now to a discussion of the Schwinger–Dyson equations, which provide a four-dimensional formulation of the scheme. It follows directly from the Langevin system (2) that

$$\langle\dot{\Phi}\rangle = \left\langle \int(\mathrm{d}x)\left(-\frac{\delta S}{\delta A^a_\mu(x)} + D^{ab}_\mu(x)Z^b(x)\right.\right.$$
$$\left.\left. + \int(\mathrm{d}y)\,(\mathrm{d}z)\,R^{ba}_{yx}\frac{\delta}{\delta A^c_\mu(z)}R^{ba}_{yx}\right)\frac{\delta\Phi}{\delta A^a_\mu(x)}\right\rangle, \tag{6}$$

where $\langle\Phi[A(\ )]\rangle$ is the average of any equal fifth-time functional of the gauge field. At equilibrium, therefore, we have the regularized Schwinger–Dyson equations

Volume 165B, number 1,2,3 PHYSICS LETTERS 19 December 1985

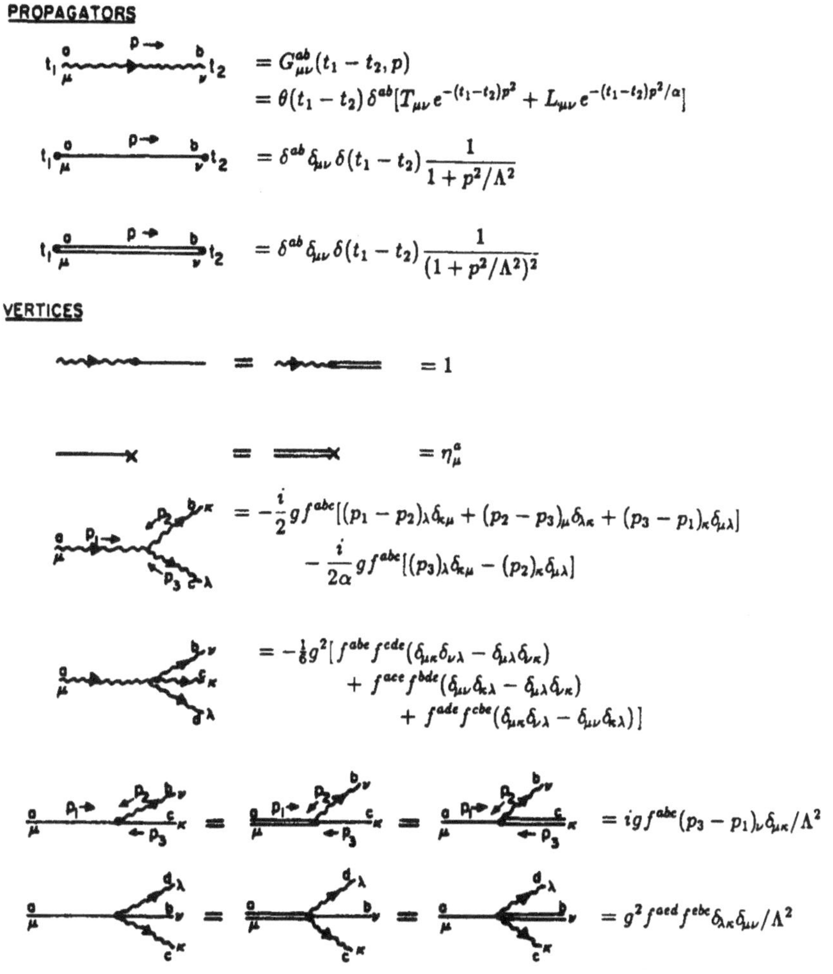

Fig. 2. Langevin Feynman rules.

$$0 = \left\langle \int (\mathrm{d}x) \left(-\frac{\delta S}{\delta A^a_\mu(x)} + D^{ab}_\mu(x) Z^b(x) + \int (\mathrm{d}y)\,(\mathrm{d}z)\, R^{bc}_{yz} \frac{\delta}{\delta A^c_\mu(z)} R^{ba}_{yx} \right) \frac{\delta \Phi}{\delta A^a_\mu(x)} \right\rangle . \qquad (7)$$

We have also developed a set of four-dimensional regularized Schwinger–Dyson Feynman rules [9] for the weak-coupling expansion of the Schwinger–Dyson equations (7). The resulting Schwinger–Dyson diagrams, equivalent to the equilibrium Langevin diagrams after all fifth-time integrations are performed, appear to provide the natural language for the discussion of perturbative renormalization [9]. The

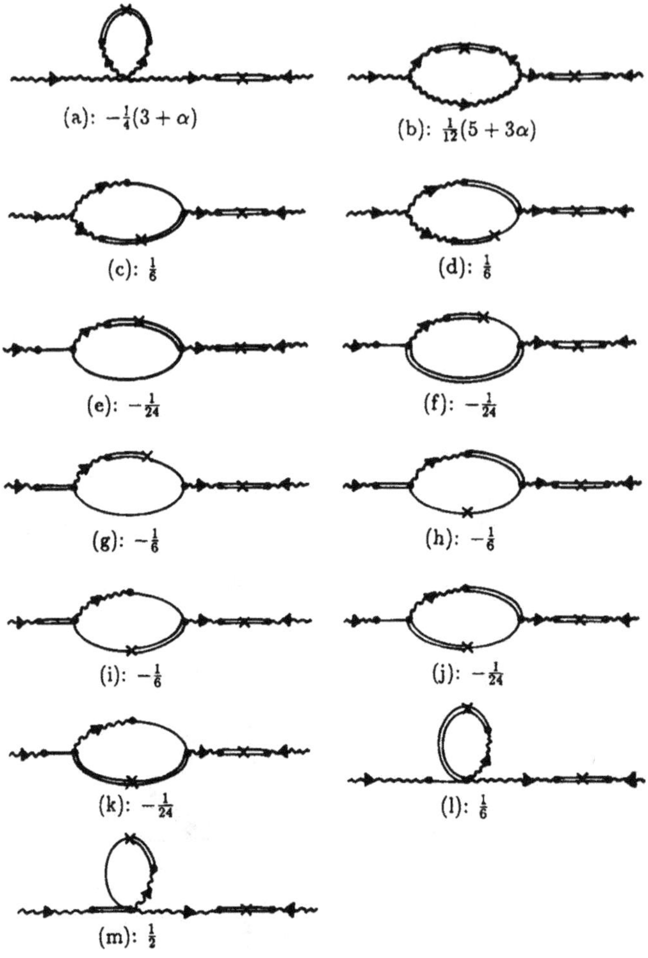

Fig. 3. One-loop contributions to $\Pi_{\mu\nu}(0)$.

Schwinger–Dyson equations will yield gauge-invariant results because the Zwanziger term in eq. (7) vanishes when Φ is any gauge-invariant observable, such as a Wilson loop $W[C]$,

$$0=\left\langle\int(\mathrm{d}x)\left(-\frac{\delta S}{\delta A_\mu^a(x)}\right.\right.$$

$$\left.\left.+\int(\mathrm{d}y)\,(\mathrm{d}z)\,R_{yz}^{bc}\,\frac{\delta}{\delta A_\mu^c(z)}\,R_{yx}^{ba}\right)\frac{\delta W[C]}{\delta A_\mu^a(x)}\right\rangle. \qquad (8)$$

Volume 165B, number 1,2,3 PHYSICS LETTERS 19 December 1985

Finally we discuss the Fokker–Planck formulation, which leads to the effective four-dimensional action. The Fokker–Planck equation may be derived by standard methods, either as the dual to the observable equation (6), or directly from eq. (2). At equilibrium,

$$\int (\mathrm{d}x) \frac{\delta}{\delta A^a_\mu(x)} [\mathcal{L}^a_\mu(x)\rho] = 0,$$

$$\mathcal{L}^a_\mu(x) \equiv \frac{\delta S}{\delta A^a_\mu(x)} - D^{ab}_\mu(x) Z^b(x) + \int (\mathrm{d}y)(\mathrm{d}z) R^{ab}_{xy} \frac{\delta}{\delta A^c_\mu(z)} R^{cb}_{zy}, \qquad (9)$$

the Fokker–Planck density ρ is the effective four-dimensional Boltzmann factor of the theory. The functional derivative in the last term of $\mathcal{L}^a_\mu$ operates on everything on the right. As we shall discuss further in ref. [9], the simpler local equation $\mathcal{L}^a_\mu(x)\rho = 0$ is not integrable. The Zwanziger term alone is known to prevent integrability [7], and our regulator doubles this effect, preventing integrability even in the scalar prototype. It follows that the local divergences found in $\delta \mathcal{L}^a_\mu(x)/\delta A^a_\mu(x)$ of eq. (9) cannot be avoided. Moreover, these divergences are related to explicit divergences in the effective four-dimensional action $S_{\mathrm{eff}} \equiv -\ln \rho$ [9]. Our scheme is therefore not an action regularization. It bears emphasis that no such divergences are found in the dual equations (6)–(8) for the averages, nor, indeed, to all orders in weak coupling, as discussed above. Not surprisingly, however, the phenomenon leads to a number of eccentricities in perturbative renormalization, which will be discussed in ref. [9].

We also point out that our scheme is not limited to QCD. Using Langevin–Schwinger–Dyson techniques and covariant derivatives that respect all relevant symmetries, we expect the method to succeed in general.

We wish to thank O. Alvarez, H.S. Chan, J. Glimm, A. Jaffe and W. Siegel for helpful discussions.

References

[1] K.G. Wilson, Phys. Rev. D10 (1974) 2445;
A.M. Polyakov, Phys. Lett. 59B (1975) 79, 82;
F. Wegner, J. Math. Phys. 12 (1971) 2259.
[2] G. 't Hooft and M. Veltman, Nucl. Phys. B44 (1972) 189.
[3] L.D. Faddeev and A.A. Slavnov, Gauge fields: Introduction to quantum theory (Benjamin/Cummings, Reading, 1980).
[4] M. Asorey and P.K. Mitter, Commun. Math. Phys. 80 (1981) 43;
I.M. Singer, Phys. Scr. 24 (1981) 817.
[5] J.D. Breit, S. Gupta and A. Zaks, Nucl. Phys. B233 (1984) 61.
[6] Z. Bern and M.B. Halpern, LBL-19714, UCB-PTH-85/24.
[7] D. Zwanziger, Nucl. Phys. B192 (1981) 259;
E.G. Floratos, J. Iliopoulos and D. Zwanziger, Nucl. Phys. B241 (1984) 221.
[8] G. Parisi and Wu Yong-Shi, Sci. Sin. 24 (1981) 483.
[9] Z. Bern, M.B. Halpern, L. Sadun and C. Taubes, in preparation.
[10] H. Hüffel and P.V. Landshoff, CERN-TH.4120/85.

A Rigorous Construction

Commun. Math. Phys. 101, 409–436 (1985)

Communications in Mathematical Physics

On the Stochastic Quantization of Field Theory

G. Jona-Lasinio[1,2,†] and P. K. Mitter[2]

1* Dipartimento di Fisica, Università di Roma "La Sapienza," Piazzale Aldo Moro 2, I-00185 Roma, Italy
2 Laboratoire de Physique Théorique et Hautes Energies**, Université Pierre et Marie Curie, Paris VI, F-75230 Paris Cedex 05, France

Abstract. We give a rigorous construction of a stochastic continuum $P(\phi)_2$ model in finite Euclidean space-time volume. It is obtained by a weak solution of a non-linear stochastic differential equation in a space of distributions. The resulting Markov process has continuous sample paths, and is ergodic with the finite volume Euclidean $P(\phi)_2$ measure as its unique invariant measure. The procedure may be called stochastic field quantization.

Introduction

Ever since the original work of Glauber [24], there has been much interest in stochastic statistical mechanical models. Such models have been rigorously studied by Holley, Stroock, Faris, Wick, and others [25–30]. The fundamental aim in these works is to obtain (and study properties of) Gibbs states of classical statistical mechanics as limiting distributions of stochastic processes. These processes are sometimes obtained as solutions of non-linear stochastic differential equations of the Langevin type (see later). Let e^{tL} be the associated semi-group, and, starting from an initial state (probability measure) μ_0, let μ_t be the evolved state under the action of the adjoint semigroup acting on the space of measures equipped with the weak * topology. If $\mu_t \to \mu$ in this topology, then μ is the unique equilibrium (invariant) measure. (Sometimes only a subsequence μ_{t_K} converges using weak commpactness criteria.) Let μ be an invariant measure. Then μ is a Gibbs state iff e^{tL} is a selfadjoint contraction on $L^2(d\mu)$. If the invariant measure is unique, the process is ergodic.

In [18], Parisi and Wu proposed such a program for Euclidean quantum field theory. We may call this the method of *stochastic quantization.* This is natural because of the analogy between Euclidean quantum field theory and classical statistical mechanics. Euclidean quantum field theory is described by a probability

* Permanent address
** Laboratoire Associé 280 au CNRS
† Supported in part by GNSM and INFN

measure which is formally Gibbsian and satisfies some important properties (reflection positivity, Euclidean invariance and some technical growth conditions), [1–3], which ensure the existence of its relativistic counterpart. However aside from Euclidean quantum mechanics (the so-called $P(\phi)_1$ models of which a stochastic version was constructed [19] by Faris and Jona-Lasinio) all non-trivial quantum field theories are plagued by ultraviolet (UV) divergences which necessitate renormalization. Under such singular circumstances it is not a priori obvious that the Parisi-Wu program can be rigorously pushed through. If it could, then the main interest of this program would be in its application to non-abelian gauge theories [18, 20–23].

In this paper we make a start on the rigorous *non-perturbative* analysis of this program. We consider the simplest non-trivial Euclidean quantum field theory which has, of course, ultraviolet divergence and hence necessitates renormalization, albeit of the simplest kind. This is the continuum $(\phi^4)_2$ model [1, 2, 3] which we consider (as a first step) in a finite spacetime volume $\Lambda \subset R^2$, since we wish to concentrate on ultraviolet aspects. We consider Λ to be a square, and the scalar fields ϕ, with values in R, are given, for definiteness, Dirichlet boundary conditions. Such a theory is then specified by the finite measure [1–3] μ on $H_{-1}(\Lambda)$,

$$d\mu = d\mu_c e^{-\frac{\lambda}{4}\int_\Lambda d^2x :\phi^4:(x)}, \tag{0.1}$$

$$\underset{\text{formally}}{=} \int \prod_{x\in\Lambda} d\phi(x) e^{-S(\phi)}, \tag{0.2}$$

$$S(\phi) = \frac{1}{2}(\phi, (-\Delta + 1)\phi)_{L^2(\Lambda)} + \frac{\lambda}{4}\int_\Lambda d^2x :\phi^4:(x). \tag{0.3}$$

Here $H_{-1}(\Lambda)$ is a Sobolev space of distributions, μ_c is the Gaussian measure of mean 0 and covariance $C = (-\Delta + 1)^{-1}$ with Dirichlet boundary conditions. : : denotes Wick ordering [1] with respect to C. Without Wick ordering the exponent in (0.1) would not be a μ_c measurable random variable. The existence of (0.1) is proved in [1–3].

In this paper we will prove the existence of an ergodic, continuous, Markov process $\hat{\phi}_t$ with values in $H_{-1}(\Lambda)$ and having the measure μ of (0.1) as its invariant measure. The process $\hat{\phi}_t$ is obtained as a *weak* solution of a generalized Langevin equation in the space of distributions $H_{-1}(\Lambda)$. Thus we have accomplished the *stochastic quantization* of finite volume $P(\phi)_2$ field theory.

In order to motivate the paper let us first consider a ferromagnetic statistical mechanical spin system (with values in R) in finite volume $\Lambda \subset Z^2$, with Gibbs distribution μ:

$$d\mu = \prod_{i\in\Lambda} d\phi(i) e^{-S(\phi)}, \tag{0.4}$$

$$S(\phi) = \frac{1}{2}(\phi, (-\Delta + 1)\phi) + \frac{\lambda}{4}\sum_{i\in\Lambda}\phi^4(i), \tag{0.5}$$

with $\lambda > 0$. Δ is the Lattice laplacian.

A symmetric diffusion process $\hat{\phi}_t$ which has (0.4) as its equilibrium measure will have a formal differential generator L symmetric with respect to $L^2(d\mu)$. The most general $L^2(d\mu)$ symmetric second order differential operator L is then of the form:

$$L=\frac{1}{2}\sum_{i,j\in\Lambda}\tilde{K}(i,j)\frac{\delta^2}{\delta\phi(i)\delta\phi(j)}-\frac{1}{2}\sum_{i,j\in\Lambda}\tilde{K}(i,j)\frac{\delta S}{\delta\phi(j)}\frac{\delta}{\delta\phi(i)}, \tag{0.6}$$

where the operator K: $L^2(\Lambda)\to L^2(\Lambda)$,

$$Kf(i)=\sum_j \tilde{K}(i,j)f(j), \tag{0.7}$$

is any bounded, positive, selfadjoint operator on $L^2(\Lambda)$. We shall restrict the choice of K so that it commutes with Δ.

In that case, formally, the dual semigroup (the Fokker-Planck semigroup) e^{tL^*} leaves μ invariant, $e^{tL^*}\mu=\mu$, i.e. μ is an equilibrium distribution. The usual choice is $\tilde{K}(i,j)=\delta_{ij}$.

The diffusion process with L [as in (0.6)] as its formal differential generator can be obtained by solving the Ito stochastic differential equation [also known as a (generalized) Langevin equation]

$$\begin{aligned} d\hat{\phi}_t(i)&=dW_t(i)-\frac{1}{2}\sum_j\tilde{K}(i,j)\frac{\delta S}{\delta\phi(j)}(\hat{\phi}_t)dt,\\ \hat{\phi}_0(i)&=\phi(i), \end{aligned} \tag{0.8}$$

and $W_t(i)$ is a Wiener process with covariance

$$E(W_t(i)W_s(j))=\tilde{K}(i,j)\min(t,s). \tag{0.9}$$

For finite lattice systems $\hat{\phi}_t$ can be rigorously constructed by standard methods. It can be shown to be ergodic, with μ of (0.4) as its invariant measure, when the spectrum of the positive self-adjoint operator $K(-\Delta+1)^{-1}$ is bounded away from zero. The latter condition assures that the gaussian process ϕ_t [corresponding to the case $\lambda=0$ in (0.5)] has a mass gap. The existence of a solution of (0.8) when $\Lambda\uparrow\mathbb{Z}^2$, and the existence of reversible invariant measures (Gibbs states) can also be proved [30].

We now turn to the continuum $P(\phi)_2$ model with (0.1)–(0.3) replacing (0.4)–(0.5). We write down the analogue of (0.8)–(0.9) directly in the continuum. We choose for the operator K [the analogue of (0.7)] on $L^2(\Lambda, d^2x)$,

$$K=C^{1-\varepsilon}=(-\Delta+1)^{-(1-\varepsilon)} \tag{0.10}$$

with $\varepsilon\leqq 1$. *Further restrictions will be imposed on* ε *presently*, for reasons to be explained below. Then the analogue of (0.8)–(0.9) reads, in operator notation,

$$\begin{aligned} d\hat{\phi}_t&=dW_t-\tfrac{1}{2}(C^{-\varepsilon}\hat{\phi}_t+\lambda C^{1-\varepsilon}:\hat{\phi}_t^3:)dt,\\ \hat{\phi}_0&=\phi, \end{aligned} \tag{0.11}$$

and W_t is a Wiener process in $H_{-1}(\Lambda)$ with covariance

$$E(W_t(f)W_s(g))=(f, C^{1-\varepsilon}g)\min(t,s), \tag{0.12}$$

where f, g are test functions.

Let ϕ_t be the unique solution of the linear stochastic differential equation in $H_{-1}(\Lambda)$,

$$d\phi_t = dW_t - \tfrac{1}{2}C^{-\varepsilon}\phi_t\,, \qquad \phi_0 = \phi\,, \tag{0.13}$$

ϕ_t is an Ornstein-Uhlenbeck (O.U.) process. Then (0.11) can be written as an integral equation

$$\hat{\phi}_t = \phi_t - \frac{\lambda}{2}\int_0^t ds\, e^{-\frac{1}{2}(t-s)C^{-\varepsilon}} C^{1-\varepsilon} :\hat{\phi}_s^3:. \tag{0.14}$$

Note that : : is Wick ordering with respect to the covariance C.

Now for each starting point ϕ, and time $t>0$, the transition probability of the O.U. process is a Gaussian measure $p_t(\phi, d\phi')$ on $H_{-1}(\Lambda)$. In order that (0.14) be free of UV divergence it is necessary and sufficient that the Gaussian measures $p_t(\phi,\cdot)$ and μ_c be equivalent. *This imposes the condition* $\varepsilon>0$, *strictly.*

Now the standard method of solving a non-linear equation like (0.11, 0.14) (Picard's method or contraction mapping) cannot be applied here since the perturbing drift [the non-linear term in (0.11)] has no continuity properties. $:\phi^3:(f)$ exists only in $L^p(d\mu_c)$, $1 \leqq p < \infty$. Instead we may attempt to construct a *weak* solution.[1]

The semi-group corresponding formally to the evolution governed by (0.11) can be written as:

$$(e^{tL}f)(\phi) = E^{(W)}_{\phi_0=\phi}\left(f(\phi_t) e^{-\frac{\lambda}{2}\int_0^t (:\phi_s^3:, dW_s) - \frac{\lambda^2}{8}\int_0^t ds(:\phi_s^3:, C^{1-\varepsilon}:\phi_s^3:)} \right), \tag{0.15}$$

where f is a bounded μ_c measurable function and ϕ_t is the O.U. process satisfying (0.13), and is considered as a functional of the Wiener process W_t, and $E^{(W)}$ is Wiener expectation. For (0.15) to be well defined, it is necessary that the exponent is a well defined random variable. *This imposes the further restriction* $\varepsilon<1$ *strictly.*

In the range $0<\varepsilon<1$ one shows straightforwardly that

$$E^{(W)}_\phi(e^{\xi_t}) \leqq 1\,, \tag{0.16}$$

where ξ_t is the exponent in (0.15). Thus e^{tL} is a contraction on $L^\infty(d\mu_c)$. However in order to construct a weak solution we must have

$$E^{(W)}_\phi(e^{\xi_t}) = 1\,. \tag{0.17}$$

To prove this we first obtain an Ito formula which permits us to rewrite (0.15) in Feynman-Kac form. Exploiting the fact that the perturbing drift in (0.11) is a gradient, one obtains

$$\int_0^t (:\phi_s^3:, dW_s) = \tfrac{1}{4}\int_\Lambda :\phi_t^4: d^2x - \tfrac{1}{4}\int_\Lambda :\phi_0^4: d^2x + \tfrac{1}{2}\int_0^t ds :(\phi_s^3, C^{-\varepsilon}\phi_s):. \tag{0.18}$$

---

1 By a weak solution of (0.14) we mean a Markov family of measures $\hat{P}_\phi$ on path space, such that $\hat{\phi}_t$ minus the second term in (0.14) has the same distributions as that of ϕ_t

In order that this formula makes sense in $L^2(dP^{(W)}, \Omega)$, where $P^{(W)}$ is Wiener measure and $\Omega = C^0([0, \infty), H_{-1}(\Lambda))$ is path space, *we must impose the further restriction:* $0 < \varepsilon < \frac{1}{2}$, (otherwise the last term in (0.18) is not a random variable). Equation (0.18) enables us to eliminate the stochastic integral (0.15). *Under the further and final restriction:*

$$0 < \varepsilon < \tfrac{1}{10} \tag{0.19}$$

(which is not necessarily optimal), we show, using estimates of the type encountered in constructive field theory [1–3], that e^{tL} is a strongly continuous self-adjoint contractive semi-group on $L^2(d\mu)$ with 1 as its unique ground state,

$$e^{tL}1 = 1\,, \tag{0.20}$$

which implies (0.17). The Markov process $\hat{\phi}_t$ is constructed, in a standard way, using the transition probabilities given by e^{tL}. There exists a continuous version of the process corresponding to a Markov family of measures $\hat{P}_\phi$ supported on $\Omega = C^0([0, t), H_{-1}(\Lambda))$. It gives a weak solution of (0.11) or (0.14) in the sense that the process

$$\hat{Z}_t = \hat{\phi}_t + \frac{\lambda}{2}\int_0^t ds\, e^{-\frac{1}{2}(t-s)C^{-\varepsilon}}C^{1-\varepsilon} :\hat{\phi}_s^3: \tag{0.21}$$

has the same joint probability distributions (with respect to $\hat{P}_\phi$) as those of the O.U. process ϕ_t. Thus $\hat{Z}_t$ can be identified as an O.U. process which is however not necessarily measurable with respect to ϕ_t.

The Markov process $\hat{\phi}_t$ is ergodic, and also mixing (because of the existence of a mass gap). This ensures that

$$\lim_{t\to\infty} E_\phi(\hat{\phi}_t(f_1)\ldots\hat{\phi}_t(f_n)) = \frac{1}{Z}\int d\mu\,\phi(f_1)\ldots\phi(f_n)\,, \qquad \mu \text{ a.e.},$$

where the f_i are test functions and $Z = \int d\mu$.

The rest of this paper gives the technical details of the above outline. We have gone into some pedagogic details as the subject is of interest to both physicists and mathematicians.

1. Preliminary: The $(\phi^4)_2$ Finite Volume Euclidean Measure

Let $\Lambda \subset R^2$ be a square with Dirichlet boundary conditions. The Laplacian

$$\Delta = \frac{\partial^2}{\partial x_1^2} + \frac{\partial^2}{\partial x_2^2} \tag{1.1}$$

with the above boundary conditions is a self-adjoint operator on $L^2(\Lambda)$. $H_1(\Lambda)$ is the Sobolev Hilbert space of functions with norm $\|\cdot\|_1$,

$$\|f\|_1^2 = \int_\Lambda d^2x |(-\Delta + 1)^{1/2} f(x)|^2\,, \tag{1.2}$$

obtained by completing $C^\infty(\Lambda)$ functions in this norm. $H_{-1}(\Lambda) \equiv E$ is the Sobolev-Hilbert space of distributions with norm $\|\cdot\|_{-1}$,

$$\|\phi\|_{-1}^2 = \int_\Lambda d^2x |(-\Delta+1)^{-1/2}\phi(x)|^2 . \tag{1.3}$$

Then the injection:

$$i: H_1(\Lambda) \hookrightarrow H_{-1}(\Lambda) \equiv E \tag{1.4}$$

is Hilbert-Schmidt. A theorem of Sazonov, Minlos [11] and Gross [8, 9] assures us that there exists a Gaussian probability measure on E with mean 0 and covariance C, denoted μ_c,

$$\int_E d\mu_c \phi(f)\phi(g) = (f, Cg), \tag{1.5}$$

where $f, g \in C^\infty(\Lambda)$ are test functions, and

$$C = (-\Delta+1)^{-1} . \tag{1.6}$$

: : denotes Wick ordering [1–3] with respect to covariance C. It is defined recursively via

$$\begin{aligned} &\phi(f_n) : \phi(f_1) \dots \phi(f_{n-1}) : \\ &= : \phi(f_1) \dots \phi(f_{n-1}) \phi(f_n) : + \sum_{j=1}^{n-1} (f_n, Cf_j) : \phi(f_1) \dots \widecheck{\phi(f_j)} \dots \phi(f_{n-1}) : \end{aligned} \tag{1.7}$$

where $\vee$ means omission. Wick ordering gives a symmetric function. Let $\{e_n, \lambda_n\}$ be the spectral basis of $(-\Delta+1)$ on $L^2(\Delta)$.

Because we are in two dimensions, $\sum_{n=1}^\infty \frac{1}{\lambda_n}$ is logarithmically divergent and

$$\sum_{n=1}^\infty \lambda_n^{-1-\delta} < \infty , \quad \delta > 0 . \tag{1.8a}$$

This will be used repeatedly. For any $\phi \in E$,

$$\phi = \sum_{n=1}^\infty (\phi, e_n) e_n = \sum_{n=1}^\infty \phi_n e_n , \tag{1.8b}$$

and

$$\phi^{(N)} = \sum_{n=1}^N \phi_n e_n$$

is the N-finite mode approximation. We can define recursively $:\phi^{(N)}:$ by

$$\begin{aligned} &\phi^{(N)}(x) : (\phi^{(N)}(x))^n : = : (\phi^{(N)}(x))^{n+1} : + nC^{(N)}(x) : (\phi^{(N)}(x))^{n-1} : , \\ &C^{(N)}(x) = \int d\mu_c (\phi^{(N)}(x))^2 . \end{aligned} \tag{1.9}$$

Then we have [1–3], the following facts:

$$:(\phi^{(N)})^n:(f) \xrightarrow[N\to\infty]{} :\phi^n(f) \quad \text{in} \quad L^p(d\mu_c, E) , \quad 1 \leqq p < \infty , \tag{1.10}$$

$$e^{-\lambda \int_\Lambda d^2x :\phi^4:(x)} \in L^1(d\mu_c, E) . \tag{1.11}$$

This leads to the definition of the finite volume Euclidean $(\phi^4)_2$ measure [1–3]

$$d\mu = d\mu_c e^{-\frac{\lambda}{4}\int_\Lambda :\phi^4:(x)d^2x}. \tag{1.12}$$

Because of Nelson's estimate (1.11), (1.12) is a well defined finite measure supported on $H_{-1}(\Lambda)$.

In the following we also need the space of vector valued (E-valued) functions on E, denoted V_E. The corresponding L^p spaces, $L^p(d\mu_c, V_E)$ are defined with the norm:

$$\|\mathscr{V}\|_p = \left(\int_E d\mu_c(\phi)\, \|\mathscr{V}(\phi)\|_{-1}^p\right)^{1/p}. \tag{1.13}$$

It is easy to check $:\phi^n:$ belongs to $L^p(d\mu_c, V_E)$, $1 \leqq p$.

2. Ornstein-Uhlenbeck Process Associated to the Euclidean Free Field, and Some Ito Functionals Thereof

Throughout this section we hold ε in the range $0<\varepsilon<\frac{1}{2}$.

Let t be a positive real number, and define:

$$C_t = tC^{1-\varepsilon} = t(-\Delta+1)^{-(1-\varepsilon)}. \tag{2.1}$$

Let μ_{C_t} be a family of Gaussian measures supported on

$$E = H_{-1}(\Lambda) \tag{2.2}$$

of mean 0 and covariance C_t obtained by the Sazonov-Minlos-Gross construction of the previous section.

For any $t>0$, $\phi \in E$, we define a family of probability measures $p_t^{(W)}(\phi, \cdot)$ on E by the formula:

$$p_t^{(W)}(\phi, B) = \mu_{C_t}(B-\phi), \tag{2.3}$$

where B is a Borel set in E.

The Wiener process W_t in E is defined by giving a family of probability measures $P_\phi^{(W)}$ with $\phi \in E$ on the path space $(\Omega, \mathbb{B}(\Omega))$, where $\Omega = C^0([0,\infty), E)$ with $\mathbb{B}$ its Borel algebra, satisfying

(i) $$P_\phi^{(W)}\{W_t \in B\} = p_t^{(W)}(\phi, B)$$

the transition probability,

(ii) for $s<t$,

$$P_\phi^{(W)}\{(W_t - W_s) \in B\} = p_{t-s}^{(W)}(0, B),$$

(iii) $$P_\phi^{(W)}\{W_s \in A, (W_t - W_s) \in B\} = p_s^{(W)}(\phi, A)\, p_{t-s}^{(W)}(0, B). \tag{2.4}$$

(i), (ii), and (iii) guarantee that $P_\phi^{(W)}$ is a Markov process supported on C^0 $([0,\infty), E)$. The covariance is given by:

$$E_0^{(W)}(W_t(f) W_s(g)) = (f, C^{1-\varepsilon} g) \min(t, s), \tag{2.5}$$

where $E_\phi^{(W)}$ is integration with respect to $dP_\phi^{(W)}$ and f, g are test functions.

The Wiener process W_t plays only an auxiliary role in the following. Our basic reference process willl be an Ornstein-Uhlenbeck (O.U.) process $\phi_t \in E$, obtained as the unique solution of the linear stochastic differential equation,

$$d\phi_t = dW_t - \tfrac{1}{2}C^{-\varepsilon}\phi_t dt. \tag{2.6}$$

The unique solution is

$$\phi_t = e^{-\frac{t}{2}C^{-\varepsilon}}\phi + \int_0^t e^{-\frac{1}{2}(t-s)C^{-\varepsilon}} dW_s, \tag{2.7}$$

which describes ϕ_t as an explicit functional $\phi_t(W)$ of the Wiener process.

Remark on the Stochastic Integral in (2.7). The stochastic integral of the type

$$I_t = \int_0^t e^{-\frac{1}{2}(t-s)C^{-\varepsilon}} dW_s \tag{2.8}$$

encountered in (2.7) is a continuous square integrable martingale in E, with zero Wiener expectation, and the martingale property is with respect to σ-algebras B_t engendered by W_s, $\forall s \leqq t$.

To see this we introduce a finite dimensional approximation $W_s^{(N)}$ to W_s:

$$W_s^{(N)} = \sum_{n=1}^{N} \lambda_n^{-\frac{1}{2}(1-\varepsilon)} \beta_s^{(n)} e_n, \tag{2.9}$$

where $\{\lambda_n^{-1}, e_n\}$ is an eigenbasis of C on $L^2(\Lambda)$, and $\beta_s^{(n)}$ is normalized Brownian motion on the line:

$$E(\beta_s^{(n)}\beta_t^{(m)}) = \delta_{nm} \min(t,s). \tag{2.10}$$

Then

$$I_t^{(N)} = \int_0^t \sum_{n=1}^{N} e^{-\frac{1}{2}(t-s)\lambda_n^{\varepsilon}} \lambda_n^{-\frac{1}{2}(1-\varepsilon)} d\beta_s^{(n)} e_n \tag{2.11}$$

as an Ito Stochastic Integral in R^N, [13–16], is a continuous square integrable martingale. We have

$$E_0^{(W)}(\|I_t^{(N+p)} - I_t^{(N)}\|_E^2) = \sum_{n=N}^{N+p} \frac{1}{\lambda_n^2}(1 - e^{-\lambda_n^{\varepsilon} t}) \xrightarrow[N\to\infty]{} 0, \tag{2.12}$$

since $\sum_{n=1}^{\infty} \lambda_n^{-2} < \infty$. Hence $\{I_t^{(N)}\}$ is a mean square convergent sequence, and its limit defines $I_t(f)$. To extract a continuous version we have by the martingale inequality [13],

$$P_0^{(W)}\left\{\sup_{0 \leqq t \leqq T} \|I_t^{(N+p)} - I_t^{(N)}\|_E > \delta\right\} \leqq \delta^{-2} E_0^{(W)}(\|I_T^{(N+p)} - I_T^{(N)}\|_E^2)$$

$$= \delta^{-2} \sum_{n=N}^{N+p} \frac{1}{\lambda_n^2}(1 - e^{-t\lambda_n^{\varepsilon}}) \xrightarrow[N\to\infty]{} 0. \tag{2.13}$$

By a standard application of the Borel-Cantelli lemma [13] we can extract from $I_t^{(N)}$ an a.e. convergent (uniformly in t) subsequence. Its limit defines the continuous version of I_t. □

The O.U. process $\phi_t^{(W)}$ starting at ϕ given by (2.7) is a continuous Markov process, which is *not* equivalent to the Wiener process W_t (starting at ϕ). P_ϕ is the probability measure on $C^0\ ([0,\infty), E) \equiv \Omega$ induced by the process. E_ϕ is the corresponding expectation.

The transition probabilities $p_t(\phi, \cdot)$

$$p_t(\phi, B) = P_\phi(\phi_t \in B) = P_\phi^{(W)}(W : \phi_c(W) \in B), \tag{2.14}$$

with B a Borel set in E can be obtained as follows.

Let

$$\phi = \sum_{n=1}^{\infty} (\phi, e_n) e_n = \sum_{n=1}^{\infty} \phi_n e_n \tag{2.15}$$

with $\{\lambda_n^{-1}, e_n\}$ the eigenbasis of C on $L^2(\Lambda)$. Then in components (2.7) reads:

$$\phi_t^{(n)} = e^{-\frac{1}{2}\lambda_n^\varepsilon}\phi^{(n)} + \int_0^t e^{-\frac{1}{2}(t-s)\lambda_n^\varepsilon}\lambda_n^{-\frac{1}{2}(1-\varepsilon)} d\beta_s^{(n)}, \tag{2.16}$$

where $\beta_s^{(n)}$ is as in (2.9), (2.10).

The stochastic integral on the right-hand-side of (2.16) is also Brownian motion with time change [13], $t \to \tau_n(t) = \dfrac{1}{\lambda_n}(1 - e^{-\lambda_n^\varepsilon t})$. Noting also the translation by $e^{-(t/2)\lambda_n^\varepsilon}\phi^{(n)}$, the component process $\phi_t^{(n)}$ has transition probabilities

$$p_{\tau_n(t)}^{(W)}(e^{-\frac{1}{2}t\lambda_n^\varepsilon}\phi^{(n)}, \cdot), \tag{2.17}$$

where $p_t^{(W)}(x, \cdot)$ is the standard Wiener transition probability in R^1.

It follows that the O.U. process ϕ_t of (2.7) has transition probabilities $p_t(\phi, \cdot)$ given by:

$$p_t(\phi, B) = \mu_{\tilde{C}_t}\left(B - e^{-\frac{t}{2}C^{-\varepsilon}}\phi\right), \tag{2.18}$$

where

$$\tilde{C}_t = (1 - e^{-tC^{-\varepsilon}})C, \tag{2.19}$$

and B is a Borel set in E.

Note that

$$p_t(\phi, \cdot) \xrightarrow[t \to \infty]{} \mu_c(\cdot) \tag{2.20}$$

in the sense of weak convergence of measures. It follows that the O.U. process is ergodic with μ_c as its unique invariant measure.

If we use the Ito calculus, appropriate to W_t,

$$\begin{aligned} E^{(W)}(dW_s(f)) &= 0, \\ E^{(W)}(dW_s(f)dW_s(g)) &= ds(f, C^{1-\varepsilon}g), \end{aligned} \tag{2.21}$$

where f, g are test functions. We have directly from (2.7)

$$\begin{aligned} E_\phi(\phi_t(f)\phi_t(g)) &= E^{(W)}(\phi_t(W)(f)\phi_t(W)(g)) \\ &= \phi(e^{-t/2C^{-\varepsilon}}f)\phi(e^{-t/2C^{-\varepsilon}}g) + (f, (1-e^{-tC^{-\varepsilon}})Cg) \xrightarrow[t\to\infty]{} (f, Cg) \\ &= \int_E d\mu_c \phi(f)\phi(g). \end{aligned} \tag{2.22}$$

Remark. The above construction of P_ϕ, our equivalently of $\phi_t(W)$, may be called the "stochastic quantization" of the Euclidean free field in the sense of [18]. It should be remarked however that the white noise in [18] has unit covariance, whereas our white noise ($\dot{W}_t$ formally) has covariance $C^{1-\varepsilon}$.

The drifts differ correspondingly. The advantage of our procedure will be the easy mathematical control of the stochastic quantization of the interacting $(\phi^4)_2$ Euclidean theory in the subsequent sections. □

The O.U. Semigroup e^{tL_0} and Its Properties

The continuous Markov process ϕ_t gives rise to a semigroup e^{tL_0}, the O.U. semigroup

$$(e^{tL_0}f)(\phi) = E_\phi(f(\phi_t)) = \int_E p_t(\phi, d\phi')f(\phi'), \tag{2.23}$$

where $f: E \to R$ is a bounded measurable function (to begin with). The following proposition summarises well known properties of the O.U. semigroup, which suffice to control the $(\phi^4)_2$ stochastic quantization.

Proposition. *The* O.U. *semigroup* e^{tL_0} *satisfies*

(i) e^{tL_0} *is positivity preserving and* $e^{tL_0}1 = 1$,
(ii) e^{tL_0} *is a contraction on all* $L^p(d\mu_c)$, $1 \leqq p < \infty$,
(iii) e^{tL_0} *is a strongly continuous, contractive, self-adjoint semigroup on* $L^2(d\mu_c)$,
(iv) e^{tL_0} *is hypercontractive:* $\exists T > 0$ *such that for* $t > T$

$$\|e^{tL_0}f\|_{L^4(d\mu_c)} \leqq \|f\|_{L^2(d\mu_c)},$$

(v) 1 *is the unique ground state.* e^{tL_0} *is positivity improving,* i.e. *if* $f, g \geqq 0$ a.e. *are non zero vectors in* $L^2(d\mu_c)$, *then* $(f, e^{tL_0}g) > 0$.

Note that (i) follows from (2.23) and that $p_t(\phi, \cdot)$ is a probability measure (2.18). (ii) follows from (i) and the Markov property, see [12, Chap. XIII]. (iii) follows from (ii), symmetry of the transition probabilities and stochastic continuity: $p_t(\phi, B)$ is continuous in t, a.e. in ϕ. (iv) was first isolated in the context of constructive field theory [2–7]. (v) follows from ergodicity, (i) [and selfadjointness in $L^2(d\mu_c)$], see [2].

An Ito Formula

We shall now derive an Ito formula which will play a key role in the subsequent sections.

By property (iii) of the previous proposition, if L_0 is the infinitesimal generator, then $-L_0$ is a non-negative selfadjoint operator. It can be realized as a second order differential operator on the subspace of twice differentiable cylindrical functions which with their derivatives are also in $L^2(d\mu_c)$. This subspace is dense in $\mathscr{D}(L_0)$.

Let
$$\phi^{(N)} = \sum_{i=1}^{N} e_i(\phi, e_i) = \sum_{i=1}^{N} \phi_i e_i\,,$$
the (finite) mode expansion of (2.15).

Then a cylindrical function is of the form
$$F(\phi^{(N)}) = F(\phi_1, \ldots, \phi_N)\,,$$
and on a twice differentiable cylindrical function which, with its derivatives, is in $L^2(d\mu_c)$,
$$L_0 F(\phi^{(N)}) = \frac{1}{2} \sum_{n=1}^{N} \left(\lambda_n^{-(1-\varepsilon)} \frac{\partial^2}{\partial \phi_n^2} - \lambda_n^{\varepsilon} \phi_n \frac{\partial}{\partial \phi_n} \right) F(\phi^{(N)})\,. \tag{2.24}$$

Define
$$\begin{aligned} F(\phi) &= \tfrac{1}{4} \int_{\Lambda} d^2x : \phi^4 : (x)\,, \\ H(\phi) &= (:\phi^3:, C^{1-\varepsilon} : \phi^3 :)\,, \\ G(\phi) &= :(\phi^3, C^{-\varepsilon}\phi):\,, \end{aligned} \tag{2.25}$$
and we recall that we restrict ε *to the range* $0 < \varepsilon < \frac{1}{2}$. Then $F, G, H \in L^p(d\mu_c)$, $1 \leqq p < \infty$, and $F(\phi^{(N)})$, $G(\phi^{(N)})$, $H(\phi^{(N)})$ converge respectively to $F(\phi)$, $G(\phi)$, $H(\phi)$ in $L^p(d\mu_c)$. By explicit calculation,
$$L_0 F(\phi^{(N)}) = -\tfrac{1}{2} G(\phi^{(N)})\,. \tag{2.26}$$

We now note the following: the transition probabilities $p_t(\phi, \cdot)$ of the O.U. process ϕ_t given by (2.18) are absolutely continuous with respect to $\mu_c(d\phi')$ for each $\phi \in E$, $t > 0$ provided $\varepsilon > 0$. The Radon-Nikodym derivative
$$\varrho_{t,\phi}(\phi') = \frac{p_t(\phi, d\phi')}{\mu_c(d\phi')} \tag{2.27}$$
is in $L^2(d\mu_c)$. Indeed by a straightforward calculation
$$\begin{aligned} a(t,\phi)^2 &\equiv \int |\varrho_{t,\phi}(\phi')|^2 d\mu_c(\phi') \\ &= [\det(I - e^{-2tC^{-\varepsilon}})]^{-1/2} \exp\{(\phi, C^{-1} e^{-tC^{-\varepsilon}} (1 + e^{-tC^{-\varepsilon}})^{-1} \phi)\}\,, \end{aligned} \tag{2.28}$$
which is well defined for $\varepsilon > 0$, because then $e^{-2tC^{-\varepsilon}}$ is of trace class.

Let $h \in L^{2p}(d\mu_c)$. Then,
$$\begin{aligned} E_\phi(|h(\phi_t)|^p) &= \int p_t(\phi, d\phi')|h(\phi')|^p \\ &= \int \varrho_{t,\phi}(\phi')|h(\phi')|^p d\mu_c(\phi') \leqq a(r,\phi) \|h\|^p_{L^{2p}(d\mu_c)}\,. \end{aligned} \tag{2.29}$$

As a consequence, $h \in L^{2p}(d\mu_c)$, $1 \leqq p < \infty \Rightarrow h(\phi_t) \in L^p(dP_\phi, \Omega)$, $1 \leqq p < \infty$.

Finally let $h(\phi) \in L^{2p}(d\mu_c)$ be obtained as the convergent limit of $h(\phi^{(N)})$, $N \to \infty$, where $h(\phi^{(N)})$ is a continuous function of $\phi^{(N)}$. Then since ϕ_s is continuous, P_ϕ a.e., in s
$$\int_0^t ds\, h(\phi^{(N)})$$

exists P_ϕ a.e. as a Riemann integral. Then $\int_0^t ds\, h(\phi_s)$ is defined as the convergent limit as $N\to\infty$ in $L^p(dP_\phi, \Omega)$ of the sequence $\int_0^t ds\, h(\phi^{(N)})$. This follows using (2.29) (2.28) in the region $s>0$ and

$$E_\phi(|h(\phi_s)|^p)\xrightarrow[s\to 0]{} |h(\phi)|^p ,$$

μ_c a.e. in ϕ. In particular, let h stand for any of the functions F, G, H of (2.25). Then

$$h(\phi_t)\in L^p(dP_\phi, \Omega), \qquad 1\leqq p<\infty ,$$

$$E_\phi\left(\int_0^t ds|h(\phi_s)|\right) <\infty .$$

We will define the stochastic integral

$$I_t=\int_0^t (:\phi^3:, dW_s), \tag{2.30}$$

where $\phi_s(W)$ is given by (2.7), as the covergent limit in $L^2(dP_\phi^{(W)}, \Omega)$ of the sequence of Ito stochastic integrals

$$I_t^{(N)}=\int_0^t (:(\phi_s^{(N)})^3:, dW^{(N)}),$$

where $W_s^{(N)}$ is given by (2.9). Indeed

$$E_\phi^{(W)}(|I_t^{(N+p)}-I_t^{(N)}|^2)$$
$$=E_\phi\left(\int_0^t ds\|C^{\frac{1-\varepsilon}{2}}:(\phi_s^{(N+p)})^3:-C^{\frac{1-\varepsilon}{2}}:(\phi_s^{(N)})^3:\|_{L^2(\Lambda)}^2\right)\xrightarrow[N\to\infty]{} 0,$$

since

$$E_\phi\left(\int_0^t ds(:\phi^3:, C^{1-\varepsilon}:\phi^3:)\right) <\infty$$

by the above.

Now by finite dimensional Ito calculus [13–16],

$$F(\phi_t^{(N)}(W))=F(\phi^{(N)})+\int_0^t (:(\phi_s^{(N)})^3:, dW_s)+\int_0^t L_0F(\phi_s^{(N)}(W))ds$$
$$=F(\phi^{(N)})+\int_0^t (:(\phi_s^{(N)})^3:, dW_s)-\tfrac{1}{2}\int_0^t ds\, G(\phi_s^{(N)}(W)),$$

where we have used (2.25), (2.26). All terms in the above equation converge in $L^2(dP_\phi^{(W)}, \Omega)$. Hence taking the limit $N\to\infty$ in this space,

$$\tfrac{1}{4}\int_\Lambda d^2x:\phi_t^4:-\tfrac{1}{4}\int_\Lambda d^2x:\phi^4:=\int_0^t (:\phi_s^3:, dW_s)-\tfrac{1}{2}\int_0^t ds:(\phi_s^3, C^{-\varepsilon}\phi_s):. \tag{2.31}$$

Finally let us define the probability measures $Q_{\mu_c}^{(W)}, Q_{\mu_c}$ on $E\times\Omega$ by

$$dQ_{\mu_c}^{(W)}=d\mu_c(\phi)dP_\phi^{(W)} \quad\text{and}\quad dQ_{\mu_c}=d\mu_c(\phi)dP_\phi .$$

Here $P_\phi^{(W)}, P_\phi$ are respectively the Wiener and O.U. measures on Ω. We have of course

$$\int dQ_{\mu_c} h(\phi_t) = \int dQ_{\mu_c}^{(W)} h(\phi_t(W)),$$

where $\phi_t(W)$ is given by (2.7).

It is easy to check that if $h \in L^p(d\mu_c)$, $1 \leqq p < \infty$, then

$$h(\phi_t(W)) \in L^p(dQ_{\mu_c}^{(W)}), \quad 1 \leqq p < \infty .$$

As a consequence $F(\phi_t)$, $G(\phi_t)$, $H(\phi_t)$, where F, G, H are given by (2.25), exist in $L^p(dQ_{\mu_c}^{(W)})$. Moreover the stochastic integral (2.30) as well as the Ito formula (2.31) is valid in $L^2(dQ_{\mu_c}^{(W)})$.

3. Markov Process Associated to the $(\phi^4)_2$ Euclidean Field Theory in Finite Volume Λ

We consider the stochastic differential equation in $E = H_{-1}(\Lambda)$,

$$d\hat{\phi}_t = dW_t - \tfrac{1}{2}(C^{-\varepsilon}\hat{\phi}_t + \lambda C^{1-\varepsilon} : \hat{\phi}_t^3 :) dt, \quad \hat{\phi}_0 = \phi, \tag{3.1}$$

which can also be written as an integral equation

$$\hat{\phi}_t = \phi_t - \frac{\lambda}{2} \int_0^t ds\, e^{-\frac{1}{2}(t-s)C^{-\varepsilon}} C^{1-\varepsilon} : \hat{\phi}_s^3 :, \tag{3.2}$$

where ϕ_t is the O.U. process of Sect. 2.

The vector

$$\sigma = C^{1-\varepsilon} : \phi^3 :$$

belongs to $L^p(d\mu_c, V_E)$, $1 \leqq p < \infty$ (see the end of Sect. 1 for the notation), but has no continuity prooerties. Hence the contraction mapping principle (Picard's method) cannot be exploited to solve (3.1) or (3.2).

Instead we will shoot for a *weak* solution. Namely we will construct a Markov family of measures $\hat{P}_\phi$, $\phi \in E$, on $\Omega = C^0([0, \infty), E)$ such that the process

$$\hat{Z}_t = \hat{\phi}_t + \frac{\lambda}{2} \int_0^t ds\, e^{-\frac{1}{2}(t-s)C^{-\varepsilon}} C^{1-\varepsilon} : \hat{\phi}_s^3 : \tag{3.3}$$

(where $\hat{\phi}_t$ are paths in Ω) has as its $\hat{P}_\phi$ joint probability distributions those of the O.U. process ϕ_t of Sect. 2. In other words under the law $\hat{P}_\phi$, $\hat{Z}_t$ is indistinguishable from ϕ_t, but not necessarily measurable with respect to it.

In this sense $\hat{P}_\phi$ solves (3.2). *This will be done under the restriction* $0 < \varepsilon < \frac{1}{10}$.

In this section we will construct the Markov family of measures $\hat{P}_\phi$ on Ω, and prove ergodic and mixing properties. Some technical estimates are relegated to the appendix. The unique invariant measure associated to this process is the $(\phi^4)_2$ measure of Sect. 1, and in this sense we have "stochastically quantized" Euclidean $(\phi^4)_2$ theory in finite volume.

In Sect. 4 we will verify that this family $\hat{P}_\phi$ actually gives the claimed weak solution.

To understand our strategy we first consider the finite dimensional approximation $\phi^{(N)} \to \phi$, of Sect. 2. Then, in this approximation, with $\phi^{(N)} = \sum_{n=1}^{N} \phi_n e_n$ identified also with the vector $(\phi_1, \dots, \phi_N)$ in R^N we have

$$d\hat{\phi}_t^{(N)} = dW_t^{(N)} - \tfrac{1}{2}(C^{-\varepsilon}\hat{\phi}_t^{(N)} + \lambda C^{1-\varepsilon} : (\hat{\phi}_t^{(N)})^3 :)dt\,, \qquad \hat{\phi}_0^{(N)} = \phi^{(N)}\,. \tag{3.4}$$

Note that,

$$:(\phi^{(N)}(x))^3: = (\phi^{(N)}(x))^3 - 3C^{(N)}(x)\phi^{(N)}(x)$$

and $C^{(N)}(x)$ is finite. As a consequence, the drift in (3.4) is a C^∞ function on R^N. Now we can apply the *standard method* to construct a unique strong solution to (3.4), [13], because we can find an increasing nested sequence of compacts in R^N on each of which the drift is bounded and Lipschitz. To make sure that there is no explosion, i.e. the solution is defined for all times, it suffices, [14], to construct a C^∞ function ϱ on R^N, such that it is non-negative and

(i) $$\varrho_R = \inf_{\|\phi^{(N)}\| > R} \varrho(\phi^{(N)}) \xrightarrow[R\to\infty]{} \infty\,,$$

where $\|\phi^{(N)}\|^2 = (\phi^{(N)}, \phi^{(N)}) = \sum_{n=1}^{N} \phi_n^2$,

(ii) there exists a constant $C > 0$, such that

$$L^{(N)}\varrho \leqq C\varrho\,,$$

where $L^{(N)}$ is the differential generator associated to (3.4).

In our case it is easy to see that the choice

$$\varrho(\phi^{(N)}) = \|\phi^{(N)}\|_{1-\varepsilon}^2 + \text{const}\,, \tag{3.5}$$

where

$$\|\phi^{(N)}\|_{1-\varepsilon}^2 = (\phi^{(N)}, C^{-(1-\varepsilon)}\phi^{(N)}) = \sum_{n=1}^{N} \lambda_n^{1-\varepsilon}\phi_n^2 \geqq \|\phi^{(N)}\|^2$$

does the job, using the explicit expression,

$$\begin{aligned} L^{(N)} = {} & \frac{1}{2} \int_{\Lambda\times\Lambda} d^2x\, d^2y\, C^{(N)1-\varepsilon}(x-y) \frac{\delta^2}{\delta\phi^{(N)}(x)\delta\phi^{(N)}(y)} \\ & - \frac{1}{2} \int_{\Lambda\times\Lambda} d^2x d^2y (C^{-\varepsilon}(x-y)\phi^{(N)}(y) + \lambda C^{1-\varepsilon}(x-y) : (\phi^{(N)})^3 : (y)) \frac{\delta}{\delta\phi^{(N)}(x)}\,. \end{aligned} \tag{3.6}$$

Thus there is a unique, non-explosive solution $\hat{\phi}_t^{(N)}$ of (3.4) which is a diffusion process, and the associated semigroup $e^{tL^{(N)}}$

$$(e^{tL^{(N)}}f)(\phi) = E_\phi^{(W)}(f(\hat{\phi}_t^{(N)}))$$

can be expressed by the Cameron-Martin-Girsanov formula [14] as

$$\begin{aligned} & (e^{tL^{(N)}}f)(\phi) = E_\phi^{(W)}(f(\phi_t^{(N)})e^{\xi_t^{(N)}(W)})\,, \\ & \xi_t^{(N)}(W) = -\frac{\lambda}{2}\int_0^t (:(\phi_s^{(N)})^3:, dW_s^{(N)}) - \frac{\lambda^2}{8}\int_0^t ds(:(\phi_s^{(N)})^3:, C^{1-\varepsilon}:(\phi_s^{(N)})^3:)\,. \end{aligned} \tag{3.7}$$

In the above $\phi_t^{(N)}(W)$ is the unique solution of (3.4), for $\lambda=0$, and is thus the finite dimensional approximation to the O.U. process ϕ_t of Sect. 2.

We also have,

$$e^{tL^{(N)}}1=E_\Phi^{(W)}(e^{\xi^{(N)}})=1\,, \tag{3.8}$$

because of the absence of explosion.

We now proceed to the infinite dimensional system (3.1), (3.2) by simply defining the semigroup e^{tL} by:

$$(e^{tL}f)(\phi)=E_\Phi^{(W)}(f(\phi_t(W))e^{\xi_t(W)})\,, \tag{3.9}$$

where

$$\xi_t(W)=-\frac{\lambda}{2}\int_0^t(:\phi_s^3:,dW_s)-\frac{\lambda^2}{8}\int_0^t(:\phi_s^3:,C^{1-\varepsilon}:\phi_s^3:) \tag{3.10}$$

and ϕ_s is the O.U. process of Sect. 2.

In Sect. 2 we showed that each term in (3.10) belongs to $L^2(dP_\Phi^{(W)},\Omega)$. Moreover $\xi_t^{(N)}\to\xi_t$ in $L^2(dP_\Phi^{(W)},\Omega)$. Thus ξ_t is a well defined non-anticipating random variable. Theorem 1, proved below, shows that e^{tL} as defined by (3.9), (3.10) exists for f a bounded measurable function, and moreover $e^{tL}1=1$. This shows in particular that e^{tL} is a contraction on $L^\infty\,(d\mu_c)$. Theorem 2, proved below, shows that e^{tL} is a strongly continuous, contractive selfadjoint semi-group on $L^2(d\mu)$ with 1 as the *unique* ground state.

First we need

Lemma 1. *Let* $0<\varepsilon<\frac{1}{10}$. *Then,*

$$\text{(i)}\qquad E_\Phi^{(W)}(e^{p\xi_t})<\infty\,,\qquad 1\leqq p<\infty\,,$$

$$\text{(ii)}\qquad E_\Phi^{(W)}(e^{p\xi_t^{(N)}})<C_{t,\phi,|\Lambda|}\,,$$

where $C_{t,\phi,|\Lambda|}$ *is a generic constant depending on* $t,\phi,|\Lambda|$ *and independent of* N.

Proof.

$$E_\Phi^{(W)}(e^{p\xi_t})\leqq E_\Phi^{(W)}\left(e^{-p\frac{\lambda}{2}\int_0^t(:\phi_s^3:,dW_s)}\right).$$

Now use the Ito formula (2.31). Hence

$$\begin{aligned}E_\Phi^{(W)}(e^{p\xi_t})&\leqq e^{\frac{p\lambda}{8}\int_\Lambda d^2x:\phi_t^4:}E_\Phi\left(e^{-\frac{p\lambda}{8}\int_\Lambda d^2x:\phi_t^4:-\frac{p\lambda}{4}\int_0^t ds:(\phi^3,C^{-\varepsilon}\phi_s):}\right)\\&\leqq C_{\phi,|\Lambda|}\left\{E_\Phi\left(e^{-\frac{p\lambda}{4}\int_\Lambda:\phi_t^4:d^2x}\right)\right\}^{1/2}\left\{E_\Phi\left(e^{-\frac{p\lambda}{2}\int_0^t ds:(\phi_s^3,C^{-\varepsilon}\phi_s):}\right)\right\}^{1/2}.\end{aligned} \tag{3.11}$$

Since $e^{-\frac{p\lambda}{4}\int_\Lambda:\phi^4:d^2x}$ belongs to $L^q(d\mu_c)$, $1\leqq q<\infty$ by Nelson's estimate (Sect. 1), it follows by virtue of (2.29) that

$$E_\Phi\left(e^{-\frac{p\lambda}{4}\int_\Lambda:\phi_t^4:d^2x}\right)<\infty\,.$$

Define

$$M(\phi)=:(\phi^3, C^{-\varepsilon}\phi):.$$

It is shown in the appendix that $\exp(-M(\phi))\in L^q(d\mu_c)$, $1\leqq q<\infty$, and also that

$$\exp\left(-\int_0^t ds\, M(\phi_s)\right)\in L^q(dP_\phi,\Omega), \qquad 1\leqq q<\infty.$$

Hence

$$E_\phi^{(W)}(e^{p\xi_t})<\infty, \qquad 1\leqq p<\infty,$$

which is (i).

Moreover, as stated in the appendix we have (by the method of [31]) not only integrability but also stability bounds for each factor in (3.11): replacing ϕ by $\phi^{(N)}$, each of the two factors in (3.11) is uniformly bounded above, independent of N. This gives (ii). □

Now we proceed to Theorem 1.

Theorem 1. *The semi-group* e^{tL} *is well defined (*i.e. *the right-hand side of* (3.9) *exists) for f, bounded and measurable. Moreover*

$$e^{tL}1=1.$$

As a consequence, e^{tL} *is a contraction on* $L^\infty(d\mu_c)$.

Proof. That the right-hand side of (3.9) is finite follows from (i) of Lemma 1. Now we prove the next statement

$$\begin{aligned}E_\phi^{(W)}(e^{\xi_t})&=E_\phi^{(W)}(e^{\xi_t^{(N)}})+E_\phi^{(W)}(e^{\xi_t}-e^{\xi_t^{(N)}})\\&=1+E_\phi^{(W)}(e^{\xi_t}-e^{\xi_t^{(N)}}),\end{aligned}$$

where we have used (3.8).

Hence

$$\begin{aligned}|E_\phi^{(W)}(e^{\xi_t})-1|&\leqq E_\phi^{(W)}(|e^{\xi_t}-e^{\xi_t^{(N)}}|)\leqq E_\phi^{(W)}(|\xi_t-\xi_t^{(N)}||e^{\xi_t}+e^{\xi_t^{(N)}}|)\\&\leqq\{E_\phi^{(W)}(|\xi_t-\xi_t^{(N)}|^2)\}^{1/2}(\{E_\phi^{(W)}(e^{2\xi_t})\}^{1/2}+\{E_\phi^{(W)}(e^{2\xi_t^{(N)}})\}^{1/2}).\end{aligned}$$

As $N\to\infty$, the first factor tends to zero, whereas the second factor is uniformly bounded above by Lemma 1. Hence, letting $N\to\infty$

$$E_\phi^{(W)}(e^{\xi_t})=1. \quad \square$$

We now turn to Theorem 2.

Theorem 2. *Let* e^{tL} *be defined by* (3.9). *Then* e^{tL} *is a bounded self-adjoint, strongly continuous semi-group on* $L^2(d\mu)$, $d\mu=d\mu_c\exp\left(-\frac{\lambda}{4}\int_\Lambda d^2x:\phi^4:(x)\right)$, e^{tL} *is positivity preserving and improving, and* 1 *is the unique ground state provided* $0<\varepsilon<\frac{1}{10}$.

Remark. Because, e^{tL} is symmetric in $L^2(d\mu)$ it follows that

$$\int d\mu e^{tL}f=\int d\mu f$$

for $f \in L^2(d\mu)$. It follows [12, Chap. XIII] that e^{tL} is not only bounded on $L^2(d\mu)$ but also contractive. Hence the infinitesimal generator L is a self-adjoint operator and $-L \geqq 0$.

Proof of the Theorem. Let

$$U : L^2(d\mu) \to L^2(d\mu_c) \tag{3.12}$$

be the unitary map given by

$$f \to \tilde{f} = Uf = e^{-\frac{\lambda}{8}\int_\Lambda d^2x :\phi^4:(x)} f .$$

Let e^{tL} be defined by (3.9); then

$$(f, e^{tL}g)_{L^2(d\mu)} = \int d\mu_c(\phi) e^{-\frac{\lambda}{4}\int_\Lambda d^2x :\phi^4:(x)} f(\phi) E_\phi^{(W)}$$

$$\cdot \left(g(\phi_t(W)) e^{-\frac{\lambda}{2}\int_0^t (:\phi_s^3(W):, dW_s) - \frac{\lambda^2}{8}\int_0^t ds \, \| C^{\frac{1}{2}(1-\varepsilon)} :\phi_s^3(W): \|^2} \right)$$

We now use the Ito formula (2.31) of Sect. 2. Let $\tilde{f}, \tilde{g}$ be the unitary transform of f, g into $L^2(d\mu_c)$ given by (3.12). Then we obtain:

$$(f, e^{tL}g)_{L^2(d\mu)} = (\tilde{f}, e^{t\tilde{L}}\tilde{g})_{L^2(d\mu_c)} , \tag{3.13}$$

where

$$(e^{t\tilde{L}}\tilde{g})(\phi) = E_\phi\left(\tilde{g}(\phi_t) e^{-\int_0^t ds \tilde{V}(\phi_s)}\right), \tag{3.14}$$

$$\tilde{V}(\phi) = \frac{\lambda}{4} :(\phi^3, C^{-\varepsilon}\phi): + \frac{\lambda^2}{8}(:\phi^3:, C^{1-\varepsilon} :\phi^3:) . \tag{3.15}$$

ϕ_t is the O.U. process of Sect. 2 and we recall that $0 < \varepsilon < \frac{1}{10}$. First note that $\tilde{V}(\phi) \in L^p(d\mu_c)$, $1 \leqq p < \infty$. We show in the appendix that $\exp - \frac{\lambda}{4} :(\phi^3, C^{-\varepsilon}\phi):$ is in $L^1(d\mu_c)$ for any $\lambda > 0$.

This fact, together with the fact that the second term in (3.9) is positive, implies that $\exp(-\tilde{V}) \in L^1(d\mu_c)$. Moreover the O.U. process ϕ_t is hyper-contractive (Sect. 2). It follows [2–6], [6] gives the abstract setting used in this paper, that $e^{t\tilde{L}}$ is a strongly continuous bounded self-adjoint semi-group on $L^2(d\mu_c)$.

Moreover $\exists T > 0$ such that for $t \geqq T$, $e^{t\tilde{L}}$ is a bounded map from $L^2(d\mu_c)$ to $L^4(d\mu_c)$, see e.g. [6]. Hence, by virtue of [10], $e^{t\tilde{L}}$ has a ground state with finite multiplicity. It now follows, because of (3.13), that e^{tL} is a strongly continuous bounded self-adjoint semi-group on $L^2(d\mu)$. Moreover e^{tL} has a ground state with finite multiplicity.

The representation (3.9) shows that e^{tL} is positivity preserving. We now prove that e^{tL} is positivity improving: i.e. if $f, g, \geqq 0$ are positive vectors in $L^2(d\mu)$ then

$$(f, e^{tL}g)_{L^2(d\mu)} > 0 \tag{3.16}$$

(by a positive vector we mean non-negative and not identically zero).

If f, g are positive vectors in $L^2(d\mu)$, then $\tilde{f}, \tilde{g}$ their unitary transforms via (3.12) are positive vectors in $L^2(d\mu_c)$. Then (3.16) will follow, because of (3.13), from

$$(\tilde{f}, e^{t\tilde{L}}\tilde{g}) > 0\,. \tag{3.17}$$

Inequality (3.17) follows from a standard argument [2]. Namely, recall Q_{μ_c} the measure on $E \times \Omega$ given at the end of Sect. 2. Then if $f, g \in L^2(d\mu_c)$, are positive vectors, then

$$\int_{E\times\Omega} dQ_{\mu_c} \tilde{f}(\phi)\tilde{g}(\phi) = (\tilde{f}, e^{tL_0}\tilde{g})_{L^2(d\mu_c)} > 0\,,$$

since the O.U. semigroup e^{tL_0} is positivity improving. It follows that $\tilde{f}(\phi)\tilde{g}(\phi_t)$ is a positive vector in $\Omega \times E$. Moreover

$$e^{-\int_0^t \tilde{V}(\phi_s)ds} > 0\,, \qquad Q_{\mu_c}\ a.e.$$

Hence

$$(\tilde{f}, e^{t\tilde{L}}\tilde{g})_{L^2(d\mu_c)} = \int_{\Omega\times E} dQ_{\mu_c} \tilde{f}(\phi)\tilde{g}(\phi_t) e^{-\int_0^t \tilde{V}(\phi_s)ds} > 0\,. \tag{3.18}$$

We conclude that e^{tL} is positivity improving. Hence by the Perron-Frobenius argument [1, 2] it follows that e^{tL} has a unique ground state. That 1 is the ground state

$$e^{tL}1 = 1$$

follows from Theorem 1. □

Remark. Theorem 2 implies by standard results [12] that the semi-group e^{tL} is ergodic. In other words, if $f \in L^2(d\mu)$, and $Z = \int d\mu$,

$$\lim_{T\to\infty} \frac{1}{T}\int_0^T ds (e^{sL}f)(\phi) = \frac{1}{Z}\int d\mu(\phi) f(\phi)\,, \qquad \mu\ a.e. \tag{3.19}$$

However, in our case we have a stronger result, namely that e^{tL} is mixing:

$$\lim_{t\to\infty} (f, e^{tL}g)_{L^2(d\mu)} = \frac{1}{Z}\left(\int d\mu f\right)\left(\int d\mu g\right). \tag{3.20}$$

This follows from the fact that e^{tL} has a mass gap.

The existence of the mass gap follows from the following facts (for details see [4, 6]): a) the resolvent of the O.U. generator $R_0 = (\lambda - L_0)^{-1}$ in a finite box is compact; b) the resolvent $\tilde{R} = (\lambda - \tilde{L})^{-1}$ is also compact as $\tilde{V}$ is an almost semibounded perturbation of L_0 and L_0 generates a hypercontractive semigroup; c) $e^{t\tilde{L}}$ has a ground state. Hence e^{tL} has a mass gap.

The Markov Process $\hat{\phi}_t$

The previous Theorems 1 and 2 and the representation (3.5) enables us to construct the desired ergodic Markov process $\hat{\phi}_t$ which gives the stochastic quantization of continuum $(\phi^4)_2$ Euclidean field theory in finite volume.

We define transition probabilities $\hat{p}_t(\phi, d\phi')$ as probability measures on $E = H_{-1}(\Lambda)$ by:

$$\hat{p}_t(\phi, B) = E_\phi^{(W)}\left(\chi\{\phi_t(W) \in B\} e^{-\frac{\lambda}{2}\int_0^t (:\phi_s^3(W):, dW_s) - \frac{\lambda^2}{8}\int_0^t ds \|C^{\frac{1}{2}(1-\varepsilon)} : \phi_s^3(W): \|^2}\right), \quad (3.21)$$

where χ is the characteristic function of an event and B is a Borel set in E.

The exponential in (3.21) is integrable by virtue of Theorem 1. Moreover $e^{tL}1 = 1$. From the countable additivity of $E_\phi^{(W)}$ it now follows that $\hat{p}_t(\phi, \cdot)$ is a family of countably additive probability measures on E. Moreover the theorems assure us that $\hat{p}_t(\phi, B)$ is μ_c measurable in ϕ, and dt measurable in t.

Hence $\hat{p}_t(\phi, B)$ are "stochastic kernels" and qualify as transition probabilities of a Markov process $\hat{\phi}_t$ with values in E. Its joint probability distributions are *defined* in the standard way: $(0 < t_1 < \ldots < t_n)$

$$\begin{aligned}
&\hat{p}_\phi\{\hat{\phi}_{t_1} \in B_1, \ldots, \hat{\phi}_{t_n} \in B_n\} \\
&= \int_{B_1} \ldots \int_{B_n} \hat{p}_{t_1}(\phi, d\phi_1) \hat{p}_{t_2 - t_1}(\phi_1, d\phi_2) \ldots \hat{p}_{t_n - t_{n-1}}(\phi_{n-1}, d\phi_n) \\
&= \hat{p}_{t_1 \ldots t_n}(B_1 \times B_2 \times \ldots \times B_n), \quad (3.22)
\end{aligned}$$

which defines $\hat{p}_{t_1 \ldots t_n}$ as a consistent family of probability measures on $(E)^n$. We shall show later that $\hat{P}_\phi$ can be realized as probability measures on Ω, our path space, and then $(\Omega, \hat{p}_\phi)$ constitutes our Markov process with continuous sample paths. Let us note, from (3.22), that if $f_1, \ldots, f_n$ are test functions,

$$\begin{aligned}
\hat{E}_\phi(\hat{\phi}_{t_1}(f_1) \ldots \hat{\phi}_{t_n}(f_n)) = E_\phi^{(W)}\Bigg(\phi_{t_1}(f_1) \ldots \phi_{t_n}(f_n) \exp\Bigg\{ &-\frac{\lambda}{2}\int_0^{t_n} (:\phi_s^3(W):, dW_s) \\
&- \frac{\lambda^2}{8}\int_0^{t_n} ds \|C^{\frac{1}{2}(1-\varepsilon)} : \phi_s^3(W)\|^2 \Bigg\}\Bigg). \quad (3.23)
\end{aligned}$$

By the preceding remark after Theorem 2, e^{tL} is mixing. Hence we have:

$$\lim_{t \to \infty} \hat{E}_\phi(\hat{\phi}_{t_1}(f_1) \ldots \hat{\phi}_{t_n}(f_n)) = \frac{1}{Z}\int d\mu(\phi)\phi(f_1) \ldots \phi(f_n), \quad \mu \text{ a.e.}, \quad (3.24)$$

which is the aim of stochastic quantization of the $(\phi^4)_2$ theory.

We shall now show that the Markov process $\hat{\phi}_t$ constructed from its transition probabilities has a continuous version. In other words, there exists μ a.e. in ϕ a Markovian family of probability measures $\hat{P}_\phi$ on $\Omega = C^0([0, \infty), E = H_{-1}(\Lambda))$ of which (3.22) are the joint probability distributions. For convenience, we run off the Markov process $\hat{\phi}_t$ with initial distribution μ. We shall prove:

Proposition. *Let* $0 < s < t < T$. *Then*

$$\int_E d\mu(\phi)\hat{E}_\phi(\|\hat{\phi}_t - \hat{\phi}_s\|_E^4) \leqq C|t-s|^2 (C_T)^{t-s}. \quad (3.25)$$

Corollary. *By virtue of the estimate* (3.25), *Kolmogoroff's theorem, see* e.g. [14], *assures us that there exists a probability measure* $\hat{Q}_\mu$ *on* $\Omega = C^0([0, \infty), E)$ *such that, if* $B_1, \ldots, B_n$ *are Borel sets in E,*

$$\begin{aligned}
&\hat{Q}_\mu\{\hat{\phi}_{t_1} \in B_1, \ldots, \hat{\phi}_{t_n} \in B_n\} \\
&= \int_E d\mu(\phi) \int_{B_1} \ldots \int_{B_n} \hat{p}_{t_1}(\phi, d\phi_1) \hat{p}_{t_2 - t_1}(\phi_1, d\phi_2) \ldots \hat{p}_{t_n - t_{n-1}}(\phi_{n-1}, d\phi_n), \quad (3.26)
\end{aligned}$$

$(\hat{Q}_\mu, \Omega)$ is a stationary Markov process. We can write

$$\hat{Q}_\mu(B) = \int d\mu(\phi)\hat{P}_\phi(B),$$

where B is a Borel set in Ω. $(\hat{P}_\phi, \Omega)$ is the Markov process started off at ϕ.

Proof of the Proposition. It is straightforward to verify that the continuous O.U. process ϕ_t of Sect. 2 satisfies

$$\int d\mu_c(\phi)E_\phi(\|\phi_r - \phi_s\|_E^{2r}) \leqq C|t-s|^r \tag{3.27}$$

for any positive integer $r \geqq 1$. Next we note

$$\begin{aligned}\int d\mu(\phi)\hat{E}_\phi(\|\hat{\phi}_t - \hat{\phi}_s\|^{2r}) &= \int d\mu(\phi)\int \hat{p}_s(\phi, d\phi_1)\int \hat{p}_{t-s}(\phi_1, d\phi_2)\,\|\phi_1 - \phi_2\|_E^{2r} \\ &= \int d\mu(\phi)\int \hat{p}_{t-s}(\phi, d\phi_2)\,\|\phi - \phi_2\|_E^{2r},\end{aligned}$$

where we have used the fact that e^{tL} leaves μ invariant. Hence (ϕ_t is the O.U. process)

$$\begin{aligned}&\int d\mu(\phi)E_\phi(\|\hat{\phi}_t - \hat{\phi}_s\|_E^{2r}) = \int d\mu(\phi)E_\phi^{(W)}(\|\phi - \phi_{t-s}(W)\|_E^{2r} \\ &\quad\cdot \exp\left\{-\frac{\lambda}{2}\int_0^{t-s}(:\phi_{s_1}^3(W):, dW_{s_1}) - \frac{\lambda^2}{8}\int_0^{t-s} ds_1\|C^{\frac{1}{2}(1-\varepsilon)} :\phi_{s_1}^3(W):\|^2\right\}\Bigg) \\ &= \int d\mu_c(\phi)E_\phi^{(W)}\Bigg(\|\phi - \phi_{t-s}(W)\|_E^{2r}\exp\left\{-\frac{\lambda}{4}\int_\Lambda d^2x :\phi^4:(x)\right. \\ &\quad \left. - \frac{\lambda}{2}\int_0^{t-s}(:\phi_{s_1}^3:, dW_{s_1}) - \frac{\lambda^2}{8}\int_0^{t-s} ds_1\|C^{\frac{1}{2}(1-\varepsilon)}:\phi_{s_1}^3:\|^2\right\}\Bigg)\end{aligned}$$

and, on using the Ito formula (2.31)

$$\begin{aligned}&= \int d\mu_c(\phi)E_\phi\Bigg(\|\phi - \phi_{r-s}\|_E^{2r}\exp\left\{-\frac{\lambda}{8}\int_\Lambda d^2x :\phi^4:(x) - \frac{\lambda}{8}\int_\Lambda d^2x :\phi_{t-s}^4:(x)\right. \\ &\quad \left. - \int_0^{t-s} ds_1 \tilde{V}(\phi_{s_1})\right\}\Bigg),\end{aligned}$$

where $\tilde{V}(\phi)$ is given by (3.15). Apply Hölder's inequality

$$\begin{aligned}&\leqq \left(\int d\mu_c(\phi)E_\phi(\|\phi - \phi_{t-s}\|_E^{8r})\right)^{1/4} \cdot \left(\int d\mu_c(\phi)E_\phi\left(e^{-\frac{\lambda}{2}\int_\Lambda :\phi^4:(x)d^2x}\right)\right)^{1/4} \\ &\quad \cdot \left(\int d\mu_c(\phi)E_\phi\left(e^{-\frac{\lambda}{2}\int_\Lambda :\phi_{t-s}:(x)d^2x}\right)\right)^{1/4} \cdot \left(\int d\mu_c(\phi)E_\phi\left(e^{-4\int_0^{t-s}\tilde{V}(\phi_{s_1})ds_1}\right)\right)^{1/4}\end{aligned} \tag{3.28}$$

$$\begin{aligned}&\leqq \left(\int d\mu_c(\phi)E_\phi(\|\phi - \phi_{t-s}\|_E^{8r})\right)^{1/4} \\ &\quad \cdot \left\|e^{-\frac{\lambda}{4}\int_\Lambda d^2x:\phi^4:(x)}\right\|_{L^2(d\mu_c)} \cdot \|e^{-4(t-s)\tilde{V}}\|_{L^1(d\mu_c)}^{1/4},\end{aligned} \tag{3.29}$$

where the last factor in (3.28) has been estimated using the Riemann sum approximation for $\int_0^{t-s} ds_1 \ldots$, Hölder's inequality (see e.g. [5]) and that e^{tL_0} leaves

μ_c invariant. The latter was also used for the second factor. It is easy to check

$$\int d\mu_c(\phi)E_\phi(\|\phi_t-\phi_s\|_E^{8r})=\int d\mu_c(\phi)E_\phi(\|\phi-\phi_{t-s}\|_E^{8r}), \tag{3.30}$$

and

$$\|e^{-4(r-s)\tilde{V}}\|_{L^r(d\mu_c)}\leqq(\|e^{-\tilde{V}}\|_{L^{4T}(d\mu_c)})^{t-s}. \tag{3.31}$$

From (3.29), (3.31), and (3.27),

$$\int d\mu(\phi)\hat{E}_\phi(\|\hat{\phi}_t-\hat{\phi}_s\|^{2r})\leqq C|t-s|^r(C_T)^{t-s}, \tag{3.32}$$

choosing $r=2$ in (3.32), the proof of the proposition is complete.

4. The Process $\hat{\phi}_t$ as a Weak Solution of (3.2)

It is legitimate to ask in what sense the ergodic Markov process $(\Omega, \hat{P}_\phi)$ constructed in Sect. 3 solves Eq. (3.2), which was our starting point.

Let $\Omega_T=C^0([0,T],E)$ and $\hat{P}_\phi$ the Markov family of measures on Ω_T obtained from the joint probability distributions (3.22) by what we have shown previously. Then the answer to the question is given by the following proposition.

Proposition.

$$\hat{Z}_t=\hat{\phi}_t+\frac{\lambda}{2}\int_0^t ds\, e^{-\frac{(t-s)}{2}C^{-\varepsilon}}C^{1-\varepsilon}:\hat{\phi}_s^3: \tag{4.1}$$

has the same joint probability distributions (with respect to $(\Omega_T, \hat{P}_\phi)$) as the O.U. *process ϕ_t of Sect. 2. Thus we have a weak solution of* (3.2). *Note that $\hat{Z}_t$ which can be identified as an* O.U. *process is not necessarily measurable with respect to ϕ_t.*

Proof. We merely have to show that the transition probabilities of $\hat{Z}_t$, with respect to $\hat{P}_\phi$, are that of the O.U. process ϕ_t. Let $\hat{E}_\phi$ be the expectation with respect to $\hat{P}_\phi$ and f a bounded measurable function on E. We have from (3.5)

$$\hat{E}_\phi(f(\hat{Z}_t))=E_\phi^{(W)}(f(Z_t)e^{\xi_0^t}), \tag{4.2}$$

where

$$Z_t=\phi_t+\frac{\lambda}{2}\int_0^t ds\, e^{-\frac{(t-s)}{2}C^{-\varepsilon}}C^{1-\varepsilon}:\phi_s^3:, \tag{4.3}$$

$$\xi_s^t=-\frac{\lambda}{2}\int_s^t(:\phi_s^3:,dW_s)-\frac{\lambda^2}{8}\int_s^t ds\|C^{1/2(1-\varepsilon)}:\phi_s^3(W):\|^2, \tag{4.4}$$

and $\phi_t(W)$ is the O.U. process, starting at ϕ, given by (2.7). By virtue of Theorem 1 of Sect. 3

$$e^{\xi_0^T}\in L^1(\Omega_T, dP^{(W)}), \tag{4.5}$$

and moreover

$$E_\phi^{(W)}(e^{\xi_0^T})=1. \tag{4.6}$$

Let $\mathbb{B}_t$ be the σ-subalgebra engendered by W_s, $\forall s \leqq t$. Then it is a straightforward consequence of (4.6) (see e.g. Lemma 2.3, Chap. 7, [16]) that the conditional expectation

$$E_\phi(e^{\xi_s^t}|\mathbb{B}_s) = 1\,, \quad P^{(W)} \text{ a.s.}. \tag{4.7}$$

Define on Ω_T the probability measure $\tilde{P}^{(W)}$ by

$$d\tilde{P}_\phi^{(W)} = e^{\xi_0^T} dP_\phi^{(W)}\,, \tag{4.8}$$

and let $\tilde{E}_\phi^{(W)}$ be expectation with respect to $\tilde{P}_\phi^{(W)}$.

Note that, using (4.8) and (4.7),

$$\begin{aligned}\tilde{E}_\phi^{(W)}(f(Z_t)) &= E_\phi^{(W)}(f(Z_t)e^{\xi_0^T}) = E_\phi^{(W)}(f(Z_t)e^{\xi_0^t}e^{\xi_t^T}) \\ &= E_\phi^{(W)}(f(Z_t)e^{\xi_0^t}E_\phi(e^{\xi_t^T}|\mathbb{B}_t)) = E_\phi^{(W)}(f(Z_t)e^{\xi_0^t})\,.\end{aligned} \tag{4.9}$$

Hence from (4.2) and (4.9) we have,

$$\hat{E}_\phi(f(\hat{Z}_t)) = \tilde{E}_\phi^{(W)}(f(Z_t))\,. \tag{4.10}$$

Thus to prove our proposition we have to show that Z_t given by (4.3), is an O.U. process, with respect to the measure $\tilde{P}_\phi^{(W)}$, whose joint probability distributions coincide with that of the O.U. process $\phi_t^{(W)}$ of Sect. 2.

Because of (4.5) and (4.6) we are assured by the Girsanov theorem [valid in our context because of (4.6)], that (W_t is the Wiener process of Sect. 2)

$$\tilde{W}_t = W_t + \frac{\lambda}{2}\int_0^t ds\, C^{1-\varepsilon} : \phi_s^3(W): \tag{4.11}$$

is *also a Wiener process in* Ω_T *with respect to the measure* $\tilde{P}^{(W)}$ *with the same covariance as* W_t

$$\tilde{E}^{(W)}(\tilde{W}_t(f)\tilde{W}_s(g)) = (f, C^{1-\varepsilon}g)\min(t,s)\,. \tag{4.12}$$

Then, Z_t is just the unique solution of

$$dZ_t = d\tilde{W}_t - \tfrac{1}{2}C^{-\varepsilon}Z_t dt\,, \quad Z_0 = \phi\,. \tag{4.13}$$

Indeed, the solution of (4.13) is:

$$\begin{aligned} Z_t &= e^{-\frac{t}{2}C^{-\varepsilon}}\phi + \int_0^t ds\, e^{-\frac{(t-s)}{2}C^{-\varepsilon}} d\tilde{W}_s \\ &= e^{-\frac{t}{2}C^{-\varepsilon}}\phi + \int_0^t ds\, e^{-\frac{1}{2}(t-s)C^{-\varepsilon}} dW_s + \frac{\lambda}{2}\int_0^t ds\, e^{-\frac{1}{2}(t-s)C^{-\varepsilon}} C^{1-\varepsilon} : \phi_s^3(W): \\ &= \phi_t(W) + \frac{\lambda}{2}\int_0^t ds\, e^{-\frac{1}{2}(t-s)C^{-\varepsilon}} C^{1-\varepsilon} : \phi_s^3(W):\,, \end{aligned} \tag{4.14}$$

which is just (4.3). Comparing (4.13) and the differential equation corresponding to (2.7) we see that Z_t is with respect to $\tilde{P}^{(W)}$ an O.U. process with the same joint probability distributions as ϕ_t of Sect. 2. This proves the proposition.

Appendix

Define

$$M \equi :(\phi^3, C^{-\varepsilon}\phi):. \tag{A.1}$$

In this appendix, C will be the covariance with free boundary conditions. In fact by Theorem VII.9 in [2], Propositions 1 and 2 below imply similar estimates for our field with Dirichlet boundary conditions in Λ.

Then in this appendix we will prove the

Proposition 1. *For any* $\lambda > 0$ *and for* ε *restricted to the range* $0 < \varepsilon < \frac{1}{10}$,

$$e^{-\lambda M} \in L^1(d\mu_c). \tag{A.2}$$

This result was used in the proof of Theorems 1 and 2 in Sect. 3.

Note that $M \in L^p(d\mu_c) 1 \leqq p < \infty$, for $\varepsilon < \frac{1}{2}$. The restriction $\varepsilon > 0$ was imposed in Sect. 2, to ensure that the transition probabilities of the O.U. process are absolutely continuous with respect to μ_c. The upper bound $\varepsilon < \frac{1}{10}$ turns out to be sufficient for (A.2) to hold.

The proof of the above proposition and (A.2) is based on a series of lemmata.

Lemma 1. *For* $0 < \varepsilon < 1$,

$$(\phi^3, C^{-\varepsilon}\phi) \geqq \int_\Lambda d^2x \phi^4(x). \tag{A.3}$$

Proof. We let $\|\cdot\|_p$ denote the $L^p(\Lambda)$ norm.

Define

$$a_\varepsilon = \int_0^\infty ds \frac{s^{-1+2\varepsilon}}{1+s^2} > 0,$$

which converges for $0 < \varepsilon < 1$.

We have the representation, converging for $0 < \varepsilon < 1$,

$$(-\Delta + 1)^\varepsilon = a_\varepsilon^{-1} \int_0^\infty ds\, s^{-1+2\varepsilon}(I - s^2(-\Delta + 1 + s^2)^{-1}).$$

Hence

$$\begin{aligned}(\phi^3, (-\Delta+1)^\varepsilon\phi) &= a_\varepsilon^{-1} \int_0^\infty ds\, s^{-1+2\varepsilon}(\|\phi\|_4^4 - s^2(\phi^3, C_{s^2+1}\phi)) \\ &\geqq a_\varepsilon^{-1} \int_0^\infty ds\, s^{-1+2\varepsilon}(\|\phi\|_4^4 - s^2|(\phi^3, C_{s^2+1}\phi)|),\end{aligned} \tag{A.4}$$

where

$$C_{s^2+1} = (-\Delta + 1 + s^2)^{-1}.$$

Using Hölder's inequality,

$$|(\phi^3, C_{s^2+1}\phi)| \leqq \|\phi\|_4^3 \|C_{s^2+1}\phi\|_4. \tag{A.5}$$

Now use Young's convolution inequality

$$\|f*\phi\|_p \leqq \|f\|_q \|\phi\|_r, \quad \frac{1}{p}=\frac{1}{q}+\frac{1}{r}-1,$$

with the choice f=integral kernel of C_{s^2+1}, $p=r=4$, $q=1$.

Note that

$$\|f\|_1 \leqq \frac{1}{s^2+1}.$$

Hence

$$\|C_{s^2+1}\phi\|_4 \leqq \frac{1}{s^2+1}\|\phi\|_4. \tag{A.6}$$

From (A.5), (A.6), we have

$$|(\phi^3, C_{s^2+1}\phi)| \leqq \frac{1}{s^2+1}\|\phi\|_4^4. \tag{A.7}$$

From (A.4), (A.7)

$$(\phi^3, (-\Delta+1)^\varepsilon\phi) \geqq a_\varepsilon^{-1}\int_0^\infty ds\, s^{-1+2\varepsilon}\left(1-\frac{s^2}{s^2+1}\right)\|\phi\|_4^4 = \|\phi\|_4^4. \quad \square$$

Next we turn to Lemma 2.

Define UV cutoff fields $\phi_\kappa(x)$ by

$$\phi_\kappa(x) = \int_{|K|\leqq\kappa} \frac{d^2K}{(2\kappa)^2} e^{iKx}\tilde{\phi}(K).$$

Define

$$M_\kappa = :(\phi_\kappa^3, C^{-\varepsilon}\phi_\kappa): = :(\phi_\kappa^3, (-\Delta+1)^\varepsilon\phi_\kappa):. \tag{A.8}$$

Then for $0<\varepsilon<\frac{1}{2}$

$$M_\kappa \xrightarrow[\kappa\to\infty]{} M, \quad \text{in} \quad L^p(d\mu_c), \quad 1 \leqq p < \infty.$$

Undoing the Wick ordering,

$$\begin{aligned} M_\kappa = {} & (\phi_\kappa^3, (-\Delta+1)^\varepsilon\phi_\kappa) - 3(C_\kappa(0)(\phi_\kappa, (-\Delta+1)^\varepsilon\phi_\kappa) \\ & + C_\kappa^{1-\varepsilon}(0)(\phi_\kappa, \phi_\kappa)) + 3|\Lambda| C_\kappa(0) C_\kappa^{1-\varepsilon}(0), \end{aligned} \tag{A.9}$$

where $C_\kappa(0)=C_\kappa(0,0)$ and $C_\kappa(x,y)$ is the integral kernel of C_κ the covariance of ϕ_κ, $C_\kappa^{1-\varepsilon}(0)=C_\kappa^{1-\varepsilon}(0,0)$, and $C_\kappa^{1-\varepsilon}(0,0)$ is the integral kernel of $C_\kappa^{1-\varepsilon}$.

Now use primitive positivity, i.e. Lemma 1, and

$$(\phi_\kappa, (-\Delta+1)^\varepsilon\phi_\kappa) \leqq (\kappa^2+1)^\varepsilon(\phi_\kappa, \phi_\kappa)$$

to obtain from (A.9)

$$\begin{aligned} M_\kappa \geqq \int_\Lambda d^2x (& (\phi_\kappa(x))^4 - 3(C_\kappa(0)(\kappa^2+1)^\varepsilon + C_\kappa^{1-\varepsilon}(0))(\phi_\kappa(x))^2 \\ & + 3C_\kappa(0)C_\kappa^{1-\varepsilon}(0)). \end{aligned} \tag{A.10}$$

Now use

$$C_\kappa(0)\sim O(\ln\kappa), \qquad C_\kappa^{1-\varepsilon}(0)\sim O(\kappa^{2\varepsilon}\ln\kappa)$$

to obtain immediately:

Lemma 2.

$$M_\kappa \geqq -\mathrm{const}\, O(\kappa^{4\varepsilon}(\ln\kappa)^2). \tag{A.11}$$

Define

$$\tilde{M}_\kappa \equiv M - M_\kappa. \tag{A.12}$$

Then it is straightforward to verify via Feynman graph calculations

Lemma 3.

$$\int d\mu_c |\tilde{M}_\kappa|^{2j} \leqq (j!)^4 b^j ((\ln\kappa)^m \kappa^{-2+4\varepsilon})^j \tag{A.13}$$

for any j, some $m>0$. b is a constant independent of j and κ.

The proof of the proposition at the beginning of the appendix now follows from Lemmata 2 and 3 by Nelson's argument [1–3].

Namely, we have

$$\mu_c\{M \leqq -\mathrm{const}(\kappa^{4\varepsilon}(\ln\kappa)^2 - 1\} \leqq \mu_c\{|\tilde{M}_\kappa|^{2j} \geqq 1\} \leqq \int d\mu_c |\tilde{M}_\kappa|^{2j}$$
$$\leqq (j!)^4 b^j ((\ln\kappa)^m \kappa^{-2+4\varepsilon})^j. \tag{A.14}$$

Using Stirling's approximation for $j!$ and an optimal κ-dependent choice of j, we have

$$\mu_c\{M \leqq -\mathrm{const}(\kappa^{4\varepsilon}(\ln\kappa)^2) - 1\} \leqq e^{-\mathrm{const}\,\kappa^{\frac{2-4\varepsilon}{4}}(\ln\kappa)^{-m/4}}. \tag{A.15}$$

This estimate, together with Lemma 2, assures us that for $\varepsilon < \frac{1}{10}$ $e^{-\lambda M} \in L^1(d\mu_c)$, $\lambda > 0$, and the proposition has been proved. □

Remark. It can also be shown by methods very similar to [31], that we have a uniform bound

$$\int d\mu_c e^{-\lambda M_\kappa} \leqq e^{\mathrm{const}\,|\Lambda|}. \tag{A.16}$$

In Sect. 2, we verified that

$$\int_0^t ds\, M(\phi_s)$$

with M defined by (A.1) is a random variable in $L^p(dP_\phi, \Omega)$, $1 \leqq p < \infty$ and P_ϕ is the O.U. measure.

Now $p_t(\phi, d\phi')$, the transition probability of the O.U. process, is absolutely continuous with respect to μ_c, for every $t>0$ and starting point ϕ, and its Radon-Nikodym derivative is in $L^2(d\mu_c)$. This fact, and the proof of Proposition 1 of this appendix leads to,

Proposition 2. *For any $\lambda>0$, and provided $0<\varepsilon<\frac{1}{10}$,*

$$E_\phi\left(e^{-\lambda\int_0^t ds\, M(\phi_s)}\right)<\infty,$$

where E_ϕ is expectation with respect to the O.U. *measure P_ϕ.*

Proof. Define

$$M^t=\int_0^t ds\, M(\phi_s), \qquad M_\kappa^t=\int_0^t ds\, M_\kappa(\phi_s),$$

M_κ as in (A.8). Then from Lemma 2,

$$M_\kappa^{(t)}\geqq -\text{const}\, t O(\kappa^{4\varepsilon}(\ln\kappa)^2). \tag{A.17}$$

Analogous to the step before Lemma 3, define

$$\tilde{M}_\kappa^{(t)}=M^{(t)}-M_\kappa^{(t)}=\int_0^t ds\, \tilde{M}_\kappa(\phi_s). \tag{A.18}$$

Lemma 4.

$$E_\phi(|\tilde{M}_\kappa^{(t)}|)^{2j}\leqq C_\phi^t((2j)!)^2 b_t^j((\ln\kappa)^m\kappa^{-2+4\varepsilon})^j \tag{A.19}$$

for any j, some $m>0$, and large κ.

Note that for large j, by Stirling approximation, $((2j)!)^2\sim(2j)^{4j}$ whereas $(j!)^4\sim j^{4j}$ so that (for large j) Lemma 3 and Lemma 4 are the same, as are Lemma 2 and (A.17), (t, is fixed).

This Suffices to Prove Proposition 2. It remains to prove Lemma 4.

We have

$$\begin{aligned}E_\phi(|\tilde{M}_\kappa^{(t)}|^{2j})&\leqq t^{2j-1}\int_0^t ds\, E_\phi(|\tilde{M}_\kappa(\phi_s)|^{2j})\\ &\leqq t^{2j-1}\int_0^t ds\int p_s(\phi,d\phi')|\tilde{M}_\kappa(\phi')|^{2j}\\ &\leqq t^{2j-1}C_\phi^t\left(\int_0^t ds\int p_s(0,d\phi')|\tilde{M}_\kappa(\phi')|^{4j}\right)^{1/2},\end{aligned} \tag{A.20}$$

where we have used the fact that $p_s(\phi,d\phi')$ is absolutely continuous with respect to $p_s(0,d\phi')$ and the radon-Nikodym derivative is in $L^2(p_s(0,d\phi'))$, and Schwarz inequality.

Now $p_t(0,d\phi')$ is a gaussian measure with mean 0 and covariance

$$C_t=(1-e^{-tC^{-\varepsilon}})C, \tag{A.21}$$

and we have for integral kernels

$$\lim_{x\to y}(C_t(x,y)-C(x,y))=\delta C_t<\infty. \tag{A.22}$$

$\tilde{M}_\kappa(\phi)$ has Wick ordering with respect to C. By virtue of (A.22) we can change the Wick ordering in $\tilde{M}_\kappa(\phi)$ to C_s-Wick ordering at the cost of introducing *lower order terms* with *finite* coefficients [1].

Now, because of (A.21), we have by Feynman graph calculations exactly as leading to Lemma 3, for large κ

$$\int_0^t ds \int p_s(0, d\phi')|\tilde{M}_\kappa(\phi')|^{4j} \leqq ((2j)!)^4 b_t^{2j}((\ln\kappa)^m \kappa^{-2+4\varepsilon})^{2j}. \tag{A.23}$$

(A.20) and (A.23) imply (A.19), which prove the lemma. □

Remark. Analogous to (A.16) we also have the uniform bound

$$E_\phi\left(e^{-\lambda\int_0^t ds M_\kappa(\phi_s)}\right) \leqq e^{C_\phi^t|\Lambda|}. \tag{A.24}$$

Acknowledgements. We thank Erhard Seiler for stimulating conversations on Stochastic quantization. G. J.-L. thanks S. Albeverio for a very useful discussion. P. K. M. thanks A. Martin, E. Seiler, W. Zimmerman and the members of the Werner Heisenberg Institut, Max Planck Institut für Physik, Munich and the Theory Division, CERN for their generous hospitality. G. J.-L. would like to express his gratitude to L. Streit, D. Elworthy, and A. Truman for their warm hospitality at the University of Bielefeld and the University of Swansea.

References

1. Glimm, J., Jaffe, A.: Quantum physics. Berlin, Heidelberg, New York: Springer 1981
2. Simon, B.: The $P(\phi)_2$ Euclidean (quantum) field theory. Princeton, NJ: Princeton University Press 1974
3. Nelson, E.: In: Constructive quantum field theory. Lecture Notes in Physics. Vol. 25, Velo, G., Wightman, A. (eds.). Berlin, Heidelberg, New York: Springer 1973
4. Glimm, J., Jaffe, A.: Quantum field models in statistical mechanics and field theory, Les Houches (1970), De Witt, C., Stora, R. (eds.). New York: Gordon and Breach 1971
5. Glimm, J.: Boson fields with nonlinear selfinteraction in two dimensions. Commun. Math. Phys. **8**, 12–25 (1968)
6. Simon, B., Høegh-Krohn, R.: J. Funct. Anal. **9**, 121–180 (1972)
7. Gross, L.: Am. J. Math. **97**, 1061–1083 (1975)
8. Gross, L.: Harmonic analysis on Hilbert space. Memoirs AMS, Vol. 26
9. Gross, L.: Abstract Wiener spaces. In: Proceedings of the 5th Berkeley Symposium on mathematical statistics and probability. Berkeley, CA: University of California Press 1968
10. Gross, L.: J. Funct. Anal. **10**, 52–109 (1972)
11. Gelfand, I., Vilenkin, N.: Generalized functions. Vol. 4, New York: Academic Press 1964
12. Yosida, K.: Functional analysis. Berlin, Heidelberg, New York: Springer 1966
13. McKean, H.: Stochastic integrals. New York, London: Academic Press 1969
14. Stroock, D., Varadhan, S.R.S.: Multidimensional diffusion processes. Berlin, Heidelberg, New York: Springer 1979
15. Simon, B.: Functional integration and quantum physics. New York: Academic Press 1979
16. Friedman, A.: Stochastic differential equations, Vol. 1. New York: Academic Press 1975
17. Yor, M., Priouret, P.: Asterisque No. 22
18. Parisi, G., Wu, Yong-Shi: Sci. Sin. **24**, 483 (1981)
19. Faris, W., Jona-Lasinio, G.: Large fluctuations for a nonlinear heat equation with noise. J. Phys. A **15**, 3025 (1982)
20. Zwanzinger, D.: Covariant quantization of gauge fields without Gribov ambiguity. Nucl. Phys. B **192**, 259 (1981)
   Baulieu, L., Zwanziger, D.: Equivalence of stochastic quantization and the Faddeev-Popov ansatz. Nucl. Phys. B **193**, 163 (1981)

21. Floratos, E., Iliopoulos, J.: Equivalence of stochastic and canonical quantization in perturbation theory. Nucl. Phys. B **214**, 392 (1983)
22. Floratos, E., Iliopoulos, J., Zwanziger, D.: A covariant ghost-free perturbation expansion for Yang-Mills theories. Nucl. Phys. B **241**, 221 (1984)
23. Seiler, E.: 1984, Schladming Lectures. Max-Planck-Institut für Physik München, Preprint
24. Glauber, R.: Time-dependent statistics of the Ising model. J. Math. Phys. **4**, 294–307 (1963)
25. Holley, R.: Free energy in a Markovian model of a lattice spin system. Commun. Math. Phys. **23**, 87 (1971)
26. Holley, R., Stroock, D.: In one and two dimensions, every stationary measure for a stochastic Ising model is a Gibbs state. Commun. Math. Phys. **55**, 37 (1977); Applications of the stochastic Ising model to the Gibbs states. Commun. Math. Phys. **48**, 249 (1976); Z. Wahrscheinlichkeitstheorie Verw. Geb. **35**, 87 (1976)
27. Holley, R., Stroock, D.: Ann. Probab. Vol. **4**, No. 2, 195; J. Funct. Anal. **42**, 29 (1981)
28. Faris, W.: J. Funct. Anal. **32**, 342 (1979); Trans. Am. Math. Soc. **261**, 579 (1980)
29. Wick, W.: Convergence to equilibrium of the stochastic Heisenberg model. Commun. Math. Phys. **81**, 361 (1981)
30. Doss, H., Royer, G.: Z. Wahrscheinlichkeitstheorie Verw. Geb. **46**, 107 (1978)
31. Dimock, J., Glimm, J.: Adv. Math. **12**, 58 (1974)

Communicated by G. Mack

Received March 26, 1985

Large-N limit

Nuclear Physics B211 (1983) 343–368

QUENCHED MASTER FIELDS

J. GREENSITE[1]

Physics Department, University of California, Berkeley, CA 94720, USA

M.B. HALPERN

Physics Department and Lawrence Berkeley Laboratory, University of California, Berkeley, CA 94720, USA

Received 27 August 1982

We derive an exact algebraic (master) equation for the euclidean master field of any large-N matrix theory, including quantum chromodynamics. The master equation is the quenched Langevin equation. The master field, a translationally covariant function of (uniform) random momenta and (gaussian) random noise, is easily constructed in perturbation theory.

1. Introduction

The concept of the large-N euclidean master field was originally introduced by Witten [1] who observed that large-N factorization (A, B singlet)

$$\langle AB\rangle \underset{N}{=} \langle A\rangle\langle B\rangle \tag{1.1}$$

was suggestive of saturation of the functional integral by a single configuration, the master field. If the generic matrix variable is A, it is the job of the master field A^* to reproduce the singlet time-ordered products of the theory

$$\langle \mathrm{Tr}\,(A_1\cdots A_n)\rangle \underset{N}{=} \mathrm{Tr}\,(A_1^*\cdots A_n^*)\,. \tag{1.2}$$

It was soon realized [2] that two distinct kinds of master field might exist: (a) the euclidean master field described above, which represents time-ordered products, and (b) the "large-N classical solution" or hamiltonian master field, which represents (the ground-state energy and) the Wightman functions [3]. The need to represent time-ordering rules out the possibility of the euclidean master field being a classical solution.

The existence of the large-N classical solution was first verified by Bardakci [4]. An alternate approach by Halpern [5] provided a complete physical interpretation: The large-N classical solution is the set of (reduced) matrix elements among the ground state and the set of generalized-adjoint eigenstates necessary to saturate

[1] Address after September 15: Niels Bohr Institute, Blegdamsvej 17, DK-2100 Copenhagen Ø, Denmark.

the large-N singlet Wightman functions. As a consequence, the large-N classical solution is known to be of the translationally covariant form [5]

$$A_{ab}(x) = e^{i(p_a - p_b)\cdot x} A_{ab}, \tag{1.3}$$

where $p_{\mu a,b}$ are the momenta of the *eigenstates a* and *b*, and $A_{ab} = A_{ab}(0)$. This is interesting, and apparently *universal*, because our euclidean master fields below are also of this form.

Large-N classical solutions have been constructed explicitly for quadratic building-block [6] quantum mechanics [2, 5], quadratic building block field theories [2, 5], free matrix quantum mechanics [7], matrix quantum mechanics [8], and the one-polygon quantum mechanics [9]. Euclidean master fields have not (until now) enjoyed an existence proof, but have been constructed explicitly for quadratic building-block theories [3], zero-dimensional matrix models [10], and quantum chromodynamics in two dimensions [11].

Apparently independent of the question of master field, large-N theory has recently enjoyed two further major advances: (a) reduction [12] and (b) quenching [13, 14, 11]. In a properly quenched theory [11], n-point functions are constructed with (1.3) as an ansatz, remaining integrations being only over $A_{ab}(0)$ and the quenched (random) momenta.

It is the purpose of this paper to bring together the idea of quenching, and ideas about the Langevin equation, put forward some time ago by Parisi and Wu [15] – in order to find the equations satisfied by quenched master fields. In this we are successful, and thereby prove the existence of (at least) our kind of large-N euclidean master field.

Our derivation of the master field equation begins with the strategy that the master field of a large-N quantum theory or statistical system must be any one of a very large class of *equilibrium* configurations associated with the system. It is well known, for example in Monte Carlo applications, that any thermal average (or vacuum expectation value, VEV) defined by

$$\langle Q \rangle = Z^{-1} \int \mathcal{D}\phi Q[\phi] e^{-S[\phi]}, \tag{1.4}$$

can be accurately represented by simply averaging over a relatively small number of important "equilibrium" configurations $\{\phi_i^{eq}\}$

$$\langle Q \rangle \simeq \frac{1}{n_{eq}} \sum_{i=1}^{n_{eq}} Q[\phi_i^{eq}]. \tag{1.5}$$

The question is how to obtain these equilibrium configurations.

As a simple physical example, consider a system of gas molecules confined to a box at fixed temperature. Suppose that initially all the molecules are in a corner of the box. In the course of time, the system evolves toward equilibrium, in which the molecules are distributed more or less uniformly throughout the box. An equilibrium configuration is a "snapshot" at large time. The probability of finding any particular equilibrium configuration at time t is proportional to the Boltzmann factor. Such quantities as density, pressure, density correlations, etc., can be well approximated by averaging over a relatively small sample of the equilibrium configurations.

The point of these elementary remarks is that one way of generating a set of equilibrium configurations is to start from an arbitrary initial configuration, and simulate the dynamical evolution of the system towards equilibrium in time.

There are various ways to do this. Given a euclidean field theory in D-dimensions, say

$$S[\phi]=\int d^D x\,\{\tfrac{1}{2}[(\partial\phi)^2+m^2\phi^2]+V(\phi)\}, \tag{1.6}$$

introduce an extra fictitious time-like dimension t in which the fields are supposed to evolve, i.e., $\phi=\phi(xt)$. Discretizing the D-dimensional space (in a finite volume) and the fictitious time, we can apply one of the standard Monte Carlo algorithms (Metropolis or heat-bath) to generate a sequence of configurations [16]. A Monte Carlo iteration corresponds to a discrete step in the timelike direction. After a finite number of steps, the system reaches equilibrium, and we begin the averaging in eq. (1.5).

An alternate approach is to solve a certain stochastic evolution equation loosely called the Langevin equation [15]*

$$\begin{aligned}\frac{\partial}{\partial t}\phi &= -\frac{\delta S}{\delta\phi}+\eta(xt)\\ &=(\partial^2-m^2)\phi-\frac{\delta V}{\delta\phi}+\eta\,,\end{aligned} \tag{1.7}$$

which also simulates the evolution toward equilibrium (at large t). Here $\eta(xt)$ is a gaussian random variable, which varies independently at each space–time point (x,t). The Langevin equation has found application in the study of critical phenomena [17], and has recently been used by Parisi and Wu to formulate perturbative non-abelian gauge theory without gauge fixing**.

However, for the problem of constructing equilibrium configurations, the Langevin equation, like the Monte Carlo algorithms, can only be applied numerically. The problem of integrating a large set of random variables over the fictitious time frustrates any analytic approach.

* The dimension of the "fifth" time t is $(\text{length})^2$. Such could be avoided by rescaling $t\to\lambda t$ with λ a dimensional constant. This is a matter of taste, and makes no difference in final results.

** A similar fifth time approach to equilibrium configurations was worked out by Greensite (unpublished) for free field theory and mean field theory.

This is where the large-N limit, and quenching in particular comes to the rescue. Consider a Langevin equation for a large-N system

$$\frac{\partial}{\partial t}A_{ab} = -\frac{\delta S}{\delta A_{ba}} + \eta_{ab}(xt)\,. \tag{1.8}$$

In the large-N limit, the Langevin noise can be quenched:

$$\eta_{ab}(xt) = \mathrm{e}^{i(p_{5a}-p_{5b})t}\,\mathrm{e}^{-i(p_a-p_b)x}\eta_{ab}\,. \tag{1.9}$$

Here the p's are quenched uniform random momenta*, and η_{ab} is a single gaussian noise matrix. Because of the reduction of degrees of freedom in the quenched noise, it becomes possible to characterize the large t equilibrium configuration precisely and algebraically. As we shall show below, the equilibrium configuration is also translationally covariant,

$$A_{ab}(xt) = \mathrm{e}^{i(p_{5a}-p_{5b})t}\,\mathrm{e}^{-i(p_a-p_b)x}A_{ab}\,, \tag{1.10}$$

where the single matrix A_{ab} satisfies the algebraic *quenched Langevin equation*:

$$i(p_{5a}-p_{5b})A_{ab} = -\frac{\delta S_Q}{\delta A_{ba}} + \eta_{ab}\,. \tag{1.11}$$

Here S_Q is the quenched action of the D-dimensional theory. The explicit equilibrium configuration (1.10) is a function of the p's and the noise matrix η_{ab}.

It was pointed out in ref. [11] that it is not necessary to integrate over the quenched momenta in the large-N limit. $(D+1)N$ uniform random numbers $\bar{p}^*$ will do as well, and one can think of the quenched momenta as "*master momenta*". In the same sense, we will show that it is not necessary to integrate over the noise matrix: A set of N^2 gaussian random numbers η^*_{ab} will do as well (η^* is the *master noise matrix*). Thus the *single* equilibrium configuration

$$\begin{aligned} A^*_{ab}(xt) &= \mathrm{e}^{i(p^*_{5a}-p^*_{5b})}\,\mathrm{e}^{-i(p^*_a-p^*_b)\cdot x}A^*_{ab}\,,\\ i(p^*_{5a}-p^*_{5b})A^*_{ab} &= -\frac{\delta S_Q}{\delta A^*_{ba}} + \eta^*_{ab}\,, \end{aligned} \tag{1.12}$$

is the master field for the theory.

The plan of the paper is as follows. Sect. 2 recalls the Langevin formulation of Parisi and Wu, focussing on the gaussian noise $\eta(xt)$. The transition from noise to quenched noise to master noise is detailed. In sect. 3, we solve the Langevin equation with quenched noise, showing that the quenching propagates uniformly to the equilibrium configuration; this is the quenched master field.

Sect. 4 is a selection of perturbative examples and construction of the master field. In sect. 5, we make some non-perturbative remarks, including as a toy, the

* The "fifth" momenta p_5, conjugate to t, could also be labelled p_{D+1}. Fifth momenta have dimension $(\mathrm{length})^{-2}$.

exact solution of the $N = 2$ master equation for the matrix model. Sect. 6 is an introduction to quantum chromodynamics. In our approach it may be possible to proceed directly in terms of a master A_μ field, without gauge fixing or constraint.

2. Noise, quenched noise, and master noise

Consider a large-N matrix theory with quartic self-interactions in D euclidean dimensions, which has an action

$$S[M] = \int d^D x \, \mathrm{Tr}\left\{\tfrac{1}{2}[(\partial M)^2 + m^2 M^2] + \frac{g}{4N} M^4\right\}, \tag{2.1}$$

where $M_{ab}(x)$ is a hermitian matrix-valued field. This action is obviously invariant under a global U(N) symmetry group. We are interested in computing U(N)-invariant expectation values of the form

$$\langle \mathrm{Tr}\, M(x_1) M(x_2) \cdots M(x_n)\rangle = Z^{-1} \int \mathscr{D}M \, \mathrm{Tr}\{M(x_1) \cdots M(x_n)\}\, e^{-S[M]}. \tag{2.2}$$

Other invariant VEV's involving products of traces, simply factorize in the large-N limit.

Introducing a fictitious timelike dimension t, the Langevin equation associated with this matrix theory is given by

$$\begin{aligned} \partial_t M_{ab} &= -\frac{\delta S}{\delta M_{ba}(x,t)} + \eta_{ab}(x,t) \\ &= (\partial^2 - m^2) M_{ab} - \frac{g}{N}(M^3)_{ab} - \eta_{ab}\,. \end{aligned} \tag{2.3}$$

Starting from a given initial configuration, e.g., $M_{ab}(x, 0) = 0$, equilibrium is reached as $t \to \infty$. Each equilibrium configuration is, of course, a functional of the stochastic noise term η_{ab}, a fact which we will emphasize by writing $M_{ab} = M^{\eta}_{ab}(x, t)$.

Now define the stochastic expectation value of any functional $F(\eta)$ of the η_{ab} fields by

$$\langle F[\eta]\rangle_S \equiv \frac{\int \mathscr{D}\eta F[\eta] \exp\left(-\frac{1}{4}\int d^D x\, dt\, \mathrm{Tr}[\eta(x,t)\eta(x,t)]\right)}{\int \mathscr{D}\eta \exp\left(-\frac{1}{4}\int d^D x\, dt\, \mathrm{Tr}[\eta(x,t)\eta(x,t)]\right)}, \tag{2.4}$$

so that

$$\begin{aligned} \underbrace{\eta_{ab}(x,t)\eta_{cd}}(x',t') &\equiv \langle \eta_{ab}(x,t)\eta_{cd}(x',t')\rangle \\ &= 2\delta_{ad}\delta_{bc}\delta^D(x-x')\,\delta(t-t')\,. \end{aligned} \tag{2.5}$$

Then the connection of the Langevin equation to ordinary field theory is expressed in the large-t limit,

$$\langle \mathrm{Tr}\,[M^{\eta}(x_1 t)\cdots M^{\eta}(x_n, t)]\rangle_S \underset{t}{=} \langle \mathrm{Tr}\,[M(x_1)\cdots M(x_n)]\rangle\,. \tag{2.6}$$

The condition of equilibrium is that the left-hand side goes to a t-independent limit.

Parisi and Wu [15] have shown how eq. (2.6) reproduces the results of ordinary perturbation theory. As a simple example, consider the free-field case $g=0$. Then

$$M^{\eta}_{ab}(x,t)=\int_0^{\infty}\mathrm{d}\tau\int \mathrm{d}^D y\, G(x-y, t-\tau)\eta_{ab}(y,\tau)\,, \tag{2.7}$$

where

$$G(x,t)=\int\frac{\mathrm{d}^D k}{(2\pi)^D}\,\mathrm{e}^{-t(k^2+m^2)}\,\mathrm{e}^{ik\cdot x}\theta(t)\,, \tag{2.8}$$

and therefore

$$\begin{aligned}\langle \mathrm{Tr}\, M^{\eta}(x,t)M^{\eta}(x',t)\rangle_S &= \int_0^{\infty}\mathrm{d}\tau\,\mathrm{d}\tau'\int \mathrm{d}^D y\,\mathrm{d}^D y' \\ &\quad\times G(x-y,t-\tau)G(x'-y',t-\tau')\langle \mathrm{Tr}\,\eta(y,\tau)\eta(y',\tau')\rangle_S \\ &= 2N^2\int_0^{\infty}\mathrm{d}\tau\int \mathrm{d}^D y\, G(x-y,t-\tau)G(x'-y,t-\tau) \\ &\underset{t}{=} N^2\int\frac{\mathrm{d}^D k}{(2\pi)^D}\,\frac{1}{k^2+m^2}\,\mathrm{e}^{ik\cdot(x-x')}\,,\end{aligned} \tag{2.9}$$

which is the usual free-field propagator.

In the interacting case, $g\neq 0$, it is necessary to solve the integral form of the Langevin equation,

$$M^{\eta}_{ab}(x,t)=\int_0^{\infty}\mathrm{d}\tau\int \mathrm{d}^D y\, G(x-y,t-\tau)\left\{\eta_{ab}(y,\tau)-\frac{g}{N}[M^3(y,\tau)]_{ab}\right\}. \tag{2.10}$$

In perturbation theory, the expansion of M^{η}_{ab} in powers of the coupling will involve products of η-fields of the form

$$\eta_{aa_1}(y_1,\tau_1)\eta_{a_1a_2}(y_2,\tau_2)\cdots\eta_{a_{m-1},b}(y_m,\tau_m)\,. \tag{2.11}$$

It is easy to see that the stochastic expectation values that occur on the l.h.s. of (2.6) are all of the form

$$\langle \mathrm{Tr}\,\eta(y_1,\tau_1)\eta(y_2\tau_2)\cdots\eta(y_n,\tau_n)\rangle_S\,. \tag{2.12}$$

Now the stochastic expectation values in (2.12) can be evaluated by Wick's theorem, with contractions defined in eq. (2.5). It is clear that the random noise theory represented by eq. (2.4) is itself a large-N, free-field matrix model. Terms in the Wick expansion can therefore be divided into planar and non-planar categories, where planar terms involve only nested Wick contractions, e.g.,

$$\mathrm{Tr}\,[\eta\,\eta\,\eta\,\eta\,\eta\,\eta\,\eta\,\eta] \quad \text{or} \quad \mathrm{Tr}\,[\eta\,\eta\,\eta\,\eta\,\eta\,\eta\,\eta\,\eta]\,, \tag{2.13}$$

while non-planar terms involve overlapping contractions, e.g.,

$$\mathrm{Tr}\,[\eta\,\eta\,\eta\,\eta\,\eta\,\eta]\,, \tag{2.14}$$

and are suppressed relative to the planar terms by powers of $1/N$.

Now suppose we can find a "master noise" field $\eta^*(\bar{x})$ $[\bar{x} \equiv (x, t)]$ such that

$$\langle \mathrm{Tr}\,\eta(\bar{x}_1) \cdots \eta(\bar{x}_n)\rangle_{\mathrm{S}} \underset{N}{=} \mathrm{Tr}\,\eta^*(\bar{x}_1) \cdots \eta^*(\bar{x}_n)\,. \tag{2.15}$$

Denote the corresponding solution of the Langevin equation by $M^{\eta^*}(xt)$. Then from (2.6), (2.10) and (2.15)

$$\mathrm{Tr}\,\{M^{\eta^*}(x_1t) \cdots M^{\eta^*}(x_nt)\} \underset{N,t}{=} \langle \mathrm{Tr}\,\{M(x_1) \cdots M(x_n)\}\rangle\,, \tag{2.16}$$

which means that the large-t limit of $M^{\eta^*}(xt)$ is the desired master field for the matrix model in (2.1).

The solution of the trivial η-theory by a master noise field is essential to solving the non-trivial models, so we will focus, for the rest of this section, on finding $\eta^*(xt)$.

The first step in constructing $\eta^*(xt)$ is to reduce the degrees of freedom in the stochastic system of eq. (2.4) by applying the Gross–Kitazawa quenching prescription [11]. This prescription, which eliminates the space–time variables in (2.4), works as follows: Write

$$\eta_{ab}(x, t) = \mathrm{e}^{i(p_{5a} - p_{5b})t}\,\mathrm{e}^{-i(p_a - p_b)\cdot x}\,\eta_{ab}\,, \tag{2.17}$$

where η_{ab} (noise matrix) is the stochastic variable of the reduced gaussian model; i.e., a model with no space-time dependence. We define reduced stochastic expectation values in this model by

$$\langle \eta_{a_1b_1}\eta_{a_2b_2} \cdots \eta_{a_nb_n}\rangle_{\mathrm{RS}} = \frac{\displaystyle\int \mathrm{d}\eta\,(\eta_{a_1b_1} \cdots \eta_{a_nb_n})\exp\left[-\frac{1}{4}\left(\frac{2\pi}{\Lambda}\right)^D\left(\frac{2\pi}{\Lambda_5}\right)\mathrm{Tr}\,\eta\eta\right]}{\displaystyle\int \mathrm{d}\eta\,\exp\left[-\frac{1}{4}\left(\frac{2\pi}{\Lambda}\right)^D\left(\frac{2\pi}{\Lambda_5}\right)\mathrm{Tr}\,\eta\eta\right]}\,, \tag{2.18}$$

so that

$$\underbrace{\eta_{ab}\eta_{cd}} \equiv \langle \eta_{ab}\eta_{cd}\rangle_{\mathrm{RS}} = 2\delta_{ad}\delta_{bc}\left(\frac{\Lambda}{2\pi}\right)^D\left(\frac{\Lambda_5}{2\pi}\right), \tag{2.19}$$

in analogy to (2.5). The quenched prescription says that

$$\langle \mathrm{Tr}\{\eta(\bar{x}_1)\cdots\eta(\bar{x}_n)\rangle_S \underset{N}{=} \int (\mathrm{d}\bar{p}) \sum_{a_1\cdots a_n} \exp[i(\bar{p}_{a_1}-\bar{p}_{a_2})\cdot\bar{x}_1]$$
$$\times\exp[i(\bar{p}_{a_2}-\bar{p}_{a_3})\cdot\bar{x}]\cdots\exp[i(\bar{p}_{a_n}-\bar{p}_{a_1})\cdot\bar{x}_n]$$
$$\times\langle\eta_{a_1a_2}\eta_{a_2a_3}\cdots\eta_{a_na_1}\rangle_{RS}\,, \tag{2.20}$$

where we have introduced the notation

$$\bar{x}\equiv(x_\mu,t)\,,\qquad \bar{p}\equiv(p_\mu,p_S)\,,\qquad \bar{p}\cdot\bar{x}\equiv p_St-p\cdot x\,,$$
$$\int(\mathrm{d}\bar{p})=\int_{-\Lambda/2}^{\Lambda/2}\prod_{a=1}^{N}\frac{\mathrm{d}^Dp_a}{\Lambda^D}\int_{-\Lambda_S/2}^{\Lambda_S/2}\prod_{a=1}^{N}\frac{\mathrm{d}p_{Sa}}{\Lambda_S}\,. \tag{2.21}$$

As an example, consider the 4-point function

$$\langle \mathrm{Tr}\,\eta(\bar{x}_1)\eta(\bar{x}_2)\eta(\bar{x}_3)\eta(\bar{x}_4)\rangle_S=4N^3\delta(\bar{x}_1-\bar{x}_2)\delta(\bar{x}_3-\bar{x}_4)+4N^3\delta(\bar{x}_1-\bar{x}_4)\delta(\bar{x}_2-\bar{x}_3)$$
$$+4N\delta(\bar{x}_1-\bar{x}_3)\delta(\bar{x}_2-\bar{x}_4)\,. \tag{2.22}$$

According to the quenched prescription in (2.20) we find

$$\langle \mathrm{Tr}\{\eta(\bar{x}_1)\eta(\bar{x}_2)\eta(\bar{x}_3)\eta(\bar{x}_4)\}\rangle_S$$
$$\underset{N}{=}\sum_{a_1a_2a_3a_4}\int(\mathrm{d}\bar{p})\exp[i(\bar{p}_{a_1}-\bar{p}_{a_2})\cdot\bar{x}_1]$$
$$\times\exp[i(\bar{p}_{a_2}-\bar{p}_{a_3})\cdot\bar{x}_2]\exp[i(\bar{p}_{a_3}-\bar{p}_{a_4})\cdot\bar{x}_3]\exp[i(\bar{p}_{a_4}-\bar{p}_{a_1})\cdot\bar{x}_4]$$
$$\times4\left(\frac{\Lambda}{2\pi}\right)^{2D}\left(\frac{\Lambda_S}{2\pi}\right)^2[\delta_{a_1a_3}+\delta_{a_2a_4}+\delta_{a_1a_4}\delta_{a_2a_3}\delta_{a_1a_2}\delta_{a_3a_4}]$$
$$=4\left(\frac{\Lambda}{2\pi}\right)^{2D}\left(\frac{\Lambda_S}{2\pi}\right)^2\int(\mathrm{d}\bar{p})\Big\{\sum_{a_1a_2a_4}[\exp[i(\bar{p}_{a_1}-\bar{p}_{a_2})\cdot(\bar{x}_1-\bar{x}_2)]$$
$$\times\exp[i(\bar{p}_{a_1}-\bar{p}_{a_4})\cdot(\bar{x}_3-\bar{x}_4)]]$$
$$+\sum_{a_1a_2a_3}\exp[i(\bar{p}_{a_1}-\bar{p}_{a_2})\cdot(\bar{x}_1-\bar{x}_4)]\exp[i(\bar{p}_{a_2}-\bar{p}_{a_3})\cdot(\bar{x}_2-\bar{x}_3)]+\sum_{a_1}1\Big\}$$
$$=4N^3\delta_c(\bar{x}_1-\bar{x}_2)\delta_c(\bar{x}_3-\bar{x}_4)+4N^3\delta_c(\bar{x}_1-\bar{x}_4)\delta_c(\bar{x}_2-\bar{x}_3)+4N\left(\frac{\Lambda}{2\pi}\right)^{2D}\left(\frac{\Lambda_S}{2\pi}\right)^2, \tag{2.23}$$

where

$$\delta_c(\bar{x})=\int(\mathrm{d}\bar{p})\,\mathrm{e}^{i\bar{p}_a\cdot x}$$
$$\underset{\Lambda,\Lambda_S}{=}\delta(\bar{x})\,. \tag{2.24}$$

Apart from a high-momentum cutoff at finite Λ, Λ_S, the value of $\langle \mathrm{Tr}\, \eta(x_1) \cdots \eta(x_n) \rangle$ obtained from the quenched prescription [eq. (2.23)] agrees, in the large-N limit, with the actual value (2.22). Note that to ensure the dominance of planar terms in (2.23), the $N \to \infty$ limit must precede the momentum cutoff $\Lambda \to \infty$ limit. Further, comparing eqs. (2.22), (2.23) we see the fact that quenched models can be trusted only over distance scales

$$d < N/\Lambda . \tag{2.25}$$

Since the quenched noise model will feed (below) into any quenched model, the relation (2.25) is universal.

We will now show two things: first, that there exists a *master noise matrix* η^* in the reduced theory, such that

$$\begin{aligned} &\langle \mathrm{Tr}\,[\eta(\bar{x}_1) \cdots \eta(\bar{x}_n)] \rangle_S \\ &\underset{N}{=} \sum_{a_1 \cdots a_n} \int (\mathrm{d}\bar{p}) \exp[i(\bar{p}_{a_1} - \bar{p}_{a_2}) \cdot \bar{x}_1] \cdots \exp[i(\bar{p}_{a_n} - \bar{p}_{a_1}) \cdot \bar{x}_n] \\ &\quad \times (\eta^*_{a_1 a_2} \eta^*_{a_2 a_3} \cdots \eta^*_{a_n a_1}) , \end{aligned} \tag{2.26}$$

and secondly, that there exists a set of *master momenta* $\{\bar{p}_a\}$ such that the master noise *field* of the *full* $\eta(\bar{x})$ theory of eq. (2.4) is just

$$\eta^*_{ab}(\bar{x}) = \mathrm{e}^{i(\bar{p}^*_a - \bar{p}^*_b) \cdot \bar{x}} \eta^*_{ab} , \tag{2.27}$$

which satisfies

$$\langle \mathrm{Tr}\,[\eta(\bar{x}_1) \cdots \eta(\bar{x}_n)] \rangle_S \underset{N}{=} \mathrm{Tr}\, \eta^*(\bar{x}_1) \cdots \eta^*(\bar{x}_n) . \tag{2.28}$$

The master noise matrix η^* is a random hermitian matrix, whose components are gaussian-distributed random numbers, while the "master momenta" $\{\bar{p}^*_a\}$ is a set of random momenta uniformly distributed in $\bar{p}$-space. In order to actually construct η^* and p^*, we require algorithms for generating gaussian random numbers r_g according to a probability density distribution

$$P(r_g) = \frac{1}{\sqrt{2\pi}} \mathrm{e}^{-r_g^2} , \tag{2.29}$$

and also for generating uniformly distributed random numbers r_u in the interval [0, 1]. Various numerical algorithms exist for generating these types of random (or, more accurately, pseudorandom) numbers [18], simple examples of which will be given below. To construct η^*, the instructions are to generate, for each pair of indices $a > b$, a pair of gaussian-distributed random numbers $r_g^{(1)}$ and $r_g^{(2)}$, and set

$$\eta^*_{ab} = \sqrt{2} \left(\frac{\Lambda}{2\pi} \right)^{D/2} \left(\frac{\Lambda_S}{2\pi} \right)^{1/2} [r_g^{(1)} + i r_g^{(2)}] , \tag{2.30}$$

while for each diagonal component η^*_{aa}, generate a single random number r_g and set

$$\eta^*_{aa} = 2\left(\frac{\Lambda}{2\pi}\right)^{D/2}\left(\frac{\Lambda_5}{2\pi}\right)^{1/2} r_g . \tag{2.31}$$

In this way we have a prescription for generating random hermitian matrices η^* with a probability distribution given by the Boltzmann factor

$$P(\eta^*) \propto \exp\left[-\frac{1}{4}\left(\frac{2\pi}{\Lambda}\right)^D\left(\frac{2\pi}{\Lambda_5}\right)\operatorname{Tr}\eta^*\eta^*\right]. \tag{2.32}$$

To construct the master momenta $\{\bar{p}^*_a\}$, simply generate, for each $a = 1, \ldots, N$, a set of $D+1$ uniform random numbers $r_u^{(1)} \cdots r_u^{(D+1)}$, and set

$$p^*_{\mu a} = \begin{bmatrix} r_u^{(1)} - \frac{1}{2} \\ \vdots \\ r_u^{(D)} - \frac{1}{2} \end{bmatrix} \cdot \Lambda , \qquad p_{5a} = (r_u^{(D+1)} - \tfrac{1}{2})\Lambda_5 , \tag{2.33}$$

Then the master field $\eta^*(xt)$ is found from (2.27). We will show that any η^* and $\eta^*(xt)$ constructed in this way will satisfy the master-field conditions (2.26) and (2.28) respectively.

The strategy for proving (2.26) is to note that

$$T = \sum_{a_1 \cdots a_n} \int (\mathrm{d}\bar{p}) \exp[i(\bar{p}_{a_1} - \bar{p}_{a_2}) \cdot \bar{x}_1] \cdots \exp[i(\bar{p}_{a_n} - \bar{p}_{a_1}) \cdot \bar{x}_n](\eta^*_{a_1 a_2} \cdots \eta^*_{a_n a_1})$$
$$= \sum_{a_1 \cdots a_n} r_{a_1 \cdots a_n} , \tag{2.34}$$

being the sum of a set of random variables $\{r_{a_1 \ldots a_n}\}$, is itself a random variable, whose mean value is precisely the l.h.s. of (2.26). What remains to be shown is that the expected deviation of T from its mean value is suppressed by at least one power of $1/N$ relative to the planar contributions to $\langle \operatorname{Tr} \eta \cdots \eta \rangle_S$, which would verify eq. (2.26). To prove this, we will need to make use of the central limit theorem of statistics, which states that if $w_1, w_2, \ldots, w_n$ are independent random variables having the same distribution with mean μ and variance σ^2, then the probability distribution $P(w)$ of the sum

$$w = \sum_{i=1}^{n} w_i , \tag{2.35}$$

tends, as $n \to \infty$, to the normal distribution

$$P(w) \propto \exp[-\tfrac{1}{2}(w - n\mu)^2/n\sigma^2] . \tag{2.36}$$

A well-known application of this theorem is that if we flip a coin n times, the difference between the number of heads and the number of tails is on the order of $\sqrt{n}$.

In using the central limit theorem, the sum over indices $a_1 \cdots a_n$ on the r.h.s. of (2.34) will first be decomposed into a set of sums, such that the terms within each sum are random variables with the same probability distribution. This means that we must pay special attention to products $\eta^*_{a_1a_2}\eta^*_{a_2a_3} \cdots \eta^*_{a_na_1}$ in which some of the indices coincide, since, e.g., the random variable $\eta^*_{ab}\eta^*_{ba}$ is positive definite, while $\eta^*_{ab}\eta^*_{cd}$ $(a \neq d)$ is not. Let us therefore introduce a notation for restricted sums in (2.34):

$$
\begin{aligned}
&(\eta_{a_1a_2}(\bar{x}_1) \cdots \eta_{a_na_1}(\bar{x}_n))^{\mathrm{R}} \\
&\quad \equiv \sum_{a_1\cdots a_n} \left\{ \int (\mathrm{d}\bar{p}) \exp[i(\bar{p}_{a_1} - \bar{p}_{a_2}) \cdot \bar{x}_1] \cdots \exp[i(\bar{p}_{a_n} - \bar{p}_{a_1}) \cdot \bar{x}_n] \right\} \\
&\qquad \times (\eta^*_{a_1a_2} \cdots \eta^*_{a_na_1}) \Delta_{\{\Phi\}}(a_1 \cdots a_n) ,
\end{aligned}
\tag{2.37}
$$

and index contractions

$$
\begin{aligned}
&(\eta_{a_1a_2} \cdots \eta_{a_la_{l+1}} \cdots \eta_{a_ma_{m+1}} \cdots \eta_{a_na_1})^{\mathrm{R}} \\
&\quad = \sum_{a_1\cdots a_n} \left\{ \int (\mathrm{d}\bar{p}) \exp[i(\bar{p}_{a_1} - \bar{p}_{a_2}) \cdot \bar{x}_1] \cdots \exp[i(\bar{p}_{a_n} - \bar{p}_{a_1}) \cdot \bar{x}_n] \right\} \\
&\qquad \times (\eta^*_{a_1a_2} \cdots \eta^*_{a_na_1}) \delta_{a_la_{m+1}} \delta_{a_ma_{l+1}} \Delta_{\{l,m\}}(a_1 \cdots a_n) ,
\end{aligned}
\tag{2.38}
$$

where

$$
\Delta_{\{m_1,m_2\cdots\}}(a_1 \cdots a_n) = \begin{cases} 1 , & \text{if } \forall_{i \neq \{m_1,m_2\cdots\}}, \forall_{j \neq \{m_1,m_2\cdots\}}, (a_i, a_{i+1}) \neq (a_{j+1}, a_j) , \\ 0 , & \text{otherwise} \end{cases}
\tag{2.39}
$$

blocks all index contractions apart from indices in the set $\{m_1, m_2 \cdots\}$. Evidently the sum T in (2.34) can be expanded as

$$
\begin{aligned}
T &= (\eta^*(\bar{x}_1) \cdots \eta^*(\bar{x}_n))^{\mathrm{R}} \\
&\quad + \{(\underbrace{\eta^*\eta^*} \cdots \eta^*)^{\mathrm{R}} + (\underbrace{\eta^*\eta^*\eta^*} \cdots \eta^*)^{\mathrm{R}} \cdots\}_{1\text{-contraction terms}} \\
&\quad + \{(\underbrace{\eta^*\eta^*}\underbrace{\eta^*\eta^*} \cdots \eta^*)^{\mathrm{R}} + (\underbrace{\eta^*\underbrace{\eta^*\eta^*}\eta^*} \cdots \eta^*)^{\mathrm{R}} + \cdots\}_{2\text{-contraction terms}} \\
&\quad + \cdots \\
&\quad + \{(\underbrace{\eta^*\eta^*}\underbrace{\eta^*\eta^*} \cdots \underbrace{\eta^*\eta^*})^{\mathrm{R}} + (\underbrace{\eta^*\underbrace{\eta^*\eta^*}\eta^*} \cdots \underbrace{\eta^*\eta^*})^{\mathrm{R}} + \cdots\}_{n/2 \text{ contraction terms.}}
\end{aligned}
\tag{2.40}
$$

The only terms in (2.40) whose mean values are non-zero are the fully contracted terms, with $\frac{1}{2}n$ index contractions. The mean value of each of these terms is identical to a corresponding Wick contracted term on the r.h.s. of (2.21) (when the expectation value $\langle\eta \cdots \eta\rangle_{\mathrm{RS}}$ is expanded according to Wick's Theorem). For a planar index contraction [see (2.13)] there are remaining sums over $\frac{1}{2}n+1$ indices, and therefore

$$\text{mean value}\{(\underbrace{\eta\,\eta}\,\underbrace{\eta\,\eta}\cdots\underbrace{\eta\,\eta})^{\mathrm{R}}\}$$

$$=\sum_{a_1\cdots a_n}\int(\mathrm{d}\bar{p})\exp[i(\bar{p}_{a_1}-\bar{p}_{a_2})\cdot\bar{x}_1\cdots\exp[i(\bar{p}_{a_n}-\bar{p}_{a_1})\cdot\bar{x}_n]$$

$$\times(\underbrace{\eta_{a_1a_2}\eta_{a_2a_3}}\underbrace{\eta\,\eta}\cdots\underbrace{\eta\,\eta})$$

$$=\mathrm{O}(N^{n/2+1}), \tag{2.41}$$

and, according to the central limit theorem, the deviation is only $\mathrm{O}[N^{(n/2+1)/2}]$.

Let us now consider the terms in (2.40) whose mean values are zero. The term with the largest deviation from zero is the term involving a sum over the largest number of indices, and this is clearly the totally uncontracted term $(\eta\cdots\eta)^{\mathrm{R}}$ which contains a sum over n indices. Again by the central limit theorem, the deviation of $(\eta\cdots\eta)^{\mathrm{R}}$ from 0 is of order $N^{n/2}$. But this is down by one power of N compared to the values of the planar, fully contracted terms as in (2.41). We can therefore conclude that the values of T in (2.34), and $\langle \mathrm{Tr}\,\eta\cdots\eta\rangle_{\mathrm{S}}$ coincide to leading order in $1/N$, which verifies (2.26).

Eq. (2.28) can be proven in just the same way. This time we consider the sum

$$T^*=\sum_{a_1\cdots a_n}\exp[i(\bar{p}^*_{a_1}-\bar{p}^*_{a_2})\cdot\bar{x}_1]\cdots\exp[i(\bar{p}^*_{a_n}-\bar{p}^*_{a_1})\cdot\bar{x}_n]\eta^*_{a_1a_2}\cdots\eta^*_{a_na_1}$$

$$=\mathrm{Tr}\,(\eta^*(\bar{x}_1)\cdots\eta^*(\bar{x}_2)), \tag{2.42}$$

and note that the mean value of the random variable T^* is again equal to $\langle \mathrm{Tr}\,\{\eta(x_1)\cdots\eta(x_n)\}\rangle_{\mathrm{S}}$. Following exactly the same steps as before, we can show that the expected deviation of $\mathrm{Tr}\,(\eta^*\cdots\eta^*)$ from its mean value $\langle \mathrm{Tr}\,\eta^*\cdots\eta^*\rangle$ is down by a factor of $1/N$, which verifies (2.28).

The completes the construction of the master noise field $\eta^*(\bar{x};\bar{p}^*\eta^*)$ in terms of the master noise matrix η^* and the master momenta $\bar{p}^*$. That master momenta p^* existed was noted in ref. [11].

The procedure for constructing $\eta^*(\bar{x})$ requires algorithms for generating both uniformly and normally distributed random numbers. Of course, truly random numbers can only be obtained from nature (one could imagine building some contraption involving Geiger counters and radioactive cesium), but for our purposes it is sufficient to use a numerical algorithm to generate pseudorandom numbers. Such algorithms are discussed in detail in ref. [18], but for the sake of completeness we will mention two of them here. The formula [19]

$$\rho_k=(7^{9k}\ \mathrm{mod}\ 10^s)/10^s, \tag{2.43}$$

for example, will generate a sequence (of length $(5.1)^s$) of pseudorandom numbers $\{\rho_k\}$ distributed uniformly in the interval [0, 1]. Further, a simple way of generating (approximately) gaussian distributed random numbers r_g, as in (2.29) is: Generate for each r_g a set of 12 uniformly distributed random numbers $\rho^{(1)}, \rho^{(2)}, \ldots, \rho^{(12)}$, and set [19]

$$r_g = \sum_{m=1}^{12} \rho^{(m)} - 6 . \tag{2.44}$$

The rationale for this method is based on the central limit theorem. More sophisticated methods for generating gaussian distributed pseudorandom numbers can be found in ref. [18].

3. Master noise to master field

In this section we wish to show that the Langevin equation allows us to construct the master field for any matrix model out of the master momenta and master noise. It will suffice to consider only the Langevin equation in the presence of quenched noise:

$$\dot{M}_{ab}(xt) = -\frac{\delta S}{\delta M_{ba}(xt)} + \eta_{ab}(xt) ,$$

$$S = \int d^D x \, \mathrm{Tr}\, \{\tfrac{1}{2}[(\partial M)^2 + m^2 M^2] + NV(MN)^{-1/2})\} , \tag{3.1}$$

$$\eta_{ab}(x, t) = e^{i(p_{5a} - p_{5b})t} e^{-i(p_a - p_b)x} \eta_{ab} .$$

After we isolate the equilibrium configuration $M(x, t; \bar{p}, \eta)$, it is trivial to construct from it

$$M^* = M(xt; \bar{p}\eta)|_{\substack{\bar{p} = \bar{p}^* \\ \eta = \eta^*}} . \tag{3.2}$$

Choosing $M(x, t) = 0$, the system is inverted as above, to

$$M_{ab}(xt) = \int_0^\infty d\tau \int d^D y \, G(x - y, t - \tau)[\eta_{ab}(y\tau) - N^{1/2} V'_{ab}(MN^{-1/2})] ,$$

$$V'_{ab}(z) \equiv \frac{\partial}{\partial z_{ba}} V(z) . \tag{3.3}$$

Doing a few iterations in powers of the quenched noise is enough to be convinced that the D-dimensional part of the quenching propagates through to the full solution $M_{ab}(x, t)$. Indeed with the non-perturbative (spatial-quenched) ansatz

$$M_{ab}(xt) = e^{-ip_{ab} \cdot x} M_{ab}(0t) ,$$

$$p_{ab} \equiv p_a - p_b , \tag{3.4}$$

it follows easily from (3.3), that

$$M_{ab}(0t) = e^{-tD_{ab}} \int_0^t d\tau\, e^{\tau D_{ab}} [e^{ip_{5ab}\tau}\eta_{ab} + N^{1/2} V'_{ab}(M(0t)N^{-1/2})],$$
$$D_{ab} \equiv p_{ab}^2 + m^2, \qquad p_{5ab} \equiv p_{5a} - p_{5b}. \tag{3.5}$$

We may call (3.5) the "spatially quenched Langevin equation", but it is also the Langevin equation for the ordinary quenched model

$$\dot{M}_{ab} = -\frac{\delta S_Q}{\delta M_{ba}} + \eta_{ab},$$
$$S_Q \equiv \tfrac{1}{2} \sum_{ab} D_{ab} M_{ab} M_{ba} + NV(MN^{-1/2}). \tag{3.6}$$

We are impressed with the power of the large-N Langevin approach. From the present point of view, we have "derived" the arbitrary quenched model simply by quenching the noise in the Langevin equation. In any case, colorfully phrased, we see that the operations "Langevin" and "quench" commute.

We still face the task of extracting the large-time equilibrium configuration from (3.5). Studying the large-t behavior of the first few iterations reveals not only that

$$M_{ab}(0t) \underset{t}{=} e^{ip_{5ab}t} M_{ab}, \tag{3.7}$$

plus exponentially decreasing corrections, but also that the leading term at each order comes precisely from the large-τ part of the integration, where the form (3.7) dominates an exponential increase in τ. [The fifth time quenched part of the master field is "self-supporting".] Therefore it is consistent to substitute ansatz (3.7) directly in (3.5). As is conventional in the passage from time-dependent to time-independent scattering theory, we also insert the large-τ convergence factor $e^{-\varepsilon\tau}$. The result of the final τ integration, at large t, is simply

$$M_{ab} = \hat{D}_{ab}^{-1}[\eta_{ab} - N^{1/2} V'_{ab}(MN^{-1/2})],$$
$$\hat{D}_{ab} \equiv p_{ab}^2 + m^2 + ip_{5ab} - \varepsilon. \tag{3.8}$$

This is the promised algebraic equation for the translationally covariant equilibrium configuration

$$M_{ab}(xt) \underset{t}{=} e^{ip_{5ab}t} e^{-ip_{ab}\cdot x} M_{ab},$$
$$\hat{D}_{ab} M_{ab} + N^{1/2} V'(MN^{-1/2}) = \eta_{ab}. \tag{3.9}$$

The result is easy to remember. It is precisely the (5-dimensionally) quenched Langevin equation for the theory

$$i[P_5, M] + [P_\mu, [P_\mu, M]] + N^{1/2} V'(MN^{-1/2}) = \eta, \tag{3.10}$$

where the P's are the conventional diagonal matrices and we have included the mass term in V. We remind the reader that the trivial substitution $\bar{p} \to \bar{p}^*$, $\eta \to \eta^*$ gives the master field M^*, as in (3.2),

$$M^*(xt; \bar{p}^*, \eta^*) = e^{i\bar{P}^* \cdot \bar{x}} M(\bar{p}^* \eta^*) e^{-i\bar{P}^* \cdot \bar{x}} . \tag{3.11}$$

The n-point functions of the theory are now constructed in the quenched form,

$$\begin{aligned} &\langle \mathrm{Tr}\,[M(x_1 t) \cdots M(x_n t)] \rangle \\ &\underset{N}{=} \int (\mathrm{d}\bar{p})\, \mathrm{Tr}\,[M(x_1 t) \cdots M(x_n t)]_\eta \\ &\underset{t}{=} \int (\mathrm{d}\bar{p}) \sum_{a_1 \cdots a_n} \exp\,[ip_{a_2} \cdot (x_1 - x_2)] \cdots \exp\,[ip_{a_1} \cdot (x_n - x_1)] \langle M_{a_1 a_2} \cdots M_{a_n a_1} \rangle_\eta , \end{aligned} \tag{3.12}$$

or in the master field form

$$\begin{aligned} &= \mathrm{Tr}\,[M^*(x_1 t) \cdots M^*(x_n, t)] \\ &= \sum_{a_1 \cdots a_n} \exp\,[ip^*_{a_2} \cdot (x_1 - x_2)] \cdots \exp\,[ip^*_{a_1} \cdot (x_n - x_1)] M^*_{a_1 a_2} \cdots M^*_{a_n a_1} . \end{aligned} \tag{3.13}$$

Note that the equilibrium configuration is *not* time independent, but five-dimensionally translationally covariant. Nevertheless, the equilibrium property

$$\frac{\partial}{\partial t} \mathrm{Tr}\,[M(x_1 t) \cdots M(x_n t)] \underset{t}{=} 0 \tag{3.14}$$

follows trivially from the translational covariance.

In the following sections, we will simplify notation by dropping stars, and referring to $M_{ab}(xt)$ at large t as the master field.

A final remark on the translation covariance of the master field (3.11) is in order. As mentioned in the introduction, this feature is now common to both euclidean and hamiltonian master fields, but in the latter case it is well understood: The large-N classical solution is a (reduced) transition matrix element among the ground state and certain "adjoint" energy-momentum eigenstates [5]. It may therefore be enlightening to consider euclidean quenching in the context of the Hilbert space formulation of large-N euclidean theories [20].

4. Perturbation theory and examples

For definiteness we illustrate with some simple computations in the theory $NV(MN^{-1/2}) = \frac{1}{3} g N^{-1/2}\, \mathrm{Tr}\, M^3$. The master field is easily computed by iteration of

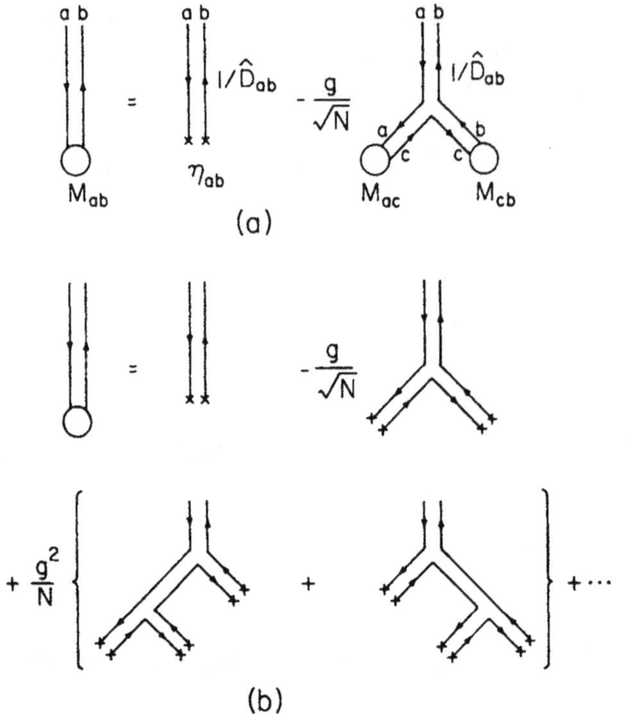

Fig. 1. Diagrammatic representation of the master equation of a matrix model with a Tr (M^3) interaction: (a) the master equation; (b) solution for M_{ab} to second order in the coupling.

(3.8) to any desired order in g. For example, through second order in g,

$$M_{ab} = r_{ab} - gN^{-1/2}\hat{D}_{ab}^{-1} \sum_c r_{ab}r_{cb}$$
$$+ g^2N^{-1}\hat{D}_{ab}^{-1} \sum_{cd} (r_{ac}\hat{D}_{cb}^{-1} r_{cd}r_{db} + \hat{D}_{ac}^{-1} r_{ad}r_{dc}r_{cb})$$
$$+ O(g^3), \qquad r_{ab} \equiv \eta_{ab}\hat{D}_{ab}^{-1} . \tag{4.1}$$

In fact this series is easily described in terms of "Feynman graphs", as illustrated in figs. 1a, b. For ease of computation, we compute η contractions from the quenched gaussian expectation value (3.18). [Indeed, having computed the master field as a function of η, we can *always* either (a) set $\eta = \eta^*$; or (b) do the gaussian integration.] Similarly $\bar{p}^*$ sums will be treated as integrals.

Two-point function to zeroth order:

$$\langle 0|\mathrm{Tr}\, M(x)M(y)|0\rangle \underset{N}{=} \int (\mathrm{d}\bar{p}) \sum_{ab} |M_{ab}|^2\, \mathrm{e}^{i(p_b - p_a)\cdot(x-y)}$$
$$= \int (\mathrm{d}\bar{p}) \sum_{ab} |\eta_{ab}|^2 |\hat{D}_{ab}|^{-2}\, \mathrm{e}^{i(p_b - p_a)\cdot(x-y)} + O(g^2) . \tag{4.2}$$

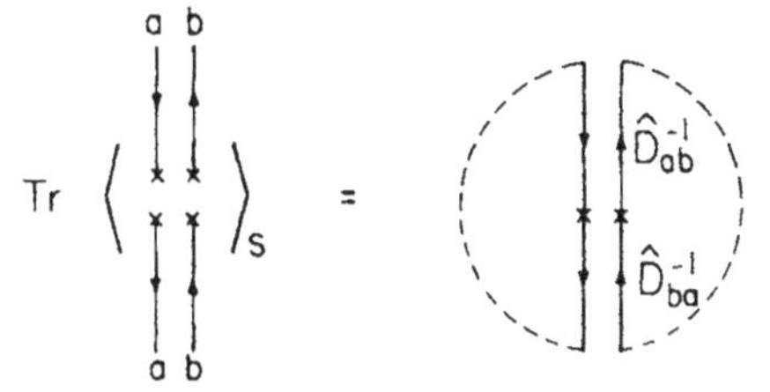

Fig. 2. Graph for the zeroth order matrix model propagator. The stochastic expectation value $\langle\ \rangle_S$ ties together the ends of tree diagrams for M_{ab}, M_{ba}, to produce a Feynman-like graph. Dotted lines indicate that the indices a, b are to be summed from 1 to N. In any such graph, there is a power of N associated with each loop formed by a solid, or solid and dotted, line.

Here we have used eq. (4.1) to zeroth order. This is equivalent to the Feynman graph of fig. 2. Further,

$$(4.2) = 2\left(\frac{\Lambda}{2\pi}\right)^D \left(\frac{\Lambda_5}{2\pi}\right) \int (\mathrm{d}\bar{p}) \sum_{ab} |\hat{D}_{ab}|^{-2}\, \mathrm{e}^{i(p_b-p_a)\cdot(x-y)}$$

$$= 2N^2\left(\frac{\Lambda}{2\pi}\right)^D \left(\frac{\Lambda_5}{2\pi}\right) \int \frac{\mathrm{d}^D p_1}{\Lambda^D} \int \frac{\mathrm{d}^D p_2}{\Lambda^D} \int \frac{\mathrm{d}^D p_{51}}{\Lambda_5} \int \frac{\mathrm{d}^D p_{52}}{\Lambda_5} |\hat{D}_{12}|^{-2}\, \mathrm{e}^{i(p_2-p_1)\cdot(x-y)}$$

$$= 2N^2 \int \frac{\mathrm{d}^D p}{(2\pi)^D}\, \mathrm{e}^{-ip\cdot(x-y)} \int \frac{\mathrm{d}p_5}{2\pi} |\hat{D}_p|^{-2}$$

$$= N^2 \int \frac{\mathrm{d}^D p}{(2\pi)^D}\, \mathrm{e}^{-ip\cdot(x-y)} D_p^{-1}\,. \tag{4.3}$$

Here we used $\hat{D}_p \equiv p^2 + m^2 + ip_5$, $D_p = p^2 + m^2$, and did the p_5 integration by residues.

Three-point function through order g:

$$\langle 0|\mathrm{Tr}\, M(x)M(y)M(z)|0\rangle \underset{N}{=} \int (\mathrm{d}\bar{p}) \sum_{abc} M_{ab}M_{bc}M_{ca}\, \mathrm{e}^{ip_b\cdot(x-y)}\, \mathrm{e}^{ip_c\cdot(y-z)}\, \mathrm{e}^{ip_a\cdot(z-x)}\,. \tag{4.4}$$

Using eq. (4.1) through $O(g)$, and dropping tadpole terms and non-leading terms in N, we obtain (i.e., the graph of fig. 3).

$$(4.4) = 4\left(\frac{\Lambda}{2\pi}\right)^{2D} \left(\frac{\Lambda_5}{2\pi}\right)^2 (-gN^{-1/2}) \int (\mathrm{d}\bar{p})\, \mathrm{e}^{ip_b\cdot(x-y)}\, \mathrm{e}^{ip_c\cdot(y-z)}\, \mathrm{e}^{ip_a\cdot(z-x)}$$

$$\times \{\hat{D}_{ab}^{-1}|\hat{D}_{bc}|^{-2}|\hat{D}_{ca}|^{-2} + |\hat{D}_{ab}|^{-2}\hat{D}_{bc}^{-1}|\hat{D}_{ca}|^{-2} + |\hat{D}_{ab}|^{-2}|\hat{D}_{bc}|^{-2}\hat{D}_{ca}^{-1}\}$$

$$= -4gN^{5/2} \int \frac{\mathrm{d}^D p_1}{(2\pi)^D} \int \frac{\mathrm{d}^D p_2}{(2\pi)^D}\, \mathrm{e}^{ip_2(z-y)}\, \mathrm{e}^{ip_1(z-x)} \{D_1 + D_2 + D_{1+2}\} I(p_1 p_2)\,, \tag{4.5}$$

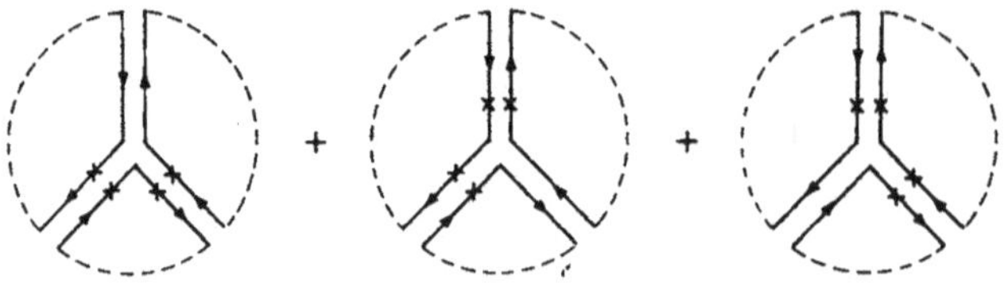

Fig. 3. Graphs contributing to the matrix 3-point function.

where $D_{1+2} \equiv D_{p_1+p_2}$; then by residues

$$I(p_1 p_2) = \int \frac{dp_{s1}}{2\pi} \int \frac{dp_{s2}}{2\pi} |\hat{D}_1|^{-2} |\hat{D}_2|^{-2} |\hat{D}_{1+2}|^2$$

$$= \tfrac{1}{4}(D_1 D_2 D_{1+2})^{-1}(D_{1+2} + D_1 + D_2)^{-1} . \tag{4.6}$$

The final result is then correct for the large-N limit,

$$-gN^{5/2} \int d^D w \, (m^2 + \partial^2)^{-1}_{xw} (m^2 + \partial^2)^{-1}_{yw} (m^2 + \partial^2)^{-1}_{zw} . \tag{4.7}$$

Two-point function through one loop. This computation corresponds to the Feynman diagrams of fig. 4. Alternately it may be performed directly from eq. (4.1), dropping tadpole graphs and non-leading terms in N. After some algebra we obtain

$$\langle 0|\mathrm{Tr}\, M(x) M(y)|0\rangle = (\text{zeroth order})$$

$$+ g^2 4N^2 \int \frac{d^D p_1}{(2\pi)^D} \int \frac{dp_{s1}}{2\pi} \int \frac{d^P p_2}{(2\pi)^D} \int \frac{dp_{s2}}{2\pi} e^{ip_1 \cdot (x-y)} |\hat{D}_1|^{-2}$$

$$\times \{ |D_{1+2}|^{-2} |\hat{D}_2|^{-2} + \hat{D}_1^{*-1} [|\hat{D}_{1+2}|^{-2} \hat{D}_2^{-1} + |\hat{D}_2|^{-2} \hat{D}_{1+2}^{*-1}]$$

$$+ \hat{D}_1^{-1} [|\hat{D}_{1+2}|^{-2} \hat{D}_2^{*-1} + |\hat{D}_2|^{-2} \hat{D}_{1+2}^{-1}] \} , \tag{4.8}$$

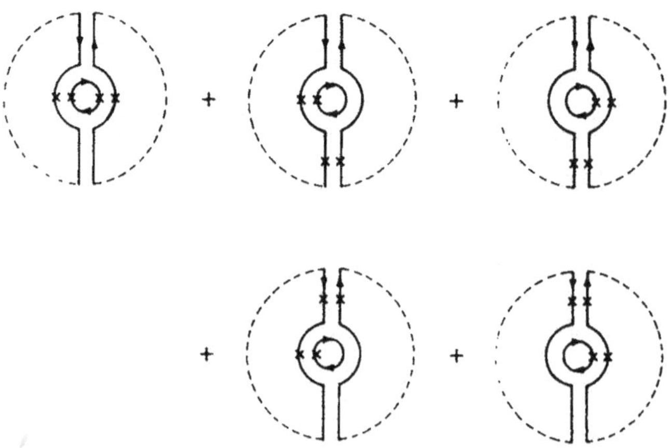

Fig. 4. 1-loop diagrams for the matrix model propagator.

and finally, after the p_5 residue algebra,

$$(4.8) = (\text{zeroth order}) + g^2N^2 \int \frac{d^Dp_1}{(2\pi)^D} \int \frac{d^Dp_2}{(2\pi)^D} e^{ip_1\cdot(x-y)} D_1^{-2} D_2^{-1} D_{1+2}^{-1}, \qquad (4.9)$$

which is the correct large-N result.

Finally, we make a remark on tadpole graphs. The master field does reproduce them correctly, proportional to $D_{aa}^{-1} = m^{-2}$. In a zero mass theory*, one may wish to eliminate them (normal order). This can be accomplished simply by using only the $a \neq b$ part of the master equation, with $M_{aa} = 0$. In the large-N limit (perturbatively) this has no effect beyond eliminating the tadpole graphs.

5. Some non-perturbative remarks

It may be possible to go beyond perturbation theory. As a toy, we solve the $N = 2$ master equation for the quartic theory $NV(MN^{-1/2}) = \frac{1}{4}gN^{-1}\,\mathrm{Tr}\,M^4$ (read $N = 2$). For simplicity, we normal order, taking

$$M = \begin{pmatrix} 0 & R^{1/2}e^{i\phi} \\ R^{1/2}e^{-i\phi} & 0 \end{pmatrix}, \qquad \eta = \begin{pmatrix} 0 & \gamma e^{i\chi} \\ \gamma e^{-i\chi} & 0 \end{pmatrix}. \qquad (5.1)$$

The master equation (3.9) reduces immediately to

$$\begin{aligned} &\tan(\chi - \phi) = p_5(D + gN^{-1}R)^{-1}, \\ &[(D + gN^{-1}R)^2 + p_5^2]R = \gamma^2, \end{aligned} \qquad (5.2)$$

where $p_5 \equiv p_{51} - p_{52}$, $p \equiv p_1 - p_2$ and $D \equiv p^2 + m^2$. What might have been two coupled cubics is a single cubic. Define

$$\begin{aligned} &\alpha \equiv (\tfrac{1}{3}N)^3 D(D^2 + gp_5^2) + \tfrac{1}{2}(N\gamma)^2 g, \\ &\beta \equiv (\tfrac{1}{3}N)^2(3p_5^2 - D^2). \end{aligned} \qquad (5.3)$$

Then since $\beta^3 + \alpha^2 > 0$, only one of the three solutions is real. The (unique) master field is then

$$\begin{aligned} &R = \frac{1}{g}[S_+ + S_- - \tfrac{2}{3}ND], \\ &S_\pm \equiv [\alpha \pm (\beta^3 + \alpha^2)^{1/2}]^{1/3}. \end{aligned} \qquad (5.4)$$

It is not difficult to check that this configuration reproduces perturbation theory.

The two-point function is now

$$G(x, y) = \int \frac{d^Dp}{\Lambda^D}(e^{ip(x-y)} + e^{ip(y-x)}) \int \frac{dp_5}{\Lambda_5} \langle R(p, p_5, \gamma)\rangle_\gamma. \qquad (5.5)$$

* In the Parisi-Wu Langevin approach zero-mass tadpoles do not equilibrate, undergoing a random walk in t. In our quenched Langevin approach, tadpoles *are* equilibrated (proportional to ϵ^{-1}).

It is easy to see here a phenomenon which is generic at any N, and which may be important in a non-perturbative weak coupling approach. We remark that the fifth momentum poles, at $p_5^2+D^2=0$, of perturbation theory, are not present when $g\neq 0$. Instead, one finds growing peaks for small g at these locations. A simple computation in the $N=2$ model gives

$$R(p_5^2+D^2=0) \underset{g\to 0}{\sim} -\frac{4D}{g}. \tag{5.6}$$

We suspect, but have not verified, that these peaks dominate (say after a euclidean p_5 rotation) at small g.

In the $N=2$ toy, we saw that, out of the multiplicity of solutions, the master field is uniquely selected. In more physical cases (higher N, double well potentials etc.) one suspects spontaneous breakdown and phase transitions may be associated with switching from one solution to another.

The case of simple spontaneous breakdown, e.g.,

$$NV=\mathrm{Tr}\left(-\tfrac{1}{2}m^2M^2+\frac{g}{4N}M^4\right),$$

is illustrative. Here one defines, following Parisi and Wu,

$$M_{ab}=\bar{M}_{ab}+\delta_{ab}\sqrt{\frac{NM^2}{g}}. \tag{5.7}$$

The Langevin equation for $\bar{M}(xt)$ can be shown, as in sect. 3, to give uniform time quenching with the choice $\bar{M}(x,0)=0$. The resulting quenched Langevin equation is

$$i[P_5,\bar{M}]+[P_\mu,[P_\mu,\bar{M}]+2m^2\bar{M}+3m\sqrt{\frac{g}{N}}\bar{M}^2+\frac{g}{N}\bar{M}^3=\eta. \tag{5.8}$$

But this is precisely the original quenched Langevin equation (3.10), with the identification (5.7). In other words, the quenched Langevin equation contains *both* the metastable solution and the stable solution.

We mention also that the quenched Langevin equation (3.9) may be put in the form

$$M_{ab}=[\hat{D}_{ab}+(M_{ab})^{-1}N^{1/2}V'_{ab}(MN^{-1/2})]^{-1}\eta_{ab}, \tag{5.9}$$

which suggests an approximation scheme based on continued fractions [21]. This method appears to lead to an "improved" perturbation theory.

6. Quantum chromodynamics

The Langevin equation is

$$\dot{A}_\mu=D_\nu F_{\nu\mu}+\eta_\mu, \tag{6.1}$$

where

$$F_{\mu\nu} = \partial_\mu A_\nu - \partial_\nu A_\mu + ig(A_\mu, A_\nu)\,,$$
$$D_\nu F_{\nu\mu} = \partial_\nu F_{\nu\mu} + ig(A_\nu, F_{\nu\mu})\,, \tag{6.2}$$
$$A_\mu = A_\mu^a \frac{\lambda^a}{\sqrt{2}}\,, \qquad \mathrm{Tr}\,(\lambda^a \lambda^b) = 2\delta_{ab}\,.$$

Further

$$\langle \eta_\mu^{ab}(xt)\eta_\nu^{cd}(y\tau)\rangle = 2\delta_{bc}\delta_{ad}\delta_{\mu\nu}\delta^D(x-y)\delta(t-\tau)\,, \tag{6.3}$$

etc.

Parisi and Wu [15] have indicated that this system describes quantum chromodynamics (at large t) without gauge fixing.

Taking the initial condition to be an arbitrary longitudinal function $A_\mu(x,0)$, we convert (6.1) to the integral equation

$$\dot{A}_\mu(xt) = A_\mu(x,0) + \int_0^\infty \mathrm{d}\tau \int \mathrm{d}^D y\, G_{\mu\nu}(x-y, t-\tau)[\eta_\nu + ig\partial_\lambda(A_\lambda, A_\nu) + ig(A_\lambda, F_{\lambda\nu})]_{y\tau}\,, \tag{6.4}$$

where

$$G_{\mu\nu}(xt) = \theta(t) \int \frac{\mathrm{d}^D k}{(2\pi)^D} \mathrm{e}^{ik\cdot x} [T_{\mu\nu}(k)\,\mathrm{e}^{-k^2 t} + L_{\mu\nu}(k)]\,,$$
$$T_{\mu\nu}(k) = \delta_{\mu\nu} - \frac{k_\mu k_\nu}{k^2}\,, \qquad L_{\mu\nu}(k) = \frac{k_\mu k_\nu}{k^2} \tag{6.5}$$
$$(\delta_{\mu\nu}\partial^2 - \partial_\mu \partial_\nu)A_\nu(x,0) = 0\,.$$

As in sect. 2, we go to the large-N limit by quenching the noise,

$$\eta_\mu^{ab}(xt) = \mathrm{e}^{ip_{5ab}t}\,\mathrm{e}^{-ip_{5ab}\cdot x}\eta_\mu^{ab}\,,$$
$$\langle \eta_\mu^{ab}\eta_\nu^{cd}\rangle = 2\left(\frac{\Lambda}{2\pi}\right)^D \left(\frac{\Lambda_5}{2\pi}\right)\delta_{bc}\delta_{ad}\delta_{\mu\nu}\,, \tag{6.6}$$

and so on.

To zeroth order in g, we compute the master field from (6.4),

$$A_\mu^{ab}(xt) \underset{t}{=} A_\mu^{ab}(x,0) + \mathrm{e}^{-ip_{ab}\cdot x}\eta_\nu^{ab}[\mathrm{e}^{ip_{5ab}t}T_{\mu\nu}(p_{ab})\hat{D}_{ab}^{-1}$$
$$+ L_{\mu\nu}(p_{ab})(\mathrm{e}^{ip_{5ab}t} - 1)(ip_5 - \varepsilon)^{-1}] + O(g)\,, \tag{6.7}$$

where now $\hat{D}_{ab} = ip_{5ab} + p_{ab}^2 - \varepsilon$. This may be used to compute the zeroth order propagator. Choosing $a_\mu(x,0)=0$, with Parisi and Wu, we obtain*

$$\langle 0|\mathrm{Tr}\, A_\mu(xt)A_\nu(yt)|0\rangle \underset{N}{=} \int (\mathrm{d}\bar{p})\, \mathrm{Tr}\, A_\mu(xt)A_\nu(yt)$$

$$\underset{t}{=} N^2 \int \frac{\mathrm{d}^D p}{(2\pi)^D}\, \mathrm{e}^{ip(x-y)}\left[T_{\mu\nu}(p)\frac{1}{p^2}+2L_{\mu\nu}(p)t\right]. \quad (6.8)$$

This result is in agreement with Parisi and Wu: The gauge-invariant part of the system is coming nicely to equilibrium at large t, but the gauge part is undergoing a random walk in time. As we mentioned in sect. 3, any time dependence at large t is due, in the quenched language, to a lack of complete time translation covariance in $A_\mu(t)$ [see eq. (6.7)].

Parisi and Wu argued that gauge-invariant quantities will always come to equilibrium at large t, so for them this phenomenon is not a problem. We desire, however, to characterize a simple, algebraic, equilibrium master A_μ to all orders, and so must change the scheme slightly.

The trick is to choose $A_\mu(x,0)$ (say, in perturbation theory) such that the time translation covariance propagates uniformly. E.g., choosing

$$A_\mu(x,0) \underset{g^0}{=} \mathrm{e}^{-ip_{ab}\cdot x}\eta_\nu^{ab} L_{\mu\nu}(p_{ab})(ip_{5ab}-\varepsilon)^{-1} \quad (6.9)$$

results in the uniformly quenched master field

$$A_\mu^{ab}(xt) \underset{g^0,t}{=} \mathrm{e}^{ip_{5ab}t}\,\mathrm{e}^{-ip_{ab}\cdot x}\eta_\nu^{ab}[T_{\mu\nu}(p_{ab})\hat{D}_{ab}^{-1}+L_{\mu\nu}(p_{ab})(ip_5-\varepsilon)^{-1}]. \quad (6.10)$$

Notice that $A_\mu(x,0)$ is indeed longitudinal, as specified in (6.5).

According to Parisi and Wu, a change of initial condition can only affect the gauge-dependent part of the system at large t. This is easily verified in computing the propagator from (6.10),

$$\langle 0|\mathrm{Tr}\, A_\mu(xt)A_\nu(yt)|0\rangle \underset{N}{=} N^2 \int \frac{\mathrm{d}^D p}{(2\pi)^D}\, \mathrm{e}^{ip(x-y)}\left[T_{\mu\nu}(p)\frac{1}{p^2}+2cL_{\mu\nu}(p)\right], \quad (6.11)$$

where

$$c=\int \frac{\mathrm{d}p_5}{2\pi}(p_5^2+\varepsilon^2)^{-1}=\varepsilon^{-1}$$

is a divergent constant. This is an inconsequential price to pay for having isolated the equilibrium A_μ master field (which itself has no such difficulties), eq. (6.10).

* A useful identity is

$$\int \frac{\mathrm{d}p_5}{2\pi}\frac{(\mathrm{e}^{itp_5}-1)(\mathrm{e}^{-itp_5}-1)}{p_5^2+\varepsilon^2}=|t|.$$

As in sect. 3, we have done an order by order analysis of the integral equation (6.4). It can easily be shown, much as we have done here at zeroth order, that if at nth order $A_\mu^{(n)}(xt)$ = totally quenched + exponentially decreasing corrections, then $A_\mu^{(n+1)}(x, 0)$ may be chosen so that $A_\mu^{(n+1)}(xt)$ = totally quenched + exponentially decreasing corrections. Therefore, by induction, we may look for a totally quenched master field at large time. Again, as in sect. 3, the totally quenched part is "self-supporting". The upshot, as in sect. 3, is the master equation for the large-t quenched master field

$$A_\mu^{ab}(xt) \underset{t}{=} e^{ip_{5ab}t} e^{-ip_{ab}\cdot x} A_\mu^{ab}(xt) ,$$

$$A_\mu^{ab} = R_{\mu\nu}^{ab}(\eta_\nu + g[P_\rho, [P_\rho, A_\nu]] + ig[A_\rho, F_{\rho\nu}])_{ab}, \tag{6.12}$$

where

$$F_{\mu\nu} = -i[P_\mu, A_\nu] + i[P_\nu, A_\mu] + ig(A_\mu, A_\nu) , \tag{6.13}$$

$$R_{\mu\nu}^{ab} = T_{\mu\nu}(p_{ab})\hat{D}_{ab}^{-1} + L_{\mu\nu}(p_{ab})(ip_{5ab} - \varepsilon)^{-1}$$

$$= \hat{D}_{ab}^{-1}\left[\delta_{\mu\nu} + \frac{p_{ab\mu}p_{ab\nu}}{ip_{5ab} - \varepsilon}\right] . \tag{6.14}$$

P_μ are the usual diagonal matrices of quenched momenta. Note also that $R_{\mu\nu}^{ab}$ satisfies the Ward identity

$$p_{ab\mu}R_{\mu\nu}^{ab} = p_{ab\nu}(ip_{5ab} - \varepsilon)^{-1} , \tag{6.15}$$

emphasizing that p_5 is playing the role of a (variable, imaginary) mass.

Having laboriously derived the master equation (6.12) from Parisi and Wu, we remark that, as in the matrix case of sect. 3, the master equation is nothing but the *totally quenched Langevin equation*

$$i[P_5, A_\mu] = [\mathcal{D}_\nu, F_{\nu\mu}] + \eta_\mu ,$$

$$\mathcal{D}_\nu = -iP_\nu + igA_\nu . \tag{6.16}$$

This is an apparently universal result. Note that (6.16) is slightly more general than (6.12). The general inversion of (6.16) corresponds to adding (arbitrary) diagonal terms of the form $\delta_{ab}k_a$ to the right-hand side of (6.12). This corresponds to adding such terms to $A_\mu(x, 0)$ in the procedure described above. Whether such constants can be added consistently is presently under investigation.

From the point of view of previous work on quenching [11], our result (6.12) or (6.16) is a surprise. Given the correctness of Parisi and Wu, we apparently have a quenched formulation of quantum chromodynamics without constraints (or Bars variables [22]). Alternative formulations with constraints such as that of ref. [11] may also be possible. Jiggling the constants mentioned in the previous paragraph may be important. Can these be used to constrain A_{aa}^μ, perhaps to zero at each order? We will return to these alternative versions elsewhere, concentrating here on the simplest, unconstrained version (6.12).

Without the constraint, "naive" [11] quenched quantum chromodynamics is a trivial theory. Our unconstrained formulation is not trivial. We have already computed the propagator above. As in the matrix model, the master equation (6.12) may be iterated to any desired order, and/or expanded Feynman diagrammatically. We give here the master field through order g:

$$A_\mu^{ab} + A_\mu^{(0)ab} + gR_{\mu\nu}^{ab}\{(P_\rho, (A_\rho^{(0)}, A_\nu^{(0)})) + i(A_\rho^{(0)}, F_{\rho\nu}^{(0)})\}_{ab} + O(g^2),$$
$$A_\mu^{(0)ab} = R_{\mu\nu}^{ab}\eta_\nu^{ab}, \tag{6.17}$$
$$F_{\mu\nu}^{(0)ab} = -i[p_{ab\mu}\eta_\nu^{ab} - p_{ab\nu}\eta_\mu^{ab}]\hat{D}_{ab}^{-1}.$$

A second function of the constraint in ref. [11] is the taming of (μ^{-2}) divergent gluonic tadpoles. In our formulation the tadpoles have a vanishing factor before quenched momentum integration:

$$\langle \mathrm{Tr}\, A_\mu \rangle \underset{N}{=} \int (\mathrm{d}\bar{p}) \sum_a A_\mu^{aa},$$
$$\sum_a A_\mu^{aa} \underset{g}{=} gR_{\mu\nu}^{aa}\,\mathrm{Tr}\,\{[P_\rho, [A_\rho^{(0)}, A_\nu^{(0)}]] + i[A_\rho^{(0)}, F_{\rho\nu}^{(0)}]\}. \tag{6.18}$$

$R_{\mu\nu}^{aa}$ is proportional to ε^{-1}, but the trace factor vanishes. The momentum integration would kill each term again. We tentatively ignore the tadpole.

The crucial test of the unconstrained master field is the computation of loop graphs at second order, to ascertain whether or not the would-be quadratic divergence cancels before momentum integration. A positive portent is that the quenched Langevin equation enjoys the gauge invariance,

$$A_\mu \to SA_\mu S^+ - \frac{1}{g}S(P_\mu, S^+),$$
$$\eta_\mu \to S\eta_\mu S^+, \qquad (P_5, S) = 0 \tag{6.19}$$

and even the generalized gauge invariance,

$$A_\mu \to UA_\mu U^+ - \frac{1}{g}U(P_\mu, U^+),$$
$$\eta_\mu \to U\eta_\mu U^+ - \frac{i}{g}(P_\mu, UP_5U^+), \tag{6.20}$$
$$P_5 \to UP_5U^+$$

with no restriction on U. The computation is apparently best approached through the Feynman diagram expansion. For this purpose, a useful form of the master

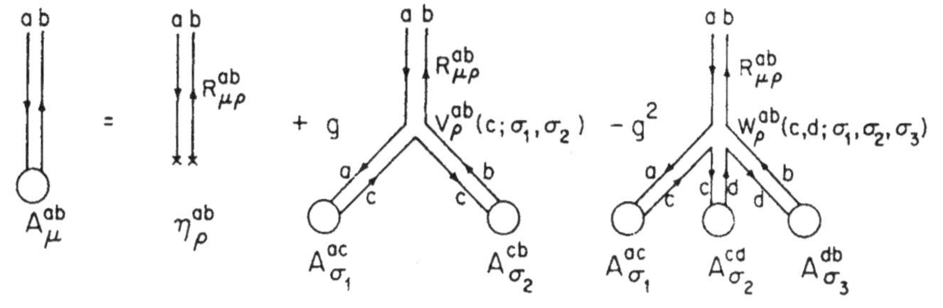

Fig. 5. The master equation for non-abelian gauge theory.

equation is $[P_\mu^{ab} \equiv (p_a - p_b)_\mu]$

$$A_\mu^{ab} = R_{\mu\rho}^{ab}[\eta_\rho^{ab} + gV_\rho^{ab}(c\sigma_1\sigma_2)A_{\sigma_1}^{ac}A_{\sigma_2}^{cb} - g^2 W_\rho^{ab}(cd, \sigma_1\sigma_2\sigma_3)A_{\sigma_1}^{ac}A_{\sigma_2}^{cd}A_{\sigma_3}^{db}]. \qquad (6.21)$$

where

$$V_\rho^{ab}(c\sigma_1\sigma_2) \equiv P_{[\sigma_1,}^{ab}\delta_{\sigma_2]\rho} + P_{[\rho,}^{ac}\delta_{\sigma_2]\sigma_1} + P_{[\sigma_1,}^{cb}\delta_{\rho]\sigma_2},$$

$$W_\rho^{ab}(cd, \sigma_1\sigma_2\sigma_3) \equiv \delta_{\sigma_1\sigma_2}\delta_{\sigma_3\rho} - 2\delta_{\sigma_1\sigma_3}\delta_{\sigma_2\rho} + \delta_{\sigma_1\rho}\delta_{\sigma_2\sigma_3}, \qquad (6.22)$$

$$P_{[\sigma_1,}^{ab}\delta_{\sigma_2]\rho} \equiv P_{ab\sigma_1}\delta_{\sigma_2\rho} - p_{ab\sigma_2}\delta_{\sigma_1\rho}, \qquad (6.23)$$

etc. This is illustrated in fig. 5. The explicit computation unfortunately involves on the order of 50 terms, not to mention the p_5 integrations, and we have not yet been able to complete it. If the computation is positive, we prefer the unconstrained approach. If negative, the "alternative versions" mentioned above must be studied.

Research was supported in part by the National Science Foundation under grant number PHY-81-18547 and the Director, Office of Energy Research, Office of High Energy and Nuclear Physics, Division of High Energy Physics, of the US Department of Energy, under contract DE-AC03-76SF-00098.

Note added in proof

Using our ideas together with those of Parisi and Sourlas [23], it is not hard to derive a "six" (or $D+2$) dimensional quenched master field. It has full 6-dimensional translational covariance, and

$$[\bar{P}_\mu, [\bar{P}_\mu, M]] + N^{1/2}V'(MN^{1/2}) = \eta,$$

$$\langle \eta_{ab}\eta_{cd}\rangle = 4\pi\left(\frac{\Lambda}{2\pi}\right)^{D+2} \delta_{ad}\,\delta_{bc},$$

where μ is summed over all $D+2$ dimensions. D-dimensional Green functions are evaluated at equal 5th and 6th momenta. There may be a hierarchy of such dimensional reduction and master fields.

References

[1] E. Witten, 1979 Cargese Lectures, recent developments in gauge theories, ed. G. 't Hooft (Plenum Press, 1980)
[2] A. Jevicki and N. Papanicolaou, Nucl. Phys. B171 (1980) 362
[3] M.B. Halpern, Nuovo Cim. 61A (1981) 207
[4] K. Bardakci, Nucl. Phys. B178 (1981) 263
[5] M.B. Halpern, Nucl. Phys. B188 (1981) 61
[6] M.B. Halpern, Nucl. Phys. B173 (1980) 504
[7] A. Jevicki and H. Levine, Phys. Rev. Lett. 44 (1980) 1443
[8] M.B. Halpern and C. Schwartz, Phys. Rev. D24 (1981) 2146
[9] M.B. Halpern, Nucl. Phys. B204 (1982) 93
[10] E. Brézin, C. Itzykson, G. Parisi and J.B. Zuber, Comm. Math. Phys. 59 (1978) 35
[11] D.J. Gross and Y. Kitazawa, Princeton preprint (1982)
[12] T. Eguchi and H. Kawai, Phys. Rev. Lett. 48 (1982) 1063
[13] G. Bhanot, U. Heller and H. Neuberger, IAS preprint (1982)
[14] G. Parisi, Phys. Lett. 112B (1982) 463
[15] G. Parisi and Wu Yongshi, Sci. Sin. 24 (1981) 483
[16] M. Creutz, L. Jacobs and C. Rebbi, Phys. Rev. D20 (1979) 1915
[17] P.C. Hohenberg and B.I. Halperin, Rev. Mod. Phys. 49 (1977) 435
[18] D.E. Knuth, The art of computer programming, vol. 2 (Addison-Wesley, Menlo Park, 1969)
[19] F. Scheid, Numerical Analysis, Schaum's outline series (McGraw-Hill, New York, 1968)
[20] O. Haan, Z. Phys. C6 (1980) 345
[21] P. Cvitanović, P.G. Lauwers and P.N. Scharbach, Nucl. Phys. B203 (1982) 385
[22] I. Bars, CERN preprint TH 3318 (1982)
[23] G. Parisi and N. Sourlas, Phys. Rev. Lett. 43 (1979) 744

Volume 121B, number 5 PHYSICS LETTERS 10 February 1983

DERIVATION OF QUENCHED MOMENTUM PRESCRIPTION BY MEANS OF STOCHASTIC QUANTIZATION ☆

J. ALFARO and B. SAKITA

Department of Physics, City College of City University of New York, New York, NY 10031, USA

Received 21 June 1982

We use the stochastic quantization method of Parisi and Wu to understand the quenched momenta prescription for large N theories. The main advantage of our procedure is its simplicity. It leads to the prescription in a straightforward manner without the explicit use of the perturbation expansion.

Recently Eguchi and Kawai [1] noted an effective reduction of degrees of freedom in the large N limit of Wilson's $U(N)$ lattice gauge theory. They proposed a reduced model on a single hyper-cubic lattice with a periodic boundary condition, which should be equivalent to the standard Wilson theory in the limit $N \to \infty$. Subsequently it was pointed out by Bhanot et al. [2] that this equivalence does not hold in the weak coupling region because of $U(1)^N$ symmetry breaking, and they suggested a modification of the model using a quenching prescription which enforces $U(1)^N$ symmetry.

Parisi [3] then noticed that the matrix model with a twisted boundary condition has a volume independent free energy per unit volume in the large N limit if the average is made over the parameters in the boundary condition. Noting that Parisi's reduction of degrees of freedom can be viewed as representing translations in space time in the internal symmetry group, a number of authors [4–7] arrived independently at the quenched momentum prescription for large N theories. These authors have based their discussion on a detailed analysis of planar Feynman graphs for all order of perturbation theory. We believe that the large N reduction phenomenon is so general that it is likely to have more transparent explanations.

The purpose of this paper is to report on a study of the reduction of degrees of freedom in the large N limit in a different theoretical setting, the stochastic quantization of Parisi and Wu [8].

Stochastic quantization. Let us consider a euclidean field theory in d dimenisons and Green's functions defined by

$$\langle \phi_{l_1}(x_1)\phi_{l_2}(x_2)\ldots\phi_{l_n}(x_n)\rangle \equiv \int D\phi\, \phi_{l_1}(x_1)\phi_{l_2}(x_2)\ldots\phi_{l_n}(x_n)\exp(-S[\phi]) \Big/ \int D\phi \exp(-S[\phi]) , \quad (1)$$

where $S[\phi]$ is the action and the l's represent internal indices. According to Parisi and Wu [8], the evaluation of Green's function can also be achieved by the stochastic method which consists of the following steps:

(i) Introduce a fictitious time t and solve the following Langevin equation

$$\partial\phi_l(x,t)/\partial t = -\delta S/\delta\phi_l(x,t) + \eta_l(x,t) , \quad (2)$$

where $\eta_l(x,t)$ is a random source with gaussian distribution, namely the random average over η has the following Wick decomposition property:

☆ Work supported by grant from National Science Foundation PHY-78-24888 and CUNY-PSC BHE Faculty Research Award.

Volume 121B, number 5 PHYSICS LETTERS 10 February 1983

$$\langle \eta_{l_1}(x_1,t_1)\eta_{l_2}(x_2,t_2)\dots\eta_{l_{2n}}(x_{2n},t_{2n})\rangle_\eta = \sum_{\substack{\text{possible}\\ \text{comb.}}}\prod_{\text{pair}}\langle \eta_{l_i}(x_i,t_i)\eta_{l_j}(x_j,t_j)\rangle_\eta\,, \tag{3}$$

$$\langle \eta_l(x,t)\eta_{l'}(x',t')\rangle_\eta = 2\delta_{ll'}\delta(x-x')\delta(t-t')\,.$$

An explicit representation of the average is given by the following functional average:

$$\langle\dots\rangle_\eta = \int D\eta(\dots)\exp\left(-\int dx\,dt\sum_l \eta_l^2(x,t)\right)\Big/\int D\eta\exp\left(-\int dx\,dt\sum_l \eta_l^2(x,t)\right). \tag{4}$$

(ii) Calculate

$$\langle \phi_{l_1}(x_1,t)\dots\phi_{ln}(x_n,t)\rangle_\eta\,.$$

(iii) Then take the limit $t\to\infty$ to obtain Green's functions

$$\langle \phi_{l_1}(x_1)\dots\phi_{l_n}(x_n)\rangle = \lim_{t\to\infty}\langle \phi_{l_1}(x_1,t)\dots\phi_{ln}(x_n,t)\rangle_\eta\,. \tag{5}$$

Namely, the equilibrium values of the classical system (2) with a gaussian random source are precisely the quantum Green's functions (statistical average) given by (1).

We emphasize the following two aspects of the stochastic method for our purpose: First, the Langevin equation can be viewed as a classical equation of motion for a system with an external source function. Second, the random source $\eta_l(x,t)$ can be quite arbitrary in its x, t dependence as far as the random average has the gaussian distribution property given by (3). These aspects are exploited to reduce the number of degrees of freedom for large N.

Reduction of degrees of freedom in the large N limit. We illustrate our method with an example of hermitian matrix model defined by the action:

$$S[\phi] = \int d^dx\,\mathrm{tr}\,[\tfrac{1}{2}(\partial_\mu\phi)^2 + \tfrac{1}{2}m^2\phi^2 + (g/4!N)\phi^4]\,, \tag{6}$$

where ϕ is a hermitian $N\times N$ matrix field. This model has a global U(N) symmetry and a reflection symmetry:

$$\phi(x)\to u\phi(x)u^+,\quad u\in U(N);\qquad \phi(x)\to-\phi(x)\,. \tag{7, 8}$$

Accordingly we consider a set of invariant Green's functions defined by

$$\langle \mathrm{tr}(\phi(x_1)\phi(x_2)\dots\phi(x_n))\rangle\quad (n=\text{even})\,. \tag{9}$$

The Langevin equation of this model is given by

$$\partial\phi_{ij}(x,t)/\partial t = (\Box - m^2)\phi_{ij}(x,t) - (g/3!N)(\phi^3(x,t))_{ij} + \eta_{ij}(x,t)\,, \tag{10}$$

where $\eta_{ij}(x,t)$ is an hermitian $N\times N$ random source with gaussian distribution.

By solving the Langevin equation by iteration and by inserting the solution into (5) and (9) we obtain a formal power series expansion of Green's function:

$$\langle \mathrm{tr}(\phi(x_1)\dots\phi(x_n))\rangle = \lim_{t\to\infty}\sum_{m=0}^{\infty}(g/N)^m\int\dots\int dy_1\,dt_1\dots dy_{2m+n}\,dt_{2m+n}\,K_m(x_1,\dots,x_n,t;y_1t_1,\dots,y_{2m+n}t_{2m+n})$$
$$\langle \mathrm{tr}(\eta(x_1,t_1)\dots\eta(x_{2m+n},t_{2m+n}))\rangle_\eta\,, \tag{11}$$

where K_m are scalar functions. A very nice point of this expression is that all the U(N) indices appear through η

so that one can study the large N limit by examining only the random average. Of course, as shown by Parisi and Wu [8], the expression (11) generates the standard Feynman–Dyson expansion if one inserts the Wick decomposition expression (3).

Next we prove the following proposition.

Proposition. Let $\bar\eta_{ij}(t)$ $(i,j=1,...,N)$ be a random source with gaussian distribution and p_i^α $(i=1,...,N;\alpha=1,...,d)$ random numbers with uniform distribution in the hypercube $[-\frac{1}{2}\Lambda,\frac{1}{2}\Lambda]^d$. As long as one restricts oneself to invariant expectation values of the form $\langle \mathrm{tr}(\eta(x_1,t_1)\ldots\eta(x_{2m},t_{2m}))\rangle_\eta$, in the large N limit

$$\eta_{ij}(x,t)=(\Lambda/2\pi)^{d/2}\exp\left[\mathrm{i}(p_i-p_j)x\right]\bar\eta_{ij}(t), \tag{12}$$

serves as variables with gaussian distribution.

The proof reduces to show that (12) gives Wick decomposition property. Let us first examine the case of $m=1$.

$$\sum_{ij}\langle\eta_{ij}(x,t)\eta_{ji}(x',t')\rangle\equiv\sum_{ij}\left(\frac{\Lambda}{2\pi}\right)^d\int\prod_{\alpha,k}\frac{\mathrm{d}p_k^\alpha}{\Lambda}\exp\left[\mathrm{i}(p_i-p_j)(x-x')\right]\langle\bar\eta_{ij}(t)\bar\eta_{ji}(t')\rangle_{\bar\eta}$$

$$=\left(\frac{\Lambda}{2\pi}\right)^d\left(2(N^2-N)\delta(t-t')\int\prod_{\alpha,k}\frac{\mathrm{d}p_k^\alpha}{\Lambda}\exp[\mathrm{i}(p_i-p_j)(x-x')]+2N\delta(t-t')\int\prod\frac{\mathrm{d}p}{\Lambda}\right).$$

In leading order in N we can neglect the last term. We then obtain

$$\langle\mathrm{tr}(\eta(x,t)\eta(x',t'))\rangle\sim 2N^2\delta(x-x')\delta(t-t'), \tag{13}$$

which is the desired result. The neglection of the second term is justified only when N is so large that the following inequality is satisfied:

$$N^{-1}(\Lambda/2\pi)^d L^d \ll 1, \tag{14}$$

where L^d is the space–time volume. This shows the degree of large N we are talking about.

Next let us examine the $m=2$ case.

$$\sum_{ijkl}\langle\eta_{ij}(x_1,t_1)\eta_{jk}(x_2,t_2)\eta_{kl}(x_3,t_3)\eta_{li}(x_4,t_4)\rangle$$

$$\equiv\left(\frac{\Lambda}{2\pi}\right)^{2d}\sum_{ijkl}\int\prod\frac{\mathrm{d}p}{\Lambda}\exp\{\mathrm{i}\,[(p_i-p_j)x_1+(p_j-p_k)x_2+(p_k-p_l)x_3+(p_l-p_i)x_4]\}$$

$$\times\langle\eta_{ij}(t_1)\eta_{jk}(t_2)\eta_{kl}(t)\eta_{li}(t_4)\rangle_{\bar\eta}.$$

Since $\bar\eta$ has gaussian distribution, we have

$$\langle\bar\eta_{ij}(t_1)\bar\eta_{jk}(t_2)\bar\eta_{kl}(t_3)\bar\eta_{li}(t_4)\rangle_{\bar\eta}$$

$$=2\delta_{ik}\delta(t_1-t_2)\delta(t_3-t_4)+2\delta_{jl}\delta(t_1-t_4)\delta(t_2-t_3)+2\delta_{il}\delta_{jk}\delta_{ji}\delta_{kl}\delta(t_1-t_3)\delta(t_2-t_4).$$

Thus

$$\langle \mathrm{tr}(\eta(x_1,t_1)\ldots\eta(x_4,t_4))\rangle$$

$$= \left(\frac{\Lambda}{2\pi}\right)^{2d} \sum_{ijkl} \int \prod \frac{\mathrm{d}p}{\Lambda} \exp[\mathrm{i}(p_i - p_j)(x_1 - x_2)] \exp[\mathrm{i}(p_i - p_l)(x_3 - x_4)]\, 2\delta_{jk}\delta(t_1 - t_2)\, 2\delta(t_3 - t_4) + \ldots$$

$$= 4\delta(t_1 - t_2)\delta(t_3 - t_4)\left(\frac{\Lambda}{2\pi}\right)^{2d}\left((N^3 - [3N^2 - N]) \int_{\substack{i\neq j, i\neq l\\ l\neq j}} \prod \frac{\mathrm{d}p}{\Lambda} \exp[\mathrm{i}(p_i - p_j)(x_1 - x_2) + \mathrm{i}(p_i - p_l)(x_3 - x_4)]\right.$$

$$+ N^2 \int \prod \frac{\mathrm{d}p}{\Lambda} \exp[\mathrm{i}(p_i - p_l)(x_3 - x_4)] + N^2 \int \prod \frac{\mathrm{d}p}{\Lambda} \exp[\mathrm{i}(p_i - p_j)(x_1 - x_2)]$$

$$\left. + N(N-1) \int \prod \frac{\mathrm{d}p}{\Lambda} \exp[\mathrm{i}(p_i - p_j)(x_1 - x_2 + x_3 - x_4)]\right) + \ldots$$

$$\approx 4\delta(t_1 - t_2)\delta(t_3 - t_4) N^3 \left[\delta(x_1 - x_2)\delta(x_3 - x_4) + \frac{1}{N}\left(\frac{\Lambda}{2\pi}\right)^d \delta(x_1 - x_2 + x_3 - x_4)\right] + \ldots .$$

We neglect the second term in the square bracket in the large N limit, using (14), and obtain

$$\langle \mathrm{tr}(\eta(x_1,t_1)\eta(x_2,t_2)\eta(x_3,t_3)\eta(x_4,t_4))\rangle \approx \mathrm{tr}[\langle\eta(x_1,t_1)\eta(x_2,t_2)\rangle\langle\eta(x_3,t_3)\eta(x_4,t_4)\rangle$$
$$+ \langle\eta(x_2,t_2)\eta(x_3,t_3)\rangle\langle\eta(x_4,t_4)\eta(x_1,t_1)\rangle] . \qquad (15)$$

It is clear now how the arguments continue. An important point to notice is that if N were large but finite η's given by (12) cease to be gaussian for $m \gtrsim N$ because we do not have enough p's to produce independent δ-functions.

The Langevin equation (10) with the random source given by (12) has a solution given by

$$\phi_{ij}(x,t) = \exp[\mathrm{i}(p_i - p_j)x]\,\bar{\phi}_{ij}(t) , \qquad (16)$$

where $\bar{\phi}_{ij}(t)$ is a solution of the reduced Langevin equation:

$$\partial\bar{\phi}_{ij}(t)/\partial t = -[(p_i - p_j)^2 + m^2]\bar{\phi}_{ij} - (g/3!N)(\bar{\phi}^3)_{ij} + \bar{\eta}_{ij}(t)(\Lambda/2\pi)^d , \qquad (17)$$

Inserting (16) into (5) and using the proposition we can show in a straightforward manner the large N equivalence between the model defined by (6) and the reduced model with quenching prescription:

$$\langle \mathrm{tr}(\phi(x_1)\ldots\phi(x_n))\rangle = \lim_{t\to\infty} \langle \mathrm{tr}(\phi(x_1,t_1)\ldots\phi(x_n,t_n))\rangle_\eta$$

$$= \lim_{t\to\infty} \int \prod \frac{\mathrm{d}p}{\Lambda} \sum_{ijk\ldots} \exp[\mathrm{i}(p_i - p_j)x_1]\ldots\langle\bar{\phi}_{ij}(t_1)\bar{\phi}_{jk}(t_2)\ldots\rangle_{\bar{\eta}}$$

$$= \int \prod \frac{\mathrm{d}p}{\Lambda} \sum_{ijk\ldots} \exp[\mathrm{i}(p_i - p_j)x_1]\ldots\langle\bar{\phi}_{ij}\bar{\phi}_{jk}\ldots\rangle_{\bar{S}} , \qquad (18)$$

where $\langle\ldots\rangle_{\bar{S}}$ is defined by

$$\langle\ldots\rangle_{\bar{S}} \equiv \int \mathrm{D}\bar{\phi}(\ldots)\exp(-\bar{S}[\phi]) \Big/ \int \mathrm{D}\bar{\phi}\exp(-\bar{S}[\phi]) , \qquad (19)$$

$$\bar{S} = (2\pi/\Lambda)^d \left(\sum_{ij} \tfrac{1}{2}[(p_i - p_j)^2 + m^2]\bar{\phi}_{ij}\bar{\phi}_{ji} + (g/4!N)\,\mathrm{tr}\,\bar{\phi}^4\right) . \qquad (20)$$

Volume 121B, number 5 PHYSICS LETTERS 10 February 1983

This is the Gross–Kitazawa version [4] of quenched momentum prescription. If we derive the Langevin equation associated with this action using (2), we obtain (17) in which t is replaced by $(\Lambda/2\pi)^d t$. Since $t \to \infty$ is taken to derive the equality (18) this scale difference does not affect the conclusion.

Remarks. The distribution (12) can be understood in terms of a change of variables as follows. Consider the generating functional of the gaussians Green's function:

$$Z[J] \equiv \int D\eta \exp\left(-\int dx\, dt\, [\tfrac{1}{2}\,\mathrm{tr}\,\eta^2 + \eta_{ij}(x,t) J_{ji}(x,t)]\right) . \tag{21}$$

We first multiply

$$I = \int \prod_{\alpha,k} \frac{dp_k^\alpha}{\Lambda} , \tag{22}$$

then change variables in the path integral

$$\eta_{ij}(x,t) \to \exp\,[i(p_i - p_j)x]\, \eta_{ij}(x,t) , \tag{23}$$

we obtain

$$Z[J] = \int D\eta \exp\left(-\int dx\, dt\, \tfrac{1}{2}\,\mathrm{tr}\,\eta^2\right) \int \prod \frac{dp}{\Lambda} \exp\left(-\int dx\, dt \sum_{ij} \eta_{ij}(x,t) \exp\,[i(p_i - p_j)x] J_{ji}(x,t)\right)$$

$$= \int D\eta \exp\left(-\int dx\, dt\, \tfrac{1}{2}\,\mathrm{tr}\,\eta^2\right) \langle e^{-F} \rangle , \tag{24}$$

where F is given by

$$F[J] = \int dx\, dt \sum_{ij} \eta_{ij}(x,t) \exp\,[i(p_i - p_j)x] J_{ji}(x,t) , \tag{25}$$

and $\langle\ \rangle$ denotes a uniform momentum average. Define

$$\bar{F}[J] = \int dt \sum_{ij} \eta_{ij}(0,t) \int dx \exp[i(p_i - p_j)x] J_{ji}(x,t) , \tag{26}$$

and use the cumulant expansion:

$$\langle e^{-F} \rangle = \exp \sum_{n=0}^{\infty} \frac{(-)^n}{n!} \langle F^n \rangle_c , \quad \langle F \rangle_c = \langle F \rangle , \quad \langle F^2 \rangle_c = \langle F^2 \rangle - \langle F \rangle^2 , \text{ etc.} \tag{27}$$

we obtain

$$Z[J] = \int \prod \frac{dp}{\Lambda} \int D\eta \exp\left(-\int dx\, dt\, \tfrac{1}{2}\,\mathrm{tr}\,\eta^2\right) e^{-\bar{F}} \exp\left(\sum_{n=0}^{\infty} \frac{(-)^n}{n!} [\langle F^n \rangle_c - \langle \bar{F}^n \rangle_c]\right) . \tag{28}$$

For the large N limit we can set effectively

$$\langle F^n \rangle_c - \langle \bar{F}^n \rangle_c \approx 0 ,$$

and get

$$Z[J] \approx \int \prod \frac{dp}{\Lambda} \int D\eta \exp\left(-\int dx\, dt\, \tfrac{1}{2}\,\mathrm{tr}\,\eta^2(x,t)\right) \exp\left(-\int dt\, \eta_{ij}(0,t) \int dx \exp\,[i(p_i - p_j)x] J_{ji}(x,t)\right) . \tag{29}$$

If we integrate out the contribution of $\eta(x,t)\, x \neq 0$ by using

$$\int dx\, dt\, \tfrac{1}{2}\,\mathrm{tr}\,\eta^2(x,t) = \left(\frac{2\pi}{\Lambda}\right)^d \sum_n \int dt\, \tfrac{1}{2}\,\mathrm{tr}\,\eta^2(x_n,t) , \tag{30}$$

we obtain

$$Z[J] = \text{const} \int \prod \frac{\mathrm{d}p}{\Lambda} \int \mathrm{D}\eta(0,t) \exp\left[-\left(\frac{2\pi}{\Lambda}\right)^d \int \mathrm{d}t \, \tfrac{1}{2} \operatorname{tr}(\eta^2(0,t))\right]$$

$$\times \exp\left(-\int \mathrm{d}t \, \eta_{ij}(0,t) \int \mathrm{d}x \exp\left[\mathrm{i}(p_i - p_j)x\right] J_{ji}(x,t)\right). \tag{31}$$

It is easy to see that $Z[J]$ so defined generates distribution (12).

References

[1] T. Eguchi and H. Kawai, Phys. Rev. Lett. 48 (1982) 1063.
[2] G. Bhanot, U. Heller and H. Neuberger, Phys. Lett. 115B (1982) 237.
[3] G. Parisi, Frascati preprint (1982).
[4] D.J. Gross and Y. Kitazawa, Princeton University preprint (1982).
[5] S. Das and S. Wadia, University of Chicago preprint (1982).
[6] T. Eguchi and H. Kawai, Tokyo University preprint (1982).
[7] G. Parisi and Zheng Yi-Cheng, Frascati preprint (1982).
[8] G. Parisi and Wu Yong-Shi, Sci. Sin. 24 (1981) 483.

Volume 131B, number 4,5,6 PHYSICS LETTERS 17 November 1983

ON COMPLEX PROBABILITIES

Giorgio PARISI

Dipartimento di Fisica, Università di Roma II "Tor Vergata", Rome, Italy

Received 8 July 1983

We show how the formalism of the Langevin equation may be used for evaluating the averages of quantities which have a complex valued distribution.

It is fashionable to use stochastic techniques for computing the expectation values in statistical mechanics, especially if the number of degrees of freedom is very high: for example if a configuration of the system is denoted by φ, we may be interested to compute

$$\int d\mu[\varphi]\, f(\varphi) \equiv \langle f \rangle ,$$

$$d\mu[\varphi] = d[\varphi] \exp\{H[\varphi]\}/Z ,$$

$$Z = \int d[\varphi] \exp\{-H[\varphi]\}, \quad \int d\mu[\varphi] = 1 . \qquad (1)$$

A very efficient method is the Monte Carlo method [‡1] in which we use explicitly the fact that $d\mu[\varphi]$ is a probability measure: generally speaking one constructs a recursive random algorithm which generates the configurations of the system according to the probability measure $d\mu[\varphi]$; more precisely we have that:

$$\lim_{N\to\infty} \frac{1}{N} \sum_{n=1}^{N} f(\varphi_n) = \langle f \rangle , \qquad (2)$$

where the φ_n are the configurations generated according to the algorithm.

An other method for continuous systems is based on the Langevin equation:

$$\dot\varphi = -\partial H/\partial\varphi + \eta(t), \quad \overline{\eta(t)\eta(t')} = 2\delta(t-t') . \qquad (3)$$

Standard mathematical manipulations shows that the probability distribution for the field φ evolves according to the Fokker–Planck equation [2]:

[‡1] For a review see ref. [1].

$$\dot P(\varphi, t) = (\partial/\partial\varphi)(-p\partial H/\partial\varphi + \partial p/\partial\varphi) . \qquad (4)$$

The reduced probability distribution $p(\varphi, t)$ will evolve according to the equation:

$$p(\varphi, t) = P(\varphi, t)u(\varphi), \quad u(\varphi) = \exp[\tfrac{1}{2}H(\varphi)] ,$$

$$\dot p(p, t) = -\mathcal{H}p \equiv \tfrac{1}{4}(\partial H/\partial\varphi)^2 - \tfrac{1}{2}\partial^2 H/\partial\varphi^2 , \qquad (5)$$

which is a Schrödinger type equation. If the number of degrees of freedom is finite and the hamiltonian H is not singular, the associated Fokker–Planck hamiltonian $\mathcal{H}$ is a self adjoint positive operator (provided that $Z < \infty$) which has only one zero eigenvalue $u(\varphi)$. These properties imply that in the large time limit $\exp(-t\mathcal{H})$ becomes the projection over $u(\varphi)$ and

$$p(\varphi, t) \xrightarrow[t\to\infty]{} \exp\{-H[\varphi]\}/Z ,$$

$$\frac{1}{t}\int_0^t f(\varphi(t'))\, dt' \xrightarrow[t\to\infty]{} \langle f \rangle . \qquad (6)$$

Up to now everything is well known.

The problem we are interested here is to study what happens if H becomes complex valued [‡2]. We argue that, while the Monte Carlo method cannot be used in this situation, nothing

[‡2] While I was writing this paper I received a very interesting paper [3] in which a similar suggestion is put forward; he shows that the existence of logarithmic cuts for the function $H(\varphi)$ does not destroy the existence of solutions of the Langevin equation. In view of this result the case in which $\partial H/\partial\varphi$ is a meromorfic function of φ with only single poles and for large $|\varphi|\partial H/\partial\varphi \propto A \cdot \varphi$ where A is a positive definite operator, seems the easiest case for obtaining rigorous results.

Volume 131B, number 4,5,6 PHYSICS LETTERS 17 November 1983

forbids to write a Langevin equation also for complex H. In this case (let us assume that H is an analytic function) eq. (3) gives an evolution law for φ in the complex plane; a real valued probability distribution on the complex plane can be defined $p(\varphi_R, \varphi_I, t)$ $(\varphi = \varphi_R + i\varphi_I)$ in the usual way; it would satisfy a generalized Fokker–Planck equation:

$$\dot{P}(\varphi_R, \varphi_I, t) = (\partial^2/\partial\varphi_R^2)p + (\partial/\partial\varphi_R)[V_R(\varphi)p] + (\partial/\partial\varphi_I)[V_I(\varphi)p]\,, \tag{7}$$

whose solution is no more as simple as before. Our main observation is that we can define a complex valued function on the real axis $P(\varphi_R, t)$, such that

$$\int d[\varphi_R] P(\varphi_R, t)\varphi_R^n = \int d[\varphi_R]\, d[\varphi_I]\, P(\varphi_R, \varphi_I, t)(\varphi_R + i\varphi_I)^n\,. \tag{8}$$

$P(\varphi_R, t)$ satisfies the Fokker–Planck equations (4), (5). In other words a complex probability distribution may be emulated by taking the expectation values of complex quantities [‡3].

It is thus natural to ask the following questions:

(a) Does eq. (3) have a solution for all times (with probability one)?

(b) Do the probabilities $P(\varphi_R, t)$, $P(\varphi_R, \varphi_I, t)$ (if they exist) have a limit when t goes to infinity?

(c) Do the averages $t^{-1}\int_0^t \varphi(t')^n\, dt'$ exist (in some sense, e.g. as distributions in time) for finite t and do they have a limit when t goes to infinity?

(d) Finally, does the following relation hold?

$$\lim_{t\to\infty} \frac{1}{t}\int_0^t \varphi^n(t')\, dt' = \lim_{t\to\infty} \int d\varphi_R P(\varphi_R, t)\varphi_R^n = \langle\varphi^n\rangle\,. \tag{9}$$

[‡3] The argument is formal and it is based on the implicit assumption that the moment problem (8) has a solution. If this happens the usual Ito differential calculus relation $d/dt\langle f\rangle = \langle \partial f/\partial\varphi\, \partial H/\partial\varphi + \partial^2 f/\partial\varphi^2\rangle$, $f(\varphi)$, being a generic function implies that $\int d\varphi P(\varphi, t)\,[\partial f/\partial t - \partial f/\partial\varphi\, \partial H/\partial\varphi - \partial^2 f/\partial\varphi^2] = 0$. By integration by part we obtain the Fokker–Planck equation as stated in text.

The answer to these questions is not evident, indeed something must go wrong for those complex hamiltonian such that $Z = 0$; in this case $\langle\varphi^n\rangle$ would be infinite. Moreover for a generic class of potentials there will be choices of the noise η such that the trajectory $\varphi(t)$ arrives to infinity in a finite time; we can only hope (it is likely to be true, provided that certain conditions are satisfied) that a solution of the differential equation exists with probability one. To be definite let us consider the case in which the hamiltonian $\mathcal{H}$ is an analytic function of a parameter g. The eigenvalues and the eigenvectors of $\mathcal{H}$ (which is supposed to be a positive selfadjoint operator for real non-negative g) are analytic functions of g which may be continued at complex g. A necessary condition for the existence of the limit at large time of $P(\varphi_R, t)$ is that all the eigenvalues of $\mathcal{H}$ have a non-negative real part. If we close our eyes on the possibility that $\int_0^t \varphi^n(t')\, dt'$ may be not defined because the field $\varphi(t)$ has a finite probability to be as large as we want, it is easy to see that if the lhs of eq. (9) exists, eq. (9) must be satisfied, as soon as the conditions on the eigenvalues of $\mathcal{H}$ are satisfied.

It is obvious that if we study the problem by assuming that the imaginary part of H is small and by using perturbation theory all points (a)–(d) are satisfied, so that there is nothing obvious against the conjecture that points (a)–(d) hold in the region of positive eigenvalues of $\mathcal{H}$: however it is quite likely that the properties of H for complex φ will play a relevant role in a rigorous proof.

In order to obtain some ideas on the possible relevance of thise proposal I have done some numerical experiments in the very simple case $H = g\varphi^4$. The solutions of the Langevin equation have been simulated on the computer for different values of g ($|g| = 1$); roughly speaking in the region where the real part of g were positive. I have checked that both

$$\frac{1}{t}\int_0^t dt'\, \varphi^2(t'), \quad \frac{1}{t}\int_0^t dt\, \varphi^4(t')\,, \tag{10}$$

seemed to have a limit when the time was going

to infinity, at least with a precision of a few percent; within the same precision they where equal, respectively, to:

$$\int d\varphi \exp(-g\varphi^4)\varphi^2/Z, \quad \int d\varphi \exp(-g\varphi^4)\varphi^4/Z. \tag{11}$$

For purely imaginary g the situation was rather confused and no evidence for the existence of the limit at large time was found. Quite remarkably in the region of the nonzero imaginary part but negative real part of g, the limits of (10) at large time seem to exist, however the integrals in eq. (11) are divergent in this region and the limits estimated for (10) did not coincide with the analytic continuation of the integrals in (11) from their convergence region.

We see that the Langevin equation may be a useful tool for studying a system with complex hamiltonians H in the region where the imaginary part of H is small (certainly far from the region of zeros of the partition function Z). It is unclear if this method may be used to study the Feynman path integral for nearly real times, not at imaginary times as it is done nowadays.

References

[1] K. Binder, ed., Monte Carlo methods (Springer, Berlin, 1979).
[2] See for example: I. Guikhman and A. Skorokhod, Introduction à la Théorie des Processus Aléatoires (Mir, Moscow, 1980);
G. Parisi and Wu Yong-shi, Sci. Sinica 24 (1981) 483.
[3] J.R. Klauder, Bell Lab. preprint, Lectures given XXIIth Schladming school).

Reprinted from JOURNAL OF STATISTICAL PHYSICS

Vol. 39, Nos. 1/2, April 1985
Printed in Belgium

Spectrum of Certain Non-Self-Adjoint Operators and Solutions of Langevin Equations with Complex Drift

John R. Klauder[1] **and Wesley P. Petersen**[1]

Received July 6, 1984; revised October 29, 1984

As part of a program to evaluate expectations in complex distributions by long-term averages of solutions to Langevin equations with complex dirft, a simple one-dimensional example is examined in some detail. The validity and rate of convergence of this scheme depends on the spectrum of an associated non-self-ad:oint Hamiltonian which is found numerically. In the regime where the stochastic evaluation should be accurate numerical solution of the Langevin equation shows this to be the case.

KEY WORDS: Diffusion; complex drift; non-self-adjoint Hamiltonian; Langevin equation; numerical solutions.

1. INTRODUCTION

1.1. Preliminaries

Consider the Fokker–Planck equation (forward Kolmogorov equation)[1] given by

$$\frac{\partial F(x, t)}{\partial t} = \frac{1}{2} \frac{\partial}{\partial x} \left[\frac{\partial}{\partial x} + \frac{\partial S(x)}{\partial x} \right] F(x, t) \tag{1}$$

for $t \geqslant 0$ subject to the initial condition $F(x, 0) = F_0(x)$. For $S(x)$ a smooth real function with the property that $e^{-S(x)}$ is integrable, it follows that

$$\lim_{t \to \infty} F(x, t) = Ce^{-S(x)} \tag{2}$$

[1] AT&T Bell Laboratories, Murray Hill, New Jersey 07974.

0022-4715/85/0400-0053$04.50/0

for all smooth initial distribution F_0. Here C is a constant determined so that

$$\int F(x,t)\,dx=\int F_0(x)\,dx=C\int e^{-S(x)}\,dx \tag{3}$$

The validity of (2) follows readily if we introduce

$$G(x,t)\equiv F(x,t)e^{(1/2)S(x)} \tag{4}$$

and note that

$$\begin{aligned}\frac{\partial G(x,t)}{\partial t}&=-HG(x,t)\\&=\frac{1}{2}\left(\frac{\partial}{\partial x}-\frac{1}{2}\frac{\partial S}{\partial x}\right)\left(\frac{\partial}{\partial x}+\frac{1}{2}\frac{\partial S}{\partial x}\right)G(x,t)\\&=-\left[-\frac{1}{2}\frac{\partial^2}{\partial x^2}+V(x)\right]G(x,t)\end{aligned} \tag{5}$$

where

$$V(x)\equiv\frac{1}{8}\left[\frac{\partial S(x)}{\partial x}\right]^2-\frac{1}{4}\frac{\partial^2 S(x)}{\partial x^2} \tag{6}$$

It is evident that H is a self-adjoint operator with a nonnegative spectrum. Indeed if $S(x)\geqslant \text{const}+\alpha x^2$ for some $\alpha>0$, as we shall assume, then the spectrum of H is purely discrete. Consequently,

$$\begin{aligned}G(x,t)&=\sum_{n=0}^{\infty}a_n\psi_n(x)e^{-\lambda_n t}\\&=Ce^{-(1/2)S(x)}+\sum_{n=1}^{\infty}a_n\psi_n(x)e^{-\lambda_n t}\end{aligned} \tag{7}$$

where

$$H\psi_n(x)=\lambda_n\psi_n(x) \tag{8}$$

and we have noted that $e^{-(1/2)S}$ is an eigenfunction of zero energy, $\lambda_0=0$. Since $e^{-(1/2)S}$ is nowhere vanishing it is the (nondegenerate) ground state of H, and thus $\lambda_n>0$, for all $n\geqslant 1$. Indeed we mean by this that $\lambda_n\geqslant\lambda_{\min}>0$ for all $n\geqslant 1$. As $t\to\infty$ it follows that

$$\lim_{t\to\infty}G(x,t)=Ce^{-(1/2)S(x)} \tag{9}$$

and therefore

$$\lim_{t \to \infty} F(x, t) = Ce^{-S(x)} \tag{10}$$

as was to be shown.

We note in addition that associated to every Fokker–Planck equation is a Langevin equation (stochastic differential equation)[1] given in the present case by

$$\dot{x}(t) = -\frac{1}{2} \frac{\partial S(x)}{\partial x}\bigg|_{x = x(t)} + \xi(t) \tag{11}$$

Here ξ denotes a generalized, standard Gaussian white noise process for which

$$\langle \xi(t) \rangle = 0$$
$$\langle \xi(t_1)\, \xi(t_2) \rangle = \delta(t_1 - t_2)$$

where $\langle \cdot \rangle$ denotes an average over the white noise ensemble. If the distribution of initial values of $x(0)$ for (11) is given by F_0, then it follows for any suitable function A that

$$\langle A(x(t)) \rangle = \frac{\int A(x)\, F(x, t)\, dx}{\int F(x, t)\, dx} \tag{12}$$

Thus if

$$\bar{A} \equiv \frac{\int A(x) e^{-S(x)}\, dx}{\int e^{-S(x)}\, dx} \tag{13}$$

we see that

$$\lim_{t \to \infty} \langle A(x(t)) \rangle = \bar{A} \tag{14}$$

Moreover, the convergence criterion (2) also ensures that the ensemble is ergodic in the sense that

$$\lim_{T \to \infty} \frac{1}{T} \int_0^T A(x(t))\, dt = \bar{A} \tag{15}$$

1.2. Can *S* Be Complex?

In this paper we raise the question: Provided $\int e^{-S(x)}\, dx \neq 0$ and $\int |e^{-S(x)}|\, dx < \infty$, how much of the foregoing scenario remains true if S is *complex*? That is, when S is complex is it possible that a general solution to the complex Fokker–Planck equation satisfies (2), and, as a consequence thereof, that solutions to the complex Langevin equation fulfill (14) and (15)? We are led to ask this question as it has recently arisen in the course of studying complex Langevin equations as a means to calculate statistical averages.[2,3]

When S is complex, the operator H defined by (5) and (6) is non-self-adjoint and nonnormal. There is a conspicuous absence of general spectral theorems in this case. If we assume that $V_i \equiv \text{Im}\, V(x)$ is a Kato-tiny perturbation[4] of $H_r \equiv H - iV_i$, namely, for general ψ that

$$\left(\int |V_i \psi|^2\, dx\right)^{1/2} \leqslant a \left(\int |\psi|^2 dx\right)^{1/2} + b\left(\int |H_r \psi|^2\, dx\right)^{1/2} \tag{16}$$

for some positive a and b, $b < 1$, then it follows from properties of the resolvent operator that eigenfunctions exist and are complete, and that both eigenfunctions and eigenvalues are analytic in the coefficient of V_i about zero. This desirable situation may always be arranged, at least for polynomial S, by analytic continuation in the independent variable x if need be. The solutions to the eigenvalue equation (again taken discrete),

$$H\psi_n(x) = \lambda_n \psi_n(x) \tag{17}$$

involve complex eigenvalues λ and eigenfunctions ψ, which need not be mutually orthogonal. Orthogonality in the form $\int \psi_n(x)\, \psi_m(x)\, dx = 0$ whenever $\lambda_n \neq \lambda_m$ does hold, however, in the present case since $\int \psi_n(x)\, H\psi_m(x)\, dx = \int \psi_m(x)\, H\psi_n(x)\, dx$ holds by integration by parts. By construction, $e^{-(1/2)S(x)}$ is always an eigenfunction with eigenvalue zero ($\lambda_0 = 0$). Consequently, the convergence criterion (2) will be fulfilled provided

$$\text{Re}\, \lambda_n \geqslant c > 0, \qquad n \geqslant 1 \tag{18}$$

This condition would follow immediately from the Feynman–Kac formula,[5] for example, if only

$$H_r \equiv -\frac{1}{2}\frac{\partial^2}{\partial x^2} + \text{Re}\, V(x) \tag{19}$$

was nonnegative, but this is generally not the case.

Choose S complex and assume (18) is true so that the asymptotic behavior (2) holds. Despite appearances the solutions of (11) diverge on only a set of measure zero (see the example below). Moreover, it is straightforward to show that (1) and (11) are still linked by (12). However, the left side of (12) involves cancellation among large numbers (see below) that could be a source of error in a numerical solution of (11). In our calculation we have taken a simple approach to this problem by cutting from a given sample those paths which are growing very large in absolute value. Clearly such cutting could affect the averaging procedure, but we find that if done symmetrically about the mean of x, the results are quite stable and relatively independent of the cutting size.

To illustrate our ideas and to study these general questions further, let us specialize to a "typical" example where

$$S(x) = \sigma x^2 + \tfrac{1}{2}x^4 \tag{20}$$

and σ is a complex parameter. In this case

$$H = -\frac{1}{2}\frac{\partial^2}{\partial x^2} + \frac{1}{2}\sigma^2 x^2 - \frac{3}{2}x^2 + \sigma x^4 + \frac{1}{2}x^6 - \frac{1}{2}\sigma \tag{21}$$

In the next section we discuss the numerical results for the spectrum of the operator in (21), and show that (18) holds whenever (i) $\mathrm{Re}\,\sigma > 0$, or (ii) $\mathrm{Re}\,\sigma < 0$ provided $|\mathrm{Im}\,\sigma|$ is sufficiently small. The associated Langevin equation for $t \geqslant 0$ reads in this case

$$\dot{z}(t) = -\sigma z(t) - z^3(t) + \xi(t) \tag{22}$$

subject to the real initial value $z(0) = x_0$, where x_0 has a suitable distribution. Despite appearances the solutions to this equation diverge in a finite time for only a set of measure zero. If at $t = t_0$, z is large in absolute magnitude then (22) reduces essentially to $\dot{z}(t) = -z^3(t)$ the solution of which is given for $t \geqslant t_0$ by

$$z(t) \equiv x(t) + iy(t) = \pm 1/(2t - \alpha - i\beta)^{1/2} \tag{23}$$

where α and β are real. If $\beta = 0$ and $2t_0 - \alpha > 0$, then $z(t)$ approaches the origin along the x axis, while if $2t_0 - \alpha < 0$, then $z(t)$ diverges in a finite time along the y axis. However, when σ is complex, $\beta = 0$ occurs with probability zero. When $\beta \neq 0$ the solution (23) remains *finite* while traversing the outline of one of the leaves of a four-leaf clover pattern, outward from the origin when near the y axis and then inward toward the origin when near the x axis. For $\alpha = T$ and $0 < |\beta| \ll T$ the contribution to $|T^{-1}\int_0^T z^2(t)\,dt|$ of the solution (23) is given approximately by $(\pi/2T)$, while the contribution to $T^{-1}\int_0^T |z(t)|^2\,dt$ is given approximately by $(\ln(T/|\beta|)/T$. This estimate illustrates the cancellation among large numbers that is involved in (12), and higher moments show that an ever increasing

Table I. Comparison of Long-Time Average (15) to Numerical integration of (13) for $A(x) = x^2$ [a]

	Long time SDE vs. numerical integration	
σ	$\frac{1}{T}\int_0^T x^2(t)\,dt$	$\frac{\int_{-\infty}^{\infty} x^2 e^{-S(x)}\,dx}{\int_{-\infty}^{\infty} e^{-S(x)}\,dx}$
2	(0.195 ± 0.015)	0.197
$2+(1/2)i$	$(0.194 \pm 0.013) - (0.032 \pm 0.003)i$	$0.193 - 0.032i$
$2+i$	$(0.177 \pm 0.014) - (0.061 \pm 0.007)i$	$0.179 - 0.061i$
$2+2i$	$(0.134 \pm 0.009) - (0.097 \pm 0.009)i$	$0.136 - 0.099i$
$2+3i$	$(0.091 \pm 0.007) - (0.104 \pm 0.011)i$	$0.091 - 0.108i$
$2+4i$	$(0.059 \pm 0.006) - (0.099 \pm 0.010)i$	$0.059 - 0.100i$
1	(0.285 ± 0.023)	0.290
$1+(1/2)i$	$(0.280 \pm 0.020) - (0.061 \pm 0.006)i$	$0.279 - 0.062i$
$1+i$	$(0.247 \pm 0.019) - (0.114 \pm 0.012)i$	$0.248 - 0.114i$
$1+2i$	$(0.160 \pm 0.014) - (0.167 \pm 0.018)i$	$0.160 - 0.166i$
$1+3i$	$(0.079 \pm 0.014) - (0.150 \pm 0.021)i$	$0.079 - 0.155i$
$1+4i$	$(0.044 \pm 0.013) - (0.133 \pm 0.023)i$	$0.040 - 0.126i$
1/2	(0.359 ± 0.030)	0.366
$1/2+(1/4)i$	$(0.361 \pm 0.027) - (0.046 \pm 0.005)i$	$0.362 - 0.045i$
$1/2+(1/2)i$	$(0.349 \pm 0.027) - (0.087 \pm 0.009)i$	$0.349 - 0.089i$
$1/2+(3/4)i$	$(0.320 \pm 0.028) - (0.125 \pm 0.013)i$	$0.328 - 0.129i$
$1/2+i$	$(0.297 \pm 0.030) - (0.160 \pm 0.019)i$	$0.300 - 0.163i$
$1/2+2i$	$(0.164 \pm 0.025) - (0.228 \pm 0.036)i$	$0.160 - 0.223i$
0	(0.467 ± 0.039)	0.478
$0+(1/4)i$	$(0.472 \pm 0.037) - (0.068 \pm 0.008)i$	$0.471 - 0.067i$
$0+(1/2)i$	$(0.449 \pm 0.035) - (0.129 \pm 0.016)i$	$0.451 - 0.132i$
$0+(3/4)i$	$(0.408 \pm 0.038) - (0.186 \pm 0.022)i$	$0.418 - 0.190i$
$0+i$	$(0.367 \pm 0.045) - (0.234 \pm 0.034)i$	$0.374 - 0.240i$
$0+2i$	$(0.242 \pm 0.032) - (0.407 \pm 0.071)i$	$0.150 - 0.308i$
$-1/2$	(0.627 ± 0.051)	0.645
$-1/2+(1/4)i$	$(0.638 \pm 0.054) - (0.101 \pm 0.014)i$	$0.635 - 0.101i$
$-1/2+(1/2)i$	$(0.620 \pm 0.047) - (0.197 \pm 0.028)i$	$0.605 - 0.198i$
$-1/2+(3/4)i$	$(0.588 \pm 0.057) - (0.307 \pm 0.044)i$	$0.554 - 0.288i$
$-1/2+i$	$(0.574 \pm 0.048) - (0.408 \pm 0.065)i$	$0.485 - 0.365i$
$-1/2+2i$	$(0.522 \pm 0.066) - (0.973 \pm 0.119)i$	$0.107 - 0.441i$

[a] Initial sample sizes for (15) are 64 paths and were cut down if $|x| > 10^6$. In this table $T = 100$, and statistical errors within the 64 paths showed $1/\sqrt{t}$ behavior.

degree of cancellation is required. The magnitude of the problem caused by such cancellations varies with the parameter σ which in turn has a strong influence on the distribution of β values in various solutions.

In the next section we illustrate the aproximate equality of (13) and

(15) (for a large finite time T) for the example of $A(x) = x^2$ by comparing the numerical integration[6] of (13) to the long-time average (15). The numerical solution of the stochastic differential equation (11) was computed by the method of Helfand.[7] We find agreement to a very satisfactory precision (see Table I).

2. NUMERICAL PROCEDURE AND RESULTS

We have diagonalized the eigenvalue equation $H\psi = \lambda\psi$ by computing the eigenvalues of the discretized problem

$$\sum_{j=0}^{N} H_{ij} a_j = \lambda a_i, \qquad 0 \leqslant i \leqslant N \tag{24}$$

where the Rayleigh–Ritz expansion

$$\psi = \sum_{j=0}^{N} a_j \phi_j(x) \tag{25}$$

is in eigenvectors of the harmonic oscillator. The matrix elements

$$H_{ij} = \int dx\, \phi_i(x) H \phi_j(x)$$

were computed from the ladder operators

$$a = \frac{1}{\sqrt{2}}\left(x + \frac{\partial}{\partial x}\right), \qquad a^+ = \frac{1}{\sqrt{2}}\left(x - \frac{\partial}{\partial x}\right)$$

We find that for the first few eigenvalues ($|\lambda|$ small), a few hundred terms in (25) are adequate provided $\operatorname{Re}\sigma > -1$. For $\operatorname{Re}\sigma \ll -1$, (25) becomes unreliable and N is prohibitively large.

In Fig. 1 we plot the modulus $|a_k|$ vs. k for the numerical solutions of the two lowest lying eigenstates when $\sigma = -\frac{1}{2}$ and $\sigma = -\frac{1}{2} + 5i$. For these figures $N \leqslant 50$ is clearly sufficient.

A modified version of H. R. Swarz's algorithm BANDR[8] was used to reduce $[H_{ij}]$ to tridiagonal form. After reduction to tradiagonal form, a complex version of TQL1[9] yields the eigenvalues. It is known that complex versions of BANDR can be unstable for some problems.[10] We did careful comparisons of our results with those from the LR method[11] and are quite comfortable with the CBANDR procedure. The even and odd parts of matrix $[H_{ij}]$ each taken separately have bandwidth = 7; hence, an operation count proportional to $7N^2$ is clearly an advantage over LR with an operation count $O(N^3)$.

MODULUS OF EVEN COEFFICIENTS

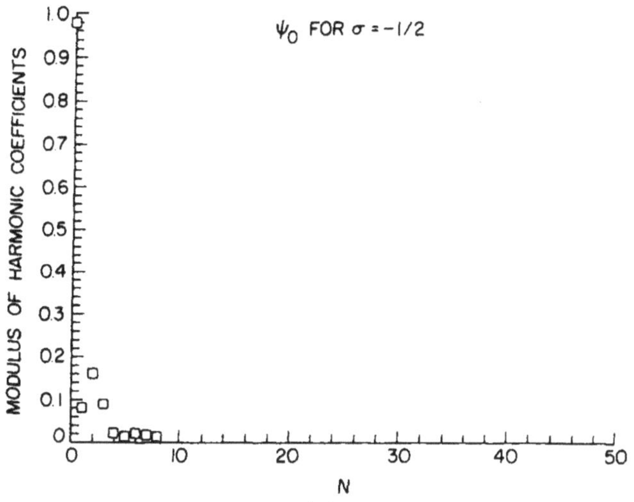

(a)

MODULUS OF ODD COEFFICIENTS

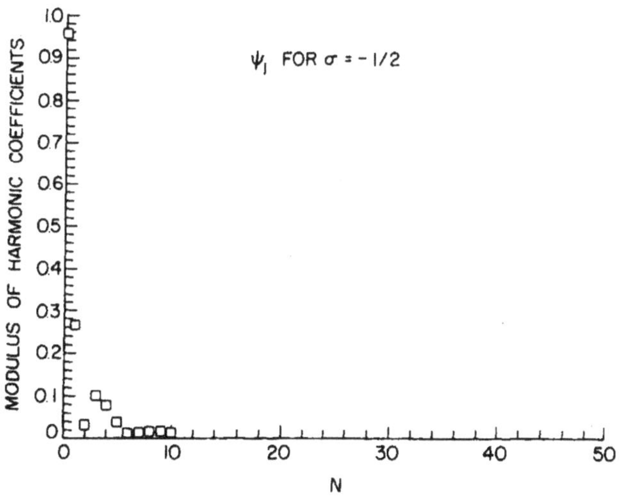

(b)

Fig. 1. Plots of modulus $|a_K|$ of harmonic oscillator coefficients in expansion (25), vs. k. Coefficients $|a_K| < 10^{-3}$ were ignored.

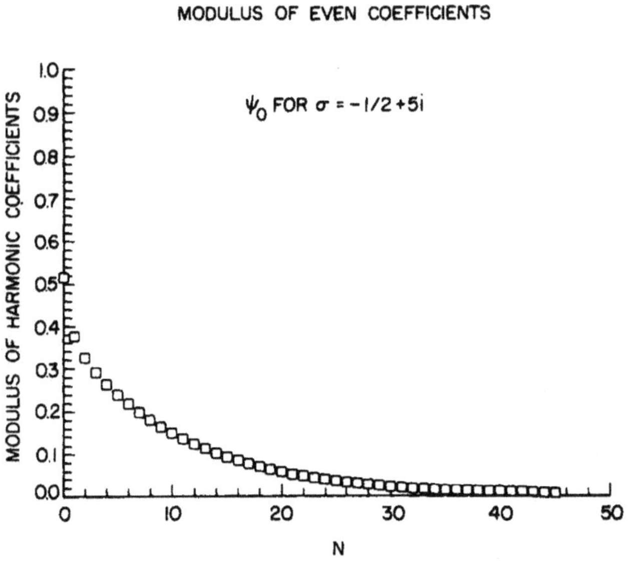

(c)

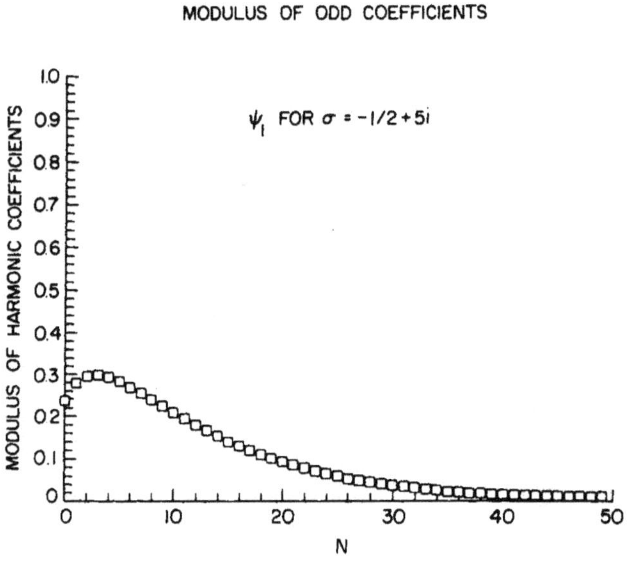

(d)

Fig. 1 (*continued*)

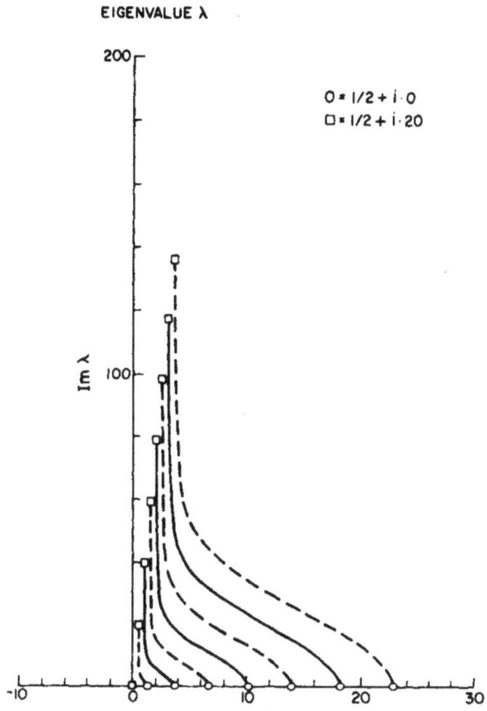

(a)

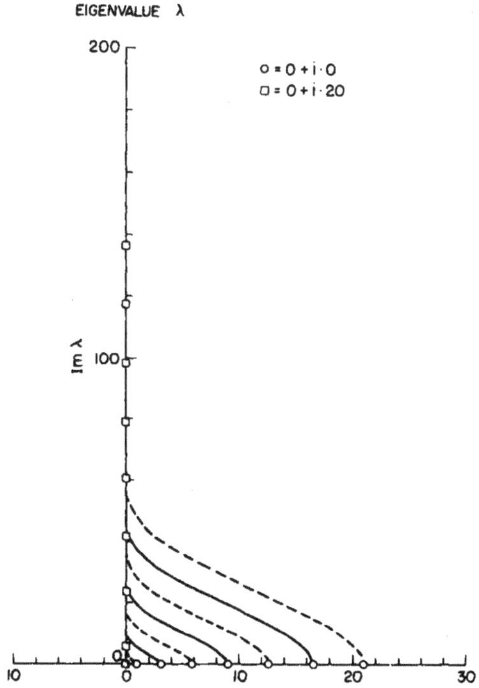

(b)

Fig. 2. Trajectories of the first eight eigenvalues of operator (23), for $0 \leqslant \mathrm{Im}\, \sigma \leqslant 20$. The even solutions are solid lines, odd solutions are dashed.

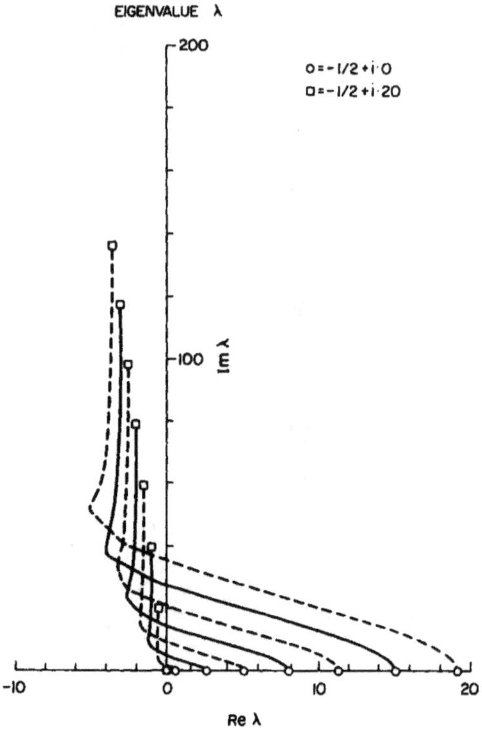

(c)

Fig. 2 (*continued*)

By scaling x in (23), $x \to \xi x$, the scaled eigenvalue equation

$$H_\xi \psi = \lambda_\xi \psi$$

can be made more stable than the original. Here $\lambda_\xi = \xi^2 \lambda$ and $H_\xi = \xi^2 H(\xi x)$. The choice $\xi^2 = 1/|\sigma|$ seemed best.

The trajectories of the first eigenvalues of (24) for $0 \leqslant \text{Im}\,\sigma \leqslant 20$ are shown in Figs. 2. Note that for large Im σ that the successive eigenvalues fall on a straight line. In the following section we show that this "complex frequency" harmonic oscillator result is not surprising.

The coefficients a_j were computed by finding the null vector of $[H_{ij}] - \lambda$. For this computation the condition number estimator CGBCO from Linpack[22] was used. We found condition numbers greater than 10^{14}.

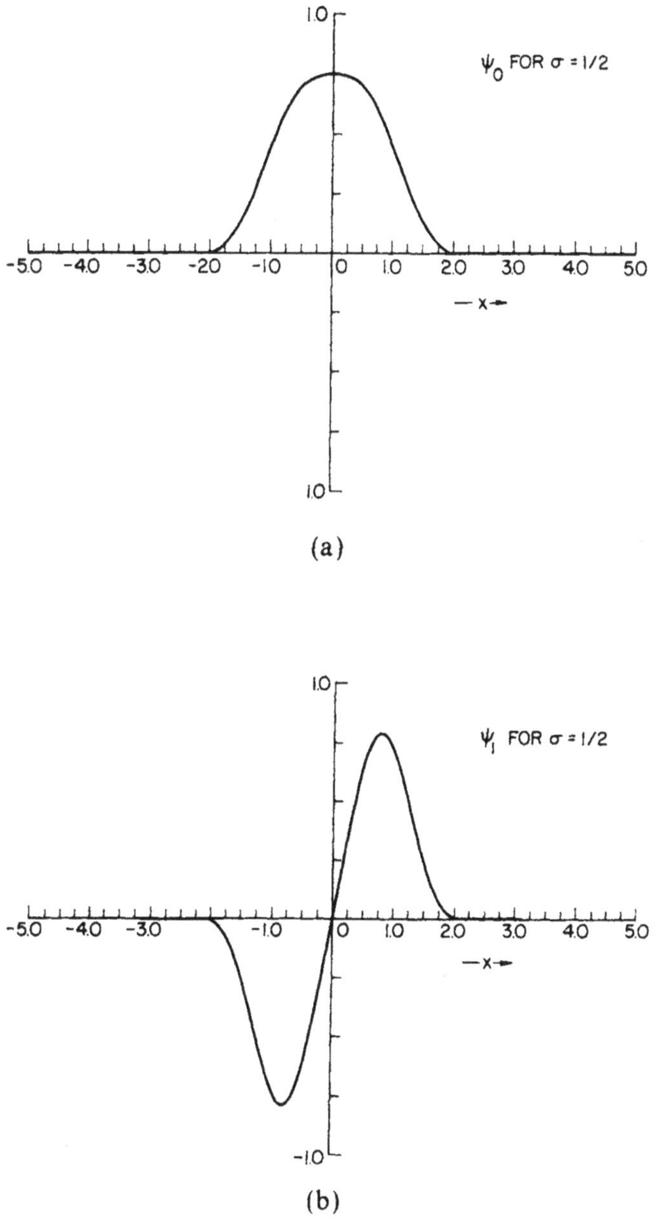

Fig. 3. Plots of the first two eigenfunctions of operator (23) in x space when Re $\sigma = \frac{1}{2}, 0, -\frac{1}{2}$. The real parts are plotted as solid lines, imaginary parts as dashed lines. The phase convention is described in the text of Section 2.

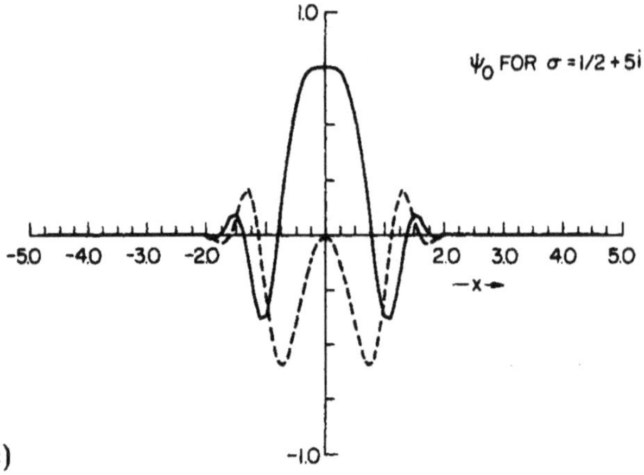

(c)

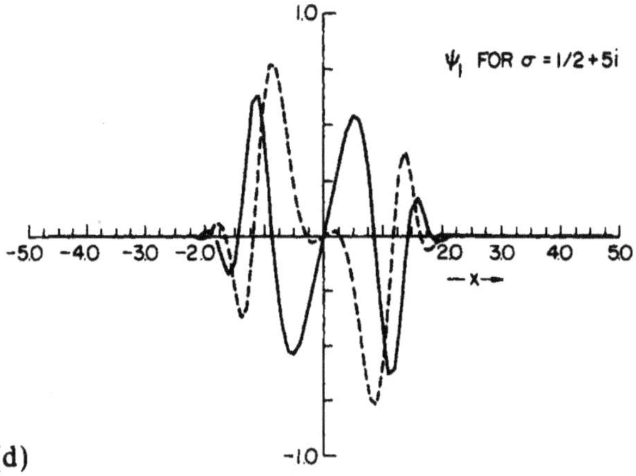

(d)

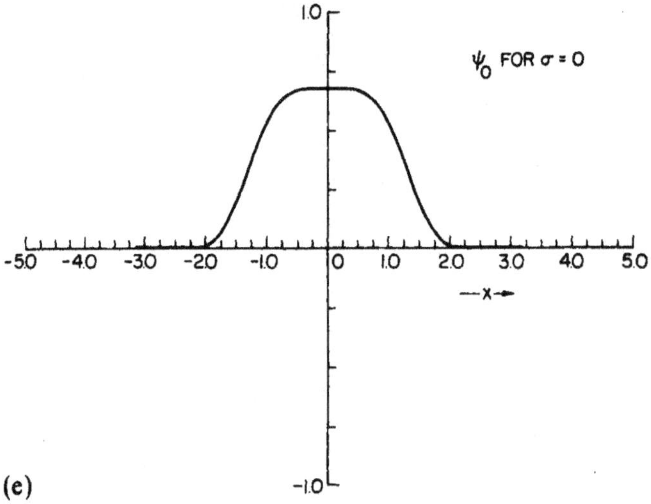

(e)

Fig. 3 (*continued*)

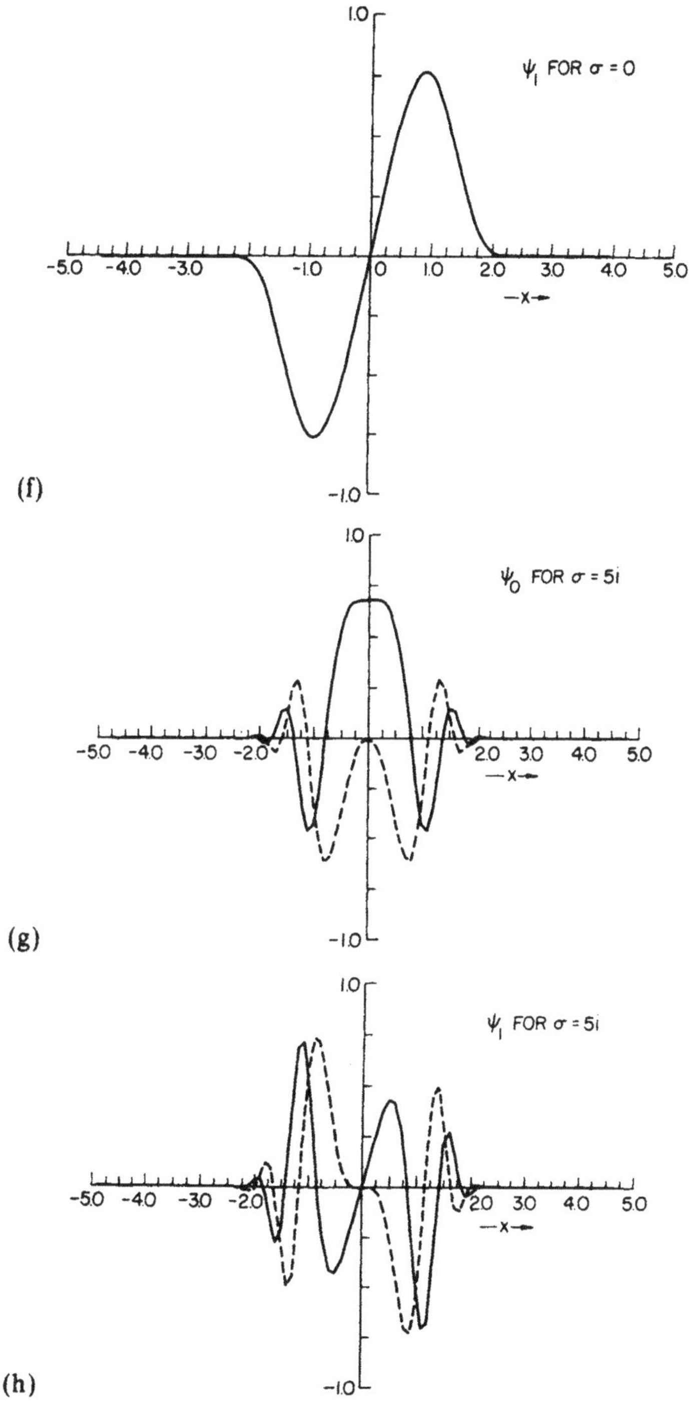

Fig. 3 (*continued*)

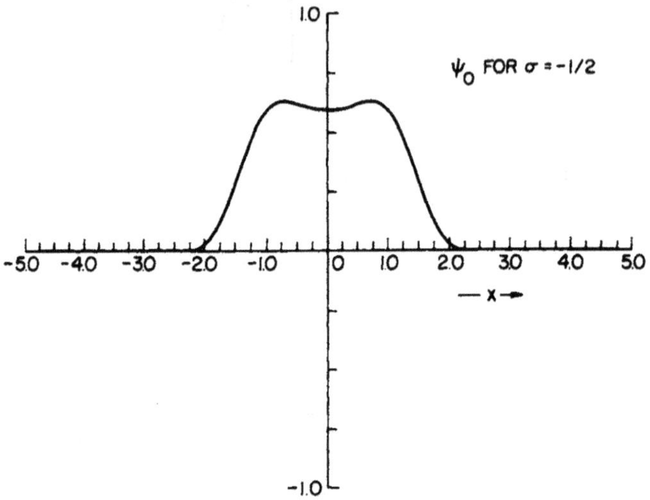

(i)

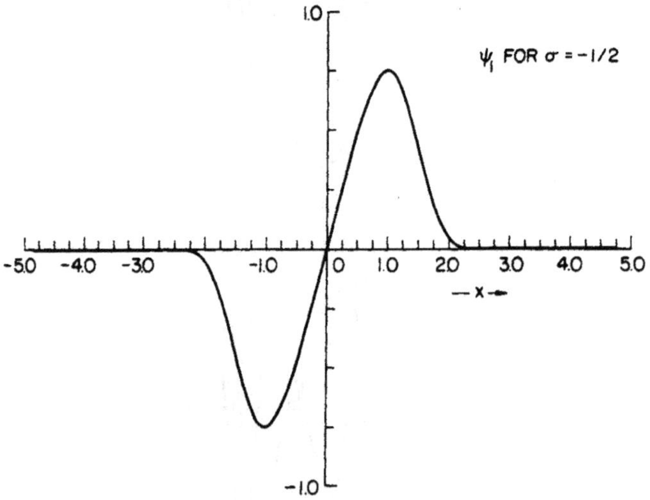

(j)

Fig. 3 (*continued*)

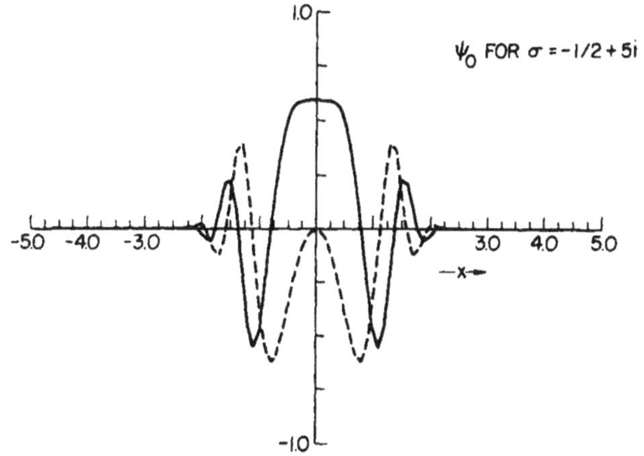

(k)

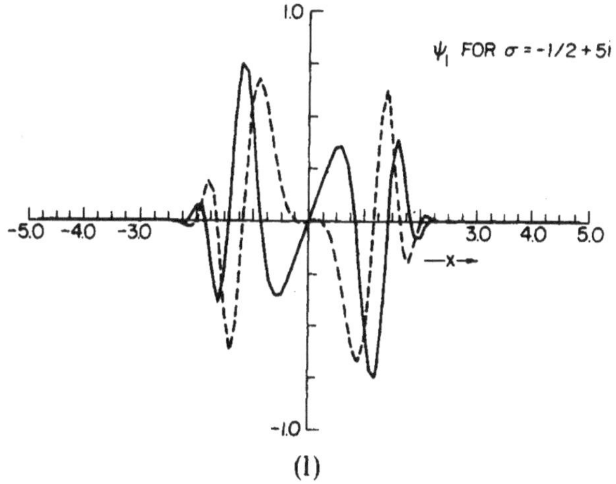

(l)

Fig. 3 (*continued*)

Furthermore, the eigenvalues from TQL1 were compared with inner products

$$\sum_{i,j} a_i^* H_{ij} a_j \Big/ \sum_i a_i^* a_i$$

and gave excellent agreement.

In Figs. 3 we illustrate the wave functions for the first two eigenvalues. Comparing 3^a, 3^e, and 3^i, note the bifurcation in the ground state at the critical point $\mathrm{Re}\,\sigma = 0$.[13] The phase of these wave functions were chosen such that $\mathrm{Im}\,\psi_0(x=0)=0$. Namely, a phase α was chosen such that

$$\mathrm{Im}(e^{i\alpha}\psi_0(x=0))=0$$

and every other ψ_λ was multiplied by this same phase factor $e^{i\alpha}$. The real parts of ψ are shown in solid lines, the imaginary parts are dashed.

2.1. Large-σ Results

In Fig. 2 we have shown numerical evidence for $\mathrm{Re}\,\sigma \geqslant 0$ that $\mathrm{Re}\,\lambda \geqslant 0$. Apparently for large $|\sigma|$ the spectrum is given by the "complex frequency" harmonic oscillator result $\lambda = n\sigma$. Equation (23) takes the form of a complex frequency harmonic oscillator when $|\sigma|$ is large. If $\mathrm{Re}\,\sigma \leqslant 0$, however, the solutions are not obviously square integrable. Our example is

$$H\psi = -\frac{1}{2}\left(\frac{\partial}{\partial x} - g\right)\left(\frac{\partial}{\partial x} + g\right)\psi = \lambda\psi \tag{26}$$

where $g=(1/2)(\partial S/\partial x)=\sigma x + x^3$. The ground state ψ_0 satisfies $(\partial_x + g)\psi_0 = 0$ and is

$$\psi_0(x) = \exp\left(\frac{-\sigma x^2}{2} - \frac{x^4}{4}\right)$$

Now, for large x, (26) is approximately

$$\frac{1}{2}\left(-\frac{\partial^2}{\partial x^2} + x^6\right)\psi \cong 0$$

Asymptotically $|\psi| \sim \exp(-x^4/4)$. We are therefore tempted to write all solutions of (26) as

$$\psi(x) = \phi(x)\,\psi_0(x)$$

with the requirement that ϕ be regular. The eigenvalue problem for ϕ is

$$-\frac{1}{2}\frac{\partial^2\phi}{\partial x^2} + (\sigma x + x^3)\frac{\partial\phi}{\partial x} = \lambda\phi \tag{27}$$

For large $|\sigma|$, noting from Figs. 2 that $\lambda(\sigma)$ is growing, (27) becomes

$$\sigma x \frac{\partial\phi}{\partial x} = \lambda\phi$$

Table II. The Ratio $|\int e^{-s}dx|/\int |e^{-s}|\,dx$ Illustrating Cancellation in the Complex Measures

σ	Contributions from complex measures $\dfrac{\lvert\int_{-\infty}^{\infty} e^{-S(x)}\,dx\rvert}{\int_{-\infty}^{\infty} \lvert e^{-S(x)}\rvert\,dx}$
2	1.0
$2+i$	0.970
$2+2i$	0.892
$2+3i$	0.803
$2+4i$	0.723
1	1.0
$1+i$	0.942
$1+2i$	0.813
$1+3i$	0.690
$1+4i$	0.600
0	1.0
$0+i$	0.880
$0+2i$	0.657
$0+3i$	0.503
$0+4i$	0.423
$-1/2$	1.0
$-1/2+i$	0.824
$-1/2+2i$	0.531
$-1/2+3i$	0.379
$-1/2+4i$	0.318

having solution $\phi \propto x^{\lambda/\sigma}$. For ϕ to be regular along the negative x axis requires $\lambda = n\sigma$, n a positive integer.

2.2. Stochastic Estimate of $\bar{A}$

In Table I we tabulate the evaluation of (13) and (15) for a finite T for the special case of $A = x^2$. The stochastic data were obtained for $T = 10^2$ and by averaging the results over as many as 64 different sample paths. Any path that diverged was thereafter dropped from the sample. For $\operatorname{Re}\sigma \geqslant \frac{1}{2}$ no more than one path diverged, for $\operatorname{Re}\sigma = 0$ as many as half the paths diverged, and for $\operatorname{Re}\sigma = -\frac{1}{2}$ only a few paths survived.

From our numerical results about the spectrum of (8), we expect that for $\operatorname{Re}\sigma > 0$ the results of (13) and (15) for a large finite T should be approximately equal. Indeed, they seem to be so. Errors in the stochastically determined results fell as $T^{-1/2}$ and showed a remarkable independence of sample size. Additional accuracy may be attained for $\mathrm{Re}\,\sigma \geqslant \frac{1}{2}$ simply by increasing T which is easily done since almost no path diverges in that case; in some examples we have used $T = 10^3$ or even 10^4 with correspondingly improved accuracy and with almost no path divergences. These data are not quoted for lack of an across-the-board comparison.

As one measure of the cancellation occurring in the complex integrands we list in Table II the ratio of complex measures given by

$$\frac{|\int dx\, e^{-S(x)}|}{\int dx\, |e^{-S(x)}|}$$

3. CONCLUSIONS

In summary, we have demonstrated that reasonably accurate numerical estimates of averages such as (13) may be computed by long-time averages involving solutions of an associated Langevin equation (11) even when the distribution (e^{-S}) is complex, provided the convergence criterion (2) is satisfied. The accuracy that may be attained with this method increases the more rapidly the convergence criterion (2) is obeyed. Although the example we have considered here is only one-dimensional $(x \in \mathbb{R})$ the method is applicable to N-dimensional examples $(x \in \mathbb{R}^N)$; for example, the model treated in Ref. 3 took N as great as 2048. Our purpose in this paper has been to give a more in-depth study of the principles of the complex Langevin method rather than deal with a complicated many-dimensional model. Finally, we observe that the property $\mathrm{Re}\,\lambda_n > 0$, $n \geqslant 1$ for $\mathrm{Re}\,\sigma > 0$—as is strongly evident from the numerical studies for the example of this paper—appears sufficiently simple in character to admit an analytic proof. Unfortunately we have been unsuccessful in that quest.

ACKNOWLEDGMENTS

The authors would like to thank L. C. Kaufman for the use of CBANDR and for discussions about using the condition number estimates for computing selected eigenvectors. We would also like to thank E. Balslev, A. Devinatz, A. Friedman, J. Jerome, L. Shepp, and L. E. Thomas for their interest in this problem.

REFERENCES

1. N. G. Van Kampen, *Stochastic Processes in Physics and Chemistry* (North-Holland, Amsterdam, 1981); P. T. Soong, *Random Differential Equations in Science and Engineering* (Academic Press, New York, 1973).
2. J. R. Klauder, in *Recent Developments in High-Energy Physics*, H. Mitter and C. B. Lang, eds. (Springer-Verlag, Vienna, 1983), p. 251.
3. J. R. Klauder, *Phys. Rev. A* **29**:2036 (1984).
4. T. Kato, *Perturbation Theory for Linear Operators* (Springer, New York, 1966).
5. B. Simon, *Functional Integrals and Quantum Physics* (Academic, New York, 1979).
6. R. Piessens, E. De Doncker-Kapenga, C. W. Uberhuber, and D. K. Kahaner, *QUADPACK—A Subroutine Package for Automatic Integration* (Springer-Verlag, Berlin, 1983), routines QAWF and QAGI.
7. E. Helfand, *Bell Syst. Tech. J.* **58**(10):2289 (1979); see also H. S. Greenside and E. Helfand, *Bell. Syst. Tech. J.* **60**(8):1927 (1981).
8. H. R. Schwarz, *Numer. Math.* **12**:231–241 (1968); J. H. Wilkinson and C. Reinsch, *Handbook for Automatic Computation, Vol. II, Linear Algebra* (Springer, New York, 1971), p. 273.
9. H. Bowdler, R. S. Martin, C. Reinsch, and J. H. Wilkinson, *Numer. Math.* **11**:293–306 (1968); J. H. Wilkinson and C. Reinsch, *Numer. Math.* **11**:227 (1968).
10. J. Cullem and R. Willoughby, *J. Comput. Phys.* **44**(2):329–358 (December 1981).
11. R. S. Martin and J. H. Wilkinson, *Numer. Math.* **12**:349–368 (1968); J. H. Wilkinson and C. Reinsch, *Numer. Math.* **12**:339 (1968); Numerical Algorithms Group (NAG version 9) library routines FO1AMF and FO2ANF.
12. J. J. Dongarra, C. B. Moler, J. R. Bunch, and G. W. Stewart, *LINPACK Users Guide* (SIAM Publication, Philadelphia, 1979).
13. E. Knobloch and K. A. Wiesenfeld, *J. Stat. Phys.* **33**(3):611 (1983).

Volume 165B, number 1,2,3 PHYSICS LETTERS 19 December 1985

NUMERICAL PROBLEMS IN APPLYING THE LANGEVIN EQUATION TO COMPLEX EFFECTIVE ACTIONS

Jan AMBJØRN and S.-K. YANG [1]

The Niels Bohr Institute, University of Copenhagen, Blegdamsvej 17, DK-2100 Copenhagen Ø, Denmark

Received 16 July 1985; revised manuscript received 20 August 1985

We analyze the Langevin equation for complex effective actions and the numerical problems associated with the equation.

1. In quantum mechanics and in field theory one often has to work with effective actions that are complex [1,2]. In these cases standard Monte Carlo methods like the Metropolis or heatbath updating algorithms cannot be used in a straightforward manner. The Langevin equation (which should of course also be included among the standard Monte Carlo techniques) admits a formal solution even when the effective action S is complex.

This was first realized by Klauder [1] and Parisi [3]. Also Gozzi investigated the problem of convergence of the complex Langevin equation [4]. In ref. [2] the complex Langevin equation was applied to the U(1)-lattice gauge theory in two dimensions. Including the static charges in the action, which then turned complex, it was possible to measure the string tension between the charges at distances which are inaccessible by conventional Monte Carlo techniques.

However, when applied to more complicated systems than U(1)-lattice systems in two dimensions, the potential very powerful method is hampered by numerical instabilities which makes extensive computer simulations of large systems quite difficult [5]. The purpose of this letter is to explain the reason for these "instabilities" (which are merely an inadequate implementation of the stochastic differential equations in discretized form on the computer) and to point to a possible way of circumventing the problem of instabilities.

[1] Supported by a Nishina Memorial Foundation fellowship.

2. The Langevin equation for a system with real euclidean action $S(x)$ is given by

$$\dot{x}(t) = -\partial S/\partial x + \eta(t), \tag{2.1}$$

where $\eta(t)$ is a gaussian random noise, appropriately normalized. Eq. (2.1) can be used to evaluate expectation values of the form

$$\bar{f}(x) \equiv N_s \int \mathrm{d}x\, f(x) \exp[-S(x)], \tag{2.2a}$$

$$N_s^{-1} \equiv \int \mathrm{d}x \exp[-S(x)]. \tag{2.2b}$$

Indeed, defining the stochastic average of $f(x(t))$ as

$$\langle f(x(t))\rangle_\eta \equiv N_\eta \int \mathcal{D}\eta(\tau) f(x_\eta(t)) \times \exp\left(-\frac{1}{4}\int_0^\infty \eta^2(\tau)\,\mathrm{d}\tau\right), \tag{2.3a}$$

$$N_\eta^{-1} \equiv \int \mathcal{D}\eta(\tau) \exp\left(-\frac{1}{4}\int_0^\infty \eta^2(\tau)\,\mathrm{d}\tau\right), \tag{2.3b}$$

where $x_\eta(t)$ is a solution to eq. (2.1) [‡1] corresponding to $\eta(t)$, it follows from standard treatments of the Langevin equation that

$$\bar{f}(x) = \lim_{t\to\infty} \langle f(x(t))\rangle_\eta. \tag{2.4}$$

Ergodicity then leads to

[‡1] We should specify the starting point $x_\eta(0) = x_0$, but omit it for notational simplicity.

0370-2693/85/$ 03.30 © Elsevier Science Publishers B.V. (North-Holland Physics Publishing Division)

$$\bar{f}(x) = \lim_{T\to\infty} \frac{1}{T}\int_0^T dt\, f(x_\eta(t)), \tag{2.5}$$

except for a class of η's of measure zero.

In the following we will discuss eq. (2.1) when $S(x)$ is *complex*. By complex actions $S(x)$ we always have in mind analytic functions where x is a *real* variable, but some parameters in $S(x)$ are complex. Further, we assume that $\exp[-S(x)]$ in eq. (2.2b) is integrable. In the following we will consider two examples in more detail:

$$S_1(x) = \tfrac{1}{2}\sigma x^2 + \tfrac{1}{4}\lambda x^4, \tag{2.6a}$$

$$\sigma = \sigma_R + i\sigma_I \tag{2.6b}$$

and

$$S_2(\theta) = -(\beta\cos\theta + i\theta), \tag{2.7}$$

where θ is an angular variable. S_1 represents the simplest non-trivial quantum mechanical action [6], while S_2 is a typical action one encounters in the context of lattice gauge theories [2].

The noise term $\eta(t)$ in eq. (2.1) is *real*, but for complex actions like S_1 and S_2 the solutions $x_\eta(t)$ are complex and we denote then $z_\eta(t) = x_\eta(t) + iy_\eta(t)$. Eq. (2.1) can now be written as

$$\dot{x}_\eta(t) = -\mathrm{Re}(\partial S(z)/\partial z)_{z=x+iy} + \eta(t), \tag{2.8a}$$

$$\dot{y}_\eta(t) = -\mathrm{Im}(\partial S(z)/\partial z)_{z=x+iy}. \tag{2.8b}$$

It is somewhat misleading to view $y_\eta(t)$ as deterministic as it couples to η through eq. (2.8a).

Eq. (2.8) is an ordinary Langevin equation in two real variables (x, y) with drift term

$$(v_x, v_y) = -(\mathrm{Re}(\partial S/\partial z), \mathrm{Im}(\partial S/\partial z))_{z=x+iy}. \tag{2.9}$$

It now follows from standard treatments of the Langevin equation [7] that the conditional probability distribution $P(x, y, t|x_0, y_0)$, defined by

$$P(x,y,t|x_0,y_0) \equiv \langle\delta(x - x_\eta(t;x_0,y_0))\,\delta(y - y_\eta(t;x_0,y_0))\rangle_\eta \tag{2.10}$$

satisfies the corresponding Fokker–Planck equation

$$\dot{P} = (\partial^2/\partial x^2)P - (\partial/\partial x)v_x P - (\partial/\partial y)v_y P, \tag{2.11}$$

and by construction we have

$$\langle f(z(t))\rangle_\eta = \int dx\, dy\, f(x + iy)\, P(x,y,t|x_0y_0). \tag{2.12}$$

For *analytic* functions, where the extension $f(x) \to f(z)$ makes sense, eq. (2.12) can be viewed as a *definition* of the stochastic average of eq. (2.1) with complex S. Corresponding to eq. (2.2) we can write

$$\langle f(z(t))\rangle_\eta = \int dx\, f(x)P(x, t|z_0). \tag{2.13}$$

$P(x, t)$ is defined from (2.12) by taking moments. For instance, taking $f_\theta(x)$ in (2.12) as $e^{i\theta x}$ leads to

$$P(x, t|z_0) = \frac{1}{2\pi}\int_{-\infty}^{\infty} d\theta\, e^{-ix\theta}\langle e^{i\theta z(t)}\rangle_\eta.$$

Again it follows from the standard arguments [7], by applying the Langevin equation on the LHS of eq. (2.13), that

$$0 = \int dx\, f(x)\{\dot{P}(x,t) - (\partial^2/\partial x^2)P(x,t) - (\partial/\partial x)[(\partial S/\partial x)P(x,t)]\}. \tag{2.14}$$

We recognize the Fokker–Planck equation *but with S complex*. The corresponding Fokker–Planck hamiltonian is

$$H_{FP} = (p_x + \tfrac{1}{2}i\, dS/dx)(p_x - \tfrac{1}{2}i\, dS/dx) = Q_2 Q_1, \tag{2.15a}$$

$$p_x = (1/i)\, d/dx. \tag{2.15b}$$

It is no longer an hermitian and quasi positive operator as $Q_2 \neq Q_1^\dagger$ because $S(x)$ is complex. However, by analyticity, the eigenfunctions still form a complete (but not orthogonal) set, and the real part of the eigenvalues are $\geqslant 0$ if the imaginary part of the action is sufficiently small [6,8]. This means that

$$P(x, t) \underset{t\to\infty}{\longrightarrow} e^{-S(x)}, \tag{2.16}$$

as desired. Further eq. (2.16) means that (2.10) converges to an equilibrium distribution $P_{eq}(x, y)$ and corresponding to the complex "probability" distribution $e^{-S(x)}$ is an ordinary probability distribution $P_{eq}(x,y)$ such that

$$N_S \int dx\, f(x)\, e^{-S(x)} = \int dx\, dy\, f(x+iy) P_{eq}(x,y) \tag{2.17}$$

(N_S defined by (2.2b)) for all analytic functions where the LHS is well defined.

3. The complex Langevin equation (2.1) or (2.8) using actions S_1 or S_2 have been carefully studied numerically [6,2] and eq. (2.17) verified for a considerable range of coupling constants σ and β. However, especially when $\sigma_R < 0$ there were numerical stability problems. For many choices of η, $|z_\eta(t)| \to \infty$ at finite times and had to be discarded in the average (2.3). The same phenomena have now been observed when the method is applied to multidimensional lattice gauge systems with static quarks included in the action [5].

The explanation is quite simple [‡2]. When solving eq. (2.8) with complex S the drift term (2.9) will no longer confine $z_\eta(t)$ to move in the neighbourhood of the classical vacuum. The action $S(z)_{z=x+iy}$ is no longer bounded from below when (x,y) are allowed to explore the whole complex plan and there exist solutions to the deterministic equation ($\eta = 0$)

$$\dot{z}(t) = -\partial S/\partial z, \tag{3.1}$$

where $|z(t)| \to \infty$ for finite t. Eq. (3.1) can be solved to give

$$t = -\int_{z_0}^{z} dz/(\partial S/\partial z). \tag{3.2}$$

For the actions S_1 and S_2 given by eqs. (2.6) and (2.7), (3.2) is easily inverted. In the case of S_1 we have

$$z(t) = z_0/[e^{2\sigma t} + (\lambda z_0^2/\sigma)(e^{2\sigma t} - 1)]^{1/2},$$
$$t \geqslant 0, \tag{3.3}$$

and it is seen that for $\sigma_R > 0$, $z = 0$ is a point of attraction for all points z_0 in the complex plane except the ones on the curves starting at the two points of repulsion (the "Higgs vacua") $\pm(-\sigma/\lambda)^{1/2}$ and given by

$$z_0(\tau) = \pm[(-\sigma/\lambda)/(1 - e^{-2\sigma/\tau})]^{1/2}. \tag{3.4}$$

For starting points $z_0 = z_0(\tau)$ in (3.3) $|z(t)| \to \infty$ for $t \to 1/\tau$. For $\sigma_R < 0$ the role of $z = 0$ and the "Higgs vacua" $\pm(-\sigma/\lambda)^{1/2}$ are interchanged. $z = 0$ is now the point of instability. The flow patterns in the case of $\sigma_R > 0$ and $\sigma_R < 0$ are shown in figs. 1a and 1b.

In the case of S_2 a similar analysis can be performed and the qualitative features are the same. We confine ourselves to show the flow pattern (fig. 2).

If we return to the stochastic equation (2.8) the "runaway" solutions corresponding to $z_0(\tau)$ in (3.4) pose no problem in principle. If $z_\eta(t)$ gets close to $z_0(\tau)$ the drift term (2.9) is large and with the wrong sign and $z_\eta(t)$ will wander a long way from the stability point(s). However, the time spend there is small exactly because the drift term $\partial S/\partial z$ is large as is seen from (3.2) and the contribution to the time average (2.5) not *necessarily* important.

From a numerical point of view there is a problem with these long journeys where the drift term is large and of wrong sign. In order to keep control over the discretized version of (2.8) the step length Δt has to be small when $|\partial S/\partial z|$ is large. If the step size Δt is not taken small in these regions one is no longer solving (2.8) but (for S_1 in a first-order implementation) iterating

$$z_{n+1} \approx -\lambda z_n^3 \Delta t, \tag{3.5}$$

and $|z_n| \to \infty$. Therefore the computer time needed in order to get good averages can be very large. It is instructive to consider the action S_1 with $\sigma_R > 0$ and $\sigma_R < 0$. Two solutions $z_\eta(t)$ to (2.8) are shown in figs. 3a and 3b. The computer time used in the two choices of σ corresponding to fig. 3 in order to get the same precision for $\langle x^2 \rangle$ differs by a factor of 50 [‡3]. The difference between the two cases can be understood from the flow patterns shown in fig. 1 if we remember that the noise term $\eta(t)$ is real and therefore only kicks $z_\eta(t)$ in the x-direction. If $\sigma_R < 0$ the stability points are $\pm(-\sigma/\lambda)^{1/2}$, but $z_\eta(t)$ is easily kicked into regions of instability, while for $\sigma_R > 0$ $Z = 0$ is surrounded by a large area of attraction and the iteration of (2.8) can proceed for long periods with relative large Δt.

As described in ref. [2] the method of complex Langevin equations seemed very promising and power-

[‡2] Similar observations are presented in ref. [6]. When writing this article we only were aware of the preprint version. We thank the referee for drawing our attention to the published version.

[‡3] We have used both ordinary first order and second order (in Δt) Langevin equations and a modified first order equation where Δt can vary such that $\Delta x < \epsilon$ is always satisfied.

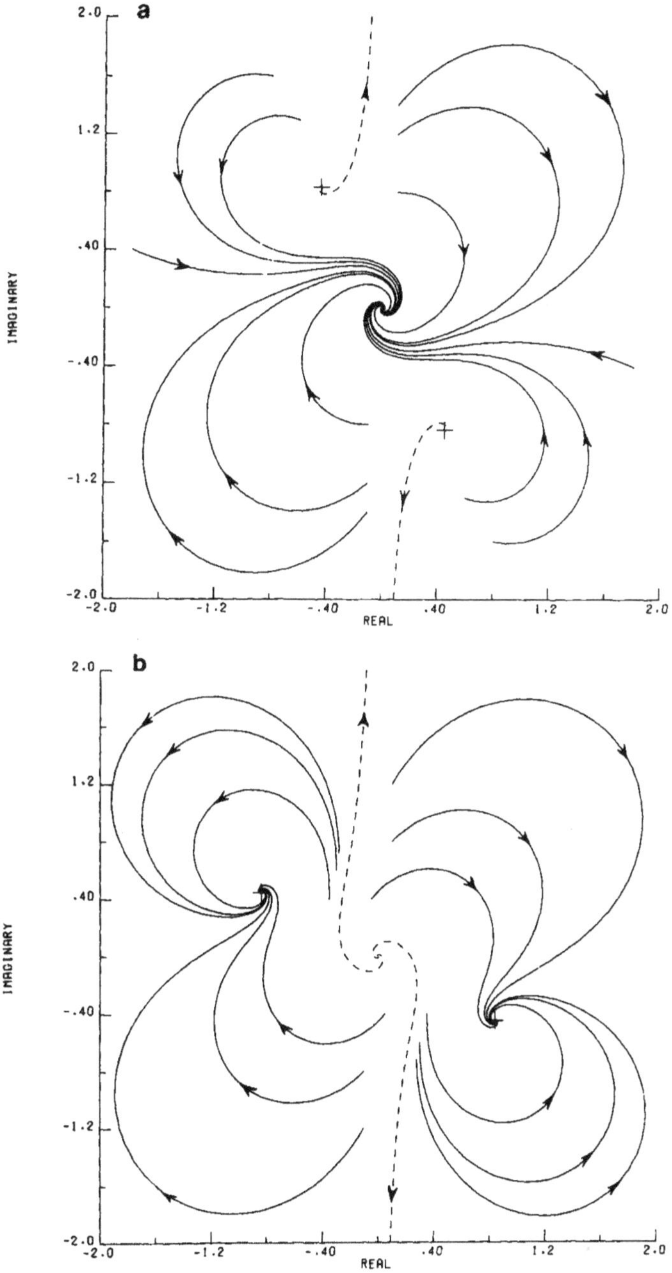

Fig. 1. (a) Flow pattern for eq. (3.1) with action $S_1 = \sigma x^2 + \frac{1}{2}x^4$, $\sigma = 0.5 + 0.75i$. (b) The same as (a), but with $\sigma = -0.5 + 0.75$ i.

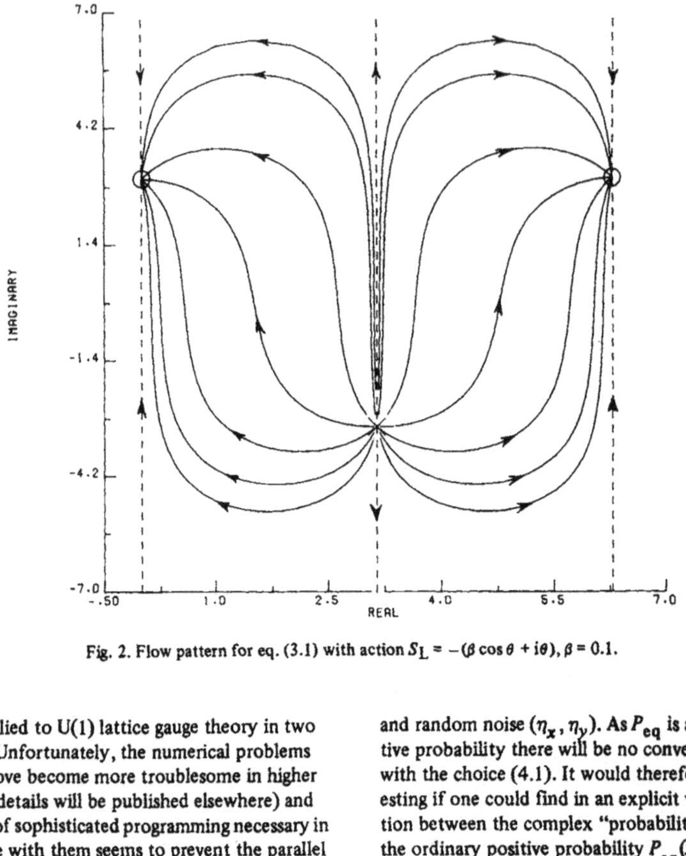

Fig. 2. Flow pattern for eq. (3.1) with action $S_L = -(\beta \cos\theta + i\theta)$, $\beta = 0.1$.

ful when applied to U(1) lattice gauge theory in two dimensions. Unfortunately, the numerical problems described above become more troublesome in higher dimensions (details will be published elsewhere) and the amount of sophisticated programming necessary in order to cope with them seems to prevent the parallel processing on fast computers necessary for large lattices.

4. As we have discussed it may not always be useful from a numerical point of view to use eq. (2.8) to generate the probability distribution $P_{eq}(x,y)$ in (2.17) because of the pathologies connected with the drift term $\partial S/\partial z$. Once we know $P_{eq}(x,y)$, however, we are free to use any standard method to generate configurations. One could use Metropolis, or even the Langevin equation, but with drift term

$$(v_x, v_y) = ((\partial/\partial x)\log P_{eq}, (\partial/\partial y)\log P_{eq}), \qquad (4.1)$$

and random noise (η_x, η_y). As P_{eq} is an ordinary positive probability there will be no convergence problem with the choice (4.1). It would therefore be very interesting if one could find in an explicit way the connection between the complex "probability" $e^{-S(x)}$ and the ordinary positive probability $P_{eq}(x,y)$ given by (2.17).

In the case of the gaussian action $S_1(\lambda = 0)$:

$$\tilde{S}_1 = \tfrac{1}{2}\sigma x^2, \quad \sigma = \sigma_R + i\sigma_I \qquad (4.2)$$

it can be done because the stationary Fokker–Planck equation (2.11) ($\dot{P} = 0$) can be solved. We get

$$P_{eq}(x,y) = N \times \exp\{-\sigma_R [x^2 + (1 + 2\sigma_R^2/\sigma_I^2)y^2 + 2(\sigma_R/\sigma_I)xy]\}. \qquad (4.3)$$

It is now an easy exercise to prove (2.17). At this point it should be noted that (2.17) alone does not

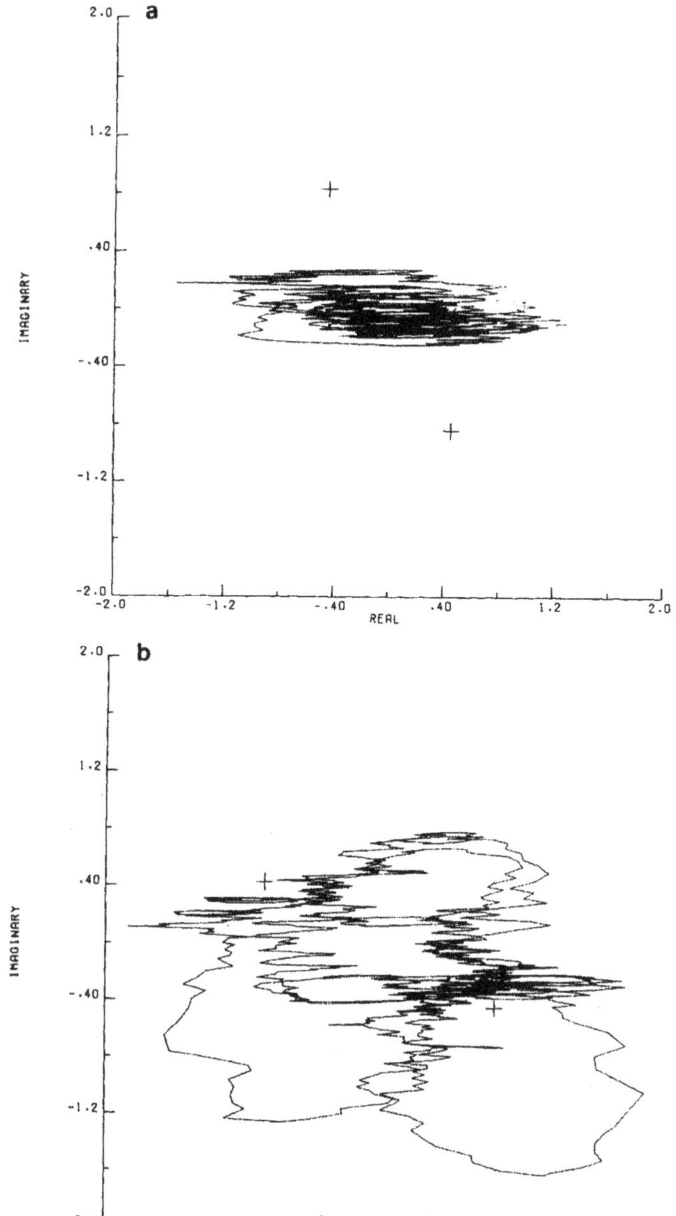

Fig. 3. (a) A solution $z_\eta(t)$ to eq. (2.8) with action $S_1 = \sigma x^2 + \frac{1}{2}x^4$, $\sigma = 0.5 + 0.75i$. (b) A solution $z_\eta(t)$, when $\sigma = -0.5 + 0.75i$. Both solutions are generated using a first-order implementation of the Langevin equation, a step size 0.01 and a total of 1000 steps.

determine $P_{eq}(x,y)$ uniquely in any way. In the present case, for instance, there is a one parameter family of gaussian solutions like (4.3) such that (2.17) is satisfied. Of course $P_{eq}(x,y)$ is unique if we require it is the stationary solution to the Fokker–Planck equation [9] (2.11) as done for eq. (4.3).

The explicit construction (4.3) of P_{eq} can of course be generalized to multidimensional cases and gaussian field theory without problems. Also in the large N limit of various vector models is the construction (4.3) valid. The only point where the nonlinearity from a term like $\lambda[\Sigma_{i=1}^{N} x_i^2]^2$ enters, is in a gap equation for $\sigma = \sigma_R + i\sigma_I$ [10]. The gap equation can be determined by the methods of, for instance, ref. [10] [‡4]

We have not yet been able to deduce the general rules of going from $e^{-S(x)}$ to $P_{eq}(x,y)$. Apart from being of general interest in the context of representing nonpositive kernels by positive ones, these rules would be extremely useful in lattice gauge theories. Knowing $P_{eq}(x,y)$ would allow us to simulate a number of problems which have so far been almost inaccessible by ordinary Monte Carlo methods. Let us only mention simulations where the effective actions contain topological terms like θ-vacua terms and Wess–Zumino terms.

‡4 We thank J. Alfaro for very useful discussion of the work in ref. [10].

It is a pleasure to thank J. Alfaro and E. Gozzi for many stimulating discussions and B. Nilsson for generous help when preparing our computer programs.

References

[1] J.R. Klauder, Phys. Rev. A29 (1984) 2036;
G. Bhanot, E. Rabinovici, N. Seiberg and P. Woit, Nucl. Phys. B230 [FS10] (1984) 291;
G. Bhanot, R. Dashen, N. Seiberg and H. Levine, Phys. Rev. Lett. 53 (1984) 519.

[2] J. Ambjorn, M. Flensburg and C. Petersen, Phys. Lett. 159B (1985) 335.

[3] G. Parisi, Phys. Lett. 131B (1983) 393.

[4] E. Gozzi, Phys. Lett. 150B (1985) 119.

[5] J. Ambjørn, M. Flensburg and C. Peterson, to be published.

[6] J.R. Klauder and W.P. Petersen, J. Stat. Phys. 39 (1985) 53.

[7] F. Langouche, D. Roekaerts and E. Tirapegui, Prog. Theor. Phys. 61 (1979) 1617.

[8] T. Kato, Perturbation theory for linear operators (Springer, Berlin, 1966).

[9] A.H. Gray Jr., J. Math. Phys. 6 (1965) 644.

[10] J. Alfaro, On the stochastic approach to the large N limit, LPTENS 85/8.

VOLUME 55, NUMBER 21 PHYSICAL REVIEW LETTERS 18 NOVEMBER 1985

Complex Langevin Simulation of the SU(3) Spin Model with Nonzero Chemical Potential

F. Karsch and H. W. Wyld
Department of Physics, University of Illinois at Urbana-Champaign, Urbana, Illinois 61801
(Received 20 September 1985)

We study the effective three-dimensional SU(3) spin model with nonzero chemical potential. This model describes the strong-coupling, large-fermion-mass limit of QCD at finite temperature and baryon density. The results obtained with a complex Langevin algorithm are encouraging for a future simulation of QCD with nonzero chemical potential. The Langevin simulations converge even for large values of the chemical potential, and excellent agreement with exact solutions in the extreme strong-coupling ($\beta=0$) limit has been found.

PACS numbers: 11.15.Ex, 05.50.+q

Only recently has it been possible to study the thermodynamics of non-Abelian gauge theories in the presence of dynamical fermions in large-scale Monte Carlo simulations.[1] These calculations gave first indications on the phase structure of QCD at finite temperature,[1-3] the mass dependence of the deconfinement and chiral transitions,[2] and the dependence on the number of flavors.[3] For a quantitative understanding of the equation of state of strongly interacting matter, it would be of considerable interest to analyze the phase diagram of QCD in the whole temperature-chemical-potential (T-μ) plane. However, although the formalism for dealing with finite chemical potentials in lattice gauge theories has been developed,[4] standard numerical simulation techniques are not applicable at $\mu\neq 0$. The reason is that for $\mu\neq 0$ the fermion determinant is complex and thus leads to a complex Euclidean action. [Here and in the following we are concerned with a SU(3) theory. In the case of SU(2) gauge theory the fermion determinant is real and exploratory simulations at nonzero μ have been performed.[5]] The integrand of the Euclidean path integral defining the partition function thus does not have an immediate probability interpretation.

It has been noticed by Parisi[6] and Klauder[7] that, while standard Monte Carlo techniques fail in the case of complex actions, it is still possible to write down a Langevin equation. In the case of a real action, time averages computed from the Langevin equation converge to equilibrium averages computed from the path integral. However, not much is known about the convergence properties of the Langevin equation for complex actions and the conditions under which it will yield results equivalent to those computed from the equilibrium distribution of the Euclidean path integral. Nonetheless, this approach has recently attracted much attention[8-11] and has been tested in the case of some simple models with complex actions which could be compared with exact solutions. These results are encouraging and show that at least for "not too large" contributions of the imaginary part in the action the Langevin equation will still resemble the distribution of the Euclidean path integral.

In the present Letter we will study the suitability of the complex Langevin algorithm for a simulation of QCD at finite chemical potential. As a preliminary step in this direction, we study the effective three-dimensional SU(3) spin model,[12-14] which approximates the full SU(3) gauge theory with fermions in the strong-coupling, large-fermion-mass limit.

The partition function is given by

$$Z=\int \prod_x dU_x\, e^{-S}, \tag{1}$$

with action

$$S=-\beta \sum_{x,l}(\mathrm{Tr}U_x \mathrm{Tr}U_{x+l}^\dagger+\mathrm{Tr}U_x^\dagger \mathrm{Tr}U_{x+l})-h\sum_x(e^{\mu}\mathrm{Tr}U_x+e^{-\mu}\mathrm{Tr}U_x^\dagger). \tag{2}$$

Here $U_x\in SU(3)$, μ is the chemical potential, and β and h are effective couplings related to the original parameters g^2, T, and m (fermion mass) of the full gauge theory.

At $\mu=0$ this model has been analyzed by use of mean-field techniques, and the phase structure in the temperature-mass plane has been studied.[12-14] It has been found that the model has a line of first-order phase transitions in the (β,h) plane starting at (0.13,0) and ending in a second-order phase transition at $(\beta_c,h_c)=(0.12,0.059)$.[14] (In the heavy-fermion-mass limit the external-field parameter h can be related to the fermion mass, $h\sim e^{-m}$, while β is related to the bare coupling g^2 and T, $\beta\sim T$.) For $\mu\neq 0$ the action, Eq. (2), is complex. We may separate the imaginary part in the action by introducing the new couplings

$$\hat h = h\cosh\mu, \quad \hat g = h\sinh\mu. \tag{3}$$

The action then reads

$$S=-\beta\sum_{x,l}(\mathrm{Tr}U_x\mathrm{Tr}U_{x+l}^\dagger+\mathrm{Tr}U_x^\dagger\mathrm{Tr}U_{x+l})-2\hat h\sum_x \mathrm{Re\,Tr}U_x-2i\hat g\sum_x \mathrm{Im\,Tr}U_x. \tag{4}$$

TABLE I. Comparison of results for $\langle \mathrm{Tr}U \rangle$ and $\langle \mathrm{Tr}U^{-1} \rangle$ at $\beta = 0$ obtained from a Langevin simulation with a discrete time step $\epsilon = 0.005$ with exact results.

		$\langle \mathrm{Tr}U \rangle$		$\langle \mathrm{Tr}U^{-1} \rangle$	
h	μ	Langevin	Exact	Langevin	Exact
0.1	0.0	0.1089(24)	0.1050	0.1089(24)	0.1050
	0.5	0.0907(14)	0.0742	0.1815(10)	0.1667
	1.0	0.0763(50)	0.0735	0.2736(43)	0.2723
	1.5	0.1297(45)	0.1217	0.4515(39)	0.4467
	2.0	0.2807(72)	0.2769	0.7268(32)	0.7271
	2.5	0.6533(93)	0.6548	1.1331(56)	1.1404
1.0	0.0	1.2682(38)	1.2676	1.2682(38)	1.2676
	0.2	1.2633(41)	1.2578	1.3441(34)	1.3401
	0.4	1.3226(47)	1.3162	1.4645(35)	1.4613
	0.6	1.4401(28)	1.4364	1.6120(21)	1.6119
	0.8	1.5989(31)	1.5986	1.7695(25)	1.7725
	1.0	1.7774(22)	1.7771	1.9264(10)	1.9287

The phase diagram at $\mu \neq 0$ has been analyzed in the mean-field approach for the case where the fields U_x are restricted to the Z(3) center of the SU(3) group.[15] There it has been found that the location of the phase transition is only very little influenced by the value of $\hat{g}$. The influence of the imaginary part thus seems to be weak, at least as far as the location of the phase transition is concerned.

The simulation of the SU(3) spin model is simplified by noting that the U's can be diagonalized simultaneously for all sites of the lattice. The eigenvalues are given by $\exp(i\theta_l)$, $l = 1, 2, 3$, with $\theta_3 = -(\theta_1 + \theta_2)$. The partition function then reads

$$Z = \int \prod_x d\theta_{1x}\, d\theta_{2x}\, e^{-S_{\mathrm{eff}}}, \tag{5}$$

where the effective action S_{eff} now contains an additional contribution from the Haar measure:

$$S_{\mathrm{eff}} = S(\theta_{1x}, \theta_{2x}) - \sum_x \ln\left[\sin^2\left(\frac{\theta_{1x} - \theta_{2x}}{2}\right)\sin^2\left(\frac{2\theta_{1x} + \theta_{2x}}{2}\right)\sin^2\left(\frac{\theta_{1x} + 2\theta_{2x}}{2}\right)\right]. \tag{6}$$

The Langevin equation for the effective action is then given by

$$\frac{d}{dt}\theta_{i,x} = -\frac{\partial S_{\mathrm{eff}}}{\partial \theta_{i,x}} + \eta_{i,x}(t), \tag{7}$$

with η being a random-Gaussian-noise term,

$$\langle \eta_{i,x}(t) \rangle = 0,$$
$$\langle \eta_{i,x}(t)\eta_{j,y}(t') \rangle = 2\delta(t - t')\delta_{x,y}\delta_{i,j}. \tag{8}$$

In the case of a complex action the phases θ_i will not stay real. The Langevin equation describes their time evolution in the complex plane. This corresponds to an analytic continuation of the original action, where $U^\dagger$ has been replaced by U^{-1}. Notice that from the original SU(3) variables the property $\det U = 1$ is preserved in this analytic continuation.

To simulate Eq. (7), we discretize the time derivative using a second-order Runge-Kutta scheme. This introduces systematic errors which are $O(\epsilon^3)$, ϵ being the discrete time step. To judge the reliability of the complex Langevin simulation, we first studied the limit $\beta = 0$ of the SU(3) spin model. In this case the partition function factorizes, and we are left with a single-site problem which is just the well-known SU(3) one-link integral.[16]

We thus can compare the results obtained in the simulation with an exact solution. In Table I we show results obtained for $\langle \mathrm{Tr}U \rangle$ and $\langle \mathrm{Tr}U^{-1} \rangle$ for two different values of h and various chemical potentials. The Langevin averages are based on simulations with 1.5×10^6 iterations. We used a discrete time step $\epsilon = 0.005$. Obviously the expectation values agree well with the exact results, although they are systematically larger, which is probably because of the finite discrete time step used to integrate the Langevin equation. Even for the largest values of μ, where $\hat{g}/\hat{h} \approx 1$, we did not observe any instabilities. Thus even when the coupling strength of the complex part is comparable with that of the real part in the action the Langevin equation gives reliable results.

For $\beta \neq 0$ the results of the simulation cannot be compared with exact solutions. However, from the mean-field analysis and the limit $\beta = 0$ we expect that the results will differ only little from those of a con-

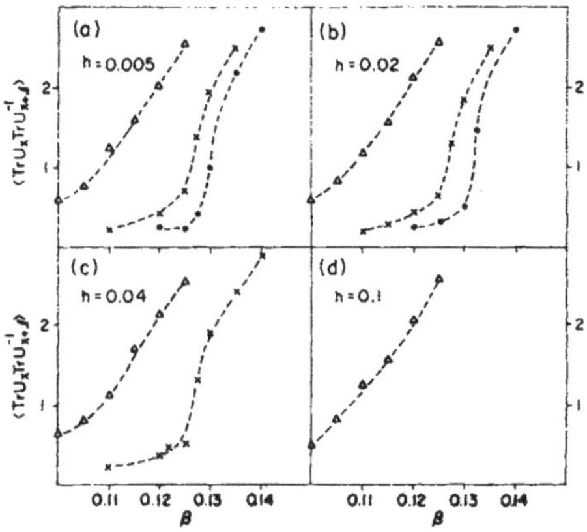

FIG. 1. Expectation value of the nearest-neighbor spin-spin correlation at nonzero μ for $h=$ (a) 0.005, (b) 0.02, (c) 0.04, and (d) 0.1. The symbols shown for the data correspond to fixed values of $\hat{h}=0.025$ (circles), 0.05 (crosses), and 0.1365 (triangles). Curves are drawn to guide the eye. The data shown have been obtained on a 10^3 lattice averaging over 3000–6000 iterations.

ventional simulation with the same value of $\hat{h}$ but $\hat{g}=0$. We thus selected values of h and μ such that $\hat{h}$ remains constant and compare these results with a conventional simulation with $\hat{g}=0$, i.e., a real action. The results of the complex Langevin simulation are shown in Fig. 1, and those of a Langevin simulation with real action are shown in Fig. 2. Clearly for all three values of $\hat{h}$ (0.025, 0.05, and 0.1365), graphs with the same value of $\hat{h}$ agree within errors, and the change due to different values of $\hat{g}$ is indeed small. They also agree well with the simulation of the "real action." The results shown in Figs. 1 and 2 have been obtained on a 10^3 lattice with 3000–6000 iterations per

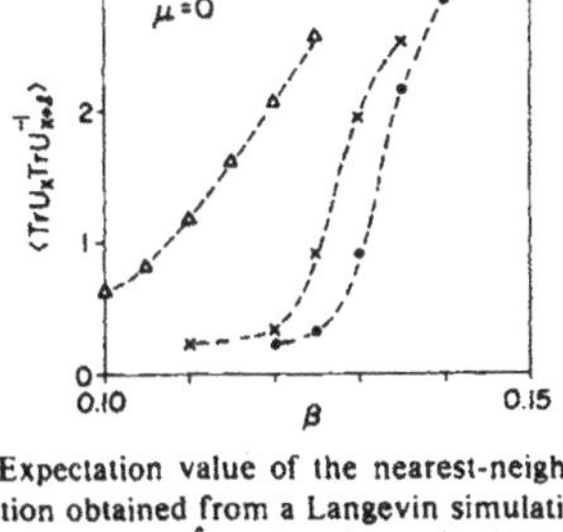

FIG. 2. Expectation value of the nearest-neighbor spin-spin correlation obtained from a Langevin simulation with a real action ($\hat{g}=0$) for $\hat{h}=0.025$ (circles), 0.05 (crosses), and 0.1365 (triangles) on a 10^3 lattice. Curves are drawn to guide the eye.

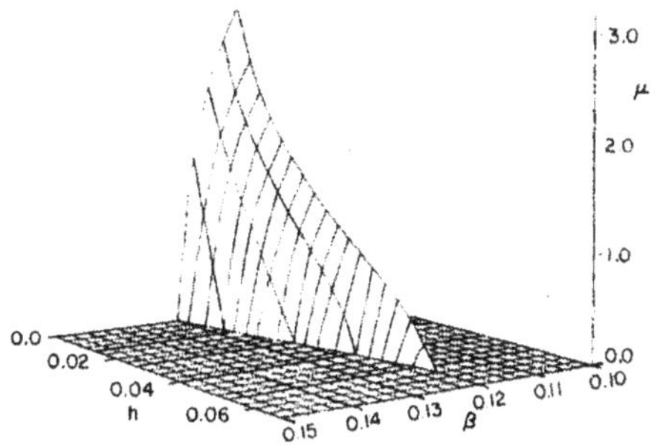

FIG. 3. Phase diagram of the SU(3) spin model at nonzero μ in the (β, h, μ) coupling space.

data point. Although we did not attempt to establish the existence of a first-order transition and the critical parameters for a second-order end point, our data are certainly consistent with the mean-field prediction. [As the transition in the SU(3) spin model, as in the Z(3) spin model, is only weakly first order, much larger lattices and better statistics would be necessary to determine the exact location of the transition.] The small deviations of the $\mu\neq 0$ simulation from those at $\mu=0$ with the same value of $\hat{h}$ indicate that the critical surface in the (β, h, μ) coupling space is well described by

$$(\beta_c, h_c, \mu_c)=(\beta_c, \hat{h}_c, 0), \tag{9}$$

where $\hat{h}_c$ is the critical coupling at $\mu=0$ and is related to h_c and μ_c through

$$\hat{h}_c=h_c \cosh\mu_c. \tag{10}$$

This phase diagram is shown in Fig. 3.

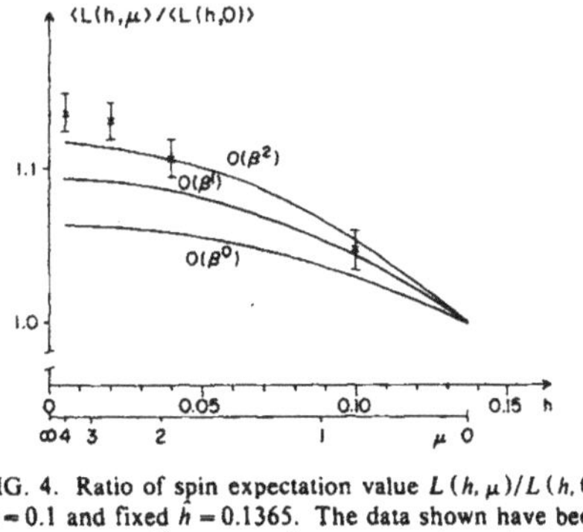

FIG. 4. Ratio of spin expectation value $L(h,\mu)/L(h,0)$ at $\beta=0.1$ and fixed $\hat{h}=0.1365$. The data shown have been obtained on a 10^3 lattice. The curves show the three lowest-order contributions of a strong-coupling expansion for $L(h,\mu)/L(h,0)$.

TABLE II. Strong-coupling expansion coefficients for $L(h,\mu)=\frac{1}{2}\langle \mathrm{Tr}U_x+\mathrm{Tr}U_x^{-1}\rangle$ for a fixed value of $\hat{h}=0.1365$. The five values of (h,μ) tabulated correspond to the data points shown in Fig. 4.

h	μ	$L_0(h,\mu)$	$L_1(h,\mu)$	$L_2(h,\mu)$
0.005	4	0.1550	1.1684	14.0440
0.02	2.5083	0.1548	1.1651	13.9983
0.04	1.8984	0.1542	1.1561	13.8750
0.1	0.8303	0.1500	1.0935	12.9981
0.1365	0	0.1457	1.0289	12.0812

To see the influence of the complex action on expectation values, we performed a high-statistics calculation at $\hat{h}=0.1365$ and $\beta=0.1$. In Fig. 4 we show results for the average spin

$$L(h,\mu)=\tfrac{1}{2}\langle \mathrm{Tr}U_x+\mathrm{Tr}U_x^{-1}\rangle, \tag{11}$$

normalized with the result for $\mu=0$. The data are based on 20000 iterations on a 10^3 lattice. The results are compared with a strong-coupling expansion for the ratio $L(h,\mu)/L(h,0)$, where

$$L(h,\mu)=L_0(h,\mu)+L_1(h,\mu)\beta+L_2(h,\mu)\beta^2+\theta(\beta^3). \tag{12}$$

The expansion coefficients for some values of h and μ are given in Table II. Although the convergence of the strong coupling series is poor at $\beta=0.1$ and $\hat{h}=0.1365$, we find that it describes quite well the trend seen for the dependence on $\hat{g}$ at fixed $\hat{h}$ in the simulation. The influence of the imaginary part of the action also reflects itself in the difference of $\langle \mathrm{Tr}U\rangle$ and $\langle \mathrm{Tr}U^{-1}\rangle$. While in the $\mu=0$ case $\langle \mathrm{Tr}U\rangle=\langle \mathrm{Tr}U^{-1}\rangle$, for $\mu\neq 0$ we find that $\langle \mathrm{Tr}U^{-1}\rangle>\langle \mathrm{Tr}U\rangle$. This is similar to what has been found from the analysis of the $\beta=0$ limit (see Table I). $\langle \mathrm{Tr}U\rangle$ ($\langle \mathrm{Tr}U^{-1}\rangle$) is related to the free energy F_q ($F_{\bar{q}}$) of static fermion (antifermion) sources:

$$\langle \mathrm{Tr}U\rangle\sim e^{-\beta F_q},\qquad \langle \mathrm{Tr}U^{-1}\rangle\sim e^{-\beta F_{\bar{q}}}. \tag{13}$$

Thus the fact that $\langle \mathrm{Tr}U^{-1}\rangle$ is larger than $\langle \mathrm{Tr}U\rangle$ for $\mu\neq 0$ indicates that the free energy of static antifermions is smaller than that for static fermion sources, i.e., it is easier to screen the charge of the antifermion in a background of fermions created as a result of nonzero values of μ.

To summarize, we find that the SU(3) spin model with nonzero chemical potential can be well simulated with a complex Langevin algorithm. Even for large values of the chemical potential the algorithm converges. This is encouraging for attempts to simulate QCD at finite density by this approach. Work in this direction is in progress. The phase diagram that we obtain from the simulation of a complex action with $\mu\neq 0$ is obviously closely related to that of the $\mu=0$ real action with a suitably adjusted external-field parameter h. This is expected to be the case also for lattice QCD in the strong-coupling, large-fermion-mass limit, where the SU(3) spin model is a good approximation. Attempts to reinterpret existing $\mu=0$ data under this assumption have been undertaken recently.[17]

We thank S. Duane and J. Polonyi for many valuable discussions. This work was supported in part by the National Science Foundation through Grant No. PHY82-01948.

[1]F. Fucito, C. Rebbi, and S. Solomon, Nucl. Phys. **B248**, 615 (1984), and Phys. Rev. D **31**, 1461 (1985); J. Polonyi, H. W. Wyld, J. B. Kogut, J. Shigemitsu, and D. K. Sinclair, Phys. Rev. Lett. **53**, 644 (1984); R. V. Gavai, M. Lev, and B. Peterson, Phys. Lett. **140B**, 367 (1984), and **149B**, 492 (1984); R. V. Gavai and F. Karsch, University of Illinois Report No. ILL-(TH)-85-#19, April 1985 (to be published).

[2]F. Fucito, R. Kinney, and S. Solomon, California Institute of Technology Report No. CALT-68-1189, October 1984 (to be published).

[3]J. B. Kogut, J. Polonyi, H. W. Wyld, and D. K. Sinclair, Phys. Rev. Lett. **54**, 1475 (1985).

[4]P. Hasenfratz and F. Karsch, Phys. Lett. **125B**, 308 (1983); J. Kogut, H. Matsuoka, M. Stone, H. W. Wyld, S. Shenkar, J. Shigemitsu, and D. K. Sinclair, Nucl. Phys. **B225[FS9]**, 93 (1983); N. Bilic and R. V. Gavai, Z. Phys. C **23**, 77 (1984).

[5]A. Nakamura, Phys. Lett. **149B**, 391 (1984).

[6]G. Parisi, Phys. Lett. **131B**, 393 (1983).

[7]J. R. Klauder, "Stochastic Quantization," Lectures given at the XXII Schladming School, March 1983 (unpublished).

[8]J. R. Klauder, Acta Phys. Austriaca, Suppl. **25**, 251 (1983), and Phys. Rev. A **29**, 2036 (1984); J. R. Klauder and W. R. Petersen, J. Stat. Phys. **39**, 53 (1985).

[9]E. Gozzi, Phys. Lett. **150B**, 119 (1985).

[10]H. W. Hamber and Hai-cang Ren, Phys. Lett. **159B**, 330 (1985).

[11]J. Ambjorn, M. Flensburg, and C. Peterson, Phys. Lett. **159B**, 335 (1985).

[12]T. Banks and A. Ukawa, Nucl. Phys. **B225[FS9]**, 145 (1983).

[13]J. Bartholomew, D. Hochberg, P. H. Damgaard, and M. Gross, Phys. Lett. **133B**, 218 (1983).

[14]F. Green and F. Karsch, Nucl. Phys. **B238**, 297 (1984).

[15]C. DeTar and T. DeGrand, Nucl. Phys. **B225[FS9]**, 590 (1983).

[16]F. Green and S. Samuel, Nucl. Phys. **B190[FS3]**, 113 (1981); J. B. Kogut, M. Snow, and M. Stone, Nucl. Phys. **B200[FS4]**, 211 (1982); K. E. Eriksson, N. Svartholm, and B. S. Skagerstam, J. Math. Phys. **22**, 2276 (1981).

[17]J. Engels and H. Satz, Phys. Lett. **159B**, 151 (1985).

Minkowski Space

Volume 148B, number 1,2,3 PHYSICS LETTERS 22 November 1984

STOCHASTIC QUANTIZATION IN MINKOWSKI SPACE

H. HÜFFEL and H. RUMPF
Institut für Theoretische Physik, Universität Wien, A-1090 Vienna, Austria

Received 15 June 1984

We propose a generalization of the euclidean stochastic quantization scheme of Parisi and Wu that is applicable to fields in Minkowski space. A perturbative proof of the equivalence of the new method to ordinary quantization is given for the self-interacting scalar field. It is argued furthermore non-perturbatively that the method generally implies the Schwinger–Dyson equations.

Parisi and Wu [1] introduced a stochastic quantization scheme which is based on the Langevin equation of non-equilibrium statistical mechanics. The method can be applied to the quantization of scalar, Dirac, and gauge fields in euclidean space–time [2].

In this paper we propose a modification of the original Parisi–Wu approach allowing to quantize scalar fields in Minkowski space right away: We introduce a Langevin equation with complex drift term; the Green functions in Minkowski space are obtained as a "weak equilibrium limit" in the distributional sense of the correlation functions of the complex process defined by the generalized Langevin equation. We give a perturbative proof of the equivalence of the quantization method for the case of a self-interacting scalar field by using a diagrammatical technique. In addition we supply a non-perturbative argument that the Schwinger–Dyson equations will hold in general for the Green functions obtained by the weak limiting procedure.

We propose to replace the Langevin equation of Parisi and Wu [1] for the euclidean scalar field Φ

$$\partial\Phi(x,t)/\partial t = -\delta S_{\mathrm{E}}[\Phi]/\delta\Phi(x,t) + \eta(x,t) \tag{1}$$

by the following generalized Langevin equation for the field Φ in Minkowski space:

$$\partial\Phi(x,t)/\partial t = \mathrm{i}(\delta S[\Phi]/\delta\Phi(x,t)) + \eta(x,t). \tag{2}$$

Here S_{E} and S denote the action in euclidean and Minkowski space, respectively; the fictitious time parameter t should not be confused with the physical time x^0. The random source η is a gaussian white noise with correlation function

$$\langle \eta(x,t)\eta(x',t)\rangle_\eta = 2(2\pi)^4\delta(t-t')\delta^{(4)}(x-x'). \tag{3}$$

Note that the process $\Phi(x,t)$ defined by eq. (2) is complex and that the existence of an equilibrium for $t\to\infty$ constitutes a more subtle problem than in the euclidean case. Actually what we are going to show is that the equilibrium limit exists in the distributional sense, namely that the correlation functions converge to the quantum Green functions for $t\to\infty$ only when interpreted as tempered distributions.

Volume 148B, number 1,2,3 PHYSICS LETTERS 22 November 1984

As an example we discuss the stochastic perturbation theory of a self-interacting scalar field with lagrangian

$$L = \tfrac{1}{2}(\partial\Phi)^2 - (g/3!)\Phi^3 \tag{4}$$

(we have put $m^2 = 0$ only for convenience; moreover all our results can be generalized for any polynomial self-interaction). The corresponding generalized Langevin equation is

$$\dot{\Phi} = -\mathrm{i}(\Box\Phi + \tfrac{1}{2}g\Phi^2) + \eta. \tag{5}$$

We consider first the free case $g = 0$. In this case the unique solution of (5) with initial condition $\Phi(t = 0) = 0$ is simply

$$\Phi^{(0)}(k,t) = \int_0^t \mathrm{d}\tau \exp\left[\mathrm{i}k^2(t-\tau)\right]\eta(\tau). \tag{6}$$

Owing to the white noise correlation function (3) of η the correlation function of the free field $\Phi^{(0)}$ is

$$\left\langle \Phi^{(0)}(k,t)\Phi^{(0)}(k',t')\right\rangle_\eta = \int_0^{\min(t,t')} \mathrm{d}\tau \exp\left[\mathrm{i}k^2(t+t'-2\tau)\right] 2(2\pi)^4\delta^{(4)}(k+k') \tag{7}$$

$$= \mathrm{i}(2\pi)^4\delta^{(4)}(k+k')k^{-2}\left\{\exp(\mathrm{i}k^2|t-t'|) - \exp\left[\mathrm{i}k^2(t+t')\right]\right\}. \tag{8}$$

Now although the limit $t = t' \to \infty$ of (8) does not exist in the ordinary sense, the expression converges if interpreted as a tempered distribution (see e.g. ref. [3]) of k, k', which is the proper interpretation in quantum field theory (all states being defined in terms of smooth wave packets). Starting from the definitions [3] it is easy to prove the following relations that hold,in the theory of tempered distributions:

$$\lim_{s\to\infty} \exp(\mathrm{i}xs) = 0,$$

$$\lim_{s\to\infty} \mathrm{P}(1/x)\exp(\mathrm{i}xs) = \mathrm{i}\pi\delta(x) \tag{9}$$

where P denotes the principal value. Therefore

$$\lim_{t=t'\to\infty} \left\langle \Phi^{(0)}(k,t)\Phi^{(0)}(k',t')\right\rangle_\eta = \mathrm{i}(2\pi)^4\delta^{(4)}(k+k')\left[\mathrm{P}(1/k^2) - \mathrm{i}\pi\delta(k^2)\right]$$

$$= \mathrm{i}(2\pi)^4\delta^{(4)}(k+k')\frac{1}{k^2+\mathrm{i}0}. \tag{10}$$

We have thus obtained the Feynman propagator from the generalized stochastic quantization. An alternative, and less sophisticated, way of arriving at (10) is to add a negative imaginary mass term $-\tfrac{1}{2}\mathrm{i}\epsilon\Phi^2$ to the lagrangian (4) and let ϵ tend to zero after all calculations have been performed. Then the second term in (8) is exponentially damped, since k^2 is replaced by $k^2 + \mathrm{i}\epsilon$.

Let us now turn to first order perturbation theory. The first order correction to (6) is

$$\Phi^{(1)}(k,t) = -\mathrm{i}\int_0^t \mathrm{d}\tau \exp\left[\mathrm{i}k_3^2(t-\tau)\right]\frac{g}{2}\int\frac{\mathrm{d}^4p}{(2\pi)^4}\Phi^{(0)}(p,\tau)\Phi^{(0)}(k_3-p,\tau). \tag{11}$$

In order to calculate the three-point Green function we consider the three-point correlation function which

Volume 148B, number 1,2,3 PHYSICS LETTERS 22 November 1984

is given to first order by

$$\left\langle \Phi^{(0)}(k_1,t)\Phi^{(0)}(k_2,t)\Phi^{(1)}(k_3,t)\right\rangle_\eta + 2 \text{ similar terms.} \tag{12}$$

A typical contribution to (12) is

$$-\mathrm{i}\int_0^t \mathrm{d}\tau \exp\left[\mathrm{i}k_3^2(t-\tau)\right]\frac{g}{2}\int\frac{\mathrm{d}^4p}{(2\pi)^4}\left\langle \Phi^{(0)}(k_1,t)\Phi^{(0)}(p,\tau)\right\rangle_\eta \left\langle \Phi^{(0)}(k_2,t)\Phi^{(0)}(k_3-p,t)\right\rangle_\eta$$

$$= \tfrac{1}{2}\mathrm{i}g(2\pi)^4\delta^{(4)}(k_1+k_2+k_3)\frac{1}{k_1^2k_2^2}\int_0^t \mathrm{d}\tau \exp\left[\mathrm{i}(k_1^2+k_2^2+k_3^2)t\right]\left\{\exp\left[-\mathrm{i}(k_1^2+k_2^2+k_3^2)\tau\right]\right.$$

$$\left. -\exp\left[-\mathrm{i}(k_1^2-k_2^2+k_3^2)\tau\right]-\exp\left[-\mathrm{i}(-k_1^2+k_2^2+k_3^2)\tau\right]+\exp\left[-\mathrm{i}(-k_1^2-k_2^2+k_3^2)\tau\right]\right\} \tag{13}$$

which tends (upon addition of the small imaginary mass term to the lagrangian) for $t \to \infty$ to

$$-\tfrac{1}{2}g(2\pi)^4\delta^{(4)}(k_1+k_2+k_3)\frac{1}{k_1^2+\mathrm{i}0}\,\frac{1}{k_2^2+\mathrm{i}0}\,\frac{1}{k_1^2+k_2^2+k_3^2+\mathrm{i}0}. \tag{14}$$

Adding finally the terms arising from all the possible permutations in k_1, k_2, k_3 one obtains

$$\left\langle \Phi(k_1,t)\Phi(k_2,t)\Phi(k_3,t)\right\rangle_\eta^{(1)} \xrightarrow{t\to\infty} -g(2\pi)^4\delta^{(4)}(k_1+k_2+k_3)\frac{1}{k_1^2+\mathrm{i}0}\,\frac{1}{k_2^2+\mathrm{i}0}\,\frac{1}{k_3^2+\mathrm{i}0}, \tag{15}$$

which complies exactly with the result of ordinary perturbation theory.

We proceed to show the equivalence of our proposed quantization scheme to the usual Minkowski space quantization for all orders of perturbation theory. We shall use a diagrammatic method that was employed in a similar proof [4] for stochastic quantization in euclidean space. For more details on the common features of that proof and the present one, the reader is referred to ref. [4].

Transforming the Langevin equation (2) into an integral equation and solving the equation by iteration we arrive at a power series expansion of the field Φ in the coupling constant(s), expressing $\Phi(x,t)$ as a functional of the white noise η. We consider an N-point correlation function $\langle \Phi(x_1,t)\dots\Phi(x_N,t)\rangle_\eta$ and substitute for Φ its perturbative expression. When the random average over η is taken diagrams are obtained which we call stochastic diagrams. They have the form of an ordinary Feynman diagram apart from crosses on the lines where two η's have been joined together (these crossed lines represent the two-point correlation function). To every Feynman diagram there exists a number of stochastic diagrams of the same topology. We are going to show by induction that the sum of all stochastic diagrams corresponding to a given Feynman diagram yields just this Feynman diagram in the weak limit $t \to \infty$.

Assume, then, that we have proved this equivalence for all numbers of vertices smaller than N with any number of external lines L (the case $N = 1$ was already demonstrated above). Consider a stochastic diagram S_F belonging to some Feynman diagram F with N vertices and L external lines, each vertex carrying a (fictitious) time τ_i which has to be integrated over. As in ref. [4] we introduce time orderings of the vertices and perform the time integrations starting with the lowest vertex time and going successively to higher ones. Leaving out the momentum denominators of the two-point correlation functions, and dropping the second term in the bracket in (8) (which does not contribute as $t \to \infty$), we get due to the ordering of vertex times for the kth time integration

$$(-\mathrm{i}g)\int_0^{\tau_{k+1}} \mathrm{d}\tau_k \exp\Big(-\mathrm{i}\tau_k \sum_{W_k} p^2\Big) = (-\mathrm{i}g)\mathrm{i}\Big(\sum_{W_k}(p^2+\mathrm{i}0)\Big)^{-1} \exp\Big(-\mathrm{i}\tau_{k+1}\sum_{W_k} p^2\Big) + \cdots, \tag{16}$$

where the elements of the index set W_k label all those momenta which attach to the vertices corresponding to the time $\tau_1, \tau_2, \ldots, \tau_k$ but which do not connect any two of them. The other momenta do not show up as their contribution has been cancelled by the preceding τ_i integrations. Strictly speaking this is true only if the contributions of the lower integration boundaries can be neglected. But this is indeed the case for $t \to \infty$ (weak limit). The same remark applies to the lower integration boundary of the τ_k integration itself, whose asymptotically vanishing contribution is indicated by the dots on the right-hand side of (16). The asymptotic suppression of these contributions occurs through multiplication with the Langevin Green function (appearing in the integrand on the right-hand side of (6)) or the two-point correlation function corresponding to every uncrossed or crossed external line of the diagram, respectively. Another consequence of these multiplications is that the whole exponential

$$\exp\left(-\mathrm{i} \sum_{\substack{\text{external}\\ \text{momenta}}} p^2\right)$$

(that remains after the τ_N integration) is reduced to unity.

Now given a fixed time ordering of vertices with, say, τ_N maximal (hence the corresponding vertex being external), we define a stochastic diagram $S_{F'}$, which is obtained from S_F by dropping the τ_N vertex with its attached external line(s); similarly we define the Feynman diagram F'. The essential ingredient of the proof is now the fact (following from (16) and the accompanying remarks) that the N-fold τ integration of S_F equals (up to a simple combinatorial factor) the $(N-1)$-fold τ integration of $S_{F'}$ times a factor K which is given by

$$K = (-\mathrm{i}g)\mathrm{i}\left(\sum_{\substack{\text{external}\\ \text{momenta}}} (p^2 + \mathrm{i}0)\right)^{-1}. \tag{17}$$

As $S_{F'}$ contains $N-1$ vertices, the induction assumption may be used so that upon summing and integrating over all time-ordered stochastic diagrams S_F (keeping τ_N as the largest time) we obtain essentially KF' in the limit $t \to \infty$. To complete the proof one has to show that the sum over all possible places for the largest vertex time τ_N just gives the Feynman diagram F. For the case of single external lines attached to the external vertices this follows immediately from the identity

$$\frac{1}{2}\sum_{\text{loc}} F' \cdot K = \frac{1}{2}\sum_{\text{loc}} p_N^2 \left(\sum_{\substack{\text{external}\\ \text{momenta}}} (p^2 + \mathrm{i}0)\right)^{-1} (-\mathrm{i}g)\frac{\mathrm{i}}{p_N^2 + \mathrm{i}0} F'$$

$$= \sum_{\text{loc}} p_N^2 \left(\sum_{\substack{\text{external}\\ \text{momenta}}} (p^2 + \mathrm{i}0)\right)^{-1} F = F, \tag{18}$$

where Σ_{loc} is the sum over all positions of the Nth vertex and p_N is the external momentum attached to it. The factor 1/2 on the left-hand side of (18) is the exact result of the sum over stochastic diagrams. The exact result for this sum in the case of double external lines is

$$-\left(\frac{1}{p_{N,1}^2 + \mathrm{i}0} + \frac{1}{p_{N,2}^2 + \mathrm{i}0}\right) KF' = -\frac{p_{N,1}^2 + p_{N,2}^2}{(p_{N,1}^2 + \mathrm{i}0)(p_{N,2}^2 + \mathrm{i}0)} KF' = (p_{N,1}^2 + p_{N,2}^2) KF. \tag{19}$$

Summation over all locations again yields F. This completes the diagrammatic proof.

Volume 148B, number 1,2,3 PHYSICS LETTERS 22 November 1984

In the euclidean case the validity of the Parisi–Wu method has been shown also non-perturbatively for non-gauge theories [2]; for gauge theories see ref. [5]. All these non-perturbative proofs are based on the Fokker–Planck equation associated with the stochastic process under consideration. In minkowskian field theory the Fokker–Planck equation for the positive definite probability distribution describing the complex process defined by (2) apparently ceases to be a useful tool: Due to the complicated structure of the equation no obvious candidate for an equilibrium distribution could be found. However it was observed recently [6] that one can define a complex valued probability distribution on the real axis such that

$$\langle F[\Phi,t]\rangle_\eta = \int D[\Phi_R]\, P[\Phi_R,t]\, F[\Phi_R], \tag{20}$$

where $P[\Phi_R,t]$ satisfies the Fokker–Planck equation (of the usual form) for the real field Φ_R subject to the complex drift force $i\delta S[\Phi_R]/\delta\Phi_R$. We expect from the perturbative treatment of the Langevin equation that (20) makes sense and that $P[\Phi_R,t]$ approaches indeed its formal equilibrium limit (i.e. the stationary solution of the Fokker–Planck equation)

$$P^{eq}[\Phi_R] = \exp(iS[\Phi_R]) \Big/ \int D[\Phi_R] \exp(iS[\Phi_R]). \tag{21}$$

We do not attempt in this paper to perform a rigorous treatment of the existence and the convergence properties of $P[\Phi_R,t]$. Rather we formulate the *relaxation hypothesis* by postulating that

$$\lim_{t\to\infty} \langle \dot{\Phi}(x,t) F[\Phi]\rangle_\eta = 0 \tag{22}$$

for any functional $F[\Phi]$. In the euclidean case (22) can be derived from the Fokker–Planck equation and the existence of an equilibrium limit [2].

In the following we prove that (22) implies the Schwinger–Dyson equations for minkowskian field theories.

The proof generalizes elements of the equivalence proofs given by Nakano [7] and Alfaro and Sakita [8] for the euclidean case. Consider the functional $Z_t[J]$ defined by

$$Z_t[J] = \left\langle \exp\left(i\int d^4x J(x)\Phi(x,t)\right)\right\rangle_\eta \equiv \langle 1\rangle_{J,t}. \tag{23}$$

We have

$$0 = \int D[\eta] \frac{\delta}{\delta\eta(x,t)} \exp\left(-\frac{1}{4}\int_0^\infty dt \int d^4x'\,\eta^2 + i\int d^4x'\,J(x')\Phi(x',t)\right) \tag{24}$$

$$= \left\langle -\tfrac{1}{2}\eta(x,t) + i\int d^4x'\,J(x')\frac{\delta\Phi(x',t)}{\delta\eta(x,t)}\right\rangle_{J,t}. \tag{25}$$

Upon using the Langevin equation and [8]

$$\delta\Phi(x',t)/\delta\eta(x,t) = \tfrac{1}{2}\delta^{(4)}(x-x'), \tag{26}$$

(25) becomes

$$\langle -\tfrac{1}{2}\dot{\Phi}(x,t) + \tfrac{1}{2}i[\delta S/\delta\Phi(x,t)] + \tfrac{1}{2}i\,J(x,t)\rangle_{J,t}. \tag{27}$$

Now (22) implies

$$\langle \dot{\Phi}(x,t) \rangle_{J,t} \stackrel{t\to\infty}{\to} 0. \tag{28}$$

Hence

$$\left(\mathrm{i}(\delta S[\chi]/\delta\chi)|_{\chi=(1/\mathrm{i})\delta/\delta J} + \mathrm{i}J \right) Z_t[J] \to 0, \tag{29}$$

i.e. $Z_t[J]$ obeys in the limit $t \to \infty$ the same functional differential equation as the conventional generating functional

$$Z[J] = \int \mathrm{D}[\Phi] \exp\left(\mathrm{i}S[\Phi] + \mathrm{i}\int \mathrm{d}^4x J(x)\Phi(x) \right) \tag{30}$$

of Minkowski space quantum field theory. As is well known [9], this functional differential equation implies the Schwinger–Dyson equations for the Green functions of the theory. Note that for linear field theories (29) implies formally

$$\lim_{t\to\infty} Z_t[J] = Z[J], \tag{31}$$

owing to the normalization condition $Z_t[0] = 1$. The general validity of (31) requires e.g. the existence of the functional Fourier transform of $\lim Z_t[J]$ which is not obvious to us.

The above proof may be extended even to gauge theories, if "stochastic gauge fixing" [10] is introduced. As an example we consider scalar electrodynamics. If we transform $A_a(x,t)$, $\phi(x,t)$ and $\phi^*(x,t)$ to

$$B_a(x,t) = A_a(x,t) - \partial_a\chi(x,t), \quad \Phi(x,t) = \exp[-\mathrm{i}e\chi(x,t)]\phi(x,t),$$
$$\Phi^*(x,t) = \exp[\mathrm{i}e\chi(x,t)]\phi^*(x,t), \tag{32}$$

then the Langevin equation for B_a picks up an additional drift term due to the dependence of χ on the fictitious time t:

$$\dot{B}_a(x,t) = \mathrm{i}\left(\delta S[B,\Phi,\Phi^*]/\delta B_a(x,t)\right) - \partial_a\dot{\chi}(x,t) + \mathrm{i}^{1/2}\eta_a(x,t). \tag{33}$$

$S[B,\Phi,\Phi^*]$ is the action of the Maxwell field (without gauge fixing term), coupled minimally to the charged scalar field; η_a denotes a gaussian "quantum variable" with formal path integral measure $\mathrm{D}[\eta]\exp(\frac{1}{4}\mathrm{i}\eta^2)$. It is important to note that one may disregard the additional drift terms $-\mathrm{i}e\dot{\chi}\Phi$, $\mathrm{i}e\dot{\chi}\Phi^*$ in the Langevin equations for Φ and Φ^*, so that we have

$$\dot{\Phi}(x,t) = \mathrm{i}\left(\delta S[B,\Phi,\Phi^*]/\delta\Phi(x,t)\right) + \eta(x,t). \tag{34}$$

This fact can most easily be understood by considering a generic element

$$F = \int \mathrm{d}x\,\mathrm{d}y\,\mathrm{d}z\, [\phi(x)\phi^*(y)]^n A(z)^m \tag{35}$$

of a gauge invariant functional of the fields. Under infinitesimal gauge transformations $\delta\phi$, $\delta\phi^*$ one has

$$(\delta F/\delta\phi)\delta\phi + (\delta F/\delta\phi^*)\delta\phi^* = 0. \tag{36}$$

Therefore the time evolution of gauge invariant quantities is not affected by the additional terms under discussion, so that they may be disregarded. The new drift term for the gauge field B_a renders possible the existence of a stationary solution of the Fokker–Planck equation. Indeed if one chooses

$$\dot{\chi} = \mathrm{i}\alpha^{-1}\partial_a B_a \tag{37}$$

then eq. (29) implies for $t \to \infty$ just the Schwinger–Dyson equations of ordinary quantum field theory with the covariant gauge fixing term $(2\alpha)^{-1}(\partial B)^2$ added to the action $S[B, \Phi, \Phi^*]$.

We conclude with the following remark. It may look superficial to insist on a Minkowski space formulation of quantum field theory with all its mathematical complications, as the existence of a well-defined euclidean counterpart is usually taken for granted. However this is not true in the important case of the gravitational field. The main motive of this paper was to provide the prerequisites for the stochastic qûantization of the gravitational field. The latter will be the subject of a forthcoming paper.

We acknowledge helpful discussions with B. Baumgartner, G. Ecker, H. Grosse, and H. Rupertsberger. H.H. is also very grateful to M. Mintchev and G. Parisi for valuable discussions.

References

[1] G. Parisi and Wu Yong-Shi, Sci. Sin. 1 (1981) 483.
[2] J.D. Breit, S. Gupta and A. Zaks, Nucl. Phys. B233 (1984) 61, and references cited therein.
[3] M.J. Lighthill, Introduction to Fourier analysis and generalized functions (Cambridge U.P., London, 1958).
[4] W. Grimus and H. Hüffel, Z. Phys. C18 (1983) 129.
[5] L. Baulieu and D. Zwanziger, Nucl. Phys. B193 (1981) 163.
[6] G. Parisi, Phys. Lett. 131B (1983) 393.
[7] Y. Nakano, Prog. Theor. Phys. 69 (1983) 361.
[8] J. Alfaro and B. Sakita, in: Gauge theory and gravitation, Proc. Intern. Symp. (Nara, Japan, 1982) (Springer, Berlin, 1983).
[9] C. Itzykson and J.-B. Zuber, Quantum field theory (McGraw-Hill, New York, 1980).
[10] D. Zwanziger, Nucl. Phys. B192 (1981) 259.

Volume 150B, number 1,2,3 PHYSICS LETTERS 3 January 1985

LANGEVIN SIMULATION IN MINKOWSKI SPACE?

E. GOZZI [1]
Department of Physics, City College of the City University of New York, New York, NY 10031, USA

Received 11 June 1984

We present in this paper a "tentative", "formal" proof that the Langevin simulation for relativistic systems can be performed directly in Minkowski space without rotating to euclidean time. The proof bypasses the difficult task of studying the spectrum of the non-self-adjoint Fokker–Planck hamiltonian and it is based on a dimensional-reduction argument recently proposed by this author.

In recent years numerical simulation of relativistic quantum field theory (RQFT) models has become very popular. The procedure is usually an adaptation of the Metropolis algorithm known under the name of Monte Carlo method [1]. As everybody knows, in RQFT we are interested in calculating averages of the type

$$\int d\mu^{M}[\varphi]\, G[\varphi] \equiv \langle G\rangle\,, \quad d\mu^{M}[\varphi] = \mathcal{D}\varphi \exp(iS[\varphi])/Z\,, \quad Z = \int \mathcal{D}\varphi \exp(iS[\varphi])\,, \quad \int d\mu^{M}[\varphi] = 1$$

(M for Minkowski). (1)

The usual trick is to rotate the system to euclidean time $t \to it$ so that the measure $\int d\mu^{M}[\varphi]$ in (1) becomes

$$d\mu^{E}[\varphi] = \mathcal{D}\varphi \exp(-S[\varphi])/Z \qquad \text{(E for euclidean)}. \tag{2}$$

The Monte Carlo method explicitly uses the fact that $d\mu^{E}[\varphi]$ is a probability measure: generally speaking one constructs [1] an algorithm which generates configurations of the systems with a frequency given by (2). More precisely: what is done is

$$\lim_{N\to\infty} \frac{1}{N}\sum_{l=1}^{N} G[\varphi_l] = \langle G\rangle\,, \tag{3}$$

where the φ_l are the configurations generated according to the algorithm.

An alternative method to the Monte Carlo has been proposed some time ago by Parisi [2] and it consists in postulating a stochastic dynamics for the system

$$\dot\varphi(x^{\mu},\tau) = -\delta S[\varphi]/\delta\varphi(x^{\mu},\tau) + \eta(x^{\mu},\tau)\,, \quad \overline{\eta(x^{\mu},\tau)\,\eta(y^{\mu},\tau')} = 2\delta(\tau-\tau')\delta(x^{\mu}-y^{\mu})\,, \quad \tau = \text{computer "time"}. \tag{4}$$

Discretizing the time τ_k, the computer generates at every "instant" τ_k a gaussian noise $\eta(\tau_k)$ and then eq. (4) is solved in steps ϵ

$$\varphi(\tau_{k+1}) - \varphi(\tau_k) = -\epsilon\, \delta S/\delta\varphi(\tau_k) + \sqrt{\epsilon}\,\eta(\tau_k)\,. \tag{5}$$

After a sufficient number of configurations $\varphi(\tau_k)$ have been generated, the "time" average is taken and

[1] Work supported in part by NSF grant no. 82-1536 and a CUNY-PSC-BHE Faculty research award.

Volume 150B, number 1,2,3 PHYSICS LETTERS 3 January 1985

$$\frac{1}{\tau}\int_0^{\tau} G[\varphi(\tau')]\,\mathrm{d}\tau' \xrightarrow[\tau'\to\infty]{} \langle G\rangle\,. \tag{6}$$

The usual proof of this relation goes through the Fokker–Planck-(FP) equation [3] associated to (4). This equation gives the "time" evolution of the probability $P(\varphi,\tau)$ for the configuration φ_η solution of eq. (4)

$$\dot{\Psi}[\varphi,\tau] = -2H^{\mathrm{FP}}\Psi[\varphi,\tau]\,,\quad \Psi[\varphi,\tau] = P(\varphi,\tau)\exp(\tfrac{1}{2}S[\varphi])\,,\quad H^{\mathrm{FP}} = -\tfrac{1}{2}\delta^2/\delta\varphi^2 + \tfrac{1}{8}(\partial S/\partial\varphi)^2 - \tfrac{1}{4}\partial^2 S/\partial\varphi^2\,. \tag{7}$$

We see that this is a Schrödinger-type eq. and, if S is real, then H^{FP} is a self-adjoint operator and positive-semidefinite [4]

$$H^{\mathrm{FP}} = \tfrac{1}{2}QQ^{+}\,,\quad Q = (\delta/\delta\varphi + \tfrac{1}{2}\partial S/\partial\varphi)\,.$$

Because of these properties of H^{FP}, the system relaxes for $\tau\to\infty$ on the "ground-state" of H^{FP} that is $\Psi_0 = \exp(-\frac{1}{2}S)$, or in terms of $P(\varphi,\tau)$

$$\lim_{\tau\to\infty} P(\varphi,\tau) = \exp(-S[\varphi])\ \int \mathcal{D}\varphi \exp(-S[\varphi])\,.$$

This basically proves (6).

Some potential advantages of the Langevin method have been pointed out in ref. [2]; principally the control over convergence which is exponential, and the possibility of arranging for cancellation of statistical fluctuations in measuring correlation functions.

We want to point out in this paper another potential advantage: the possibility to simulate the systems directly in Minkowski space. The Monte Carlo algorithm cannot be applied in this case because the measure of integration in Minkowski space

$$\mathrm{d}\mu^{\mathrm{M}}[\varphi] = \exp(\mathrm{i}S)\Big/\int \mathcal{D}\varphi \exp(\mathrm{i}S)\,,$$

is not real anymore so it looses its meaning of probability. On the contrary, nothing forbids us to use the Langevin equation (4) in Minkowski space

$$\dot{\varphi}(x^\mu,\tau) = +\mathrm{i}\delta S[\varphi]/\delta\varphi + \eta(x^\mu,\tau)\,. \tag{8}$$

The computer can still solve this equation in steps and calculate the LHS of (6)

$$\frac{1}{\tau}\int_0^{\tau} G[\varphi(\tau')]\,\mathrm{d}\tau'\,, \tag{9}$$

but we have to prove that, for $\tau\to\infty$, the "time" average (9) converges to

$$\lim_{\tau\to\infty}\frac{1}{\tau}\int_0^{\tau} G[\varphi(\tau')]\,\mathrm{d}\tau' = \int \mathcal{D}\varphi\, G\exp(\mathrm{i}S)\Big/\int \mathcal{D}\varphi \exp(\mathrm{i}S)\,. \tag{10}$$

This is difficult to prove [5] because we now have that the corresponding FP operator

$$H^{\mathrm{FP}} = -\tfrac{1}{2}\delta^2/\delta\varphi^2 - \tfrac{1}{8}(\partial S/\partial\varphi)^2 + \tfrac{1}{4}\mathrm{i}\partial^2 S/\partial\varphi^2$$

is not self-adjoint anymore, so we do not know if eq. (7) has eigenfunctions and what the eigenvalues are. We only know that $\exp(\mathrm{i}S/2)$ is an eigenfunction with eigenvalue zero. The authors of ref. [5] have analyzed recently this problem in the general case when the drift force $\partial S/\partial\varphi$ is complex and not simply imaginary as in our case. They [5] have tried to prove the positivity of the real part of the eigenvalues λ_n^{FP} of the non-self-adjoint

Volume 150B, number 1,2,3 PHYSICS LETTERS 3 January 1985

$$H^{FP} = -\tfrac{1}{2}\delta^2/\delta\varphi^2 + \tfrac{1}{8}(\partial S/\partial\varphi)^2 - \tfrac{1}{4}\partial^2 S/\partial\varphi^2\,, \quad S \text{ complex}\,, \quad \eta \text{ real}\,. \tag{11}$$

[Associated with the Langevin equation (4) when S is complex and η is kept real.] In fact, if Re $\lambda_n^{FP} \geqslant 0$, then the system is going to relax, for $\tau \to \infty$, to $\exp(-S)$, or in our case to $\exp(iS)$.

Of course, to prove Re $\lambda_n^{FP} \geqslant 0$ is not easy because so little is known of the spectrum of non-hermitian operators. A rigorous proof only exists for the H^{FP} associated with the Minkowski action of the harmonic oscillator (see Klauder [6]). This proof is made by directly diagonalizing H^{FP}. Of course this is prohibitive for more complicated S.

Stimulated by the result of Klauder, we had a strong belief that the result had to be valid for any action. In this paper we want to present a proof of this. Our analysis will bypass the difficult task of analyzing the spectrum of H^{FP} and it will be based on a path-integral approach [4,7]. We feel that the functional approach, because of its "global" character [in contrast to the local or differential character of (7)], is more suitable to prove the relation (10) that is, after all, an asymptotic relation and therefore more sensible to the global properties than to the local ones of the Fokker–Planck dynamics.

The path-integral formalism for the Langevin dynamics has been developed in ref. [4] and we are going to skip the details here. The main idea is that the correlation functions $\langle\varphi_\eta(\tau_1)\ldots\varphi_\eta(\tau_l)\rangle_\eta$ of fields φ_η solutions of eq. (4) can be derived from a generating functional

$$\langle\varphi_\eta(\tau_1)\ldots\varphi_\eta(\tau_l)\rangle_\eta = \delta^l Z_{SS}[J]/\delta J^l|_{J=0}\,, \tag{12}$$

with

$$Z_{SS}[J] = \int \tilde{\mathcal{D}}\varphi\tilde{\mathcal{D}}\eta\,\delta(\varphi-\varphi_\eta)\exp\left(-\int_{-\infty}^{+\infty} \tfrac{1}{4}\eta^2\,d\tau + \int_{-\infty}^{+\infty} J\varphi\,d\tau\right),$$

$\tilde{\mathcal{D}}\varphi \equiv \lim_{N\to\infty} \mathcal{D}\varphi(\tau_1), \mathcal{D}\varphi(\tau_i)$ is the usual path integration in x^μ.

This generating functional can be brought to a supersymmetric [4] form (SS), integrating away η and introducing Grassmann fields $\psi, \bar\psi$

$$Z_{SS} = \int \tilde{\mathcal{D}}\varphi\tilde{\mathcal{D}}\psi\tilde{\mathcal{D}}\bar\psi\exp\left(-\int_{-\infty}^{+\infty}[\mathcal{L}_{SS} + J\varphi]\,d\tau\right), \quad \mathcal{L}_{SS} = \tfrac{1}{2}\dot\varphi^2 + \tfrac{1}{8}(\partial S/\partial\varphi)^2 + \bar\psi(\partial_\tau + \tfrac{1}{2}\partial^2 S/\partial\varphi^2)\psi\,. \tag{13}$$

For details of the derivation see ref. [4].

In (12) we will restrict the current $J(x^\mu, \tau)$ to the form $J(x,\tau) = \tilde{J}(x^\mu)\delta(\tau)$, so that we automatically get equal "time" correlations at $\tau = 0$. The choice of having the fields, in the correlations, at $\tau = 0$ is correct because in (13) we prepared the system at $\tau = -\infty$, so that at $\tau = 0$ it should already be relaxed to equilibrium. Essentially what we have to prove is that

$$\langle\varphi_\eta(0)\ldots\varphi_\eta(0)\rangle_\eta = \int\mathcal{D}\varphi[\varphi\ldots\varphi]\exp(-S[\varphi])\Big/\int\mathcal{D}\varphi\exp(-S[\varphi])\,, \tag{14}$$

where S is complex.

Before proceeding, some observations have to be made. Our original field φ was real, but now, solving the Langevin equation with S complex, we have to allow φ to get an imaginary part even if we hold η real. The integration in (13) is now over the complex field φ, the lagrangian $\mathcal{L}_{SS}$ is not real anymore and the corresponding hamiltonian

$$H_{SS} = +\tfrac{1}{2}(\delta/\delta\varphi + \tfrac{1}{2}\delta S/\delta\varphi)\psi(-\delta/\delta\varphi + \tfrac{1}{2}\delta S/\partial\varphi)\bar\psi + (\psi \leftrightarrow \bar\psi)\,, \tag{15}$$

is not hermitian anymore, but $\mathcal{L}_{SS}$ is still invariant under the following SS [4]

$$\delta\varphi = \epsilon\psi + \bar{\epsilon}\bar{\psi}, \quad \delta\psi = \bar{\epsilon}(\dot{\varphi} + \tfrac{1}{2}\partial S/\partial\varphi), \quad \delta\bar{\psi} = \epsilon(\dot{\varphi} - \tfrac{1}{2}\partial S/\partial\varphi), \quad \delta\varphi^* = 0,$$

ϵ, $\bar{\epsilon}$ infinitesimal anticommuting parameters. (16)

The generators of this symmetry are

$$Q_\psi = (\delta/\delta\varphi + \tfrac{1}{2}\partial S/\partial\varphi)\psi, \quad Q_{\bar{\psi}} = \bar{\psi}(-\delta/\delta\varphi + \tfrac{1}{2}\partial S/\partial\varphi), \quad [Q_\psi, H_{SS}] = 0 = [Q_{\bar{\psi}}, H_{SS}], \quad \tfrac{1}{2}\{Q_\psi, Q_{\bar{\psi}}\} = H_{SS}. \tag{17}$$

As $\mathcal{L}_{SS}$ (13) is not real, the probability function itself is not real, but this is something we expect as the equilibrium probability itself $\exp(-S)$, when S is complex, is not real. The fact that the probability is complex does not bother us at all, because the Langevin algorithm does not use the probability, but just solves the eq. (4).

Somebody might even be bothered because we have a path-integral formulation (13) for non-hermitian operators, but we should remember that for the path-integral approach, what we need is just the usual Smoulukowsky property [3] for the transition probabilities $P(\varphi 0|\varphi_1\tau_1)$.

$$\int P(\varphi 0|\varphi_1\tau_1)P(\varphi_1\tau_1|\varphi_2\tau_2)\,\mathrm{d}\varphi_1 = P(\varphi 0|\varphi_2\tau_2),$$

and this is a reflection of the markoffian character of the stochastic process (4) independent of the fact that $P(\varphi)$ might be complex. The careful reader might have noticed the "strange" invariance (16) of $\mathcal{L}_{SS}$: the field φ "rotates" in ψ, $\bar{\psi}$, but φ^* is invariant. Nevertheless, this is still a supersymmetry transformation in any respect as the algebra (17) closes. Of course, it has lost the hermitian character we are accustomed to for supersymmetric algebra, but this is something we expect for a non-hermitian H_{SS}. The different transformations of φ and φ^* are due to the fact that we kept η real, so that only the real part of φ has a stochastic dynamics, while the imaginary part has a deterministic one [see eq. (4)]. The supersymmetry (16) is a reflection of this stochastic character [4] of the variables φ and so it applies only to half of them.

Having clarified these points, we proceed to prove (14). We will follow ref. [7] and we are going to repeat here all the steps for completeness.

First of all, let us rewrite (13) using superfields Φ

$$\Phi = \varphi + \bar{\alpha}\psi + \alpha\bar{\psi} + \alpha\bar{\alpha}\omega, \tag{18}$$

where ω is an auxiliary field, α, $\bar{\alpha}$ are elements of a Grassmann algebra,

$$\int \mathrm{d}\alpha = \int \mathrm{d}\bar{\alpha} = 0, \quad \int \mathrm{d}\alpha\,\alpha = 1 = \int \mathrm{d}\bar{\alpha}\,\bar{\alpha}, \quad \int \mathrm{d}\tau\,\mathcal{L}_{SS} = \int \mathrm{d}\tau\,\mathrm{d}\alpha\,\mathrm{d}\bar{\alpha}\,(K[\Phi] - S[\Phi]),$$

with $K[\Phi] = D_\alpha\Phi\, D_{\bar{\alpha}}\Phi$, $D_\alpha = \partial_\alpha - \bar{\alpha}\,\partial\tau$, and $S[\Phi]$ is the usual complex action of our system, but where the place of the scalar field φ is taken by the superfield Φ. Even if S is complex, Φ is a scalar superfield under the transformation (16). As usual this supersymmetry can be seen as a transformation on the components fields φ, ψ, $\bar{\psi}$, ω induced by the supertranslations in the superspace

$$\delta\alpha = \epsilon, \quad \delta\bar{\alpha} = \bar{\epsilon}, \quad \delta\tau = -(\bar{\alpha}\epsilon - \bar{\epsilon}\alpha). \tag{19}$$

In contrast to the usual relativistic case, the superspace here is built up only of the spurious variables τ, α, $\bar{\alpha}$.

Using (18) we can still write the Z_{SS} formally as

$$Z_{SS} = \int \tilde{\mathcal{D}}\Phi \exp\left(\int \mathrm{d}\tau \int \mathrm{d}\alpha \int \mathrm{d}\bar{\alpha}\,(K[\Phi] - S[\Phi] + \tilde{J}\Phi)\right),$$

where $\tilde{\mathcal{D}}\Phi = \tilde{\mathcal{D}}\varphi\,\tilde{\mathcal{D}}\omega\,\tilde{\mathcal{D}}\psi\,\tilde{\mathcal{D}}\bar{\psi}$ and $\tilde{J}$ is restricted to $\tilde{J}(x, \alpha, \bar{\alpha}, \tau) = J(x)\delta(\tau)\delta(\alpha)\delta(\bar{\alpha})$, because we are only interested in correlations of φ fields. Let us now introduce [7] the interpolating generating functional

$$Z_{SS}^{\lambda}[J] = \int \tilde{\mathcal{D}}\Phi \exp\left(-\int \mathrm{d}\tau \int \mathrm{d}\alpha \int \mathrm{d}\bar{\alpha}\,\{[\lambda + (1-\lambda)\delta(\tau)\delta(\bar{\alpha})\delta(\alpha)]S[\Phi] - K[\Phi]\}\right). \tag{20}$$

Volume 150B, number 1,2,3 PHYSICS LETTERS 3 January 1985

It is easy to see that for $\lambda = 1$ this is the supersymmetric generating functional (13), while for $\lambda = 0$ we get the usual quantum one (14) $Z = \int \mathcal{D}\varphi \exp(-S[\varphi])$. In fact

$$Z_{SS}^{\lambda=0} = \int \widetilde{\mathcal{D}}\Phi \exp\left(-S[\varphi(\tau=0)] + \int d\tau\, d\alpha\, d\bar{\alpha}\, K[\Phi]\right) = \mathcal{N} \int \mathcal{D}\varphi(\tau=0) \exp\{-S[\varphi(\tau=0)]\}, \tag{21}$$

with $\mathcal{N}$ a normalizing constant.

In the last step above, we have just performed [remembering the form of the path integration measure (12)] $N-1$ integrations for φ (avoiding the one in $\tau = 0$) and all the N path-integrations for ψ, $\bar{\psi}$, ω. These last integrations can be easily done as ψ, $\bar{\psi}$, ω enter only in $K[\Phi]$ and this is just the free lagrangian

$$\int d\tau\, d\alpha\, d\bar{\alpha}\, K[\Phi] \propto \int (\omega^2/2 + \omega\dot{\varphi} - \bar{\psi}\dot{\psi})\, d\tau .$$

We should also notice that Z_{SS}^{λ} is still invariant under the transformation (16), because $S[\Phi]$ and $K[\Phi]$ are separately invariant, and moreover $\int d\tau \int d\alpha \int d\bar{\alpha}\, \delta(\tau)\delta(\bar{\alpha})\delta(\alpha)$ can be written as $\int d\tau \int d\alpha\, d\bar{\alpha}\, \theta(\tau + \bar{\alpha}\alpha)$ that is also invariant. The next step is to show that the correlation functions derived from Z_{SS}^{λ} are independent of λ. Let us discuss for simplicity the two-point function

$$C^{\lambda}(x, \tau = 0, \alpha = 0 = \bar{\alpha}) = \langle \Phi(x, \tau = 0 = \alpha = \bar{\alpha})\Phi(0)\rangle^{\lambda} .$$

Its derivative with respect to λ is

$$\frac{\partial C^{\lambda}}{\partial \lambda} = -\int dx' \int d\tau\, d\alpha'\, d\bar{\alpha}' \,[1 - \delta(\tau')\delta(\bar{\alpha}')\delta(\alpha')]\langle \Phi(x,0)\Phi(0) S[x', \tau', \alpha', \bar{\alpha}']\rangle^{\lambda} . \tag{22}$$

Because of the invariance (16), $\langle \Phi(x,0)\Phi(0)\delta\,[x', \tau', \alpha', \bar{\alpha}']\rangle^{\lambda}$ can only have the form $f(x', x, \tau' + \bar{\alpha}'\alpha)$; in fact the only combination invariant under (19) is $(\tau + \bar{\alpha}\alpha)$. Making use now of the well known property

$$\int dx \int_a^b d\tau \int d\alpha \int d\bar{\alpha}\, g(x, \tau + \bar{\alpha}\alpha) = \int dx \int_a^b d\tau\, g'(x,\tau) = \int dx\, [g(x,b) - g(x,a)] ,$$

we obtain for (22) [‡1]

$$\frac{\partial C^{\lambda}}{\partial \lambda} = -\int dx' \int_{-\infty}^{+\infty} d\tau'\, d\alpha'\, d\bar{\alpha}' \,[1 - \delta(\tau')\delta(\bar{\alpha}')\delta(\alpha')]\,\theta(-\tau')\, f(x, x', \tau' + \bar{\alpha}'\alpha')$$

$$= -\int dx' \int_{-\infty}^{+\infty} d\tau'\, \theta(-\tau') f'(x, x', \tau') + \int dx'\, f(x, x', 0)$$

$$= -\int dx' \int_{-\infty}^{0} d\tau'\, f'(x, x', \tau') + \int dx'\, f(x, x', 0) = \int dx'\, f(x, x', -\infty) .$$

The fields φ at $\tau = -\infty$ can always be chosen to be of any value [because of the markoffian character of the Langevin eq. (4)], even zero so that $f(x', x, -\infty) = 0$ and thus

$$\partial C^{\lambda}/\partial \lambda = 0 .$$

This means that the correlations functions are independent of λ, so the ones derived from $Z_{SS}^{\lambda=1}$ [that is (13)] are the same as the ones derived from $Z_{SS}^{\lambda=0}$ (that is the quantum one). This basically proves (14).

‡1 The $\theta(-\tau)$, in the first step of this derivation, appears because we choose, as in ref. [2], the advanced Green functions of the Langevin process.

We want to stress that *this proof relies only on the supersymmetric invariance of Z_{SS} and this invariance is also there for complex S.* The only other ingredient we used was the markoffian character of the stochastic process.

The reader might be puzzled why supersymmetry did the job for us. The answer lies in the "global" or "topological" properties of Z_{SS} [8], that allowed the deformation of Z_{SS} to Z_{SS}^{λ}. After all, as we said earlier, we needed to exploit the "global" character of the Fokker–Planck dynamics to prove the asymptotic relation (10).

Indirectly, our proof also confirms that Re $\lambda_n^{FP} \geqslant 0$; in fact if just one λ_n^{FP} had Re $\lambda_n^{FP} < 0$, the probability would have increased without any limit and not relaxed to $\exp(-S)$.

Of course our argument does not provide us with any estimate of how big Re λ_n^{FP} are, and so, of how fast the system goes to the equilibrium. Even with this point [9] still unanswered, we feel that it might be very important to have a tool to simulate relativistic systems directly in Minkowski space, without having to go every time to euclidean time and then having to prove all the properties [10] (reflection positivity, etc.) that are required to have equivalence between the two formulations.

Moreover, simulating directly in Minkowski space we can numerically evaluate scattering amplitude, and this was impossible in the usual Monte Carlo–euclidean simulation. The proof presented here is only a formal one and it can be confirmed or disconfirmed (and this is the reason for the question mark in the title) only by numerical analysis, work is in progress in this direction [9].

I wish to thank G. Marchesini who triggered my interest in this problem. Many illuminating discussions with B. Sakita and J. Likken are gratefully acknowledged.

Note added. Of course the argument that we have given here fails when the supersymmetry cannot be invoked. This is the case when supersymmetry is spontaneously broken (S negative and unbounded). Supersymmetry cannot be invoked even in the case that the Langevin equation (4) has more than one solution, that is when the jacobian $\|\delta\eta/\delta\varphi\|$ develops a zero. We feel that this is the case in one of the examples (Re $\sigma < 0$) recently studied in ref. [11]. This would also happen in our Minkowski simulation if, instead of adding $i\epsilon\varphi^2$ to the action to get the Feynman propagator, we would add $-i\epsilon\varphi^2$.

For a clear discussion of the case of multiple solutions of the Langevin equation, see Parisi and Sourlas [4].

References

[1] K. Binder, Monte Carlo Methods (Springer, Berlin, 1979).
[2] G. Parisi and Y. Wu, Sci. Sin. 24 (1981) 483;
G. Parisi, Nucl. Phys. B180 (1981) 378; B205 (1982) 337.
[3] M.C. Wang and G.E. Uhlenbeck, Rev. Mod. Phys. 17 (1945) 323.
[4] G. Parisi and N. Sourlas, Nucl. Phys. 206B (1982) 321;
C. Bender et al., Nucl. Phys. 219B (1983) 61;
E. Gozzi, Phys. Rev. D28 (1983) 1922; Phys. Lett. 130B (1983) 183; CCNY-HEP 83/16, Phys. Rev. D, to be published.
[5] J.R. Klauder, Lectures given at the 22th Schladming School (Austria, 1983);
G. Parisi, Phys. Lett. 131B (1983) 393.
[6] J.R. Klauder, Bell Laboratory preprint.
[7] E. Gozzi, Phys. Lett. 143B (1984) 183;
R. Kirschner, Karl Marx University preprint.
[8] S. Ceccotti and L. Girardello, Phys. Lett. 110B (1982) 39.
[9] E. Gozzi, work in progress.
[10] K. Osterwalder and R. Schrader, Commun. Math. Phys. 31 (1973) 83; 42 (1975) 281.
[11] J. Klauder and P. Petersen, Bell Laboratory preprint (July 1984).

Nuclear Physics B180 [FS2] (1981) 378–384

CORRELATION FUNCTIONS AND COMPUTER SIMULATIONS

G. PARISI

INFN, Laboratori Nazionali di Frascati, Frascati, Italy

Received 11 November 1980

If the equilibrium properties of a statistical system are obtained by solving numerically the associated Langevin equation describing the approach to equilibrium, the connected correlation functions can be computed directly with small effort and high precision.

1. Introduction

Computer simulations of statistical systems are starting to be popular also among high energy physicists [1]; they are mostly interested in computing the correlation length $\xi = m^{-1}$. If $A(x)$ is a local operator one can prove that

$$\langle A(x)A(y)\rangle_c \equiv \langle A(x)A(y)\rangle - \langle A(x)\rangle\langle A(y)\rangle$$
$$\to \exp(-m_A|x-y|)/|x-y|^{(D-1)/2}\,, \tag{1}$$

where m_A is the minimum mass of the state created from the vacuum by the operator A; the power law in the prefactor is correct only if the minimum mass state is a one-particle state*.

In order to extract the value of the mass from the connected correlation functions and to compute the mass spectrum, it is necessary to control the correlation function at large $|x-y|$, i.e. in the region where it is small. This fact is particularly sad; a typical Monte Carlo simulation [1, 2] has a random statistical error proportional to $N_T^{-1/2}$, N_T being the total number of extractions; in many physically interesting cases, in particular in gauge theories, it is impossible to give reasonable estimates of the mass, also using very high statistics.

In this paper I present a new algorithm which allows the direct computation of connected correlation functions for continuous systems, the algorithm being based on the Langevin equation. I have not done a systematic study of the advantages presented by this new algorithm: after the description of the method (sect. 2) I present the results which I have obtained in some simple cases (sect. 3). The difficulties in extending this method to more interesting cases are briefly discussed in sect. 4.

* For a recent review of the connections between statistical mechanics and field theory, see ref. [2].

2. The Langevin equation for correlation functions

Let us consider a statistical system with hamiltonian $H[\phi]$, $\phi_i (i = 1, N)$ being N continuous variables. The equilibrium expectation value of a function $f[\phi]$ is given by

$$\langle f[\phi]\rangle = \int d[\phi] \exp(-H[\phi]) f[\phi] \Big/ \int d[\phi] \exp(-H[\phi]) . \qquad (2)$$

For simplicity we have set $\beta = (kT)^{-1} = 1$.

A practical way of computing $\langle f[\phi]\rangle$ is based on the Langevin equation*:

$$\dot{\phi}_i = -\frac{\partial H}{\partial \phi_i} + \eta_i(t) , \qquad \langle \eta_i(t) \eta_j(t)\rangle = 2\delta_{ij}\delta(t - t') , \qquad (3)$$

$\phi_i(t)$ being a function of the "time" t and $\eta_i(t)$ being a random gaussian variable. Indeed, we have

$$\langle f[\phi]\rangle = \lim_{T\to\infty} \frac{1}{T} \int_0^T dt\, f[\phi(t)] , \qquad (4)$$

where $\phi(t)$ is the solution of eq. (3) for a given random choice of η.

Eqs. (3), (4) are a practical tool to compute the statistical expectation value, alternative to the standard Monte Carlo technique.

Before actually solving eq. (3) we must discretize the time:

$$\phi_i^{k+1} = -\varepsilon \frac{\delta H}{\partial \phi_i}[\phi^k] + \sqrt{2\varepsilon} R_i^k , \qquad (5)$$

where we have set $t = \varepsilon k$ and the R_i^k are random gaussian variables:

$$\langle R_i^k R_{i'}^{k'}\rangle = \delta^{kk'} \delta_{ii'} . \qquad (6)$$

The computer time is proportional to ε^{-1} and the errors on the final result are proportional to ε. If accurate results are needed, this procedure is slower than the standard Monte Carlo: the time discretized Langevin equation is very similar to the Monte Carlo method with small steps proportional to $\varepsilon^{1/2}$.

From the fluctuation-dissipation theorem we know that:

$$\langle \phi_i \phi_j\rangle_c = \frac{d}{d\lambda} \langle \phi_i\rangle_\lambda |_{\lambda = 0} , \qquad (7)$$

* The properties of the Langevin equation have been recently studied in connection with the critical slowing down near a second-order phase transition [3]: for a direct application of the Langevin equation to gauge theories see ref. [4].

where $\langle\phi\rangle_\lambda$ is computed with the hamiltonian $H_\lambda = H - \lambda\phi_j$. We can introduce the λ-dependent Langevin equation*

$$\dot{\phi}_i(t,\lambda) = -\frac{\partial H}{\partial \phi_i} + \lambda\delta_{ij} + \eta_i(t)\,. \tag{8}$$

Expanding $\phi_i(t,\lambda)$ in powers of $\lambda\,(\phi_i(t,\lambda) = \phi_i^{(0)} + \lambda\phi_i^{(1)} + \frac{1}{2}\lambda^2\phi_i^{(2)} + \cdots)$ and identifying the terms, we get

$$\begin{aligned} &\dot{\phi}_i^{(0)} = -\frac{\partial H}{\partial \phi_i} + \eta_i(t)\,, \qquad \dot{\phi}_i^{(1)} = -\frac{\partial^2 H}{\partial\phi_i\,\partial\phi_1}\phi_1^{(1)} + \delta_{ij}\,, \\ &\dot{\phi}_i^{(2)} = -\frac{\delta^2 H}{\partial\phi_i\,\partial\phi_1}\phi_1^{(2)} - \frac{\partial^3 H}{\partial\phi_i\,\partial\phi_1\,\partial\phi_n}\phi_1^{(1)}\phi_n^{(1)}\,, \end{aligned} \tag{9}$$

where H and its derivatives are computed as functions of $\phi^{(0)}$ only.

We finally get

$$\begin{aligned} \langle\phi_i\phi_j\rangle_c &= \lim_{T\to\infty}\frac{1}{T}\int_0^T \phi_i^{(1)}(t)\,dt\,, \\ \langle\phi_i\phi_j\phi_j\rangle_c &= \lim_{T\to\infty}\frac{1}{T}\int_0^T \phi_i^{(2)}(t)\,dt\,. \end{aligned} \tag{10}$$

Similar expressions can be obtained for higher order correlation functions.

We have found a direct method to computing the correlation functions; this method cannot be implemented in the standard Monte Carlo approach: we have used in an essential way the fact that the trajectory is a continuous (analytic) function of the force; that property does not hold in Monte Carlo simulations, where the force controls a probability factor which gives a yes-no answer. In sect. 3 we shall see the application of this method to simple examples.

3. Some examples

The simplest example we can think about is free field theory: the hamiltonian is

$$H = \tfrac{1}{2}\int d^D x[m^2\phi^2(x) + (\partial_\mu\phi)^2]\,, \tag{11}$$

the field being defined on a D-dimensional space.

The equations for $\phi^{(0)}$, $\phi^{(1)}$ and $\phi^{(2)}$ are respectively

$$\begin{aligned} &\dot{\phi}^{(0)}(x,t) = -(-\Delta + m^2)\phi^{(0)}(x,t) + \eta(x,t)\,, \\ &\langle\eta(x,t)\eta(x't')\rangle = 2\delta^D(x-x')\delta(t-t')\,, \\ &\dot{\phi}^{(1)}(x,t) = -(-\Delta + m^2)\phi^{(1)}(x,t) + \delta^D(x)\,, \\ &\dot{\phi}^{(2)}(x,t) = -(-\Delta + m^2)\phi^{(2)}(x,t)\,. \end{aligned} \tag{12}$$

* This approach is very similar to the technique of ref. [5] for computing time-dependent correlation functions in the molecular dynamics approach; their method fails to compute static correlation functions, the time evolution of the system being ergodic.

The solution of the last two equations is trivial in momentum space:

$$\begin{aligned}\phi^{(1)}(k,t)&=[1-\exp(-t(k^2+m^2))]/(k^2+m^2)\,,\\ \phi^{(2)}(k,t)&=0\,.\end{aligned}\tag{13}$$

The boundary conditions $\phi^1(x,0)=\phi^2(t,0)$ have been imposed. The same asymptotic result for $\tau\to\infty$ is also obtained if we discretize the time.

Generally speaking, in the method proposed here the equations for the connected correlation functions are deterministic if the hamiltonian is gaussian. The convergence is therefore very fast; no random errors are present.

Let us consider an interacting case:

$$H=\int \mathrm{d}x^D[\tfrac{1}{2}m^2\phi^2(x)+\tfrac{1}{2}(\partial_\mu\phi)^2+\tfrac{1}{4}g\phi^4]\,,$$

$$\begin{aligned}\dot\phi^{(0)}(x,t)&=-(-\Delta+m^2)\phi^{(0)}(x,t)+g(\phi^0(x,t))^3+\eta(x,t)\,,\\ \dot\phi^{(1)}(x,t)&=-(-\Delta+m^2)\phi^{(1)}(x,t)+3g(\phi^0(x,t))^2\phi^{(1)}(x,t)+\delta^D(x)\,.\end{aligned}\tag{14}$$

The equations are very simple if we substitute $\langle\phi^{(0)2}\rangle$ for $(\phi^{(0)}(x,t))^2$ in the last equation: we get the Hartree-Fock approximation.

We notice that for m^2 positive, the second derivative $\delta^2H/\delta\phi(x)\,\delta\phi(y)$ is a positive operator and an absolute bound on $\phi^{(1)}(x,t)$ can easily be obtained. The method may fail (although I do not believe that it fails) when m^2 is negative; $\phi^{(1)}(x,t)$ may oscillate and become larger and larger with time and the limit $\tau\to\infty$ in eq. (10) may not exist; this problem deserves a more accurate study.

In order to verify the efficiency of the method, I have done a very simple test. I have considered a one-dimensional system, the hamiltonian of which is

$$H=\sum_{i=1}^{N}(\tfrac{1}{4}\phi_i^4+4(\phi_i-\phi_{i+1})^2)\,,\tag{15}$$

where periodic boundary conditions have been imposed, and N has been taken equal to 98.

Computer simulations of eqs. (5), (6) have been done in two cases:

$$\begin{aligned}&\text{(a)}\quad \varepsilon=0.01\,, &&\tau=500\,,\\ &\text{(b)}\quad \varepsilon=0.0025\,, &&\tau=125\,.\end{aligned}\tag{16}$$

The results for the correlation function as a function of distance are shown in fig. 1 for case b. The typical statistical error is a few percent; the systematic error is proportional to ε. Going from case a to case b the correlation fuñctions change by about 10%, so the systematic error can be also estimated in the few percent range.

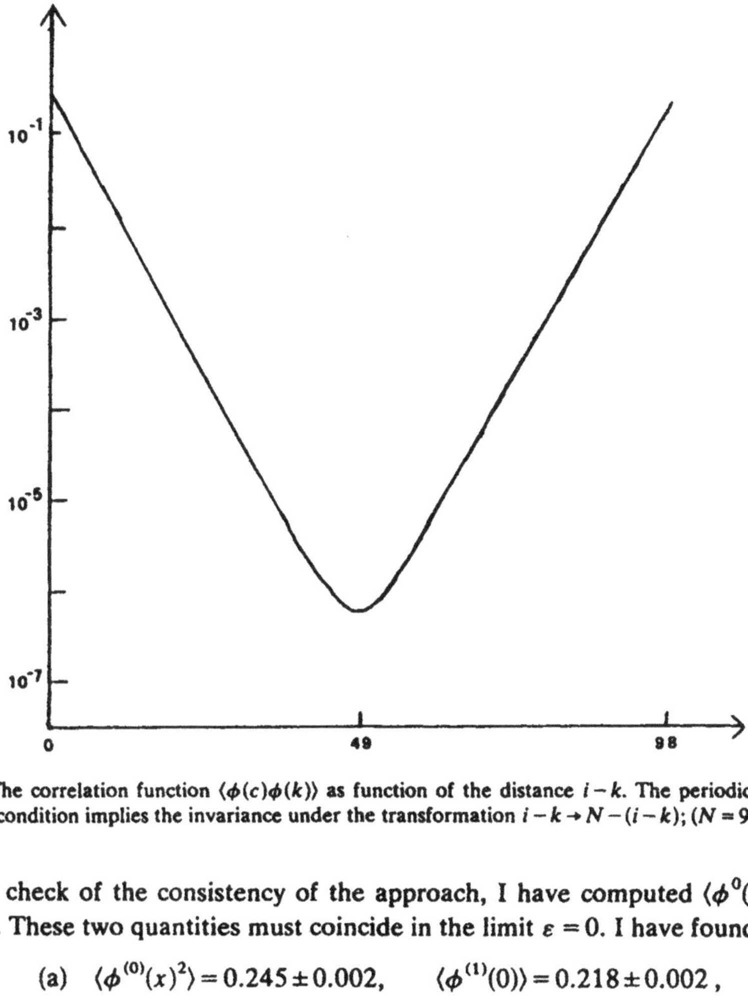

Fig. 1. The correlation function $\langle\phi(c)\phi(k)\rangle$ as function of the distance $i-k$. The periodic boundary condition implies the invariance under the transformation $i-k \to N-(i-k)$; $(N=98)$.

As a check of the consistency of the approach, I have computed $\langle\phi^0(x)^2\rangle$ and $\langle\phi^1(0)\rangle$. These two quantities must coincide in the limit $\varepsilon=0$. I have found:

$$\begin{aligned} &\text{(a)} \quad \langle\phi^{(0)}(x)^2\rangle = 0.245 \pm 0.002, \qquad \langle\phi^{(1)}(0)\rangle = 0.218 \pm 0.002\,, \\ &\text{(b)} \quad \langle\phi^{(0)}(x)^2\rangle = 0.234 \pm 0.004, \qquad \langle\phi^{(1)}(0)\rangle = 232 \pm 0.006\,, \end{aligned} \tag{17}$$

The error being purely statistical. We clearly see that in case a the two results do not agree, while in case b good agreement is found.

In cases a and b the mass is estimated to be 0.285 and 0.289, respectively.

4. Open problems

If the method I propose also gives good results for higher dimensional systems, it would be rather interesting to extend it both to discrete systems, such as the Ising model, and to constrained systems like the non-linear σ-model or gauge theories.

Generally speaking it is impossible to extend the method to a general discrete system, the continuity and differentiability of the time evolution being a prerequisite for applying this method. However, in simple cases, like the Ising model, something can be done: indeed the partition function of the Ising model can be written as [6]

$$\sum_{\{0=\pm 1\}} \exp\left[\tfrac{1}{2}\beta \sum_{i,k} \sigma_i \sigma_k V_{ik}\right]$$

$$= \int \mathrm{d}\phi_i \exp\left[-\tfrac{1}{2}\beta \sum_{i,k} \phi_i \phi_k V_{ik} + \sum_i \ln \mathrm{ch}\left(\sum_k \beta V_{ik}\phi_k\right)\right],$$

$$\langle \sigma_i \rangle = \left\langle \mathrm{th}\left(\sum_k \beta V_{ik}\phi_k\right)\right\rangle, \tag{18}$$

$$\langle \sigma_i \sigma_j \rangle = \left\langle \mathrm{th}\left(\sum_k \beta V_{ik}\phi_k\right) \mathrm{th}\left(\sum_l \beta V_{jl}\phi_l\right) + \delta_{ih}\, \mathrm{ch}^{-2}\left(\sum_k \beta V_{ik}\phi_k\right)\right\rangle .$$

The correlation functions of the σ spins can be easily reconstructed from the correlation functions of ϕ fields, which may be computed using the method proposed here.

The evolution equation for ϕ is

$$\dot{\phi}_i = -\sum_k V_{ik}\left[\phi_k - \mathrm{th}\left(\sum_l \beta V_{kl}\phi_l\right)\right] + \eta_i ,$$

$$\langle \eta_i(t)\eta_k(t')\rangle = 2\delta_{ik}\delta(t-t')/\beta . \tag{19}$$

If we neglect the noise term (η_i), we recover the mean field equations.

Unfortunately this operation of substituting a discrete variable for a continuous one can be easily done only if the interaction is quadratic. Serious difficulties would be present in an Ising model with a 4-spin interaction.

For constrained systems, if the above described trick does not work, we have many possibilities, whose respective advantages and disadvantages are unclear to me, the simplest one consisting in replacing the delta function of the constraint with a strongly peaked function. This question deserves more investigation.

Another problem comes from the Langevin equation itself: in order to simulate it on a computer we must discretize the time, introducing a non-zero ε. In the algorithm described in sect. 3 the error is proportional to ε; there are algorithms for which the error is of order ε^2 but they are rather complicated. In reality we need only that the error on the asymptotic behaviour is small: we are not interested to know with high precision the time dependence of the solution of the Langevin equation. In order to reach thermodynamic equilibrium we need to enforce the detailed balance principle for the transition probabilities $P(x, x')$:

$$\exp[-H(x)]P(x, x') = P(x', x)\exp[-H(x')] , \tag{20}$$

as can be seen from the corresponding master equation [7].

For example, in the one-dimensional case we can write the equation

$$\phi^{(k+1)}-\phi^{(k)}=-\varepsilon\frac{\partial H}{\partial\phi}C(\phi)+S(\phi)\sqrt{2\varepsilon}R^{(k)}, \tag{21}$$

where the correction factors $C(\phi)$ and $S(\phi)$ are chosen in such a way as to enforce the detailed balance at order ε^2.

Possible choices are

$$C(\phi)=1+\tfrac{1}{2}\varepsilon\frac{\partial^2 H}{\partial\phi^2}, \qquad S(\phi)=1, \tag{22a}$$

or

$$C(\phi)=1, \qquad S(\phi)=1-\tfrac{1}{4}\varepsilon\frac{\partial^2 H}{\partial\phi^2}. \tag{22b}$$

In the quadratic case ($H=\frac{1}{2}m^2\phi^2$) the recurrence equation would be

$$\phi^{(k+1)}-\phi^{(k)}=-A\phi^{(k)}+\sqrt{2\varepsilon}BR^{(k)}, \tag{23}$$

where A and B are given by the condition

$$A-\tfrac{1}{2}A^2=m^2\varepsilon B^2. \tag{24}$$

No solutions are found for $B=1$ if $m^2\varepsilon>\frac{1}{2}$. It would be interesting to extend this procedure to the general case in an efficient way, bypassing the condition $m^2\varepsilon<\frac{1}{2}$. This step may be crucial if we want to replace the constraints by a stiff potential.

It is also possible to avoid the introduction of the constraint and to write directly the Langevin equation on the constrained system.

The author is grateful to G. Ciccotti, C. de Dominicis, L. Peliti and K. Wilson for illuminating discussions and comments.

References

[1] K. Wilson, Talk at 1977 Crete Summer School, Cargese Lecture Notes (1979);
M. Creutz, L. Jacobs and C. Rebbi, Phys. Rev. Lett. 42 (1979) 1390; Phys. Rev. D20 (1979) 1915;
M. Creutz, Phys. Rev. Lett. 45 (1980) 313
[2] J. Kogut, Rev. Mod. Phys. 51 (1979) 791;
G. Parisi, Proc. 20th Conf. on High-energy physics, Madison 1980, to be published
[3] B. Halperin and P.C. Hoemberg, Rev. Mod. Phys. 49 (1977) 435
[4] G. Parisi and Wu Wong-shi, ASITP preprint 80-004 (1980), Sci. Sin., to be published
[5] G. Ciccotti, G. Jacucci and I. R. McDonald, J. Stat. Phys. 21 (1979) 1
[6] A. M. Polyakov, JETP (Sov. Phys.) 28 (1969) 533
[7] K. Binder, Monte Carlo methods, ed. K. Binder (Springer, 1979)

Nuclear Physics B225 [FS9] (1983) 475–496

CONSIDERATIONS ON NUMERICAL ANALYSIS OF QCD

H.W. HAMBER

The Institute for Advanced Studies, Princeton, NJ08540, USA

E. MARINARI

Service de Physique Théorique, CEN-SACLAY, 91191 Gif sur Yvette, France

G. PARISI

Universitá di Roma II, Tor Vergata, Roma, Italy

C. REBBI

Brookhaven National Laboratory, Upton, NY 11973, USA

Received 13 June 1983

We discuss the strategy to remove the quenched approximation and to minimize systematical and statistical errors occurring in the numerical simulation of lattice QCD.

We suggest a way to compute the flavour singlet sector of the mass spectrum, and comment about the relation between the fluctuations of the mass values and the scattering amplitudes.

1. Introduction

Recently many efforts have been devoted to the analysis of the low-energy part of the hadronic spectrum and of the pattern of chiral symmetry breaking in lattice QCD. While most of the numerical work has been done using the so-called quenched approximation [9, 2–7] (i.e. neglecting the quark vacuum polarization diagrams), some recent results have been obtained for the full theory [1]. Although the results of these computations are affected by systematical errors (due for example to the small size of the lattice [8]) that are very difficult to control, they suggest (and make somehow concrete) the possibility of more precise investigations, to be done on larger lattices, which would determine the hadronic mass spectrum in a picture where systematic errors are under control and statistical errors are reasonably small.

In this paper we discuss the strategy to remove the quenched approximation. The claim that these computations can be performed by using an amount of CPU time comparable with the time needed for a quenched computation is supported by the results of the numerical simulations [1] we mentioned above.

In sect. 2 we review the general formalism [9–11, 2–4] for numerical simulations of lattice gauge theories with fermions. In sect. 3 we study the sources of systematic errors affecting the evaluation of the effects of quark loops, and we suggest the computational methods to be preferred. In sect. 4 we outline how the computation can be organized.

Three appendices focus on specific problems: in appendix A we show how to compute the flavour singlet ($I=0$) sector of the mass spectrum. In appendix B we discuss the relation between the fluctuation of the mass values and the scattering amplitudes, and about the formation of many quark bound states. In appendix C we give some details about our computer codes, and we suggest efficient computational methods.

2. Dealing with fermionic theories: the general formalism

2.1. THE LATTICE FERMIONIC THEORY: EFFECTIVE ACTION AND SECOND-ORDER FORMALISM

Let us consider a lattice theory, characterized by the following euclidean action:

$$S[A, \bar{\psi}, \psi] = S_G[A] + \sum_{f=1}^{n_f} \bar{\psi}_i^{(f)} \Delta_{ik}[A] \psi_k^{(f)}, \qquad (2.1)$$

where A stands for the set of bosonic fields defined on the links of the lattice, ψ for the set of fermionic fields defined on sites, i and k denote the lattice sites, n_f is the number of flavours. Although the action of eq. (2.1) has a quite general form, in the following discussion we will assume that the A fields are gauge fields and that the ψ fields are quarks.

Using the Matthews–Salam formula [12] we can rewrite the mesonic and the baryonic propagator in the following way (for the sake of simplicity we will drop color, flavour and spinor indices):

$$\langle \bar{\psi}(x)\psi(x)\bar{\psi}(0)\psi(0)\rangle_{I\neq 0} = \int \mathrm{d}\mu_{\mathrm{eff}}[A]\, \Delta_{x,0}^{-1}[A]\Delta_{0,x}^{-1}[A], \qquad (2.2a)$$

$$\langle \bar{\psi}(x)\psi(x)\bar{\psi}(0)\psi(0)\rangle_{I=0} = \int \mathrm{d}\mu_{\mathrm{eff}}[A]\, \Delta_{x,0}^{-1}[A]\Delta_{0,x}^{-1}[A] - n_f \int \mathrm{d}\mu_{\mathrm{eff}}[A]\, \Delta_{0,0}^{-1}[A]\Delta_{x,x}^{-1}[A], \qquad (2.2b)$$

$$\langle \bar{\psi}(x)\bar{\psi}(x)\bar{\psi}(x)\psi(0)\psi(0)\psi(0)\rangle = \int \mathrm{d}\mu_{\mathrm{eff}}[A]\, (\Delta_{x,0}^{-1}[A])^3, \qquad (2.2c)$$

where

$$\mathrm{d}\mu_{\mathrm{eff}}[A] \propto \mathrm{d}[A]\, \mathrm{e}^{-S_{\mathrm{eff}}[A]} = \mathrm{d}[A]\, \mathrm{e}^{-S_G[A]} \det{}^{n_f} \Delta[A]$$
$$= \mathrm{d}[A] \exp\{-[S_G[A] - n_f \operatorname{Tr}(\ln \Delta[A])]\}, \qquad (2.3)$$
$$\int \mathrm{d}\mu_{\mathrm{eff}}[A] = 1 .$$

By $I=0$ we denote the flavour singlet sector; we will assume that flavour symmetry is unbroken.

Fermions on the lattice are usually described by the Wilson [13] or the Kogut–Susskind [14–16] (KS) formalism: in both schemes the determinant of Δ is a real number, and it is positive in the KS picture. In both cases the operator Δ is not hermitian; however we can use the following relation*

$$\det(\Delta)\in \mathbb{R}^+ \Rightarrow \det(\Delta)=[\det(\Delta)\det(\Delta^*)]^{1/2}$$
$$\Rightarrow \det(\Delta)=\det(\Delta\Delta^*)^{1/2}\equiv \det(\tilde{\Delta})^{1/2},$$

in order to deal with a hermitian operator. Substituting $\tilde{\Delta}$ for Δ in (2.3) yields a theory with $2n_f$ flavours; for the continuum this procedure corresponds to the use of a second-order formalism theory [17, 9], i.e. to quantize the action

$$\bar{\psi}(-D_\mu D^\mu+\sigma_{\mu\nu}F^{\mu\nu}+m^2)\psi,$$

where the ψ are parafermions of order $\frac{1}{2}$. Substituting at the same time $\tilde{\Delta}$ for Δ and $\frac{1}{2}n_f$ to n_f we will get a theory with the same number of flavours as the original one. Now for KS fermions $d\mu_{\text{eff}}[A]$ can be thought as a probability measure; the same holds for Wilson fermions if one is in conditions such to avoid a change in sign of $\det(\Delta)$, or restricts oneself to even n_f.

The computational procedure we have in mind is the following: by using some probabilistic algorithm we produce configurations of the gauge fields thermalised with respect to the measure $d\mu_{\text{eff}}[A]$, and then we measure expectation values such as those considered in equations (2.2). The inverse operators needed for this last step (fermionic Green function) Δ^{-1}_{ik} can be obtained, with the required high precision, for the given configuration of the gauge fields by using a fast algorithm, suitable to invert sparse matrices (relaxation, Gauss–Seidel, conjugate gradient). A difficult goal to achieve is to measure the mass differences between $I=0$ and $I\neq 0$ states; we will comment about this point in appendix A.

It should be noticed that in eq. (2.3) n_f is not necessarily an integer number. This observation can be used to compensate for the species doubling of the KS fermions. The KS action gives rise, in the continuum limit, to

$$n_f^{(\text{KS})}=2^{d/2}$$

fermionic species (d is the number of space-time dimensions), and setting the parameter n_f contained in the effective action to be equal to (wanted number of species)/$n_f^{(\text{KS})}$ we will get the quark-loop corrections for the wanted number of flavours.

* We denote by A^* the adjoint of the operator A (hermitian conjugation)

$$(A^*)_{ik}=\text{Re}(A_{ki})-i\,\text{Im}(A_{ki}).$$

2.2. GENERATING THE GAUGE FIELD CONFIGURATIONS

We discuss now the algorithm which can be used to generate configurations of the gauge fields A with the measure $d\mu_{\rm eff}[A]$. We will temporarily forget the unitarity constraint on A. These will be considered in sect. 4.

Let us outline two possible computational methods. The first possibility consists in writing an equation "à la Langevin" for the fields $A_i(t)$ (t is in this case the Langevin time):

$$\begin{aligned}\frac{dA_i}{dt} &= -\frac{\delta}{\delta A_i}(S_{\rm eff}) + \eta_i(t) \\ &= -\frac{\delta}{\delta A_i}(S_G) + n_f \,\mathrm{Tr}\left\{\Delta^{-1}\frac{\delta}{\delta A_i}(\Delta)\right\} + \eta_i(t) \\ &= -\frac{\delta}{\delta A_i}(S_G) + \tfrac{1}{2}n_f \,\mathrm{Tr}\left\{\tilde{\Delta}^{-1}\frac{\delta}{\delta A_i}(\tilde{\Delta})\right\} + \eta_i(t)\,, \end{aligned} \tag{2.4}$$

where $\tilde{\Delta} = \Delta\Delta^*$, and $\eta_i(t)$ is a gaussian white noise with

$$\overline{\eta_i(t)\eta_j(t')} = 2\delta_{ij}\delta(t-t')\,. \tag{2.5}$$

The other possibility would be to use the Metropolis algorithm [18]. One A field at once is updated each time: the new tentative field $A_i^{\rm T}$ is given by

$$A_i^{\rm T} = A_i^0 + \rho r_i\,, \tag{2.6}$$

where A_i^0 is the old value of the field, r_i is a random 3×3 matrix, with elements such that

$$\langle r_i^{ab}\rangle = 0\,, \qquad \langle |r_i^{ab}|^2\rangle = 1\,,$$

$(a, b = 1, 2, 3)$, and ρ is a control parameter. We should remind the reader that in this analysis we are neglecting the constraint $A_i \in SU(3)$, that is implemented in the actual computation. Now we will select the new value for our field A_i (we call $A_i^{\rm N}$) by setting $A_i^{\rm N} = A_i^{\rm T}$ with probability p (exchange is performed) and $A^{\rm N} = A^0$ with probability $(1-p)$ (no change). The probability p is given by

$$\begin{aligned} p &= \min\{\exp\{-(S_{\rm eff}[A^{\rm T}] - S_{\rm eff}[A^0])\}, 1\} \\ &\equiv \min\{p_{\rm T}, 1\}\,. \end{aligned} \tag{2.7}$$

For small ρ we can write

$$\begin{aligned} p_{\rm T} &= \exp\left\{-\frac{\delta S_{\rm eff}}{\delta A_i}\rho r_i + O(\rho^2)\right\} \\ &= \exp\left\{-\left[(S_G[A_{\rm T}] - S_G[A_0]) - \tfrac{1}{2}n_f\rho r_i\,\mathrm{Tr}\left\{\tilde{\Delta}^{-1}\frac{\delta}{\delta A_i}\tilde{\Delta}\right\}\right]\right\}, \end{aligned} \tag{2.8}$$

where terms of order ρ^2 are neglected. Here ρ is a free parameter, that we can choose small enough for $O(\rho^2)$ in (2.8) to be negligible; we will have to however pay the price in the time of approach to equilibrium which grows as ρ^{-2}.

The two methods we have proposed can be related through a Kolmogorov result [19]: if we identify

$$t=\rho^2 n\,,$$

n being the number of times we upgrade "à la Metropolis" each of our link variables, in the limit where ρ goes to zero at fixed time t the Kolmogorov equation for the evolution of the transition probability for the A fields will tend to the Fokker–Planck equation associated to the Langevin eq. (2.4). In simpler words a Monte Carlo simulation with small up-dates is equivalent to a Langevin equation. The Monte Carlo procedure for continuous variables can be considered as a wise (the asymptotic equilibrium distribution is preserved) discretization in time (introducing a mesh for the time variable) of the Langevin equation.

It seems to us that, if the lattice size is not too small to get sensible physical results, the only possibility of avoiding the need of enormous computer time and/or of an unacceptably large quantity of computer memory lies in performing small average up-dates of the gauge fields, independently of the upgrading method chosen.

Using p_T as in eq. (2.8) for a non-zero ρ we expect to find an error of order ρ on the equilibrium distribution $d\mu_{\text{eff}}[A]$.

The limit $\rho\to 0$ can be taken in different ways: we can compute the physical quantities we are interested in for a few different small values of ρ, and extrapolating the result to $\rho=0$ (after verifying that we are in a zone in which a linear extrapolation makes sense). The other possibility is setting ρ to such a small value $\bar{\rho}$ that $\langle 0\rangle_{\rho=\bar{\rho}}-\langle 0\rangle_{\rho=0}$ is negligible. To this respect we want to remark that the contribution to the RHS of eqs. (2.4) and (2.8) due to the fermion fields should in principle be updated every time that a single link variable is substituted. In the $\rho\to 0$ limit updating this contribution, for example, just after one full sweep on all the gauge fields A, will induce on the equilibrium probability an error vanishing with ρ.

2.3. THE FERMIONIC CONTRIBUTION TO THE ACTION

We come now to the crucial point: how to compute $\operatorname{Tr}\{\tilde{\Delta}^{-1}(\delta/\delta A)\tilde{\Delta}\}$, appearing in (2.4) and (2.8). The computation of $((\delta/\delta A)\tilde{\Delta})_{ik}$ is straightforward: using the lattice equivalent of the second-order formalism only elements for which i and k are at most second-nearest-neighbour sites will be non-zero. So we are left with the computation of the elements of $(\tilde{\Delta}^{-1})$. Let us analyze some efficient techniques.

The first possibility (defermionization) consists in building an auxiliary Monte Carlo procedure for some auxiliary bosonic fields φ_i (lying on sites) that we call pseudofermions [10, 9]. We start from the identity

$$(\tilde{\Delta}^{-1})_{ik} = \int d\mu[\varphi]\varphi_i^* \varphi_k ,$$

$$\int d\mu[\varphi] = 1 ,$$

$$d\mu[\varphi] \propto d[\varphi] \exp\left\{ -\sum_{lm} \varphi_l^* \tilde{\Delta}_{lm} \varphi_m \right\} , \tag{2.9}$$

which suggests to us that we compute the two-point function of the φ field (with measure $d\mu[\varphi]$) by a Monte Carlo procedure consisting of n_{PF} sweeps on all the φ fields. We can now precisely expose the full recipe for constructing a configuration of A fields in equilibrium with $d\mu_{eff}[A]$. Each complete cycle will consist of two phases. First one does n_{PF} Monte Carlo sweeps for the φ fields (using the probability measure of eq. (2.9)), and afterward updates once the A fields (on the whole lattice) using, in (2.4) or (2.8),

$$\begin{aligned} \mathrm{Tr}\left\{ \tilde{\Delta}^{-1} \frac{\delta}{\delta A_i} \tilde{\Delta} \right\} &= (\tilde{\Delta}^{-1})_{lm} \frac{\delta}{\delta A_i} \{\tilde{\Delta}_{ml}\} \\ &\simeq \frac{\delta}{\delta A_i} \{\tilde{\Delta}_{ml}\} \overline{\varphi_m^* \varphi_l} \\ &= 2\,\mathrm{Re}\left[\Delta_{mk}^* \frac{\delta}{\delta A_i} \{\Delta_{kl}\} \overline{\varphi_m^* \varphi_l} \right] , \end{aligned} \tag{2.10}$$

where the bar denotes the average over the last $(n_{PF} - n_D)$ iterations over the pseudofermions, where n_D stands for the number of iterations used at the beginning of any pseudofermionic cycle for bringing the pseudofermions to equilibrium, and not used in computing average values (discarded). The error intrinsic to this approximation vanishes in the limit $1/n_{PF} \to 0$. It should be noticed that the limit of small ρ implements the limit of small $1/n_{PF}$: when the dynamics of the gauge fields slows down, the relative speed of the pseudofermionic one becomes greater.

It is clear that the second-order pseudofermionic formalism we have built up can be used independently from the probabilistic algorithm we choose to update the pseudofermions: a good choice to construct the equilibrium probability $d\mu[\varphi]$ may be to use the heat bath method [20]. The main reason which makes this procedure very convenient for the upgrading of pseudofermions is that the φ equilibrium probability is gaussian. Other advantages can be seen in the heat bath formulation: first there are no free parameters to be adjusted by hand (like $\overline{\delta\varphi^2}$ in the Metropolis algorithm), the procedure being optimized by itself. This is a very nice feature, because tuning parameters and trying to reach the maximum efficiency is always very demanding timewise. Secondly the heat bath is slightly faster (of a factor of order 2) than the Monte Carlo procedure, even if in the latter multiple hitting is used [21]. Finally the computer code does not contain in this case "logical IF" statements; in other words there are no discontinuities in the evolution of the φ_i fields with respect to the variables $\Delta_{ik}[A]$. This fact will be shown to be useful when computing the splitting between $I=0$ and $I\neq 0$ masses (see appendix A).

In both methods (Metropolis and heat bath) the use of the second-order formalism is compulsory: the pseudofermion action has to contain the hermitian operator $\tilde{\Delta}=\Delta\Delta^*$. On the contrary if we want to perform a direct evaluation of $(\Delta^{-1})_{ij}$ using the first-order formalism we can consider the following set of Langevin-type equations:

$$\dot{\varphi}_1^i = -\Delta_{ij}\varphi_1^j + \eta_i(t),$$
$$\dot{\varphi}_2^i = -(\Delta^{\mathrm{T}})_{ij}\varphi_2^j + \eta_i(t), \tag{2.11}$$

where

$$\overline{\eta_i(t)\eta_k(t')} = 2\delta_{ik}\delta(t-t'),$$

Δ^{T} is the transposed Δ and in the two equations (2.11) the stochastic noise is the same. Assuming that Δ_{ij} is a positive operator, in the sense that

$$\lim_{t\to\infty} \mathrm{e}^{-\Delta t} = 0,$$

we get

$$\lim_{T\to\infty} \frac{1}{T}\int_0^T \varphi_1^i(\tau)\varphi_2^k(\tau)\,\mathrm{d}\tau = (\Delta^{-1})_{ik}, \tag{2.12}$$

or

$$\lim_{t\to\infty} \overline{\varphi_1^i(t)\varphi_2^k(t)} = (\Delta^{-1})_{ik}, \tag{2.13}$$

where the bar denotes the average over the noise η. We can easily prove eq. (2.13): let us formally integrate eq. (2.11) by writing

$$\varphi_1^i(t) = \int_0^t (\mathrm{e}^{-\Delta\tau})_{ij}\eta_j(t-\tau)\,\mathrm{d}\tau,$$

$$\varphi_2^k(t) = \int_0^t (\mathrm{e}^{-\Delta^{\mathrm{T}}\tau})_{kl}\eta_l(t-\tau)\,\mathrm{d}\tau$$

$$= \int_0^t \eta_l(t-\tau)(\mathrm{e}^{-\Delta\tau})_{lk}\,\mathrm{d}\tau. \tag{2.14}$$

Now

$$\lim_{t\to\infty} \overline{\varphi_1^i(t)\varphi_2^k(t)} = \lim_{t\to\infty} \int_0^t\int_0^t (\mathrm{e}^{-\Delta\tau'})_{ij}(\mathrm{e}^{-\Delta\tau''})_{lk} 2\delta_{jl}\delta(\tau''-\tau')\delta\tau'\delta\tau''$$

$$= \lim_{t\to\infty} 2\int_0^t (\mathrm{e}^{-2\Delta\tau})_{ik}\,\mathrm{d}\tau = (\Delta^{-1})_{ik}, \tag{2.15}$$

and eq. (2.13) is proved.

Practically we will estimate $(\Delta^{-1})_{ik}$ from eq. (2.13) using finite time: we will integrate our equation not for an infinite time but up to $t=t_{PF}$ (here t_{PF} plays the role of n_{PF} in the second-order algorithms), with a t_{PF} long enough for the error to be negligible.

Now we want to study numerically the Langevin system (2.11): so we will have to discretize the time. The most naive expression is

$$\varphi(n+1)=\varphi(n)+\varepsilon O\varphi(n)+\sqrt{2\varepsilon}\,r(n)\,, \tag{2.16}$$

where we omitted the euclidean space-time indices, and where O is $(-\Delta)$ for φ_1 and $(-\Delta^T)$ for φ_2, n labels the discretized Langevin time, and the $r(n)$ are random numbers with

$$\langle r(n)\rangle=0\,,\qquad \langle r(n)^2\rangle=1\,.$$

This naive transcription has been tested on a variety of different physical systems; it turns out that it succeeds in being a good approximation of the corresponding continuous equation only for very small integration steps ε, such that the computer time needed for the integration is prohibitively long. The way out is to consider a more accurate discrete transcription (that will ensure realistic computing time):

$$\varphi_1(n+1)=\varphi_1(n)-\varepsilon\Delta\varphi_1(n)+\tfrac{1}{2}\delta\varepsilon^2\Delta^2\varphi_1(n)+\sqrt{\varepsilon}\{(1+\tfrac{1}{2}\delta\Delta)r(n)+s(n)\}\,, \tag{2.17}$$

and the same for φ_2, with $\Delta\to\Delta^T$. Here $r(n)$ and $s(n)$ are two independent random variables. The case in which $\delta=1$ corresponds to a second-order Runge-Kutta approximation (for the deterministic part): this is the only case we have considered [1], although it is not clear whether a value of δ slightly larger than 1 could be better.

To end this section we note that, in the same way that we proved eq. (2.13), one can show from eq. (2.11) that, if the operator Δ is normal ($[\Delta,\Delta^*])=0$),

$$\lim_{t\to\infty}\overline{\varphi_1^i(t)\varphi_1^k(t)^*}=2[(\Delta+\Delta^*)^{-1}]_{ik}\,. \tag{2.18}$$

We will use this relation in sect. 3 in order to estimate the error implied from considering a finite t_{PF}.

3. Minimizing the errors

We have seen in sect. 2 that in the method we are proposing for simulating the feedback of quark loops over the gauge field dynamics there is an intrinsic systematic error of order ρ (or order $t^{1/2}$ in the Langevin formulation, (2.4)). Assuming that ρ is small we want to estimate the error induced by the fact that n_{PF} (or t_{PF}, in the Langevin pseudofermionic formalism) is finite.

Consider the complete Monte Carlo cycle for constructing $d\mu_{eff}[A]$ that we described in sect. 2: although not compulsory when one begins a new cycle it is convenient to use as a starting point (initial conditions for the φ-fields) the n_{PF}th value of the pseudofermionic variables one computed in the previous cycle. Indeed they will be off-equilibrium by a quantity proportional to ρ. For n_{PF} large enough we can write [10].

$$\frac{1}{n_{PF}}\sum_{n=1}^{n_{PF}} \varphi_k^{(n)*}\varphi_i^{(n)} = (\tilde{\Delta}^{-1})_{ik} + S_{ik} + R_{ik}\,, \tag{3.1}$$

where S_{ik} is a systematic effect proportional to $1/n_{PF}$ and R_{ik} is a noise proportional to $(1/n_{PF})^{1/2}$. The S_{ik} term would eventually be exponentially small in n_{PF} if one could approximate (Δ^{-1}) with

$$\frac{1}{n_{PF}(1-x)}\sum_{n=xn_{PF}}^{n_{PF}} \varphi_k^{(n)*}\varphi_i^{(n)}\,, \tag{3.2}$$

where x $(0 \leq x < 1)$ is such that (xn_{PF}) is an integer ((xn_{PF}) is called n_D in sect. 2). But one should be aware that this requires very large values of n_{PF}.

In the Langevin framework (and, by using the equivalence of the small-step Monte Carlo procedure with a Langevin-like procedure, also for the Metropolis updating scheme) it is very easy to see that the global systematical error done in computing expectation values, over $d\mu_{eff}[A]$ will be of the form (n_f is the number of fermionic species)

$$\frac{\rho^2}{n_{PF}}(An_f + Bn_f^2) + O\left(\frac{1}{n_{PF}^2}\right), \tag{3.3}$$

where the term proportional to n_f^2 comes from the noise contribution (R_{ik}) and the n_f part originates from the systematic effects (S_{ik}). It follows that the choice of the algorithm to be used depends on the number of fermionic species one wants to consider: if we recall that in our formulation n_f is not necessarily an integer number, and that the x parameter of eq. (3.2) has the role of making small the systematic error term appearing in eq. (3.1), S_{ik}, it becomes clear that x should be chosen close to one for a small n_f $(n_f \to 0)$, while a small x value is convenient for large n_f.

The last point we want to discuss here concerns the advantages presented by a heat bath (or Monte Carlo) defermionization with respect to a Langevin one, and vice versa. Let us assume we are using Kogut–Susskind fermions. We have seen that the Monte Carlo approach is based on the second-order formalism: so we expect that the time for equilibrating the pseudofermionic system, t_{eq}, will be proportional, when the bare quark mass m_q goes to zero, to m_q^{-2}. On the other hand the noise contribution to the error should not be too big. On the contrary the Langevin-like algorithm is based on a first-order formalism. In this case the diagonal term of the operator we want to equilibrate will be m_q, and $t_{eq} \sim m_q^{-1}$. One pays for this nice feature by the fact that

$$\overline{|\varphi_1^i|^2} = \frac{1}{m_q}\,,$$

as can be seen from the relation (2.18), and from the fact that in the KS approach [16]

$$\Delta + \Delta^* = 2m_q, \tag{3.5}$$

This means that there is a noise on $\overline{\varphi_1\varphi_2}$ proportional to m_q^{-1}; it is a stronger effect than the one found in the Monte Carlo approach (numerical experiments show that the noise on the equivalent quantity stays quite small for small m_q).

We can now formulate our conclusions, by using the expression (3.3). The use of the Langevin equation is suitable for small number of flavours, while the Monte Carlo and heat bath are to be preferred for large n_f. Moreover, the value n_f^c for which it is convenient to flip from the first-order formalism to the second-order one, is a function of m_q.

As a last remark we note that in the expression (3.3) there is an overall ρ^2 factor, and that the global error is given by this expression plus the effect due to the finiteness of ρ. So it seems to us to be convenient to fix n_{PF} to a value reasonably large (for the values of the coupling constant, of the lattice size and of m_q used in ref. [1] $n_{PF} \simeq 50$ appeared to be an acceptable choice), and to extrapolate ρ to zero. This procedure is, as we noticed in subsects. 2.2, 2.3, a suitable way of avoiding an extrapolation in a two-variables space.

4. The structure of the numerical experiment

In the previous sections we have enumerated a frightening quantity of possible options, ways of extrapolating, practical differences in implementing the suggested algorithms: we have now to be clear about which are the optimal ones.

Let us start by discussing which lattice formulation we have to choose for the fermionic contribution to the effective action. In our opinion the job is best done by the KS fermions. Several advantages over Wilson fermions can be easily seen: there is manifest chiral symmetry (see the last of the references in [16] and [9]), the fermionic determinent is positive, the bare quark mass is defined on an absolute scale (due to chiral invariance the critical point of the theory has to stay at $m_q = 0$; there is not yet a value as the Wilson K_c, critical K, to be found). Moreover the corrections to the continuum limit are proportional to a^2 (where a is the lattice spacing), while on the contrary they are proportional to a in the Wilson action (in this case, due to the presence of an explicit chiral invariance breaking term, operators of dimension-5 like $\bar{\psi}D_\mu D^\mu\psi$ or $\bar{\psi}\sigma_{\mu\nu}F^{\mu\nu}\psi$ are allowed. These operators are not present in KS formulation, which preserves chiral symmetry). Last but not least in the KS formulation just one variable for the lattice site is present ("staggered fermions"), in contrast with the four spin components in the Wilson formulation; so the computer KS program turns out to be faster by a factor of order 4 than the Wilson fermion program, and occupies much less memory. On the other hand the identification of quantum numbers (spin, isospin) is straightforward for Wilson fermions, and less simple (in particular for the baryonic sector) for KS fermions [16]. So a possibility could be to represent the *external* quark lines (Δ^{-1} in eq. (2.2)) in the Wilson picture. For the mesons, however, it costs only a slight increase of CPU time to perform the computation also using KS external lines. A separate discussion is needed for the $I=0$ part of the spectrum, and we will present it in appendix A.

Let us turn now to the practical way in which the mass spectrum computation (of a theory with $n_f=2$ or 3) should be performed. For obtaining reasonably accurate results one may have to analyze to the order of 10^2 statistically independent gauge field configurations (on a lattice with something like $10^3\cdot 20$ sites, at squared coupling constant $g^2\sim 1$). The number of Metropolis time steps (with optimized efficiency) needed to destroy the correlation between the estimated masses associated with two gauge field configurations is unclear. In ref. [4] masses were found to be correlated up to ~1000 steps; this effect is however mainly due to the influence of the boundary conditions on the fermionic Green functions in the Wilson formulation [8]. Anyhow if in a quenched computation one computes the Wilson quark correlation functions on one configuration over one hundred, the pure gauge Metropolis upgrading only takes an order of 10% of the total computer time [4, 5]; therefore if we loose a factor 10 for generating the gauge field configurations using $d\mu_{\text{eff}}[A]$, instead of the pure gauge measure $d\mu[A]$, we just need a factor 2 more of total computer time.

We can go further by using the following strategy: we upgrade the whole system (gauge fields and pseudofermions) one hundred times, using a large value of ρ, we call ρ_L (let us assume ρ_L is such that the pure gauge Metropolis procedure is optimized). This way we will produce a configuration which is slightly off-equilibrium (with respect to the measure $d\mu_{eH}[A]$): now we will equilibrate it by performing another hundred sweeps at a small ρ value, ρ_S. We will compute the Green functions only on the last configuration (of the 200 we have produced in the way we have described).

This was just a very naive sketch of a possible efficient scheme, but we can try now to improve it. Our defermionization algorithm can be thought as characterized by the three parameters ρ, $\beta=6/g^2$ and m_q: so let us suppose our final configurations to have been obtained with the choice of the parameters $\bar{\rho}$, $\bar{\beta}$ and $\bar{m}_q$ (where $\bar{\rho}$ is small enough to make negligible the error induced by its finiteness, $\bar{m}_q$ goes as close to zero as the size of our lattice allows it and $\bar{\beta}$ is in the scaling region). We can think now of performing our 100 "fast" steps using the parameters ρ_F, β_F and m_q^F chosen such that

$$E(\rho_F,\beta_F,m_q^F)=E(\bar{\rho},\bar{\beta},\bar{m}_q)\,,$$

$$\langle\bar{\psi}\psi\rangle(\rho_F,\beta_F,m_q^F)=\langle\bar{\psi}\psi\rangle(\bar{\rho},\bar{\beta},\bar{m}_q)\,. \tag{4.1}$$

In this way the configuration we get at the end of our 100 fast sweeps (done with

ρ_F, β_F and m_q^F) will be as close as possible to the thermal equilibrium with $d\mu_{eff}[A]$ $(\tilde{\rho}, \tilde{\beta}, \tilde{m}_q)$. We should however stress that the number 100 + 100 we are quoting here is arbitrary enough; the optimal number of sweeps needed between two configurations on which the averages are done is at the moment quite unclear, and will depend on the size of the lattice.

What will be the effect of internal quark-loops on the masses? One plausible possibility is that, for the $I \neq 0$ particles and the lowest-lying baryons, it is negligible, while for the heavier states it could produce some large widths (small widths have the nice feature of simplifying the analysis of the correlation functions). Let us consider the hypothetical situation in which there already exists a very accurate computation done in the quenched approximation. This computation has been done on a large lattice, using an improved action [22] (in order to minimize the effects of the non-zero lattice spacing), and the masses have been computed with high precision (of a few percent). Now if we consider a typical mass ratio, for example

$$R = \frac{m_p}{m_\rho},$$

it is possible to expand it in powers of n_f:

$$R(n_f) = R_0 + n_f R_1 + n_f^2 R_2 + O(n_f^3).$$

Phenomenological arguments suggest that R_1 is much smaller than R_0, and that the R_2 term is negligible for $n_f = 2$; in this situation a computation of R_1 with a not-so-high accuracy (and on a not-so-large lattice) would be by fair enough. It would be very convenient to compute R_1 in a direct way by comparing the results obtained for $n_f = 0$ and $n_f = 0.1$. We should recall now what we explicitly claimed in sect. 2: our approach to the simulation of fermionic internal loops is valid also for non-integer n_f. Of course this cannot be done by comparing the results of two different Monte Carlo simulations: the statistical error would overcome the mass difference. It seems to us that what is needed is an algorithm $\mathscr{A}$ (thermalizing according to the measure $d\mu_{eff}[A]$) such that the gauge field configuration obtained by applying T times $\mathscr{A}$ to the system is a smooth function of n_f. This situation can be realized by avoiding the MC procedure, i.e. by upgrading the gauge field in the Langevin approach, and the pseudofermions by the heat bath or Langevin methods. Working at small n_f also presents the advantage that the noise contribution to the error (the R_{ik} term in eq. (3.1)), which is proportional to $\rho^2 n_f^2 n_{PF}^{-1}$ (eq. (3.3)), is in this limit negligible, and one can concentrate one's efforts in minimizing the systematic contributions to the error.

These arguments suggest that we exploit in a more systematic way the possibility of upgrading the gauge fields with a Langevin-like equation. We see two different possibilities: the first one is based on relaxing the condition that the gauge fields A are unitary, and adding to the action a term

$$\mathrm{Tr}\,\{(AA^* - 1)^2 + |\det(A) - 1|^2\},$$

which enforces on average the unitarity constraint. The second consists in writing a Langevin equation directly on the group manifold. In short, all we need is an upgrading algorithm producing in the phase space a trajectory that is a smooth function of the force, and satisfying the detailed balance (at least in an approximate way). We can write

$$A_\mu = \exp\{i\varepsilon^{1/2}\rho_\alpha\lambda_\alpha + \varepsilon P_A[FA_0^*]\}A_0, \tag{4.2}$$

with

$$F = \frac{\delta}{\delta A}(S_{\text{eff}}).$$

where the ρ_α are gaussian random numbers, the λ_α are the SU(3) Gell–Mann matrices and P_A is the projector onto the SU(3) algebra (i.e. it picks up the traceless antihermitian part of FA_0^*). For practical purposes it is convenient to truncate the Taylor expansion of the exponential and to renormalize the new matrix to SU(3). Eq. (4.2) satisfies the detailed balance condition:

$$P(A_0, A_N) = P(A_N, A_0)\exp\{-(S_{\text{eff}}[A_N] - S_{\text{eff}}[A_0])\}, \tag{4.3}$$

(where $P(A, B)$ is the probability of transition from A to B) with an accuracy of order ε^2. Since the number of iterations grows as ε^{-1}, the effective deviations from the detailed balance turn out to be of order ε. In the limit of small ε eq. (4.2) provides an algorithm for which the trajectories are smooth functions of the parameters which appear in S_{eff}. Multiple hitting, (updating a link more than once any time we touch it) can be used to increase the efficiency of the algorithm.

The method we are suggesting is sensible if two nearby trajectories do not separate exponentially in time. In order to see under which conditions this happens let us consider the one-variable case

$$\dot{x}(t) = -\frac{\partial U}{\partial x} + \eta(t). \tag{4.4}$$

We are interested in understanding what happens to two nearby trajectories $x_1(t)$ and $x_2(t)$, which evolve under the same noise $\eta(t)$. If we set $\delta = x_2 - x_1$ we get

$$\dot{\delta} = -\left.\frac{\partial^2 U}{\partial x^2}\right|_{x=x_1(t)} \delta + O(\delta^2), \tag{4.5}$$

and in the limit of small δ we see that if $\partial^2 U/\partial x^2 > 0$ the two trajectories converge, and they diverge if $\partial^2 U/\partial x^2 < 0$. It is, most of the time, in the region where $\partial^2 U/\partial x^2$ is positive (as suggested by naive arguments) the trajectories should not diverge.

The situation is completely analogous in the many-variable case:

$$\dot{x}_i(t) = -\frac{\partial S}{\partial x_i} + \eta_i(t), \tag{4.6}$$

where now the crucial condition is that the average in time of the hessian matrix

$$H = \frac{\partial^2 S}{\partial x_i \partial x_j},$$

should be positive.

A serious difficulty arises if, by the symmetry of the problem, the hessian has a zero eigenvalue; in this case δ may start growing up in the direction of the correspondant eigenvector, and this could result in trajectories exponentially separating in time. Due to gauge invariance this condition is realized in gauge theories. We see three possible remedies: one is to add to the action a small term which violates gauge invariance. The second one is to fix the gauge (axial gauge) without producing ghosts. The third one consists in making after every few iterations an explicit gauge transformation over the gauge fields, in such a way as to bring them as close as possible to the identity. It seems to us that the third method should be preferred; adding a gauge breaking term to the action could badly modify the expectation value of large-distance correlation functions, and the simulation in the axial gauge converges too slowly towards thermal equilibrium. This procedure also affects the pseudofermionic sector of our cycle: we think that using the third option, also the systematic contribution in (3.1), will be diminished.

This is the computational procedure we think to be the more suitable; we obtain in this way two slightly different configurations of the gauge fields (in the sense that the trajectory in time of the values of the interesting operators stay close), respectively, let's say, with $n_f = 0$ and $n_f = 0.1$. The splitting of the masses due to the quark loops will be now free from most of the statistical errors.

We wish to thank O. Napoly for a critical reading of the manuscript and interesting discussions, and Mrs. S. Zaffanella for the very careful typing of a very tricky manuscript. The work of H.W.H. was supported by the US Department of Energy under grant no. DE-AC02-76ER02220.

Appendix A

THE $I = 0$ SECTOR

One of the goals one can try to reach by computer simulations is to obtain results about the masses of the $I = 0$ mesons; this is clearly a problem of great physical interest. It is, unluckily enough, also rather hard: one has to evaluate in this case

$$\overline{G_{ii}(A) G_{kk}(A)},$$

(where G are the fermionic Green functions) and that is very difficult. In fact to estimate the correlations one has to do an explicit average over the A field configurations, and also if i is kept fixed one has to do the computation for all k.

The simplest possibility is to use the Metropolis or Langevin to compute $G_{kk}(A)$ for all k in the quenched limit; if we set

$$M_{l=1} = m_{l=0} + \Delta m n_f + O(n_f^2),$$

for n_f small we get

$$\lim_{n_t = |i-k| \to \infty} \frac{G_{ii}[A] G_{kk}[A]}{G_{ik}[A] G_{ki}[A]} = \Delta m n_t . \tag{A.1}$$

This object is very difficult to evaluate for all channels except the pseudoscalar one, where the signal is rather high. A second possible way out would be to compute $G_{kk}[A]$ by means of the hopping-parameter expansion [6]. The common bottle-neck of these methods is that in practice they do not allow us to measure the signal at a distance larger than $n_t \sim 3$: this is due to the presence of a large statistical error.

Let us describe now the method we think could be applied to solve this problem*. We introduce two different pseudofermionic fields φ_1 and φ_2 (splitting the n_f fermionic species contribution to the action into two $\frac{1}{2}n_f$ equal parts), and consider the operator

$$O_\alpha(t) = \sum_x \varphi_\alpha^*(x, t) \varphi_\alpha(x, t) , \qquad \alpha = 1, 2 , \tag{A.2}$$

where the sum over the spatial part x is done over the tth hyperplane of dimension 3. We add now to the action a term $\varepsilon O_1(t=0)$, setting ε to a small value ε_0, and measuring the quantities

$$R_\alpha(t) = \frac{1}{\varepsilon} \{ \langle O_\alpha(t) \rangle_{\varepsilon=\varepsilon_0} - \langle O_\alpha(t) \rangle_{\varepsilon=0} \} . \tag{A.3}$$

We get in this way

$$R_1(t) \sim e^{-m_{l=0} t_{l=0}} + e^{-M_{l=1} t_{l=1}}$$

$$R_2(t) \sim e^{-m_{l=0} t_{l=0}} - e^{-M_{l=1} t_{l=1}} . \tag{A.4}$$

If the mass splitting is large enough we can measure in this way $m_{l=0}$ and $M_{l=1}$. But the most important remark (and we built all this machinary just for this goal) is that if the splitting is small we get

$$\frac{R_2(t)}{R_1(t)} \sim -\Delta m t . \tag{A.5}$$

In this way it should be possible to measure also a very small splitting. The price we will have to pay is that we will have to perform a separate computation for every choice of the particle quantum numbers; but this seems to us to be the only realistic possibility.

* For the application of an analogous method to the evaluation of the glueball mass see ref. [23].

Appendix B

SCATTERING AMPLITUDES VIA MASSES FLUCTUATIONS

The way in which the particles' masses are computed in a numerical simulation is quite straightforward. For the $I \neq 0$ mesons, for example, one first computes

$$G_{\Gamma}(\boldsymbol{x}, t|A) \equiv \langle \psi\Gamma\psi(\boldsymbol{x}, t)\bar{\psi}\Gamma\psi(\boldsymbol{0}, 0)\rangle_A \tag{B.1}$$

and sets his operators to zero spatial momentum by defining

$$G_{\Gamma}(t, A) = \sum_{\boldsymbol{x}} G_{\Gamma}(\boldsymbol{x}, t|A) . \tag{B.2}$$

This procedure is repeated for each configuration $A^{(n)}$. Finally one computes

$$G_{\Gamma}(t) = \int d\mu_{\rm eff}[A] G_{\Gamma}(t|A) , \tag{B.3}$$

and the interesting mass will be given from the large-time behavior

$$\lim_{t\to\infty} \left\{ -\frac{1}{t} \ln G_{\Gamma}(t) \right\} = m . \tag{B.4}$$

We can also think of defining a value of the mass for each $A^{(n)}$th configuration, by the relation

$$\lim_{t\to\infty} \left\{ -\frac{1}{t} \ln G_{\Gamma}(t|A^{(n)}) \right\} = m_{\Gamma}[A^{(n)}] . \tag{B.5}$$

Now we ask ourselves if the following relation is true:

$$\int d\mu_{\rm eff}[A] m_{\Gamma}[A] \equiv \tilde{m}_{\Gamma} \stackrel{?}{=} m_{\Gamma} . \tag{B.6}$$

The answer is easily given, and it is no [25]. We will now discuss this point in some detail. Let us consider the case of a fairly large box, so that the fluctuations induced in the masses by the boundary conditions can be neglected (these fluctuations should go to zero as a high power of L), and look at the behavior of

$$G_{\Gamma}^{(n)}(t) = \int d\mu_{\rm eff}[A][G_{\Gamma}(t|A)]^n \tag{B.7}$$

in a finite large box of size L. Now we expect $G_{\Gamma}^{(n)}(t)$ to behave like

$$\lim_{t\to\infty} G_{\Gamma}^{(n)}(t) = e^{-m_{\Gamma}^{(n)} t} , \tag{B.8}$$

where $m_{\Gamma}^{(n)}$ is the mass of the lightest state composed from n particles present in the box, in which each quark has different spin or flavor quantum numbers. If $m_{\Gamma}^{(n)}$

is an analytic function of n we have

$$\frac{\mathrm{d}}{\mathrm{d}n} m_{\Gamma}^{(n)}|_{n=0} = \tilde{m}_{\Gamma}. \tag{B.9}$$

The interaction between particles of different flavors can be expected to be attractive (van der Waals forces are generally attractive); this also follows by convexity arguments (e.g. $\overline{G^2} > \bar{G}^2$). There are two possible scenarios:

(a) particles do not form a bound state, and have an attractive scattering length L;

(b) they do form a bound state with binding energy m_B.

A simple ansatz for $m^{(n)}$ is therefore for the two cases

$$m_{\Gamma}^{(n)} = nm_{\Gamma} - \frac{n(n-1)}{2V}\frac{l}{m_{\Gamma}} \tag{B.10a}$$

$$m_{\Gamma}^{(n)} = nm_{\Gamma} - \tfrac{1}{2}n(n-1)m_B. \tag{B.10b}$$

These two equations follow from assuming the existence of two-body forces, and small l and Δm; the constant can be computed in the framework of non-relativistic potential models [24]. We will just use them at a very qualitative level. Using eq. (B.9) we get

$$\tilde{m}_{\Gamma} = m_{\Gamma} + \frac{1}{2V}\frac{l}{m_{\Gamma}}, \tag{B.11a}$$

$$\tilde{m}_{\Gamma} = m_{\Gamma} + \tfrac{1}{2}m_B. \tag{B.11b}$$

This means that there is a non-zero well-defined difference between the most probable mass (i.e. the average mass) and the true mass. In the case (a) the difference disappears when the volume of the box grows to infinity, while in case (b) there is a finite gap also for large volumes, i.e. fluctuations in the mass values do *not* go to zero when $V \to \infty$. This last one is certainly the case of the proton; moreover m_B is corresponding to the binding energy of the Ξ with a nucleon, which can be extracted from measurements on hypernuclei (hyperfragments). m_B can be estimated to be in this case less than 100 MeV, so we expect an error not greater than ~50 MeV (5 per cent) rising from computing $\tilde{m}_{\mu}$ instead of m_{μ}.

The situation can be further clarified if we write the probability distribution for the effective mass at distance t. Using the relation

$$G_{\Gamma}(t) = \int \mathrm{d}p_t(\mu)\, \mathrm{e}^{-\mu t} \tag{B.12}$$

and (B.10) we get

$$\mathrm{d}p_t(\mu) \propto \sqrt{\frac{t}{\Delta m \cdot \pi}} \exp\left\{-t\frac{(\mu - \tilde{m}_{\Gamma})^2}{2\Delta m}\right\} \mathrm{d}\mu, \tag{B.13}$$

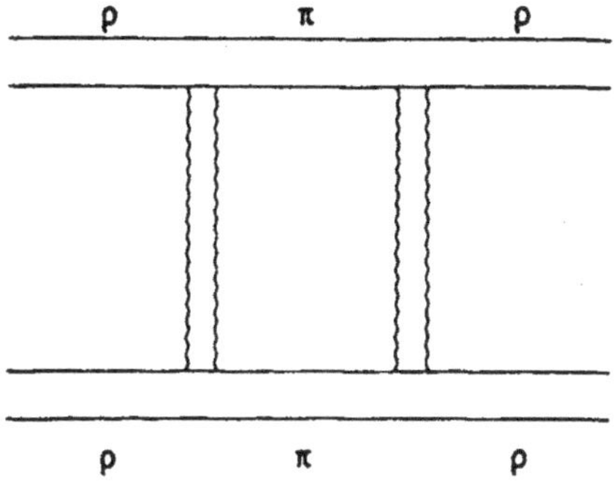

Fig. 1. The diagram dominating the two-ρ propagator large time behavior.

where for respectively the cases (a) and (b)

$$\Delta m = \frac{l}{V m_\Gamma}, \tag{B.14a}$$

$$\Delta m = m_B, \tag{B.14b}$$

It is easily seen that when $t \to \infty$ the distribution of the mass becomes very sharply peaked around the wrong value. This phenomenum is very well-known [25]. The value of m which maximizes the integral has a probability of the order $e^{-t\Delta m}$. For the protons (and likely for the other hadrons) the effect does not seem to be very important; the time t needed to resolve different resonances having the same quantum numbers is of the order of $(150\text{–}200\ \text{MeV})^{-1}$, so that in the useful region $t\Delta m$ is normally a small quantity (only for unnecessarily long lattices does this effect become important). This also means that the integral (B.12) is dominated by the contributions from the center of the distribution and not from the tail, implying that it is not necessary to consider too many configurations.

Unfortunately the situation is not so nice in mesonic channels: two ρ have the same quantum numbers of two pions, so that the diagram of fig. 1 dominates the two-ρ propagator large-time behavior. This approximately means

$$G_\rho^2(t) \sim \frac{1}{V} e^{-2m_\pi t} + e^{-2m_\rho t}. \tag{B.15}$$

In other words in the ρ propagator there is a small component which behaves as $V^{-1/2} e^{-m_\pi t}$ and not as $e^{-m_\rho t}$. This effect can also be expected from naive arguments: in the presence of a background field $\bar\psi\gamma_\mu\psi$ excites from the vacuum a ρ and also a π with a random phase. Therefore the total probability of creating a π from the vacuum by acting with $\int_V \bar\psi\gamma_\mu\psi\, d_x^D$ is proportional to $V^{-1/2}$. This effect can be dangerous, but the diagrams of fig. 1 are likely to be small at the threshold.

It is clear then that a study of mass fluctuations can be seen as the analysis of the contributions of some diagrams to the scattering amplitude. Eqs. (B.11) are particularly interesting in the case of the pion; from chiral symmetry arguments it turns out that the scattering length-squared of the pion is proportional to $m_\pi^2 f_\pi^{-4}$ so that we get

$$\tilde{m}_\pi \simeq m_\pi + \frac{1}{V f_\pi^2},$$

implying that the fluctuations of the pion mass do not increase when m_π goes to zero (on-shell pions at zero momentum are free when m_π goes to zero). This is not the case for Wilson fermions; in this formulation chiral symmetry is explicitly broken and the squared scattering length-squared is different from zero (proportional to a) for zero mass pions. Larger fluctuations are present in this case.

While the pattern of small mass fluctuations has been "experimentally" observed in the KS scheme, the opposite situation seems to hold for the Wilson formulation on small lattices (for $r = 1$, see ref. [4]). The main cause can be found in the non-zero expectation value acquired from the space loops and in the effect of the boundary conditions on the K_c [8]. Although this effect asymptotically disappears for large L, it is dominating on medium size lattices (0.5–0.8 fm*). A drastic solution in the quenched approximation would be to choose zero boundary conditions (or to sum over different Z_3 "gauge" transformations). This problem is likely to be alleviated by choosing a value of r (the Wilson chiral symmetry breaking parameter) less than one (e.g. $r = \frac{1}{2}$), and by using an improved action for the quark fields. Moreover, if the vacuum polarization quarks satisfy antiperiodic boundary conditions in space and time, the configuration in which the space and time loops are close to the identity are likely to be preferred in the $\beta \to \infty$ limit. We can conclude that it could be possible in this way to reach together the two goals of reducing the mass fluctuations to their natural size and to have interesting information about low-energy scattering amplitudes.

Appendix C

THE PSEUDOFERMIONS: COMPUTER CODES

In this appendix we want to give some details about our computer program for inserting in the Monte Carlo simulation the contribution of the fermionic determinant. We will treat here the numerical implementation of eqs. (2.9) and (2.10).

Let us remind the reader of some general features. We have defined a cycle on our system to be composed by an updating of the full lattice of link variables (gauge fields), other than by n_{PF} (number of pseudofermionic steps) sweeps on the site variables φ (pseudofermions), using as a starting point the last $\{\varphi\}$ configuration computed in the last cycle. This second part of our updating cycle has eventually

* Lattices of ~1 fm show that Z_3 effects are dramatically reduced [26].

the duty of producing the contribution to the action of the fermionic fields. When starting the next cycle (and updating the gauge fields again) we will fetch this contribution in the action we will use to update the gauge fields.

The part of the program used to update the gauge fields is different from the normal pure gauge program in one single line, in which the fermionic contribution is added to the pure gauge action: so we will describe here only the fundamental points of the computer code used to update the pseudofermions.

Let us start from a general description of our code, that is basically made up from three different phases. In the first phase, given the configuration of the gauge fields $\{A\}$ (that is the output of the first part of our cycle), and the configuration of the pseudofermions $\{\varphi\}$ (we will use as the "initial condition" the last φ-fields computed in the last cycle), we will compute the fields

$$H_i \equiv (\not{D}+m)_{ij}\varphi_j, \tag{C.1}$$

the utility of which will become clearer in the following.

The second phase has the rôle of updating the fermions, while in the third phase the contribution to the gauge action is computed (together with $\langle\bar{\varphi}\varphi\rangle$). The second phase is repeated $n_{\rm PF}\times\tilde{n}$ number of times, the third one is repeated $n_{\rm PF}$ times (that means that the contribution of the φ fields to the bosonic action is computed just from every $\tilde{n}$th step).

The computer memory we need is basically twice the one needed for a pure gauge simulation. If we indicate with N_c the number of colors (3), with d the number of dimensions (4), with N the number of sites per dimension (6–8), we need to store the fields

$$A(2dN^4N_c^2)\,,\qquad \varphi(2N^4N_c)\,,$$
$$G(2dN^4N_c^2)\,,\qquad H(2N^4N_c)\,. \tag{C.2}$$

The value in parentheses is the amount of real numbers we need to store for every field (the 2 is just due to the fact that the fields are complex). The A's are the gauge fields, the G's are the contributions to be summed to the pure gauge action, the φ are the pseudofermions and the H's have been defined in (C.1).

The upgrading phase is very simple: one random increment $\delta\varphi^a$ is chosen, where

$$\langle\delta\varphi^a\rangle = 0,\qquad \langle\delta\varphi^{a2}\rangle = \delta\,, \tag{C.3}$$

and δ is turned in such a way as to optimize the convergence toward equilibrium. We note that eq. (2.8) can be read as

$$e^{-\bar{\varphi}\Delta\varphi} = e^{\bar{H}H}\,. \tag{C.4}$$

So

$$\Delta S = (2\,\mathrm{Re}\,H + \mathrm{Re}\,(\Delta H))\,\mathrm{Re}\,(\Delta H) + (2\,\mathrm{Im}\,H + \mathrm{Im}\,(\Delta H))\,\mathrm{Im}\,(\Delta H)\,. \tag{C.5}$$

A variation in the φ-field in one site will induce $(2d+1)$ variations in the H-field.

The computation of these contributions (and the knowledge of $H \forall$ the sites we have from the first phase) allows us to compute ΔS in a very straightforward way.

Now if for example we apply the Metropolis procedure, given the random number r uniformly distributed in the interval (0, 1), if $\exp\{-\Delta S\} < r$ we do not change the φ, and try again on the same site (n times) or on the next site. Otherwise we set

$$\varphi(n) \to \varphi(n) + \delta\varphi(n),$$
$$H(n) \to H(n) + \delta H(n),$$
$$H(n \pm n_\mu) \to H(n \pm n_\mu) + \delta H(n \pm n_\mu). \tag{C.6}$$

where the fields H are changed in $2d+1$ locations.

The last duty of the code is to compute the contribution to the bosonic action (that will be averaged over n_{pF} added terms). If we write

$$S_{\mathrm{eH}} = S_{\mathrm{G}} + \tfrac{1}{4} n_{\mathrm{f}} \sum_{ij} G_{ij} U_{ij},$$

we will get

$$G_{ij} = \mathrm{Re}[\bar{H}^i(n+n_\mu)\varphi^j(n) - \bar{\varphi}^i(n+n_\mu)H^j(n)],$$

where μ is the direction of the link U_{ij}.

References

[1] H. Hamber, E. Marinari, G. Parisi and C. Rebbi, Phys. Lett. 124B (1983) 99
[2] H. Hamber and G. Parisi, Phys. Rev. Lett. 47 (1981) 1792;
E. Marinari, G. Parisi and C. Rebbi, Phys. Rev. Lett. 47 (1981) 1795;
H. Hamber, E. Marinari, G. Parisi and C. Rebbi, Phys. Lett. 108B (1982) 314
[3] D. Weingarten, Phys. Lett. 109B (1982) 57
[4] H. Hamber and G. Parisi, Phys. Rev. D27 (1983) 208;
F. Fucito, G. Martinelli, C. Omero, G. Parisi, R. Petronzio and F. Rapuano, Nucl. Phys. B210 [FS6] (1982) 407;
D. Weingarten, Indiana University preprint IUHET-82 (1982)
[5] C. Bernard, T. Draper and K. Olynyk, Hadron mass calculations in QCD, UCIA preprint 82/TEP/10 (June 1982);
K.C. Bowler, E. Marinari, G.S. Pawley, F. Rapuano and D.J. Wallace, Nucl. Phys. B220 [FS8] (1983) 137;
R. Gupta and A. Patel, Nucl. Phys. B226 (1983) 152
[6] A. Hasenfratz, Z. Kunszt, P. Hasenfratz and C.B. Lang, Phys. Lett. 110B (1982) 289;
A. Hasenfratz, P. Hasenfratz, Z. Kunszt and C.B. Lang, Phys. Lett. 117B (1982) 81;
P. Hasenfratz and I. Montvey, Phys. Rev. Lett. 50 (1983) 309
[7] J. Kogut, M. Stone, H.W. Wyld, J. Shigemitsu, S.H. Shenker and D.K. Sinclair, Phys. Rev. Lett. 48 (1982) 1140;
J. Kogut, M. Stone, H.W. Wyld, W.R. Gibbs, J. Shigemitsu, S.H. Shenker and D.K. Sinclair, Phys. Rev. Lett. 50 (1983) 393
[8] G. Martinelli, G. Parisi, R. Petronzio and F. Rapuano, Boundary effects and hadron masses in lattice QCD, CERN preprint TH3456 (Nov. 1982)
[9] E. Marinari, G. Parisi and C. Rebbi, Nucl. Phys. B190 [FS3] (1981) 734;
S. Otto and M. Randeira, Nucl. Phys. B220 [FS8] (1983) 479

[10] F. Fucito, E. Marinari, G. Parisi and C. Rebbi, Nucl. Phys. B180 [FS2] (1981) 369;
F. Fucito and E. Marinari, Nucl. Phys. B190 [FS3] (1981) 266
[11] D.H. Weingarten and D.N. Petcher, Phys. Lett. 99B (1981) 333
[12] T. Matthews and A. Salam, Nuovo Cim. 12 (1954) 563; 2 (1955) 120
[13] K.G. Wilson, Phys. Rev. D10 (1974) 2445; in New phenomena in subnuclear physics (Erice 1975), ed. A. Zichichi (Plenum, New York, 1977)
[14] T. Banks, S. Raby, L. Susskind, J. Kogut, D.R.T. Jones, P.N. Scherbach and D.K. Sinclair, Phys. Rev. D15 (1977) 1111;
L. Susskind, Phys. Rev. D16 (1977) 3031
[15] N. Kawamoto and J. Smit, Nucl. Phys. B192 (1981) 10
[16] H. Kluberg-Stern, A. Morel, O. Napoly and B. Peterson, Nucl. Phys. B190 [FS3] (1981) 504;
H. Kluberg-Stern, A. Morel and B. Peterson, Phys. Lett. 114B (1982) 152; Nucl. Phys. B215 [FS7] (1983) 527;
H. Kluberg-Stern, A Morel, O. Napoly and B. Peterson, Susskind fermions in configuration space, Saclay preprint DPhT/82/56; Flavours of lagrangian Susskind fermions, Nucl. Phys. B220 [FS8] (1983) 447.
[17] R.P. Feynman, Phys. Rev. 84 (1951) 108;
R.P. Feynman and M. Gell-Mann, Phys. Rev. 109 (1958) 193
[18] N. Metropolis, A.W. Rosenbluth, A.M. Teller and E. Teller, J. Chem. Phys. 21 (1953) 1087;
Monte Carlo Methods, ed. K. Binder (Springer-Verlag, Berlin 1979)
[19] I. Guikhman and A. Skorokhod, Introduction à la théorie des processus aléatoires (Mir, Moscow, 1980)
[20] M. Creutz, L. Jacobs and C. Rebbi, Phys. Rev. Lett. 42 (1979) 1390;
M. Creutz, Phys. Rev. Lett. 43 (1979) 553; Phys. Rev. D21 (1980) 2308
[21] N. Cabibbo and E. Marinari, Phys. Lett. 119B (1982) 387
[22] K. Symanzik, Mathematical problems in theoretical physics, eds. R. Schrader et al., Conf. Berlin 1981 (Springer Verlag 1982) Lecture Notes 153;
G. Martinelli, G. Parisi and R. Petronzio, Phys. Lett. 114B (1982) 251;
P. Weisz, DESY preprint 82-044 (1982)
[23] M. Falcioni, E. Marinari, M.L. Paciello, G. Parisi, B. Taglienti and Zhang-Yi chen, Nucl. Phys. B215 [FS7] (1983) 265
[24] L.D. Landau and E.M. Lifshitz, Quantum mechanics (Pergamon, London–Paris, 1958)
[25] B. Derrida and H. Hilhorst, J. Phys. C14 (1981) L539
[26] H. Lipps, G. Martinelli, R. Petronzio and F. Rapuano, Phys. Lett. 126B (1983) 250.

PHYSICAL REVIEW D VOLUME 29, NUMBER 5 1 MARCH 1984

Glueball-mass estimates in lattice QCD

Herbert W. Hamber and Urs M. Heller
The Institute for Advanced Study, Princeton, New Jersey 08540
(Received 21 March 1983)

We show how the Langevin equation for SU(3) gauge fields can be used to compute glueball correlation functions at large separation. On a $6\times6\times6\times6$ lattice we estimate the mass of the lowest 0^{++} glueball by determining the corresponding correlation function at separations 0, 1, 2, and 3. Several values for the coupling constant are investigated which lie in a region where scaling behavior for the string tension is observed. Our preliminary study indicates $m_{0^{++}}=(240\pm70)\Lambda_0$ or alternatively $m_{0^{++}}=(1.1\pm0.2)m_\rho$.

I. INTRODUCTION

The lattice gauge theory presents a well defined framework in which nonperturbative effects in QCD can be studied. One of the more fruitful techniques in this respect has been the Monte Carlo method. Several attempts have been made to extract the mass of the lightest state, the scalar glueball, in the pure gauge theory.[1-8] An appropriate connected two-point correlation function of operators that have the quantum numbers of the glueball is evaluated numerically, and from its large-distance exponential falloff the mass of the lowest state is extracted.

In the currently studied coupling-constant regime the lowest masses are of order one in lattice units, which implies that the correlation functions themselves become rapidly very small as the separation is increased. The statistical fluctuations in a numerical simulation are of order $N^{-1/2}$, where N is the number of Monte Carlo sweeps per variable over which the averaging is done. Because of this signal-to-noise-ratio problem one is limited in practice to rather short distances over which the correlation functions can be evaluated, if machine time is to be kept within reasonable limits. Also, as the gauge coupling becomes weaker it is necessary to determine the correlation functions at larger distances in order to separate the exponential tail from the uninteresting short-distance power behavior.

Glueball-mass estimates have been limited in the past to a study of correlations at distances 1 and 2, and in some rare instances 3.[2-7] An exception is Ref. 8 in which for the group SU(2) the glueball correlation function was determined up to separation 5. On the other hand, there is in general no reason to believe that the true asymptotic behavior of the correlation function is reached at such short distances. In Ref. 9 it was suggested that the Langevin equation could be used for computing connected glueball correlation functions, with an increase in accuracy of several orders of magnitude over the Monte Carlo method, and some calculations were performed for the group SU(2). The increase in accuracy is achieved by allowing for a coherent cancellation of statistical errors between two highly correlated stochastic processes.

In this paper the analysis is extended to the group SU(3). After introducing the Langevin equation for the group SU(3), we calculate the connected glueball correlation function at $\beta=5.4$, 5.5, 5.6, 5.7, and 5.8 and extrapolate the results for the glueball mass to the continuum limit using the renormalization group. Our results are in reasonable agreement with previous results (for a list of references to previous work we refer the reader to refs. 2 and 4, though they tend to indicate slightly lower values for the mass of the lowest 0^{++} glueball state.

The plan of the paper is as follows. In Sec. II we introduce the Langevin equation on the group manifold of SU(3) and discuss its time-discretized form. Then it is shown how the connected two-point correlation function can be evaluated by setting up two closely correlated processes with slightly different parameters in the actions. Section III goes into the details of the numerical simulation and presents our results, together with a discussion of statistical and systematic errors. In Sec. IV we discuss our results and compare to previous similar calculations. After pointing out the importance of extraneous effects such as the presence of a peak in the specific heat in the region where the masses are calculated and the spin-wave behavior of the correlation function at short distances, we discuss the extrapolation of the masses to the continuum limit.

II. THE LANGEVIN EQUATION FOR SU(3)

Let us first establish some notation. The gauge degrees of freedom are defined on the links of a four-dimensional periodic hypercubic lattice of spacing a and linear size L, and are elements of the group SU(3). We use the Wilson form of the action[10]

$$S_G=-\frac{\beta}{6}\sum_{n,\mu<\nu}\mathrm{Tr}U_{n,\mu}U_{n+\mu,\nu}U^\dagger_{n+\nu,\mu}U^\dagger_{n,\nu}+\mathrm{c.c.} \quad (2.1)$$

with $\beta=6/g^2$.

It is known that the Langevin equation[9] represents a useful alternative to the Monte Carlo method for generating a set of equilibrium configurations.[11-13] Given a field $\varphi(x)$ in the continuum, one introduces an extra time t and writes down the evolution equation

$$\dot{\varphi}(x,t) = -\frac{\delta S}{\delta \varphi} + \eta(x,t) , \quad (2.2)$$

where $\eta(x,t)$ is a Gaussian white noise

$$\langle \eta(x,t) \rangle = 0 , \quad \langle \eta(x,t)\eta(x',t') \rangle = 2\delta(x-x')\delta(t-t') . \quad (2.3)$$

In order to solve the above equation numerically, one needs to discretize the time. With a time step ϵ and $t=\epsilon k$, Eq. (2.3) becomes

$$\varphi^{(k+1)}(x) = \varphi^{(k)}(x) - \epsilon \frac{\delta S}{\delta \varphi}\bigg|_{\varphi=\varphi^{(k)}} + (2\epsilon)^{1/2}\eta^{(k)}(x) \quad (2.4)$$

with

$$\langle \eta^{(k)}(x) \rangle = 0 , \quad \langle \eta^{(k)}(x)\eta^{(k')}(x') \rangle = \delta(x-x')\delta_{kk'} . \quad (2.5)$$

This procedure introduces an error of order ϵ in the averages, which can in principle be reduced by going to small enough ϵ. It is possible to write down equations such that the detailed balance condition for the transition probabilities $P(c,c')$

$$e^{-S(c)}P(c,c') = P(c',c)e^{-S(c')} \quad (2.6)$$

is enforced to higher order in ϵ. Algorithms exist for which the error is of order ϵ^2, but they are rather complicated.[14]

The Langevin equation for Dirac fermions is also known and is discussed in Refs. 15 and 16 in the context of lattice gauge theories. Here we are interested in writing evolution equations for elements of the group SU(3). It is easy to show that the correct equilibrium distribution is recovered, up to order ϵ, if the matrices $U_{n\mu}$ evolve according to the stochastic equation[17,18]

$$U_{n\mu}^{(k+1)} = \exp\left\{ i(2\epsilon)^{1/2}\omega_{n\mu}^{a}\frac{\lambda_a}{2} - \epsilon P_{AT}\left[\frac{\delta S}{\delta U_{n\mu}^{(k)}}U_{n\mu}^{(k)}\right]\right\} U_{n\mu}^{(k)} . \quad (2.7)$$

Here the λ_a's are the Gell-Mann matrices, generators of the group SU(3), and the $\omega_{n\mu}^{a}$'s are random real numbers with zero mean and unit variance

$$\langle \omega_{n\mu}^{a} \rangle = 0 , \quad \langle \omega_{n\mu}^{a}\omega_{n'\mu'}^{\beta} \rangle = \delta_{nn'}\delta_{\mu\mu'}\delta^{\alpha\beta} . \quad (2.8)$$

The operator P_{AT} projects out the anti-Hermitian traceless part of the operator in parentheses. Its effect is to constrain the new element to lie still on the group manifold. The force term $\delta S/\delta U_{n\mu}$ contains the effect of the $6(d-1)$ neighboring links, and, in the case when fermion degrees of freedom are present, will include the contribution from the fermion currents.

In practice we prefer not to take the exponential of an operator, such as written in Eq. (2.7). The random SU(3) matrices

$$R_{n\mu}(\epsilon) = \exp\left[i(2\epsilon)^{1/2}\omega_{n\mu}^{a}\frac{\lambda_a}{2}\right] \quad (2.9)$$

are computed by expanding the exponential to fourth order (which is justified for small ϵ) and projecting them uniformly on the group. This is achieved, for example, by choosing at random a row or column of R, and orthonormalizing the remaining rows (viz., columns) with respect to the chosen one.[19] We have checked that the error introduced by this procedure is negligible (we typically use ϵ of order 10^{-3}). A table of 200 random R matrices (containing always both R and R^+ to ensure detailed balance) is updated every time a full sweep through the lattice is completed.

Because of the smallness of ϵ we have also chosen to expand the force contribution in Eq. (2.7) to lowest order in ϵ. The deviations from unitarity of the new matrices $U_{n\mu}^{(k+1)}$, which arises because we neglect higher-order contributions in ϵ, are corrected by projecting the matrices back on the group in the same way as described above for the $R_{n\mu}$ matrices.

In order to speed up the approach toward thermal equilibrium each link matrix $U_{n\mu}$ is updated ten times before we proceed to the next link. As we will show in the next section, we have checked that our procedure reproduces the correct energy density (average plaquette) with an error of order ϵ.

Let us now discuss the determination of the correlation functions.[9] In order to compute the connected correlation function of two glueball operators, we set up two correlated stochastic processes. We consider two initially identical systems E_0 and E_1, which are in thermal equilibrium at a temperature β. E_0 is then allowed to further evolve in time according to the evolution equations (2.7). In system E_1 the action is changed to

$$S \rightarrow S - \delta\tilde{O}(t_0) , \quad (2.10)$$

where the operator $\tilde{O}(t_0)$ is the zero-spatial-momentum operator

$$\tilde{O}(t_0) = \sum_{\vec{x}} O(t_0,\vec{x}) \quad (2.11)$$

and $O(t\vec{x})$ is a (not necessarily local) operator function of the $U_{n\mu}$ fields, which is summed over a fixed "time slice" t_0. The parameter δ is chosen to be small, and the system E_1 is then allowed to evolve in time with the *same* noise distribution $\{\omega_{n\mu}^{a}\}$ as system E_0. Note that because of the logical statements present in a Metropolis Monte Carlo procedure, phase coherence and ensuing cancellation of errors cannot be achieved for long runs. If the same operator $\tilde{O}(t)$ is then averaged over a different time slice t_1, one has for system E_0

$$\langle \tilde{O}(t_1) \rangle_{E_0} = \frac{\int [dU]\tilde{O}(t_1)e^{-S[U]}}{\int [dU]e^{-S[U]}} \quad (2.12)$$

while for system E_1 one obtains

$$\langle \tilde{O}(t_1)\rangle_{E_1} = \frac{\int [dU]\tilde{O}(t_1)e^{-S[U]+\delta\tilde{O}(t_0)}}{\int [dU]e^{-S[U]+\delta\tilde{O}(t_0)}}$$

$$= \langle \tilde{O}(t_1)\rangle_{E_0} + \delta\frac{\int [dU]\tilde{O}(t_1)\tilde{O}(t_0)e^{-S[U]}}{\int [dU]e^{-S[U]}} - \delta\frac{\int [dU]\tilde{O}(t_1)e^{-S[U]}\int [dU]\tilde{O}(t_0)e^{-S[U]}}{\left[\int [dU]e^{-S[U]}\right]^2} + O(\delta^2)\,. \quad (2.13)$$

From this one realizes that the connected correlation function of the operators $\tilde{O}(t)$ at distances $|t_1-t_0|$ is given by

$$\langle \tilde{O}(t_1)\tilde{O}(t_0)\rangle_c = \delta^{-1}[\langle \tilde{O}(t_1)\rangle_{E_1} - \langle \tilde{O}(t_1)\rangle_{E_0}] + O(\delta)\,. \quad (2.14)$$

For convenience we have chosen the measured operator $[\tilde{O}(t_1)]$ and the one in the action $[\tilde{O}(t_0)]$ to be the same, but the method of course allows for the possibility of having different operators $\tilde{O}_\alpha(t_1)$ and $\tilde{O}_\beta(t_0)$ on different times slices, and study their mixing. In this framework one can also gain information about the mixing between glueballs and operators containing fermion fields. One considers, for example, a glueball operator $\tilde{O}(t_0)$ added to the action and measures on the remaining time slices the expectation value of the meson operator

$$\sum_{\vec{x}} \bar{\psi}(\vec{x},t)\Gamma\psi(\vec{x},t)\,, \quad (2.15)$$

where Γ is a Dirac gamma matrix. After formal integration over the Fermi fields the above operator is replaced by the matrix element of the inverse of the lattice Dirac operator

$$\sum_{\vec{x}} \mathrm{Tr}\left\langle \vec{x},t\left|\frac{\Gamma}{\not{D}+m}\right|\vec{x},t\right\rangle \quad (2.16)$$

evaluated in a background gauge configuration.

In the present study we limited ourselves to the one-plaquette operator summed over spatial orientations

$$\tilde{O}(t) = \tfrac{1}{6}\sum_{\vec{x}} \mathrm{Tr}[U_P(\vec{x},t) + \mathrm{H.c.}]\,. \quad (2.17)$$

The change in the action induced on one time slice amounts in this case to an increase in the inverse gauge coupling $\beta\to\beta+\delta$ on the same slice. The operator in (2.17) has the correct spin-parity assignment for the $J^{PC}=0^{++}$ glueball state.

III. NUMERICAL RESULTS

In this section we present the results of a numerical simulation using the discretized Langevin equation (2.7). First we want to show that this provides an alternative way to create equilibrium configurations of the SU(3) lattice gauge theory. To this end we did runs on a $4\times4\times4\times4$ lattice from ordered starts at $\beta=5.6$ with $\epsilon=0.001$ and 0.0025. Figure 1 shows a graph of the average action per plaquette as a function of the number of sweeps through the lattice. As one can see it converges to the right value known from Monte Carlo simulations. The magnitude of ϵ is chosen as a compromise between two requirements: it has to be small since every step introduces an error of order ϵ. It should not be too small, because smaller ϵ provides slower convergence to equilibrium (see Fig. 1). We have found that $\epsilon=0.001$ to 0.0025 works reasonably well.

The update of a single link with the Langevin method takes about twice as long as in conventional Monte Carlo algorithms. The main reason for this is the need to project out the anti-Hermitian traceless part of $(\partial S/\partial U_{n\mu})U_{n\mu}$ in Eq. (2.7). However, the advantage of the Langevin formalism is that one can measure connected correlation functions accurately as described in the previous section. We have done this for various couplings: $\beta=5.4$, 5.5, 5.6, 5.7, and 5.8. These values of β were chosen because the crossover from strong to weak-

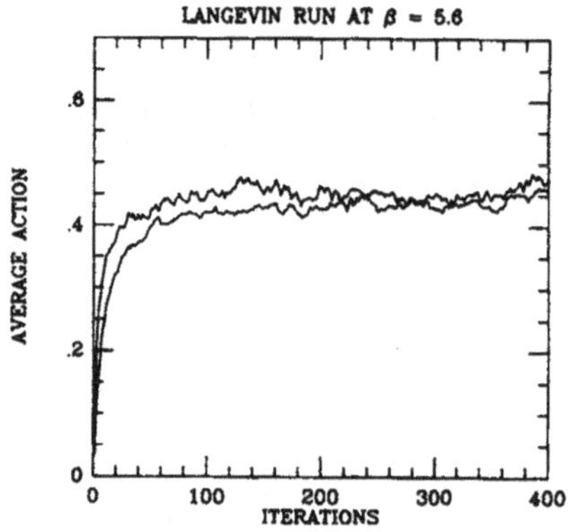

FIG. 1. Comparison of two Langevin runs at $\beta=5.6$ for $\epsilon=0.001$ (lower curve) and $\epsilon=0.0025$ (upper curve).

coupling behavior in the string tension is observed at $\beta=5.3-5.4$, which roughly coincides with the observed peak in the specific heat, and since the string tension scales according to the asymptotic-freedom formula for $\beta=5.4-6.0$. The lattice used should be big enough to accurately represent long-wavelength fluctuations, which means that it should at least be somewhat larger than the correlation length. We always worked on a $6\times6\times6\times6$ symmetric lattice. Monte Carlo data for the string tension show that such a lattice should be large enough for the values of β considered here.

As explained in Sec. II, at each β we had to perform two highly correlated runs (E_0 and E_1). We started them from configurations which were previously brought to equilibrium. At $\beta=5.6$ we used a configuration which we had on tape. It was thermalized from an ordered start by 2400 Monte Carlo iterations. The starting configurations for the other couplings were obtained from this by another 500 Monte Carlo sweeps. In the Langevin runs for computing the correlation functions we used both $\epsilon=0.001$ and 0.0025 for $\beta=5.5$, 5.6, and 5.7, while at $\beta=5.4$ and 5.8 we limited ourselves to $\epsilon=0.001$. We had to stop our runs after $\sim$400 iterations, because at that point the phase coherence between the two correlated systems was lost due to the accumulated roundoff errors. At $\beta=5.7$ we also did a run in which we set ϵ to 0.0025 for the first 100 iterations (in both systems E_0 and E_1) to speed up the approach to equilibrium, and then changed it to 0.001 to reduce the statistical fluctuations. We used $\delta=\delta\beta=0.05$ in system E_1 throughout. In all cases we were able to reproduce the correct average action per plaquette (see Table I). Figure 2 shows a plot of the average action per plaquette for the two correlated Langevin runs at $\beta=5.6$ and for $\epsilon=0.0025$. It clearly shows that the two systems E_0 and E_1 fluctuate together.

Because of our use of periodic boundary conditions we were allowed to average over forward and backward correlations. We computed

TABLE I. Comparison of the average action per plaquette. For the Langevin runs the average is over all 400 iterations of system E_0, while for the Monte Carlo runs the average is over the last 400 iterations at each β.

β	ϵ	Langevin	Monte Carlo
5.4	0.001	0.5231±0.0035	0.5239±0.0030
5.5	0.001	0.5006±0.0058	0.5017±0.0041
	0.0025	0.5009±0.0054	
5.6	0.001	0.4690±0.0095	0.4761±0.0011
	0.0025	0.4687±0.0065	
5.7	0.001	0.4442±0.0042	
	0.0025	0.4488±0.0041	0.4477±0.0021
	0.0025/1	0.4456±0.0034	
5.8	0.001	0.4278±0.0024	0.4347±0.0033

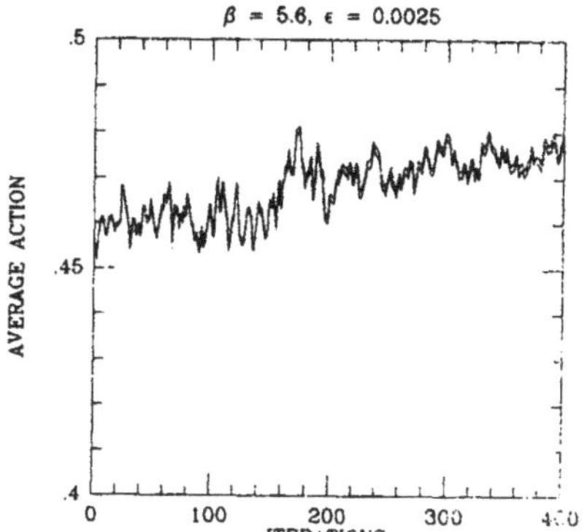

FIG. 2. Average action at $\beta=5.6$ for $\epsilon=0.0025$. In the run corresponding to the lower curve the temperature β has been raised on one time slice by an amount $\delta\beta=0.05$.

$$G(t)=\tfrac{1}{2}[\langle\tilde{O}(t_0+t)\tilde{O}(t_0)\rangle_c+\langle\tilde{O}(t_0-t)\tilde{O}(t_0)\rangle_c] \tag{3.1}$$

with $t\leq L/2$ and L the linear size of the lattice in the time direction. In Fig. 3 we show the correlation functions $G(t)$ for $t=0$, 1, 2, and 3 at $\beta=5.6$ and for $\epsilon=0.0025$. As expected, the noise in the correlation functions increases with distance. While the result for the average plaquette is not noticeably dependent on ϵ in the range we investigated, the approach to equilibrium for correlations at longer distances is considerably slower for smaller ϵ (0.001 as compared to 0.0025). This phenomenon becomes more acute when the mass gap m_G gets smaller. For large (Langevin) times we expect the relaxation time τ that governs the approach to equilibrium to scale as

$$\tau\sim m_G^{-z}, \tag{3.2}$$

where z is a dynamical critical exponent.[11] No estimate for this exponent is known to us for gauge theories in four dimensions. For a simple φ^4 theory near four dimensions one finds $z=2$ whereas for the O(3) Heisenberg model near $d=6$ one gets $z=4$. The rather large dynamical critical exponent for the O(3) model is connected to the presence of the global continuous symmetry and Goldstone modes.

We would expect the index z to be rather large for gauge theories as well since motions along gauge orbits cost no energy and do not drive the system toward equilibrium. In principle this difficulty could be overcome by doing longer runs. However, we had to limit our runs to 400 iterations, because after that the phase coherence between the two correlated runs was lost. Some of those problems could probably be avoided by using double precision or a 64-bit machine (we worked on a 32-bit machine). We found a somewhat improved convergence

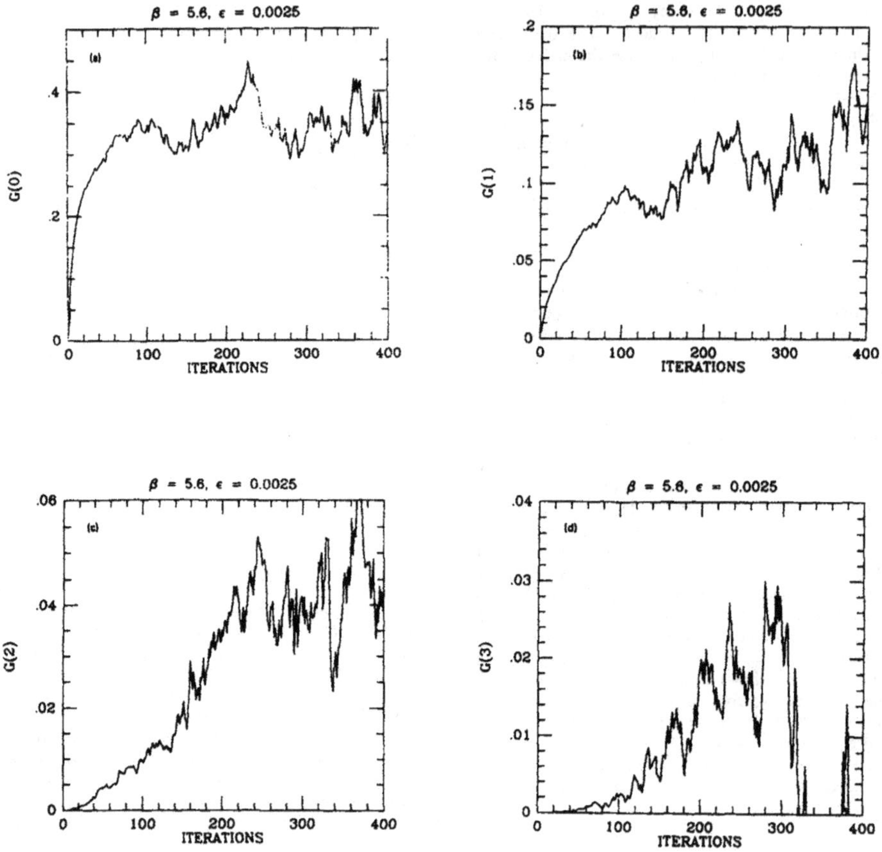

FIG. 3. Glueball correlation function for $\beta=5.6$ and $\epsilon=0.0025$ at separation 0 (a), separation 1 (b), separation 2 (c), and separation 3 (d).

rate when a bigger value of ϵ was taken at the beginning of a run and later was reduced in order to improve on the errors. The study of an optimal choice of ϵ's is beyond the scope of the present investigation.

Since the operator $\tilde{O}(t)$ we used in Eq. (3.1) is summed over all sites of the time slice t it projects out the zero-spatial-momentum part of the correlation function. When r labels the physical states that couple $\tilde{O}$ we assume that $G(t)$ behaves as

$$G(t)=\sum_r A_r \cosh\left[m_r\left(\frac{L}{2}-t\right)\right] . \tag{3.3}$$

(Periodic boundary conditions have been used.) At large separations in the time direction the state with the lowest mass dominates the correlation function (3.3). For the operator used [Eq. (2.17)] this is presumably the $J^{PC}=0^{++}$ glueball. We were not able to measure $G(t)$ for long enough times to make a fit of the form (3.3). Instead we used a one-mass parametrization of $G(t)$ and extracted a distance-dependent mass $m(t)$ from

$$\frac{G(t)}{G(t-1)} \equiv \frac{\cosh[m(t)(L/2-t)]}{\cosh\{m(t)[L/2-(t-1)]\}} . \tag{3.4}$$

The lowest mass m_0 is then found as

TABLE II. Correlation functions $G(t)$ ($t=0,\ldots,3$) for various values of β and ϵ, averaged over clusters of 50 iterations with standard deviation in each cluster.

	Cluster number	Correlation functions $t=0$	$t=1$	$t=2$	$t=3$
$\beta=5.4$	1	0.1581±0.0086	0.0233±0.0015	0.0002±0.0000	0.0000±0.0000
$\epsilon=0.001$	2	0.2687±0.0030	0.0484±0.0008	0.0010±0.0000	0.0000±0.0000
	3	0.3250±0.0017	0.0662±0.0006	0.0017±0.0000	0.0000±0.0000
	4	0.3705±0.0015	0.0812±0.0010	0.0033±0.0002	0.0003±0.0001
	5	0.3815±0.0013	0.0901±0.0007	0.0029±0.0002	0.0009±0.0000
	6	0.3646±0.0017	0.1011±0.0008	0.0013±0.0004	−0.0018±0.0001
	7	0.4414±0.0037	0.1126±0.0008	0.0055±0.0004	0.0015±0.0004
	8	0.4634±0.0030	0.1254±0.0023	0.0235±0.0015	0.0095±0.0004
$\beta=5.5$	1	0.1617±0.0087	0.0236±0.0015	0.0002±0.0000	0.0000±0.0000
$\epsilon=0.001$	2	0.2622±0.0018	0.0501±0.0007	0.0012±0.0000	0.0001±0.0000
	3	0.3241±0.0026	0.0713±0.0009	0.0024±0.0001	0.0003±0.0000
	4	0.3620±0.0022	0.0914±0.0009	0.0033±0.0001	0.0007±0.0000
	5	0.4034±0.0036	0.0942±0.0005	0.0043±0.0002	0.0020±0.0001
	6	0.4364±0.0016	0.1009±0.0008	0.0046±0.0003	0.0048±0.0002
	7	0.4175±0.0021	0.0985±0.0014	−0.0002±0.0006	0.0013±0.0003
	8	0.3811±0.0043	0.1087±0.0024	−0.0017±0.0011	−0.0078±0.0004
$\beta=5.5$	1	0.2416±0.0108	0.0433±0.0025	0.0013±0.0001	0.0000±0.0000
$\epsilon=0.0025$	2	0.3607±0.0037	0.0857±0.0014	0.0048±0.0002	0.0010±0.0001
	3	0.3909±0.0020	0.1047±0.0014	0.0085±0.0004	0.0043±0.0003
	4	0.3754±0.0033	0.1082±0.0032	0.0039±0.0011	0.0068±0.0005
	5	0.3599±0.0078	0.0939±0.0079	−0.0085±0.0021	−0.0012±0.0016
	6	0.5601±0.0133	0.2054±0.0096	−0.0206±0.0040	−0.0705±0.0060
	7	0.3534±0.0266	0.2401±0.0203	−0.0604±0.0092	−0.1001±0.0102
	8	0.2684±0.0467	0.7452±0.0276	0.2585±0.0392	0.1304±0.0298
$\beta=5.6$	1	0.1714±0.0089	0.0254±0.0016	0.0003±0.0000	0.0000±0.0000
$\epsilon=0.001$	2	0.2665±0.0017	0.0508±0.0008	0.0016±0.0001	0.0001±0.0000
	3	0.3023±0.0010	0.0666±0.0007	0.0040±0.0002	0.0003±0.0000
	4	0.3376±0.0024	0.0809±0.0008	0.0068±0.0001	0.0008±0.0000
	5	0.3622±0.0007	0.0914±0.0003	0.0077±0.0002	0.0004±0.0000
	6	0.3978±0.0031	0.1090±0.0009	0.0141±0.0003	0.0008±0.0001
	7	0.4332±0.0022	0.1239±0.0006	0.0160±0.0002	−0.0012±0.0001
	8	0.4418±0.0013	0.1273±0.0008	0.0214±0.0008	0.0042±0.0004
$\beta=5.6$	1	0.2306±0.0094	0.0421±0.0024	0.0014±0.0002	0.0001±0.0000
$\epsilon=0.0025$	2	0.3312±0.0018	0.0802±0.0012	0.0068±0.0003	0.0011±0.0001
	3	0.3249±0.0025	0.0863±0.0009	0.0134±0.0004	0.0043±0.0003
	4	0.3417±0.0026	0.1039±0.0017	0.0269±0.0008	0.0105±0.0005
	5	0.3887±0.0040	0.1221±0.0013	0.0419±0.0008	0.0182±0.0005
	6	0.3238±0.0024	0.1053±0.0014	0.0389±0.0007	0.0197±0.0009
	7	0.3435±0.0026	0.1187±0.0018	0.0384±0.0011	0.0030±0.0016
	8	0.3624±0.0047	0.1438±0.0025	0.0479±0.0011	−0.0132±0.0016
$\beta=5.7$	1	0.1628±0.0083	0.0245±0.0015	0.0004±0.0000	0.0000±0.0000
$\epsilon=0.001$	2	0.2582±0.0021	0.0471±0.0007	0.0019±0.0001	0.0002±0.0000
	3	0.2835±0.0013	0.0598±0.0004	0.0035±0.0001	0.0006±0.0000
	4	0.3012±0.0004	0.0669±0.0003	0.0042±0.0001	0.0008±0.0000
	5	0.3016±0.0005	0.0710±0.0002	0.0054±0.0001	0.0011±0.0000
	6	0.3001±0.0016	0.0776±0.0004	0.0075±0.0002	0.0010±0.0001
	7	0.3271±0.0011	0.0790±0.0007	0.0101±0.0001	0.0003±0.0001
	8	0.3326±0.0011	0.0822±0.0005	0.0097±0.0002	−0.0002±0.0001

TABLE II. (*Continued.*)

	Cluster number	Correlation functions $t=0$	$t=1$	$t=2$	$t=3$
$\beta=5.7$	1	0.2269±0.0089	0.0417±0.0023	0.0018±0.0002	0.0001±0.0000
$\epsilon=0.0025$	2	0.3140±0.0024	0.0738±0.0009	0.0051±0.0001	0.0012±0.0001
	3	0.3305±0.0017	0.0829±0.0006	0.0082±0.0003	0.0002±0.0001
	4	0.3289±0.0017	0.0897±0.0010	0.0120±0.0004	−0.0001±0.0003
	5	0.3413±0.0022	0.0809±0.0018	0.0089±0.0011	0.0017±0.0007
	6	0.3165±0.0032	0.0845±0.0018	0.0250±0.0018	0.0115±0.0014
	7	0.4644±0.0140	0.1282±0.0083	0.0060±0.0025	0.0229±0.0039
	8	0.5082±0.0075	0.1932±0.0056	−0.0338±0.0034	−0.0347±0.0040
$\beta=5.7$	1	0.2269±0.0089	0.0417±0.0023	0.0018±0.0002	0.0001±0.0000
$\epsilon=0.0025$	2	0.3140±0.0024	0.0738±0.0009	0.0051±0.0001	0.0012±0.0001
−0.001	3	0.3366±0.0010	0.0869±0.0004	0.0077±0.0002	0.0010±0.0001
	4	0.3431±0.0009	0.0836±0.0004	0.0080±0.0001	−0.0007±0.0001
	5	0.3294±0.0013	0.0887±0.0005	0.0080±0.0001	0.0036±0.0003
	6	0.2993±0.0018	0.0855±0.0007	0.0174±0.0007	0.0079±0.0003
	7	0.3345±0.0016	0.0998±0.0007	0.0188±0.0004	0.0082±0.0005
	8	0.3411±0.0031	0.1103±0.0008	0.0206±0.0010	0.0011±0.0005
$\beta=5.8$	1	0.1581±0.0080	0.0242±0.0015	0.0004±0.0000	0.0000±0.0000
$\epsilon=0.001$	2	0.2486±0.0017	0.0473±0.0006	0.0015±0.0000	0.0000±0.0000
	3	0.2685±0.0008	0.0551±0.0003	0.0035±0.0001	0.0004±0.0000
	4	0.2794±0.0007	0.0627±0.0005	0.0054±0.0001	0.0005±0.0000
	5	0.2892±0.0014	0.0660±0.0003	0.0065±0.0000	0.0010±0.0000
	6	0.2878±0.0007	0.0697±0.0004	0.0071±0.0001	0.0018±0.0001
	7	0.2994±0.0009	0.0767±0.0003	0.0090±0.0001	0.0023±0.0001
	8	0.3010±0.0013	0.0774±0.0005	0.0079±0.0001	0.0008±0.0002

$$\lim_{\substack{t\to\infty \\ L\gg t}} m(t)=m_0 \,. \tag{3.5}$$

We assume that the gap between m_0 and higher masses that contribute to $G(t)$ in (3.3) is large enough that $m(t)$ gives a good estimate for m_0 already at small distances.

In order to extract the correlation functions and mass estimates we chose to average the correlation functions over 8 clusters of 50 iterations. The result, including the standard deviation in each cluster, is presented in Table II. Since the equilibrium values of the correlation functions and mass estimates are reached at large Langevin time (number of iterations), we have to extrapolate our results. We can proceed in two different ways.

(i) Extrapolation of the correlation functions. We fit the cluster averages of Table II as a function of $1/N$ where N is the number of the cluster, and extrapolate to $1/N=0$. For this we made a polynomial fit

$$G\left(t,\frac{1}{N}\right)=G(t,0)+a(t)\frac{1}{N}+b(t)\frac{1}{N^2} \,. \tag{3.6}$$

TABLE III. Mass estimates for various value of β and ϵ. The mass values are the average of the two extrapolation procedures (see text) and the error estimates their differences.

β	ϵ	$m(1)$	$m(2)$	$m(3)$
5.4	0.001	1.16±0.06	2.2±0.9	1.7±1.0
5.5	0.001	1.26±0.05	2.8±0.2	1.3±0.4
	0.0025	0.95±0.10	2.7±0.8	1.3±0.9
5.6	0.001	1.11±0.05	1.4±0.4	2.7±0.6
	0.0025	0.91±0.05	0.8±0.3	1.0±0.2
5.7	0.001	1.25±0.05	1.9±0.2	2.3±0.5
	0.0025	1.20±0.15	1.6±0.6	1.3±0.5
	0.0025/1	1.12±0.05	1.45±0.2	1.4±0.5
5.8	0.001	1.26±0.05	1.75±0.3	1.6±0.3

The masses $m(t)$ are then extracted from $G(t,0)$ as in Eq. (3.4).

(ii) Extrapolation of the mass estimates. For each cluster calculate the masses $m(t,1/N)$ from (3.4), fit them as a function of $1/N$, and extrapolate to $1/N=0$. Again we used a polynomial fit similar to Eq. (3.6).

The two methods provide us with two estimates for the distance-dependent masses. In Table III we list the average of the two for the various values of β and ϵ. The error induced by the extrapolation procedure can be estimated as the difference between the two values for the masses obtained in (i) and (ii). It is also quoted in Table III.

At longer distances the mass estimates from the runs with $\epsilon=0.001$ tend to be larger than the ones from runs with $\epsilon=0.0025$. This is probably due to the fact that relaxation for $\epsilon=0.001$ takes much longer and has not been achieved in those runs. In all cases we have observed that, not unexpectedly, the approach to equilibrium for the correlation functions (and hence the masses) at longer distances is much slower. The mass estimates at distances 2 and 3 come therefore with larger errors than at distance 1. In Fig. 4 we show these mass estimates as a function of β. At $\beta=5.4$ and 5.8, $\epsilon=0.001$ was used, whereas at the other β's the results from the runs with $\epsilon=0.0025$ are plotted. Also shown in Fig. 4 are the spin-wave estimates for weak coupling (for $t=2$ and 3) and the scaling curves predicted from asymptotic freedom.

IV. DISCUSSION

Let us now come to a discussion of our results. In order to extrapolate the glueball mass to the continuum limit ($\beta=\infty$) we use the renormalization group. Define the lattice scale parameter for SU(3):

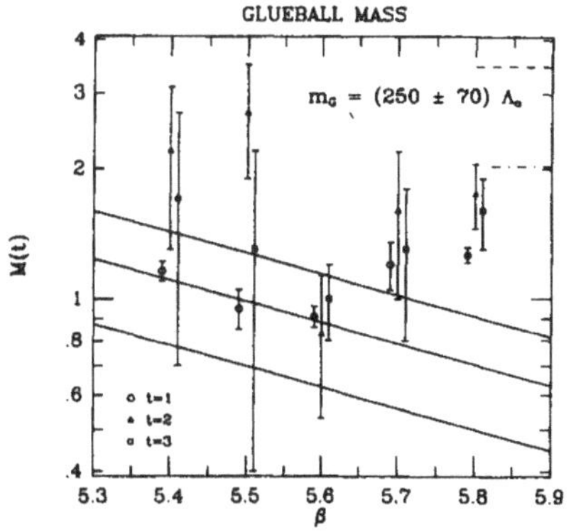

FIG. 4. 0^{++}-glueball-mass estimates at different couplings for different separations. The continuous lines represent the expected renormalization-group behavior $m_G=(240\pm70)\Lambda_0$. The dashed and dashed-dotted lines represent the expected weak-coupling spin-wave behavior of the distance-dependent masses [Eq. (3.4)] $m(1)$ and $m(2)$.

$$\Lambda_0=a^{-1}\left[\frac{8\pi^2}{33}\beta\right]^{51/121} e^{-(4\pi^2/33)\beta}[1+O(\beta^{-1})]\,. \tag{4.1}$$

Then for large β we expect the glueball mass to scale as

$$m_G=c\Lambda_0\,. \tag{4.2}$$

It is not clear from our data to what extent the values for m_G seem to follow this prediction. The mass estimates at $\beta=5.4$, 5.5, and 5.6 are consistent with the expected behavior, although the values at larger separation have large error bars due to the statistical fluctuations. (For larger mass the correlation functions decrease more rapidly and are therefore more difficult to measure.)

The situation, however, is not so nice if one considers what happens to the specific heat in the same region. By the Monte Carlo method we have computed the energy density between $\beta=4.5$ and 6.5 using an increment $\delta\beta=0.03$ and 80 iterations for each value of β on a $5\times5\times5\times5$ lattice, averaging over the last 40 iterations. An eight-order polynomial fit to this data was then performed, and from it the derivative was estimated. From this analysis we find a peak in the specific heat at $\beta=5.5$–5.6, which can probably be ascribed to a nearby complex zero of the partition function. If a singularity were to appear for a real $\beta=\beta_c$, then a scaling argument would suggest

$$\frac{\partial E}{\partial\beta}\underset{\beta\to\beta_c}{\sim} A\,|\beta-\beta_c|^{-\alpha}$$
$$\sim Bm^{(dv-2)/v}\,, \tag{4.3}$$

where α and v are critical exponents, m is the inverse of the correlation length, and A and B are regular functions for $\beta\to\beta_c$. This would in turn imply that m goes to zero as

$$m\underset{\beta\to\beta_c}{\sim}\left|B^{-1}\frac{\partial E}{\partial\beta}\right|^{v/(dv-2)}\,. \tag{4.4}$$

In the case of SU(3) lattice gauge theories with the Wilson action it seems unlikely that the peak in the specific heat will not have some effect on the glueball mass (a dip) around and before $\beta=5.6$, a behavior that is unrelated to the asymptotic freedom scaling expected for large β. A similar behavior is also observed in the string tension and the hadron masses. Therefore we conclude that our estimates for $\beta<5.6$ are likely to be affected by at present uncontrollable systematic effects due to nearby spurious singularities.

On the other hand, at large β ($\beta\geq5.7$) we see that our data clearly deviates from the asymptotic-freedom prediction. Indeed it is known that for weak enough coupling the spin-wave result

$$G(t)\sim t^{-5} \tag{4.5}$$

should be recovered for small enough distances, $t\ll m_G{}^{-1}$. In Fig. 4 we have included the spin-wave prediction for the distance-dependent mass estimates $m(2)$ and $m(3)$ defined in the preceding section. Since we are limited, because of the size of our $6\times6\times6\times6$ lattice, to separations of up to three lattice spacings, we are at the present unable to penetrate more deeply into the scaling region. It would seem therefore that our most believable point, as far as the extrapolation to $\beta\to\infty$ is concerned, is $\beta=5.6$. Our conclusion is thus

$$m_G=(240\pm70)\Lambda_0 . \tag{4.6}$$

An alternative way of presenting our results would be as a ratio of the 0^{++} glueball mass to a known physical quantity also accessible by the Monte Carlo method. Using recent high-statistics data for the string tension on an $8\times8\times8\times8$ lattice[20] we find at $\beta=5.6$

$$\frac{m_G}{\sqrt{T}}=\frac{(0.9\pm0.1)a^{-1}}{(0.35\pm0.05)a^{-1}}=2.6\pm1.0 . \tag{4.7}$$

The string tension is not known directly, but if we use $T=(420\text{ MeV})^2$ we get $m_G=1080\pm400$ MeV.

In Ref. 16 the ρ mass at $\beta=5.6$ was estimated to be $m_\rho=(0.8\pm0.1)a^{-1}$ using the Wilson fermion action[10] on a lattice of maximum size $6\times6\times6\times12$. From this estimate we obtain

$$\frac{m_G}{m_\rho}=\frac{(0.9\pm0.1)a^{-1}}{(0.8\pm0.1)a^{-1}}=1.1\pm0.2 \tag{4.8}$$

which would give, using the physical ρ mass as input, for the mass of the lowest glueball about 850 ± 200 MeV.

Let us finally briefly compare our results with previous estimates. Our values for the masses at different coupling constants are in reasonable agreement with previous results, although they usually tend to be slightly lower. In the first of Ref. 15 the finite-size scaling method was used to estimate the mass of the scalar glueball at $\beta=6.0$, giving the large value $(1.2\pm0.1)a^{-1}$ at separations 2 and 3 on lattices of maximum size $7\times7\times7\times7$, an estimate that now would seem to be contaminated by the spin-wave behavior of Eq. (4.5). The variational estimate of Ref. 6 at $\beta=5.7$ on a $4\times4\times4\times8$ lattice is about 30% above our best result in terms of Λ_0. However, at the coupling our masses seem already to be influenced by the spin-wave behavior as well. The same technique was used to estimate the glueball mass in Ref. 7 at several values of β ranging from 5.0 to 5.8 on a $4\times4\times4\times8$ lattice. Their latest estimate is about 15% higher than ours, but the agreement seems good in view of the different techniques used.

Several explanations for the discrepancies are possible. Among these we mention the slightly larger lattice used in the present calculation and the influence of the peak in the specific heat, which might be size dependent. It now seems important that longer runs on larger lattices be done. Work in this direction is in progress and will be reported elsewhere.

ACKNOWLEDGMENTS

One of us (H.H.) would like to thank E. Marinari and G. Parisi for many fruitful discussions on the Langevin equation. We have written together part of the code that was later adapted for the present study. This research was supported by the U.S. Department of Energy under Grant No. DE-AC02-76ER02220. U.H. wishes to express his thanks for a grant by the Federal Republic of Germany.

[1] J. Kogut, D. K. Sinclair, and L. Susskind, Nucl. Phys. B114, 199 (1976).

[2] C. Rebbi, in *Proceedings of the 21st International Conference on High Energy Physics, Paris, 1982*, edited by P. Petiau and M. Porneuf [J. Phys. (Paris) Colloq. 43, (1982)].

[3] G. Bhanot and C. Rebbi, Nucl. Phys. B180 [FS2], 469 (1981).

[4] B. Berg, in *Lattice Gauge Theories, Supersymmetry, and Grand Unification*, proceedings of the 6th Johns Hopkins Workshop on Current Problems in Particle Theory, Florence, 1982 (Physics Department, Johns Hopkins University, Baltimore, 1982).

[5] M. Falcioni *et al.*, Phys. Lett. 110B, 295 (1982).

[6] K. Ishikawa, G. Schierholz and M. Teper, Phys. Lett. 110B, 399 (1982); 116B, 429 (1982); 120B, 387 (1983); DESY Report No. 83-04 (unpublished).

[7] B. Berg and A. Billoire, Phys. Lett. 113B, 65 (1982); 114B, 324 (1982); Nucl. Phys. B221, 109 (1983).

[8] K. H. Mütter and K. Schilling, Phys. Lett. 117B, 75 (1982); 121B, 267 (1983).

[9] M. Falcioni *et al.*, Nucl. Phys. B215 [FS7], 265 (1983).

[10] K. G. Wilson, Phys. Rev. D 10, 2445 (1974); in *New Phenomena in Subnuclear Physics*, Proceedings of the 14th Course of the International School of Subnuclear Physics, Erice, 1975, edited by A. Zichichi (Plenum, New York, 1976).

[11] B. Halperin and P. C. Hohenberg, Rev. Mod. Phys. 49, 435 (1977).

[12] G. Parisi and Wu Yong-Shi, Sci. Sin. 24, 483 (1980).

[13] G. Parisi, Nucl. Phys. B180 [FS2], 378 (1981); B205 [FS5], 337 (1982).

[14] I. Drummond, S. Duane, and R. Horgan, Cambridge University Report No. DAMPT 82/19, 1982 (unpublished).

[15] H. Hamber and G. Parisi, Phys. Rev. Lett. 47, 1795 (1981); E. Marinari, G. Parisi, and C. Rebbi, *ibid.* 47, 1798 (1981).

[16] H. Hamber and G. Parisi, Phys. Rev. D 27, 208 (1983).

[17] G. Parisi (unpublished).

[18] J. Alfaro and B. Sakita, CCNY Report No. HEP-82/16, 1982 (unpublished); A. Guha and S. C. Lee, Phys. Rev. D 27, 2412 (1983).

[19] We thank E. Marinari for suggestions on this point.

[20] M. Creutz, R. Ardill, and K. Moriarty, Brookhaven Report No. BNL-32377, 1982 (unpublished).

Nuclear Physics B239 (1984) 201–208

NUMERICAL EVIDENCE FOR A BARRIER AT THE GRIBOV HORIZON*

E. SEILER
Max-Planck-Institut für Physik und Astrophysik, München, FRG

I. O. STAMATESCU
CERN, Geneva, Switzerland

D. ZWANZIGER
University of New York, New York, NY 10003, USA

Received 29 August 1983

We demonstrate numerically that in stochastically quantized euclidean Yang–Mills theory with a non-conservative gauge fixing force, the Gribov horizon represents a barrier that keeps the system inside a bounded region of field space.

1. Introduction

In a series of papers, stochastic quantization of euclidean Yang–Mills theory was studied [1–3]. A central point is the replacement of the Faddeev–Popov procedure by a non-conservative but local gauge fixing force that does not require the introduction of ghost fields. In ref. [3], it is argued that in the continuum, the system in the limit of an infinitely strong gauge fixing force will be confined to a convex region in field space which is bounded in each direction.

In ref. [4] (see especially sect. 2.2), a lattice version of the system (with finite gauge fixing force) was studied. The non-conservative character of the gauge fixing force K_{GF} leads to the presence of steady currents in the equilibrium measure, as is pointed out in that work.

Furthermore, the gauge fixing force K_{GF} shows some peculiar features related to the Gribov ambiguity: while it vanishes everywhere on the gauge fixing surface given by the Landau condition $\partial\cdot A=0$ or by the background gauge condition $\partial_{\bar{A}}(A-\bar{A})=0$, it is attractive to that surface only within a bounded convex subregion Ω; the boundary of Ω is the Gribov horizon [5]. Outside the Gribov horizon, K_{GF} will have "unstable modes", i.e. there will be directions along which K_{GF} is repelled from the gauge fixing surface. See ref. [4] for details.

On the other hand, it was shown in ref. [4] that K_{GF} is always a restoring force in the sense that it tends to reduce the size of the gauge field A. More precisely, it

* This research was supported in part by the National Science Foundation grant no. PHY-8116102.

was shown in ref. [4] that under the motion $\dot{A} = K_{GF}(A)$, we have the relation

$$\frac{d}{dt}|A-\bar{A}|^2 = -2\alpha|\partial_{\bar{A}}\cdot(A-\bar{A})|^2 . \tag{1}$$

The interplay of these features leads to a peculiar behaviour of the system once it manages to diffuse along the gauge fixing surface beyond the Gribov horizon. The gauge fixing force K_{GF} will drive it away from the gauge fixing surface, but at the same time back towards $\bar{A}$, so after completing a kind of loop, the system will end up again essentially inside the Gribov horizon close to the gauge fixing surface. "Instability stabilizes". These "loops" are closely related to the existence of a steady current. This mechanism is illustrated in fig. 2 of ref. [4].

In this note, we study these effects numerically and confirm the picture outlined above. The effects are most clearly seen if we drop completely the classical force due to the Yang–Mills action (i.e. if we go to infinite coupling). It is a remarkable fact that the system remains stable in this limit; this may be called "self-compactification".

Let us now briefly recall the main features of the model. The configuration space (field space) is described by variables $A_\mu^a(m)$ living on the links (m, μ) of a four-dimensional simple cubic lattice with periodic boundary conditions; a is a colour index. Stochastic quantization introduces a time-dependent probability density $P(A, t)$ obeying the diffusion equation

$$\dot{P}(A,t) = -\nabla\cdot\boldsymbol{J}, \tag{2}$$

where the current $\boldsymbol{J}$ is given by

$$\boldsymbol{J} = -\hbar\nabla P + \boldsymbol{K}P, \tag{3}$$

and $\boldsymbol{K}$ is the drift force consisting of two parts:

$$\boldsymbol{K} = \alpha\boldsymbol{K}_{GF} + \beta\boldsymbol{K}_{CL}, \tag{4}$$

$$K^a_{GF_\mu}(m) = D^{ab}_\mu(A)(\partial_{\bar{A}}\cdot(A-\bar{A}))^b(m), \tag{5}$$

$$K^a_{CL_\mu}(m) = -\frac{\partial S}{\partial A^a_\mu(m)}, \tag{6}$$

where S is the lattice version of the classical Yang–Mills action (see ref. [4]). As remarked, we will mostly set $\beta = 0$ in this note; since we also found essentially no effect due to the background field $\bar{A}$, we have also put $\bar{A} = 0$ to save computing time.

To study the system numerically, we discretize both time and field space. All fields $A_\mu^a(m)$ will be integer multiples of a basic unit η, called step size. The diffusion eqs. (2), (3) are then simulated by a random walk in field space; the transition probability from a configuration to a neighbouring one (changing one of the components $A_\mu^a(m)$ by a unit η in $\pm$ direction) is given by $\frac{1}{2}(1 \pm \frac{1}{2}\eta K^a_\mu(m))$ for small ηK—for large ηK some modification is necessary in order to avoid negative probabilities. Details may be found in ref. [6].

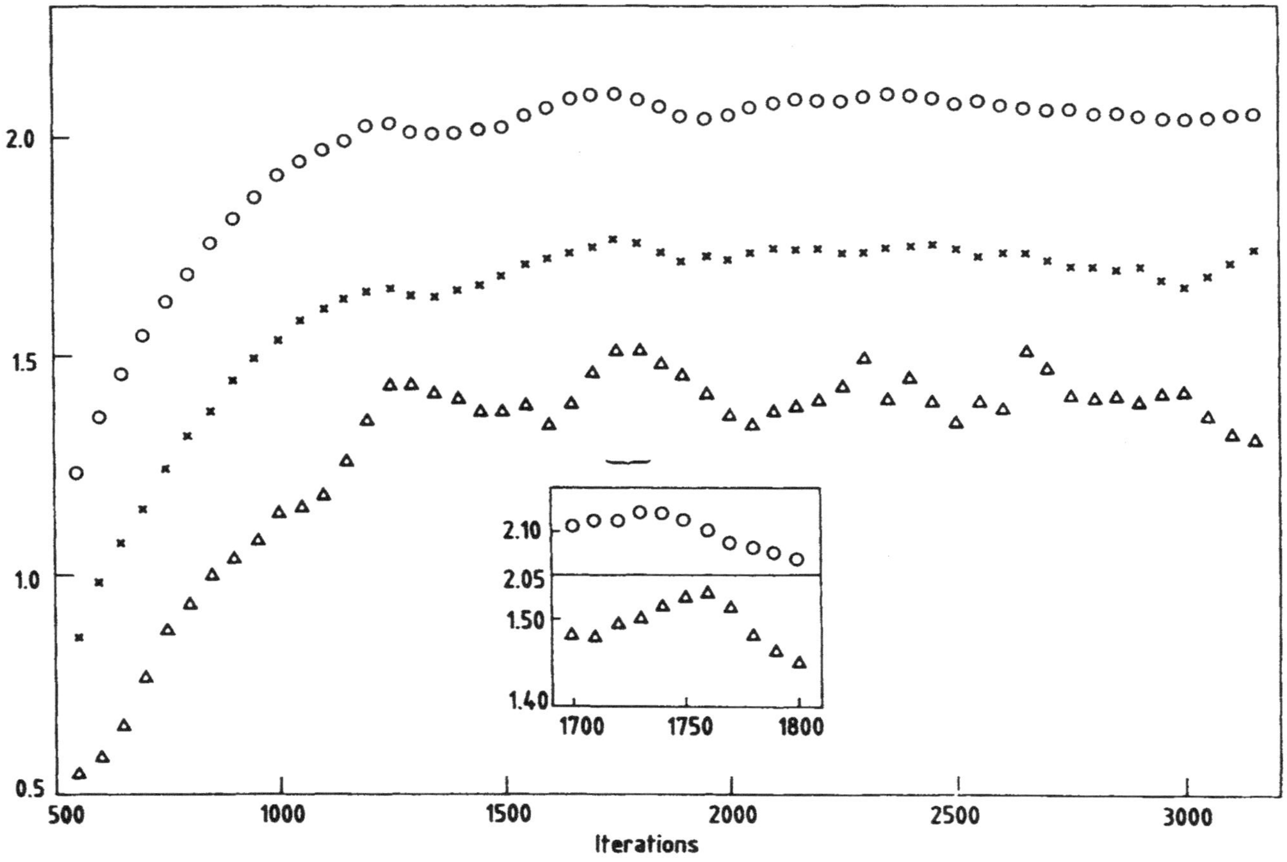

Fig. 1. Action ($\times\frac{1}{100}$): crosses, $A_\mu^2(\times\frac{1}{10})$: circles and $(D_\mu A_\mu)^2(\times\frac{1}{10})$: triangles. Each point represents an average over the 50 sweeps (respectively 10, in the box) between it and the preceding point. Here $\beta = 0.0001$, $\alpha = 1$ and the step size $\eta = \frac{1}{8}$ on a 6^4 lattice. All the averages are given per link.

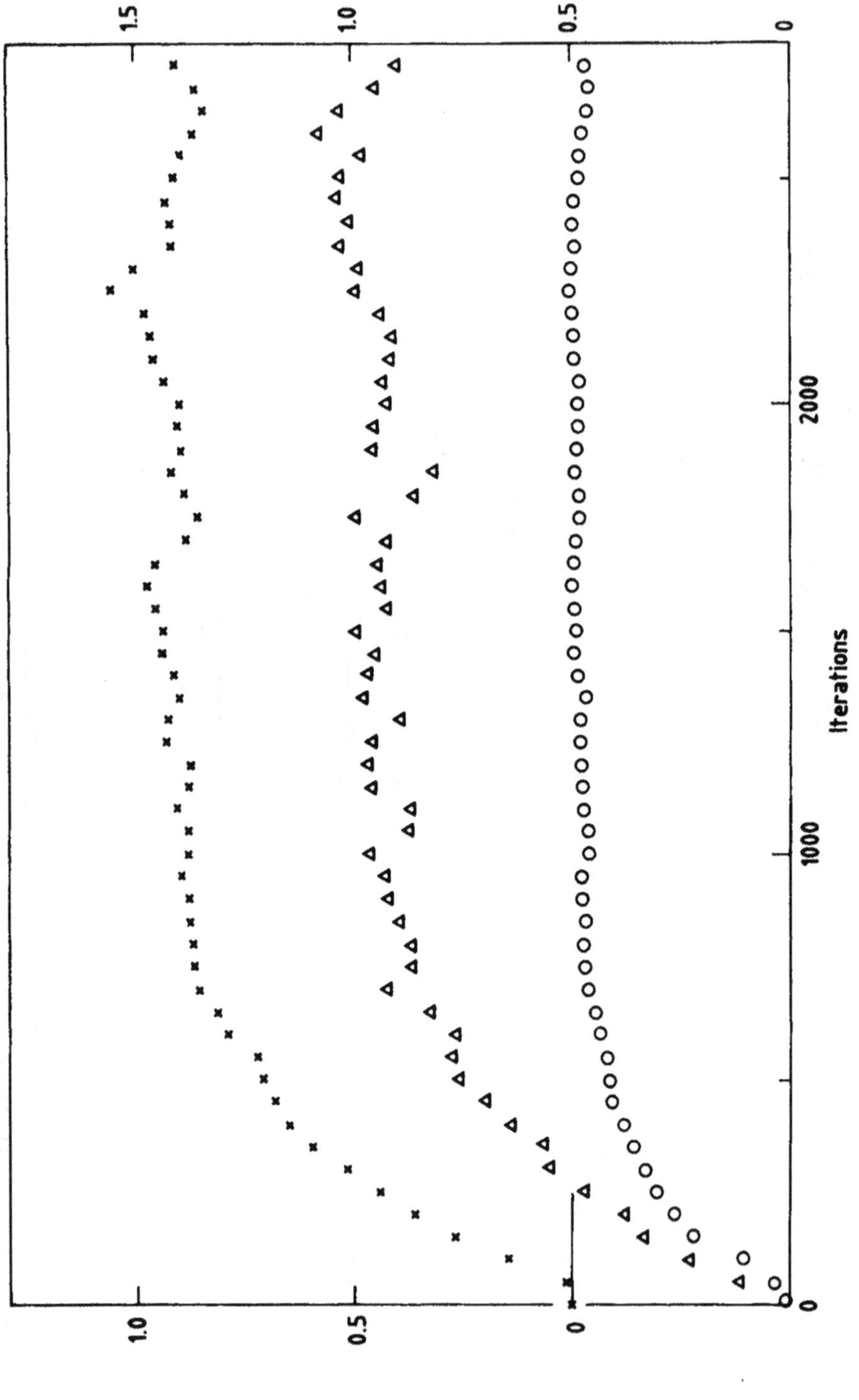

Fig. 2. Action $(\times\frac{1}{1000})$, (left-hand scale), $A_\mu^2 \times (\frac{1}{100})$ and $(D_\mu A_\mu)^2(\times\frac{1}{100})$, (right-hand scale). Averages over 50 sweeps; $\beta=0$, $\alpha=0.15$ and $\eta=\frac{1}{4}$ on a 4^4 lattice. Same symbols as before.

2. Stability of the system started at $A=0$

We started the system "cold", i.e. at $A \equiv 0$ on a 6^4 lattice with step size $\eta=\frac{1}{8}$, $\alpha=1$, $\beta=0.0001$; another run was done on a 4^4 lattice with $\eta=\frac{1}{4}$, $\alpha=0.15$, $\beta=0.0$. The results are presented in figs. 1 and 2.

Both runs show eventual stabilization after a slow and long heating period. Even after equilibrium has been reached, low frequency fluctuations are clearly visible; they reflect the presence of a steady current in equilibrium. Presumably what is happening is this: once in a while, the system diffuses a little bit outside the Gribov horizon; then it picks up an unstable mode of the gauge force, driving it away from the surface $\partial \cdot A=0$. On the other hand, as is shown by eq. (1), non-negligible values of $\partial \cdot A$ will tend to reduce $|A|$ and bring the system back near the region Ω inside the Gribov horizon. This phenomenon will be seen in a much more dramatic way if we start the system far outside the Gribov horizon, as described in the next section.

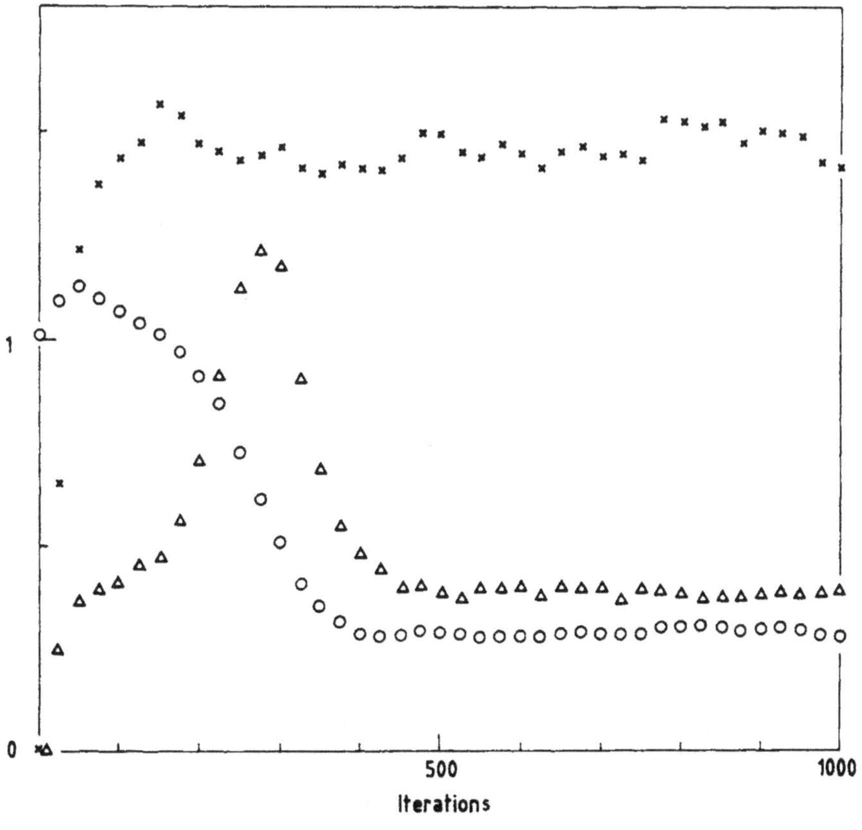

Fig. 3. Action, $A_\mu^2(\times\frac{1}{5})$ and $(D_\mu A_\mu)^2(\times\frac{1}{2})$. Averages over 25 sweeps; $\beta=0.25$, $\alpha=1$ and $\eta=\frac{1}{8}$ on a 4^4 lattice started outside the Gribov horizon. Same symbols as before.

3. Return from outside the Gribov horizon

The simplest field configurations outside the Gribov horizon are given by sufficiently large constant colour vectors

$$A^a_\mu(m) = c\delta_{a3}\delta_{\mu 4}\,, \tag{7}$$

(see ref. [4]).

So, by starting the system at such a configuration, we can induce a return cycle. We present three examples.

The first one (fig. 3) was done in the presence of the classical force but at $\alpha > \beta$ (at $\alpha \leqslant \beta$ the cycles can hardly be seen). We chose $\beta = 0.25$, $\alpha = 1$, $\eta = \frac{1}{8}$ and a starting configuration

$$\mathring{A}^a_\mu = 36\eta\delta_{\mu 4}\delta_{a3} = 4.5\delta_{\mu 4}\delta_{a3}$$

on a 4^4 lattice.

Figs. 4 and 5 show runs done with $\beta = 0$ on 6^4 lattices; in fig. 4, $\alpha = 0.15$, $\eta = \frac{1}{6}$, $\mathring{A}^a_\mu = 108\eta\delta_{\mu 4}\delta_{a3} = 18\delta_{\mu 4}\delta_{a3}$, whereas in fig. 5, $\alpha = 5$, $\eta = \frac{1}{20}$, $\mathring{A}^a_\mu = 144\eta\delta_{\mu 4}\delta_{a3} =$

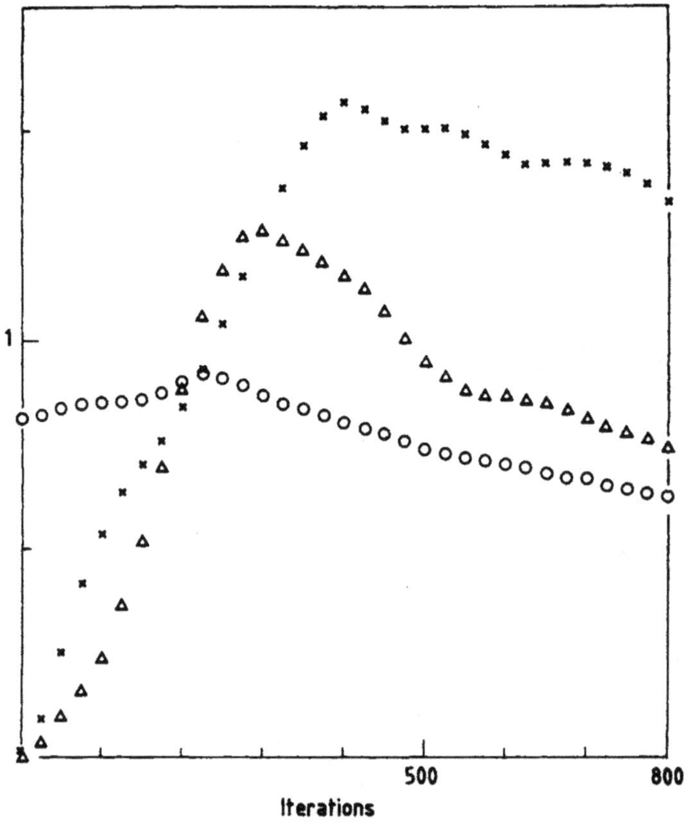

Fig. 4. Action ($\times\frac{1}{1000}$), $A^2_\mu(\times\frac{1}{100})$ and $(D_\mu A_\mu)^2(\times\frac{1}{200})$. Averages over 25 sweeps; $\beta = 0$, $\alpha = 0.15$ and $\eta = \frac{1}{6}$ on a 6^4 lattice started outside the Gribov horizon. Same symbols as before.

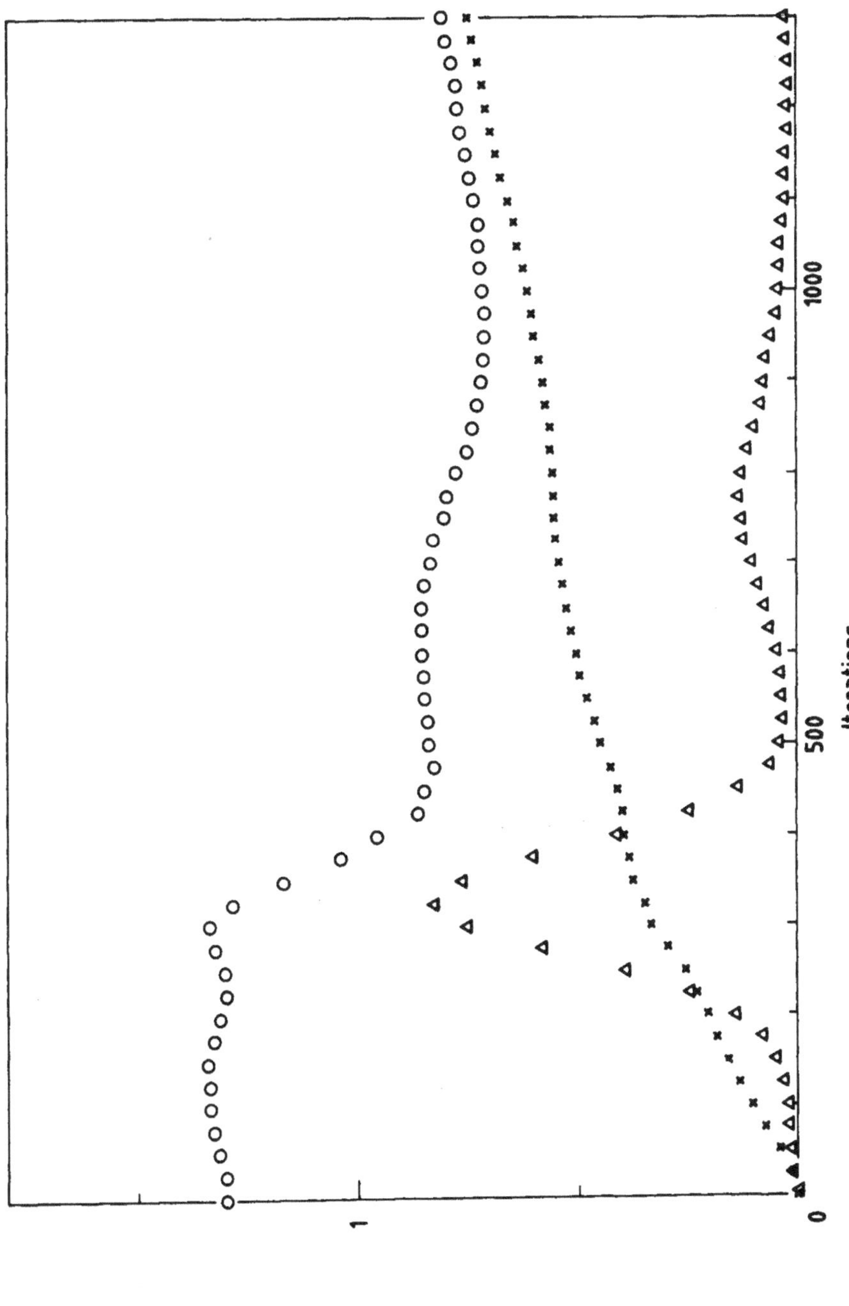

Fig. 5. Action $(\times\frac{1}{100})$, $A_\mu^2(\times\frac{1}{10})$ and $(D_\mu A_\mu)^2(\times\frac{1}{50})$. Averages over 25 sweeps; $\beta=0$, $\alpha=5$, $\eta=\frac{1}{20}$ on a 6^4 lattice started outside the Gribov horizon. Same symbols as before.

7.2 $\delta_{\mu 4}\delta_{a3}$. The return cycles are very clearly visible; the approximate validity of eq. (1) can also be seen. Note that $|A|$ always reaches its maximum before $|\partial \cdot A|$. Of course, we cannot expect eq. (1) to hold exactly because it neglects the fluctuating force and comes from a purely classical treatment.

4. Conclusions

The numerical simulations confirm the intuitive picture developed in ref. [4] and here. It is very striking that the system "self-compactifies" and allows the limit $\beta \to 0$ without showing any instability.

Of course, this "self-compactification" should be understood in an average sense. If α is not infinite, arbitrary large values of A are possible, but their probability decreases rapidly with $|A|$. In equilibrium, the system will apparently mostly drift within the Gribov horizon, performing small cycles; only occasionally will it venture out further, performing a larger cycle. From the numerical point of view, these cycles are, of course, a nuisance, since they mean slow convergence. So it is gratifying that for $\beta \gtrsim \alpha$ they seem to be already pretty much suppressed (see ref. [4]). However, we think it is remarkable that we can "see" the Gribov horizon in the way shown here.

References

[1] D. Zwanziger, Phys. Lett. 114B (1982) 337
[2] L. Baulieu and D. Zwanziger, Nucl. Phys. B193 (1981) 163
[3] D. Zwanziger, Nucl. Phys. B209 (1982) 336
[4] E. Seiler, I.O. Stamatescu and D. Zwanziger, Nucl. Phys. B239 (1984) 177
[5] V.N. Gribov, Nucl. Phys. B139 (1978) 1
[6] I.O. Stamatescu, U. Wolff and D. Zwanziger, Nucl. Phys. B225 [FS9] (1983) 377

VOLUME 55, NUMBER 18 PHYSICAL REVIEW LETTERS 28 OCTOBER 1985

Langevin Simulation Including Dynamical Quark Loops

A. Ukawa
Institute of Physics, University of Tsukuba, Ibaraki 305, Japan

and

M. Fukugita
Research Institute for Fundamental Physics, Kyoto University, Kyoto 606, Japan
(Received 9 April 1985)

A Langevin method is proposed for simulation of full QCD including dynamical quark loops. It is shown that the method works well once a proper discretization of the fictitious time is made. A realistic simulation with this method is perfectly feasible on vector computers presently available.

PACS numbers: 12.35.Cn, 02.50.+s, 02.70.+d, 11.15.Ha

Recent applications of Monte Carlo techniques to lattice gauge theories[1] have greatly facilitated our understanding of hadron dynamics. It has enabled us to calculate a variety of important physical quantities including, albeit within the quenched approximation, the hadron spectrum.[2] For a full treatment of lattice QCD, however, one needs efficient techniques for incorporating the effect of vacuum quark loops. The majority of the attempts so far utilized the pseudofermion technique.[3] More recently the microcanonical method[4] has been extended to include dynamical quarks.[5] Both methods involve approximations and assumptions which require a careful scrutiny for their justification: In the pseudofermion case, the ratio of the quark determinant is approximated by the leading term in the variation of the gauge variables. The validity of the microcanonical procedure hinges on the assumption of ergodicity besides the apparent drawback that the coupling constant has to be calculated through the simulation itself.

We point out that the Langevin formulation of field theories[6] provides an interesting alternative for solving full QCD including dynamical quark loops, which seems free from these potential sources of trouble. Inclusion of dynamical quark loops is achieved by expressing the quark determinant in terms of effective bosonic variables.[7] The solution of the corresponding Langevin equation generates an ensemble having the measure of full QCD. There is no approximation involved in the framework itself.

We have undertaken a study of the practical problems that arise in solving numerically the Langevin equation. Our study has revealed the following: (i) A conventional discretization of the fictitious time derivative leads to large systematic errors, especially for correlations at large distances, which practically invalidates the procedure. (ii) With a second-order improved discretization, however, the Langevin method works beautifully, both without and with dynamical quarks. Our extensive calculation with the SU(2) gauge group shows that the effect of quark loops substantially modifies the result obtained in the quenched approximation. A preliminary study for SU(3) indicates that a simulation on a 10^4 lattice or larger with several thousand sweeps is feasible with vector computers available today. We have also examined the microcanonical method. A comparison will be made with the Langevin method toward the end of this note.

We begin the consideration of systematic errors in the Langevin simulation for pure gauge system. Let $S(\mathbf{U})$ be the action with U_l the SU(N) gauge variable on the link l, and let $\nabla_l = t^a\nabla_l^a$ be the right derivative with $\nabla_l^a = \sum_{ij}(U_l t^a)_{ij}\partial/\partial U_{lij}$ and $\mathrm{tr}(t^a t^b) = \delta^{ab}$. The Langevin equation is given by[8]

$$-iU_l(\tau)^{-1}\frac{d}{d\tau}U_l(\tau) = -i\nabla_l S(\mathbf{U}(\tau)) + \eta_l(\tau), \qquad (1)$$

with τ the fictitious time and $\eta_l(\tau) = t^a\eta_l^a(\tau)$ the Gaussian noise satisfying

$$\langle \eta_l^a(\tau)\eta_{l'}^b(\tau')\rangle = 2\delta_{ll'}\delta^{ab}\delta(\tau-\tau').$$

Let us discretize the time τ in steps of $\Delta\tau$ and write $U_l^{(n)} \equiv U_l(n\Delta\tau)$. The simplest discretization of (1) preserving the unitarity constraint on U_l is given by

$$U_l^{(n+1)} = U_l^{(n)}\exp[iX_l(\mathbf{U}^{(n)},\eta^{(n)})], \qquad (2a)$$

$$X_l(\mathbf{U}^{(n)},\eta^{(n)}) = -i\Delta\tau\nabla_l S(\mathbf{U}^{(n)}) + (\Delta\tau)^{1/2}\eta_l^{(n)}, \qquad (2b)$$

and $\eta_l^{(n)} = t^a\eta_l^{a(n)}$ satisfies

$$\langle \eta_l^{a(n)}\eta_{l'}^{b(n')}\rangle = 2\delta_{ll'}\delta^{ab}\delta^{nn'}.$$

In Fig. 1, we show by open circles the SU(3) Wilson-loop averages for the standard single-plaquette action from the iterative solution of (2) with $\Delta\tau = 0.01$ on 4^4 lattice and compare them with the result of the Monte Carlo simulation (crosses). The Langevin result is systematically smaller and the difference between the two is substantial, especially for the 2×2 Wilson loop.

We found that this is a systematic error arising from

the discretization of time (2). To show this, we define the distribution function $\rho^{(n)}(\mathrm{U})$ by

$$\rho^{(n)}(\mathrm{U}) = \langle \prod_l \delta(U_l; U_l^{(n)}) \rangle, \tag{3}$$

where $\langle \cdots \rangle$ denotes average over the noise $\eta^{(0)}, \ldots, \eta^{(n-1)}$. Substituting (2) and expanding in $\Delta\tau$, one finds that the Fokker-Planck equation for (2) takes the form

$$(1/\Delta\tau)(\rho^{(n+1)} - \rho^{(n)}) = -\mathscr{D}^{(n)}\rho^{(n)} - \Delta\tau\mathscr{D}_1^{(n)}\rho^{(n)} + O(\Delta\tau^2), \tag{4}$$

where the operators $D^{(n)}$ and $D_1^{(n)}$ are defined by

$$\mathscr{D}^{(n)}\rho^{(n)} = \sum_l \{\nabla_l^a(\nabla_l^a S^{(n)}\rho^{(n)}) + \nabla_l^a\nabla_l^a\rho^{(n)}\}, \tag{5}$$

$$\mathscr{D}_1^{(n)}\rho^{(n)} = -\sum_{l,l'}\{\tfrac{1}{2}\nabla_l^a\nabla_{l'}^b(\nabla_l^a S^{(n)}\nabla_{l'}^b S^{(n)}\rho^{(n)}) + \tfrac{1}{3}(\nabla_l^2\nabla_{l'}^b + \nabla_{l'}^b\nabla_l^2 + \nabla_l^a\nabla_{l'}^b\nabla_l^a)(\nabla_{l'}^b S^{(n)}\rho^{(n)})$$
$$+ \tfrac{1}{6}(\nabla_l^2\nabla_{l'}^2 + \nabla_l^a\nabla_{l'}^2\nabla_l^a + \nabla_l^a\nabla_{l'}^b\nabla_l^a\nabla_{l'}^b)\rho^{(n)}\}, \tag{6}$$

with $S^{(n)} = S(\mathrm{U}^{(n)})$ and $\nabla_l^2 = \nabla_l^a\nabla_l^a$.

Equation (4) shows that the limiting distribution for the first-order discretization (2) deviates from the desired form $\exp(-S)$. The stationary solution ρ_∞ of (4) takes the form

$$\rho_\infty = \exp[-S - \Delta\tau\delta S_1 + O(\Delta\tau^2)], \tag{7a}$$

$$\delta S_1 = -\tfrac{1}{2}\sum_l \nabla_l^2 S + \tfrac{1}{12}c_2 S + \tfrac{1}{4}\sum_l \nabla_l^a S\nabla_l^a S, \tag{7b}$$

where c_2 represents the quadratic Casimir invariant in the adjoint representation. For the standard single-plaquette action the first two terms of (7b) shift β to

$$\beta_{\text{eff}} = \{1 - [2(N^2-1)/N - \tfrac{1}{6}N]\Delta\tau\}\beta, \tag{8}$$

while the third term generates an additional action containing two plaquettes. Both these terms favor disordering of the gauge configuration. Hence we understand the trend in Fig. 1 that the Langevin result is smaller than the Monte Carlo value. The magnitude of the difference, estimated by ignoring the $(\nabla S)^2$ term for simplicity, is in rough accord with the data shown in Fig. 1.

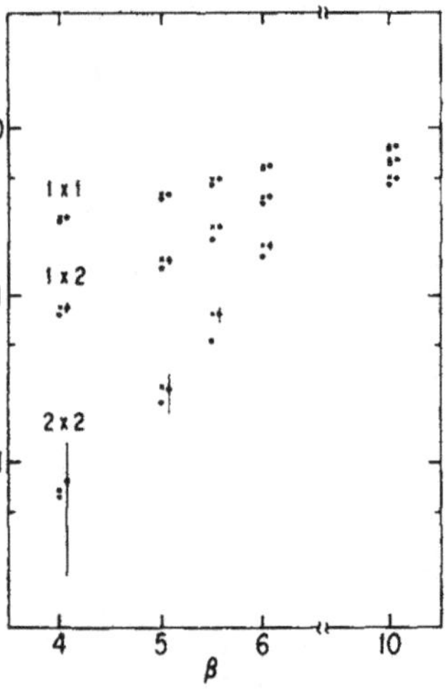

FIG. 1. Wilson loop for SU(3) pure gauge theory. Langevin results are represented by open and filled circles for naive and the second-order discretization, respectively. Crosses are Monte Carlo results. The error bars denote mean square deviation.

The systematic error could, in principle, be reduced by choosing $\Delta\tau$ sufficiently small. Unfortunately, 1% accuracy in $\beta_{\text{eff}}(\Delta\tau \lesssim 0.002)$ does not guarantee the same accuracy in the correlation functions because their β derivative is large at large distances, especially in the crossover region. Furthermore, the number of iterations needed to generate an independent configuration is proportional to $\Delta\tau^{-1}$, and hence so is the computer time. This makes the simulation including quarks practically impossible, for the computer time per iteration for full QCD is at least an order of magnitude more than that of the pure gauge case.

Clearly, reducing $\Delta\tau$ is not the way to diminish the systematic error. One rather needs an accurate discretization algorithm which ensures that the limiting distribution agrees with $\exp(-S)$ up to order $\Delta\tau$.[9] Our algorithm is given by

$$\begin{aligned} X_{0l}^{(n)} &= -i\Delta\tau\nabla_l S(\mathrm{U}^{(n)}) + (\Delta\tau)^{1/2}\eta_l^{(n)}, \\ U_l^{(n+1/2)} &= U_l^{(n)}\exp iX_{0l}^{(n)}, \\ X_{1l}^{(n)} &= -i\Delta\tau\nabla_l S(\mathrm{U}^{(n+1/2)}) + (\Delta\tau)^{1/2}\eta_l^{(n)}, \\ U_l^{(n+1)} &= U_l^{(n)}\exp\{i(\beta X_{0l}^{(n)} + \gamma X_{1l}^{(n)})\}. \end{aligned} \tag{9}$$

The requirement on the corresponding Fokker-Planck equation spelled out above is met if the parameters β

and γ satisfy

$$\beta=\tfrac{1}{2}+\Delta\tau\beta_1,\quad \gamma=\tfrac{1}{2}+\Delta\tau\gamma_1,$$
$$\beta_1+\gamma_1=\tfrac{1}{12}c_2, \tag{10}$$

and if the Gaussian noise $\eta_l^{(n)}=t^a\eta_l^{a(n)}$ is renormalized by

$$\langle\eta_l^{a(n)}\eta_{l'}^{b(n')}\rangle=2(1-\Delta\tau\tfrac{1}{4}c_2)\delta_{ll'}\delta^{ab}\delta^{nn'}. \tag{11}$$

In Fig. 1 we plot by filled circles the Wilson-loop averages from the second-order algorithm [(9)–(11)] with $\Delta\tau=0.01$ and $\beta_1=\gamma_1=c_2/24$. The very nice agreement with the Monte Carlo results demonstrates the effectiveness of this algorithm. With $\Delta\tau=0.01$, the systematic error is reduced by two orders of magnitude by repeating essentially twice the first-order algorithm and hence by using only twice the computer time. (We also carried out runs with $\Delta\tau$ up to 0.1. The systematic error exhibited a quadratic dependence for large $\Delta\tau$ as expected for the second-order algorithm. No noticeable deviation from the Monte Carlo values is found up to $\Delta\tau\sim 0.05$.)

We now describe the extension of the method to full QCD. The effect of vacuum quark loops arises from the determinant $\det D(U)$. If $D(U)$ is positive definite, $\det D(U)$ equals the Gaussian integral of a complex scalar field Y_s on site s with the action $S_f(U,Y)=Y_s^\dagger D^{-1}(U)_{ss'}Y_{s'}$. Thus, one may take

$$S_{\rm eff}(U,Y)=S_f(U,Y)+S(U) \tag{12}$$

as the effective action for the full QCD.[7] The Langevin equation

$$-iU_l(\tau)^{-1}\frac{d}{d\tau}U_l(\tau)$$
$$=-i\nabla_l S_{\rm eff}(U(\tau),Y(\tau))+\eta_l(\tau), \tag{13}$$

$$\frac{d}{d\tau}Y_s(\tau)=-D^{-1}(U(\tau))_{ss'}Y_{s'}(\tau)+\xi_s(\tau), \tag{14}$$

with η_l and ξ_s the Gaussian noise generates the distribution $\exp(-S_{\rm eff})$ as $\tau\to\infty$.[10]

In discretizing (13) and (14), a second-order algorithm is certainly needed to avoid systematic error. Using the Fokker-Planck equation, we found that the standard Runge-Kutta algorithm with $\langle\xi_s^{(n)}\xi_{s'}^{(n')\dagger}\rangle=2\delta_{ss'}\delta^{nn'}$ for (14), together with the discretization procedure [(9)–(11)] for (13) (with $S\to S_{\rm eff}$), gives the limiting distribution that deviates only by terms of $O(\Delta\tau^2)$ from $\exp(-S_{\rm eff})$. Compared with the pure gauge case, the additional complication is the calculation of $D^{-1}Y$ from Y which could be handled by standard methods such as conjugate gradient.

We have tested the above framework for SU(2) using Wilson's fermion action

$$D_0=1-K\{\textstyle\sum(1-\gamma_\mu)U_l+(1+\gamma_\mu)U_{l'}^\dagger$$

with two flavors ($D=D_0^\dagger D_0$). In Fig. 2(a) we show the K dependence of the Wilson loop and in 2(b) $\langle\bar\psi\psi\rangle$ at $\beta=4/g_0^2=2.0$ on a 4^4 lattice, the latter being calculated by the identity $\langle\bar\psi\psi\rangle=-\langle{\rm tr}(D^{-1}YY^\dagger D^{-1}\times D_0^\dagger)\rangle$. At each K, we carried out 2500 second-order iterations using the last 1500 iterations for the average ($\Delta\tau=0.01$).

The dynamical quark loops make the gauge configuration more ordered than in the pure gauge case. This is clearly visible in Fig. 2(a) beyond $K\sim 0.12$. (The points at $K=0$ are the Monte Carlo results for the pure gauge system.) The 4^4 lattice is probably too small to observe the effects of quark-pair creation in the static potential. It is nonetheless suggestive that the Wilson loop at $K\gtrsim 0.15$ plotted against area exhibits a concave shape in contrast to an exponential decrease for $K\gtrsim 0.12$. The loop effect is also apparent in the chiral order parameter [Fig. 2(b)] which deviates from the quenched value (cross) from $K\sim 0.12$.

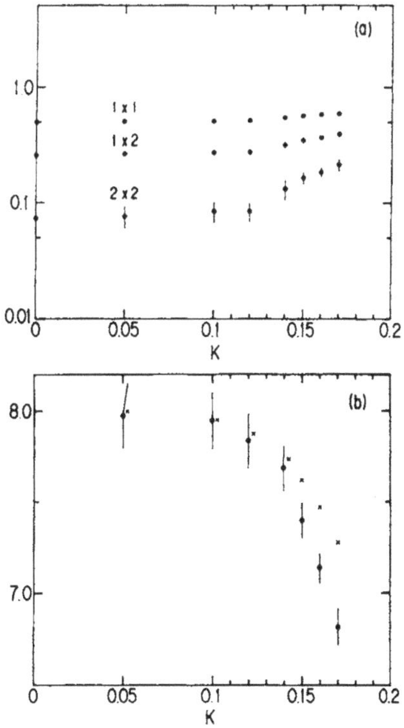

FIG. 2 (a) Wilson loop for full SU(2) theory with Wilson fermion (two flavors) at $\beta=2.0$ on 4^4 lattice. (b) $\langle\bar\psi\psi\rangle$ at $\beta=2.0$ on 4^4 lattice. Crosses represent the quenched value.

These trial results show that the effect of quark loops could be quite large so that the quenched results for properties of hadrons might be substantially modified.

A check on the systematic error is provided by the equality $\langle S_f \rangle = 4\times 3\times V$ with V the number of sites. The deviation of the measured value from this was less than 1%. We also repeated the calculation with a smaller time step, $\Delta\tau = 0.002$, and aside from the statistical fluctuation did not find any deviation from those of Fig. 2.

We examined the convergence by repeating the calculation with a different sequence of random numbers and a different type of initial configurations. The convergence was particularly slow around $K \sim 0.15$ compared with smaller or larger values of K. A plausible explanation is that at $K \sim 0.15$ the effective value of β reaches 2.2 at which a sharp crossover occurs in the pure gauge case.

The Langevin simulation is very much suited to vector computers. Our preliminary study shows that SU(3) simulation (Wilson fermion) on a 10^4 lattice can be done with about fifty megabytes of storage and about 30 h of computer time for 2500 iterations.

The microcanonical method[5] also tries to generate an ensemble governed by the effective action (12). In order to compare with the Langevin method, we have examined it for SU(2), paying particular attention to the questions associated with ergodicity.[11] We note that the ergodicity assumption fails in the limit of weak coupling and/or large quark mass since the system (or part of it) becomes integrable in these limits and hence by the Kolmogorov-Arnold-Moser theorem[12] invariant tori exist on the energy surface.

Our microcanonical formalism is slightly different from that of Ref. 5 in that $-U_l^{-1}dU_l/d\tau$ is identified with the gauge momentum and the gauge degree of freedom is fixed by the complete axial gauge. Our study on a 4^4 lattice shows that the method works for the pure gauge system; the Wilson loop averages agree with Monte Carlo values and the kinetic energy distribution coincides with that calculated from the microcanonical partition function. With quarks included, however, a delicate problem arises. On a 4^4 lattice (Wilson action) with $\Delta\tau = 0.01$, a large number of iterations $\sim (1-2)\times 10^4$ (to be compared with 2500 for the Langevin case) were necessary before the value of β obtained from the quark and gauge kinetic energies by the equipartition theorem agreed within a few percent. The situation was worse on a 2^4 lattice for which the two estimates of β in some runs differed by (10–20)% even after 2×10^4 iterations. This troublesome behavior was most prominent for β around 2.0 and for small K. While it is not quite clear if these results reflect some nonergodic behavior, it shows at least that the microcanonical method is less efficient than the Langevin in exploring the phase space of the system. Furthermore, though the values of Wilson loops and $\langle \bar\psi\psi \rangle$ obtained near $\beta \sim 2.0$ on a 4^4 lattice are consistent with those of the Langevin simulation, it is not clear to us how one could reliably estimate the error associated with the mismatch of β.

The Langevin method is apparently free from these problems of the microcanonical formalism, and we have shown above that it, indeed, works beautifully. We conclude by stressing that a realistic SU(3) simulation (on a 10^4 lattice or larger) with the Langevin algorithm presented in this note is prefectly feasible on presently available vector computers with regards to both storage and time consumption.

We are grateful to the Theory Division of The National Laboratory for High Energy Physics for warm hospitality and generous support for our work, to M. Okawa for informative discussions, and to Y. Oyanagi for advice on programming. One of us (A.U.) thanks Aspen Center for Physics for hospitality, where part of this work was done, and N. Christ, Y. Iwasaki, J. Kogut, O. Martin, and J. Polonyi for discussions.

[1]M. Creutz, Phys. Rev. D **21**, 2308 (1980); K. G. Wilson, in *Recent Progress in Gauge Theories, Cargèse Lectures 1979*, edited by G. 't Hooft *et al.* (Plenum, New York, 1980).

[2]D. Weingarten, Phys. Lett. **109B**, 57 (1982); H. Hamber and G. Parisi, Phys. Rev. Lett. **47**, 1972 (1981); A. Hasenfratz *et al.*, Phys. Lett. **110B**, 289 (1982).

[3]F. Fucito *et al.*, Nucl. Phys. **B180**, 369 (1981).

[4]D. J. E. Callaway and A. Rahman, Phys. Rev. Lett. **49**, 613 (1982).

[5]J. Polonyi and H. W. Wyld, Phys. Rev. Lett. **51**, 2257 (1983); J. Polonyi *et al.*, Phys. Rev. Lett. **53**, 644 (1984).

[6]G. Parisi and Y.-S. Wu, Sci. Sin. **14**, 483 (1981).

[7]D. Weingarten and D. Petcher, Phys. Lett. **99B**, 333 (1981).

[8]A. Guha and S.-C. Lee, Phys. Rev. D **27**, 2412 (1983).

[9]One can develop a second-order discretization by modifying the X in (2b) with addition of terms of order $(\Delta\tau)$ [I. T. Drummond *et al.*, Nucl. Phys. **B200**, 119 (1983)]. The additional term, however, involves higher derivatives of the action of complicated form. Even worse, when quarks are included, this method requires an inversion of quark matrices ($D^{-1}\cdot\partial D/\partial U_l\cdot D^{-1}Y$, see text below for notation) for every link l on the lattice in each iteration, which is practically impossible [N. Christ, private communication].

[10]D. Zwanziger [Phys. Rev. Lett. **50**, 1886 (1983)] also proposed a Langevin method. Our method differs from his in one important aspect. His method uses $Y^\dagger DY$ for the effective quark action. In order to recover the correct distribution $(\det D)\exp(-S)$, this necessitates an extrapolation in the results whose magnitude seems difficult to bring under control, and hence should be avoided.

[11]M. Fukugita and A. Ukawa, to be published.

[12]See, for example, J. Moser, *Stable and Random Motions in Dynamical Systems* (Princeton Univ. Press, Princeton, N. J., 1973), and references therein.

PHYSICAL REVIEW D VOLUME 32, NUMBER 10 15 NOVEMBER 1985

Langevin simulations of lattice field theories

G. G. Batrouni, G. R. Katz, A. S. Kronfeld, G. P. Lepage, B. Svetitsky,* and K. G. Wilson
Newman Laboratory of Nuclear Studies, Cornell University, Ithaca, New York 14853
(Received 17 May 1985)

We present a new analysis of Langevin simulation techniques for lattice field theories, including a general discussion of errors and algorithm speed. We introduce Fourier techniques that greatly accelerate simulations on large lattices. We also introduce a new technique for including quark vacuum-polarization corrections that admits any number of flavors, odd or even, without the need for nested Monte Carlo calculations. Our analysis is supported by a variety of numerical experiments.

I. INTRODUCTION

Considerable progress has been made over the past few years in the numerical simulation of quantum chromodynamics (QCD). With rapid developments in computing technology we can expect major new advances in the near future, leading to definitive tests of the full theory. To realize this goal the next generation of QCD simulations will have to address several critical issues:

the efficient inclusion of effects due to quark vacuum polarization;
the critical slowing down of existing simulation algorithms with larger lattices, resulting in computing needs that can grow quadratically or worse with the lattice volume;
the need for economical methods for high-statistics measurements of such things as the spectrum of the light hadrons, the glueball spectrum, or the heavy quark potential;
the need for a range of checks on the systematic errors in such measurements.

In this paper we demonstrate how the first two problems—those relating to the generation of field configurations—can be tackled using stochastic quantization via the Langevin equation.[1] Langevin updating proves to be ideal here, largely because the entire lattice is updated simultaneously, rather than link-by-link or site-by-site. We will discuss strategies for high statistics measurements on these configurations in future papers.

The first problem, that of including fermion determinants, can be solved by adding a new nonlocal noise term to the Langevin equation for gauge fields. As we demonstrate in Sec. IV, the only price paid for the nonlocality is that linear equations of the form

$$M\psi=\phi \tag{1.1}$$

must be solved for each update, where $M=\gamma_5(D\cdot\gamma+m)$ is the inverse fermion propagator. This is a reasonable price if by "each update" one means an update of the entire field configuration, as in Langevin updating. By contrast, a heat-bath or Metropolis algorithm demands that similar equations be solved once for the update of each separate link and site—an extremely costly venture. The algorithm we describe in Sec. IV has several other attractive features; most notably, it allows the simulation of any number of quark flavors, even or odd, without the need for a Monte Carlo calculation within a Monte Carlo calculation.

Langevin updating also offers a new approach to the problem of critical slowing-down. A crucial weakness in most updating algorithms is that short-wavelength structure tends to evolve far more quickly than structure at long wavelengths—typically faster by a factor of ξ^2, where ξ is the longest correlation length of the theory (measured in units of the lattice spacing). Since one must limit the size of a single update to maintain stability at short distances, the evolution of large-scale features is greatly slowed. Indeed the number of sweeps of the lattice required for appreciable change at these wavelengths grows as ξ^2, which tends to infinity in the continuum limit. In Sec. III we show how this problem can be remedied by updating field configurations in momentum space rather than in coordinate space.[2] Fast Fourier transforms (FFT's) are used to relate the two descriptions, and again the procedure is only feasible when the entire configuration is updated simultaneously (two FFT's are required per update). Using momentum-space updating for several two-dimensional spin models, we found that simulation speed can be increased by factors of 20—50 on 16×16 lattices. Similar gains seem attainable with QCD.

Momentum-space updating of gauge fields is only useful if calculations are done in a smooth gauge such as $A^0=0$ gauge; otherwise, the gauge freedom obscures the relationship between momentum and wavelength. With Langevin updating the cost of such gauge fixing is almost insignificant, once again because it is required only once for each update of the entire field.

Using Fourier transforms one can also accelerate the convergence of the iterative matrix methods needed to solve Eq. (1.1) or to compute fermion propagators. The problems here are very similar to those just discussed for updating. For instance, large-scale structure develops $O(\xi)$ times more slowly than short-wavelength structure when solving Eq. (1.1) with the conjugate gradient algorithm. The remedy is also very similar, as we illustrate in Sec. IV. Fourier acceleration makes possible QCD algorithms for which the amount of computation needed grows only linearly (up to logarithms) with the number of lattice points as the mesh is refined.

An analysis of algorithm speed requires an understanding of the errors inherent in Langevin updating. Errors arise because the differential Langevin equation must be replaced by a finite-difference equation for the purposes of a computer simulation. In scalar field theory, for example, the differential Langevin equation is[1]

$$\frac{\partial\phi(x,\tau)}{\partial\tau}=-\frac{\delta S[\phi]}{\delta\phi(x,\tau)}+\bar{\eta}(x,\tau), \quad (1.2)$$

where τ is a fictitious time labeling successive field configurations and $\bar{\eta}$ is a Gaussian random noise normalized by

$$\langle\bar{\eta}(x,\tau)\bar{\eta}(x',\tau')\rangle_{\bar{\eta}}=2\,\delta(x'-x)\,\delta(\tau'-\tau).$$

Once τ is sufficiently large, the field configurations $\{\phi(x,\tau)\}$ for different τ's are distributed with probability density $e^{-S[\phi]}$, as desired for a simulation of the field theory. A finite-difference form of this equation, suitable for computer simulations, can be obtained by making τ discrete with step size $d\tau=\epsilon$, e.g.,

$$\phi(x,\tau_{n+1}) = \phi(x,\tau_n) - f_x[\phi,\eta]\,, \quad (1.3a)$$

where

$$f_x = \bar{\epsilon}\frac{\delta S}{\delta\phi(x,\tau_n)}+\sqrt{\bar{\epsilon}}\,\eta(x,\tau_n) \quad (1.3b)$$

and where now η is normalized by

$$\langle\eta(x_i,\tau_i)\,\eta(x_j,\tau_j)\rangle_\eta = 2\,\delta_{x_i,x_j}\delta_{\tau_i,\tau_j}\,.$$

Discrete equations of this type need not be regarded solely as approximations to Eq. (1.2), valid only as $\bar{\epsilon}\to 0$. Rather, they are in themselves viable stochastic equations for any value of $\bar{\epsilon}$. However, for τ sufficiently large, the field configurations generated by the discrete equation (1.3) are distributed with a probability density $e^{-\bar{S}[\phi]}$, where now the equilibrium action $\bar{S}$ differs from S in $O(\bar{\epsilon})$ and beyond:

$$\bar{S}[\phi] = S[\phi]+\bar{\epsilon}S_1[\phi]+\cdots\,. \quad (1.4)$$

The question of errors centers upon the significance of these new additions to the action.

In Sec. II we examine the leading correction, $\bar{\epsilon}S_1$, for local field theories. We find that these corrections simply renormalize existing couplings in S, either directly or through quantum fluctuations. This pattern seems likely to persist in $O(\epsilon^2)$ and beyond, indicating that $\bar{S}$ and S represent the same continuum theory, at least if $\bar{\epsilon}$ is not too large. There is no error associated with using (1.3) in place of (1.2).

More generally, the step size $\bar{\epsilon}$ in Eq. (1.3b) can be replaced by a matrix $\epsilon_{x,y}$.[3] In fact, the key to Fourier acceleration is the proper choice of this matrix. The errors in this case are not merely renormalization effects, as above. Nevertheless, they are easily understood and bounded, as we illustrate in Sec. II. We find that algorithm speed (see Sec. III) depends somewhat upon the accuracy desired, but most importantly, it is independent of the correlation length ξ when $\epsilon_{x,y}$ is well chosen. Furthermore, one can reduce such errors substantially by employing higher-order difference equations, like the Runge-Kutta scheme of Sec. II C, in place of (1.3). None of these conclusions is altered when fermion loops are included as in Sec. IV.

This paper concludes, in Sec. V, with a brief summary of our findings, particularly as they apply to simulations of QCD.

II. DISCRETE LANGEVIN PROCESSES

In this section, we analyze the errors associated with the use of a discrete Langevin equation in place of the differential equation. Although most of our results are quite general, we focus here on theories for which the continuum action is local. Simulations involving the nonlocal effects of fermion loops are discussed in Sec. IV.

A. The effective equilibrium action—scalar fields

When the step size is replaced by a matrix Eq. (1.3) becomes

$$\begin{aligned}\phi(x,\tau_{n+1}) &= \phi(x,\tau_n)-f_x[\phi,\eta]\,,\\ f_x &= \sum_y\left[\epsilon_{x,y}\frac{\delta S}{\delta\phi(y,\tau_n)}+\sqrt{\epsilon_{x,y}}\,\eta(y,\tau_n)\right],\end{aligned} \quad (2.1)$$

where η is Gaussian noise with $\langle\eta^2\rangle=2$. As $n\to\infty$, this equation generates an ensemble of field configurations from the ensemble of random variables $\eta(x,\tau_n)$. Here we wish to study the equilibrium distribution of these asymptotic field configurations.

The Fokker-Planck equation determines the evolution of the probability density, $P(\{\phi\},\tau_n)$, for obtaining a configuration $\{\phi\}$ after Langevin time τ_n:

$$P(\tau_{n+1})-P(\tau_n)=\sum_{n=1}^{\infty}\sum_{x_1\cdots x_n}\frac{\delta}{\delta\phi(x_1)}\cdots\frac{\delta}{\delta\phi(x_n)}\times[\Delta_{x_1\cdots x_n}P(\tau_n)]\,, \quad (2.2a)$$

where

$$\Delta_{x_1\cdots x_n}=\frac{1}{n!}\langle f_{x_1}\cdots f_{x_n}\rangle_\eta\,. \quad (2.2b)$$

This equation is readily derived from the evolution equation

$$P(\{\phi\},\tau_{n+1})=\left\langle\int d\phi\,P(\{\phi\},\tau_n)\times\prod_x\delta(\phi(x,\tau_{n+1})-\phi(x,\tau_n)+f_x)\right\rangle_\eta \quad (2.3)$$

by Taylor expanding the δ functions in powers of f and averaging over the noise, η. Equation (2.2a) can be analyzed order-by-order in ϵ since only $2n$ terms on the right-hand side contribute through $O(\epsilon^n)$. The left-hand side vanishes as $\tau\to\infty$, leading to an equation for the asymptotic distribution $\bar{P}(\{\phi\})\equiv P(\{\phi\},\tau\to\infty)$. To $O(\epsilon)$, at least, Eq. (2.2a) then becomes[4]

$$-\frac{\delta}{\delta\phi(x)}\left[\frac{\delta\bar{P}}{\delta\phi(x)}+\bar{S}_x\bar{P}\right]=0, \tag{2.4}$$

where the effective Langevin force $\bar{S}_x$ can be written as a gradient, $\delta\bar{S}/\delta\phi(x)$. Consequently the asymptotic distribution is

$$\bar{P}(\{\phi\})=e^{-\bar{S}[\phi]}, \tag{2.5a}$$

where the equilibrium action is

$$\bar{S}[\phi]=S[\phi]+\tfrac{1}{4}\sum_{x,y}\epsilon_{x,y}\left[2\frac{\delta^2S[\phi]}{\delta\phi(x)\delta\phi(y)}-\frac{\delta S[\phi]}{\delta\phi(x)}\frac{\delta S[\phi]}{\delta\phi(y)}\right]+\cdots. \tag{2.5b}$$

It is $\bar{S}$ rather than S that determines the physics of the discrete-time simulation.

The impact of the $O(\epsilon)$ terms in $\bar{S}$ is most transparent when $\epsilon_{x,y}$ is completely local:

$$\epsilon_{x,y}=\bar{\epsilon}\delta_{x,y}\,. \tag{2.6}$$

Then the correction terms in $\bar{S}$ are local—in a $\lambda\phi^4$ theory, for example, the $O(\epsilon)$ corrections include terms like $\bar{\epsilon}\lambda\phi^2$, $\bar{\epsilon}(\partial^2\phi)^2$, and $\bar{\epsilon}\lambda^2\phi^6$. Since the new interactions generally respect the internal symmetries of the original action, they either renormalize existing couplings in S, or introduce nonrenormalizable interactions. In either case the continuum limit of the theory is completely unaffected. One need only shift the bare couplings in S, by amounts of relative order $\bar{\epsilon}$, to compensate for all effects due to the new terms in $\bar{S}$. Of course, if $\bar{\epsilon}$ is too large (i.e., ~ 1), such shifts might drastically alter the nature of S (e.g., $\lambda\to\bar{\lambda}<0$), thereby destabilizing the simulation. Assuming that this is not the case, $\bar{S}$ is no less fundamental or correct than S as a lattice approximation to the continuum action.

The correction terms in $\bar{S}$ are more complicated when $\epsilon_{x,y}$ is nonlocal. In $\lambda\phi^4$ theory, for example, one obtains new interaction terms like

$$\lambda^2\sum_{x,y}\phi(x)^3\epsilon_{x,y}\phi(y)^3 \tag{2.7a}$$

and

$$\lambda\sum_{x}\phi(x)^2\epsilon_{x,x}. \tag{2.7b}$$

These are terms of the kind one would obtain in S if there were another scalar "particle" ζ, with propagator $\epsilon_{x,y}$, coupled to the ϕ field via interactions like $\lambda\phi(x)^3\zeta(x)$ and $\lambda\phi(x)^2\zeta(x)^2$. In this picture, the local $\epsilon_{x,y}$ considered above [Eq. (2.6)] is just the infinite-mass limit for the ζ, and so it is not surprising that the ζ corrections in effect simply renormalize the bare couplings of S. As we shall see in Sec. III, a better choice for $\epsilon_{x,y}$ is

$$\epsilon_{x,y}=\bar{\epsilon}G_{x,y}\,, \tag{2.8}$$

where $G_{x,y}$ is the propagator for a free scalar particle. In this case, the fictitious ζ particle behaves very much like the ϕ particle, and its contribution to physical measurables typically will be $\bar{\epsilon}$ times that from the ϕ field. Therefore to simulate S accurately to 10%, for example, one requires $\bar{\epsilon}\sim 0.1$.

Curiously, if one works only to $O(\epsilon)$, all of the new nonrenormalizable and nonlocal interactions in $\bar{S}$ can be removed, leaving behind terms that can be completely absorbed into S simply by shifting the coupling constants by calculable amounts. This is accomplished by changing field variables to

$$\bar{\phi}(x)=\phi(x)-\tfrac{1}{4}\sum_{y}\epsilon_{x,y}\frac{\delta S}{\delta\phi(y)}. \tag{2.9}$$

Then the action can be reexpressed in terms of $\bar{\phi}$,

$$S[\phi]=S[\bar{\phi}]+\tfrac{1}{4}\sum_{x,y}\frac{\delta S}{\delta\phi(x)}\epsilon_{x,y}\frac{\delta S}{\delta\phi(y)}+\cdots, \tag{2.10}$$

while the measure becomes

$$\begin{aligned}\prod_x d\phi(x)&=\prod_x d\bar{\phi}(x)\left[1+\tfrac{1}{4}\sum_{y}\epsilon_{x,y}\frac{\delta^2S}{\delta\phi(x)\delta\phi(y)}+\cdots\right]\\&=\left[\prod_x d\bar{\phi}(x)\right]\exp\left[\tfrac{1}{4}\sum_{x,y}\epsilon_{x,y}\frac{\delta^2S}{\delta\phi(x)\delta\phi(y)}+\cdots\right].\end{aligned} \tag{2.11}$$

With these changes the equilibrium action is

$$\bar{S}=S[\bar{\phi}]+\tfrac{1}{4}\sum_{x,y}\epsilon_{x,y}\frac{\delta^2S}{\delta\phi(x)\delta\phi(y)}+\cdots \tag{2.12}$$

which, for most theories, differs from $S[\phi]$ only by calculable shifts of the bare couplings already present in S. In $\lambda\phi^4$ theory with $\epsilon_{x,y}=\bar{\epsilon}\delta_{x,y}$, for example, the only modification is $m^2\to m^2+\bar{\epsilon}\lambda/4$. Moreover, when ϵ is local like this the change of variables has no effect on the matrix elements of any operator linear in the $\phi(x)$'s, i.e.,

$$\langle f[\bar{\phi}]\rangle=\langle f[\phi]\rangle+O(\epsilon^2)\,, \tag{2.13}$$

where $f[\phi]$ is any multilinear functional. This result follows directly from the equation of motion

$$\int d\phi\, e^{-S[\phi]} \frac{\delta S}{\delta\phi(x)} \frac{\delta f}{\delta\phi(x)} = 0. \tag{2.14}$$

This technique for removing $O(\epsilon)$ corrections is an effective way of speeding up simulations with a nonlocal $\epsilon_{x,y}$, as in Eq. (2.8), since larger $\bar{\epsilon}$'s can be employed with no loss of accuracy. The added cost of shifting couplings and fields is generally negligible.

We have not yet investigated the $O(\epsilon^2)$ corrections that arise in the discrete formulation of the Langevin equation. The significance of these corrections can be assessed by analyzing their effect on ordinary perturbation theory. A preliminary study of this sort indicates that corrections in higher order are similar in nature to those in $O(\epsilon)$.

B. Non-Abelian fields

The analysis for scalar fields can be generalized readily to include non-Abelian variables such as the link field in QCD or the meson field in a chiral model.[5] All that one needs is to define differentiation with respect to the non-Abelian variable, or, more precisely, differentiation within its group manifold. Assuming that the non-Abelian variable, U, is an element of a Lie group in some representation with generators T_i (where $[T^i,T^j]=ic_{ijk}T^k$ and $\mathrm{tr}(T^iT^j)=\delta_{ij}/2$), a standard definition for this derivative is obtained from

$$f(e^{i\delta\cdot T}U) = f(U) + \delta^i\partial_i f + O(\delta^2). \tag{2.15}$$

This choice allows integration by parts—the critical operation in passing from Eq. (2.3) to Eq.(2.2)—under the invariant Haar measure of the group. Given this definition, the most convenient discrete Langevin equation is

$$U(\tau_{n+1}) = e^{-if\cdot T}U(\tau_n), \tag{2.16}$$

where $f_j=f_j(U(\tau_n),\eta)$ is the driving function. The simplest choice for the driving function is to take $\epsilon_{x,y}$ local,

$$f_j(U(\tau_n),\eta) = \bar{\epsilon}\partial_j S[U] + \sqrt{\bar{\epsilon}}\eta_j, \tag{2.17}$$

leading to results very similar to those obtained from Eqs. (1.3) for scalar fields. In manipulating these equations one must be careful because the derivatives no longer commute. Rather, they obey the algebra of the Lie group:

$$[\partial_i,\partial_j] = -c_{ijk}\partial_k. \tag{2.18}$$

Consequently, the equilibrium action is

$$\bar{S}[U] = \left[1+\frac{\bar{\epsilon}}{12}C_A\right]S[U] + \frac{\bar{\epsilon}}{4}\sum_j \{2\partial_j^2 S[U] - (\partial_j S[U])^2\} + \cdots \tag{2.19}$$

which differs from that for a scalar theory in the term involving C_A, the Casimir invariant of the Lie group's adjoint representation [$C_A=n$ for SU(n)].

These equations generalize in the obvious way when $\epsilon_{x,y}$ is nonlocal, and the significance of the $O(\epsilon)$ terms here is much the same as for the scalar field theories discussed above. If $\epsilon_{x,y}$ is local, the main effect of these terms is to renormalize the bare couplings in the original theory. Corrections involving a nonlocal $\epsilon_{x,y}$ resemble terms one might obtain from a new particle with propagator $\epsilon_{x,y}$ and with the quantum numbers of the non-Abelian field.

The use of a nonlocal $\epsilon_{x,y}$ in simulations of gauge theories (such as QCD) is complicated by the need to preserve local gauge invariance. Any breaking of this invariance tends to generate a large mass for the gauge bosons, either directly or through quadratic (in four dimensions) ultraviolet divergences in the quantum fluctuations. In QCD, this problem can be avoided by introducing gauge-invariant field variables. This is easily done through a gauge transformation. After each update of a configuration one applies the gauge transformation

$$U_\mu(n) \rightarrow G(n)U_\mu(n)G^\dagger(n+\hat{\mu}), \tag{2.20}$$

where $G(n)$ is a product of links along some path connecting site n to the origin. Such a transformation completely fixes the gauge, leaving only a global symmetry which is preserved for any choice of $\epsilon_{x,y}$. The axial gauge $A^0=0$ is an example of such a gauge. (Note that if $\epsilon_{x,y}=\bar{\epsilon}\delta_{x,y}$, the gauge fixing has no effect whatsoever on the evolution, in Langevin time, of gauge invariant quantities.) An alternative to the use of such gauges is to make $\bar{\epsilon}$ very small—$O(1/\xi^2)$ in four dimensions—so that the boson mass and other symmetry-breaking effects are negligible. Another possibility, rather similar to gauge fixing, is to reformulate the theory directly in terms of gauge-invariant fields—e.g., the corner variables of Ref. 6.

To gain experience with Langevin algorithms we applied them to SU(3) gauge theory in four dimensions, using the standard plaquette action

$$S[U] = -\frac{\beta}{2n}\sum_p \mathrm{tr}(U_p+U_p^\dagger), \tag{2.21}$$

where U_p is the plaquette operator. For this theory, the discrete Langevin equations (2.16) and (2.17) take the form

$$U_\mu(\tau_{n+1}) = e^{-F(\tau_n)}U_\mu(\tau_n), \tag{2.22a}$$

where

$$F(\tau_n) = \frac{\bar{\epsilon}\beta}{4n}\sum_{U_p\supset U_\mu}(U_p-U_p^\dagger) - \frac{\bar{\epsilon}\beta}{4n^2}\sum_{U_p\supset U_\mu}\mathrm{tr}(U_p-U_p^\dagger) + i\sqrt{\bar{\epsilon}}H(\tau_n), \tag{2.22b}$$

and $H(\tau_n)$ is a traceless Hermitian noise matrix obeying

$$\langle H_{ik}(x,\tau)H_{lm}(x',\tau')\rangle_H = \left[\delta_{il}\delta_{km} - \frac{1}{n}\delta_{ik}\delta_{lm}\right]\delta_{x,x'}\delta_{\tau,\tau'}. \tag{2.22c}$$

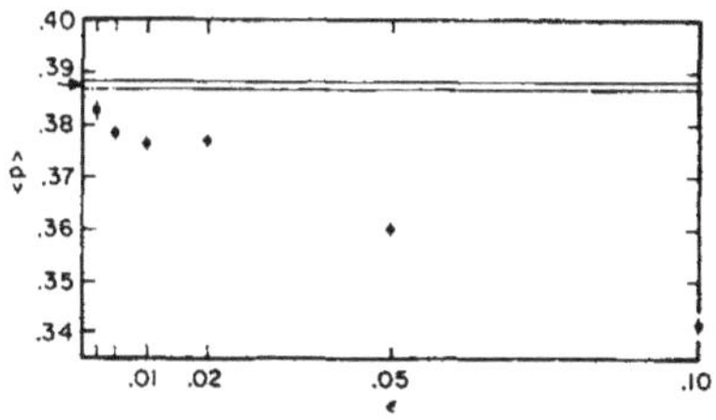

FIG. 1. Plot of the average plaquette vs ϵ at $\beta=4.9$ for the simple Euler algorithm. The horizontal lines near $\langle p\rangle=0.39$ indicate the Monte Carlo values. 1200 sweeps at $\epsilon=0.1$.

We iterated these equations with $\beta=4.9$ on a 2×5^3 lattice. The data for the average plaquette are plotted versus $\bar\epsilon$ in Fig. 1. (Note that the plaquette is dominated by ultraviolet contributions, and thus is quite sensitive to the differences between S and $\bar S$.) As $\bar\epsilon$ increases, the links tend to become more random, giving a smaller value for the average plaquette. At $\bar\epsilon=0.1$ our result differs from the Monte Carlo value by about 11%, as might have been expected from our error analysis.

One can remove $O(\epsilon)$ terms in $\bar S$ for non-Abelian variables, as for scalar variables, by a change of variables,

$$U \to \bar U = e^{-i(\bar\epsilon/4)\partial_j S T^j}\, U, \tag{2.23}$$

leaving an equilibrium action

$$\bar S[U] = \left[1+\frac{\bar\epsilon}{12}C_A\right]S[U] + \frac{\bar\epsilon}{4}\sum_j \partial_j^2 S[U] + \cdots . \tag{2.24}$$

(This works for a nonlocal $\epsilon_{x,y}$, as well.) For the plaquette action Eq. (2.21), $\bar S$ is then identical to S, but with β replaced by

$$\bar\beta = \beta[1-\bar\epsilon(C_l - C_A/12)], \tag{2.25}$$

where C_l is the Casimir invariant for the representation in which the U matrices are taken [e.g., $C_F=(n^2-1)/2n$ for the fundamental representation in SU(n)]. We redid the simulations described above, but now adjusting β so that $\bar\beta=4.9$ for all $\bar\epsilon$. The results, plotted in Fig. 2, are significantly closer to the Monte Carlo result, suggesting that $O(\epsilon)$ terms have indeed been removed. In computing the plaquette expectation values, we used the original U's, rather than the shifted $\bar U$'s, since in general

$$\langle f(\bar U)\rangle = (1+\bar\epsilon C_l/4)^{\#\,\mathrm{links}}\langle f(U)\rangle + O(\bar\epsilon^2) \tag{2.26}$$

when f is any linear functional of the U's. Consequently the additional effort required to remove all $O(\bar\epsilon)$ effects was negligible.

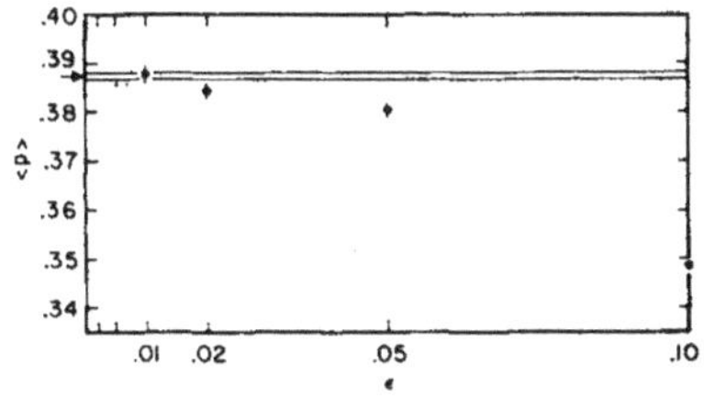

FIG. 2. Plot of the average plaquette vs ϵ at $\beta=4.9$ for the simple Euler algorithm with shifted coupling. 3200 sweeps at $\epsilon=0.1$.

C. Runge-Kutta algorithms

Equation (1.3) represents the simplest of an infinite number of discrete Langevin equations. In particular, there is a group of "Runge-Kutta" algorithms that model the differential Langevin equation (1.2) more closely. For example, one can model the action $S[\phi]$ correctly up to $O(\epsilon^2)$ by using the first-order Runge-Kutta equation[7,8]

$$\phi(\tau_{n+1}) = \phi(\tau_n) - \frac{\bar\epsilon}{2}\left[\frac{\delta S[\phi(\tau_n)]}{\delta\phi(\tau_n)} + \frac{\delta S[\tilde\phi(\tau_{n+1})]}{\delta\tilde\phi(\tau_{n+1})}\right] + \sqrt{\bar\epsilon}\,\eta(\tau_n), \tag{2.27}$$

where $\tilde\phi$ is a tentative update using the lowest-order algorithm,

$$\tilde\phi(\tau_{n+1}) = \phi(\tau_n) - \bar\epsilon\frac{\delta S[\phi(\tau_n)]}{\delta\phi} - \sqrt{\bar\epsilon}\,\eta(\tau_n).$$

The analogous equation for non-Abelian variables is obtained by replacing f_j in Eq. (2.17) with

$$f_j = \frac{\bar\epsilon}{2}\left[1+\frac{C_A\bar\epsilon}{6}\right]\{\partial_j S[U(\tau_n)] + \partial_j S[\tilde U(\tau_{n+1})]\} + \sqrt{\bar\epsilon}\,\eta_j(\tau_n), \tag{2.28}$$

where, as in the scalar case, $\tilde U(\tau_{n+1})$ is obtained from $U(\tau_n)$ using the leading order equations (2.16) and (2.17) with the same random noise variable $\eta_j(\tau_n)$. Such higher-order algorithms allow a much larger step size $\bar\epsilon$ before the equilibrium action $\bar S$ deviates much from the original action S. (They also generalize for nonlocal $\epsilon_{x,y}$.)

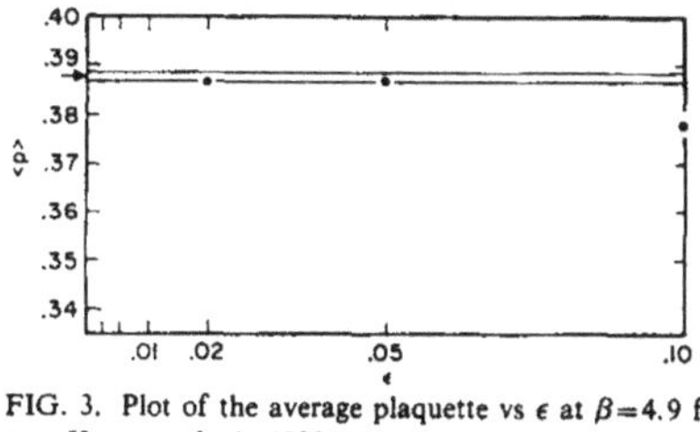

FIG. 3. Plot of the average plaquette vs ϵ at $\beta=4.9$ for the Runge-Kutta method. 1000 sweeps at $\epsilon=0.1$.

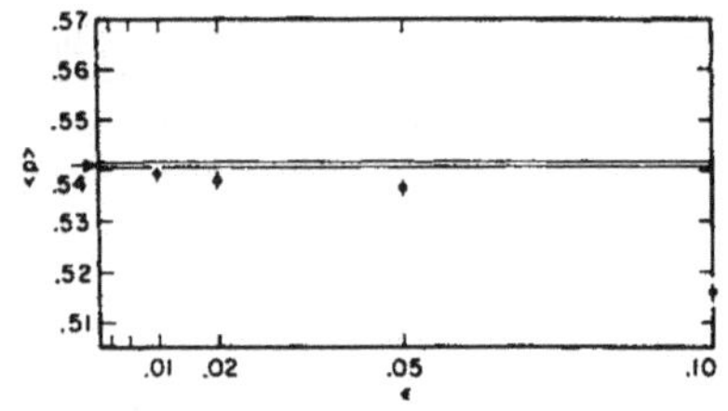

FIG. 4. Plot of the average plaquette vs ϵ at $\beta=5.5$ for the Runge-Kutta method. 1000 sweeps at $\epsilon=0.1$.

On the other hand, there is a penalty of roughly a factor of 2 in both computation and data storage for these particular Runge-Kutta algorithms.

We again simulated the pure gauge theory on a 2×5^3 lattice with $\beta=4.9$ and 5.5, this time using the first order Runge-Kutta scheme just described. Our data for the plaquette are plotted for different $\bar{\epsilon}$'s in Figs. 3 and 4. This algorithm ran 20–50 times faster than the simplest algorithm [Eqs. (2.16) and (2.17)] for similar precision. It also outperforms the algorithm with shifted couplings [Eq. (2.25)].

III. ALGORITHM SPEED AND FOURIER ACCELERATION

A. Free field theory

A critical figure of merit related to the speed of simulation algorithms is the correlation "time" N_c, which is the number of updates of the (whole) field configuration required to obtain a new, statistically uncorrelated configuration. Making the step size $\epsilon_{x,y}$ [Eq. (2.1)] larger reduces N_c, and is desirable. However, one is limited to $|\epsilon_{x,y}| \lesssim 1$ if the simulation is to remain stable (Sec. II A). If $\epsilon_{x,y}$ is local [Eq. (2.6)], this usually means that the correlation time grows quadratically,

$$N_c \sim \xi^2, \tag{3.1}$$

as ξ, the longest correlation length of the theory (in lattice units), increases (see below). With such behavior, the amount of computation required to achieve a given statistical accuracy grows as $V^{3/2}$ when the number of lattice points V is increased. (This assumes that the *physical* volume of space-time covered by the lattice remains constant, as will usually be the case.)

This behavior is evident in free field theory, with the quadratic action $S=\frac{1}{2}\phi M\phi$ where $M=-\partial^2+m^2$. The discrete Langevin equation is

$$\phi(x,\tau_{n+1}) = (1-\bar{\epsilon}M)\phi(x,\tau_n)+\sqrt{\bar{\epsilon}}\,\eta(\tau_n), \tag{3.2}$$

and it leads to an equilibrium action

$$\bar{S} = \tfrac{1}{2}\phi M\left[1-\frac{\bar{\epsilon}}{2}M\right]\phi + \text{constants}. \tag{3.3}$$

This equilibrium theory is obviously unstable unless

$$|\bar{\epsilon}M| < 2, \tag{3.4}$$

which is true if

$$\bar{\epsilon} \lesssim \frac{1}{p_{\max}^2+m^2}, \tag{3.5}$$

where $p_{\max}\sim 1$ is the maximum momentum (in lattice units) on the lattice. Thus, as expected, we require that $\bar{\epsilon}\lesssim 1$ for a stable simulation as $\xi\to\infty$ $(m\to 0)$.

To see how this restriction affects the correlation time N_c, we can solve the evolution equation (3.2) in momentum space, where[9] $M\to p^2+m^2$. With the initial condition $\phi(p,\tau=0)=0$, the solution is

$$\phi(p,\tau_n) = \sqrt{\bar{\epsilon}}\sum_{j=1}^{n}(1-\bar{\epsilon}[p^2+m^2])^{n-j}\eta(p,\tau_{j-1}). \tag{3.6}$$

Since $|1-\bar{\epsilon}[p^2+m^2]| < 1$ from Eq. (3.5), $\phi(p,\tau_n)$ depends less and less upon $\eta(p,\tau_j)$ as $n-j$ grows. Roughly then, we can define the correlation time $N_c(p)$ for a given $\bar{\epsilon}$ and p by the relation

$$\log_e\{1-\bar{\epsilon}[p^2+m^2])^{N_c(p)}\}\sim -1$$

which implies

$$N_c(p) \sim \frac{1}{\bar{\epsilon}[p^2+m^2]}. \tag{3.7}$$

Choosing $\bar{\epsilon}\sim\bar{\epsilon}_{\max}\sim[p_{\max}^2+m^2]^{-1}$ implies that the high-momentum components of the field will evolve with a very short correlation time: $N_c(p_{\max})\sim 1$. However, the correlation time for the low momentum structure usually will be much longer:

$$N_c(p\sim 0) \sim \frac{1}{m^2} \sim \xi^2. \tag{3.8}$$

Thus $O(\xi^2)$ times as many updates of the field configuration are required to obtain the same statistics as for high momenta. Unfortunately, it is these long-wavelength components that are usually the most important physically.

This analysis illustrates a general problem with the local Langevin algorithm, and indeed with most other Monte Carlo algorithms used today. The evolution of long-wavelength structure is held back by the need for stability at short wavelengths. Unfortunately, the problem becomes more and more severe as the grid is refined and the correlation length ξ increases.

Such critical slowing down can only be remedied by resolving different length scales and dealing with them separately. Our analysis of free field theory suggests that this can be accomplished by updating the fields in momentum space, using different ϵ's for different momenta in such a way as to speed up the evolution at low momenta without destabilizing the high momenta.[2] This means employing a nonlocal $\epsilon_{x,y}$:

$$\epsilon_{x,y} = \sum_p e^{ip\cdot(x-y)}\epsilon(p).$$

Equations (3.5) and (3.7) indicate that the choice

$$\epsilon(p) \sim \frac{1}{p^2+m^2} \tag{3.9}$$

is optimal, at least for free field theory. Then the stability criterion [Eq. (3.5)] is satisfied, while the correlation time [Eq. (3.7)] is $N_c \sim 1$ for all momenta, not just for $p \sim p_{max}$. Consequently, the number of field updates required for a simulation is reduced by a factor of $1/\xi^2$, and computing power need only grow linearly with the size of the lattice, rather than as $V^{3/2}$.

B. Interacting field theories

The basic strategy for accelerating simulations of an interacting field theory should be similar to that for free field theory. In particular, working in momentum space provides a way of separating the different length scales. Introducing a momentum-dependent step size is both straightforward and not particularly expensive with Langevin updating: Eqs. (1.3) are replaced by

$$\Delta\phi(x) = -\hat{F}\left[\epsilon(p)\hat{F}^{-1}\frac{\delta S}{\delta\phi(x)} + \sqrt{\epsilon(p)}\eta(p)\right], \tag{3.10}$$

where $\hat{F}$ is an operator representing a Fourier transform. The computation required for fast Fourier transforms is roughly comparable to that needed to evaluate $\delta S/\delta\phi$, since only two transforms are required for each update of the entire field configuration. A similar strategy would be impossible with Metropolis or heat-bath updating, where a new Fourier transform would be necessary for each hit at each site.

The optimal $\epsilon(p)$ for an interacting theory again should be

$$\epsilon(p) = \frac{\bar{\epsilon}}{p^2+m^2}, \tag{3.11}$$

where $\bar{\epsilon}$ determines the accuracy of the simulation (Sec. II) and where m will usually[10] be of order $1/\xi$.

In some theories ultraviolet and infrared behavior are strongly coupled. For example, the bare and renormalized masses in a four-dimensional scalar field theory differ by $O(\xi^2)$ due to ultraviolet fluctuations. One might worry about speeding up evolution at low momenta for such a theory, since the high-momentum structure must be sufficiently well developed to allow a very accurate estimate of the $O(\xi^2)$ renormalizations. However, such effects are largely determined by structure at distances of order a lattice spacing, and thus one has $O(\xi^4)$ independent samples in a single configuration from which they can be determined. Consequently the mass renormalization, for example, is specified to an accuracy of $\xi^2/(\xi^4)^{1/2} \sim 1$ by a single configuration—just adequate for our purposes. Such linkage between high momenta and low momenta generally does not seem to be a problem.

This discussion of scalar field theories applies to non-Abelian theories as well. In particular, Fourier transforms can be used just as they are for scalar theories. The transformations used in Eq. (3.10) can be applied without change to Eq. (2.17), used in updating non-Abelian fields. In gauge theories it is also important that the gauge be fixed as otherwise long-wavelength physics can be disguised by high-frequency fluctuations due to crinkly gauges. The $A^0=0$ gauge mentioned in Sec. II is sufficiently smooth.

C. A numerical example

To investigate the utility of Fourier acceleration we used it in simulating the two-dimensional XY model:[11]

$$S = -\beta\sum_{n,\mu}\cos(\theta_n - \theta_{n+\mu}) - h\sum_n \cos(\theta_n). \tag{3.12}$$

We shifted β, h, and the field θ_n so that all $O(\epsilon)$ corrections due to discrete Langevin updating cancelled [Eqs. (2.9)–(2.12)]. We then ran on a 16×16 lattice with both a p-dependent and a p-independent step size:

$$\epsilon(p) = \begin{cases} 0.02 \\ 0.02\dfrac{\left[\beta\sum_\mu[1-\cos(2\pi p_\mu/N)]+h\right]_{max}}{\beta\sum_\mu[1-\cos(2\pi p_\mu/N)]+h}, \end{cases} \tag{3.13}$$

where the p dependence in the second case is that of the free propagator for θ. Our results for the momentum-space Green's function $\langle|\theta(p)|^2\rangle$, with $\beta=1.5$ and $h=0.23$, are given in Table I. As expected, performance is greatly improved by Fourier acceleration. At low momenta 400 sweeps with Fourier acceleration give results

TABLE I. $\langle|\theta(p)|^2\rangle$ for the XY action of Eq. (3.12). Values of the momentum-space propagator $\langle|\theta(p)|^2\rangle$ for the two-dimensional XY model on a 16^2 lattice. Columns 1 and 2 are results from perturbation theory, 3 and 4 are Langevin results with Fourier acceleration, 5 and 6 are the same but without Fourier acceleration. Runs were done at $\beta=1.5$ and $h=0.23$.

	Perturbation		FFT		No FFT	
p^2/p^2_{max}	0th order	1 loop	400 steps	6000 steps	400 steps	6000 steps
$\frac{1}{32}$	2.093	2.512	2.372(172)	2.568(36)	2.373(710)	2.654(180)
$\frac{1}{16}$	0.798	0.958	0.971 (49)	0.973(11)	1.023(127)	0.998 (35)
$\frac{1}{8}$	0.338	0.405	0.443 (17)	0.421 (6)	0.356 (37)	0.402 (10)
$\frac{1}{4}$	0.190	0.228	0.257 (18)	0.239 (3)	0.263 (19)	0.232 (4)
$\frac{1}{2}$	0.105	0.126	0.131 (7)	0.129 (2)	0.125 (8)	0.131 (2)
1	0.082	0.098	0.095 (18)	0.097 (4)	0.067 (19)	0.098 (4)

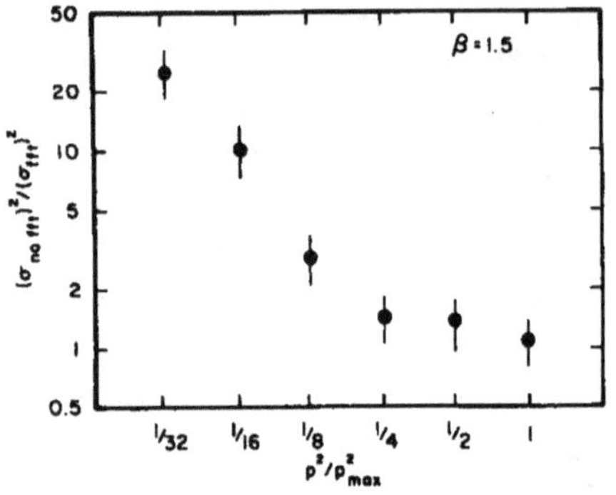

FIG. 5. Improvement due to Fourier acceleration in measurements of the momentum Green's functions for the *XY* model at $\beta = 1.5$ and $h = 0.23$. The squares of the standard deviation with and without acceleration are compared for different momenta.

superior to those from 6000 sweeps without acceleration. In Fig. 5 we plot the square of the ratio of the standard deviations with and without FFT's versus momentum. The standard deviations were estimated by measuring fluctuations among 16 samples of 400 measurements each. At high momenta there is little difference, but the simulation at $p \sim 0$ was 20–25 times faster with Fourier acceleration. We also measured the correlation time N_c by binning our data in various sized bins and measuring correlations between successive bins. The results, in Fig. 6, indicate a correlation time of just a few sweeps of the lattice for *any* momentum with Fourier acceleration, while without FFT's the correlation time grows dramatically at low momenta. The values we obtained for the

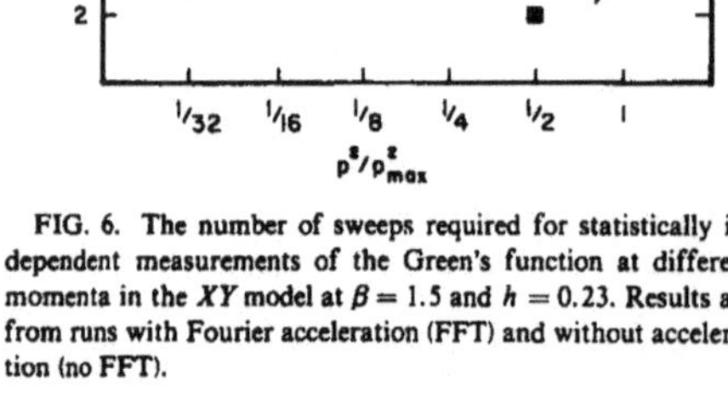

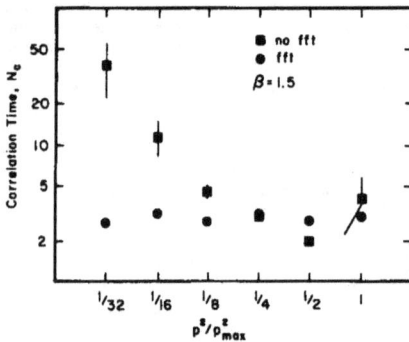

FIG. 6. The number of sweeps required for statistically independent measurements of the Green's function at different momenta in the *XY* model at $\beta = 1.5$ and $h = 0.23$. Results are from runs with Fourier acceleration (FFT) and without acceleration (no FFT).

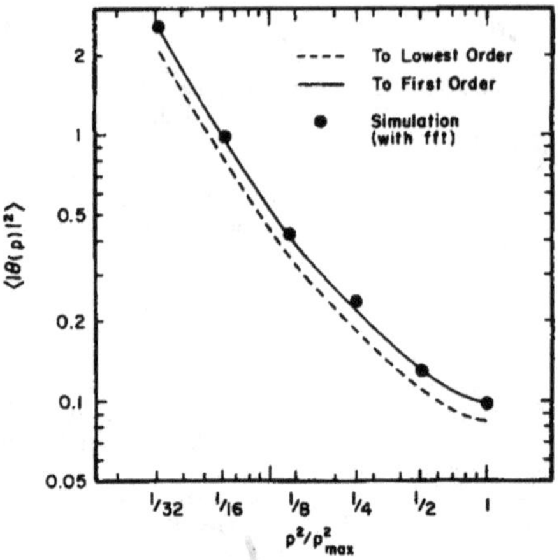

FIG. 7. The Green's function $\langle |\theta(p)|^2 \rangle$ for the *XY* model with $\beta = 1.5$ and $h = 0.23$. Results from lowest and first order perturbation theory are compared with results from our simulation (with Fourier acceleration).

propagator agree well with perturbation theory when one-loop corrections are included—these shift β to 1.25 and h to 0.2. Our data for the propagator are shown in Fig. 7.

We have also simulated the *XY* model in its nonperturbative phase, running at $\beta = 1.0$ and 0.7. Again Fourier acceleration speeds the simulation by a factor of order ξ^2,

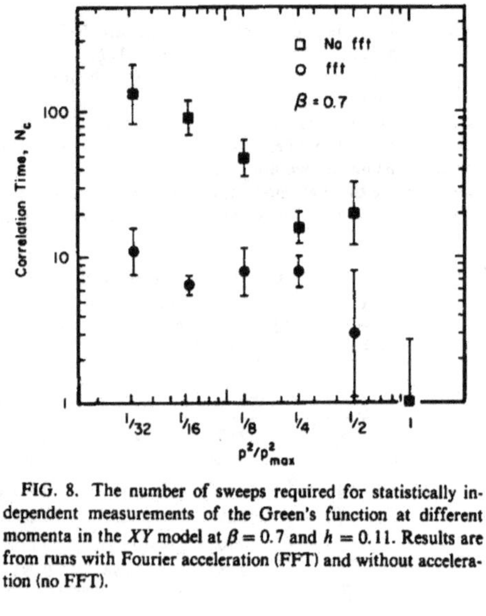

FIG. 8. The number of sweeps required for statistically independent measurements of the Green's function at different momenta in the *XY* model at $\beta = 0.7$ and $h = 0.11$. Results are from runs with Fourier acceleration (FFT) and without acceleration (no FFT).

even though many vortex-antivortex pairs are present. Lacking perturbation theory, one can determine the optimal p dependence of $\epsilon(p)$ from measurements of the correlation time for different momenta (cf. Fig. 6); $\epsilon(p)$ should be roughly proportional to $N_c(p)$. We show the correlation times, with and without Fourier acceleration, for $\beta=0.7$ in Fig. 8. The improvement due to Fourier acceleration is substantial, though smaller than at $\beta=1.5$ since the correlation length is shorter. Clearly, Fourier acceleration is effective whenever different wavelengths evolve at very different speeds, whether or not the theory is in a perturbative domain.

IV. LANGEVIN SIMULATIONS OF FERMIONIC THEORIES

A. Fermion loops

One of the major hurdles in simulating QCD has been the inclusion of effects due to quark vacuum polarization. In principle, quark loops can be included by adding a term $-\operatorname{Tr}\ln M[U]$ for each flavor to the gauge boson action $S_g[U]$. Here, $M=\gamma_5(D\cdot\gamma+m)$ is the inverse propagator of the quark, together with an additional factor of γ_5 which makes M Hermitian without affecting the fermion determinant. Thus for a single flavor, the full action is

$$S = S_g[U] - \operatorname{Tr}\ln M[U]. \tag{4.1}$$

Remarkably, such an action is readily simulated using the Langevin equation. The fermion term can be generated simply by adding a new bilinear noise term:

$$U(\tau_{n+1}) = e^{-if\cdot T}U(\tau_n), \tag{4.2a}$$

$$f_j = \bar{\epsilon}\left[\partial_j S_g - \tfrac{1}{2}\mathrm{Re}\left[\eta_q^\dagger \frac{1}{M}\partial_j M\eta_q\right]\right] + \sqrt{\bar{\epsilon}}\eta_j, \tag{4.2b}$$

where η_q and η_j are Gaussian random numbers (with $\langle|\eta|^2\rangle=2$). That this equation yields fields distributed according to the action (4.1) is evident from the Fokker-Planck equation. To $O(\epsilon)$ in the Fokker-Planck equation, f_j is always averaged over the bilinear noise η_q:

$$\begin{aligned}\langle f_j\rangle &= \bar{\epsilon}\left[\partial_j S_g - \operatorname{Tr}\left[\frac{1}{M}\partial_j M\right]\right] + \sqrt{\bar{\epsilon}}\eta_j \\ &= \bar{\epsilon}\partial_j(S_g - \operatorname{Tr}\ln M) + \sqrt{\bar{\epsilon}}\eta_j .\end{aligned}$$

As this is just the f_j one would use for the fermionic action (4.1), the fermion contribution is properly included. Note that the derivative of $\operatorname{Tr}\ln M$ has no imaginary part, and so we can drop the Re from (4.2b) when averaging. The only significant cost for including the fermions is that $\psi=M^{-1}\eta_q$ must be determined once for each update by solving the linear equations

$$M[U]\psi = \eta_q; \tag{4.3}$$

in terms of ψ, the driving function f_j is local:

$$f_j = \bar{\epsilon}[\partial_j S_g - \mathrm{Re}(\psi^\dagger\partial_j M\eta_q)] + \sqrt{\bar{\epsilon}}\eta_j . \tag{4.2c}$$

The generalization for nonlocal $\epsilon_{x,y}$ is obvious.

This procedure has several attractive features. Most importantly, it does not involve a Monte Carlo calculation within a Monte Carlo calculation, as do most pseudofermion methods. Also, any number of flavors, odd or even, may be simulated by including a bilinear noise term for each quark, or if the masses are the same, by multiplying the existing term by n_f, the number of flavors.[12] This follows because our pseudofermions enter only as noise, not as dynamical fields. Finally, the analysis developed in the preceding sections provides tight control over errors.

The $O(\epsilon)$ corrections here are similar to those for local theories (Sec. II). In deriving the Fokker-Planck equation one averages over both η_q and η_j. The asymptotic probability density then satisfies

$$\partial_i(\partial_i\bar{P} + \bar{S}_i P) = 0, \tag{4.4a}$$

where

$$\bar{S}_i = \partial_i\bar{S} + \frac{\bar{\epsilon}}{8}(\partial_j - \partial_j S)\operatorname{Tr}\left[\frac{1}{M^2}\partial_i M^2 \frac{1}{M^2}\partial_j M^2\right] + \cdots \tag{4.4b}$$

and $\bar{S}$ is given by Eq. (2.19) with S as in Eq. (4.1). For a general $\epsilon_{x,y}$ the new terms in $\bar{S}$ resemble corrections one might expect from a new gauge field with propagator $\epsilon_{x,y}$. For example, the term

$$\sum_{x,y}\operatorname{Tr}\left[\frac{1}{M}\partial_{i,x}M\frac{1}{M}\partial_{i,y}M\right]\epsilon_{x,y}$$

is just a self-energy correction to the quark propagator (in a background U field). Thus, as before, these terms introduce no new problems.

The second term in Eq. (4.4b) is not an exact differential, and therefore cannot be absorbed easily into a redefinition of $\bar{S}$. One might worry that this term introduces unwanted infrared sensitivity due to the rather singular factors of $1/M^2$ (as opposed to $1/M$) that it contains. However, we have examined contributions from this term to various n-point Green's functions in perturbation theory, and we find that almost all such contributions cancel. (This follows from the equations of motion: $\langle(\partial_j-\partial_j S)f[U]\rangle=0$ for any functional f of the fields.) The few terms that remain seem no more singular than contributions from ordinary perturbation theory, and are therefore suppressed by $\bar{\epsilon}$. Consequently, the performance of the QCD algorithms discussed in Sec. II is unaffected by the inclusion of fermion loops.

We have tested algorithm (4.2) on a 2^4 lattice for QCD with Wilson fermions, for which

$$\begin{aligned}\gamma_5 M\psi = \psi(n) - \kappa\sum_\mu[&(1-\gamma_\mu)U_\mu(n)\psi(n+\hat{\mu}) \\ &+(1+\gamma_\mu)U_\mu^\dagger(n-\hat{\mu})\psi(n-\hat{\mu})].\end{aligned} \tag{4.5}$$

We tried several values of β, κ, and n_f. Our results for $\beta=4$ and $\kappa=0.15$ with two flavors are displayed in Table II, together with recent results by Weingarten.[13] While our algorithm requires roughly the same number of sweeps over the lattice as Weingarten's Metropolis algorithm, it is much faster than his because the matrix equation (4.3) is solved only once per sweep, not once per link-update. Our algorithm is also far superior to the pseudofermion method he analyzes. Note that we use skew boundary conditions for three of the four dimensions and consequently our results should differ slightly from Weingarten's. On large lattices our algorithm's performance can be further enhanced by Fourier acceleration.

We have also investigated a number of related algorithms for including fermion loops. One is to simulate a theory with a new auxiliary scalar field ϕ,

$$S[U,\phi] = S_g[U] + \phi^\dagger \frac{1}{M^2}\phi, \tag{4.6}$$

that introduces fermion loops for two flavors. An update of the gauge fields for this theory is identical to Eqs. (4.2), but with the bilinear noise term replaced by $\psi^\dagger(\partial_j M^2)\psi$, where $M^2\psi=\phi$. The scalar field is updated using the Langevin equation[14]

$$\Delta\phi = -\bar{\epsilon}_q\phi - (\bar{\epsilon}_q)^{1/2} M\,\eta_q. \tag{4.7}$$

This procedure is comparable in speed to the one discussed above (in fact, they become identical when $\bar{\epsilon}_q \to 1$), but it is limited to even numbers of flavors since M is not positive definite.

B. Fermion propagators

Another important consideration relating to fermions is the speed with which linear equations like Eq. (4.3) can be solved. Such computations are critical both for including fermion loops and for studying fermionic observables.

Most iterative matrix methods work better when the matrix is positive definite, and so we consider the equation

$$M^2\psi = \phi \tag{4.8}$$

TABLE II. Data from three different algorithms for simulating QCD: Langevin [(Eq. 4.2)], Metropolis, and pseudofermion (Ref. 13). Runs were done on a 2^4 lattice, $\beta=4$, $\kappa=0.15$.

	Number of sweeps	Tr[line]	Tr[plaquette]
Eqs. (4.2)			
$\epsilon = 0.05$	6000	−0.15(2)	0.29(1)
$\epsilon = 0.01$	12000	−0.15(2)	0.31(1)
$\epsilon = 0.005$	24000	−0.15(2)	0.30(1)
Ref. 13			
Metropolis	2400	−0.16(2)	0.36(2)
pseudofermion	85000	−0.15(1)	0.31(3)

in place of (4.3). The speed with which these algorithms converge is intimately related to the spectrum of M^2. This is illustrated by the simplest of iterative procedures, the Jacobi algorithm, where the $(n+1)$st estimate is obtained from

$$\psi^{(n+1)} = (1-\delta M^2)\psi^{(n)} + \delta\phi. \tag{4.9}$$

The parameter δ controls the speed and stability of the algorithm. This equation is easily solved formally:

$$\psi^{(n)} = \delta\sum_{j=0}^{n}(1-\delta M^2)^{j-1}\phi. \tag{4.10}$$

Note the similarity of these equations to the stochastic equations used for updating scalar fields [Eqs. (3.2) and (3.6)]. As in that case, this algorithm converges provided δ is chosen small enough,

$$\delta \lesssim \frac{1}{|M^2|_{\max}} \sim \frac{1}{p^2_{\max}+m^2}, \tag{4.11}$$

while the number of iterations required to obtain a fixed accuracy is [cf. Eq. (3.7)]

$$N_{\rm inv} \sim \frac{1}{\delta\,|M^2|_{\min}} \sim \xi^2. \tag{4.12}$$

Again one must wait for the long wavelength components of ψ to develop; the short wavelength components converge after only a few iterations.

The solution to this problem is identical to that for the updating: i.e., we solve the equations in momentum space, making the step size δ momentum dependent. For the Jacobi method, this means replacing Eq. (4.9) by

$$\Delta\psi = \hat{F}\,\delta(p)\hat{F}^{-1}(-M^2\psi^{(n)}+\phi), \tag{4.13}$$

where again the optimal $\delta(p)$ is something like

$$\delta(p) = \frac{\bar{\delta}}{p^2+m^2}. \tag{4.14}$$

With this sort of Fourier acceleration, requiring only two FFT's per iteration, all wavelengths evolve at roughly the same rate. Then, only $N_{\rm inv}\sim 1$ iterations are required for a complete solution ψ.

Fourier acceleration can be used with any matrix algorithm that, like the Jacobi algorithm, updates the whole field ψ simultaneously. In particular, it is compatible with the conjugate gradient algorithm.[15] On the other hand, it is unlikely that this acceleration technique can be used with methods like the Gauss-Seidel algorithm that update one site at a time.

When using this method for gauge theories, it is again essential to fix the gauge so that high momentum does in fact correspond to short distances (and not just a crinkly gauge function). Gauges like the $A^0=0$ gauge discussed in Sec. II B are suitable. However, if for some reason one is using Fourier acceleration only in matrix inversion and not for updating the gauge fields, other gauges can be employed. For example, one can apply the gauge transformation

$$U_\mu(n) \to G(n)\, U_\mu(n)\, G(n+\hat\mu),$$
$$G(n) = \frac{1}{\alpha}\sum_\nu \left[U_\nu(n-\hat\nu) - U_\nu(n) - \frac{1}{n}\mathrm{Tr}[U_\nu(n-\hat\nu) - U_\nu(n)] \right] \quad (4.15)$$

before each update. In an Abelian gauge theory, this is equivalent to adding a gauge fixing term $-(\partial\cdot A)^2/\bar\alpha$ to the continuum action. The effect is similar for non-Abelian fields, except that these gauges bear no simple relation to common perturbative gauges (witness the absence of ghost fields here).

We tested Fourier acceleration of the conjugate gradient algorithm for QCD on a small lattice (4^4 sites). Using Wilson fermions with $n_f=2$ and $\beta=6$ we solved Eq. (4.3) for ψ as part of the updating algorithm for the gauge fields. [Actually, we multiply Eq. (4.3) by M on each side since the conjugate gradient method requires a positive matrix.] Near the critical κ, $\kappa_c \sim 0.158$, Fourier acceleration reduced the number of sweeps needed by a factor of $\frac{1}{3}$, roughly as expected for such small lattices. The algorithm functioned well for all κ's, both above and below κ_c. We also examined the residual vector, $M^2\psi - \phi$, as a function of momentum. Again as expected, the error was distributed uniformly in momentum with Fourier acceleration, while it was dominated by low momenta without FFT's. We also found that Fourier acceleration had no effect without gauge fixing: there was then no correlation between momentum and error size. We are currently testing Fourier acceleration on larger lattices where the gain should be substantially larger, particularly with small quark masses.

V. SUMMARY

In this paper we have introduced a variety of new algorithms designed to speed the simulation of field theories like QCD. Fourier acceleration, both for updating and for matrix inversion, promises to reduce or even to eliminate critical slowing down. Introducing fermions as a Langevin noise not only avoids the need for a Monte Carlo calculation within a Monte Carlo calculation, but allows simulations with any number of flavors (even, odd, fractional, negative,...). Runge-Kutta versions of the discrete Langevin equation allow much larger steps to be taken for each update. In addition, the Langevin formulation opens up new possibilities such as the use of complex actions, or the direct calculation of connected Green's functions. These may be very important in dealing with both systematic and statistical errors.

The tight control we have over errors is one of the most attractive features of these algorithms. The nature of the errors is readily apparent from the equilibrium action, and their significance is easily estimated, both analytically and numerically. This remains true even with Fourier acceleration and fermion loops.

Our theoretical analysis is amply supported by a variety of numerical experiments, on small lattices or in low dimensions. Needless to say, however, the extent to which our techniques affect QCD simulations will become apparent only after high-statistics tests on large lattices. Results from such tests will be forthcoming.

ACKNOWLEDGMENTS

We are indebted to many of our colleagues for their comments and suggestions throughout the course of this work. In particular, we want to thank Ian Drummond, Ron Horgan, Giorgio Parisi, Pietro Rossi, and Don Weingarten. G.P.L. thanks the Department of Applied Mathematics and Theoretical Physics at Cambridge for their hospitality during the early stages of this work. His work is also partially supported by the Alfred P. Sloan Foundation. B.S. thanks the I.B.M. Corporation for support.

*Present address: Center for Theoretical Physics, Laboratory for Nuclear Science, Massachusetts Institute of Technology, Cambridge, MA 02139.

[1]G. Parisi and Wu Yongshi, Sci. Sin. 24, 483 (1981).

[2]Similar ideas are discussed by G. Parisi, in *Progress in Gauge Field Theory*, edited by G. 't Hooft *et al.* (Plenum, New York, 1984).

[3]This is also true for the Langevin equation in the continuous time limit Eq. (1.2). See, for example, P. H. Damgaard and K. Tsokos, Nucl. Phys. B235, 75 (1984).

[4]This result is obvious to lowest order in ϵ. One can show that it still applies in first order by using the lowest-order solution to simplify the right-hand side of Eq. (2.2a). Note that the discrete-time Fokker-Planck equation then takes the form $\Delta P = -Q^\dagger Q\, P$, where the operator Q has a zero eigenvalue. The analysis of this equation for $\tau \to \infty$ is quite similar to that used in the continuous-time Langevin formalism.

[5]M. B. Halpern, Nucl. Phys. B228, 173 (1983); I. T. Drummond, S. Duane, and R. R. Horgan, *ibid.* B220 [FS8], 119 (1983); A. Guha and S. C. Lee, Phys. Rev. D 27, 2412 (1983); H. Hamber and U. Heller, Phys. Rev. D 29, 928 (1984).

[6]I. Bars, Nucl. Phys. B148, 445 (1979); B149, 39 (1979); G. G. Batrouni and M. B. Halpern, Phys. Rev. D 30, 1782 (1984).

[7]This is easy to demonstrate using the Fokker-Planck equation, (2.2).

[8]E. Helfand, Bell Syst. Tech. J. 58, 2289 (1979); H. S. Greenside and E. Helfand, *ibid.* 60, 1927 (1981).

[9]We are being somewhat schematic here. On a lattice, p^2+m^2 is replaced by the inverse of a lattice propagator.

[10]Insofar as a nonlocal $\epsilon_{x,y}$ mimics a new field in the effective action, the optimal m may be somewhat different from $1/\xi$ due to renormalization.

[11]J. Tobochnik and G. V. Chester, Phys. Rev. B 20, 3761 (1979).

[12]In fact, n_f could also be fractional, which might be useful in

studying the phase structure of QCD.

[13]D. Weingarten, IBM report, 1985 (unpublished).

[14]This is the standard Langevin equation for action (4.6) but with $\epsilon_{xy} = \bar{\epsilon} M^2$. This choice minimizes errors due to finite step size while at the same time avoiding all critical slowing down in the ϕ fields. This algorithm (but with $\epsilon_{xy} = \epsilon$ rather than $\bar{\epsilon} M^2$) has also been suggested by A. Ukawa and M. Fukugita, KEK report (1985), and by O. Martin, S. Otto, and J. Flower, Caltech report (1985). The Metropolis version of this algorithm is discussed in D. Weingarten and D. Petcher, Phys. Lett. **99B**, 333 (1981).

[15]M. R. Hestenes and E. Stiefel, J. Res. Natl. Bur. Stand. **49**, 409 (1952); L. A. Hageman and D. M. Young, *Applied Iterative Methods* (Academic, New York, 1981).